Lecture Notes in Computer Science 10657

Commenced Publication in 1973
Founding and Former Series Editors:
Gerhard Goos, Juris Hartmanis, and Jan van Leeuwen

More information about this series at http://www.springer.com/series/7412

Marcelo Mendoza · Sergio Velastín (Eds.)

Progress in Pattern Recognition, Image Analysis, Computer Vision, and Applications

22nd Iberoamerican Congress, CIARP 2017
Valparaíso, Chile, November 7–10, 2017
Proceedings

Springer

Editors
Marcelo Mendoza (ID)
Universidad Federico Santa María
Santiago
Chile

Sergio Velastín (ID)
Carlos III University of Madrid
Madrid
Spain

ISSN 0302-9743   ISSN 1611-3349   (electronic)
Lecture Notes in Computer Science
ISBN 978-3-319-75192-4   ISBN 978-3-319-75193-1   (eBook)
https://doi.org/10.1007/978-3-319-75193-1

Library of Congress Control Number: 2018931880

LNCS Sublibrary: SL6 – Image Processing, Computer Vision, Pattern Recognition, and Graphics

Printed on acid-free paper

This Springer imprint is published by the registered company Springer International Publishing AG
part of Springer Nature
The registered company address is: Gewerbestrasse 11, 6330 Cham, Switzerland

Preface

This volume contains the papers presented at CIARP 2017, the 22nd Iberoamerican Congress on Pattern Recognition held at the Universidad Técnica Federico Santa María in Valparaíso, Chile. CIARP publishes original, high-quality papers related to pattern recognition, welcoming contributions on any aspect of pattern recognition theory as well as practical issues and applications. CIARP 2017 was endorsed by the IAPR, the International Association for Pattern Recognition. CIARP 2017 was the 22nd in the CIARP series of annual international conferences devoted to all aspects of pattern recognition, image processing, computer vision, multimedia, and related fields.

The conference was held during November 7–10, 2017, and consisted of four days of papers and keynotes. There were 156 submissions. Each submission was reviewed by three Program Committee members. The committee decided to accept 86 papers. We used a double-blind reviewing process, i.e., the reviewers did not know who the authors of the papers were and the authors did not know who the reviewers were. The reviewers were chosen based on their expertise ensuring that they came from different countries and institutions around the world. This reviewer process assignment worked remarkably well as indicated by the high average confidence value the reviewers gave their own reviews. We would like to thank all the members of the Program Committee for their work, which we are sure contributed to improving the quality of the selected papers.

The program also included four invited talks by Professors Aaron Courville of the University of Montreal, Canada, Gert Lanckriet, principal researcher at Amazon, USA, Ingela Nytrom of Uppsala University in Sweden, and Nello Cristianini of the University of Bristol in the UK. It was a pleasure to work with this outstanding group of invited speakers. We would like to also thank Professor José Ruiz Schulcloper for his keynote during the CIARP student consortium. We extend our special appreciation to our sponsoring organizations: Department of Informatics at Universidad Técnica Federico Santa María, the Chilean Association of Pattern Recognition, and the International Association of Pattern Recognition.

Last but not least, we thank all the people involved in this conference and the Steering Committee of CIARP. We would like to thank especially the members of our local Organizing Committee for helping us in the organization of the conference.

December 2017

Marcelo Mendoza
Sergio Velastín

Organization

General Chairs

Marcelo Mendoza   Universidad Técnica Federico Santa María, Chile
Sergio Velastín   Universidad Carlos III, Spain

Steering Committee

Marta Mejail   SARP, Argentina
Helio Cortes Vieira Lopes   SIGPR-BR, Brazil
Marcelo Mendoza   AChiRP, Chile
Andrés Gago Alonso   ACRP, Cuba
Eduardo Bayro-Corrochano   MACVNR, México
César Beltrán-Castañón   APeRP, Perú
Luis Fillipi Barbosa de Almeida   APRP, Portugal
Roberto Paredes-Palacios   AERFAI, Spain
Álvaro Pardo   APRU, Uruguay

Session Chairs

Kalman Palagyi   University of Szeged, Hungary
Bogdan Kwolek   AGH University of Science and Technology, Poland
Saha Punam   University of Iowa, USA
Bruno Motta de Carvalho   Universidad Federal do Rio Grande do Norte, Brazil
Daniel Riccio   University of Naples Federico II, Italy
William Robson Schwartz   Universidade Federal do Minas Gerais, Brazil
José Ruiz Shulcloper   University of Informatics Sciences, Cuba
Luis Baumela   Politechnica University of Madrid, Spain
Héctor Allende-Cid   Pontificia Universidad Católica de Valparaíso, Chile
Ricardo Ñanculef   Universidad Técnica Federico Santa María, Chile
César San Martín   Universidad de la Frontera, Chile

ACHIRP Organization

Héctor Allende-Cid   Pontificia Universidad Católica de Valparaíso, Chile
José Luis Jara   Universidad de Santiago, Chile
Marco Mora   Universidad Católica del Maule, Chile

Local Organization

Claudia Arancibia
Ignacio Espinoza
Margarita Bugueño
Fernanda Weiss
Francisco Mena
Yerko Cuzmar
Claudia Sanhueza

Program Committee

Luís A. Alexandre	UBI and Instituto de Telecomunicações, Portugal
Daniel Acevedo	Universidad de Buenos Aires, Argentina
Niusvel Acosta-Mendoza	Advanced Technologies Application Center (CENATAV), Cuba
Enrique Alegre	University of Leon, Spain
Héctor Allende-Cid	Pontificia Universidad Católica de Valparaíso, Chile
Arnaldo Araujo	Universidade Federal de Minas Gerais, Brazil
Mauricio Araya-López	Universidad Tecnica Federico Santa Maria, Chile
Akira Asano	Kansai University, Japan
George Azzopardi	University of Malta, Malta
Antonio Bandera	University of Malaga, Spain
Rafael Berlanga	Universitat Jaume I, Spain
Gunilla Borgefors	Uppsala University, Sweden
Mario Bruno	Universidad de Playa Ancha, Chile
Odemir Bruno	University of São Paulo, Brazil
María Elena Buemi	Universidad de Buenos Aires, Argentina
Luis Vicente Calderita Estévez	University of Extremadura, Spain
Jesus Ariel Carrasco-Ochoa	INAOE
Carlos Castillo	Universitat Pompeu Fabra, Spain
Max Chacon	Universidad de Santiago de Chile, Chile
Rama Chellappa	University of Maryland, USA
Diego Sebastián Comas	Universidad Nacional del Mar del Plata, Argentina
Mauricio Delbracio	Universidad de La República, Uruguay
Mariella Dimiccoli	Computer Vision Center (CVC)
Jefersson Alex Dos Santos	Universidade Federal de Minas Gerais (UFMG), Brazil
Jacques Facon	Pontificia Universidade Catolica do Parana, Brazil
Marcelo Fiori	Universidad de la República, Uruguay
Alejandro Frery	Universidade Federal de Alagoas, Brazil
Maria Frucci	Institute for High Performance Computing and Networking National Research Council of Italy (ICAR-CNR), Italy

Bernardete Ribeiro   University of Coimbra, Portugal
Paul Rosin   Cardiff University, UK
Jose Ruiz-Shulcloper   Advanced Technologies Applications Center
                                  (CENATAV/DATYS)
Carolina Saavedra   Valparaiso University, USA
Alessia Saggese   University of Salerno, Italy
Gabriella Sanniti di Baja   Institute for High-Performance Computing
                                  and Networking, CNR, Italy
William Robson Schwartz   Federal University of Minas Gerais, Brazil
Rafael Sotelo   Universidad de Montevideo, Uruguay
Antonio-José   Universitat Politècnica de València, Spain
   Sánchez-Salmerón
Mariano Tepper   Duke University, USA
Maria Trujillo   Universidad del Valle, Colombia
Carlos Valle   Universidad Técnica Federico Santa María, Chile
Sergio A Velastin   Universidad Carlos III de Madrid, Spain
Mario Vento   Università degli Studi di Salerno, Italy
Pablo Zegers   Universidad de los Andes, Colombia

Contents

A Citation k-NN Approach for Facial Expression Recognition

Daniel Acevedo[1,2]($\boxtimes$), Pablo Negri[3]🆔, María Elena Buemi[1],
Francisco Gómez Fernández[1,2], and Marta Mejail[1,2]

[1] Facultad de Ciencias Exactas y Naturales, Departamento de Computación,
Universidad de Buenos Aires, Buenos Aires, Argentina
dacevedo@dc.uba.ar
[2] Instituto de Investigación en Ciencias de la Computación (ICC),
CONICET-Universidad de Buenos Aires, Buenos Aires, Argentina
[3] CONICET-Universidad Argentina de la Empresa (UADE),
Buenos Aires, Argentina

Abstract. The identification of facial expressions with human emotions plays a key role in non-verbal human communication and has applications in several areas. In this work, we propose a descriptor based on areas and angles of triangles formed by the landmarks from face images. We test this descriptors for facial expression recognition by means of an adaptation of the k-Nearest Neighbors classifier called Citation-kNN in which the training examples come in the form of sets of feature vectors. Comparisons with other state-of-the-art techniques on the CK+ dataset are shown. The descriptor remains robust and precise in the recognition of expressions.

1 Introduction

Facial expressions have been studied due to the great deal of applications that are required for interpreting human communication through facial gestures, such as automatic behavior recognition, human-computer interaction, pain assessment, health-care, surveillance, deceit detection and sign language recognition [1]. In Human Computer Interaction (HCI), it can be used to supersede other forms of non-verbal communication, since the expressiveness of faces are usually linked to an emotional state [2].

We aim at recognizing human emotions from the complete set of the Extended Cohn-Kanade AU-Coded Facial Expression Database (CK+) [3]. This database captures several persons performing facial expressions, including seven basic emotions: anger, contempt, disgust, fear, happy, sadness and surprise.

In this work, a geometric descriptor that captures facial expression evolution within a video sequence is introduced. It is based on the angles and areas formed by the landmarks placed on the images of faces that perform an expression. A similar descriptor was introduced in [4] and its dynamic was modeled with a Conditional Random Field (as opposed to our Multiple Instance Learning

© Springer International Publishing AG, part of Springer Nature 2018
M. Mendoza and S. Velastín (Eds.): CIARP 2017, LNCS 10657, pp. 1–9, 2018.
https://doi.org/10.1007/978-3-319-75193-1_1

approach presented in this work). Geometric descriptors were also proposed for other applications and different contexts. In [5] geometric descriptors are used for kinship verification. Angles from triangles formed by 76 landmarks are extracted from images as well as segment lengths and its ratios. In [6] the authors propose for age invariant face recognition the use of a set of triangles proportion ratios to estimate the degree of similarity among triangles. A thorough study of different methodologies can be found at [7].

We use the geometric descriptor (angles and areas) in a variant of a supervised machine learning framework in which the descriptors are grouped in a form of bags. By means of a distance between bags (Hausdorff distance), an adaptation of the k-nearest neighbor algorithm is used as a classifier, called Citation k-NN [8].

The paper is organized as follows: Sect. 2 introduces the proposed descriptor. Section 3 presents the adaptation of our geometric descriptor to be used by the Citation k-NN classifier. Section 4 presents the experimental results. Finally Sect. 5 presents conclusions.

2   Geometrical Descriptor

Expression recognition using temporal approaches usually tracks the changes of landmarks spatial positions within the images on a video sequence. This analysis, however, is sensible to movement of the head while the expression is occurring.

Our descriptor, computes a measure independent of the spatial position of the landmarks on the image. They are computed from the triangles obtained by three landmarks, inspired on the polygon mesh representation. In this way, the approach is independent of the face pose, and the features only measure facial transformations by evaluating the changes on the angles and areas on the respective triangles between consecutive frames. Three landmarks (ℓ_a, ℓ_b, ℓ_c) define a triangle Δ. The geometry of Δ can be described by its *area* and the *internal angles*. Thus, let be: θ_Δ, the angular descriptor where ℓ_a is the landmark at the central point, ℓ_b and ℓ_c are the extreme points of the angle, and α_Δ is the area of triangle Δ. Figure 1 shows a subset of landmarks and the evolution of an

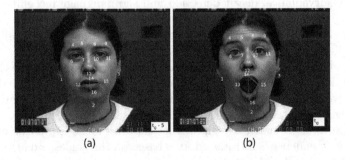

(a)   (b)

Fig. 1. Images of a surprise expression with landmarks numbered and the triangle $\Delta = (\ell_{14}, \ell_{15}, \ell_{16})$ displayed in red lines, (a) $t_0 - 5$ frame, centered at angle $\hat{\ell}_{15} = 45°$, (b) t_0 frame, centered at angle $\hat{\ell}_{15} = 96°$.

angle on a surprise expression using a temporal interval of 5 frames. Traditionally, the set of landmarks for each capture is composed of 68 points. We work with a subset of $n = 18$ landmarks that define 814 triangles describing the geometrical shape of the face and their dynamic.

3   Multiple Instance Learning and Citation-kNN

Multiple Instance Learning (MIL) is a variant of supervised machine learning in which the training examples come in the form of *bags*. A bag is a set of feature vectors, called *instances*. Each training bag has, as in traditional supervised learning, an assigned class label. The goal is to learn a model capable of predicting class labels for unseen query bags. Note how this setting considerably differs from classical single instance supervised learning, in which training data is organized as individual feature vectors, and a class label is available for each vector. A variety of approaches to the MIL task have been proposed over the years [8]. Every method relies on a specific assumption about the relationship between a bag label and the instances within that bag. Citation-kNN is a way of adapting the *k-Nearest Neighbors* (kNN) approach to the MIL task. It is based on the very loose assumption that bags that are similar according to a given distance measure are likely to have the same class label. In the single-instance variant of kNN, traditional distance functions such as L_p norms can be used to compare examples. In MIL though, proper distance functions between bags have to be used instead.

The chosen distance function in Citation-kNN is a modified version of the Hausdorff distance. Given two bags $A = \{a_1, \ldots, a_M\}$ and $B = \{b_1, \ldots, b_N\}$, and two integers S and L such that $1 \leq S \leq M$ and $1 \leq L \leq N$ the Hausdorff distance is defined as:

$$H_{SL}(A, B) = \max(h_S(A, B), h_L(B, A)) \tag{1}$$

The function $h_F(A, B)$ is known as the *directed Hausdorff distance*. Given bags A and B, and an integer F such that $1 \leq F \leq |A|$, it is defined as: $h_F(A, B) = F^{th} \min_{b \in B} \|a - b\|$, where $\|\cdot\|$ is a norm on the points of A and B and F^{th} is the F-th ranked distance in the set of distances (one for each element in A). In other words, for each point in A the distance to its nearest neighbor in B is computed, and the points of A are ranked according to the values of this distance. The F-th ranked value decides the distance $h_F(A, B)$.

The values of S and L are usually defined indirectly by specifying the fraction $0 \leq K \leq 1$ of the points to be considered. That is, given bags A and B, S and L are respectively set as: $S = \max(\lfloor |A|K \rfloor, 1)$ and $L = \max(\lfloor |B|K \rfloor, 1)$, where $\lfloor \cdot \rfloor$ is the floor function and $|\cdot|$ denotes cardinality .

The usual kNN method for selecting the label of a query bag based on the majority class of the closest neighbors may not lead to good classification results in the MIL setting. To overcome this problem, [9] incorporates the notion of *citation* into the voting mechanism used to decide the label of a query bag.

The proposed approach defines the *R-nearest references* of a bag A as the R nearest neighbors of A. Moreover, it defines the *C-nearest citers* of A to be the set that includes any given bag X if and only if A is one of the C nearest neighbors of X. For binary classification, the number of positive votes p is determined by summing the number of positive references R_p and positive citers C_p, i.e. $p = R_p + C_p$. Likewise, if the number of positive and negative references are R_n and C_n respectively, the number of negative votes is $n = R_n + C_n$. The query is predicted positive if $p > n$, and negative otherwise.

3.1 Modified Citation-kNN

The voting procedure described in Sect. 3 was thought for binary classification. However, a natural extension to the multiple class scenario can be devised for citation-kNN. Let L be the set of possible labels. For a given query bag X, assume $\{r_1, \ldots, r_{n_r}\}$ and $\{c_1, \ldots, c_{n_c}\}$ are the class labels of its n_r nearest references and its n_c nearest citers respectively. In our extension, the predicted label l^* for X is obtained by summing the votes of citers and references as:

$$l^* = \arg\max_{l \in L} \left(\sum_{i=1}^{n_c} \delta(l, c_i) + \sum_{j=1}^{n_r} \delta(l, r_j) \right) \tag{2}$$

where $\delta(a, b) = 1$ if $a = b$ and $\delta(a, b) = 0$ otherwise.

The criteria presented in Eq. (2) assigns the same importance to the vote of any citer or reference. However, it seems reasonable to pay more attention to those neighbors that are closer to the query bag. This can be achieved by using weighted voting. Formally, let X be the query example, L be the set of possible labels, and assume $\{R_1, \ldots, R_{n_r}\}$ and $\{C_1, \ldots, C_{n_c}\}$ are the sets of the n_r nearest references and the n_c nearest citers of X respectively. Further, let $\{r_1, \ldots, r_{n_r}\}$ and $\{c_1, \ldots, c_{n_c}\}$ be the class labels of those nearest references and citers respectively. Then, the weighted voting approach for predicting label l^* for X is:

$$l^* = \arg\max_{l \in L} \left(\sum_{i=1}^{n_c} \alpha_i \delta(l, c_i) + \sum_{i=1}^{n_r} \beta_i \delta(l, r_i) \right) \tag{3}$$

where α_i and β_i are the weights for citer C_i and reference R_i respectively.

Clearly, the weight for a particular example should be a decreasing function of its distance to X. In this work we set $\alpha_i = \frac{1}{H(X,C_i)}$ and $\beta_i = \frac{1}{H(X,R_i)}$, where $H(\cdot)$ is the Hausdorff distance described in Eq. (1); we omit the S and L subscripts here for clarity.

3.2 Citation kNN on Bags of Time-Stamped Facial Expression Descriptors

Our MIL approach for facial expression recognition represents descriptor sequences using bags of time-stamped descriptors and classifies new bags using

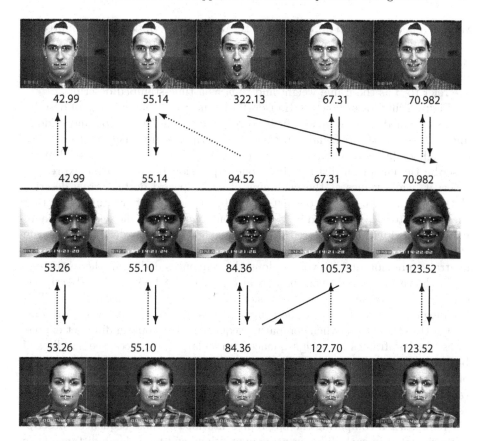

Fig. 2. Illustration of the Hausdorff distance between bags of face expression descriptors, corresponding to the *happy* (top and middle respectively) and *sadness* (bottom) emotions. Considering each pair of expressions (top-middle and middle-bottom), values at the beginning of each downward solid arrow describes the euclidean distance from one expression to the nearest expression below it, while values at the beginning of each upward dotted arrow describes the euclidean distance from one expression to the nearest above it.

Eq. (3). We now formalize the proposed representation and present a series of examples that aim at gaining further insight into the Hausdorff distance. For clarity purposes, the original sequences used in the examples were brought to a length of only 5 frames.

From the definition of areas and angles in Sect. 2, let $x_i = (\theta_{\Delta_{i1}}, s\,\alpha_{\Delta_{i1}}, \ldots,$ $\theta_{\Delta_{ih}}, s\,\alpha_{\Delta_{ih}})$ be a descriptor for each frame i, where $i = 1, \ldots, M$, M is the number of frames, h the number of considered triangles, and s a scale factor that puts into balance the influence of areas over angles. Then, the descriptor X for each sequence of M frames is defined as $X = (x_1, \ldots, x_M)$ and its time-stamped bag representation is the set $B_X = \{b_1, b_2, \ldots, b_M\}$, where $b_i = (x_i, \beta \frac{i}{M})$ is the ordered pair obtained by appending a single value to descriptor x_i. The appended value $\beta \frac{i}{N}$ adds temporal information to the original descriptor. It is computed by multiplying the relative temporal position of the descriptor (i.e. the video frame at which the descriptors is located divided by the total number of frames in the video) by a constant β, that regulates the weight of the temporal information. We refer to β as the frame weight and to b_i as a time-stamped descriptor.

Figure 2 illustrates how the Hausdorff distance between two bags of descriptors is obtained. The compared bags correspond to the *happy* (top and middle) and *sadness* (bottom) facial expressions. Descriptors are linked to their nearest neighbor by arrows. Values at the beginning of each downward solid arrow describe the euclidean distance from one expression to the nearest expression below it, while values at the beginning of each upward dotted arrow describe the euclidean distance from one expression to the nearest above it. Observe that each descriptor is represented by its corresponding face with the selected 16 landmarks marked with dots. Moreover, faces are shown in the order they appear in the original sequence. The leftmost image of each sequence corresponds to the neutral expression and it evolves along the sequences ending in the expression itself, shown in the rightmost images in Fig. 2. It can be seen that descriptors corresponding to these neutral expressions are relatively close to each other. This is an expected behavior, since the neutral face is common to all the expressions.

On the other hand, when comparing sequences that contain different expressions, the last frames for each sequence lead to larger distances, see middle and bottom sequences in Fig. 2. This example illustrates an important assumption of our approach: even though two differently labeled expressions sequences may (and most probably will) contain many similar descriptors, at least one of the sequences will contain expressions that differentiate it from the other sequence. Therefore, when comparing the bag representations of those sequences, descriptors originated in such differentiating expressions will be at large distance from their nearest neighbors in the other bag. As a consequence, the Hausdorff Distance will be large.

Recall from Sect. 3 that the modified Hausdorff Distance can be tuned by specifying the aforementioned parameter $0 \leq K \leq 1$. To illustrate the effect of this parameter, Fig. 2 (top and middle) shows the directed Hausdorff Distance between two *happy* bags. More precisely, if we use the names A and B to denote the bags from the top and the middle respectively, the figure illustrates $h_S(A, B)$ with solid lines and $h_L(B, A)$ with dotted lines. Recall from Sect. 3 that the value of S is determined by K as $S = \max(\lfloor 5K \rfloor, 1)$. Note that the third expression in A corresponds to a *surprise* expression (considered here as noise). As a consequence, its associated descriptor lies far away from every descriptor in B (distance equal to 322.13 from its nearest descriptor). Therefore, if we set $K = 1$ (and thus $S = 5$) an undesirably large Hausdorff Distance is obtained. The desired behavior can be obtained, by letting $K = \frac{4}{5}$ (and $S = 4$). It might seem that reducing K is always beneficial. However, note that setting $K = \frac{2}{5}$ (and $S = 2$) would ignore the three more discriminative expressions in A.

4 Experiments and Results

In this Section the geometric features are tested by means of the Citation-kNN classification method. We performed several experiments for evaluating the accuracy of emotion recognition. A leave-one-out subject cross validation methodology was used to assess the performance in the study: each sequence belonging to a subject in the database is classified (i.e., tested) and the rest of the subjects are used for training.

In our experiments we utilized the extended Cohn-Kanade dataset (CK+) [3], which is one of the most widely used resources in the development of expression recognition systems. The CK+ database, contains sequences of frames that begin with a neutral expression (Ne) and end in one basic emotion: anger (An), contempt (Co), disgust (Di), fear (Fe), happiness (Ha), sadness (Sa) or surprise (Su). There are 118 subjects that perform from one to six expressions each, to a total of 327 sequences which are labeled as one of the seven aforementioned expressions (but no sequence is labeled as neutral on the database). The database also provides 68 landmarks for each frame of each sequence. We utilized all the subjects (118) and all of its sequences (327), that are labeled with one of the 7 emotions provided by the database.

In Citation-kNN, the time-stamped versions of the features are used. They are grouped into bags and the classification process is as depicted in the previous section. Different parameter values are considered in the following ranges: $R \in [2, 20]$, $C \in [2, 20]$, $K \in [0.1, 1]$ with step 0.1, $\beta \in [1, 70]$ with step 5 and $s \in [2.9, 3.6]$ with step 0.1. Best results, shown in the last row of Table 2, were obtained for $R = 8$ for the R-nearest references, $C = 20$ for the C-nearest citers, $K = 0.9$, $\beta = 30$, and $s \in [2.9, 3.6]$. The overall performance achieved by this method is 89.30% (well-classified sequences from a total of 327).

We notice that the Citation-kNN method has the advantage that the location of the apex frame (i.e., the frame where the expression itself is achieved) can be determined automatically. As shown in Fig. 2, if we consider the distance between apex frames we observe that it increases considerably. On the other hand, the lowest distance between frames from different videos is the one that corresponds to the neutral expression.

Table 1. Accuracy with the different trained and tested datasets, OA: Onset-Apex, OAO: Onset-Apex-Offset.

Method	Train: OA Test: OA	Train: OAO Test: OA	Train: OA Test: OAO	Train: OAO Test: OAO
Accuracy (%)	89.30	83.18	83.79	89.30

We performed an experiment to test the robustness with respect to the position of the apex in the sequence. For that, the CK+ sequences were mirrored and appended at the end of the regular sequences in order to build a longer sequence where the apex is at the center, we call this sequence OAO: Onset-Apex-Offset.

Regular CK+ sequences, called OA (Onset-Apex), were used for training and OAO sequences were used for testing, and viceversa. Also, OAO sequences were used for training and testing.

In all the cases, a leave-one-out subject cross validation methodology was employed. We observed that when the apex location in the test sequences is different from the apex location in the training sequences (for example, one is in the middle and the other at the end, respectively), the accuracy remains similar and maintains the 89.3% accuracy when the extended sequences were used for training and testing, see Table 1.

For the sake of comparison among the existing methods in the bibliography Table 2 shows the classification rates of the seven expression of CK+ dataset. All the methodologies are sequence based multi-class recognition systems. Some of the methods compared use different number of folds when performing cross-validation. In Cov3D [10] and Interval Temporal Bayesian Network (ITBN) [11] tests were performed using 5 and 15 fold-cross validation, respectively. The Constrained Local Method (CLM) [12] and Lucey's work [3] use leave-one-out testing approach. The proposed methodology, Cit-kNN (last row in Table 2), achieves classification accuracy comparable to the state-of-the-art supervised 2D facial expression recognition techniques on CK+ database.

Table 2. Benchmark results. Best scores are in bold and 2nd best scores are underlined.

	An	Co	Di	Fe	Ha	Sa	Su
Cov3D [10]	**94,4**	**100,0**	95,5	**90,0**	96,2	70,0	**100,0**
SPTS+CAPP [3]	75.0	84.4	94,7	65,2	**100,0**	68,0	96,0
ITBN [11]	91,1	78,6	94,0	83,3	89,8	**76,0**	91,3
CLM [12]	70,1	53,4	92,5	72,1	94,2	45,9	93,6
Cit-kNN	84,4	77,8	**96,6**	44,0	98,6	75,0	98,8

5   Conclusions

In this paper we have presented a simple geometric descriptor for facial expression recognition based on angles and areas of selected landmarks corresponding to strategic locations based on motion and position. An adaptation of the k-NN approach to the Multiple Instance Learning task has been used as a classifier. In case there is no movement in the sequence, the behavior of Citation-kNN is not affected by this fact. Also, Citation-kNN can automatically obtain the expression apex frame and performs robustly when the apex frame is located at different positions in the sequence video.

References

1. Sariyanidi, E., Gunes, H., Cavallaro, A.: Automatic analysis of facial affect: a survey of registration, representation, and recognition. IEEE Trans. PAMI **37**, 1113–1133 (2015)

2. Zen, G., Porzi, L., Sangineto, E., Ricci, E., Sebe, N.: Learning personalized models for facial expression analysis and gesture recognition. IEEE Trans. Multimed. **18**, 775–788 (2016)
3. Lucey, P., Cohn, J., Kanade, T., Saragih, J., Ambadar, Z., Matthews, I.: The CK+ Dataset: a complete dataset for action unit and emotion-specified expression. In: CVPRW, pp. 94–101 (2010)
4. Acevedo, D., Negri, P., Buemi, M.E., Mejail, M.: Facial expression recognition based on static and dynamic approaches. In: ICPR, pp. 4124–4129 (2016)
5. Bottino, A., Vieira, T., Ul Islam, I.: Geometric and textural cues for automatic kinship verification. Int. J. Pattern Recogn. **29** (2015)
6. Osman Ali, A., et al.: Age-invariant face recognition using triangle geometric features. Int. J. Pattern Recogn. Artif. Intell. **29**, 1556006 (2015)
7. Corneanu, C.A., et al.: Survey on RGB, 3D, thermal, and multimodal approaches for facial expression recognition: history, trends, and affect-related applications. IEEE Trans. Pattern Anal. **38**, 1548–1568 (2016)
8. Ubalde, S., Gómez-Fernández, F., Goussies, N.A., Mejail, M.: Skeleton-based action recognition using citation-kNN on bags of time-stamped pose descriptors. In: ICIP, pp. 3051–3055 (2016)
9. Wang, J., Zucker, J.D.: Solving the multiple-instance problem: a lazy learning approach. In: Proceedings of the 17th International Conference on Machine Learning ICML 2000, pp. 1119–1126 (2000)
10. Sanin, A., Sanderson, C., Harandi, M., Lovell, B.: Spatio-temporal covariance descriptors for action and gesture recognition. In: WACV, pp. 103–110 (2013)
11. Wang, Z., Wang, S., Ji, Q.: Capturing complex spatio-temporal relations among facial muscles for facial expression recognition. In: CVPR, pp. 3422–3429 (2013)
12. Chew, S.W., Lucey, P., Lucey, S., Saragih, J., Cohn, J., Sridharan, S.: Person-independent facial expression detection using constrained local models. In: FG 2011, pp. 915–920 (2011)

Mining Generalized Closed Patterns
from Multi-graph Collections

Niusvel Acosta-Mendoza[1,2]([⊠]) [ID], Andrés Gago-Alonso[1],
Jesús Ariel Carrasco-Ochoa[2], José Francisco Martínez-Trinidad[2],
and José Eladio Medina-Pagola[1]

[1] Advanced Technologies Application Center (CENATAV),
7a # 21406 e/214 and 216, Siboney, Playa, 12200 Havana, Cuba
nacosta@cenatav.co.cu
[2] Instituto Nacional de Astrofísica, Óptica y Electrónica (INAOE),
Luis Enrique Erro No. 1, Sta. María Tonantzintla, 72840 Puebla, Mexico

Abstract. Frequent approximate subgraph (FAS) mining has become an important technique into the data mining. However, FAS miners produce a large number of FASs affecting the computational performance of methods using them. For solving this problem, in the literature, several algorithms for mining only maximal or closed patterns have been proposed. However, there is no algorithm for mining FASs from multi-graph collections. For this reason, in this paper, we introduce an algorithm for mining generalized closed FASs from multi-graph collections. The proposed algorithm obtains more patterns than the maximal ones, but less than the closed one, covering patterns with small frequency differences. In our experiments over two real-world multi-graph collections, we show how our proposal reduces the size of the FAS set.

Keywords: Approximate graph mining
Frequent multi-graph mining · Generalized frequent patterns

1 Introduction

In the last decade, frequent approximate subgraph (FAS) mining has become an important topic [9,11,18]. This topic includes different approaches for mining subgraphs which frequently appear, in an approximate way, into a graph collection. Several algorithms have been developed for FAS mining, based on different approximate graph matching techniques [2,4,11,17,18]. However, although several authors have claimed that for some practical problems the multi-graph representation can model the nature of the data in a better way than the simple-graph representation [1,4,16], only a few algorithms have been designed for mining FASs from multi-graph collections [1,2,4].

On the other hand, in several real-world applications, it is common that a large number of patterns be mined when a FAS miner is applied [3,9,11]. This happens because, into the mining process, several redundant FASs are mined;

© Springer International Publishing AG, part of Springer Nature 2018
M. Mendoza and S. Velastín (Eds.): CIARP 2017, LNCS 10657, pp. 10–18, 2018.
https://doi.org/10.1007/978-3-319-75193-1_2

being infeasible the use of these patterns. For this reason, several researchers have focused on mining only maximal[1] or closed[2] subgraphs from the whole set of FASs [6,8,12,14]. However, to the best of our knowledge, there is no FAS mining algorithm for mining maximal or closed FASs from multi-graph collections.

There are several works about closed patterns, but in our approximate context, when we apply the traditional closed condition over the FASs, usually the whole set of FASs is obtained as closed patterns. Since in the approximate graph mining approach rarely two patterns have the same frequency. For this reason, another solution, originally known as δ-tolerance or generalized closed patterns [5–7,10,15], was proposed allowing a relaxation for the strict traditional closed condition. In this paper, we introduce an algorithm for mining generalized closed frequent approximate subgraphs from multi-graph collections.

The organization of this paper is the following. In Sect. 2, some basic concepts are provided. The proposed algorithm is introduced in Sect. 3. An experimental evaluation of our proposal, over two real-world multi-graph collections, is shown in Sect. 4. Finally, our conclusions and future work directions are discussed in Sect. 5.

2   Basic Concepts

In this paper, we focused on collections of undirected labeled multi-graphs. Thus, the first concepts to be defined are labeled graph, simple-graph and multi-graph.

A *labeled graph* in the domain of all possible labels $L = L_V \cup L_E$, where L_V and L_E are the label sets for vertices and edges respectively, is a 5-tuple, $G = (V_G, E_G, \phi_G, I_G, J_G)$, where V_G is a set of vertices, E_G is a set of edges, $\phi_G : E_G \to V_G^\bullet$ is a function that returns the pair of vertices of V_G which are connected by a given edge, where $V_G^\bullet = \{\{u,v\}|u,v \in V_G\}$, $I_G : V_G \to L_V$ is a labeling function for assigning one label to one vertex in V_G and $J_G : E_G \to L_E$ is a labeling function for assigning one label to one edge in E_G.

Multi-edges are two or more edges between the same pair of vertices (i.e. e and e' are multi-edges if $e \neq e'$ and $\phi_G(e) = \phi_G(e') = \{u,v\}$ such that $u, v \in V_G$, $u \neq v$). A *loop* is an edge connecting a vertex to itself (i.e., when $\phi_G(e) = \{u\}$ since $\phi_G(e) = \{u,v\}$ with $v = u$; in a loop $|\phi_G(e)| = 1$). Then, a *multi-graph* is a graph that can contain more than one edge between a pair of vertices (multi-edges) or edges connecting a vertex to itself (loops), while a *simple-graph* does not contain multi-edges and loops.

Given two multi-graphs G_1 and G_2, a pair of functions (f, g) is an *isomorphism* between these graphs iff $f : V_{G_1} \to V_{G_2}$ and $g : E_{G_1} \to E_{G_2}$ are bijective functions, such that: $\forall u \in V_{G_1}$ where $f(u) \in V_{G_2}$ and $I_{G_1}(u) = I_{G_2}(f(u))$; $\forall e_1 \in E_{G_1}$, where $\phi_{G_1}(e_1) = \{u,v\}$, $e_2 = g(e_1) \in E_{G_2}$, $\phi_{G_2}(e_2) = \{f(u), f(v)\}$

---

[1] A maximal frequent subgraph is a frequent subgraph which is not a subgraph of any other frequent subgraph [12,14].

[2] A closed subgraph is a frequent subgraph that is not a subgraph of another frequent subgraph with the same frequency [8,15].

and $J_{G_1}(e_1) = J_{G_2}(e_2)$; and $\forall e_1 \in E_{G_1}$, where $\phi_{G_1}(e_1) = \{v\}$, $e_2 = g(e_1) \in E_{G_2}$, $\phi_{G_2}(e_2) = \{f(v)\}$ and $J_{G_1}(e_1) = J_{G_2}(e_2)$.

If there is an isomorphism between G_1 and G_2, then we say that G_1 and G_2 are *isomorphic*. Besides, if G_1 is isomorphic to a subgraph of G_2, then there is a *sub-isomorphism* between G_1 and G_2; in this case we say that G_1 is *sub-isomorphic* to G_2.

For approximate graph matching, in this paper, only variations in vertex and edge labels will be allowed but preserving the graph structure. In this scenario, a similarity function that allows performing approximate comparisons between labeled graphs preserving the structure is required.

Given two multi-graphs G_1 and G_2, the *similarity* between G_1 and G_2, preserving the graph structure, is defined as:

$$sim(G_1, G_2) = \begin{cases} \max\limits_{(f,g) \in \Upsilon(G_1,G_2)} \Theta_{(f,g)}(G_1, G_2) & if\ \Upsilon(G_1, G_2) \neq \emptyset \\ 0 & otherwise \end{cases} \quad (1)$$

where $\Upsilon(G_1, G_2)$ is the set of all possible isomorphisms between G_1 and G_2 without taking into account the labels, and $\Theta_{(f,g)}(G_1, G_2)$ is a similarity function for comparing the label information between G_1 and G_2, according to the isomorphism (f, g).

Given three multi-graphs G_1, G_2 and G_3, a similarity function $sim(G_1, G_2)$ preserving the graph structure, and a similarity threshold $\tau \in [0, 1]$, there is an *approximate isomorphism* between G_1 and G_2 if $sim(G_1, G_2) \geq \tau$. Also, if there is an approximate isomorphism between G_1 and G_2, and G_2 is a subgraph of G_3, then there is an *approximate sub-isomorphism* between G_1 and G_3, denoted as $G_1 \subseteq_A G_3$.

Since several correspondences between two multi-graphs could exist, the *maximum inclusion degree* $maxID(G_1, G_2)$ of G_1 in G_2 is defined as the maximum value of similarity at comparing G_1 with all of the subgraphs of G_2.

Given a multi-graph collection $D = \{G_1, \ldots, G_{|D|}\}$, a similarity function $sim(G_1, G_2)$, a similarity threshold τ, a minimum support threshold $minsup$ and a multi-graph G, the approximate support value of G in D is computed as: $appSupp(G, D) = \sum_{G_i \in D, G \subseteq_A G_i} maxID(G, G_i)/|D|$. Therefore, if $appSupp(G, D) \geq minsup$, then G is a *frequent approximate subgraph* (FAS) in D.

As it was mentioned in Sect. 1, we are interested in the generalized closed FASs. There are different ways for defining this kind of patterns [5–7,10,15]. In this paper, we use the δ-closed pattern condition defined in [15] because it is the one used in a graph mining context with good results.

Given a multi-graph collection D, a minimum support threshold $minsup$, a closed threshold ($\delta = [0, 1]$), and a frequent multi-graph G in D, G is a generalized closed FAS iff there is no G' where G is subgraph of G' with $appSup(G', D) \geq max([1 - \delta]appSupp(G, D), minsup)$.

According to [15], when $\delta = 0$ the traditional closed condition is obtained, and when $\delta = 1$ this definition coincides with the maximal condition. Besides, when $0 < \delta < 1$ we can obtain more patterns than the maximal ones, but less than the closed one which cover patterns with small frequency differences.

Given a support threshold, a closed threshold, a similarity function between multi-graphs and a similarity threshold, the problem addressed in this paper consists in proposing an algorithm for mining generalized closed FAS in multi-graph collections.

3   Proposed Algorithm

There are three variants for mining maximal or closed FASs: (1) directly mining only these kinds of FASs from the multi-graph collections, (2) mining all the FASs and extracting only those maximal or closed patterns from all the FASs, and (3) including some restrictions in the mining process of a FAS miner for just storing the maximal or closed FASs from all the traversed FASs. Although by using the first variant the maximal or closed FASs may be mined fastest, each algorithm proposed over this variant can mine only these specific kinds of patterns for each developed algorithm; while with the second variant we can extract different kinds of FASs from the whole set of FASs. However, the second variant includes an additional cost to obtain the final result, while with the third variant we are able to obtain the maximal or closed FASs without changing the mining process of a FAS miner.

In this paper, we follow the third variant by extending the MgVEAM algorithm proposed in [4] for mining generalized closed FASs from multi-graph collections. It is important to highlight that we chose to extend MgVEAM because, this algorithm was designed for working with multi-graph collections and the use of the generalized closed condition does not introduce additional computational complexity in the mining process.

In our proposed algorithm, for computing only the generalized closed FASs, some restrictions were included into the MgVEAM mining process. In Algorithm 1, we show the pseudo-code of our proposed algorithm, in which, from a given multi-graph collection D, all frequent approximate vertices are computed. Next, each frequent vertex is recursively extended, by calling the function *"Search"*; traversing only those multi-graphs $G_i \in D$ containing at least one occurrence of the candidate FAS.

Algorithm 1. $Main(D, \tau, minsup, \delta, F)$

Input: D : A multi-graph collection, τ : Similarity threshold, $minsup$: Support threshold,
    δ : Closed threshold.
Output: F : Generalized closed FAS set.

1 $C \leftarrow$ *The frequent approximate single-vertex graph set in D based on the approximate sub-isomorphism;*
2 $F \leftarrow C$;
3 $NF \leftarrow \emptyset$;
4 **foreach** $T \in C$ **do**
5 | $D_T \leftarrow$ *The subgraphs of D which are approximate sub-isomorphic to T;*
6 | Search($T, D_T, D, \tau, minsup, \delta, F, NF$);

In Algorithm 2, the Search function is detailed. This function, recursively performs the extension of a FAS. The candidate FASs are obtained by calling the

"GenCandidate" function; extending a FAS in all possible extensions (by adding one edge). Then, only those candidates that fulfills the support constraint and which have not been identified in previous steps are extended by a recursive call to the Search function. However, only those patterns that are identified as generalized closed are kept as output FASs (see lines $3 - 5$ of Algorithm 2).

Algorithm 2. $Search(G, D_G, D, \tau, minsup, \delta, F, NF)$

Input: G : A FAS, D_G : Occurrence set of G, D : A multi-graph collection, τ : Similarity threshold, $minsup$: Support threshold, δ : Closed threshold, F : Generalized closed FAS set, NF : A multi-graph set.
Output: F : Generalized closed FAS set.

1 $C \leftarrow$ GenCandidate(G, D_G, D, τ) ;   // following Algorithm 3
2 **foreach** $T \in C$ **do**
3   **if** $appSupp(T, D) \geq minsup$ **then**
4   **if** $appSup(T, D) \geq \lceil 1 - \delta \rceil appSupp(G, D)$ **then**
5              $F \leftarrow F \setminus \{G\}; NF \leftarrow NF \cup \{G\}$;   // based on the generalized closed condition
6   **if** $T \notin F$ *and* $T \notin NF$ **then**
7              $F \leftarrow F \cup \{T\}$;
8              $D_T \leftarrow$ *The subgraphs of D which are approximate sub-isomorphic to T*;
9   Search$(T, D_T, D, \tau, minsup, \delta, F, NF)$;

The aim of the "GenCandidate" function is to compute all candidate extensions of a given frequent graph T. In this candidate generation step, all children of T and their occurrences in the multi-graph collection D'_T are searched. Later, the CAM code for these subgraphs are computed by the "ComputeCAM" function (introduced in [4]) invoked in the line 3 of Algorithm 3. Finally, each pattern G, its corresponding CAM code, and its similarity value are stored as an output candidate in C.

Algorithm 3. $GenCandidate(T, D_T, \tau, C)$

Input: T : A FAS, D_T : Occurrence set of T, τ : Similarity threshold.
Output: C : Candidate graph set.

1 $O \leftarrow \{(G, T_e) | G$ *is a child of T and T_e is an occurrence of G in $G_k \in D_T\}$;
2 **foreach** $(G, T_e) \in O$ **do**
3      $CAM_G =$ ComputeCAM(G, e') ;   // following the proposed in [4]
4      $sim(G, T_e)$ *is inserted into C using CAM_G as identifier*;

Our proposed algorithm examines the same set of candidates as MgVEAM [4]; and the generalized closed condition verification does not introduce additional cost to the mining process. For these reasons, the computational complexity of MgVEAM is kept in our proposal, which is $O(dm(mm! + n!))$, where each multi-graph G_i in the collection D has the same size (i.e. each multi-graph has n vertices and m edges), $d = |D|$, every $G_i \in D$ is completely connected, and all labels for vertices and edges are identical.

4   Experiments

With the purpose of evaluating the performance of our proposals over some real-world multi-graph collections, in this section, we apply our proposal and MgVEAM [4] over two real-world multi-graph collections (*PROT-DB* and *WEB-DB*), which were taken from the IAM graph database repository [13]. The used multi-graph collections are available online[3]. In WEB-DB, the 600 multi-graphs are composed by 9395 vertex labels and 64 edge labels with an average size of 65 vertices and 51 edges, and an average of 43 multi-edges per graph. PROT-DB has 600 multi-graphs, which are composed by 3 vertex labels and 986 edge labels with an average size of 33 vertices and 73 edges, and an average of 69 multi-edges per graph.

Our experiment was carried out on a personal computer with an Intel(R) Core(TM) i5-3317U CPU @ 1.70 GHz with 4 GB of RAM. All the algorithms were implemented in ANSI-C on Microsoft Windows 10.

In our experiment we apply our proposal and MgVEAM [4] over both WEB-DB and PROT-DB multi-graph collections, which were previously described. We use different closed threshold values for our proposal and the same values for the support and similarity thresholds for both our proposal and MgVEAM. Then, with this settings, as we can see in Table 1, both algorithms achieved the same performance, in terms of runtime. In this way, as we expected, the generalized closed condition introduced into MgVEAM does not affect the performance of MgVEAM.

Table 1. Performance, in terms of runtime (seconds), achieved by MgVEAM and our proposal over two multi-graph collection.

(a) WEB-DB				(b) PROT-DB		
Support	**Algorithms**			**Support**	**Algorithms**	
threshold	Our proposal	MgVEAM		**threshold**	Our proposal	MgVEAM
0.01	60	60		0.02	762	762
0.02	52	52		0.03	389	389
0.03	43	43		0.04	229	229
0.04	38	38		0.05	146	146
0.05	31	31		0.06	97	97

On the other hand, we show the pattern set reduction achieved by our proposal for both WEB-DB and PROT-DB multi-graph collections. This reduction is shown in Table 2. This table is composed by two sub-tables: (a) showing the results achieved over WEB-DB collection, and (b) showing the results achieved over PROT-DB collection. The first column of each one shows the support threshold values used for the experiment. The second column shows the number of FASs mined by MgVEAM (i.e. all FASs) and the other six consecutive

---

[3] http://ccc.inaoep.mx/~ariel/MgVEAM/.

columns show the number of generalized closed FASs mined by our proposal with different closed threshold values (δ). It is important to highlight that, when $\delta = 0$ and $\delta = 1$, our proposal mines the traditional closed and maximal FASs, respectively.

Table 2. Number of FASs identified by MgVEAM (denoted by "All FASs") and by our proposal with different closed threshold δ values over two multi-graph collection.

(a) WEB-DB

Support threshold	All FASs	Our proposal					
		$\delta = 0$	$\delta = 0.2$	$\delta = 0.4$	$\delta = 0.6$	$\delta = 0.8$	$\delta = 1$
0.01	2468	1617	1512	1440	1364	1314	1277
0.02	1125	732	681	652	623	610	603
0.03	467	384	361	343	330	326	321
0.04	293	260	244	234	226	224	220
0.05	210	192	185	176	170	168	165

(b) PROT-DB

Support threshold	All FASs	Our proposal					
		$\delta = 0$	$\delta = 0.2$	$\delta = 0.4$	$\delta = 0.6$	$\delta = 0.8$	$\delta = 1$
0.02	109029	108890	107475	104101	98261	94352	94350
0.03	48890	48890	48679	47825	45861	44164	44164
0.04	25363	25363	25271	24919	24003	22868	22868
0.05	13504	13504	13458	13279	12705	11907	11907
0.06	7837	7837	7816	7701	7318	6799	6799

In Table 2(a), it can be seen that the size of the obtained FAS set is reduced by applying our proposal, where we achieve an important FAS reduction of 48 and 13% over WEB-DB and PROT-DB multi-graph collections, respectively. On the other hand, as we can see in Table 2(b), in most cases, our proposal mines all FASs when the traditional closed patterns are searched (i.e. when $\delta = 0$). In this way, we can conclude that, for reducing the size of the FAS set, the traditional closed patterns are less useful than the generalized closed patterns computed by our proposal.

5   Conclusions and Future Work

In this paper, we propose an algorithm for mining generalized closed FASs from multi-graph collections by including the condition of the generalized closed patterns into the mining step of a previously reported algorithm for FAS mining without introducing additional computational costs to this process but reducing the number of mined patterns. The proposed algorithm, to the best of our knowledge, is the first algorithm reported in the literature for mining this kind of pattern over multi-graph collections. In our experiments over real-world multi-graph collections, the performance of our proposal, in terms of runtime and the reduction of the number of FASs, was evaluated. From our results we can conclude that, by using our proposed algorithm, the size of the FAS set is reduced.

As future work, we are going to study the canonical form used by FASs miners for developing algorithmic improvements in the generalized closed pattern mining process.

Acknowledgements. This work was partly supported by the National Council of Science and Technology of Mexico (CONACyT) through the scholarship grant 287045.

References

1. Acosta-Mendoza, N., Carrasco-Ochoa, J.A., Martínez-Trinidad, J.F., Gago-Alonso, A., Medina-Pagola, J.E.: A New method based on graph transformation for FAS mining in multi-graph collections. In: Carrasco-Ochoa, J.A., Martínez-Trinidad, J.F., Sossa-Azuela, J.H., Olvera López, J.A., Famili, F. (eds.) MCPR 2015. LNCS, vol. 9116, pp. 13–22. Springer, Cham (2015). https://doi.org/10.1007/978-3-319-19264-2_2
2. Acosta-Mendoza, N., Gago-Alonso, A., Carrasco-Ochoa, J.A., Martínez-Trinidad, J.F., Medina-Pagola, J.E.: A new algorithm for approximate pattern mining in multi-graph collections. Knowl.-Based Syst. **109**, 198–207 (2016)
3. Acosta-Mendoza, N., Gago-Alonso, A., Carrasco-Ochoa, J.A., Martínez-Trinidad, J.F., Medina-Pagola, J.E.: Improving graph-based image classification by using emerging patterns as attributes. Eng. Appl. Artif. Intell. **50**, 215–225 (2016)
4. Acosta-Mendoza, N., Gago-Alonso, A., Carrasco-Ochoa, J.A., Martínez-Trinidad, J.F., Medina-Pagola, J.E.: Extension of canonical adjacency matrices for frequent approximate subgraph mining on multi-graph collections. Int. J. Pattern Recogn. Artif. Intell. **31**(7), 25 (2017)
5. Boley, M., Horváth, T., Wrobel, S.: Efficient discovery of interesting patterns based on strong closedness. Stat. Anal. Data Min. **2**(5–6), 346–360 (2009)
6. Bringmann, B., Nijssen, S., Zimmermann, A.: Pattern-based classification: a unifying perspective, p. 15. arXiv preprint arXiv:1111.6191 (2011)
7. Cheng, J., Ke, Y., Ng, W.: Effective elimination of redundant association rules. Data Min. Knowl. Discov. **16**(2), 221–249 (2008)
8. Demetrovics, J., Quang, H.M., Anh, N.V., Thi, V.D.: An optimization of closed frequent subgraph mining algorithm. Cybern. Inf. Technol. **17**(1), 1–13 (2017)
9. Emmert-Streib, F., Dehmer, M., Shi, Y.: Fifty years of graph matching, network alignment and network comparison. Inf. Sci. Int. J. **346**(C), 180–197 (2016). https://doi.org/10.1016/j.ins.2016.01.074
10. Gay, D., Selmaoui-Folcher, N., Boulicaut, J.F.: Application-independent feature construction based on almost-closedness properties. Knowl. Inf. Syst. **30**(1), 87–111 (2012)
11. Li, R., Wang, W.: REAFUM: representative approximate frequent subgraph mining. In: SIAM International Conference on Data Mining, Vancouver, BC, Canada, pp. 757–765. SIAM (2015)
12. Liu, M., Gribskov, M.: MMC-Marging: identification of maximum frequent subgraphs by metropolis Monte Carlos sampling. In: IEEE International Conference on Big Data, pp. 849–856. IEEE (2015)
13. Riesen, K., Bunke, H.: IAM graph database repository for graph based pattern recognition and machine learning. Orlando, USA, pp. 208–297 (2008)
14. Shu-Jing, L., Yi-Chung, C., Li-Don, Y., Jungpin, W.: Discovering long maximal frequent pattern. In: 8th International Conference on Advanced Computational Intelligence, Chiang Mai, Thailand, pp. 136–142. IEEE. 14–16 February 2016

15. Takigawa, I., Mamitsuka, H.: Efficiently mining δ-tolerance closed frequent subgraphs. Mach. Learn. **82**(2), 95–121 (2011)
16. Verma, A., Bharadwaj, K.K.: Identifying community structure in a multi-relational network employing non-negative tensor factorization and GA k-means clustering. Wiley Interdisc. Rev. Data Min. Knowl. Discov. **7**(1), 1–22 (2017)
17. Wang, H., Ma, W., Shi, H., Xia, C.: An interval algebra-based modeling and routing method in bus delay tolerant network. KSII Trans. Internet Inf. Syst. **9**(4), 1376–1391 (2015)
18. Wu, D., Ren, J., Sheng, L.: Uncertain maximal frequent subgraph mining algorithm based on adjacency matrix and weight. Int. J. Mach. Learn. Cybern. (2017). https://doi.org/10.1007/s13042-017-0655-y. ISSN 1868-808X

Edge Detection Based on Digital Shape Elongation Measure

Faisal Alamri[1][✉] and Joviša Žunić[2][✉]

[1] Computer Science, University of Exeter, Exeter, UK
fa269@exeter.ac.uk
[2] Mathematical Institute, Serbian Academy of Sciences, Belgrade, Serbia
jovisa_zunic@mi.sanu.ac.rs

Abstract. In this paper, we justify the hypothesis that the methods based on the tools designed to cope with digital images can outperform the standard techniques, usually coming from differential calculus and differential geometry. Herein, we have employed the shape elongation measure, a well known shape based image analysis tool, to offer a solution to the edge detection problem. The shape elongation measure, as used in this paper, is a numerical characteristic of discrete shape, computable for all discrete point sets, including digital images. Such a measure does not involve any of the infinitesimal processes for its computation. The method proposed can be applied to any digital image directly, without the need of any pre-processing.

1 Introduction

In this paper, we propose a new method for detecting edges on digital images. Because of its importance, the edge detecting problem is well studied [1, 4, 6, 7, 11, 13]. Due to the diversity of images and applications, there is no single method that would outperform all the others in all situations. This is why there are several methods in a common use - not a single one (i.e. 'the best one'). Most of the methods are based on theoretical frameworks already established, mostly in differential calculus and geometry areas. These methods accept that there must be an inherent error when dealing with digital images. Simply, infinitesimal processes cannot be performed in discrete spaces. Thus, there are reasons to assume that in some situations, methods developed to deal with discrete data can outperform the existing ones [3,8]. We will follow this concept and design a new method without using any theoretical framework that involves infinitesimal processes (e.g. gradients, curvature, etc.). The performance of the new method will be illustrated by a comparison with the Sobel and Prewitt edge detectors [2]. Those detectors use a 3×3 neighborhood for a judgment whether a certain pixel is a good candidate to be treated as an edge point or not. It is important to notice that the new method does work (without any modification) on an

J. Žunić—Work supported by the Serbian Ministry of Education, Science and Technology.

M. Mendoza and S. Velastín (Eds.): CIARP 2017, LNCS 10657, pp. 19–27, 2018.
https://doi.org/10.1007/978-3-319-75193-1_3

arbitrary pixel neighborhood. The same 3×3 neighborhood has been chosen herein, for a fair comparison, with the edge detectors mentioned.

We provide a collection of experimental illustrations[1]. Examples are selected to be illustrative, rather than to insist on a dominance of the new method.

2   Preliminaries

In a common theoretical framework, an image is given by the image function, $I(x, y)$, defined on a rectangular array of pixels, i.e. $(x, y) \in \{1, 2, \ldots, m\} \times \{1, 2, \ldots, n\}$, while a pixel (x_0, y_0), roughly speaking, is considered to be an *edge pixel* if the image function $I(x, y)$, essentially changes the value in a certain neighborhood of (x_0, y_0). From such an interpretation, it is very natural to employ the standard differential calculus, which is very powerful and theoretically well founded mathematical tool, designed to analyze the behavior of functions in arbitrary dimensions. In this approach, the image function $I(x, y)$ is initially treated as a function defined on a continuous domain, let say $[a, b] \times [c, d]$ with $a, b, c, d \in R$, and the discontinuities of $I(x, y)$, are considered as candidates for the edge points, on the image defined by $I(x, y)$. A very popular method, Sobel edge detector and its variants (e.g. Prewitt and Canny) are based on such an approach. To determine the discontinuities of the image function $I(x, y)$, the derivatives are used. However, the function derivatives assume infinitesimal processes, what brings us to a well recognised problem: *Infinitesimal processes cannot be performed in a discrete space.* Only the approximative solutions can be given, in discrete spaces (by the way, as it has been done in Sobel method). To be more precise, the first derivative $\frac{\partial I(x,y)}{\partial x}$, defined as,

$$\frac{\partial I(x, y)}{\partial x} = \lim_{\Delta x \to 0} \frac{I(x + \Delta x, y) - I(x, y)}{\Delta x} \tag{1}$$

requires the infinitesimal process '$\Delta x \to 0$', which cannot be performed in the digital images domain, because Δx is bounded below with the pixel size, i.e. Δx cannot be smaller than the pixel size. Thus, all the edge detecting methods based on differential calculus have an inherent loss of precision, since the derivatives can be approximated only[2] - not computed exactly. The question is: *How big the loss (made in the computation precision) is?* The experience has shown that, in many situations, the loss is not big and that the related methods are efficient. Of course, there are situations, where the approach lead to different performances of the related methods - in some situations some of the methods perform well, while others show a bed performance.

In this paper, we suggest another approach that can be, at least, compatible with the established methods. Our hypothesis is that when dealing with a discrete space, some of the methods developed to work in discrete spaces may have better performances than the existing methods do. If we go back to the

---

[1] Please always enlarge the images appearing, in order to provide a better viewing.

[2] Common approximation in (1) is: $\frac{\partial I(x,y)}{\partial x} \approx I(x + 1, y) - I(x, y)$, assuming $\Delta x = 1$.

initial intuition, which a good candidate for an edge point should be a point (x_0, y_0) where the image function $I(x, y)$ essentially changes its value (i.e. in a continuous space the point (x_0, y_0) is a discontinuity of $I(x, y)$), then a good candidate to indicate such a discontinuity can be *the shape elongation measure*, applied to a neighborhood of the pixel considered. Notice that there are many other shape descriptors/measures intensively studied in the literature and in a frequent use in image analysis tasks [5,9,10,12].

The rest of the paper includes: The basic definitions and statements about the elongation measure (Subsect. 2.1); The new edge detecting method (Sect. 3); Some experimental results (Sect. 4); Concluding remarks (Sect. 5).

2.1 Shape Elongation Measure

We deal with the digital images defined on a domain $\mathcal{D} = \{1, \ldots, m\} \times \{1, \ldots, n\}$, by the image function $I(x, y)$. A point $(x_0, y_0) \in \mathcal{D}$, is referred to as a pixel, while $I(x_0, y_0)$ represents the the image intensity, at the pixel (x_0, y_0).

The centroid (x_s, y_S) of S is $(x_S, y_S) = \left(\frac{\sum_{(x,y) \in S} x}{\sum_{(x,y) \in S} 1}, \frac{\sum_{(x,y) \in S} y}{\sum_{(x,y) \in S} 1} \right)$ while $\overline{m}_{p,q}(S) = \sum_{(x,y) \in S} (x - x_S)^p \cdot (y - y_S)^q$, are the centralized moments.

The shape orientation of a discrete point set S, is defined by the line that minimizes the total sum of the square distances of the points of S to a certain line. Such a line passes through the centroid of S. The line, which maximizes the sum of the squared distances of all the points of S and passes through the centroid of S, is orthogonal to the axis of the least second moment of inertia. Both these extreme quantities, minimum and maximum, are straightforward to compute and their ratio defines the shape elongation measure. The elongation of S is denoted as $\mathcal{E}(S)$, and computed as follows:

$$\mathcal{E}(S) = \frac{\overline{m}_{2,0}(S) + \overline{m}_{0,2}(S) + \sqrt{(\overline{m}_{2,0}(S) - \overline{m}_{0,2}(S))^2 + 4\overline{m}_{1,1}(S)^2}}{\overline{m}_{2,0}(S) + \overline{m}_{0,2}(S) - \sqrt{(\overline{m}_{2,0}(S) - \overline{m}_{0,2}(S))^2 + 4\overline{m}_{1,1}(S)^2}} \quad (2)$$

The measure $\mathcal{E}(S)$ is invariant to translation, rotation and scaling transformations, and $\mathcal{E}(S) \in [1, \infty)$. Several shape examples are below. The $\mathcal{E}(S)$ values (given below the shapes), are in accordance with our expectation.

$\mathcal{E}(S) = 1.57$ $\mathcal{E}(S) = 5.08$ $\mathcal{E}(S) = 13.32$ $\mathcal{E}(S) = 52.40$ $\mathcal{E}(S) = 129.57$

3   New Method for Detecting Edges

As it has been seen from (2), $\mathcal{E}(S)$ can be computed for any discrete point set S. Our hypothesis is that the $\mathcal{E}(S)$ value, calculated in a certain neighborhood of a pixel, from a grey-level image, can be used as a good indicator whether this pixel belongs to an edge or not. We will use the 3×3 pixel neighborhood, the same as the Sobel method does. Since $\mathcal{E}(S)$ definition allows multiple points, its extension to the grey-level images is straightforward. So, for a given image function, $I(x,y)$, we have the following formulas

$$(x_P, y_P) = \left(\frac{\sum_{(x,y)\in P} x \cdot I(x,y)}{\sum_{(x,y)\in P} I(x,y)}, \frac{\sum_{(x,y)\in P} y \cdot I(x,y)}{\sum_{(x,y)\in P} I(x,y)} \right) \tag{3}$$

$$\overline{m}_{p,q}(P) = \sum_{(x,y)\in P} (x - x_P)^p \cdot (y - y_P)^q \cdot I(x,y), \quad \text{and } p,q \in \{0,1,2\ldots\}. \tag{4}$$

$$\mathcal{E}(P) = \frac{\overline{m}_{2,0}(P) + \overline{m}_{0,2}(P) + \sqrt{(\overline{m}_{2,0}(P) - \overline{m}_{0,2}(P))^2 + 4\overline{m}_{1,1}(P)^2}}{\overline{m}_{2,0}(P) + \overline{m}_{0,2}(P) - \sqrt{(\overline{m}_{2,0}(P) - \overline{m}_{0,2}(P))^2 + 4\overline{m}_{1,1}(P)^2}} \tag{5}$$

for the centroid, centralized moments, and elongation of a grey-level image P, respectively. Notice that we have used the same notations for both black-white (i.e. $\mathcal{E}(S)$) and grey-level (i.e. $\mathcal{E}(P)$) images. This is because there is no conceptual difference between these two situations. Examples in Fig. 1 illustrate how $\mathcal{E}(P)$ acts on grey-level images. The obtained values suggest that the central pixel, in Fig. 1(a) is a good candidate to be an edge pixel, since its 3×3 neighborhood has the largest $\mathcal{E}(P) = 1.46$ value, comparing to all situations displayed in Fig. 1(a)–(e). The computed value $\mathcal{E}(P) = 1.00$, for the central pixel in Fig. 1(e), suggests that this pixel is not a candidate to be an edge pixel. This supports our hypothesis that $\mathcal{E}(P)$ can be efficiently employed in an edge detection task.

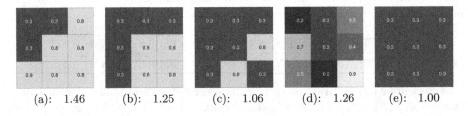

(a):   1.46   (b):   1.25   (c):   1.06   (d):   1.26   (e):   1.00

Fig. 1. $\mathcal{E}(P)$ values are displayed below the 3×3 figures. Normalized pixel intensities $I(x,y) \leq 1$ are inscribed in the corresponded pixel.

In order to present the core idea of the new method, we use the Matlab implementation of Sobel method. The only change made is that the image produced by using the gradient operator, is replaced with the image produced by using the elongation $\mathcal{E}(P)$ operator, both applied to 3×3 pixel neighborhood. Thus, basically both implementations have the following common steps: **Step-1:** Input (original) image; **Step-2:** Image produced by applying the related operator (i.e. the elongation operator, $\mathcal{E}(P)$, for the new method, and the gradient operator for Sobel method); **Step-3:** Thresholding applied to the image produced in step 2.; **Step-4:** Computation of the output image.

Original image	$\mathcal{E}(P)$ values	Elongation	Gradient	Sobel

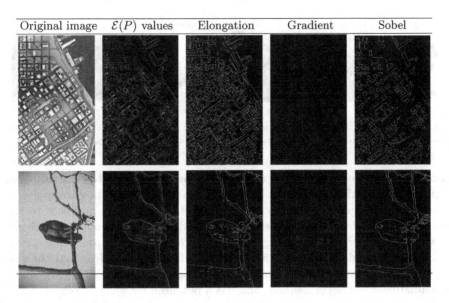

Fig. 2. Original images, images produced by the $\mathcal{E}(P)$ values and the output images, by the new method, are in the 1st, 2nd, and 3rd column, respectively. Images produced by the gradient operator and the images produced by Sobel method, are in the 4th and 5th column, respectively.

Once again, we have used Matlab code, and used $\mathcal{E}(P)$), in step 2. The threshold level (in step 3.) was not changed and adapted to the new method for a better performance achieved. These steps are illustrated in Fig. 2. The images obtained after applied elongation operator and the output images produced by the new method are in the 2nd and 3rd column, respectively. The images produced by the Sobel operator and the output images produced by Sobel method are in the 4th and 5th column, respectively. The images produced by $\mathcal{E}(P)$ and by the Sobel operator have a different structure. This causes different output images produced by these two methods.

4   Experimental Illustrations

In this section, we provide a number of experiments to illustrate the performance of the new method and compare it with the Sobel and Prewitt method.

First Experiment: Black-white images. We start with the black-white images, in Fig. 3. 1st and 4th images are original ones. The edges detected by the new method are in the 2nd and 5th image, while Sobel method produced images are 3rd and 6th ones. The new method performs better in the case of the camel image. In the case of the stick image the new method also performs better in detecting the long straight sections, while Sobel method detects a small spot in the middle of the shape.

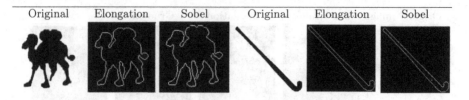

| Original | Elongation | Sobel | Original | Elongation | Sobel |

Fig. 3. 1st and 4th image: original images; 2nd and 5th image: edges detected by the new method; 3rd and 6th image: edges detected by the Sobel method.

Second Experiment: Images with the provided ground-truth images. Figure 4 includes the original images (column on the left) and the corresponding ground truth images (the column on the right). Output images produced by the new method and by Sobel method are in the 2nd and 3rd columns, respectively. At this moment, we are not going to judge which method performs better, even though the ground-truth images are provided. However, it is clear that the new method produces essentially different outputs (see the 2nd row in Fig. 4), which is compatible with Sobel method, and has a potential for applications.

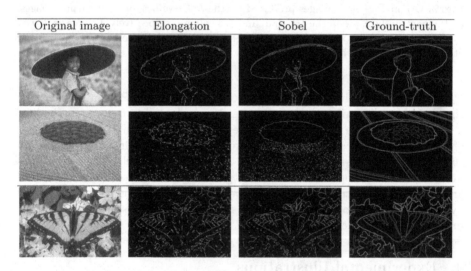

| Original image | Elongation | Sobel | Ground-truth |

Fig. 4. Original images and output images, produced by the new method and by Sobel method, and ground-truth images are in the 1st, 2nd, 3rd, and 4th column, respectively.

Third Experiment: Images without clearly enhanced edges. In this experiment we have used art painting/drawings images[3]. As such, they do not have enhanced edges. The elongation based edge detector, Prewitt, and Sobel edge detectors, have been applied. The original images and the results produced

---

[3] The authors thank Radoslav Skeledžić - Skela, a Serbian painter, who has provided these images.

Original image	Elongation	Prewitt	Sobel

Fig. 5. Art painting images are displayed in the left column. Edges detected by the new method, Prewitt and Sobel methods are in the 2nd, 3rd, and 4th column, respectively.

by the three different edge detectors are displayed in Fig. 5. The results of Prewitt and Sobel edge detectors are slightly similar. The new method is based on a different concept, and this is why its results essentially differ from the results obtained by Prewitt and Sobel edge detectors. Without a judgment which of these three edge detectors performs the best, it could be said that the new edge detector is compatible with the existing ones.

5   Concluding Remarks

We have proposed a new method for detecting edges on digital images. The method uses shape elongation measure and does not use any mathematical tools based on infinitesimal processes. Such processes involve an inherent error, whose influence varies and strongly depends on the input image. The new measure conceptually differs from popular edge detectors. Thus, it is expected to produce very different results. Our aim was to present the core idea of the new method. This is why we have followed Matlab implementation of Sobel method. We have replaced the intermediate image produced by the gradient/Sobel operator, with the image produced by the elongation operator, and have used 3×3 pixel neighborhood (as Sobel method does). Alos, the necessary threshold parameter has been computed by Matlab, directly. It could come out that on this way we have disadvantaged the new method. At this moment, our goal was to show that the new method has a good potential to be used, together with the other edge detectors. We have provided a spectrum of small, but different in their nature, experiments. It is difficult to judge which method dominates the others. However, it is clear that the methods behavior differs from situation to situation. This implies that a use of one of the methods, in a particular situation, does not exclude a use of the other methods in some different situations.

We have not used the available common methods to upgrade our method, again, in order to point out the core idea. Some additional upgrade possibilities are already visible. First of all, a variable pixel neighborhoods and adaptive thresholding. This would be investigated in our further work.

References

1. Förstner, W.: A framework for low level feature extraction. In: Eklundh, J.-O. (ed.) ECCV 1994. LNCS, vol. 801, pp. 383–394. Springer, Heidelberg (1994). https://doi.org/10.1007/BFb0028370
2. Gonzalez, R., Woods, R.: Digital Image Processing. Addison Wesley, Reading (1992)
3. Hahn, J., Tai, X.C., Borok, S., Bruckstein, A.M.: Orientation-matching minimization for image denoising. Int. J. Comput. Vis. **92**, 308–324 (2011)
4. Harris, C., Stephens, M.: A combined corner and edge detector. In: Proceedings of Fourth Alvey Vision Conference, pp. 147–151 (1988)
5. Lachaud, J.-O., Thibert, B.: Properties of Gauss digitized shapes and digital surface integration. J. Math. Imaging Vis. **54**, 162–180 (2016)
6. Köthe, U.: Edge and junction detection with an improved structure tensor. In: Michaelis, B., Krell, G. (eds.) DAGM 2003. LNCS, vol. 2781, pp. 25–32. Springer, Heidelberg (2003). https://doi.org/10.1007/978-3-540-45243-0_4
7. Lyvers, E.P., Mitchell, O.R.: Precision edge contrast and orientation estimation. IEEE Trans. Pattern Anal. Mach. Intell. **10**, 927–937 (1988)
8. Lukić, T., Žunić, J.: A non-gradient-based energy minimization approach to the image denoising problem. Inverse Prob. **30**, 095007 (2014)
9. Rosin, P.L., Žunić, J.: Measuring squareness and orientation of shapes. J. Math. Imaging Vis. **39**, 13–27 (2011)

10. Roussillon, T., Sivignon, I., Tougne, L.: Measure of circularity for parts of digital boundaries and its fast computation. Pattern Recogn. **43**, 37–46 (2010)
11. Schmid, C., Mohr, R., Bauckhage, C.: Evaluation of interest point detectors. Int. J. Comput. Vis. **37**, 151–172 (2000)
12. Stojmenović, M., Nayak, A., Žunić, J.: Measuring linearity of planar point sets. Pattern Recogn. **41**, 2503–20511 (2008)
13. Torre, V., Poggio, T.: On edge detection. IEEE Trans. Pattern Anal. Mach. Intell. **8**, 147–163 (1984)

Mining the Criminal Data of Rio de Janeiro: Analyzing the Impact of the Pacifying Police Units Deployment

Cassio Almeida⬥, Sonia F. Gonzalez⬥, Pedro C. L. Souza,
Simone D. J. Barbosa⬥, and Hélio Lopes(✉)⬥

Pontifícia Universidade Católica do Rio de Janeiro, Rio de Janeiro, RJ, Brazil
{calmeida,sgonzalez,simone,lopes}@inf.puc-rio.br, pedroclsouza@gmail.com

Abstract. Rio de Janeiro is recognized as a violent city, but the pacifying police program was established with the goal to modify this situation. This paper shows that after six years and many pacified communities, the program has had an impact on the decrease of violent deaths with effects both before and after the deployment of the pacifying police units (UPPs). We also evidence a spatial dynamic where the neighboring areas to the UPPs present a more pronounced decrease. Those results raise new issues in the analysis of public safety.

Keywords: Spatial-temporal analysis · Data mining
Public security · Difference-in-differences

1 Introduction

The city of Rio de Janeiro is widely recognized by its touristic attractions, wonderful landscapes, beaches, and vibrant cultural life. In contrast to this image that pleases and attracts everyone, the city appears in the news in episodes of violence, usually related to drug trafficking. Nowadays, with large international events, such as the World Cup in 2014 and the Olympic Games in 2016, there is much talk about the violence and safety issues of sports delegations and tourists. To change this scenario, the state government implemented a pacifying program that creates Pacifying Police Units (UPPs) in communities dominated by criminal factions, expelling them and enabling the entry of public services and other investments [7]. The first UPP was deployed in December 2008. In 2014 there were 37 in the city of Rio de Janeiro, covering over 200 communities and an estimated population of 562,691 inhabitants. The expansion of the program raised several criticisms questioning the effectiveness of the UPPs in reducing criminality. In 2015, the Public Safety Institute (ISP), the organism responsible for the official criminality statistics in the state of Rio de Janeiro, published a report which showed a decrease in criminality within the UPPs [6]. However, it did not examine the areas neighboring areas that received UPPs. A program of such nature may influence criminality in the surrounding areas as well (hereafter called non-UPP areas), and thus deserves a broader evaluation.

In this work we used the open criminality statistics data from ISP (www.ispdados.rj.gov.br) and the demographic data from the census conducted by Instituto Brasileiro de Geografia e Estatística (IBGE) in 2000 and 2010. We needed the IBGE data to estimate the population of each Police Department (DP) area.

The goal of this work is to evaluate the impact of UPPs on the occurrence of violent deaths in the city of Rio de Janeiro. We conducted the analysis in two stages. First, we made a descriptive analysis of yearly series by region. Second, we used *difference-in-differences* (DID) models [4] with the monthly series in the DP areas with UPPs and in the adjacent DP areas.

The remainder of this paper is divided in four sections. In Sect. 2, we describe the models, variables, and modeling strategy adopted. Then, in Sect. 3 we describe and discuss the obtained results. Finally, Sect. 4 summarizes the contributions and points to future work.

2   Empirical Strategy

Public safety policy in Rio de Janeiro can be analyzed for two different periods: before UPPs and after UPPs. To evaluate the effect of the Pacifying Program, we consider the smallest area of data aggregation available, the DP areas. Those areas can be classified into two groups: areas which had received at least one UPP within the study period, and areas which had not. This way, a DP that received one or more UPPs can be analyzed in its period pre- and post-UPP.

The group of DPs that had received UPPs can also be compared, for the same period, to the group of DPs that did not undergo the intervention. Figure 1(b) shows the DP areas that had received UPPs. We focus on the city of Rio de Janeiro because, during the period of study, only DPs from that city received UPP interventions.

In the deployment process of UPPs, the criminal groups were expelled and the areas were kept densely occupied by police forces. Such an occupation reduces several crimes, especially those related to violent deaths. Thus, the variable adopted in this work, named homicides, corresponds to the aggregation of data on three types of crimes related to violent deaths and available for public access: murders, manslaughter, and robbery followed by death. This variable was transformed into rate by 100,000 inhabitants[1], named homicide rate (*homic_r*). We chose this indicator because of its relevance, the loss of lives. It is also less susceptible to subnotifications and, because it is aggregated, it eliminates problems due to subtle differences in classification. Even if the same kind of fact is at one time classified as murder and at another as manslaughter, they will both be included in the homicide aggregated variable, making the differences in the classification irrelevant for this indicator [3].

When planning an experiment, a series of measures are taken, such as randomization and restrictions in the choice of groups that will receive the treatment or not. This allows us to compare statistics calculated for each group to

---

[1] $homic\_r_{it} = \frac{Homicides_{it}}{Population_{it}} * 100,000.$

((a)) Subregions of the
State of Rio de Janeiro

((b)) Subregions of the city of Rio
de Janeiro: areas with and without
UPPs

Fig. 1. Subregions of the state of Rio de Janeiro.

evidence the effects of the treatment. In a simplified way, for each group that receives a treatment, there is another one, similar and independent, that remains the same, without undergoing any treatment. This way, after a period of time, the difference between the two can be attributed to effects of the treatment. In the context of the Pacifying Program, the location of UPPs is not a planned experiment, but a deliberate choice of the designated authorities. Therefore, we need to take some measures to identify suitable control groups.

In this work, the treatment group comprises the DPs that received at least one UPP. The candidate control groups (which comprise DPs that had not received a UPP) were aggregated in four groups and shown in Fig. 1(a), and which are characterized as follows:

– Capital: aggregation of the DPs in the city of Rio de Janeiro that did not receive any UPP;
– Baixada: aggregation of the DPs in the cities of Itaguaí, Seropédica, Paracambi, Japeri, Queimados, Nova Iguaçu, Mesquita, Belford Roxo, São João de Meriti, Nilópolis, Duque de Caxias, and Magé;
– Greater Niterói: aggregation of DPs in the cities of Niterói, São Gonçalo, Itaboraí, Tanguá, Guapimirim, and Maricá;
– Interior: aggregation of DPs in the remaining cities (those that do not belong to the previous groups).

The aggregation of the subregions Capital, Baixada, and Greater Niterói, comprises the Metropolitan Region of Rio de Janeiro.

The basic idea is to compare the treatment group with each control group to identify the effect of the Pacifying Program. In a certain way, the distance between those control groups observe a question of distancing from the areas with UPPs and could be affected in a different way in the case of crime migration.

Considering that the evaluation of the impact of UPPs is not a planned experiment, some steps need to be taken: first, a parallel trends model was used to compare the treatment group with each candidate control group in the period pre-UPP to verify whether the monthly homicide rates of each DP in each group are similar, i.e., whether there is evidence that these series are parallel. If we can state that a series of homicides in a certain candidate control group has a parallel trend to the series of the treatment group in the period of pre-UPP, so there are evidences that the subregions, for this dimension, do not differ in the pattern of occurrences, and therefore that control group can be deemed suitable for further analysis. This supports the idea that the differences in these two groups in the periods pre- and post-UPPs are effects of the UPPs. The second step was the use of panels of difference-in-differences (DID) models to evaluate the homicide rates in the UPP areas and the control groups. For this analysis we used linear models for data in panels [1] implemented in R [5]. The model of parallel trends between treatment and candidate control groups pre-UPP has the Eq. 1.

$$h_{it} = \gamma_i + \sum_{t=1}^{n} \beta_t monthYear_t + \sum_{t=1}^{n} \delta_t(monthYear \cdot UPP_{it}) + e_{it}, \qquad (1)$$

where h_{it} represents the homicide rate in the DP i at time t, γ_i the fixed effect of the DPs, β_t the general effect at time t, $monthYear_t$ a month-year time dummy, δ_t the effect of the interaction between time and the DPs that will receive a UPP in the future, $monthYear \cdot UPP_{it}$ the dummy for time and UPP presence, and e_{it} the random error of DP i at time t. We want to assess the null hypothesis H_0 that $\delta_t = 0$, implying that there is no interaction effect. At this stage time is limited to all months from 2003 to 2008, the period pre-UPP.

To verify the effect of UPPs, we used two variables: UPP_{it}, which is a dummy variable indicating the existence of a UPP in DP i at time t, and $cover_{it}$, which is the population coverage of the UPP. This variable is the ratio between the population in UPPs in DP i at time t and the total population of DP i at time t.[2] To verify the effect of UPPs, we used Eq. 2:

$$h_{it} = \gamma_i + \lambda UPP_{it} + \delta.cover_{it} + \sum_{t=1}^{n} \beta_t.monthYear_t + e_{it}, \qquad (2)$$

where h_{it} represents the dependent variable homicide rate in DP i at time t, γ_i the fixed effect of DPs, β_t the general effect at time t, $monthYear_t$ a month-year time dummy, λ the effect of UPPs, δ the effect of coverage, and e_{it} the random error of DP i at time t. At this stage time refers to the entire period of study, including all months from 2003 to 2014.

To verify possible effects before and after the inauguration of a UPP, we used a *leads-lags* model, represented by Eq. 3:

$$h_{it} = \gamma_i + \sum_{d=0}^{m} \lambda_{\pm d}UPP_{i,t\pm d} + \sum_{d=0}^{m} \delta_{\pm d}cover_{i,t\pm d} + \sum_{t=1}^{n} \beta_t monthYear_t + e_{it} \quad (3)$$

with pre-deployment lags of -12 and -6 months and post-deployment leads of 6, 12 and 24 months. In the model, the sum of the m *negative lags* ($\lambda_{-1}, \ldots, \lambda_{-m}$ and $\delta_{-1}, \ldots, \delta_{-m}$) is the post-treatment effect, and the sum of the q *positive leads* ($\lambda_{+1}, \ldots, \lambda_{+q}$ e $\delta_{+1}, \ldots, \delta_{+q}$) is the pre-treatment effect.

---

[2] The UPP occupies only part of a DP area, hence the need to consider the population coverage of the UPP.

Table 1. Effects of UPPs on the homicides - results for the estimated models.

	CONTROLS													
	Grande Niterol							Interior						
	1	2	3	4	5	6	7	1	2	3	4	5	6	7
Presence of UPP (UPP1)	-0.3917*** (0.1006)		-0.2722** (0.1087)	0.2915 (0.2135)			0.3206 (0.2194)	-1.0635*** (0.1003)		-0.9262*** (0.1088)	0.2066 (0.2119)			0.2436 (0.2168)
UPP Coverage (cob)		-0.3781*** (0.0643)	-0.2809*** (0.0669)	-0.0789 (0.2136)	-0.1577** (0.0689)		-0.1775 (0.2158)		-0.7623*** (0.0797)	-0.3094*** (0.0696)	-0.1261 (0.2203)	-0.1865** (0.0707)		-0.1805 (0.2178)
Presence of UPP -12 months (UPPpre1a)				0.2516 (0.1988)			0.2839 (0.2004)				-0.1895 (0.2012)			-0.153 (0.2033)
Presence of UPP -6 months (UPPpre6m)				-0.7656*** (0.2312)		-0.3557** (0.1119)	-0.7501** (0.2349)				-0.7374** (0.2291)		-0.7994*** (0.1122)	-0.7092** (0.2329)
Presence of UPP +6 months (UPPpos6m)				-0.1205 (0.2002)			-0.0997 (0.2059)				-0.2083 (0.2036)			-0.1891 (0.2094)
Presence of UPP +12 months (UPPpos1a)				-0.2173 (0.1554)		-0.0636 (0.1132)	-0.2163 (0.1519)				-0.2877* (0.157)		-0.3315** (0.1148)	-0.301* (0.1632)
Presence of UPP +24 months (UPPpos2a)				0.0631 (0.1347)			0.162 (0.1437)				-0.1373 (0.137)			-0.0539 (0.1463)
UPP Coverage -12 months (cobpre1a)				-0.107 (0.138)			-0.1904 (0.1405)				-0.4139** (0.1361)			-0.1464 (0.1382)
UPP Coverage -6 months (cobpre6m)				-0.1039 (0.23)			0.0431 (0.225)				-0.0858 (0.2297)			0.0251 (0.2242)
UPP Coverage +6 months (cobpos6m)				0.0253 (0.1671)			0.1222 (0.192)				0.0482 (0.222)			0.1612 (0.1993)
UPP Coverage +12 months (cobpos1a)				-0.0188 (0.1807)			-0.0281 (0.1775)				-0.0236 (0.2067)			-0.0122 (0.1874)
UPP Coverage +24 months (cobpos2a)				-0.4494** (0.2145)		-0.39** (0.1776)	-0.4827*** (0.2149)				-0.7139** (0.2719)		-0.3976** (0.1854)	-0.4092** (0.2189)
Fixed Effect - DP	Yes	Yes	Yes	Yes	Yes	Yes	Yes	Yes	Yes	Yes	Yes	Yes	Yes	Yes
R-Adj	0.1453	0.1455	0.1463	0.147	0.146	0.1478	0.1482	0.0476	0.0443	0.048	0.0492	0.045	0.0495	0.0495
N	4752	4752	4752	4752	4752	4752	4752	10937	10937	10937	10937	10937	10937	10937
p-value (pre-upp)				0.17964							0.21776			

Significance level 10%(*); 5%(**) e 1%(***)

Another relevant hypothesis is that an effect of a DP with UPP may affect the neighboring DPs. To assess this, we used the Eq. 4:

$$h_{it} = \gamma_i + \alpha vupp_{it} + \sum_{t=1}^{n} \beta_t monthYear_t + e_{it},\qquad(4)$$

where h_{it} represents the homicide rate in DP i at time t, γ_i the fixed effect of the DPs, β_t the general effect at time t, $monthYear_t$ a month-year time dummy, α the effect in the non-UPP DP areas adjacent to DP areas that received UPPs, and the variable $vupp_{it}$ a dummy indicating that DP i at time t is adjacent to a DP with UPP at time t.

Based on the aforementioned equations, we developed 7 modeling strategies to access the effects of the variables. The estimated results, i.e. the variables' effects, are shown in Table 1. Models 1, 2 and 3, were developed with Eq. 2. The first model tested only Presence of UPP, the second only the UPP coverage and the third both. Models 4, 5, 6 and 7 were developed with Eq. 3. The model 4 tested only effects of UPP, the 5 only effects of cover, the 6 used only significants variables from model 7, which is the estimated results of the complete model.

3 Results

The UPP program began at the end of 2008 with one deployed UPP, and grew rapidly. The year of the biggest impact was 2012, when the program doubled the number of involved communities.

In Figs. 2(a) and (b), we observe that homicides decreased throughout the state in the period from 2003 and 2011, and after that period they relapsed into increase. We observe that in the period from 2008 to 2012, corresponding to the beginning of the UPP program, there was a great decrease of homicides in all regions. In the following period, from 2012 to 2014, there is a relapse into increase in almost all regions; we highlight Baixada, with the largest increase

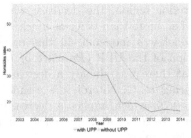
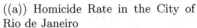

((a)) Homicide Rate in the City of Rio de Janeiro

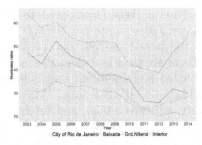

((b)) Homicide Rate in the Subregions of the State of Rio de Janeiro

Fig. 2. Evolution of the homicide rates in the subregions of Rio de Janeiro.

(39.15%). In the UPPs region, the increase was 5.58%, which calls attention for its contrast with the negative rate in the areas without UPPs (Capital) in the same city. As the regions affected are those of DPs with UPP, we can still investigate whether the increase in homicides occurred within the UPPs or outside the UPPs. The UPP balance report published by ISP [6] shows that the UPPs had a decrease of 31% in homicides during the same period. We observe in the DP areas with UPP, in the same period, an increase of 7.72% and, in the areas without UPP, a decrease of 5.02%. The difference between the rates suggests that violence has increased in the UPP margins, i.e., in the DP areas with UPP, but outside the boundaries of the UPPs themselves. This fact shows a change in the criminality dynamic. Considering the conclusions by [2,3], there was a strong positive correlation between the presence of criminal factions and violence in the community surroundings. The results of our work, however, evidence that in the UPP margins the rates relapsed into increase. In the DP areas without UPP, the rates are still decreasing, although less strongly.

In the verification of the impact of UPPs with the DID model, the first step was to evaluate the possibility of using the aforementioned regions as control, and for this we used the parallel trends model described. As can be seen in Table 1, in the last row, the *p-values (pre-UPP)* are the p-values of the F test for the coupled significance of the interactions. In the case of Greater Niterói and Interior, we have evidence that, in the pre-UPP period, the homicide trends in those regions are parallel to the UPP area and we assume that those regions are acceptable as control to the UPP areas.

The results obtained in the adjustments of the fixed effect models for DP, with the two control groups, are in Table 1. In the rows of this table we find the estimated parameters. The columns are divided in two groups, one for each control, G. Niterói and Interior. In those groups, each column refers to a model identified by a number, according to aforementioned.

In general, there is an agreement regarding the significant estimated parameters. For model 1, the estimate of the homicide rate decrease due to the presence of UPP is −0.39 homicides-month per 100,000 people when compared to G. Niteroi. When compared to the Interior region, the decrease was of −1.06 homicides-month per 100,000 people, i.e., more than twice as much.

In the tested models, our preferred model was model 6. In this model, the effects six months before the inauguration of the UPP (*Presence of UPP −6 months*) are significant and indicate that there was already a decrease in homicide rates at that time. When comparing to G. Niterói, the estimated monthly rate was −0.35, and for Interior −0.79, which implies a decrease of 13% and 34%, respectively, when compared to the 2014 rates. This effect can be explained by observing the process of UPP deployment.

In general, the tactical intervention begins one year before the deployment, and within six months the area is stabilized and with strong policing, which promotes the decrease of homicides. In practice, considering that in 2014 the average monthly rate in Interior was 2.34 homicides-month per 100,000 people, a decrease of 34% represents on average 34 homicides a month. In the case of Greater Niterói, the decrease of 12% represents an average decrease of 9 homicides per month.

In the moment of deployment, the coverage (*cover*) has significant effects, with −0.158 and −0.166 for G. Niterói and Interior, respectively. The parameters estimated for Interior represent a decrease in the monthly rate of 0.166 for a coverage of 1 (100%). Put in another way, for each 1% of coverage, the decrease is of 0.00166 homicides-month per 100,000 people. This represents an average decrease of 0.07 homicides each month for each 1% of coverage. If we used the average coverage of 2014, which was 21%, we would have an average decrease of 1.47 homicides each month. When compared to Greater Niterói, also considering the average coverage of 2014, the decrease would be, on average, 0.84 homicides-month per 100,000 people. The effect of coverage is significant and stronger two years after the inauguration of the UPP (UPP coverage +24 months). At this moment the decrease for each percentage of coverage is of 0.14% with respect to G. Niterói and of 0.17% with respect to Interior.

In the models for the group of non-UPP DPs neighboring the DPs with UPP, model 4, we also used the parallel trends model to define the control groups. The best control groups were Baixada and Greater Niterói, *p-values (pre-UPP)* and 0.29 respectively. Using the model 4, the estimated effects were significant (1%), with parameters −1.097 and −0.8315 when compared to Baixada and Niterói, respectively. This represents a decrease of 30% in the rates, or 44 fewer homicides each month, when compared to Baixada.

It calls attention that the decrease effect due to the presence of a UPP, in the UPP areas (see Table 1 model 1), is lower (the estimated parameter is higher) than the same effect in the non-UPP areas neighboring UPPs, both with relation to Greater Niterói. This shows that the neighboring areas present a stronger decrease in homicide. Knowing that approximately 80% of the DPs without UPP in the city of Rio de Janeiro are neighbors to an area with UPP, this result corroborates that areas without UPP in the city of Rio de Janeiro have kept the decrease in yearly homicide rates between 2012 and 2014.

4 Conclusions

The presented results indicate that the DPs which received at least one UPP presented a decrease in homicide rates. The estimated effects are general, obtained by monthly series from 2003 to 2017, pre- and post-UPP periods, and therefore do not show a change in trends of the yearly series (see Table 1), but the estimates are certainly affected.

We also observe that the decrease in homicide is stronger in non-UPP areas neighboring the areas with UPP. The analysis of yearly series showed a new dynamic of homicide occurrences. After a period of decrease throughout the state, the rates increased between 2012 and 2014 in almost all regions. We highlight the city of Rio de Janeiro where homicide rates increased around the UPPs, i.e., in the DP areas that have UPPs, but outside the pacified communities (i.e., in its margins). In the remaining regions of the city the homicides have decreased: in the UPP areas, 30% [6], and in the areas without UPP, 5%. The new strategy of crime behavior can be attributed to the police occupation. With the territory reclaimed and densely policed, the intervention inhibits criminal action and moves the confrontations to the margins of UPPs. The fact that the rates in the remaining regions have increased is also an evidence of a change in criminality, which deserves attention in future work.

References

1. Angrist, J.D., Pischke, J.S.: Mostly Harmless Econometrics: An Empiricist's Companion. Princeton University Press, Princeton (2008)
2. Barcellos, C., Zaluar, A.: Homicídios e disputas territoriais nas favelas do Rio de Janeiro. Revista de Saúde Pública 48(1), 94–102 (2014)
3. Cano, I., et al.: 'Os Donos do Morro': Uma Avaliação Exploratória do Impacto das Unidades de Polícia Pacificadora (UPPs) no Rio de Janeiro. Fórum Brasileiro de Segurança Pública (2012)
4. Card, D., Krueger, A.B.: Minimum wages and employment: a case study of the fast-food industry in New Jersey and Pennsylvania: reply. Am. Econ. Rev. 90(5), 1397–1420 (2000)
5. Croissant, Y., Millo, G.: Panel data econometrics in R: the PLM package. J. Stat. Softw. 27(2) (2008). http://www.jstatsoft.org/v27/i02/
6. ISP: Balanço de Indicadores da Política de Pacificação (2007–2014) (2015). http://arquivos.proderj.rj.gov.br/isp_imagens/uploads/BalancodeIndicadoresdaPol iciadePacificacao2015.pdf
7. UPP: Unidade de Polícia Pacificadora (2015). http://www.upprj.com

Going Deeper on BioImages Classification:
A Plant Leaf Dataset Case Study

Daniel H. A. Alves[1]([✉]) [iD], Luís F. Galonetti[1], Claiton de Oliveira[1],
Pedro H. Bugatti[1] [iD], and Priscila T. M. Saito[1,2] [iD]

[1] Department of Computing, Federal University of Technology - Paraná,
Cornélio Procópio, Brazil
{alvesd,galonetti}@alunos.utfpr.edu.br,
{claitonoliveira,pbugatti,psaito}@utfpr.edu.br
[2] Institute of Computing, University of Campinas, Campinas, Brazil

Abstract. In this paper, we present and evaluate the accuracy of a
Deep Convolutional Neural Network (DCNN) architecture, with other
traditional methods, to solve a bioimage classification problem. The main
contributions of this work are the application of a DCNN architecture
and the further comparison of different types of classification and feature
extraction techniques applied to a plant leaf image dataset. Furthermore,
we go deeper on the analysis of a cross-domain transfer learning approach
using a state-of-the-art deep neural network called Inception-v3. Our
results show that we manage to classify a subset of 53 species of leafs
with a notable mean accuracy of 98.2%.

Keywords: Deep learning · Transfer learning · Pattern recognition
Image classification · Bioimages

1 Introduction

The plant biodiversity study and preservation has always been an important
concern. On that sense finding automatic ways that can precisely identify and
help catalog plant is essential. Fortunately, the advances on computer vision field
through machine learning techniques can make this not only possible but also
accessible for anyone with a mobile device [1].

Machine learning is present in many different areas, from a simple recom-
mendation system [2] to a robust system for detecting fraud in financial transactions [3].

With the increase of computational power and the use of machine learning
tools new experiments have been performed, which contribute to advancing the
state-of-the-art of this research area. As an example is the ImageNet Large Scale
Visual Recognition Challenge (ILSVRC) [4], a computer vision challenge held
each year since 2010, whose data consists of approximately 1.2 million training
images and 1,000 object classes. Since then, new enhancements have been made

© Springer International Publishing AG, part of Springer Nature 2018
M. Mendoza and S. Velastín (Eds.): CIARP 2017, LNCS 10657, pp. 36–44, 2018.
https://doi.org/10.1007/978-3-319-75193-1_5

and even more complex networks were introduced showing the great capabilities of deep learning techniques in the computer vision field.

In this paper, we present and evaluate the accuracy of a Deep Convolutional Neural Network (DCNN) architecture, with other traditional methods, to solve a bioimage classification problem. The main contribution of this work is the comparison of different image classification techniques applied to a plant leaf image dataset. Furthermore, we go deeper on the analysis of a cross-domain transfer learning approach using a state-of-the-art deep neural network called Inception-v3.

In the following section, we will briefly introduce some basic concepts and works about deep learning, deep convolution neural networks and transfer learning. Next, we show our methodology, the dataset used on experiments and the feature extraction techniques that work as inputs on the traditional classification algorithms. We also describe the techniques and experiments that shown a substantial increase in the accuracy when applying a deep learning model on the proposed dataset. Finally, we bring a conclusion with thoughts for future works.

2 Background

2.1 Deep Learning

Deep learning is a sub-area of machine learning which deals with complex neural networks, that is, with a large number of layers and units of neurons. It goes far beyond a Multi-layer Perceptron [5], since it is a set of techniques designed to allow computational models to learn representations of data at levels of abstraction similar to the human brain [6].

In general, deep learning is able to solve image classification problems better than any other parametric or non-parametric method in the literature [6]. Of course, there is a trade-off between computational cost and accuracy: more complex networks tend to take more computing time (but with better results) and the availability of a large dataset (more than $5,000$ samples per class) is shown to be crucial for the positive results [7].

2.2 Deep Convolutional Neural Networks (DCNNs)

Deep Convolutional Neural Networks are the current state-of-the-art method for image classification using deep learning. As in any neural network, they are composed of layers and their respective neurons. Each neuron receives input data and performs a dot product with weights and biases and optionally followed by an activation function. One of the differences between a classical neural network and the DCNNs is the reduction in the number of connections between the neurons of each layer, due to the convolution and polling operations. Hence, the layers are not fully connected, as consequence we have a reduced computational cost.

LeNet-5 [8] was the first successful case in the application of DCNNs to classify images of handwritten characters. It is a simple architecture consisting of only 2 layers of convolution with polling (subsampling) followed by 2 full connected layers.

2.3   Transfer Learning

A typical way to do image classification is to train and test a learning model with the same domain or feature space data. However, this is not always feasible, especially when working with deep neural networks that require a large amount of labeled data to train on. *Transfer Learning* is a technique that tries to address this problem, in other words, we can see it as a reuse of a previous knowledge. It is also known that lower layers of the network can be used as simplistic descriptors (e.g. edge detectors), while higher layers (i.e. deeper) increase the abstraction by concatenating the descriptors from previous layers and are able to identify more complex shapes like a face or a car wheel for example [9].

There are a plenty of strategies for using transfer learning. In [10,11], these strategies were evaluated with deep convolution neural networks evidencing powerful generic image descriptors extracted from this process. Moreover, they pointed that this descriptors should be the primary candidates in a image classification task.

3   Proposed Methodology

Our methodology proposes two approaches for classification of plant leaf images. In the first approach we choose a group of techniques for extracting image features (detailed in the Sect. 3.2). Then, we use the data extracted as input for the classifiers defined in the Sect. 3.3. The second approach uses deep learning techniques for bioimage classification, we detail the architecture and other choices in the Sect. 3.4.

For all experiments, we use the dataset and train/test splits that are described in the Sect. 3.1. Figure 1 shows the proposed methodology pipeline.

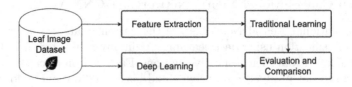

Fig. 1. Pipeline of the proposed methodology

3.1 Dataset Description

The dataset used in our experiments is a subset of Pl@ntLeaves II dataset [12]. This subset consists of 3,655 scan-like images of 53 classes (species), with more than 30 images per class to avoid dealing with an extremely unbalanced dataset. The images were divided into 80% training and 20% testing data. Figure 2 illustrates some image samples of the subset we used and Fig. 3 shows the distribution of images for each class.

Fig. 2. Image samples of the subset of Pl@ntLeaves II dataset [12]

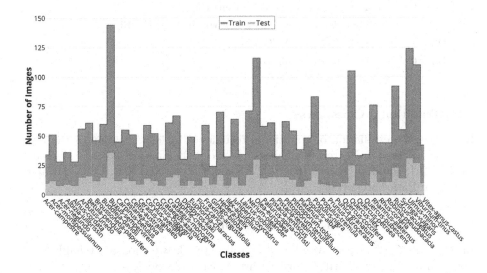

Fig. 3. Distribution of images for each class

3.2 Feature Extraction

In order to work with traditional classification algorithms, first we need to extract information (i.e. features) from images. To do so, we employ 12 popular techniques for feature extraction. Table 1 summarizes all these methods used in the experiments. The first column defines a name for each group of extractors (SD1 stands for the Shape Descriptor 1, TD1 for the Texture Descriptor 1, and so on) that will be analyzed for classification. The second column brings the type of descriptor among: Shape, Texture or Color. The third column contains the names of the techniques used to extract the features, and lastly, the fourth column shows the number of features extracted in each group of techniques. Each set of features is a numerical representation of an image.

Table 1. Techniques applied for feature extraction

Name	Type	Techniques	#Feat.
SD1	Shape	Centroid distance function signature [13]	139
		Chain code histogram [13]	
		Fourier descriptors [13]	
SD2	Shape	Pyramid histogram of oriented gradients (PHOG)	630
TD1	Texture	Quantized compound change histogram (QCCH) [14]	92
		Haralick descriptors [13]	
TD2	Texture	Edge histogram MPEG-7 [15]	80
TD3	Texture	Gabor filter [16]	60
CD1	Color	Border/interior classification (BIC) [17]	192
		Cumulative global color histogram (CGCH) [13]	
CD2	Color	ColorLayout MPEG-7 [18]	33
CD3	Color	ColorCorrelogram [19]	1024

3.3 Traditional Classifiers

Regarding the evaluation of traditional techniques for classification of the leaf dataset, we used k-Nearest Neighbors (k-NN) [20], Support Vector Machine (SVM) [21], Random Forest (RF) [22], Naive Bayes (NB) [23], C4.5 [24] and Optimum Path Forest (OPF) [25].

3.4 Inception-v3 with Transfer Learning

In the context of deep convolutional neural network, we choose a reliable and powerful model to pair with transfer learning called Inception-v3 [26].

We evaluated two strategies of transfer learning. The *complete* transfer learning, that retrains only the final layer of the network, and the *classical*, that uses all the values until the last but one layer. In both approaches, we replaced the output layer, this was necessary because the number of classes in the original pre-trained model (1,000 classes) is different from the number of classes in the leaf dataset (53 classes).

For the hyper-parameters settings, we fixed the maximum number of iterations at 10k. Furthermore, we tested the model with 2 ranges of learning rates, 1E−01 and 1E−02. The training batch size was also evaluated with values 20, 50 and 100. With these tests, we were able to fine-tune our network in order to improve accuracy even more.

Figure 4 illustrates the Inception-v3 model and the layers replaced for the complete and classical transfer learning approaches.

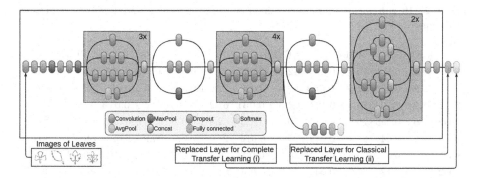

Fig. 4. Inception-v3 model and transfer learning strategies: replaced layer for (i) complete transfer learning and (ii) classical transfer learning

4   Results and Discussion

We started evaluating the traditional classifiers on the bioimage classification task. As presented on previous section, there are 6 classifiers to be tested with 8 groups of features extracted from the images. Table 2 show all the results applying traditional classifiers and feature extraction techniques. Analyzing the results, we can see that the best one was obtained with the OPF algorithm using the features extracted by the ColorCorrelogram (CD3), reaching a mean accuracy of 91.8%. This is surprisingly high, considering we did not apply any feature selection technique. Moreover, it also outperformed the accuracy obtained by a framework proposed by Brilhador [27], where they reached an accuracy of up to 87.5% with SVM on a very similar dataset.

Although 91.8% accuracy is an acceptable result, we went deeper and analyzed the performance with the Inception-v3 architecture. We choose to adopt the Transfer Learning strategy, for two main reasons: first, our dataset is relatively small, as consequence this might either overfit or not converge at all. Second, this model is really big and computational expensive to train from scratch. In fact, these are the main reasons that lead to the use of transfer learning in the literature. Not less important, to work with transfer learning it is required a pre-trained network, and since the ILSVRC 2012 dataset is considered an academic benchmark for computer vision research, this seems to be suitable to our experiments.

Considering our worst and best results with Inception-v3 Transfer Learning, using simple hyper-parameters tweaks, we reached an accuracy gain of up to 1.4%. In comparison with our best accuracy on traditional techniques (i.e. ColorCorrelogram with OPF), we achieved 98.2% of mean accuracy, a 6.4% of improvement. The best settings we found it was executing a classical transfer learning with a learning rate of 0.1 and train batch size of 20. These results can be viewed in Table 3, where we vary different hyper-parameters.

Table 2. 10 run mean accuracy with traditional classifiers and feature extraction techniques

	k-NN [20]	SVM [21]	RF [22]	NB [23]	C4.5 [24]	OPF [25]
SD1	42.6 ± 0.101	30.2 ± 0.023	**62.2 ± 0.073**	27.3 ± 0.063	42.6 ± 0.057	61.2 ± 0.001
SD2	80.9 ± 0.233	**91.1 ± 0.063**	83.9 ± 0.026	72.6 ± 0.028	44.5 ± 0.041	89.2 ± 0.001
TD1	78.9 ± 0.103	75.2 ± 0.123	**83.5 ± 0.031**	61.2 ± 0.065	**67.6 ± 0.013**	74.5 ± 0.001
TD2	67.6 ± 0.093	75.9 ± 0.087	70.6 ± 0.023	62.8 ± 0.038	39.9 ± 0.021	**80.7 ± 0.001**
TD3	22.6 ± 0.044	27.4 ± 0.045	26.0 ± 0.012	21.0 ± 0.077	25.2 ± 0.021	**58.3 ± 0.001**
CD1	75.6 ± 0.013	65.5 ± 0.073	**88.9 ± 0.004**	59.4 ± 0.043	67.6 ± 0.003	73.8 ± 0.001
CD2	65.8 ± 0.123	66.5 ± 0.006	75.7 ± 0.024	65.5 ± 0.013	60.3 ± 0.076	**81.9 ± 0.001**
CD3	86.5 ± 0.103	90.5 ± 0.087	87.6 ± 0.012	60.1 ± 0.065	57.9 ± 0.098	**91.8 ± 0.001**

Table 3. Settings, 10 run mean accuracy and standard deviation on the Inception-v3 with transfer learning

Type	Learning rate	Train batch size	Mean accuracy	Std. deviation
Complete	1.00E−02	20	97.03%	5.00E−04
Complete	1.00E−02	50	96.80%	1.15E−03
Complete	1.00E−02	100	96.88%	1.26E−03
Complete	1.00E−01	20	97.28%	1.26E−03
Complete	1.00E−01	50	97.23%	1.50E−03
Complete	1.00E−01	100	97.28%	1.26E−03
Classical	1.00E−02	20	97.35%	2.89E−03
Classical	1.00E−02	50	97.33%	2.87E−03
Classical	1.00E−02	100	97.30%	1.41E−03
Classical	1.00E−01	20	**98.20%**	1.41E−03
Classical	1.00E−01	50	98.10%	1.15E−03
Classical	1.00E−01	100	98.13%	5.00E−04

5   Conclusion

In this paper, we proposed a methodology based on Deep Convolutional Neural Network (DCNN) architecture, and compared it with other traditional methods, to solve a bioimage classification problem.

The results testify that even with a relatively small dataset we can achieve high accuracy when using deep learning for a classification problem, outperforming the state-of-the art approaches. This is mainly because we use a robust architecture (Inception-v3) that was previously trained on a much larger dataset, a process called transfer learning. Besides, further we show that fine tunning can be made to try to get even higher accuracy. Future work might be try other classifiers on top of different layers of the pre-trained network and augmenting the dataset with synthetic data generated by different deep learning techniques.

Acknowledgment. This research was supported by grants from CNPq (#431668/2016-7 and #422811/2016-5), Fundação Araucária, FAPESP, SETI and UTFPR.

References

1. Joly, A., Bonnet, P., Goëau, H., Barbe, J., Selmi, S., Champ, J., Dufour-Kowalski, S., Affouard, A., Carré, J., Jean-François, M., Boujemaa, N., Barthélémy, D.: A look inside the Pl@ntNet experience. Multimed. Syst. **22**(6), 751–766 (2016)
2. Ricci, F., Rokach, L., Shapira, B.: Introduction to recommender systems handbook. In: Ricci, F., Rokach, L., Shapira, B., Kantor, P. (eds.) Recommender Systems Handbook. Springer, Boston (2011). https://doi.org/10.1007/978-0-387-85820-3_1
3. Phua, C., Lee, V., Smith, K., Gayler, R.: A comprehensive survey of data mining-based fraud detection research. arXiv preprint arXiv:1009.6119 (2010)
4. Russakovsky, O., Deng, J., Su, H., Krause, J., Satheesh, S., Ma, S., Huang, Z., Karpathy, A., Khosla, A., Bernstein, M., Berg, A.C., Fei-Fei, L.: ImageNet large scale visual recognition challenge. IJCV **115**(3), 211–252 (2015)
5. Rosenblatt, F.: The perceptron: a probabilistic model for information storage and organization in the brain. Psychol. Rev. **65**(6), 386 (1958)
6. LeCun, Y., Bengio, Y., Hinton, G.E.: Deep learning. Nature **521**(7553), 436–444 (2015)
7. Goodfellow, I., Bengio, Y., Courville, A.: Deep Learning. MIT Press, Cambridge (2016)
8. LeCun, Y., Bottou, L., Bengio, Y., Haffner, P.: Gradient-based learning applied to document recognition. Proc. IEEE **86**(11), 2278–2324 (1998)
9. Yosinski, J., Clune, J., Bengio, Y., Lipson, H.: How transferable are features in deep neural networks? In: Advances in Neural Information Processing Systems, pp. 3320–3328 (2014)
10. Razavian, A.S., Azizpour, H., Sullivan, J., Carlsson, S.: CNN features off-the-shelf: an astounding baseline for recognition. In: CVPR, pp. 512–519 (2014)
11. Shin, H.C., Roth, H.R., Gao, M., Lu, L., Xu, Z., Nogues, I., Yao, J., Mollura, D., Summers, R.M.: Deep convolutional neural networks for computer-aided detection: CNN architectures, dataset characteristics and transfer learning. IEEE Trans. Med. Imaging **35**(5), 1285–1298 (2016)
12. Goëau, H., Bonnet, P., Joly, A., Yahiaoui, I., Barthélémy, D., Boujemaa, N., Molino, J.: The imageCLEF 2012 plant identification task. In: CLEF 2012: Conference and Labs of the Evaluation Forum, Rome, Italy, September 2012
13. Gonzalez, R.C., Woods, R.E.: Digital Image Processing, 3rd edn. Prentice-Hall Inc., Upper Saddle River (2006)
14. Huang, C.B., Liu, Q.: An orientation independent texture descriptor for image retrieval. In: ICCCAS, pp. 772–776 (2007)
15. Won, C.S., Park, D.K., Park, S.-J.: Efficient use of MPEG-7 edge histogram descriptor. ETRI J. **24**(1), 23–30 (2002)

16. Nixon, M.S., Aguado, A.S.: Feature Extraction & Image Processing for Computer Vision. Academic Press, Cambridge (2012)
17. Stehling, R.O., Nascimento, M.A., Falcão, A.X.: A compact and efficient image retrieval approach based on border/interior pixel classification. In: CIKM, pp. 102–109. ACM (2002)
18. Kasutani, E., Yamada, A.: The MPEG-7 color layout descriptor: a compact image feature description for high-speed image/video segment retrieval. In: ICIP, pp. 674–677. IEEE (2001)
19. Huang, J., Kumar, S.R., Mitra, M., Zhu, W.-J.: Spatial color indexing and applications. In: ICCV, pp. 602–607. IEEE (1998)
20. Cover, T., Hart, P.: Nearest neighbor pattern classification. IEEE Trans. Inf. Theory **13**(1), 21–27 (1967)
21. Hearst, M.A., Dumais, S.T., Osuna, E., Platt, J., Scholkopf, B.: Support vector machines. IEEE Intell. Syst. Appl. **13**(4), 18–28 (1998)
22. Breiman, L.: Random forests. Mach. Learn. **45**(1), 5–32 (2001)
23. Domingos, P., Pazzani, M.: On the optimality of the simple Bayesian classifier under zero-one loss. Mach. Learn. **29**(2), 103–130 (1997)
24. Quinlan, J.R.: C4.5: Programs for Machine Learning. Elsevier, Amsterdam (2014)
25. Papa, J.P., Falcão, A.X., Suzuki, C.T.N.: Supervised pattern classification based on optimum-path forest. Int. J. Imaging Syst. Technol. **19**, 120–131 (2009)
26. Szegedy, C., Vanhoucke, V., Ioffe, S., Shlens, J., Wojna, Z.: Rethinking the inception architecture for computer vision. In: CVPR, pp. 2818–2826 (2016)
27. Brilhador, A., Colonhezi, T.P., Bugatti, P.H., Lopes, F.M.: Combining texture and shape descriptors for bioimages classification: a case of study in imageCLEF dataset. In: CIARP, pp. 431–438 (2013)

Benchmarking Head Pose Estimation in-the-Wild

Elvira Amador[1], Roberto Valle[1], José M. Buenaposada[2], and Luis Baumela[1($\boxtimes$)]

[1] Univ. Politécnica Madrid, Madrid, Spain
{rvalle,lbaumela}@fi.upm.es
[2] Univ. Rey Juan Carlos, Móstoles, Spain
josemiguel.buenaposada@urjc.es

Abstract. Head pose estimation systems have quickly evolved from simple classifiers estimating a few yaw angles, to the most recent regression approaches that provide precise 3D face orientations in images acquired "in-the-wild". Accurate evaluation of these algorithms is an open issue. Although the most recent approaches are tested using a few challenging annotated databases, their published results are not comparable. In this paper we review these works, define a common evaluation methodology, and establish a new state-of-the-art for this problem.

Keywords: Head pose estimation · Convolutional neural networks

1 Introduction

We define head pose as the yaw, pitch and roll angles that determine the orientation of the head in the camera reference system [13]. It has attracted much research due to its relevance as a pre-processing step of many face analysis tasks such as alignment of facial landmarks [2,20] or facial expressions recognition [4]. It is also used in video-surveillance [11] and intrinsically linked with human-computer interaction in social communication [12], gaze [18] and focus of attention [1] estimation.

There are many approaches for image-based head pose estimation. Some of them use very low resolution images [11] or 3D range data [5]. In this paper we only consider methods that use 2D images of average or high resolution. Among these, *manifold embedding* and *non-linear regression* techniques are possibly the most popular ones. The former assume that separated continuous head pose subspaces exist according to appearance [14]. Non-linear regression methods learn a mapping from image features to pose angles. Random Forests [5,19] and Convolutional Neural Networks (CNNs) [6,10,15] are some of the most prevailing.

At present, the best performing approaches are based on CNNs. Yang *et al.* [20] use a small CNN for regression of yaw, pitch and roll angles with 3 convolutional layers, 3 pooling layers and 2 fully connected layers. Ranjan *et al.* [15] fuse intermediate feature layers at different resolutions, and use a multi-task approach to detect faces, estimate facial landmarks, head pose and gender.

© Springer International Publishing AG, part of Springer Nature 2018
M. Mendoza and S. Velastín (Eds.): CIARP 2017, LNCS 10657, pp. 45–52, 2018.
https://doi.org/10.1007/978-3-319-75193-1_6

The H-CNN architecture [10] uses an inception module [17] that pools and concatenate features from intermediate layers and is jointly trained on the visibility, facial landmarks and head pose estimation parameters. In Table 1 we show the performance of these approaches. Although they use the same databases, their results cannot be immediately compared. This will be further discussed in Sect. 2.

In this paper we review the problem of estimating head pose by regressing the yaw, pitch and roll head angles from medium/high resolution images acquired "in-the-wild", i.e. in realistic unrestricted conditions. Our contributions are:

- A brief survey of the best head pose estimation algorithms.
- Definition of an evaluation methodology and publicly available benchmark to precisely compare the performance of head pose estimation algorithms.
- The establishment of the state-of-the-art on this benchmark.

2 Benchmarking Head Pose

There are many public databases with face labeled data. However very few of them provide ground truth head pose, because of the difficulty in accurately estimating these angles. Traditionally, pose estimation algorithms have been evaluated with databases acquired in laboratory conditions and with imprecise angular information [13]. Later, more realistic and accurate data-sets such as AFLW [8] emerged. They have images in challenging real-world situations acquired without any position, illumination or quality restriction.

Here we propose the use of three databases:

- **AFLW** [8]. It contains a collection of 25993 faces acquired in an uncontrolled scenario with head poses ranging between ±120° for yaw and ±90° for pitch and roll angles. It provides a mean face 3D structure and manual annotations for 21 face landmarks. We compute the pose angles from the labeled landmarks using the POSIT algorithm [3] and assuming each face has the 3D structure of the mean face.
  We have found several annotations errors and, consequently, removed these faces from our benchmark. From the remaining faces we randomly choose 21074, 2068 and 1000 instances for training, validation and testing respectively. These images will be available after publication.
- **AFW** [21]. This small database has been traditionally used only for testing purposes. It has 250 images with 468 faces in quite challenging settings. It provides discrete yaw labels ranging from −90° to 90° with 15° intervals, plus the facial bounding box. These labels were manually annotated, hence often they are not very accurate.
- **300W**[1]. It includes 689 challenging faces obtained from the testing subsets of other databases (HELEN, LFPW and IBUG). This is the most popular face alignment benchmark. It provides face bounding boxes and 68 manually annotated landmarks. It does not provide any pose information.

---

[1] https://ibug.doc.ic.ac.uk/resources/300-W/.

We use again AFLW mean 3D face and the POSIT algorithm [3] to estimate the three pose angles for each face instance. This data-set will also be publicly available.

Table 1. Head pose estimation published results. For AFLW and 300 W we show the Mean Absolute Error (MAE) in degrees. For AFW we show the classification success rate.

Method	AFLW (MAE)			AFW	300 W (MAE)		
	Yaw	Pitch	Roll	Yaw	Yaw	Pitch	Roll
Peng et al. [14]	-	-	-	86.3%	-	-	-
Valle et al. [19]	12.26°	-	-	83.54%	-	-	-
Gao et al. [6]	6.60°	5.75°	-	-	-	-	-
Yang et al. [20]	-	-	-	-	4.20°	5.19°	2.42°
Ranjan et al. [15]	7.61°	6.13°	3.92°	97.7%	-	-	-
Kumar et al. [10]	6.45°	5.85°	8.75°	96.67%	-	-	-

In Table 1 we show the published results of the best head pose estimation algorithms. AFLW figures are not comparable among any of the cited works. Some select 1000 test images at random and use the rest for training [10,15]. Valle et al. [19] chose 10% of the images for testing and the rest for training. Gao et al. [6] use 15561 randomly chosen image faces for training and the remaining 7848 for testing. Moreover, none of these AFLW subsets are publicly available, hence it is impossible to make a fair comparison among any of these approaches.

Similarly, the results for AFW are not comparable. Some approaches test on the whole database [15,19]. However, each was trained on a different subset of AFLW. Moreover, Kumar et al. [10] test on the 341 images whose height is larger than 150 pixels. Peng et al. [14] test on a different set of 459 faces.

Finally, the head pose labels for 300 W are not available. Yang [20] computes them from an average face composed of 49 3D points. Unfortunately, this information is not public.

In summary, to have comparable results all algorithms should use the same train, validation and test data-sets. For our benchmark we propose to use a single train and validation data-set composed respectively by 21074 and 2068 face images randomly chosen from AFLW. For testing we have three data-sets: the AFLW test is performed on the remaining 1000 images; when testing with AFW and 300 W we use respectively all 468 and 689 faces from AFW and 300 W test sets.

Note also that our labels may also have small errors caused by the assumption that all faces have the same 3D structure.

3   Experiments

3.1   Methodology

Following the models used by the best published results [6, 10, 15, 20], we use
a distributed face representation extracted from a deep CNN. Training such a
model from scratch requires a large amount of data and computing power. The
usual approach in computer vision is to use a general architecture already trained
on a related problem and fine-tune it for the task at hand (see Fig. 1).

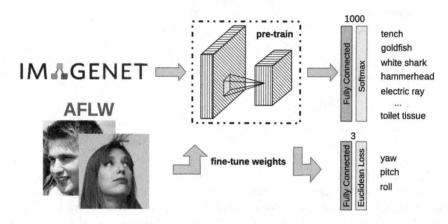

Fig. 1. Transfer learning methodology to fine-tune ImageNet generic weights.

To build our baseline regressors we use AlexNet [9], GoogLeNet [17],
VGG [16] and ResNet [7] trained architectures, top performers in the image
classification task of the ILSVRC competition. AlexNet was also used by
Ranjan *et al.* [15], GoogLeNet by Kumar *et al.* [10], and VGG-Net[2] by Gao
et al. [6]. In each architecture we change the last 1000 units Softmax classification layer with an Euclidean Loss layer with three units for modeling the yaw,
pitch and roll angles.

For fine-tuning and evaluation we use the Caffe framework with a GeForce
GTX 1080 (8 GB) graphics processor. We followed the same procedure for each
model. We use Nesterov Accelerated Gradient Descent (NVG) method, initialize
the learning rate to $\alpha = 10^{-5}$ and reduce it with $\gamma = 0.1$ factor after "step size"
iterations (see Table 2). Momentum was set to $\mu = 0.9$. Table 2 reports the
remaining optimization of parameters for each architecture. We optimize the
GPU memory occupation by setting the batch length and number of iterations
on the basis of the network size. So, large networks use a small batch and larger
number of iterations (see Table 2). The network weights used for tests are those
at the last iteration. They will be publicly available after publication.

---

[2] They used VGG-Face, a VGG-16 architecture trained on the VGG face database.

Table 2. Training parameter values for each architecture.

Model	Image size	Iterations	Weight decay	Step size	Batch
AlexNet [9]	227 × 227	25000	0.0005	10000	24
GoogLeNet [17]	224 × 224	25000	0.005	10000	24
VGG-16 [16]	224 × 224	25000	0.0005	10000	24
VGG-19 [16]	224 × 224	25000	0.005	10000	24
ResNet-50 [7]	224 × 224	63000	0.000005	21000	10
ResNet-101 [7]	224 × 224	126000	0.000005	42000	5
ResNet-152 [7]	224 × 224	252000	0.005	84000	2

It takes 8 h for fine tuning the parameters of the largest net, ResNet-152, and process test images on average at a rate of 4 FPS. In Fig. 2 we show a pair of learning curves for VGG-19 and ResNet-152 architectures. Validation curves are more stable because we always process all test images. However, depending on the batch, the training performance has a larger variance. Vertical dashed red lines mark the number of iterations required to complete an epoch.

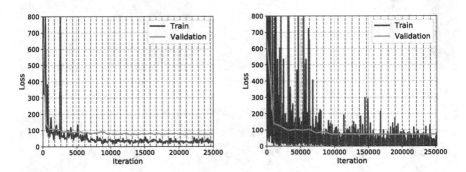

Fig. 2. Sample learning curves for VGG-19 and ResNet-152 architectures.

In Table 3 we present the results of the baseline classifiers for each network architecture. In general, these results confirm that the deeper the representation, the better the performance. This is a well-known fact in the deep learning literature [7].

In AFLW we use the Mean Absolute Error (MAE) of each angle as evaluation metric. Hence, the baseline model using AlexNet achieves better performance than Ranjan *et al.* [15]. Similarly, GoogLeNet results improve those by Kumar *et al.* [10]. For VGG-16, results are only marginally better thank those by Gao *et al.* [6], although our net was trained on the more general ImageNet data-set.

In AFW, since it provides discrete labels, we use as metric the classification success rate. Here, although again the results are also not strictly comparable, the models by Kumar *et al.* [10] and Ranjan *et al.* [15] improve those achieved by our baseline classifiers. This is surprising since in the more precise AFLW

Table 3. Head pose baseline estimation results.

Method	AFLW (MAE)			AFW	300 W (MAE)		
	Yaw	Pitch	Roll	Yaw	Yaw	Pitch	Roll
AlexNet [9]	6.28°	5.02°	3.36°	86.32%	6.86°	6.61°	5.82°
GoogLeNet [17]	6.40°	5.31°	3.74°	**95.51%**	5.71°	7.99°	6.85°
VGG-16 [16]	6.23°	4.96°	3.35°	85.68%	6.35°	7.02°	5.98°
VGG-19 [16]	5.78°	4.79°	3.20°	94.23%	5.56°	6.35°	4.65°
ResNet-50 [7]	6.00°	4.90°	3.14°	94.44%	5.71°	5.91°	3.23°
ResNet-101 [7]	**5.59°**	**4.79°**	**2.83°**	94.44%	**5.13°**	**5.87°**	**3.03°**
ResNet-152 [7]	5.61°	4.79°	3.03°	94.01%	5.52°	6.16°	3.18°

regression case, the result is the opposite. Perhaps in this case the discretization played against our models or, since AFW was manually labeled, the annotation error is higher. Hence, the MAE differences are less significant.

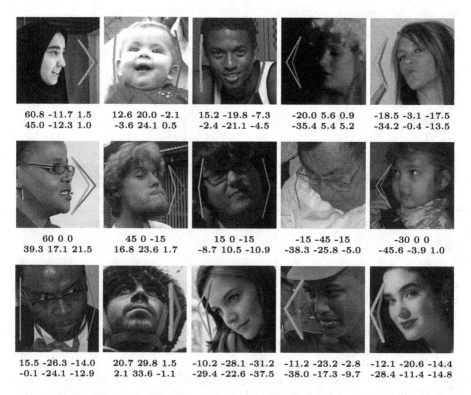

60.8 -11.7 1.5	12.6 20.0 -2.1	15.2 -19.8 -7.3	-20.0 5.6 0.9	-18.5 -3.1 -17.5
45.0 -12.3 1.0	-3.6 24.1 0.5	-2.4 -21.1 -4.5	-35.4 5.4 5.2	-34.2 -0.4 -13.5

60 0 0	45 0 -15	15 0 -15	-15 -45 -15	-30 0 0
39.3 17.1 21.5	16.8 23.6 1.7	-8.7 10.5 -10.9	-38.3 -25.8 -5.0	-45.6 -3.9 1.0

15.5 -26.3 -14.0	20.7 29.8 1.5	-10.2 -28.1 -31.2	-11.2 -23.2 -2.8	-12.1 -20.6 -14.4
-0.1 -24.1 -12.9	2.1 33.6 -1.1	-29.4 -22.6 -37.5	-38.0 -17.3 -9.7	-28.4 -11.4 -14.8

Fig. 3. Representative results with yaw errors greater than 15° for AFLW (top), AFW (middle) and 300W (bottom) databases. Below each image we display the yaw, pitch and roll angle values. Green and blue colors represent respectively estimated and ground truth angles. (Color figure online)

The MAEs of Yang *et al.* [20] in 300 W, although not strictly comparable, are better than those of our baseline classifiers. This may be caused by the fact that they train their CNN on the 300 W training data-set and, perhaps, over-fit to it.

Finally, in Fig. 3 we present some representative face images with head pose estimation errors greater than 15° obtained using ResNet-152 architecture. As can be noticed, sometimes the estimation seems to be more accurate than the annotation. This may be caused by the manual annotation error.

4   Conclusions

We have surveyed the state-of-the-art on face pose estimation "in-the-wild". Although some of the best performing approaches use the same train and test data-sets, their results are not comparable.

In this paper we have defined an evaluation procedure and benchmark data-sets with images captured in unrestricted settings. We have also trained a set of CNN-based classifiers that provide baseline results for our benchmark. The results in Table 3 represent the reproducible state-of-the-art for this problem.

The model based on the deepest network architecture, ResNet, provides the best overall performance. Hence, confirming that deeper representations have better generalization capabilities. When confronted with the best published results in the literature, although not strictly comparable, the ResNet model achieves better performance in the challenging AFLW dataset.

By making publicly available the baseline classifiers and the benchmark data-sets, we expect that future algorithms will be compared on fair grounds.

Acknowledgments. The authors gratefully acknowledge computer resources provided by the Super-computing and Visualization Center of Madrid (CeSViMa) and funding from the Spanish Ministry of Economy and Competitiveness under projects TIN2013-47630-C2-2-R and TIN2016-75982-C2-2-R.

References

1. Ba, S.O., Odobez, J.M.: Multiperson visual focus of attention from head pose and meeting contextual cues. IEEE Trans. Pattern Anal. Mach. Intell. **33**(1), 101–116 (2011)
2. Dantone, M., Gall, J., Fanelli, G., Gool, L.V.: Real-time facial feature detection using conditional regression forests. In: Proceedings Conference on Computer Vision and Pattern Recognition (2012)
3. DeMenthon, D., Davis, L.S.: Model-based object pose in 25 lines of code. Int. J. Comput. Vis. **15**(1–2), 123–141 (1995)
4. Demirkus, M., Precup, D., Clark, J.J., Arbel, T.: Soft biometric trait classification from real-world face videos conditioned on head pose estimation. In: Proceedings of Conference on Computer Vision and Pattern Recognition Workshops (2012)
5. Fanelli, G., Dantone, M., Gall, J., Fossati, A., Van Gool, L.: Random forests for real time 3D face analysis. Int. J. Comput. Vis. **101**(3), 437–458 (2013)

6. Gao, B.B., Xing, C., Xie, C.W., Wu, J., Geng, X.: Deep label distribution learning with label ambiguity. IEEE Trans. Image Process. **26**(6), 2825–2838 (2016)
7. He, K., Zhang, X., Ren, S., Sun, J.: Deep residual learning for image recognition. In: Proceedings of Conference on Computer Vision and Pattern Recognition (2016)
8. Koestinger, M., Wohlhart, P., Roth, P.M., Bischof, H.: Annotated facial landmarks in the wild: a large-scale, real-world database for facial landmark localization. In: IEEE International Workshop on Benchmarking Facial Image Analysis Technologies (2011)
9. Krizhevsky, A., Sutskever, I., Hinton, G.E.: ImageNet classification with deep convolutional neural networks. In: Proceedings of Neural Information Processing Systems (NIPS) (2012)
10. Kumar, A., Alavi, A., Chellappa, R.: KEPLER: keypoint and pose estimation of unconstrained faces by learning efficient H-CNN regressors. In: Proceedings of International Conference on Automatic Face and Gesture Recognition (2017)
11. Lee, D., Yang, M., Oh, S.: Fast and accurate head pose estimation via random projection forests. In: Proceedings of Conference on Computer Vision and Pattern Recognition (2015)
12. Marín-Jiménez, M.J., Zisserman, A., Eichner, M., Ferrari, V.: Detecting people looking at each other in videos. Int. J. Comput. Vis. **106**(3), 282–296 (2014)
13. Murphy-Chutorian, E., Trivedi, M.M.: Head pose estimation in computer vision: a survey. IEEE Trans. Pattern Anal. Mach. Intell. **31**(4), 607–626 (2009)
14. Peng, X., Huang, J., Hu, Q., Zhang, S., Metaxas, D.N.: Three-dimensional head pose estimation in-the-wild. In: Proceedings of International Conference on Automatic Face and Gesture Recognition (2015)
15. Ranjan, R., Patel, V.M., Chellappa, R.: HyperFace: a deep multi-task learning framework for face detection, landmark localization, pose estimation, and gender recognition. CoRR abs/1603.01249 (2016)
16. Simonyan, K., Zisserman, A.: Very deep convolutional networks for large-scale image recognition. CoRR abs/1409.1556 (2014)
17. Szegedy, C., Liu, W., Jia, Y., Sermanet, P., Reed, S.E., Anguelov, D., Erhan, D., Vanhoucke, V., Rabinovich, A.: Going deeper with convolutions. In: Proceedings Conference on Computer Vision and Pattern Recognition (2015)
18. Valenti, R., Sebe, N., Gevers, T.: Combining head pose and eye location information for gaze estimation. IEEE Trans. Image Process. **21**(2), 802–815 (2012)
19. Valle, R., Buenaposada, J.M., Valdés, A., Baumela, L.: Head-pose estimation in-the-wild using a random forest. In: Proceedings of Articulated Motion and Deformable Objects (AMDO) (2016)
20. Yang, H., Mou, W., Zhang, Y., Patras, I., Gunes, H., Robinson, P.: Face alignment assisted by head pose estimation. In: Proceedings of British Machine Vision Conference (2015)
21. Zhu, X., Ramanan, D.: Face detection, pose estimation, and landmark localization in the wild. In: Proceedings of Conference on Computer Vision and Pattern Recognition (2012)

Forensic Document Examination:
Who Is the Writer?

Aline Maria M. M. Amaral[1] ⓘ, Cinthia O. de Almendra Freitas[2],
Flávio Bortolozzi[3] ⓘ, and Yandre Maldonado e Gomes da Costa[1(✉)] ⓘ

[1] Departament of Informatics, State University of Maringá, Maringá, Brazil
amiotto75@gmail.com, yandre@din.uem.br
[2] Polytechnic School, Pontifical Catholic University of Paraná, Curitiba, Brazil
almendracinthia@gmail.com
[3] Departament of Informatics, UniCesumar, Maringá, Brazil
flavio.bortolozzi.53@gmail.com

Abstract. This work presents a baseline system to automatic handwriting identification based only on graphometric features. Initially a set composed of 12 features was presented and its extraction process demonstrated. In order to evaluate the efficiency of these features, a selection process was applied, and a smaller group composed only of 4 features (*GS = Goodness Subset*) present the best writer identification rates. Experiments were conducted in order to evaluate the performance, individually and in group, of the graphometric features; and to identify the number of writers that significantly affect the accuracy of the system. The accuracy of the system applied to 100 different writers taking account the *GS* features set were 84% (TOP1), 96% (TOP5) and 98% (TOP10). These results are comparable to others in the literature on graphometric features. It can be observed that gradually the relation between the number of writers and accuracy is stabilized, and with 200 writers the results are maintained.

Keywords: Graphometry · Feature extraction · Forensic letter

1 Introduction

In dispute cases, questions related to the authenticity of documents presented as evidence can be discussed in Court. The problem becomes greater when dealing with handwriting documents, since the attempts to fraud and forgeries are more easily accessible, because high technology is not necessary to do them. In most cases, a writer and a pen are sufficient to accomplish a fraud or a forgery.

Currently, the forensic handwriting identification is performed by experts using optical (optical device) and/or chemicals methods. Based on Sheikholeslami [1], the manual process of feature extraction and observation is tedious and may leave doubts about the writer identification. In addition, different Forensic Document Examiners (FDE) may extract the same features from a document in a different way. Then, the use of semi-automatic identification systems can be useful and helpful to experts.

According to Sreeraj and Idicula [2], the automatic handwriting-based writer identification is an active research arena. As it is one of the most difficult problems

© Springer International Publishing AG, part of Springer Nature 2018
M. Mendoza and S. Velastín (Eds.): CIARP 2017, LNCS 10657, pp. 53–60, 2018.
https://doi.org/10.1007/978-3-319-75193-1_7

encountered in the field of digital image processing and pattern recognition, the handwriting-based writer identification problem faces with several sub problems such as: designing algorithms to identify handwritings of different individuals; identifying and representing relevant features of the handwriting and evaluating the performance of automatic methods.

Although different approaches have been presented in researches such as [3–7] the principal difference between them is the feature set used to represent the handwriting. In this work, we present a baseline system to automatic handwriting identification based only on graphometric features, i.e., the same principles used by the FDEs during their analysis. Initially a set composed of 12 features was defined and its extraction process developed. To evaluate the efficiency of these features, a selection process was applied, and a smaller group composed only of 4 features present the best writer identification rates (considering the experiments realized).

It is worth mentioning that in a previous work [8] the features selection process was applied in a group composed by 8 features. In this work we increment this group including features to extract information related to the writer loops habits and better results were obtained.

Besides, experiments were conducted to determine the number of writers required to validate the baseline system. All experiments were realized considering TOP1, TOP5 and TOP10 choices. These classifications mean that the baseline system return a group of possible writers (one for TOP1, five for TOP5 and ten for TOP10) of a questioned document, and the correct writer is present in this group.

The paper is divided into the following sections. Section 2 presents the principles of forensic handwriting analysis used to define the proposed baseline system. Section 3 summarizes the baseline system, including the feature set defined and the feature selection process used to obtain the best feature set. Section 4 presents the experimental results and a brief discussion based on results obtained. Finally, Sect. 5 provides some considerations and indicates future investigations.

2 Forensic Handwriting Analysis

This work presents a baseline system to automatic handwriting identification based only on forensic features. Thus, in this section we present a discussion of the forensic handwriting analysis.

2.1 Forensic Principles and Concepts

According to Morris [9], the forensic handwriting identification is part of criminology and it analyses provide a great number of elements that affect a person's writing. This important area also knows the relevance of writing systems and how they influence the writer since his childhood even his graphic maturity writing.

According to Schomaker [10], contrary to biometrics with a purely physical or biophysical basis, the biometric analysis of handwriting requires a very broad knowledge at multiple levels of observation. For the identification of a writer in a large collection of known samples of handwriting, multi-level knowledge must be considered. In forensic practice, many aspects are considered, ranging from the physics of ink deposition [11] to knowledge on the cultural influences in a population [12].

Bensefia [4] point out that each writer can be characterized by his own handwriting, by the reproduction of details and unconscious practices. Handwriting identification is based on the principle that there are individual features that distinguish one person's writing from that of another.

According to Bensefia [4], the **writer identification** task concerns the retrieval of handwritten samples from a database using the handwritten sample under study as a graphical query. It provides a subset of relevant candidate documents, on which complementary analysis will be developed by the expert. Whereas, the **writer verification** task, on its own, must conclude about two samples of handwriting and determines whether they are written by the same writer or not.

2.2 Graphometry and Other Approaches

Based on Sreeraj and Idicula [2], approaches related to the feature extraction for writer identification can be divided into: global (extracted from paragraphs, lines, or just pieces of the text image); and local (extracted from characters and words).

Different approaches for handwriting identification have been presented in the literature. Many of them apply features extracted from the document image, such as texture approaches [13–17] or codebook approaches [6, 18]. These features are not considered graphometric, because they consist of complex computational transformations and procedures on the document image and do not consider the same principles used by the FDEs. The approach presented in this work uses specifically graphometric features as presented in [19–23]. These features are those observed by FDEs during their analyses.

3 Forensic Handwriting Identification Based on Graphometry

In this work, we propose a baseline system for handwriting writer identification. To conduct the experiments which validate the system, we apply documents from 200 different writers from Brazilian Forensic Letter Database [24]. This base, that is text dependent, is composed by three copies of the same letter for each writer.

During the first stage (**training stage**) it is necessary provide the model for each writer randomly selected from forensic database. Two letters from each writer was used in this stage. At second stage (**testing stage**), the baseline system compares a specific writer against the models established in the training stage applying the third letter of each writer. In the next sections we describe the preprocessing, feature extraction, classification and feature selection steps.

3.1 Preprocessing

The preprocessing consists in **five tasks**, that are: *thresholding*, that is the process of converting the 256-gray images in a binary image using the OTSU algorithm;

lines segmentation, this process consists in finding and targeting the lines in the forensic letter; *segmentation the words of each line*, this task realizes the segmentation of the words of each line for further processing it; *contours extraction*, the stroke contours were obtained through the application of morphological filters; and *document image segmentation*, this process consist in spliting the image in 24 segments (6 × 4).

3.2 Feature Extraction

Based on the study of graphometry, the set of feature used for forensic handwriting identification process in current work is: relative placement habits (f_1, f_3, f_4, f_5, f_6) relative relationship between individual words height (f_2 and f_7), axial slant (f_8) and relative loop habits (f_9, f_{10}, f_{11}, f_{12}) as presented in Table 1.

An important feature related to handwriting individuality is relative placement habits [9]. Writers can make a better use of the paper sheet and write to its physical limit.

Another important feature is related to the size of the first word of each handwriting line. When this feature had to be computed, the first word of each line was bounded by a box and its height and proportion of black pixels were computed.

The axial slant is a graphometric feature extensively used in approaches to automatic writer identification. In fact, it represents the general angle of the handwriting and has the best individual performance in the baseline system.

The relative loop habits are a set of graphometric features extracted from words and characters. These features present information about the upward and downward loops of the words (height, width, number of pixels and axial slant).

Table 1. Feature description

Group	Description
f_1 (primitive: 1)	Number of lines in each forensic letter
f_2 (primitives: 2 to 21)	Proportion of black pixels
f_3 (primitives: 22 to 41)	Right margin position
f_4 (primitive: 42)	The lower left margin position
f_5 (primitive: 43)	Upper margin position
f_6 (primitive: 44)	Bottom margin position
f_7 (primitives: 45 to 64)	Height of the first word
f_8 (primitives: 65 to 81)	Axial slant: general angle of the handwriting
f_9 (primitive: 82)	Medium height of the upward and downward loops
f_{10} (primitive: 83)	Medium width of the upward and downward loops
f_{11} (primitive: 84)	Medium size of the upward and downward loops
f_{12} (primitive: 85)	General angle of the upward and downward loops

Figures 1 and 2 presents an overview of the extraction process from a letter image of the Brazilian Forensic Letter Database [24]. The result of the extraction process is a vector containing 85 primitives (as can be observed in Table 1).

This vector is applied to SVM classifier in the training and testing stages. All features were normalized to improve the classification process.

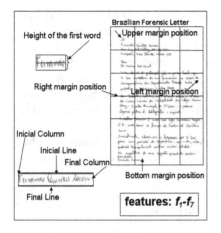

Fig. 1. Feature extraction $f_1 - f_7$ [8]

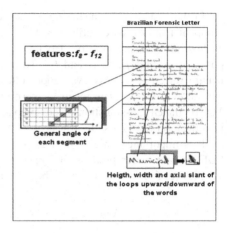

Fig. 2. Feature extraction $f_8 - f_{12}$

3.3   Classification

The classification task consists in submitting the vectors of primitives extracted from the forensic letters to the SVM classifier. We select SVM classifier based on the literature and based on some tests applying other classifiers. In this stage, the questioned document (forensic letter) is confronted with the models generated for each writer (all-against-all), and a confusion matrix is generated as result. This matrix contains the probably of each writer to be the author of the questioned document. These probabilities permit to identifying not only the correct classification (TOP1), but also the five and ten (TOP5 and TOP10 respectively) candidates to be the author of the questioned document.

3.4   Feature Selection

In order to validate our feature set, a feature selection process was applied in the entire set $(f_1, f_2, f_3, f_4, f_5, f_6, f_7, f_8, f_9, f_{10}, f_{11}, f_{12})$ and a group composed only by the features f_1, f_6, f_8 and f_{12} present the best writer identification rates. This selection process was reported by [8], in which, the group of features was composed by the features $(f_1, f_2, f_3, f_4, f_5, f_6, f_7, f_8,)$ and the selected features was composed by $(f_1 \& f_6 \& f_8)$.

According to Dy and Broadley [25], feature selection is a process that selects a subset of original features. A general feature selection process comprises four steps: subset generation, subset evaluation, stopping criterion, and result validation.

4   Experimental Results and Discussion

To validate the baseline system experiments are realized focusing on: analyze the resulting group of the feature selection process and reach a maximum number of writers used in the experiments that significantly affect the accuracy of the system.

In the first experiment the feature selection process was used to achieve the best group of features, as described in Sect. 3.4. By a sequential forward search and an evaluation criterion based on dependency, the goodness set (*GS*) obtained was composed of features f_1, f_6, f_8 and f_{12}. To ensure that the feature set, resulting from the feature selection process, was good, other sets of features empirically defined were evaluated (Table 2). The result of these experiments is also analyzed and TOP5 and TOP10 match classifications were prepared (as showed in Table 3) reaching writer identification rates close to 100%.

It is important to highlight that using TOP5 and TOP10 match classification the FDEs obtain better productivity since they can reduce the number of handwriting samples (to 5 or 10) which must be manually analyzed.

Table 2. Comparison between feature set and *GS*

Features set	Number of writers/accuracy (%)						
	40	80	120	160	200	240	300
f_6 & f_8	90	80	75	69	69	68	68
f_1 & f_6 & f_8	90	84	78	75	69	67	67
f_1 & f_2 & f_3 & f_4 & f_5 & f_6 & f_7 & f_8 & f_9 & f_{10} & f_{11} & f_{12}	60	55	53	50	49	49	48
GS = f_1 & f_6 & f_8 & f_{12}	**93**	**88**	**83**	**76**	**71**	**71**	**70**

Table 3. TOP1, TOP5 and TOP10 match classification for *GS*

GS = f_1 & f_6 & f_8 & f_{12}	Number of writers/accuracy (%)				
	20	40	60	80	100
TOP1	85	93	88	88	84
TOP5	95	95	96	96	98
TOP10	99	100	99	98	96

As mentioned before, another group of experiments was conducted to determine the number of writers which stabilizes the baseline system. To perform this task, writers randomly selected from the Brazilian Forensic Letter Database [24] were added in the group of users experimented in the baseline system, from 40 to 300 writers. The experiments were done with all the features, the best group of features (*GS*) and ensemble of features (Table 2) and the writer identification performance was computed.

It can be observed that gradually the relation between the number of writers and accuracy is stabilized, and with 200 writers the results are maintained. It is important note that applying 200 different writers represents to consider 400 letters in the training stage and 200 letters in the test stage, totaling 600 letters. Furthermore, the writer identification rate with the larger group (200 writers) was of 71%.

Table 4 presents a brief comparison of the results obtained using the baseline system and other present in the literature. Considering the number of writers and the accuracy, our results are very promising, as can be observed, with 160 writers our identification rate is 76% while other work with similar sample size [22] is 58%.

Table 4. Recent studies comparison.

Author	Number of writers	Accuracy
Zois and Anastassopoulos [19]	50	92
Hertel and Bunke [20]	50	90
Schlapbach and Bunke [21]	50	94
Pervouchine and Leedham [22]	165	58
Chen et al. [23]	60	55
$GS = f_1 \& f_6 \& f_8 \& f_{12}$	**160**	**76**

5 Conclusion and Future Works

Current paper discussed the efficiency of a graphometric feature set which can be applied to writer identification. Firstly, we have described the main features based on graphometric principles and research related to them. Thereafter, we presented the baseline system. We have demonstrated, based on experimental results and a feature selection process, that these features achieved promising results for forensic handwriting analysis. Results were improved in TOP1 classification when the *GS* was applied, and results were comparable to others in the literature (Table 4) when graphometric features were considered. Considering TOP5 and TOP10 classifications, the writer identification rates achieved was close to 100%. It is important to detach the productivity gain obtained for forensic handwriting analysis when reducing the number of handwriting samples (to 5 or 10) which must be manually analyzed.

Besides, experiments were conducted to determine the number of writers which stabilizes the baseline system performance, and with 200 different writers no significantly gain or damage was perceived in the results. As future work, new features will be studied and included in the baseline system trying to improve the results and some tests with other classifiers will be prepared.

References

1. Sheikholeslami, G., Srihari, S.N., Govindaraju, V.: Computer aided graphology. In: Proceedings of the 5th International Workshop on Frontiers in Handwriting Recognition, Essex, England, pp. 457–460 (1996)
2. Sreejaj, M., Idicula, S.M.: A survey on writer identification schemas. Int. J. Comput. Appl. **26**(2), 23–33 (2011)
3. Plamondon, R., Lorrete, G.: Automatic signature verification and writer identification: the state of the art. Pattern Recogn. **37**(2), 107–131 (1989)
4. Bensefia, A., Paquet, T., Heutte, A.: A writer identification and verification system. Pattern Recogn. Lett. **26**(13), 2080–2092 (2005)
5. Blankers, V., Niels, R., Vuurpijl, L.: Writer identification by means of explainable features: shapes of loops and lead-in strokes. In: Proceedings of the 19th Belgian-Dutch Conference on Artificial Intelligence, pp. 17–24 (2007)
6. Schomaker, L., Franke, K., Bulacu, M.: Using codebooks of fragmented connected-component contours in forensic and historic writer identification. Pattern Recogn. Lett. **28**, 719–727 (2007)

7. Luna, E.C.H., Riveron, E.M.F., Calderon, S.G.: A supervisoned algorihm with a new differentiated-weighting scheme for identifying the author of a handwritten text. Pattern Recogn. Lett. **32**, 1139–1144 (2011)
8. Amaral, A.M.M.M., Freitas, C.O.A., Bortolozzi, F.: Feature selection for forensic handwriting identification. In: Proceedings of 12th International Conference on Document Analysis and Recognition. Washington: (IAPR), vol. 1, pp. 10–15 (2013)
9. Morris, R.N.: Forensic Handwriting Identification: Fundamental Concepts and Principles. Academic Press, Cambridge (2000)
10. Schomaker, L.: Writer identification and verification. In: Ratha, N.K., Govindaraju, V. (eds.) Advances in Biometrics: Sensor, Algorithms and Systems, pp. 247–264. Springer, London (2008). https://doi.org/10.1007/978-1-84628-921-7_13
11. Franke, K., Rose, S.: Ink-deposition model: the relation of writing and ink deposition processes. In: Proceedings of the Ninth International Workshop on Frontiers in Handwriting Recognition (IWFHR 2004), pp. 173–178. IEEE Computer Society (2004)
12. Schomaker, L., Bulacu, M.: Automatic writer identification using connected-component contours and edge based features of upper-case western script. IEEE Trans. Pattern Anal. Mach. Intell. **26**(6), 787–798 (2004)
13. Said, H.E.S., Tan, T., Baker, K.: Writer identification based on handwritings. Pattern Recogn. **33**(1), 133–148 (2000)
14. Bulacu, M., Schomaker, L. Brink, A.: Text-independent writer identification and verification on offline Arabic handwriting. In: Proceedings of the 9th Conference on Document Analysis and Recognition (ICDAR) (2007)
15. He, Z., You, X., Tang, Y.: Writer identification of Chinese handrwriting documents using hidden Markov tree model. Pattern Recogn. **41**, 1295–1307 (2008)
16. Helli, B., Moghaddam, E.: A text-independent Persian writer identification based on feature relation graph (FRG). Pattern Recogn. **43**, 2199–2209 (2010)
17. Hanusiak, R.K., Oliveira, L.S., Justino, E., Sabourin, R.: Writer verification using texture-based features. Int. J. Doc. Anal. Recogn. **15**, 213–226 (2011)
18. Siddiqi, I., Vincent, N. Combining global and local features for writer identification. In: Proceedings of the 11th International Conference on Frontiers in Handwriting Recognition, pp. 48–53 (2008)
19. Zois, E., Anastassopoulos, V.: Morphological waveform coding for writer identification. Pattern Recogn. **33**(3), 385–398 (2000)
20. Hertel, C. Bunke, H.: A set of novel features for writer identification. In: Proceedings of the 4th International Conference on Audio- and Video-Based Biometric Person Authentication, pp. 679–687 (2003)
21. Schlapbach, A., Bunke, H.: Off-line handwriting identification using HMM based recognizers. In: Proceedings of 17th International Conference of the Pattern Recognition – ICPR 2004, vol. 2 (2004)
22. Pervouchine, V., Leedham, G.: Extraction and analysis of forensic document examiner features used for writer identification. Pattern Recogn. **40**, 1004–1013 (2007)
23. Chen, J., Lopresti, D., Kavallieratou, E.: The impact of ruling lines on writer identification. In: 12th International Conference on Frontiers in Handwriting Recognition (2010)
24. Freitas, C.O.A., Oliveira, L.S., Bortolozzi, F., Sabourin, R.: Brazilian forensic letter database. In: Proceedings of the 11th International Workshop on Frontiers on Handwriting Recognition (2008)
25. Dy, J.G., Broadley, C.E.: Feature subset selection and order identification for unsupervised learning. In: Proceedings of 17th International Conference on Machine Learning, pp. 247–254 (2000)

Traversability Cost Identification of Dynamic Environments Using Recurrent High Order Neural Networks for Robot Navigation

Nancy Arana-Daniel[(✉)], Julio Valdés-López, Alma Y. Alanís,
and Carlos López-Franco

Universidad de Guadalajara, Guadalajara, Mexico
nancyaranad@gmail.com

Abstract. In this paper, it is presented a neural network methodology for learning traversability cost maps to aid autonomous robotic navigation. This proposal is based on the control theory idea for dynamical system identification i.e. we solve the problem of learning and recognizing the pattern which describes the best the behavior of the cost function that represents the environment to obtain traversability cost maps as if we are identifying a dynamical system that is the rough terrain where the robot navigates. Recurrent High Order Neural Networks (RHONN) trained with Extended Kalman Filter (EKF) are used to identify rough terrain traversability costs, and besides the good results in the identification tasks, we get the advantages of using a robust machine learning method such as RHONNs. Our proposal gives the robot the capability to generalize the knowledge learned in previous navigation episodes when it is navigating on similar (but not necessarily equal) environments, so the robot can re-use learned knowledge, generalize it and also it can recognize hidden states. Experimental results show that our proposed approach can identify and learn very complex cost maps, we prove it with artificially generated maps as well as satellite maps of real land.

1 Introduction

Autonomous navigation in rough terrain is an important goal in robotics, in order to achieve it, it is necessary, among other things, a virtual representation of the environment where the robot is going to navigate (a map). This map has to contain information about the traverse cost of the regions to let the robot knows how hard is going to traverse these areas.

This paper presents the results of applying the control theory idea of dynamical system identification to the task of learning traversability cost maps. The rough and dynamic environment is then treated as a dynamical system and the measured data obtained by robot sensors are the descriptors of the environment that help to identify the mathematical model (cost map) of the terrain.

As we have said before, the type of terrain that the RHONN is learning is unstructured or rough terrain that it is found in outdoor environments such as

© Springer International Publishing AG, part of Springer Nature 2018
M. Mendoza and S. Velastín (Eds.): CIARP 2017, LNCS 10657, pp. 61–68, 2018.
https://doi.org/10.1007/978-3-319-75193-1_8

valleys, mountains, rivers, roads etc. The elements of this class of terrain cannot be defined just by traversable and non-traversable but as a continuous interval between those states that measures the difficulty that has a robot to traverse the environment.

There are some related works that deal with unstructured terrain mapping. The LEARCH (LEArning to seaRCH) is a family of imitation learning algorithms [3]. The algorithm tries to find a cost function that makes that a path planner generates the same path as an expert does. The advantage of this is that a priori knowledge of the terrain is not needed to generate the cost map. However, the accuracy of the cost map falls in cells that are far from the experts path. Also, it requires retraining on each new environment, and the training phase is computationally expensive, so it cannot be performed online.

In [2] it is proposed a probabilistic cost map where each cell in the map grid is represented by a distribution over the probable cost of the terrain rather than a single scalar value. This approach allows planning algorithms that use heuristics such as A* to be more efficient in time and memory. However the lack of an automated methodology to assign the values of each grid of the environment makes necessary the assignment of them by a human expert.

The authors in [1] define a terrain traversability gradation method using an online non-supervised learning algorithm to classify images obtained from a stereoscopic camera using the robot's previous experiences. It is a very flexible algorithm since it can adapt to new scenarios by comparing the terrain images with previously found images, although it is inconvenient that as the number of images increased, also the requirements of memory are, and the number of comparisons is computationally expensive.

In [5] a cost map is generated using the visual characteristics of the road. A SVM (Support Vector Machine) is trained to weight the pixels in the image and determine if it is a valid path. This approach starts with a small database made by a human expert and expands it from its experiences with online training. Only it is limited to the immediate local planning of the area to be crossed.

In [4] a robot dog is trained to classify the cost to cross the terrain, where each point of the terrain is modeled by a paraboloid. The dog classifies the points by which it passes as good or bad and the cost of new terrain is obtained using this classification and some parameters of paraboloids. The disadvantage of this approach is that they only use the data that the dog learned as a table, which means that it is not able to correctly classify points that are not in the database, or in their absence, it is required a very large database to be able to sort the points correctly.

Recurrent High Order Neural Network (RHONNs) as described in [6,7] are very good for modeling and identifying dynamic systems that can be both linear and nonlinear. These characteristics are useful since the environment in which a robot moves can change as there are external agents that modify its path. An unstable climate that causes rain and turns streams into flows or an earthquake that generates debris are agents that can transform a previous path that was safe in a practically impassable one. If we take these aspects into account, we can consider that any real navigation environment is a dynamic system.

This paper is organized as follows: In Sect. 2 it is introduced the RHONN's theory trained with EKF, Sect. 3 is devoted to showing the design of our solution to learn traversability cost maps using RHONNs with EKF as well as the experimental results obtained with our approach. Section 4 presents the conclusions of this work.

2   Recurrent High Order Neural Networks

Or RHONNs for short, are a discrete-time nonlinear neural network which have had a wide use in modeling dynamical systems [6,7], the model of the RHONN is described by:

$$x_i(k+1) = w_i^\top z_i(x(k), u(k)), \quad i = 1, \cdots, n, \tag{1}$$

where x_i is the state of the i-th neuron, $u = [u_1, u_2, \cdots, u_m]^\top$ is the input vector to the neural network, w_i are the adjustable weights of the network, and z_i is given by:

$$z_i(x(k), u(k)) = \left[z_{i_1} z_{i_2} \cdots z_{i_{Li}}\right]^\top = \left[\prod_{j \in I_1} \xi_{i_j}^{d_{i_j}(1)} \prod_{j \in I_2} \xi_{i_j}^{d_{i_j}(2)} \cdots \prod_{j \in I_{L_i}} \xi_{i_j}^{d_{i_j}(L_i)}\right]^\top, \tag{2}$$

$\{I_1, I_2, \cdots, I_{L_i}\}$ is a collection of unodered subsets of $\{1, 2, \cdots, m+n\}$, where L_i represents the number of high order connections, $d_{i_j}(.) \in \mathbb{Z}_{>0}$, and ξ_i defined as:

$$\xi_i = \left[\xi_{i_1} \cdots \xi_{i_n} \xi_{i_{n+1}} \cdots \xi_{i_{n+m}}\right]^\top = \left[S(x_1(k)) \cdots S(x_n(k)) u_1(k) \cdots u_m(k)\right]^\top. \tag{3}$$

$S(.)$ is a sigmoid function and is usually represented by:

$$S(\varsigma) = \frac{1}{1 + \exp(-\beta\varsigma)}, \quad \beta > 0, \text{ where } \varsigma \in \mathbb{R}. \tag{4}$$

2.1   The Extended Kalman Filter (EKF) Training Algorithm

There are a number of methods to train neural networks. The one that we will be using is the EKF algorithm [7], it has a remarkable rate of convergence. The algorithm is described by:

$$K_i(k) = P_i(k)H_i(k)[R_i(k) + H_i^\top(k)P_i(k)H_i(k)]^{-1}, \tag{5}$$

$$w_i(k+1) = w_i(k) + \eta_i K_i[y(k) - \hat{y}(k)], \tag{6}$$

$$P_i(k+1) = P_i(k) - K_i(k)H_i^\top(k)P_i(k) + Q_i(k), \tag{7}$$

where $P_i \in \mathbb{R}^{L_i \times L_i}$, $w_i \in \mathbb{R}^{L_i}$ is the weight vector of the network, L_i is the total number of neural network weights, $y \in \mathbb{R}^m$ is the desired output vector, $\hat{y} \in \mathbb{R}^m$ is the network output, η_i is a learning rate parameter, $K_i \in \mathbb{R}^{L_i \times m}$

represents the Kalman gain matrix, $Q_i \in \mathbb{R}^{L_i \times L_i}$, $R_i \in \mathbb{R}^{m \times m}$, and $H_i \in \mathbb{R}^{L_i \times m}$ is defined as:

$$H_{ij}(k) = \left[\frac{\partial \hat{y}(k)}{\partial w_{ij}(k)} \right]_{w_i(k) = \hat{w}_i(k+1)} , \quad i = 1, \cdots, n \text{ and } j = 1, \cdots, L_i. \qquad (8)$$

The matrices P_i, Q_i and R_i can be initialized as a zero matrix of their respective size.

3   Cost Maps Identification Using RHONN's

A map of the environment can be seen as $\mathcal{E} \in \mathbb{R}^{m \times n}$. Each $u_{ij} \in \mathcal{E}$ is also a vector $u \in \mathcal{F}$ where $\mathcal{F} \subseteq \mathbb{R}^l$ is the feature space, l is the number of terrain features that describe the environment. In order for a path planner to generate an optimal path in the environment it requires a transit cost map, thereby to calculate the cost map we need a cost function $C : \mathcal{F} \rightarrow t_c$ with the transit cost $t_c \in \mathbb{R}$, i.e. a function that maps describing features from the environment to traversal cost.

We propose to use RHONN to learn the cost function C due to its qualities of generalization and reuse of knowledge. The dynamical rough terrain in which the agent is going to navigate can be seen as a complex dynamical system to be identified and learned by the RHONN. The huge variety of these environments makes it necessary the use of a learning algorithm which helps to avoid to have to present each variation of the environment to the navigation system to obtain a traverse cost function for each one. So, good generalization quality of RHONNs [7] helps to minimize the issue by letting a smaller set of knowledge handle a wider variety of cases. Furthermore, using a RHONN to identify and learn a traverse cost function allows the navigation agent to re-use knowledge (cost functions) that had been learned in previous navigation episodes if the environment is similar. The above, because the RHONNs are based on Hopfield networks which are well-known by their associative memory capabilities [6].

3.1   Experimental Results

To prove using RHONNs as cost function of rough terrain, we designed the next methodology: Each patch of rough terrain was divided into a grid (as can be seen in Figs. 13 and 16). Each cell of this grid was described using four real values that represent features of the terrain: slope, the quantity of rubble present, the density of vegetation and water depth (grids of Figs. 1, 3, 5, 7, 10 and 13). Using these describing values, a human expert produces a map of traversal cost of the terrain, i.e. a human expert maps the four features of each cell of the grid into a real value that represents the difficulty for a navigation agent to cross given patches of terrain (first row of Figs. 2, 4, 6, 8, 11 and 14). We use these expert-generated maps of traversal cost as training data for the RHONN and then we compare the map of traversal cost learned by the RHONN (second row of Figs. 2,

4, 6, 8, 11 and 14) with the one generated by the human expert to obtain the precision error of our system.

For our experiments we used a RHONN with input $u = \{u_1, u_2, u_3, u_4\} \in \mathbb{R}^4$ where each feature u_i represents slope, rubble, density of vegetation and density of water in that order. Output $\hat{y} = x \in \mathbb{R}$ where x is the single state vector, and $\{I_1, I_2, \cdots, I_{20}\} = \{[0\ 1], [0\ 2], [0\ 3], [0\ 4], [0\ 5], [1\ 1], [1\ 2], [1\ 3], [1\ 4], [1\ 5], [2\ 2], [2\ 3], [2\ 4], [2\ 5], [3\ 3], [3\ 4], [3\ 5], [4\ 4], [4\ 5], [5\ 5]\}$. Figures 1 and 5 are randomly generated maps each used to train a different RHONN. Using as training data the mapping examples from experts in Figs. 2 and 6.

Then to test the generalization capabilities of the RHONNs a simulation of a dynamic environment begins by gradually randomizing up to 30% the value of the elements in each feature vector until 90% of the map is changed. Tables 1 and 2 show the mean quadratic error from the RHONNs.

Figures 4 and 8 show that the RHONN cost approximation capability is very good even if the environment is different from the training set.

Finally we test the RHONN with some real world environments as shown in Figs. 9 and 12.

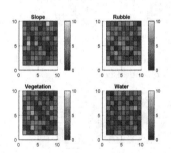

Fig. 1. Randomly generated feature map.

Table 1. Generalization test of Fig. 1.

% of changed cells	Mean $(\text{error})^2$
0%	1.4780
10%	1.3350
20%	1.1071
30%	2.0531
40%	2.1616
50%	2.2692
60%	1.7913
70%	2.4622
80%	3.9386
90%	2.8945

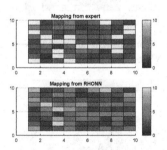

Fig. 2. Comparison of expert's and RHONN's outputs of Fig. 1.

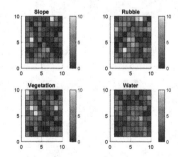

Fig. 3. Generated map from Fig. 1 after 90% changed cells

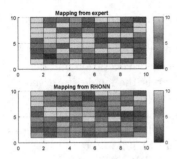

Fig. 4. Comparison of expert's and RHONN's outputs of Fig. 3.

Table 2. Generalization test of Fig. 5

% of changed cells	Mean (error)^2
0%	1.4056
10%	1.6540
20%	2.1847
30%	3.0615
40%	3.6561
50%	4.3490
60%	3.5884
70%	3.6276
80%	2.2350
90%	7.3610

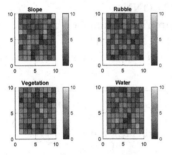

Fig. 5. Randomly generated feature map.

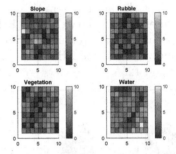

Fig. 7. Generated map from Fig. 5 after 90% changed cells.

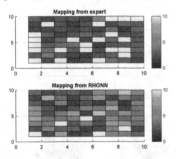

Fig. 6. Comparison of expert's and RHONN's outputs of Fig. 5.

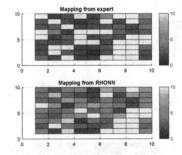

Fig. 8. Comparison of expert's and RHONN's outputs of Fig. 7.

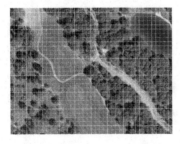

Fig. 9. Satellite map of a grove.

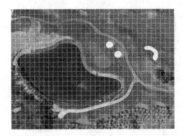

Fig. 12. A pretty view from a golf field.

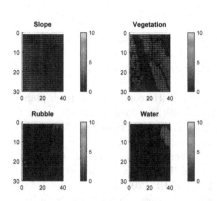

Fig. 10. Feature map of Fig. 9.

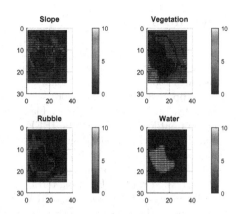

Fig. 13. Feature map of Fig. 12.

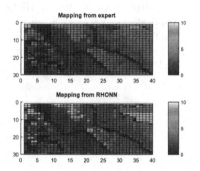

Fig. 11. Comparison of expert's and RHONN outputs of Fig. 9.

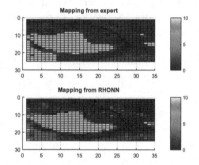

Fig. 14. Comparison of expert's and RHONN outputs of Fig. 12.

4   Conclusion

The tests applied to the RHONNs show very encouraging results. The tests show that the learning capability of the RHONNs is very good to recognize the model of the dynamic environment in which a robot could navigate even when these maps are made from an expert's criterion without additional complexity from the RHONN's model.

Furthermore, it can be seen from the experimental results that the RHONNs learn the cost function of the environment even when sudden changes have happened, i.e. the RHONNs gave back the correct traverse cost for a patch of terrain that belongs to a previously learned environment even when the descriptors of this patch have changed because the terrain has suffered a sudden change (as unforeseen landslide or flood by an example). These good results can be attributed to the capabilities of the RHONN for identifying and modeling complex dynamical systems. For future work, the identification system using the RHONN will be implemented to learn the navigation environment of a robot in real time.

References

1. Kim, D., Sun, J., Oh, S.M., Rehg, J.M., Bobick, A.F.: Traversability classification using unsupervised on-line visual learning for outdoor robot navigation. In: Proceedings 2006 IEEE International Conference on Robotics and Automation, ICRA 2006, pp. 518–525. IEEE (2006)
2. Murphy, L., Newman, P.: Risky planning on probabilistic costmaps for path planning in outdoor environments. IEEE Trans. Robot. **29**(2), 445–457 (2013)
3. Ratliff, N.D., Silver, D., Bagnell, J.A.: Learning to search: functional gradient techniques for imitation learning. Auton. Robots **27**(1), 25–53 (2009)
4. Rebula, J., Hill, G., Bonnlander, B., Johnson, M., Neuhaus, P., Perez, C., Carff, J., Howell, W., Pratt, J.: Learning terrain cost maps. In: IEEE International Conference on Robotics and Automation, ICRA 2008, p. 2217. IEEE (2008)
5. Roncancio, H., Becker, M., Broggi, A., Cattani, S.: Traversability analysis using terrain mapping and online-trained terrain type classifier (2014)
6. Rovithakis, G.A., Christodoulou, M.A.: Adaptive Control with Recurrent High-Order Neural Networks: Theory and Industrial Applications. Springer Science & Business Media, Berlin (2012). https://doi.org/10.1007/978-1-4471-0785-9
7. Sanchez, E.N., Alanis, A.Y., Loukianov, A.G.: Discrete-Time High Order Neural Control. Springer, Berlin (2008). https://doi.org/10.1007/978-3-540-78289-6

Texture Classification of Phases of Ti-6Al-4V Titanium Alloy Using Fractal Descriptors

André R. Backes[1] and Jarbas Joaci de Mesquita Sá Junior[2(✉)]

[1] Faculdade de Computação, Universidade Federal de Uberlândia,
Av. João Naves de Ávila, 2121, Uberlândia, MG 38408-100, Brazil
arbackes@yahoo.com.br
[2] Curso de Engenharia de Computação Campus de Sobral,
Universidade Federal do Ceará, Rua Estanislau Frota, S/N, Centro, Sobral,
Ceará 62010-560, Brazil
jarbas_joaci@yahoo.com.br

Abstract. Traditionally, the evaluation of metal microstructures and their physical properties is a subject of study in Metallography. Through microscopy, we obtain images of the microstructures of the material evaluated, while a human expert performs its analysis. However, texture is an important image descriptor as it is directly related to the physical properties of the surface of the object. Thus, in this paper, we propose to use texture analysis methods to automatically classify metal microstructures, more specifically, the phases of a Titanium alloy, Ti-6Al-4V. We performed texture analysis using the Bouligand-Minkowski fractal dimension method, which enables us to describe a texture image in terms of its irregularity. Experiments were performed using 3900 texture samples of 2 different phases of the titanium alloy. We used LDA (Linear Discriminant Analysis) to evaluate computed texture descriptors. The results indicated that fractal dimension is a feasibility tool for the evaluation of the microstructures present in the metal samples.

Keywords: Ti-6Al-4V · Texture · Complexity analysis
Automated visual inspection · Titanium

1 Introduction

Texture analysis is an area of intense research as it plays a major role in many disciplines. It is also a major field of development for several computer vision applications, ranging from computer assisted diagnosis, defect detection, to food inspection and remote sensing [1,2].

Texture is certainly one of the most studied image attributes. Nevertheless, it lacks a consensus regarding a formal definition. It is usually described as a repetition, exactly or with small variation, of a small pattern over a region, where the characteristics of this small pattern are directly connected to the physical properties of the surface of the object. Consequently, texture analysis enable us to

© Springer International Publishing AG, part of Springer Nature 2018
M. Mendoza and S. Velastín (Eds.): CIARP 2017, LNCS 10657, pp. 69–76, 2018.
https://doi.org/10.1007/978-3-319-75193-1_9

distinguish regions with different reflectance characteristics, and thus, different colors in a specific combination of bands [2,3].

The literature describes a great variety of methods for texture characterization, and fractal dimension is an important one. Basically, fractal dimension measures how complex an object is. For a texture image, this complexity is measured in terms of organization of pixels, i.e., fractal dimension enable us to quantify the texture image in terms of homogeneity. This allows comparison with other texture patterns [3–5].

In this paper, our goal is to use fractal dimension based descriptors to correctly discriminate samples of an inspected metal: the phases of a Titanium alloy, Ti-6Al-4V. Visual inspection and quality control is a common task of the manufacturing process. Traditionally, the evaluation of metal microstructures and their physical properties is a subject of study in Metallography. Through microscopy (optical and electron microscopes can be used), it is possible to acquire images of the microstructures of the material evaluated, while a human expert performs its analysis. Due to the human component in this process, this task is susceptible to errors. In this sense, computer vision techniques are an effective approach to minimize errors and improve our discrimination ability [6–9].

Our paper is organized as follows: Sect. 2 describes the titanium alloy properties and its microstructure. Section 3.1 presents the Bouligand-Minkowski fractal dimension method, which is used to extract texture features from the images of titanium alloy's microstructure. We perform an experiment with 3900 texture samples of 2 different phases of the titanium alloy to evaluate our proposed descriptors (Sect. 4). The results (Sect. 5) indicate that fractal dimension is a feasibility tool for the discrimination of different phases of the titanium alloy as it surpasses all the compared approaches, while Sect. 6 presents some remarks about this work.

2   Titanium Alloy Properties

In the past decades, titanium alloys have been extensively used in several applications, such as seawater, marine and industrial chemical service. This is mostly due to its high strength characteristics and excellent corrosion resistance. Titanium alloys rival directly stainless steels, nickel-based alloys and composites. When analyzing the microstructure of titanium alloys, it is possible to differentiate three different types, called phases: Alpha, Beta, and Alpha+Beta. These phases depend on the manufacturing process used, which also results in different mechanical and physical properties. These phases can also be described as a function of thermo-mechanical evolution of the alloy [6,10]. Figure 1 shows an example of each phase microstructure.

To obtain a specific phase, alloying elements (called stabilizers) must be added to the titanium. These stabilizers produce physical-chemical effects, thus resulting in the creation of an alloy with the desired microstructure. In this work, we propose to study the Ti-6Al-4V, the most commonly used titanium alloy as it is significantly stronger than commercially pure titanium while having similar thermal properties. This alloy is composed of 6% aluminum, 4% vanadium, 0.25% (maximum) iron, 0.2% (maximum) oxygen, and the remainder is titanium. At room temperature, this alloy presents 91% of Alpha phase and 9% of Beta phase in its microstructure [6, 10].

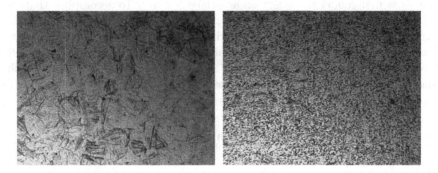

Fig. 1. Metallography of Alpha (light) and Beta (dark) microstructure. Left: Alpha+ Beta; Right: Alpha and Beta (images provided by the authors of the paper [6]). (Color figure online)

3 Texture Analysis for Texture Classification

3.1 Bouligand-Minkowski Fractal Dimension

Complexity is an important tool for texture analysis and characterization. Although its importance in many areas and its use in many image based applications [11–14], it is a term that lacks a formal definition in the literature. Complexity is commonly understood as a measure of object irregularity. For shapes, this irregularity is presented as small variations in the contour of the object; for image textures, the irregularity is characterized by variations of the gray levels in a specific region [3, 15].

There exist many different approaches to estimate the complexity of an object in the literature. Among these approaches, a successful one is the fractal dimension, D. Introduced in the 1970s by Benoit Mandelbrot, fractals are objects which present great complexity although they are generated from iteration of simple rules. Fractal geometry enables us to model many natural phenomena and surfaces (e.g. coastlines, brick, skin, rocks etc.). An interesting property of fractal objects is their non-integer dimension (non-Euclidean geometry) called fractal dimension. It is a property related to the complexity and space occupation of the fractal [4, 5].

A texture image is not a fractal object. Real fractals present self-similarity at infinite scales. Images, on the other hand, present a limited self-similarity. Nevertheless, fractal dimension can be used to characterize these objects. Many methods have been proposed over the years to compute the fractal dimension of shapes and textures. One of the most accurate methods to estimate the fractal dimension is the Bouligand-Minkowski method [3,4]. Initially used for shape analysis and later expanded to include texture analysis, this method uses the influence area of the object to compute its irregularity. This approach makes this method very sensitive to even small structural changes of the object.

To compute the Bouligand-Minkowski fractal dimension from a texture image I, we must first model the image as a surface $S \in R^3$. To accomplish that, we consider each element $s \in S$, $s = (y, x, z)$, as defined by the pixel coordinates (y and x) in the image texture I and the intensity $I(x, y) = z$ at this point. Next, we perform the dilation of the surface S using a sphere of radius r [3,4,16]. Through dilation of each point $s \in S$ by a radius r, we are able to compute the surface's influence volume, $V(r)$. Next, we estimate the fractal dimension D as the (continuous) plot of $\log(V(r))$ versus $\log(r)$:

$$D = 3 - \lim_{r \to 0} \frac{\log V(r)}{\log(r)} \tag{1}$$

with

$$V(r) = \left\{ p \in R^3 | \exists p' \in S : |p - p'| \le r \right\}. \tag{2}$$

Notice that the estimation of the fractal dimension D involves the calculus of a limit. Nevertheless, this limit can be estimated as the slope a of the line regression of the log-log curve $\log(r) \times \log(V(r))$:

$$D = 3 - a. \tag{3}$$

3.2   Error-Based Fractal Descriptors

According to Mandelbrot [5], fractal dimension is a property of fractal objects and it is related to the self-similarity of the fractal at infinite scales. Unlike fractals objects, real world objects (such images) have limited resolution and finite size. In this case, self-similarity is not present at infinite scales, but only at some few scales. As a result, the complexity measured for an image changes its value according to the visualization scale. This effect is clearly noticed when analyzing the log-log curve computed by the Bouligand-Minkowski method. Small variations in the image produce a large set of variations in the curve, resulting in a richness of details that cannot be represented by a single fractal dimension value as computed using the line regression approach, which is the common estimation process for classical fractal dimension.

In order to overcome this problem, we propose to use a set of descriptors which explores the curve details instead of a single fractal dimension value, so that we are able to perform a more accurate texture characterization, as proposed in [3]. We start by computing the log-log curve using the Bouligand-Minkowski method for a dilation radius r. Next, we compute the line regression $y = ax + b$ that best fits the log-log curve, where x and y are, respectively, the log of the radius r and the influence volume $V(r)$, a is the slope and b gives the y-intercept of the line. According to Sect. 3.1, the fractal dimension of the image is estimated as $D = 3 - a$.

Due to the richness of details in the log-log curve, the calculated line is a rough approximation of its true behavior. There is a small error between the points of log-log curve and the value predicted by the line regression. This error is due to the small variation in the log-log curve, a result of the aspect of the image. Variations in pixel position and intensity depend on image context and influence the resulting log-log curve, so that it cannot be described using a simple line regression. Because of this, we propose to use this difference between line regression and the computed curve as an image descriptor. This simple difference, e_i, is defined as

$$e_i = a \log r_i + b - \log V(r_i), \tag{4}$$

where $r_i \in [1, r]$ is a radius value which exists in the log-log curve and r is the dilation radius used to compute the log-log curve. To properly characterize a texture pattern, we select n equidistant radius values from the curve, thus composing a feature vector $\psi(n)$:

$$\psi(n) = [e_1, e_2, \ldots, e_n]. \tag{5}$$

4   Experiments

For the experiments, we built a dataset of texture samples of "pure" Alpha and Beta and "pure" Alpha+Beta phase. For this, we used 15 images of size 1079×816 pixels size for each material. Then, we divided each image into blocks of 80×80 pixels size. As the image size is not a multiple of 80, residual pixels are discarded. We chose this block size as it is "the nearest value to the size of a microstructure in an Alpha+Beta phase" [6]. This process resulted in a dataset composed of 3900 texture samples, 1950 representing the "pure" Alpha and Beta and 1950 the "pure" Alpha+Beta phase. It is important to emphasize that we discarded all color information of the images and considered only its luminance.

To improve our analysis, we also compared our error-based fractal descriptors with traditional texture analysis methods. They are: *Fourier descriptors* [17], *Co-occurrence matrices* [18], *Wavelet descriptors* [19], *Tourist walk* [20] and *Lacunarity* [21,22]. For classification, we used Linear Discriminant Analysis (LDA)[23] with the *leave-one-out cross-validation* scheme.

5   Results and Discussion

Before applying the proposed fractal approach to texture analysis, some considerations are necessary. We must define the maximum dilation radius, r, used by Bouligand-Minkowski method and the number of error based descriptors selected, n.

The dilation radius r is directly related to the computational cost and discrimination ability of the method. The Bouligand-Minkowski method requires a 3D matrix in order to compute the influence volume of a texture pattern [3]. For a $N \times N$ pixels image with L gray levels, the size of this 3D matrix will be $(N+2r) \times (N+2r) \times (L+2r+1)$. Thus, large values of r will increase the computational cost. On the other hand, small values of r result in a log-log curve with few points, thus limiting the discrimination ability of the method. During the experiments, we have decided to use $r = 10$. This radius value achieves a good balance between computational cost and discrimination ability of the method.

Next, we evaluated the number of descriptors n extracted from the log-log curve using the error based approach. Figure 2 shows the success rate achieved for different values of n. We notice an increase in the success rate as the number of selected descriptors increases. The method yields the best performance when using $n = 78$, which results in a success rate of 95.21%.

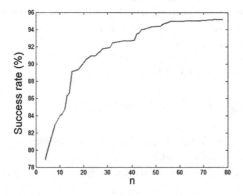

Fig. 2. Classification accuracy observed for different values of n.

Table 1. Results yielded by the proposed approach and traditional texture analysis methods.

Methods	No of descriptors	Success rate (%)
Fourier descriptors	39	92.67
Tourist walk	48	86.00
Co-occurrence matrix	16	86.26
Wavelet descriptors	18	88.90
Lacunarity	19	90.54
Proposed approach	78	95.21

After defining the parameters radius, r, and number of descriptors, n, we performed a comparison experiment with other traditional texture discrimination methods. Table 1 shows the results obtained by each method evaluated. For each compared approach, we used the configuration based on the authors' paper or the common use in the literature. The results show that our approach surpasses all the other methods in discriminating the Titanium alloy phases, i.e., our approach provides a more precise characterization of the Titanium alloy phases. Besides the ability to discriminate different phases of the Titanium alloy with considerable quality, our approach holds potential for many applications in Metallography, where physical properties are expressed in different metal microstructures with different reflectance properties.

6   Conclusion

In this paper, we proposed the use of complexity analysis methods to automatically classify metal microstructures. We investigated the potential of fractal dimension to discriminate the three different phases of a Titanium alloy, Ti-6Al-4V: Alpha, Beta, and Alpha+Beta. We used an error based approach to compute fractal descriptors from the Bouligand-Minkowski fractal dimension method. This approach enables us to explore the details in the microstructure's influence volume, thus resulting in a more precise characterization of the sample. The results indicated that our proposed approach is capable of discriminating the different phases of the Titanium alloy with considerable quality, as it surpasses traditional texture analysis methods in a comparison experiment. In this sense, our approach holds great potential for applications involving the evaluation of metal microstructures and their physical properties as it is able to distinguish regions with different reflectance characteristics.

Acknowledgments. André R. Backes gratefully acknowledges the financial support of CNPq (National Council for Scientific and Technological Development, Brazil) (Grant #302416/2015-3), FAPEMIG (Foundation to the Support of Research in Minas Gerais) (Grant #APQ-03437-15) and PROPP-UFU. Jarbas Joaci de Mesquita Sá Junior acknowledges the financial support of CNPq (Grant 453835/2017-1). We also thank the authors of the paper [6] for kindly providing the titanium alloy images used in this paper.

References

1. Xie, X.H.: A review of recent advances in surface defect detection using texture analysis techniques. Electron. Lett. Comput. Vis. Image Anal. **7**(3), 1–22 (2008)
2. Fernández, A., Álvarez, M.X., Bianconi, F.: Texture description through histograms of equivalent patterns. J. Math. Imaging Vis. **45**(1), 76–102 (2013)
3. Backes, A.R., Casanova, D., Bruno, O.M.: Color texture analysis based on fractal descriptors. Pattern Recogn. **45**(5), 1984–1992 (2012)
4. Tricot, C.: Curves and Fractal Dimension. Springer, Heidelberg (1995)
5. Mandelbrot, B.: The Fractal Geometry of Nature. Freeman & Co., New York (2000)
6. Ducato, A., Fratini, L., La Cascia, M., Mazzola, G.: An automated visual inspection system for the classification of the phases of Ti-6Al-4V titanium alloy. In: Wilson, R., Hancock, E., Bors, A., Smith, W. (eds.) CAIP 2013. LNCS, vol. 8048, pp. 362–369. Springer, Heidelberg (2013). https://doi.org/10.1007/978-3-642-40246-3_45

7. Malamas, E.N., Petrakis, E.G.M., Zervakis, M., Petit, L., Legat, J.D.: A survey on industrial vision systems applications and tools image and vision computing. Image Vis. Comput. **21**, 171–188 (2003)
8. Topalova, I., Mihailov, A., Tzokev, A.: Automated classification of heat resistant steel structures based on neural networks. In: IEEE 25th Convention of Electrical and Electronics Engineers in Israel 2008, IEEEI 2008, pp. 437–440 (2008)
9. Wejrzanowski, T., Spychalski, W.L., Rózniatowski, K., Kurzydlowski, K.J.: Image based analysis of complex microstructures of engineering materials. Appl. Math. Comput. Sci. **18**(1), 33–39 (2008)
10. Bruschi, S., Poggio, S., Quadrini, F., Tata, M.: Workability of Ti-6Al-4V alloy at high temperatures and strain rates. Mater. Lett. **58**(2728), 3622–3629 (2004)
11. Bruno, O.M., de Oliveira Plotze, R., Falvo, M., de Castro, M.: Fractal dimension applied to plant identification. Inf. Sci. **178**, 2722–2733 (2008)
12. Plotze, R.O., Falvo, M., Pádua, J.G., Bernacci, L.C., Vieira, M.L.C., Oliveira, G.C.X., Bruno, O.M.: Leaf shape analysis using the multiscale minkowski fractal dimension, a new morphometric method: a study with passiflora (passifloraceae). Can. J. Bot. **83**(3), 287–301 (2005)
13. Backes, A.R., Eler, D.M., Minghim, R., Bruno, O.M.: Characterizing 3D shapes using fractal dimension. In: Bloch, I., Cesar, R.M. (eds.) CIARP 2010. LNCS, vol. 6419, pp. 14–21. Springer, Heidelberg (2010). https://doi.org/10.1007/978-3-642-16687-7_7
14. Wu, J.M., Kuo, C.M., Chen, C.J.: Dose verification in intensity modulation radiation therapy: a fractal dimension characteristics study (2013)
15. Li, J., Sun, C., Du, Q.: A new box-counting method for estimation of image fractal dimension. In: International Conference on Image Processing, pp. 3029–3032 (2006)
16. Schroeder, M.: Fractals, Chaos, Power Laws: Minutes From an Infinite Paradise. W. H. Freeman, New York (1996)
17. Weszka, J.S., Dyer, C.R., Rosenfeld, A.: A comparative study of texture measures for terrain classification. IEEE Trans. Syst. Man Cybern. SMC **6**(4), 269–285 (1976)
18. Haralick, R.M.: Statistical and structural approaches to texture. Proc. IEEE **67**(5), 786–804 (1979)
19. Chang, T., Kuo, C.J.: Texture analysis and classification with tree-structured wavelet transform. IEEE Trans. Image Process. **2**(4), 429–441 (1993)
20. Backes, A.R., Gonçalves, W.N., Martinez, A.S., Bruno, O.M.: Texture analysis and classification using deterministic tourist walk. Pattern Recogn. **43**(3), 685–694 (2010)
21. Allain, C., Cloitre, M.: Characterizing the lacunarity of random and deterministic fractal sets. Phys. Rev. A **44**(6), 3552–3558 (1991)
22. Du, G., Yeo, T.S.: A novel lacunarity estimation method applied to SAR image segmentation. IEEE Trans. Geosci. Remote Sens. **40**(12), 2687–2691 (2002)
23. Fukunaga, K.: Introduction to Statistical Pattern Recognition, 2nd edn. Academic Press, Boston (1990)

Pyramidal Zernike Over Time: A Spatiotemporal Feature Descriptor Based on Zernike Moments

Igor L. O. Bastos[1]([⊠]) [iD], Larissa Rocha Soares[2] [iD],
and William Robson Schwartz[1] [iD]

[1] Smart Surveillance Interest Group, Department of Computer Science,
Universidade Federal de Minas Gerais, Belo Horizonte, Brazil
`igorcrexito@gmail.com, william@dcc.ufmg.br`
[2] Reuse in Software Engineering, Universidade Federal da Bahia, Salvador, Brazil
`larissars@dcc.ufba.br`

Abstract. This paper aims at presenting an approach to recognize human activities in videos through the application of Zernike invariant moments. Instead of computing the regular Zernike moments, our technique, named Pyramidal Zernike Over Time (PZOT), creates a pyramidal structure and uses the Zernike response at different levels to associate subsequent frames, adding temporal information. At the end, the feature response is associated to Gabor filters to generate video descriptions. To evaluate the present approach, experiments were performed on the UCFSports dataset using a standard protocol, achieving an accuracy of 86.05%, comparable to results achieved by other widely employed spatiotemporal feature descriptors available in the literature.

Keywords: Activity recognition · Feature extraction
Zernike moments

1 Introduction

Over the previous years, a growing interest in video surveillance applications such as elder care, home nursing and video monitoring, has been noticed, encouraging researches related to the field of activity recognition [3]. These activities are connected with the time domain and important information is denoted by changes in appearance and motion of elements over time [8]. In a simplistic way, they are sequences of primitive sub-actions [3], in which spatial and temporal features are associated to describe the whole event.

To describe activities in videos, features are extracted allowing a representation in a discriminative and compact way [2]. Due to the richness of spatial information on every frame, 2D local feature descriptors are commonly applied in this field. However, these descriptors do not consider the time contribution for this task, making room for the use of spatiotemporal descriptors. In this context, this work exploits Zernike Moments [9] and proposes a feature descriptor, called

© Springer International Publishing AG, part of Springer Nature 2018
M. Mendoza and S. Velastín (Eds.): CIARP 2017, LNCS 10657, pp. 77–85, 2018.
https://doi.org/10.1007/978-3-319-75193-1_10

Pyramidal Zernike Over Time (PZOT), that takes into account the temporal information in the description video sequences containing human activities.

This paper proposes a novel spatiotemporal method to recognize activities in videos based on a descriptor derived from the conventional Zernike invariant moments. To evaluate the method, experiments are performed on the UCFS-ports dataset, employing a standard visual recognition pipeline (Bag-of-Words [17] followed by the SVM classifier), reaching an accuracy of 86.05%, which is comparable to results achieved by other widely employed spatiotemporal feature descriptors, demonstrating the viability of the proposed approach.

2   Related Work

Due to the recent interest and large demand for activity recognition approaches, many studies have been developed for this task. The main idea of using Zernike invariant moments on the present approach regards the importance of spatial information, making this technique suitable to recognize activities.

The application of Zernike moments derives from their ability to provide information related to shapes in images. This fact was exploited by Tsolakidis et al. [18] and Newton and Hse [13], where the technique was used for leaf and digit recognition, respectively. Besides, the use of Zernike moments is desirable due to their amplitude invariance to rotation and translation, the possibility of changing their order and repetition parameters, and their ability to extract finer or coarser information regarding shapes with no overlap or redundancy [9].

The present approach tackles the extension of conventional Zernike Moments with the aim of adding temporal description. In this sense, the approaches proposed by Lassoued et al. [12] and Constantini et al. [4] deserve attention, since both designed Zernike based methods that deal with time information. Lassoued et al. [12] presented a temporal adaptation of Zernike moments to recognize human activities consisting on the extraction of 3D silhouettes over time on which a variant of Zernike moments is applied. Although it also adapts the Zernike moments to the temporal context, the work proposed in [12] differs from the present approach since it proposes a volumetric modeling of moments, changing their formulation to contemplate time. Similarly, the method proposed by Constantini et al. [4] is based on a volumetric modeling of Zernike moments computed over subsequent frames and associated to a spatial pyramid for matching.

Although presenting similarities with the aforementioned Zernike-based methods, the current approach presents a different idea since the temporal information is gathered from the combination of 2D Zernike responses. Moreover, the pyramid applied on the proposed approach is different from the one proposed in [4], since in [4] it consists of an arrangement of an image in different spatial scales, which are considered for matching descriptor responses. In turn, here, we create a pyramid by splitting the image in an increasing number of grids, used to extract the finer shape information through the Zernike responses.

PZOT and the methods proposed in [4] and [12] modify a 2D spatial descriptor to incorporate temporal information. This strategy is followed by well-known

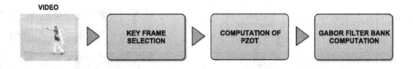

Fig. 1. Steps of the proposed approach.

approaches, producing spatiotemporal descriptors from spatial descriptors, such as HOG3D [10] or the temporal strands of SURF [20] and SIFT [15].

Nowadays, one can notice the increasing number of deep neural models used for activity recognition, as the ones proposed by Simonyan and Zisserman [16] and Feichtenhofer et al. [7]. However, this paper focuses on the proposition of a handcrafted feature descriptor. These descriptors are valuable since they can be associated to less complex classification models, which present a lower computational cost for training and require a smaller amount of training data. Hence, the validation performed in this study is based on tests and comparisons between the present method and state-of-art spatiotemporal handcrafted descriptors.

3 Proposed Approach

The present research proposes a novel spatiotemporal feature descriptor based on the usage of Zernike invariant moments to be applied to the activity recognition task. Thus, to operate correctly, some steps need to be taken, ranging from an initial selection of key frames in the video to the feature extraction with Zernike and Gabor descriptors. This section describes the steps of our approach, illustrated in Fig. 1. It is important to mention that the descriptor proposed in this work is intended to be applied over bounding boxes of the subjects performing the activity in the scene, usually provided by the video datasets.

3.1 Key Frame Selection

The first step of the approach performs a key frame selection to find relevant frames and avoid redundancy and unnecessary processing. For every video, the difference between subsequent frames is computed and those that satisfy a threshold (empirically determined) are selected as relevant. This strategy is based on the idea that similar frames do not contribute for the activity recognition. The feature extraction performed on PZOT occurs only on the key frames.

3.2 Computation of Pyramidal Zernike Over Time

To obtain information related to shapes on frames, a sequence of image processing techniques is employed before computing the Zernike moments. This step intends to produce binary masks in which the subject contours (shape) are highlighted. In addition, internal contours tend to be discarded, resulting in images

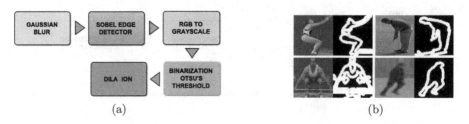

(a)   (b)

Fig. 2. Preprocessing steps prior to Zernike moments computation. (a) Sequence of image processing techniques; (b) resulting binary masks.

containing silhouettes of the subjects. With that, the Zernike response for each pair of parameters regards these silhouettes and, consequently, produces a more accurate outcome, which is more related to subject postures and is less influenced by background. This sequence of image processing techniques must be applied on every bounding box region of every frame of the videos. Figure 2(a) illustrates the processing pipeline and Fig. 2(b) exemplifies some obtained outputs. It is worth mentioning that both Sobel and Gaussian blur operators are applied over RGB images considering each channel independently. Furthermore, the dilation operation is performed over binary images using a 4×4 disk structuring element.

Instead of computing regular Zernike moments, a temporal approach, named the Pyramidal Zernike Over Time (PZOT) is assembled. This approach consists of placing every selected frame in different pyramid levels, in which every level is represented by the original image split in an increasing number of subimages. The number of subimages ranges from 1 (level one at the top of pyramid) to 4^{n-1} (at the base of pyramid), where n represents the number of levels.

At every level of the pyramid, nine Zernike moments are computed (amplitude and phase) for each subimage. These moments differ one from another according to the order and repetition parameters, making them able to represent finer and coarser details of the shapes. The Zernike moments are obtained from

$$Z_{nm} = \frac{(n+1)}{\pi} \sum_x \sum_y f(x,y)V_{nm}(x,y), \tag{1}$$

where $V_{nm}(x,y)$ represents the complex Zernike polynomial of order n and repetition m projected in polar coordinates and $f(x,y)$ represents a pixel of the input image. The phase and amplitude are obtained from each value of Z_{nm}.

The goal of splitting in an increasing number of subimages is to capture local shape information, obtaining responses to shapes of the entire image and local regions, improving the descriptiveness of the technique. For a three-level pyramid, for instance, $9 \times (1) + 9 \times (4) + 9 \times (16)$ moments are computed, resulting in 189 amplitude and 189 phase values.

Despite the increase of descriptiveness, the pyramid structure does not incorporate temporal information. To this end, it is necessary to create a relation between the frames of the video, which is made by the difference of the amplitude and phase values of consecutive frames, generating coefficients that describe distortions in the shapes (amplitude difference) and their positioning (phase difference). Therefore, the PZOT descriptor is composed by the computation of Zernike Pyramid for the first frame (representing the base shape) and the difference of every pyramid level for each subsequent pair of frames, resulting in a descriptor with 189 (base frame) $+ (k - 1) \times 189$ values for both amplitude and phase, where k represents the number of video frames. Section 4 describes a variation of this strategy that produced better results, in which Zernike responses are computed for each frame and added to this feature descriptor.

3.3   Gabor Filter Bank Computation

PZOT describes a video sequence through the base shape and its variations on subsequent frames. This strategy efficiently computes these variations, but does not capture much information to describe each frame, not exploring an important source of information for the task of activity recognition. Thus, Gabor filters have been included on the approach to describe the contours at different frequencies and orientations. These filters were intended to be applied over bounding boxes provided by video datasets (not applied on binary masks). However, instead of computing Gabor responses for every frame, we only compute for frames where the amplitude difference of Zernike moments satisfy a threshold value, indicating a strong variation on the silhouette (appearance) of the action subject.

To select the frames from which Gabor filters will be extracted, the Zernike amplitudes are used to compose a curve at the first level of pyramid for each frame. The difference of areas under these curves for a frame and the base shape frame are compared to a threshold (20% of the area under the base shape curve). Every time this value is satisfied, the frame is considered relevant (different from its predecessor) and the Gabor responses are computed. It is important to mention that the difference is computed, initially, according to the first frame (first base shape). Afterwards, this base shape is updated every time the threshold is satisfied and every subsequent comparison is performed considering this update. Figure 3(a) shows the responses for two subsequent frames for a diving activity video (red and blue curves) of the UCFSports dataset.

An 8-degree polynomial (derived from nine amplitude responses of Zernike moments) is applied to adjust the curves and after that, the difference is computed (showed in yellow in Fig. 3). The degree of the polynomial enforces a non-linear relation among amplitude values, which implicitly corresponds to responses for different frequencies. Besides that, fitting a curve emphasizes the difference between responses for subsequent frames. The degree of this polynomial was adjusted experimentally. For any degree smaller than 8, the curves were not well adjusted to the 9 Zernike amplitudes, resulting in low accuracies.

The Gabor filter bank employed consists of filters with four different orientations and four different frequencies. Those filters are weighted according to the difference obtained in the curve. Figure 3(b) shows a curve difference obtained for two non-consecutive frames of a diving activity video of the UCFSports dataset. It is possible to notice that the difference between curves (showed in light and dark yellow) is more relevant at the high order moments, which indicate

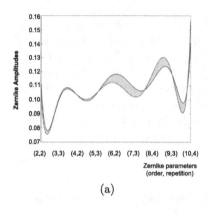 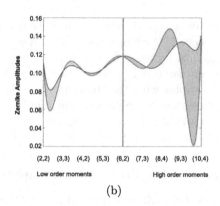

(a)   (b)

Fig. 3. Difference between Zernike curves (amplitude as a function of order/repetition). Both curves were fit by a 8-degree polynom. (a) Difference for two subsequent frames; (b) difference for two non-subsequent frames. (Color figure online)

a stronger variation on finer shapes. Such Zernike behavior may be associated to high frequency responses, as low order moments may be associated to low frequency responses. Thus, the curve difference is split into two regions and the Gabor filters with the two lowest frequencies are weighted by the low order area, while the Gabor filters with the two highest frequencies are weighted by the high order area, as shown on Fig. 3(b). The weighting coefficients are computed by the area of each region normalized by the total difference area.

4   Experimental Results

The UCFSports dataset [14] was used to evaluate the approach. This dataset is composed by 150 sequences representing 10 different sport activities. It also provides, along with videos, bounding boxes detaching the subjects involved in each activity. Furthermore, to homogenize the tests, a classification pipeline described in [2] was employed. It uses SVM classifier with a RBF kernel and a bag of 4000 visual words. The executed tests intend to evaluate the feasibility of our descriptor, even if the obtained results depend on steps that precede its use.

The first step of the evaluation of our method regards the selection of Zernike parameters, which were chosen by a preliminary study conducted on a Brazilian Sign Language recognition dataset [1]. From this study, nine pairs of parameters (order and repetition) were set, varying from lower to higher order moments: (10, 4), (9, 3), (8, 4), (7, 3), (6, 2), (5, 3), (4, 2), (3, 3), and (2, 2).

Once the Zernike parameters have been defined, we conducted a test with the application of PZOT (without Gabor filters) on the video frames, computing the base shape for the first frame and amplitude and phase distortions over time. We achieved an accuracy of 63.64%. Then, a second test was conducted. Instead of computing only the distortions regarding the base shape, the Zernike amplitude/phase responses (describing shapes of every frame) were added, reaching an accuracy of 70.54%. Finally, the complete approach was tested (adding Gabor), as described in Sect. 3.3. This test achieved an accuracy of 86.05%.

Table 1. Accuracies of different approaches applied to the UCFSports dataset.

	Approach	Acc (%)
Published results	HOG [5]	77.00
	HOF [11]	84.00
	HOG/HOF [11]	81.60
	HOG3D [10]	85.60
	MBH [6]	90.53
	DT [19]	88.20
	OFCM [2]	92.80
Our results	PZOT with base shape and only distortions	63.64
	PZOT with shapes on every frame	70.54
	Gabor with no PZOT	79.47
	PZOT + Gabor (proposed approach)	**86.05**

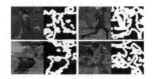

Fig. 4. Binary masks with noisy shapes detached.

According to Table 1, PZOT provides results that are comparable to previous researches applied to the UCFSports dataset. It is important to highlight that the features extracted with PZOT contribute with the ones extracted by Gabor filters in a complementary way, what is revealed by the increase of accuracy obtained with the combination of techniques. Lastly, three pyramid levels were applied, as they produced better results than two or four levels.

The *Skateboarding* and *Running* classes of UCFSports were the ones for which the method obtained the lowest accuracies: 75.09% and 70.91%, respectively. For these classes, the image processing pipeline generated noisy images, preventing the Zernike moments from being properly extracted, as shown in Fig. 4. One can notice the amount of clutter and how human shapes are not well-segmented.

5   Conclusions

This paper presented a temporal feature descriptor based on the Zernike invariant moments. Tests were conducted to evaluate this descriptor on the UCFSports dataset, reaching an accuracy of 86.05%. Despite achieving a smaller accuracy than some Spatiotemporal descriptors, the present approach shows to be valuable since it achieved comparable results using a simple and inexpensive method. For some activities of the UCFSports dataset, the number of extracted features is approximately 1/3 of the ones extracted by the OFCM [2], the state-of-art feature descriptor for this dataset. At last, studies should be directed to points such as the simplicity of the image processing pipeline, application of few Zernike pairs of parameters and evaluation of a different Gabor filter bank, representing future directions for the present approach.

Acknowledgments. The authors would like to thank the Brazilian National Research Council – CNPq, the Minas Gerais Research Foundation – FAPEMIG (Grants APQ-00567-14 and PPM-00540-17) and the Coordination for the Improvement of Higher Education Personnel – CAPES (DeepEyes Project).

References

1. Bastos, I.L.O., Angelo, M.F., Loula, A.: Recognition of static gestures applied to Brazilian sign language (libras). In: SIBGRAPI, pp. 305–312 (2015)
2. Caetano, C., Santos, J.A., Schwartz, W.R.: Optical flow co-occurrence matrices: a novel spatiotemporal feature descriptor. In: Proceedings of the 23rd ICPR (2016)
3. Chaquet, J.M., Carmona, E.J., Antonio, F.: A survey of video datasets for human action and activity recognition. Comput. Vis. Image Underst. **117**, 633–659 (2013)
4. Costantini, L., Seidenari, L., Serra, G., Capodiferro, L., Del Bimbo, A.: Space-time Zernike moments and pyramid kernel descriptors for action classification. In: Maino, G., Foresti, G.L. (eds.) ICIAP 2011, Part II. LNCS, vol. 6979, pp. 199–208. Springer, Heidelberg (2011). https://doi.org/10.1007/978-3-642-24088-1_21
5. Dalal, N., Triggs, B.: Histograms of oriented gradients for human detection. In: Proceedings of IEEE CVPR 2005, vol. 1, pp. 886–893 (2005)
6. Dalal, N., Triggs, B., Schmid, C.: Human detection using oriented histograms of flow and appearance. In: Leonardis, A., Bischof, H., Pinz, A. (eds.) ECCV 2006. LNCS, vol. 3952, pp. 428–441. Springer, Heidelberg (2006). https://doi.org/10.1007/11744047_33
7. Feichtenhofer, C., Pinz, A., Zisserman, A.: Convolutional two-stream network fusion for video action recognition. In: Proceedings of the IEEE CVPR 2016 (2016)
8. Ke, S., Thuc, H., Lee, Y., Hwang, J., Yoo, J., Choi, K.: A review on video-based human activity recognition. Computers **2**, 88–131 (2013)
9. Khotanzad, A., Hong, Y.H.: Invariant image recognition by Zernike moments. IEEE Trans. Pattern Anal. Mach. Intell. **12**(5), 489–497 (1990)
10. Klaser, A., Marszalek, M., Schmid, C.: A spatio-temporal descriptor based on 3D-gradients. In: Proceedings of 19th BMVC, pp. 275:1–10 (2008)
11. Laptev, I., Marszalek, M., Schmid, C., Rozenfeld, B.: Learning realistic human actions from movies. In: Proceedings of IEEE CVPR 2008, pp. 1–8 (2008)
12. Lassoued, I., Zagrouba, E., Chahir, Y.: An efficient approach for video action classification based on 3D Zernike moments. In: Park, J.J., Yang, L.T., Lee, C. (eds.) FutureTech 2011. CCIS, vol. 185, pp. 196–205. Springer, Heidelberg (2011). https://doi.org/10.1007/978-3-642-22309-9_24
13. Newton, A.R., Hse, H.: Sketched symbol recognition using Zernike moments. In: Proceedings of the 17th ICPR, vol. 1, pp. 367–370 (2004)
14. Rodriguez, M.D., Ahmed, J., Shah, M.: Action mach: a spatio-temporal maximum average correlation height filter for action recognition. In: Proceedings of CVPR 2008 (2008)
15. Scovanner, P., Ali, S., Shah, M.: A 3-dimensional sift descriptor and its application to action recognition. In: Proceedings of the 15th ACM ICM, pp. 357–360 (2007)

16. Simonyan, K., Zisserman, A.: Two-stream convolutional networks for action recognition in videos. In: Advances in NIPS 27, pp. 568–576 (2014)
17. Sivic, J., Zisserman, A.: Video Google: a text retrieval approach to object matching in videos. In: Proceedings of the IEEE ICCV 2003, vol. 2, pp. 1470–1477 (2003)
18. Tsolakidis, D.G., Kosmopoulos, D.I., Papadourakis, G.: Plant leaf recognition using Zernike moments and histogram of oriented gradients. In: Likas, A., Blekas, K., Kalles, D. (eds.) SETN 2014. LNCS (LNAI), vol. 8445, pp. 406–417. Springer, Cham (2014). https://doi.org/10.1007/978-3-319-07064-3_33
19. Wang, H., Klaser, A., Schmid, C., Liu, C.: Action recognition by dense trajectories. In: Proceedings of the IEEE CVPR 2011, pp. 3169–3176 (2011)
20. Willems, G., Tuytelaars, T., Van Gool, L.: An efficient dense and scale-invariant spatio-temporal interest point detector. In: Forsyth, D., Torr, P., Zisserman, A. (eds.) ECCV 2008, Part II. LNCS, vol. 5303, pp. 650–663. Springer, Heidelberg (2008). https://doi.org/10.1007/978-3-540-88688-4_48

A Multilabel Extension of LDA Based on the Gram-Schmidt Orthogonalization Procedure

Juan Bekios-Calfa[ID] and Brian Keith[(✉)][ID]

Departamento de Ingeniería de Sistemas y Computación,
Universidad Católica del Norte, Av. Angamos 0610, Antofagasta, Chile
{juan.bekios,brian.keith}@ucn.cl

Abstract. Multilabel classification is a generalization of the traditional unidimensional classification problem, the goal of multilabel classification is to learn a function that maps instances into a set of relevant labels. This article proposes an extension to linear discriminant analysis in the context of multilabel classification. The new method is based on Gram-Schmidt orthogonalization procedure. The theoretical basis and underlying assumptions of the new model are described and the method is experimentally evaluated on the Emotions data set for multilabel classification. The analysis of the empirical results support that this new method is competitive and in some instances superior to the baseline.

Keywords: Linear discriminant analysis
Gram-Schmidt orthogonalization · Multilabel classification

1 Introduction

Traditionally the classification task focuses on the estimation of a single output variable. This is done through learning from a set of examples. These instances are associated with a unique tag λ_i from a set of disjoint tags $\mathcal{L} = \{\lambda_1, \lambda_2, \ldots, \lambda_n\}$ [13]. If $|\mathcal{L}| = 2$ the learning task is called a problem of binary classification, where generally $\mathcal{L} = \{\lambda, \neg\lambda\}$ indicates if an instance belongs or not to the class indicated by the label. For the case where $|\mathcal{L}| > 2$ the problem is called multiclass classification. The set of unique tags that can be associated with an instance is also known as class variable [13].

Consider the m-dimensional input space $\mathcal{X}$, with $\mathcal{X} = \mathcal{X}_1 \times \ldots \times \mathcal{X}_m$, ($\mathcal{X}_i$; $i = 1, \ldots, m$), where $\mathcal{X}_i \in \mathbb{N}$ (nominal features), $\vee \ \Omega_{X_i} \in \mathbb{R}$ (numeric features) and $\mathcal{L}$ as the set of output tags. The learning task consists of finding a function h

$$h : \mathcal{X}_1 \times \cdots \times \mathcal{X}_m \to \mathcal{L}$$
$$(x_1, \ldots, x_m) \mapsto \lambda_i \in \mathcal{L}$$

such that h must be able to generalize correctly, in the sense of minimizing the loss of expected prediction with respect to a specific loss function.

A natural generalization of the classical classification problem is the multidimensional classification problem. In this case, the classifier is associated with

M. Mendoza and S. Velastín (Eds.): CIARP 2017, LNCS 10657, pp. 86–93, 2018.
https://doi.org/10.1007/978-3-319-75193-1_11

multiple output variables. These variables can either be binary or multiclass like in the classical problem. Depending on the *a priori* information that can be obtained from these variables and the output type the following taxonomy is used to classify these problems: multilabel, multi-dimensional or structured output. This article focuses on the multilabel problem.

A multilabel classifier can be seen as generalization of the single-output classifier. In these classifiers the instances are associated with a set of labels L, where $L \subset \mathcal{L}$. However, in contrast with the traditional approach, the labels assigned by the classifier are not mutually exclusive. Therefore, any instance in the data set can be associated with more than one label. Historically, these classifiers were motivated by problems in text classification and medical diagnosis.

In multilabel classification the learning task can be reformulated as finding the function h

$$h : \mathcal{X}_1 \times \cdots \times \mathcal{X}_m \to \mathcal{P}(\mathcal{L})$$
$$(x_1, \ldots, x_m) \mapsto L = \{\lambda_r, \ldots, \lambda_q\} \subseteq \mathcal{L}$$

where $\mathcal{X}_1 \times \cdots \times \mathcal{X}_m$ represents the input space and $\mathcal{P}(\mathcal{L})$ denotes the power set of $\mathcal{L}$ (i.e. the set of all possible subsets). This function associates each example with a set $L = \{\lambda_r, \ldots, \lambda_q\}$ of labels. The labels present in this subset of $\mathcal{L}$ are called the relevant labels. Thus, the learning task is to find the function h such that h minimizes the expected prediction loss for the label set [13].

An important assumption of this model is that the different class variables associated (either through manual or automatic labeling) for each of the instances serve to improve the efficiency of the classifiers. This is because they contain information that can be used to subdivide the space of original characteristics and enhance its discriminating power ever more [14,19,20].

The problem of multilabel classification is an active area of research, recent surveys demonstrate several challenges in multilabel classification. Some of these include dealing with high dimensionality, the exploration of label correlations in an efficient manner and the understanding of these correlations [6,17].

This article proposes a new method for multilabel classification based on Linear Discriminant Analysis (LDA) [7] and Gram-Schmidt Orthogonalization (GS) [4] procedure through QR Factorization [1]. While the idea of applying GS in the context of LDA is not new (see for example: [2,3,18]), however none of those works have been focused on solving the multilabel problem, instead they deal with issues such as computational complexity and efficiency. On the other hand, there have been various proposals of multilabel extensions for LDA [8–10,15]. None of the papers found in the reviewed literature used an approach similar to the one proposed in this article.

2   Proposed Method: ML-GS-LDA

The classical LDA method is based on the computation of the scatter matrices S_w and S_b (the within class scatter matrix and the between class scatter matrix, respectively) and the optimization of the following objective function:

$$J(W) = \mathbf{tr}\left(\frac{W^T S_b W}{W^T S_w W}\right)$$

The projection W^* which maximizes $J(W)$ is calculated through Singular Value Decomposition (SVD) [7].

However, while this approach works for multiclass problems, it is still limited to a single label. One possible extension dealing with multilabel classification is through the definition of class-wise scatter matrices [15]. Another approach centered on multilabel classification efficiency can be made through the use of two successive QR factorization instead of conventional SVD and a clever application of null spaces [10]. Both extensions empirically demonstrate that classification performance can be improved by using multiple class labels together.

In this context, this article proposes a new method for multilabel classification based on LDA. The proposed method is arguably simpler and easy to implement, though it could be slightly more expensive in terms of computational time, basing itself on iterated LDA over different labels and Gram-Schmidt orthogonalization. The method is described in Algorithm 1, the details are given in the next paragraph.

The proposed method works as follows, for all the included labels the classical LDA projection matrix is built using the input data X and the corresponding column from the matrix Y. These projections all have the same dimensionality d, given by the number of eigenvectors selected, according to the parameter Dim. In this case, since there are only two possible classes for each label, a

---

Algorithm 1. ML-GS-LDA

Input: X, matrix containing the data in rows; Y, matrix containing the labels associated with each data point in X; LabelOrder, a list indicating the order in which the model must be trained (e.g. [0, 1, 2] means to predict the class using the information from labels zero, one and two); Target, the classification target of the algorithm, indicates the label that must be predicted; Dim, value indicating the number of dimensions of each LDA projection.

Output: W, the projection matrix generated after successive LDA computation and the application of the Gram-Schmidt procedure.

```
 1: function ML-GS-LDA
 2:     W = null
 3:     for all Label in LabelOrder do
 4:         W' = LDA(X, Y[Target, :], Dim)
 5:         if W == null then
 6:             W = W'
 7:         else
 8:             W = [W, W']
 9:         end if
10:     end for
11:     return Gram-Schmidt(W)
12: end function
```

---

value of $Dim = 1$ was assigned. The conditional inside the loop considers the first iteration separately from the rest, indicating that when W is not initialized it corresponds to the classical LDA projection matrix of the first label. The matrices are concatenated in each iteration, forming the final matrix W of order $|A| \times d \cdot |C|$ where $|A|$ is the number of attributes in the original data, d is the dimensionality of each projection and $|C|$ is the number of classes given by the length of the *LabelOrder* parameter. Note that each of the individual projections is orthogonal, however, the resulting matrix W is not necessarily orthogonal. So, in order to guarantee the orthogonality (and therefore uncorrelatedness) of each axis, the Gram-Schmidt procedure is applied to the matrix W. The resulting matrix is returned and can be used to project the original data set onto the new subspace.

Some observations need to be made about the underlying assumptions in the proposed method. The first one is that orthogonality between a feature axis (in terms of the covariance dot product) implies uncorrelatedness of those same axis [11]. The second one is that the application of Gram-Schmidt orthogonalization procedure on the concatenated projection basis generates an adequate projection for this problem. And finally, the last assumption is that the generated projection will mainly encode the information of the first label while at the same time retaining the extra information from the auxiliary labels.

By using multiple information sources (i.e. a multilabel approach) and eliminating the correlation in this new subspace it is expected that there is improvement in terms of classification accuracy, similar to how the use of Gram-Schmidt in the context of classical linear regression produces some improvements in accuracy and uncorrelated regressors [5].

3   Methodology

To evaluate the proposed method the Emotions [12, 16] data set was used. This data set contains 593 instances with 72 numerical attributes. These attributes represent 30 s from a musical piece fragment. The 72 attributes are obtained through the application of different filters and transformations to each musical fragment. Also, each instance can be associated with up to six labels corresponding to different emotional states, see Table 1.

Table 1. Description of the labels contained in the Emotions data set.

Label	Description	#Examples
L1	Amazed-surprised	173
L2	Happy-pleased	166
L3	Relaxing-calm	264
L4	Quiet-still	148
L5	Sad-lonely	168
L6	Angry-fearful	189

3.1 Validation

The method was evaluated using a 5-*fold* cross validation approach. For each experiment PCA (Principal Component Analysis) was applied for a first stage dimensionality reduction. Afterwards, for further dimensionality reduction either the supervised method LDA or the proposed algorithm ML-GS-LDA were applied. To obtain the performance for each label a Naive Bayes classifier was used for each method. The parameters to define the best configuration for the PCA dimension and the label ordering selection for the ML-GS-LDA algorithm were performed using a *wrapper* method.

Two different approaches for ML-GS-LDA were evaluated with the purpose of verifying the significance of the orderings of the labels on the ML-GS-LDA algorithm and their impact on the performance of the classifier. The first one considers combinations of the labels (i.e. the ordering does not matter) and the second one considers permutations (i.e. the ordering matters). See line 3 of Algorithm 1.

4 Results and Discussion

The obtained accuracy results are shown in Table 2. The analysis of these values shows a marginal increment in the classification accuracy for each label. Also, by analyzing the labels selected for each projection some relationships and dependencies between the different labels can be inferred.

Marginal improvements in classification accuracy can be found for all the six labels, independently of the applied strategy (combinations or permutations). The analysis of these results suggests that the combination of different LDA projections, in conjunction with orthogonalization, could improve classification accuracy in the multilabel problem.

The proposed algorithm can find dependencies and relationships between the different labels, in the sense that the information obtained from their LDA projection and subsequent orthogonalization allows for prediction of another label.

Table 2. Results for the Emotions data set [13] for the PCA + LDA and PCA + ML-GS-LDA and a Naive Bayes classifier.

Classifier	Labels					
	L1	L2	L3	L4	L5	L6
PCA + LDA	80.95 ± 2.29 PCA: 15	73.20 ± 2.23 PCA: 30	73.68 ± 2.75 PCA: 45	88.87 ± 4.12 PCA: 35	83.31 ± 2.30 PCA: 55	79.26 ± 3.62 PCA: 20
PCA + ML + GS + LDA combinations	82.13 ± 3.21 PCA: 20 L: (1, 5, 6)	74.38 ± 3.00 PCA: 30 L: (1, 3, 5)	76.37 ± 5.42 PCA: 20 L: (1, 2, 3, 4, 5)	90.06 ± 4.82 PCA: 35 L: (1, 4)	84.33 ± 2.80 PCA: 55 L: (2, 4, 5)	80.45 ± 4.15 PCA: 30 L: (1, 2, 4, 6)
PCA + ML + GS + LDA permutations	82.13 ± 3.21 PCA: 20 L: (1, 5, 6)	76.24 ± 3.56 PCA: 50 L: (1, 3, 4, 2)	77.04 ± 5.27 PCA: 20 L: (1, 3, 4, 2, 6)	90.06 ± 4.02 PCA: 35 L: (5, 1, 4, 2)	84.33 ± 2.80 PCA: 55 L: (2, 4, 5)	82.64 ± 4.43 PCA: 30 L: (2, 4, 1, 6)

A quick review of the relationships is described. However, an in-depth analysis of these dependencies is beyond the scope of this work. The following relationships were found:

- **Label 1:** In this case, both the combination and permutation approach have the same optimal label ordering. According to this, L1 would be related with L5 and L6. It is interesting to note that in both cases the target label is first in the ordering.
- **Label 2:** In this case, the combination and permutation approach have different optimal label orderings. However, in both cases L1 and L3 are present, while the difference between the two is the interchanging of L5 for L4 and L2, resulting in a higher classification accuracy for the permutation approach. It is interesting to note that the predicted label was found to give better results when placed last in the ordering instead of being in the first position.
- **Label 3:** In this case, the results are similar, however L5 in the combination approach has been replaced by L6 in the permutation approach. Also, the ordering of L2, L3 and L4 has been changed for the permutation case. This results in an increase in classification accuracy for this label.
- **Label 4:** In this case, the classification accuracy is the same for both cases, however the label orderings are different. It is interesting to note that the label ordering found by the combination approach is a subset of the one found by the permutation approach. Also, while the average value is the same, by using the permutation approach a lower variance was found.
- **Label 5:** In this case, both methods produce the same results. From this it can be found that L5 is related with L2 and L4.
- **Label 6:** In this case, classification accuracy is higher for the permutation case. However, the only difference between the methods is the label ordering, since both sets use the same labels, this is interesting because it reveals the importance of the ordering for finding the optimal classifier.

Based on the obtained results and the previous analysis, apart from the accuracy gain discussed before the benefits from this proposal are two-fold:

- The search for the optimal label ordering provides a new way to study the dependencies and relationships between the different labels. While this model provides a basis for this analysis, the theoretical implications and properties of these orderings must be further evaluated, both in terms of empirical analysis and the mathematical properties associated with them.
- The final method is fast in comparison with other approaches, since the results are obtained through a simple matrix multiplication and the application of Naive Bayes. It is important to note the training phase and the search for the optimal parametrization can require plenty of computational resources, however, this is true for all methods that require finding several hyper-parameters.

It is possible that through the application of another classifying scheme, such as some variant of SVM better results could be obtained. However, the use of NB has the aforementioned benefit of being more efficient in terms of computational resources compared to SVM.

5 Conclusions

This work has explored a novel generalization of the classical LDA method for multilabel problems, this proposal is based on the Gram-Schmidt orthogonalization procedure through QR factorization. The theoretical basis for this model and its underlying assumptions have been exposed and the proposal has been validated through experimental evaluation on the Emotions data set. The analysis of the empirical results support that this new method is competitive and in some instances superior to the baseline.

One of the main benefits of the proposed algorithm is that it serves as an exploration tool for label dependencies and relationships. An in-depth analysis, both theoretical and empirical, of the relationships that this algorithm finds is proposed as future work. Another important plus of this method is that it is fast once the classifier has been trained. Since the resulting projection can be obtained through the application of a simple matrix multiplication.

On the other hand, one of the key aspects of the method is the order of the decomposition, which affects the obtained results. The election of the right order is a challenge that must still be addressed. Also, experimental results show that the results vary from label to label. The task of correctly exploiting these results to provide the best global result is still pending and is considered as future work. Comparison with other state of the art methods, such as deep learning approaches for emotion recognition, is also considered as future work.

Finally, some limitations of the proposed method in its current form correspond with the combinatorial optimization required to determine the optimal multilabel combination. To approach this, a step-wise approach like the one used in the process of finding the best regressors for linear regression or a greedy strategy based on an easy to evaluate heuristic could be useful to find better solutions, although optimality of the combinations would still be an open problem. Future work plans on dealing with the implementation of a modified version of the algorithm that includes optimization routines and exhaustive evaluation on multilabel data sets with different metrics.

References

1. Björck, Å.: Numerics of Gram-Schmidt orthogonalization. Linear Algebra Appl. **197**, 297–316 (1994)
2. Cai, D., He, X., Han, J.: SRDA: an efficient algorithm for large-scale discriminant analysis. IEEE Trans. Knowl. Data Eng. **20**(1), 1–12 (2008)
3. Cevikalp, H., Neamtu, M., Wilkes, M., Barkana, A.: Discriminative common vectors for face recognition. IEEE Trans. Pattern Anal. Mach. Intell. **27**(1), 4–13 (2005)
4. Cohen, H.: A Course in Computational Algebraic Number Theory, vol. 138. Springer Science & Business Media, Heidelberg (1993). https://doi.org/10.1007/978-3-662-02945-9
5. Farebrother, R.: Algorithm as 79: Gram-Schmidt regression. J. Roy. Stat. Soc.: Ser. C (Appl. Stat.) **23**(3), 470–476 (1974)

6. Gibaja, E., Ventura, S.: Multi-label learning: a review of the state of the art and ongoing research. Wiley Interdisc. Rev.: Data Min. Knowl. Discov. **4**(6), 411–444 (2014)
7. Izenman, A.J.: Linear discriminant analysis. In: Modern Multivariate Statistical Techniques. STS, pp. 237–280. Springer, Heidelberg (2013). https://doi.org/10.1007/978-0-387-78189-1_8
8. Ji, S., Ye, J.: Linear dimensionality reduction for multi-label classification. In: IJCAI, vol. 9, pp. 1077–1082 (2009)
9. Oikonomou, M., Tefas, A.: Direct multi-label linear discriminant analysis. In: Iliadis, L., Papadopoulos, H., Jayne, C. (eds.) EANN 2013. CCIS, vol. 383, pp. 414–423. Springer, Heidelberg (2013). https://doi.org/10.1007/978-3-642-41013-0_43
10. Park, C.H., Lee, M.: On applying linear discriminant analysis for multi-labeled problems. Pattern Recogn. Lett. **29**(7), 878–887 (2008)
11. Rodgers, J.L., Nicewander, W.A., Toothaker, L.: Linearly independent, orthogonal, and uncorrelated variables. Am. Stat. **38**(2), 133–134 (1984)
12. Trohidis, K., Tsoumakas, G., Kalliris, G., Vlahavas, I.P.: Multi-label classification of music into emotions. In: Bello, J.P., Chew, E., Turnbull, D. (eds.) ISMIR, pp. 325–330 (2008). http://dblp.uni-trier.de/db/conf/ismir/ismir2008.html#TrohidisTKV08
13. Tsoumakas, G., Katakis, I.: Multi-label classification: an overview. Int. J. Data Warehouse. Min. **3**(3), 1 (2007)
14. Wan, H., Wang, H., Guo, G., Wei, X.: Separability-oriented subclass discriminant analysis. IEEE Trans. Pattern Anal. Mach. Intell. (2017)
15. Wang, H., Ding, C., Huang, H.: Multi-label linear discriminant analysis. In: Daniilidis, K., Maragos, P., Paragios, N. (eds.) ECCV 2010. LNCS, vol. 6316, pp. 126–139. Springer, Heidelberg (2010). https://doi.org/10.1007/978-3-642-15567-3_10
16. Wieczorkowska, A., Synak, P., Raś, Z.W.: Multi-label classification of emotions in music. In: Kłopotek, M.A., Wierzchoń, S.T., Trojanowski, K. (eds.) Intelligent Information Processing and Web Mining. AINSC, vol. 35, pp. 307–315. Springer, Heidelberg (2006). https://doi.org/10.1007/3-540-33521-8_30
17. Zhang, M.L., Zhou, Z.H.: A review on multi-label learning algorithms. IEEE Trans. Knowl. Data Eng. **26**(8), 1819–1837 (2014)
18. Zheng, W., Zou, C., Zhao, L.: Real-time face recognition using Gram-Schmidt orthogonalization for LDA. In: Proceedings of the 17th International Conference on Pattern Recognition, ICPR 2004, vol. 2, pp. 403–406. IEEE (2004)
19. Zhu, M., Martinez, A.M.: Selecting principal components in a two-stage LDA algorithm. In: 2006 IEEE Computer Society Conference on Computer Vision and Pattern Recognition, vol. 1, pp. 132–137. IEEE (2006)
20. Zhu, M., Martinez, A.M.: Subclass discriminant analysis. IEEE Trans. Pattern Anal. Mach. Intell. **28**(8), 1274–1286 (2006)

Linear Modelling of Cerebral Autoregulation System Using Genetic Algorithms

Felipe-Andrés Bello Robles[1]([⊠]) [iD], Ronney B. Panerai[2],
and Max Chacón Pacheco[1]

[1] Informatics Engineering Department,
Universidad de Santiago de Chile, Santiago, Chile
{felipe.bello,max.chacon}@usach.cl
[2] Cardiovascular Sciences Department, University of Leicester, Leicester, UK
rp9@le.ac.uk

Abstract. Cerebral autoregulation (CA) represents the brain's capacity to maintain the cerebral blood flow constant, independent of the activities realized by an individual. There are pathologies like Alzheimer, vascular dementia, ischemic stroke, subarachnoid haemorrhage and severe brain injury, where a degradation of CA can be found. Despite limited understanding of its physiological basis, assessment of CA is relevant for diagnosis, monitoring and treating some of these pathologies. CA modelling is done by using mean arterial blood pressure (MABP) as input and cerebral blood flow velocity (CBFV) as output; the standard model used is transfer function analysis, although CA has been modelled with support vector machines (SVM) and other methods. In this work a resistive-capacitive model (R-C) is presented where parameters can be estimated from MABP and CBFV signals through Genetic Algorithms (GA), comparing its discrimination capacity against SVM models. Signals from 16 healthy subjects were used with 5 min of spontaneous variations (SV) and 5 min breathing oxygen with 5% of CO_2 (hypercapnia). Results show that both models can capture CA and the degradation induced by hypercapnia. Using the autoregulation index (ARI), the R-C model discriminates with a ROC area of 0.89 against 0.72 from SVM, thus representing a promising alternative to assess CA.

Keywords: Cerebral autoregulation · Genetic algorithms · R-C model

1 Introduction

Human brain is sensitive to changes in its blood supply, where low cerebral blood flow (CBF ischemia) could lead to loss of consciousness or even death after few minutes. On the other hand excessive CBF could generate damage on the vessels and cerebral tissue through intracranial hypertension or even hemorrhage. The mechanism that controls the level of blood flow in the brain (associated to the energy consumption) independently of changes in MABP is known as cerebral autoregulation (CA). In patients with Alzheimer, vascular dementia, ischemic brain stroke, and severe head injury, a degradation of CA has been reported, but detailed modelling of this system remains

© Springer International Publishing AG, part of Springer Nature 2018
M. Mendoza and S. Velastín (Eds.): CIARP 2017, LNCS 10657, pp. 94–101, 2018.
https://doi.org/10.1007/978-3-319-75193-1_12

challenging. In the 80s, the appearance of transcranial Doppler [1] opened up the possibility of CA assessment through maneuvers to induce changes in MABP of the subject to get an autoregulatory response [2–5]. These maneuvers are not always applicable therefore models that can captures CA at rest, based on spontaneous variations (SV), for example using transfer function analysis (TFA) have been widely used [3, 6, 7] other non-linear models such as neural networks, Wiener-Laguerre and SVM [3, 8, 9] had allowed improvements in performance and the comprehension of this phenomenon during SV. CA is often evaluated through the response of a theoretical inverse step of MABP, measuring the recovering of CBFV levels by using an autoregulation index ARI [10]. The lack of more physiological information related to the understanding of CA, and the high complexity of black box models such as SVM, makes it attractive to present a new model that parameterizes resistive and capacitive components (R-C), which could provide greater understanding of the underlying physiology, by using genetic algorithms (GA). To evaluate CA, the subject breathes oxygen with 5% of CO_2 which produces CA degradation by the effect of CO_2 over the vessels [2, 11]. Considering data of healthy subjects that inhaled CO_2, two hypotheses arise: first, a GA allows learning human cerebral autoregulation pattern by using a resistive-capacitive model. Second, GA applied over a resistive-capacitive model can discriminate between a normal and degraded status of cerebral autoregulation, with a ROC area greater than the one from a linear SVM.

2 Methods

2.1 CA Modelling by Using SVM

CA modelling through black box approach is implemented by using SVM under the concept of "support vector regression" in which is done an estimation of CBFV $\widehat{v}(t)$ values, based on MABP $p(t)$ using a finite impulse response (FIR) structure:

$$\widehat{v}(t) = f(p(t), \ldots, p(t - n_p)) \tag{1}$$

CBFV estimation $\widehat{v}(t)$ is given by a static SVM with n_p inputs, corresponding to the delays number of the model. Each segments signal (SV and under CO_2) is normalized and divided in two, one part for training and the other for validation (balanced cross validation). Trainings are done for each subject with hyper parameter's configurations for SVR, with its proper validation. For hyper parameter's configurations C, v and n_p a grid search is realized and using a Linear Kernel, same behaviour as TFA [6] to compare under similar conditions. Estimated $\widehat{v}(t)$ and measured $v(t)$ are compared through correlation coefficient and Mean Squared Error (MSE) per each training. Finally models are selected through normalized inverse MABP step response criteria where the model selected is the one with higher correlation coefficient in prediction that satisfies the physiological response of CA as shown in Fig. 1.

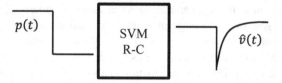

Fig. 1. Theoretical response from trained SVM/R-C models to an inverse MABP step.

Responses $\hat{v}(t)$ per each subject from models are evaluated by using an autoregulatory index ARI [10], with values from 0 (absence of autoregulation) to 9 (best autoregulation).

2.2 Proposed Model Resistive-Capacitive R-C

The standard for cerebral autoregulation modelling is based on TFA [6, 12], which behaves as high pass filter over signals, obtaining parameters such as gain, coherence and phase. Knowing this, the use of a resistive-capacitive circuit is proposed, implementing a second-order high pass filter. MABP represents the voltage input $p(t)$ and CBF corresponds to the input current $v(t)$ to C_1 capacitor as shown in Fig. 2.

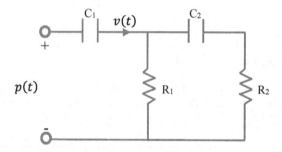

Fig. 2. Proposed model for cerebral autoregulation with resistive-capacitive components.

In this hydraulic-electric analogy, to convert CBF into velocity (CBFV), a standard diameter of 3 mm for middle cerebral artery is considered. The resistive components of the model are measured in mmHg s/ml (cerebrovascular resistance) and capacitive components in ml/mmHg (possibly compliance of the vessels). The combination of the four component's values is presented as an optimization problem, where the MSE is minimized in a frequency band from 0–0.2 Hz (where CA phenomenon is present) [12], comparing CBF estimated and measured. This is a linear model, therefore the complete signal is used for simulations (SV or CO_2), without normalizing them to estimate CBFV. Literature shows ranges of values for resistive and capacitive components for normal subjects [13], but for CA degradations using CO_2 are unknown. Numerical approaches require initial conditions near the optimum to converge and avoid local minimums, therefore the use of Genetic Algorithms is proposed (GA) since they have been used in electrical circuit design [14–16], as well as hydraulic design problems [17], similar to the R-C model proposed. Using this heuristic, a grid is implemented to establish different ranges for the initial conditions per each component of the model, where each element of the initial population of the GA is a tuple composed of random values (within feasible physiological ranges) following a uniform distribution for R_1, R_2, C_1 and C_2 with which the simulation of the circuit will get a CBFV estimated per each tuple. GA will have a finish condition given by the number of generations of the algorithm or by not finding a difference greater than the millionth part in the MSE. Per each combination of initial conditions for the model, the best model from GA is obtained. To evaluate their learning of CA pattern, an inverse MABP step is applied to the model, with its maximum and minimum values got from the subject's signal, and a model's response is obtained by measuring CBFV (Fig. 1). Following the same selection process than SVM and calculating ARI after that.

2.3 Subjects and Measurements

Sixteen healthy subjects aged 31.8 ± 8.5 years were studied. None of them had a history of hypertension, diabetes, migraine, epilepsy or any other cardiovascular or neurologic disease. The study was approved by the Leicestershire Research Ethics Committee and informed consent was obtained in all cases [11]. CBFV-measured was recorded in the right MCA with transcranial Doppler (Scimed QVL-120) using a 2 MHz transducer. ABP was measured non-invasively using arterial volume clamping of the digital artery (Finapres 2300 Ohmeda). The signals CBFV-measured and ABP were recorded for 5 min for SV followed by 5 min recording while subjects breathed a combination of oxygen and 5% of CO_2 inducing an hypercapnic status. By using average's heart beat interpolation from CBFV-measured and ABP, CBFV and MABP are obtained, sampled at 5 Hz. To evaluate the proposed hypotheses different measures are used. First an analysis of inverse MABP step response dynamic from models will be realized, followed by repeated measures ANOVA for models between SV and CO_2 conditions using ARI. Finally a ROC curve will be obtained for models RC and SVM to compare their area under the curve.

3 Results

FIR-SVM obtained from a parameter's grid search that learns physiological CA pattern gives the following results shown in Table 1.

Table 1. Statistical mode, mean values, parameter's standard deviation and SVM metrics.

Parameter/metric	SV	5% CO_2
Mode n_p	4 [1–8]	8 [1–8]
C	303.5 ± 986.67	46.25 ± 84.7
v	0.32 ± 0.59	0.41 ± 0.34
Correlation coefficient	0.59 ± 0.17	0.66 ± 0.18
ARI	6.29 ± 0.79	5.4 ± 1.05

Based on component's values under normal conditions of subjects obtained from [13], an initial condition grid is defined where the ranges for the components are $R_1 = [0.5–100]$; $R_2 = [0.01–1; 1–100]$; $C_1 = [0.02–2; 2–200]$; $C_2 = [0.01–1; 1–100]$ realizing 8 executions of GA based on the initial conditions. Calibrating the GA the size of population changed from 50 to 400, and GA generations from 25 to 175 without finding greater differences in correlation or MSE in prediction, however a population of 300 and 125 generations are selected due present greater classification power between normal and CO_2 conditions as shown in Table 2.

Table 2. Mean values and standard deviation of best R-C models per subject.

Parameter/metric	SV	5% CO_2
R_1 mmHg s/ml	40.22 ± 28.00	19.78 ± 16.03
R_2 mmHg s/ml	5.31 ± 8.8	8 ± 24.13
C_1 ml/mmHg	37.09 ± 50.19	74.07 ± 64.43
C_2 ml/mmHg	7.47 ± 22.79	6.79 ± 19.2
Correlation Coef.	0.53 ± 0.17	0.52 ± 0.19
Correlation Coef. [0–0.2 Hz]	0.53 ± 0.17	0.56 ± 0.22
ARI	5.54 ± 2.01	1.92 ± 2.21

Average Inverse MABP step response to evaluate CA is shown on Fig. 3 for R-C and SVM models, trained with SV and CO_2 signals, in which the suddenly CBFV drop happens at 5 s and stabilize at different levels depending on the signal used in training.

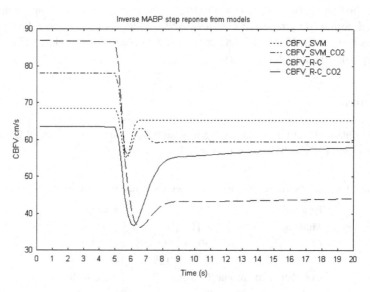

Fig. 3. Inverse average step response from R-C and SVM models adjusted under SV and CO_2 signals.

A repeated measures ANOVA, shows differences for models under both conditions (SV and CO_2) with *p-value = 0.00034*, Fig. 4.

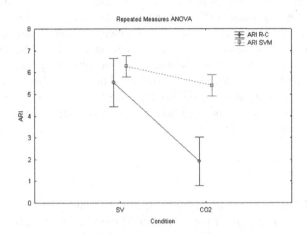

Fig. 4. Repeated measures ANOVA for R-C and SVM models between SV and CO_2

Post-hoc Tukey analysis shows differences for R-C and SVM models *p-value = 0.000186* and *p-value = 0.014* respectively.

Also the area under the ROC curve was calculated where R-C models have 0.89 against 0.73 from SVM.

4   Discussion

SVM achieves an accuracy to estimate CBFV of 0.59 in correlation during SV and slightly higher under CO_2, while R-C model presents a correlation near of 0.53 for both conditions (SV and CO_2). The dynamic of the inverse MABP step response with a drop at 5 s, applied over both models shows a normal autoregulation response for healthy subjects under normal conditions (SV). SVM responses under CO_2 show a behaviour where after the drop it tends to raise and then stabilize, while R-C response during CO_2 drops and keeps in the bottom showing a clear CA degradation.

R-C model's initial conditions grid enables feasible ranges per each component in which the combination allows valid physiological responses, decreasing the number of models to evaluate and giving clarity to the search process.

Repeated measures ANOVA tells that both models present differences among SV and CO_2 conditions. Post-hoc Tukey, shows that despite both models present differences between subject's conditions, R-C models discriminate more markedly with a p-value much lower. Therefore both models can capture CA during SV, but there are differences between models on capturing CA degradations under CO_2. ROC area from R-C model is 17% larger than SVM which validates the better classification power of the proposed model.

Limitations of this work makes it necessary to perform comparisons between R-C model and SVM trained with CBFV under the same frequency band, because the higher correlation of SVM could be given by higher frequencies than the CA phenomenon and therefore out of a range where the information of CA degradation is contained. Considering future works, R-C model could have a physiological meaning from the cerebral haemodynamics perspective, since in Table 2 can be observed a marked difference in the values of R_1 and C_1 among SV and CO_2. Therefore, the components R_1, R_2, C_1 and C_2 could be studied individually or together because of the possibility of classifying subjects in different conditions even better than ARI and improve the understanding of CA. Another approach to be investigated is the use of a non-linear RC model to be implemented and compared against non-linear models previously proposed [3, 8, 9].

5 Conclusions

An R-C model has been presented by using an hydraulic-electric analogy through genetic algorithms to learn the pattern of cerebral blood flow changes in response to fluctuations in arterial blood pressure. This model also discriminates between normal and degraded CA conditions (CO_2) better than SVM, validating both hypotheses proposed in this work. From these results, new clinical applications will allow improvements in differentiating subject's condition through simulations based on measured data from subject/patient, improving comprehension of CA phenomenon in normal as well as pathological conditions.

References

1. Newell, D.W., Aaslid, R., Lam, A., Mayberg, T.S., Winn, H.R.: Comparison of flow and velocity during dynamic autoregulation testing in humans. Stroke **25**, 305–338 (1994). https://doi.org/10.1161/01.STR.25.4.793
2. Panerai, R.: Assessment of cerebral pressure autoregulation in humans - a review of measurement methods. Physiol. Meas. **19**, 305–338 (1998). https://doi.org/10.1088/0967-3334/19/3/001
3. Panerai, R., Dawson, S., Potter, J.: Linear and nonlinear analysis of human dynamic cerebral autoregulation. Am. J. Physiol. - Heart Circ. Physiol. **277**, 698–705 (1999)
4. Tiecks, F., Lam, A., Aaslid, R., Newell, D.: Comparison of static and dynamic cerebral autoregulation measurements. Stroke **26**, 1014–1019 (1995)
5. Panerai, R., Dawson, S., Eames, P., Potter, J.: Cerebral blood flow velocity response to induced and spontaneous sudden changes in arterial blood pressure. Am. J. Physiol. **280**, 2162–2174 (2001)
6. CAR-NET webpage. http://www.car-net.org/. Accessed 11 May 2017
7. Czosnyka, M., Smielewski, P., Lavinio, A., Pickard, J., Panerai, R.: An assessment of dynamic autoregulation from spontaneous fluctuations of cerebral blood flow velocity: a comparison of two models, index of autoregulation and mean flow index. Anesth. Analg. **106**, 234–239 (2008)

8. Chacón, M., Araya, C., Panerai, R.: Non-linear multivariate modeling of cerebral Hemodynamics with autoregressive support vector machines. Med. Eng. Phys. **33**, 180–187 (2011)
9. Panerai, R.B., Chacon, M., Pereira, R., Evans, D.H.: Neural network modelling of dynamic cerebral autoregulation: assessment and comparison with established methods. Med. Eng. Phys. **26**, 43–52 (2004)
10. Tiecks, F.P., Lam, A.M., Aaslid, R., Newell, D.W.: Comparison of static and dynamic cerebral autoregulation measurements. Stroke **26**, 1014–1019 (1995)
11. Panerai, R.B., Deverson, S.T., Mahony, P., Hayes, P., Evans, D.H.: Effect of CO2 on dynamic cerebral autoregulation measurement. Physiol. Meas. **20**, 265–275 (1999)
12. Claassen, J., Meel-van den Abeelen, A., Simpson, D., Panerai, R.B.: Transfer function analysis of dynamic cerebral autoregulation: a white paper from international cerebral autoregulation research network. J. Cereb. Blood Flow Metab. **36**, 1–16 (2016)
13. Gommer, E., Martens, E., Aalten, P., Shijaku, E., Verhey, F., Mess, W., Ramakers, I., Reulen, J.: Dynamic cerebral autoregulation in subject with Alzheimer's Disease, mild cognitive impairment, and controls: evidence for Increased peripheral vascular resistance with possible predictive value. J. Alzheimer's Dis. **30**, 805–813 (2012)
14. Menozzi, R., Piazzi, A., Contini, F.: Small-signal modeling for microwave FET linear circuits based on a genetic algorithm. IEEE Trans. Circ. Syst.-I: Fundam. Theory Appl. **46**, 839–847 (1996)
15. Yap, R., Lam, K., Bugayong, R., Hernandez, E., De Guzman, J.: Hardware design and implementation of genetic algorithm for the controller of a DC to DC boost converter. J. Teknol. **78**, 117–122 (2016)
16. Khalil, N., Ahmed, R., Abul Seoud, R., Soliman, A.: An intelligent technique for generating equivalent gyrator circuits using Genetic Algorithm. Microelectron. J. **46**, 1060–1068 (2015)
17. Ko, M.: A novel design method of optimizing an indirect forced circulation solar water heating system based on life cycle cost using a genetic algorithm. Energies **8**, 11592–11617 (2015)

Knowledge Transfer for Writer Identification

Diego Bertolini[1,2]([✉]), Luiz S. Oliveira[3], Yandre M. G. Costa[2][iD],
and Lucas G. Helal[2]

[1] Federal University of Technology - Paraná (UTFPR),
Campo Mourão, PR, Brazil
diegobertolini@utfpr.edu.br
[2] State University of Maringá (UEM), Maringá, PR, Brazil
[3] Federal University of Paraná (UFPR), Curitiba, PR, Brazil

Abstract. The technical literature on writer identification usually considers the best case scenario in terms of data availability, i.e., a database composed of hundreds of writers with several documents per writer is available to train the machine learning models. However, in real-life problems such a database may not be available. In this context, learning from one dataset and transferring the knowledge to other would be extremely useful. In this paper we show how to transfer knowledge from one dataset to another through a framework that uses a writer-independent approach based on dissimilarity. Experiments on five different databases under single- and multi-script environments showed that the proposed approach achieves good results. This is an important contribution since it makes it possible do deploy the writer identification system even when no data from that particular writer are available for training.

1 Introduction

Writer identification is the task in which the goal is to identify whether the writer of a handwritten sample belongs to a set of writers or not. This matter has aroused interest once it can be used in several different applications, i.e. personalized handwriting recognition, writer retrieval, forensic document examination, and classification of ancient manuscripts. The literature shows that this problem has been attacked from several different perspectives for variated number of scripts, such as Latin [2,3,23], Arabic [6], Chinese [25], Japanese [14], among others. The amount of data used in terms of sample size may vary from several pages [3,5], paragraphs [1], to single words [2]. Recently, the writer identification problem was extended to a multi-script scenario where the rationale is recognizing an individual of a given text written in one script from the samples of the same individual written in another script [4,9].

A common trait among all the works reported in the literature is the dependence on the dataset. Usually, a dataset composed of n writers is divided into training and testing so that the writer is equally represented in both training and testing. The machine learning models are then trained on the training sets

© Springer International Publishing AG, part of Springer Nature 2018
M. Mendoza and S. Velastín (Eds.): CIARP 2017, LNCS 10657, pp. 102–110, 2018.
https://doi.org/10.1007/978-3-319-75193-1_13

and results reported on the testing set. This approach is usually known as writer-dependent (WD). An alternative approach called writer-independent (WI) for writer identification was presented in [3], which converts the k-class pattern recognition problem (where k is the number of writers) into a 2-class problem through a dichotomy transformation.

However, in both cases (WD and WI), writers used for training and testing are draw from the same dataset, which in general contains a considerable number of individuals. Just to cite a few, IAM, IFN/ENIT, BFL, and CVL contains 650, 411, 315, and 310 writers, respectively.

In real-life application problem, though, such large datasets may not be available to train the machine learning models. A possible solution in this context is to learn in one dataset (source) and transfer the knowledge to other (target) dataset. This strategy has been explored in different domains of application with several different names, such as knowledge transfer, cumulative learning, life-long learning, transfer learning, etc. A review on this subject can be found in [21].

In this work we argue that the writer-independent (WI) approach based on dissimilarity described in [3] makes knowledge transfer possible because of its main property, i.e., the writer that did not contribute for the training set can still be identified by the system. To corroborate this hypothesis we carried out a series of experiments using five different datasets largely used in the literature (IAM [18], BFL [10], CVL [16], QUWI [17] and LAMIS [8]) and two different textural descriptors, the Local Binary Pattern (LBP) [19] and Robust Local Binary Pattern (RLBP) [7]. The selected databases contain document written in different languages (English, French, German, Portuguese, and Arabic) and two different scripts (Roman and Arabic).

In the first experiment we considered only Roman scripts where two databases were used as source and other three as target. In a second experiment we evaluated if the results found for Roman script can be replicated for Arabic scripts. Finally, the third experiment, we addressed the multi-script problem approach.

2   Knowledge Transfer with Dissimilarity

As mentioned before, an interesting aspect about the dissimilarity approach is the possibility of reducing any pattern recognition problem to a 2-class problem. The idea consists in extracting the feature vectors from both questioned and reference samples and then computing what we call the dissimilarity feature vectors. In ideal conditions, it is expected that if both samples come from the same writer (genuine), then all the components of such a vector should be close to 0, otherwise, the components should be far from 0.

Given a queried handwritten document and a reference handwritten document, the goal consists in determining whether or not the two documents were produced by the same writer. Let V and Q be two vectors in the feature space, labeled l_V and l_Q respectively. Let Z be the dissimilarity feature vector resulting from the dichotomy transformation $Z = |V - Q|$, where $|\cdot|$ is the absolute value. This dissimilarity feature vector has the same dimensionality as V and Q.

In the dissimilarity space, there are two classes that are independent of the number of writers: the within class (+) and the between class (o). The dissimilarity vector Z is assigned the label l_Z,

$$l_Z = \begin{cases} + \text{ if } l_V = l_Q, \\ o \text{ otherwise} \end{cases} \tag{1}$$

The rationale of knowledge transfer using the dissimilarity approach is presented in Fig. 1. In this example, (a) depicts the source dataset in the feature space from five different writers while (b) shows the dissimilarity vectors for the source dataset, which are the results of the dichotomy transformation between the features of each pair of samples to form vectors. The same representation for the target dataset is presented in Figs. 1(c) and (d). Differently of the feature space, where multiple boundaries are necessary to discriminate the writers, in the dissimilarity space, only one boundary is necessary, since the problem is reduced to a 2-class classification problem. Besides, even writers whose specimens were not used for training can be identified by the system. This characteristic is quite attractive, since it obviates the need to train a new model every time a new writer is introduced.

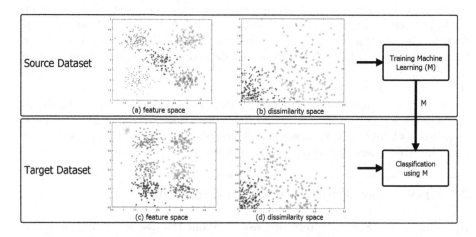

Fig. 1. Knowledge transfer with dissimilarity: (a) samples of the source dataset in the feature space, (b) samples of the source dataset in the dissimilarity space, (c) samples of the target dataset in the feature space, and (d) samples of the target dataset in the dissimilarity space. In (b) and (d), "+" stands for the vectors associated to the within class and "o" stands for the vectors associated to the between class.

In spite of the different number of writers and distributions of the feature space observed in the source and target datasets (Figs. 1(a) and (c)), the dichotomy transformation impacts in the geometry producing very similar distribution for both source and target datasets in the dissimilarity space (Figs. 1(b) and (d)). This observation leads us to investigate whether knowledge transfer is an alternative in this context, i.e., to learn the machine learning model (M) in the source dissimilarity space and deploying it to other target dataset.

In order to generate the positive samples $(+)$ to train the classifier, we computed the dissimilarity vectors among the R genuine samples (references) of each writer which resulted in $\binom{R}{2}$ different combinations. The same number of negative samples (o) is generated by computing the dissimilarity between one reference of one writer against one reference of other writers picked at random. In our experiments, the best results were found using 9 references per writer.

3   Databases and Feature Extraction

As mentioned earlier, we have selected five different databases to be used as source and target in our experiments. These databases contain handwriting in different languages and scripts.

The IAM dataset [18] is widely used in problems such as handwriting recognition and writer identification. It contains forms with handwritten English text of variable content (text-independent). A total of 650 writers have contributed to the dataset. BFL database [10] is composed of 315 writers, with three samples per writer, for a total of 945 images. This makes it suitable for text-dependent writer identification as well. In the CVL database [16] contains 310 writers with 1,604 text-dependent handwriting. Furthermore, 27 writers have six documents in English and one in German, while 283 writers have four documents in English and one in German. LAMIS [8] contains 1,200 handwritten text images from 100 different writers. To acquire the handwritten samples writers were instructed to produce 12 pages of handwriting, six of them in French and other six in Arabic, all the letters are text dependent. To the best of our knowledge, this database has not been used for writer identification. Lastly, the QUWI database [17] contains 4,068 handwritten text images from 1,017 different writers. In order to acquire the handwritten samples, volunteers were instructed to produce four pages of handwriting as follows: First and second pages in the Arabic language free-text and copied or (text-independent). The third and fourth page in English of free-text and copied, respectively.

The recent literature on writer identification shows that researchers have been investigating a great number of representations for writer identification, which can be classified into local and global. The local approach takes into account specific features of the writing and generally involve some kind of segmentation process. Such features are usually extracted from words, characters, or allographs [2,22]. The global approach, on the other hand, tries to avoid the segmentation process by representing the handwriting as a texture [3,11,12]. In these cases, several textural descriptors have been tried out such as Grey-Level Co-occurrence Matrices (GLCM) [13], Local Phase Quantization (LPQ) [20], Local Binary Patterns (LPB) [19] and its variations such as Robust LBP [7].

In this work we adopted the global approach described by Bertolini et al. [3]. In this case, a texture of the handwriting is created scanning the document top-down left-write and putting together all the connected components found in

Fig. 2. Example of the texture produced from the LAMIS database (a) Arabic and (b) English handwritings for the same writer.

the image. Small components, such as strokes, commas, and noise are discarded. Then, the texture is segmented into nine 256×256 blocks. Figure 2 shows two examples of the handwriting texture produced from English and Arabic handwritings for the same writer of the LAMIS database.

As representation, we have tested several textural descriptors but our best results were always achieved either with LBP or RLBP. Both descriptors are quite similar, but RLBP considers a more flexible concept of uniformity. For example, if there is one, and only one value in the binary code which makes the LBP non-uniform, it is possibly caused by some noise and, for this reason, it must be considered as a uniform pattern. In our experiments, both LBP and RLBP representations produce a 59-dimensional feature vector. In both cases, the feature vectors are normalized with the min-max rule. For those readers interested in these textural descriptors, please refer to [7,19].

4   Experimental Results and Discussion

In all experiments described in this work, the Support Vector Machine (SVM) was used to perform the classification [24]. The parameters of the system for training were chosen using 5-fold cross validation. The best results were achieved using Gaussian kernel. Parameters C and γ were determined through a grid search.

To corroborate the hypothesis presented in the introduction we have designed a set of three experiments. In the first experiment we considered only Roman scripts where two databases were used as source (QUWI and LAMIS) and other three as target (IAM, BFL, CVL). To better compare the results, we fixed in 100 the number of writers for both source datasets, which were picked randomly.

Table 1 summarizes the first experiment. In the first part of the table we report the performance on the three target datasets using QUWI, LAMIS, and the union of both as source data. The second part of the table allows us to better assess these results since source and target are the same. As we can observe, the results using knowledge transfer are quite similar to those where the same dataset was used for training and testing. In the case of IAM, using QUWI or LAMIS as source produced better results and using IAM itself. Besides, we notice that the source data can benefit from a larger number of writers. When both

Table 1. Performance on IAM, BFL, and CVL using the models trained on QUWI and LAMIS. The number of writers used is in parenthesis

Source	Target					
	BFL (315)		CVL (310)		IAM (650)	
	LBP	RLBP	LBP	RLBP	LPB	RLBP
QUWI (100)	97.8	99.3	91.6	93.8	93.5	92.9
LAMIS (100)	98.4	99.4	89.7	91.9	92.3	92.5
QUWI + LAMIS (200)	99.0	99.6	93.5	92.9	94.8	93.3
BFL (100)	99.0	99.0				
CVL (100)			93.5	93.8		
IAM (100)					89.0	90.0

source datasets were combined (LAMIS + QUWI), we perceived some improvement in all three target datasets. Finally, the best results were achieved in the BFL dataset, which is a text-dependent database that contains a great deal of handwriting so that we have no problems in generating the same amount of texture for all writers. In the case of CVL and IAM, some writers are represented by just one letter, hence, compromising the texture generation process. Table 1 also shows the good performance of the knowledge transfer when source and target contain handwriting in different languages, using the Roman script, though. In this case, the source datasets contain handwriting in English (QUWI) and French (LAMIS) while the target datasets are written in Portuguese (BFL), English/German (CVL). Before discussing the second experiment, it is important mentioning that the literature shows identification rates on IAM and CVL around 97% and 99%, respectively [15]. In these cases, the same datasets are used as source and target and all the same users are considered for both training and testing.

So far all the experiments have considered only the Roman script. Since both QUWI and LAMIS contain documents written in Arabic, we assessed the knowledge transfer applied for Arabic. It is noteworthy that the QUWI database contains two types of documents, free and copy text. The free-text is composed of approximately six lines of handwriting in Arabic while the copy text contains three paragraphs of Arabic handwriting. In other words, the copy text has considerably

Table 2. Performance of the knowledge transfer for Arabic scripts.

Source	Target					
	LAMIS		QUWI-free		QUWI-copy	
	LBP	RLBP	LBP	RLBP	LPB	RLBP
QUWI	97.0	94.0	77.0	79.0	97.0	97.0
LAMIS	94.0	97.0	72.0	69.0	98.0	99.0

more handwriting, which translates directly in a richer texture. In this experiment, 100 writers were considered in both source and target datasets. In the case of the QUWI, only copy documents were used.

Table 2 reports the results of the knowledge transfer for the Arabic script. Similarly to the experiments on the Roman script, the knowledge transfer achieves good performance in both cases. The discrepancy in terms of performance between both copy and free texts is related to the different amount of text available for copy and free text. Since the samples of free texts are composed of few lines, we could not create dense textures as we did on copied texts, which are composed of three paragraphs.

Table 3. Performance of the knowledge transfer for multi-script.

Script train/script test	Source	Target	LBP	RLBP
Roman/Arabic	LAMIS	QUWI-free	72.0	74.0
	LAMIS	QUWI-copy	100.0	98.0
	QUWI-copy	LAMIS	98.0	93.0
	QUWI-free	LAMIS	97.0	98.0
Arabic/Roman	LAMIS	QUWI-free	75.0	83.0
	LAMIS	QUWI-copy	97.0	98.0
	QUWI-copy	LAMIS	98.0	98.0
	QUWI-free	LAMIS	96.0	99.0

Finally, we address the problem of multi-script where the idea is to identify an individual of a given text written in one script from the samples of the same individual written in another script. In the context of knowledge transfer, the source dataset could be Arabic and the target in Roman, and vice-versa. Similarly to the previous experiments, 100 writers were considered in both source and target datasets. Table 3 presents the performance of the knowledge transfer for multi-script where LAMIS and QUWI were used as source and target. In the first part of Table 3 the model was trained on Roman while the testing was performed on Arabic. The second part of the table the scripts were inverted.

The results presented in Table 3 show the robustness of this approach. In all tests we have performed (except when QUWI-free has been used as target, as explained before) the results on multi-script compare to single-script results. This shows that this strategy is viable for both these cases.

5   Conclusion

In this work we have shown through a series of experiments on five different databases using two different textural descriptors that the writer-independent approach underpinned on the concept of dissimilarity allows knowledge transfer for writer identification. We have assessed the proposed approach single-script (Roman/Arabic) and multi-script environments and observed that in all cases one can transfer the knowledge from one dataset to another. This is an important contribution since it makes it possible do deploy the writer identification system even when no data from that particular writer are available for training.

References

1. Al-Maadeed, S.: Text-dependent writer identification for Arabic handwriting. J. Electr. Comput. Eng. (2012)
2. Bensefia, A., Paquet, T.: Writer identification based on a single handwriting word samples. EURASIP J. Image Video Process. **2016**(1), 34 (2016)
3. Bertolini, D., Oliveira, L.S., Justino, E., Sabourin, R.: Texture-based descriptors for writer identification and verification. Expert Syst. Appl. **40**(6), 2069–2080 (2013)
4. Bertolini, D., Oliveira, L.S., Sabourin, R.: Multi-script writer identification using dissimilarity. In: International Conference on Pattern Recognition (ICPR), pp. 3020–3025. IAPR (2016)
5. Brink, A., Schomaker, L., Bulacu, M.: Towards explainable writer verification and identification using vantage writers. In: 9th International Conference on Document Analysis and Recognition, Curitiba, Brazil, pp. 824–828 (2007)
6. Bulacu, M., Schomaker, L.: Text-independent writer identification and verification using textural and allographic features. IEEE Trans. Pattern Anal. Mach. Intell. **29**(4), 701–717 (2007)
7. Chen, J., Kellokumpu, V., Zhao, G., Pietikäinen, M.: RLBP: robust local binary pattern. In: Proceedings of the British Machine Vision Conference (2013)
8. Djeddi, C., Gattal, A., Souici-Meslat, L., Siddiqi, I., Chibani, Y., Abed, H.: LAMIS-MSHD: a multi-script offline handwriting database. In: ICFHR, pp. 93–97 (2014)
9. Djeddi, C., Siddiqi, I., Souici-Meslati, L., Ennaji, A.: Text-independent writer recognition using multi-script handwritten texts. Pattern Recogn. Lett. **34**, 1196–1202 (2013)
10. Freitas, C., Oliveira, L.S., Sabourin, R., Bortolozzi, F.: Brazilian forensic letter database. In: 11th International Workshop on Frontiers on Handwriting Recognition, Montreal, Canada (2008)
11. Hannad, Y., Siddiqi, I., Kettani, M.: Writer identification using texture descriptors of handwritten fragments. Expert Syst. Appl. **47**, 14–22 (2016)
12. Hanusiak, R.K., Oliveira, L.S., Justino, E., Sabourin, R.: Writer verification using texture-based features. Int. J. Doc. Anal. Recogn. (IJDAR) **15**(3), 213–226 (2012)
13. Haralick, R., Shanmugam, K., Dinstein, I.: Textural features for image classification. IEEE Trans. Syst. Man. Cybern. **3**(6), 610–621 (1973)
14. Kameya, H., Mori, S., Oka, R.: A segmentation-free biometric writer verification method based on continuous dynamic programming. Pattern Recogn. Lett. **27**(6), 567–577 (2006)
15. Khan, F.A., Tahir, M., Khelifi, F., Bouridane, A., Almotaeryi, R.: Robust off-line text independent writer identification using bagged discrete cosine transform features. Expert Syst. Appl. **71**, 404–415 (2017)
16. Kleber, F., Fiel, S., Diem, M., Sablatnig, R.: CVL-Database: an off-line database for writer retrieval, writer identification and word spotting. In: 12th International Conference on Document Analysis and Recognition, pp. 560–564. IEEE (2013)

17. Maadeed, S.A., Ayouby, W., Hassaïne, A., Aljaam, J.M.: QUWI: an Arabic and English handwriting dataset for offline writer identification. In: 2012 International Conference on Frontiers in Handwriting Recognition, pp. 746–751, September 2012
18. Marti, U.-V., Bunke, H.: The IAM-database: an English sentence database for offline handwriting recognition. Int. J. Doc. Anal. Recogn. 5(1), 39–46 (2002)
19. Ojala, T., Pietikainen, M., Maenpaa, T.: Multiresolution gray-scale and rotation invariant texture classification with local binary patterns. IEEE Trans. Pattern Anal. Mach. Intell. 24(7), 971–987 (2002)
20. Ojansivu, V., Heikkilä, J.: Blur insensitive texture classification using local phase quantization. In: Elmoataz, A., Lezoray, O., Nouboud, F., Mammass, D. (eds.) ICISP 2008. LNCS, vol. 5099, pp. 236–243. Springer, Heidelberg (2008). https://doi.org/10.1007/978-3-540-69905-7_27
21. Pan, S.J., Yang, Q.: A survey on transfer learning. IEEE Trans. Knowl. Data Eng. 22(10), 1345–1359 (2010)
22. Siddiqi, I., Djeddi, C., Raza, A., Souici-meslati, L.: Automatic analysis of handwriting for gender classification. Pattern Anal. Appl. 18(4), 887–899 (2015)
23. Siddiqi, I., Vincent, N.: Text independent writer recognition using redundant writing patterns with contour-based orientation and curvature features. Pattern Recogn. 43(11), 3853–3865 (2010)
24. Vapnik, V.: The Nature of Statistical Learning Theory. Springer, New York (1995). https://doi.org/10.1007/978-1-4757-2440-0
25. Wang, X., Ding, X.: An effective writer verification algorithm using negative samples. In: 2004 Ninth International Workshop on Frontiers in Handwriting Recognition, IWFHR-9 2004, pp. 509–513. IEEE (2004)

Real-Time Brand Logo Recognition

Leonardo Bombonato[(✉)], Guillermo Camara-Chavez, and Pedro Silva

Federal University of Ouro Preto, Ouro Preto, Minas Gerais, Brazil
leonardobombonato@gmail.com

Abstract. The increasing popularity of Social Networks makes change the way people interact. These interactions produce a huge amount of data and it opens the door to new strategies and marketing analysis. According to Instagram (https://instagram.com/press/) and Tumblr (https://www.tumblr.com/press), an average of 80 and 59 million photos respectively are published every day, and those pictures contain several implicit or explicit brand logos. The analysis and detection of logos in natural images can provide information about how widespread is a brand. In this paper, we propose a real-time brand logo recognition system, that outperforms all other state-of-the-art methods for the challenging FlickrLogos-32 dataset. We experimented with 5 different approaches, all based on the Single Shot MultiBox Detector (SSD). Our best results were achieved with the SSD 512 pretrained, where we outperform by 2.5% of F-score and by 7.4% of recall the best results on this dataset. Besides the higher accuracy, this approach is also relatively fast and can process with a single Nvidia Titan X 19 images per second.

Keywords: Computer vision · Brand logo recognition
Deep learning · CNN

1 Introduction

Brand logos are graphic entities that represent organizations, goods, *etc.* Logos are mainly designed for decorative and identification purposes. A specific logo can have several different representations, and some logos can be very similar in some aspects. Logo classification in natural scenes is a challenging problem since they often appear in various angles and sizes, making harder to extract keypoints especially due to significant variations in texture, poor illumination, and high intra-class variations (Fig. 1). The automatic classification of those logos gives to the marketing industry a powerful tool to evaluate the impact of brands. Marketing campaigns and medias can benefit with this tool, detecting unauthorized distributions of copyright materials.

Several techniques and approaches were proposed in the last decades for object classification, such as Bag of Visual Words (BoVW), Deep Convolutional Neural Networks (DCNN), feature matching with RANSAC, *etc.* The most successful approaches in logo classification were based on the BoVW model

© Springer International Publishing AG, part of Springer Nature 2018
M. Mendoza and S. Velastín (Eds.): CIARP 2017, LNCS 10657, pp. 111–118, 2018.
https://doi.org/10.1007/978-3-319-75193-1_14

Fig. 1. This figure exemplifies the challenges of classifying logos in natural scenes, such as high intra-class variation, warping, occlusion, rotation, translation and scales.

and DCNN. Most recent approaches in logo recognition use a Region-Based Convolution Neural Network (RCNN) [15], this network introduces a selective search to find candidates. This approach has shown great results comparing to others proposed in previous researches. Even though the good results achieved by RCNN, it is really hard to train and test due to the characteristics of selective search that generates several potential bounding boxes categorized by a classifier. After classification, post-processing is used to refine the bounding boxes, eliminating duplicate detection, and re-scoring the boxes.

In this paper, we propose an approach for logo detection based on a deep learning model. Our results outperform state-of-the-art approach results, achieving a higher accuracy on the FlickrLogos-32 dataset. Our proposal uses transfer learning to improve the logo image representations, being not only more accurate but also faster in logo detection, processing 19 images per second using a NVidia Titan X card.

2   Related Works

Different approaches for logo recognition have been proposed through the last years. A few years ago, only shallow classifiers have been proposed to solve this challenge [3,18,19]. But with the increasing popularity of deep learning frameworks and because of its success in image recognition, some researches started to come up [2,8].

The first successful approach in logo recognition was based on contours and shapes in images with a uniform background. Francesconi *et al.* [6] proposed an adaptive model using a recursive neural network, the authors used the area and the perimeter of the logo as features.

After 2007, with the popularization of SIFT, countless applications started to use it, due to its robustness to rotation and scale transformations and partially invariant to occlusions. Many approaches based their proposals on SIFT descriptors [3,17–20].

RANSAC became a popular learning module for object recognition since its use in Lowe's research work [13]. Lowe used this technique to compare matched descriptors and find outlines, eliminating the false positives matches and thus locating the object. In logo recognition, some researchers explored this method and achieved significant results, e.g. [3, 20].

Deep Convolutional Neural Networks are on the trends in computer vision and especially in object recognition. Recent approaches in logo recognition applied this technique, achieving impressive results like [5, 7, 8, 15].

3 Deep Convolutional Neural Networks

Artificial Neural Network is a classification machine learning model where the layers are composed of interconnected neurons with learned weights. These weights are learned by a training process. Convolutional Neural Network (CNN) is a type of feedforward artificial neural network and a variation of a multilayer perceptron. A neural network with three or more hidden layers is called deep network.

Transfer Learning. In a CNN each layer learns to "understand" specific features from the image. The first layers usually learn generic features like edges and gradients, the more we keep forwarding in the layers, the more specific the features the layer detects. In order to "understand" these features, it is necessary to train the network, adjusting the net weights according to a predefined loss function. If the network weights initiate with random values, it requires much more images and training iterations compared to using pretrained weights. The use of net weights trained with other dataset is called "fine-tuning" and it demonstrates to be extremely advantageous compared to training a network from scratch [4]. This technique is useful when the number of training images per class is scarce (e.g. 40 images for this problem), which makes it hard for the CNN to learn. Furthermore, transfer learning also speeds up the training convergence [16].

Data Augmentation. Training a DCNN requires lots of data, especially very large/deep networks. When the dataset does not provide enough training images, we can add more images using data augmentation process. This process consists of creating new synthetic images, that simulate different view angles, distortions, occlusions, lighting changes, etc. This technique usually increases the robustness of the network resulting in better results.

3.1 Single Shot MultiBox Detector

Single Shot MultiBox Detector (SSD [11, 12]) makes predictions based on feature maps taken at different stages, then it divides each one into a pre-established set of bounding boxes with different aspect ratios and scales. The bounding boxes adjust itself to better match the target object. The network generates scores using a regression technique for estimate the presence of each object category in each bounding box. The SSD increases its robustness to scale variations by concatenating feature maps from different resolutions into a final feature map. This network generates scores for each object category in each bounding box and produces adjustments to the bounding box that better match the object shape. At the end, a non-maximum suppression is applied to reduce redundant detections (Fig. 2).

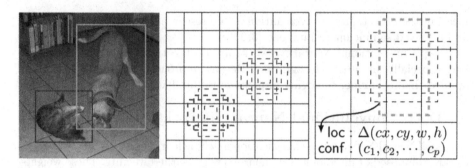

Fig. 2. (a) The final detection produced by the SSD. (b) A feature map with 8×8 grid. (c) A feature map with 4×4 grid and the output of each box, the location and scores for each class. Image extracted from [11].

SSD Variants. The SSD approach uses a base network to extract features from images and use them in detection layers. The extra layers in the SSD are responsible for detecting the object. There are some differences between SSD 300/500 and SSD 512. The SSD 512 is an upgrade of SSD 500, the improvements are presented as follows:

1. The pooling layer (*pool6*) between fully connected layers (*fc6* and *fc7*) was removed;
2. The authors added convolutional layers as extra layers;
3. A new color distortion data augmentation, used for improving the quality of the image, is also added;
4. The network populates the dataset by getting smaller training examples from expanded images;
5. Better proposed bounding boxes by extrapolating the image's boundary.

4   Our Approaches and Contributions

Logos detection can be considered a sub problem of object detection since they usually are objects with a planar surface. Our approaches are based on the SSD framework since it performs very well in object detection. We explore the performance of SDD model on logo images domain. We analyze the impact of using pretrained weights, rather than training from scratch, with the technique called transfer learning. We compare different implementations of the SSD and we also explore the impact of warping image transformations to meet the shape requirements of the SSD input layer.

4.1   Transfer Learning Methodology

To use the transfer learning technique was necessary to redesign the DCNN. This re-design remaps the last layer, adapting the class labels between two different datasets. Therefore, all convolution and pooling layers are kept the same, and the last fully-connected layers (responsible for classification) are reorganized for the new dataset. For logos detection, the fine-tuning was made over a pretrained network, trained for 160.000 iterations on PASCAL VOC2007 + VOC2012 + COCO datasets [12].

4.2 Our Proposal Approaches

We explored 5 different approaches, Table 1 shows all different setups. All networks were trained for 100.000 iterations using the Nesterov Optimizer [14] with a fixed learning rate of 0.001. The SSD 300 and SSD 500 were only explored using pretrained weights because they were easily surpassed by the SSD 512. The approach SSD 500 AR was an attempt to reduce the warp transformation of the input image since in the training and testing phase, the SSD needs to fit the input image into a square resolution.

Table 1. Our five proposal approaches

Acronym	Method	Training details	Extra
SSD 300	SSD 300	Pretrained	
SSD 500	SSD 500	Pretrained	
SSD 500 AR	SSD 500	Pretrained	Preserving aspect ratio
SSD 512 FS	SSD 512	From scratch	
SSD 512 PT	SSD 512	Pretrained	

5 Experiments

We evaluate and analyze our approaches on FlickrLogos-32 dataset [19]. Our experiments ran on the Caffe deep learning framework [9] and using 2× Nvidia Tesla K80. We first describe the dataset, then we compare the performance of our approaches and finally we compare our results to state-of-the-art methods in logo recognition.

5.1 DataSet

FlickrLogos-32 (FL32) is a challenging dataset and the most promising approaches in logo recognition experimented their proposals on it. This dataset was proposed by Romberg [19], many approaches evaluated their performances on this dataset [3,5,8,18]. Romberg also defined an experimental protocol, splitting the dataset into training, validation and testing sets. In all approaches, we strictly follow this protocol. Table 2 shows the distribution between, train, validation and test sets. We have used P1 + P2 (except no-logos) for training and P3 for testing.

Table 2. Evaluation protocol table

Subset	Description	Images	Sum
P1	Hand-picked images, single logo, clean background	10 per class	320
P2	Images showing at least a single logo under various views	30 per class	3960
	Non-logo images	3000	
P3	Images showing at least a single logo under various views	30 per class	3960
	Non-logo images	3000	
Total			8240

5.2 Comparison of Our Approaches

All the 5 different approaches are represented in the left chart of Fig. 3, while the right figure shows the metrics for our best approach, the SSD pretrained. Analyzing the figure we can see that the approaches SSD 300, SSD 500 and SSD 500 AR achieved poor results if compared to the SSD 512. We see that in all cases using pretrained weights resulted in better performance. Analyzing only the best result, SSD 512 PT, we see that we achieve our best F-score with a threshold of 90.

5.3 Comparison Against Other Researches

The comparison among other researches and our best result (the SSD 512 with pretrained weights) can be seen in the Table 3. Analyzing the results we can see that our method outperforms by 2.5% the F-score and by 7.4% the recall

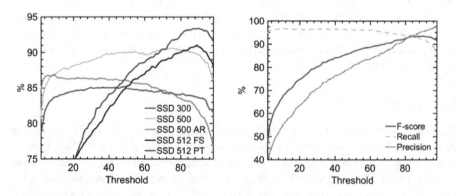

Fig. 3. The left image shows the F-score of out 5 different approaches. The right image shows the F-score, Precision and Recall of the best approach (AR - Preserving aspect ratio, FS - From scratch and PT - Fine-tuning of a pretrained model).

of the state-of-the-art. The high recall achieved is due to the fact that the SSD uses some of its extra layers to estimate the object location and also it can well generalize the object. The approach proposed by Li et al. [10] achieved such high precision due to the process of feature matching that eliminates false positive matches.

Table 3. Comparison of out best approach with the methods in the state-of-the-art (HCF - Hand Crafted Feature, DL - Deep Learning)

Method	Method	Year	Precision	Recall	F1
Romberg et al. [19]	HCF	2011	0.981	0.610	0.752
Revaud et al. [17]	HCF	2012	0.980	0.726	0.841
Romberg et al. [18]	HCF	2013	0.999	0.832	0.908
Li et al. [10]	HCF	2014	**1.000**	0.800	0.890
Bianco et al. [1]	DL	2015	0.909	0.845	0.876
Eggert et al. [5]	DL	2015	0.996	0.786	0.879
Oliveira et al. [15]	DL	2016	-	-	0.890
Bianco et al. [2]	DL	2017	0.976	0.676	0.799
Our	DL	2017	0.954	**0.919**	**0.933**

6   Conclusion

In this work, we investigated the use of DCNN, transfer learning and data augmentation on logo recognition system. The combination among them has shown that DCNN is very suitable for this task, even with relatively small train set it provides greater recall and f-score. A relevant contribution of this paper is the use of data augmentation combined with transfer learning to surpass the lower data issue and allow to use deeper networks. These techniques improve the performance of DCNN in this scenario. The results of our approach reinforce the robustness of DCNN approach, which surpasses the F1-score literature results.

Acknowledgements. The authors thank UFOP and funding Brazilian agency CNPq.

References

1. Bianco, S., Buzzelli, M., Mazzini, D., Schettini, R.: Logo recognition using CNN features. In: Murino, V., Puppo, E. (eds.) ICIAP 2015. LNCS, vol. 9280, pp. 438–448. Springer, Cham (2015). https://doi.org/10.1007/978-3-319-23234-8_41
2. Bianco, S., Buzzelli, M., Mazzini, D., Schettini, R.: Deep learning for logo recognition. arXiv preprint arXiv:1701.02620 (2017)
3. Boia, R., Florea, C.: Homographic class template for logo localization and recognition. In: Paredes, R., Cardoso, J.S., Pardo, X.M. (eds.) IbPRIA 2015. LNCS, vol. 9117, pp. 487–495. Springer, Cham (2015). https://doi.org/10.1007/978-3-319-19390-8_55

4. Chatfield, K., Simonyan, K., Vedaldi, A., Zisserman, A.: Return of the devil in the details: delving deep into convolutional nets. arXiv preprint arXiv:1405.3531 (2014)
5. Eggert, C., Winschel, A., Lienhart, R.: On the benefit of synthetic data for company logo detection. In: Proceedings of the 23rd Annual ACM Conference on Multimedia Conference, pp. 1283–1286. ACM (2015)
6. Francesconi, E., Frasconi, P., Gori, M., Marinai, S., Sheng, J.Q., Soda, G., Sperduti, A.: Logo recognition by recursive neural networks. In: Tombre, K., Chhabra, A.K. (eds.) GREC 1997. LNCS, vol. 1389, pp. 104–117. Springer, Heidelberg (1998). https://doi.org/10.1007/3-540-64381-8_43
7. Hoi, S.C., Wu, X., Liu, H., Wu, Y., Wang, H., Xue, H., Wu, Q.: Logo-net: large-scale deep logo detection and brand recognition with deep region-based convolutional networks. arXiv preprint arXiv:1511.02462 (2015)
8. Iandola, F.N., Shen, A., Gao, P., Keutzer, K.: Deeplogo: hitting logo recognition with the deep neural network hammer. arXiv preprint arXiv:1510.02131 (2015)
9. Jia, Y., Shelhamer, E., Donahue, J., Karayev, S., Long, J., Girshick, R., Guadarrama, S., Darrell, T.: Caffe: convolutional architecture for fast feature embedding. arXiv preprint arXiv:1408.5093 (2014)
10. Li, K.W., Chen, S.Y., Su, S., Duh, D.J., Zhang, H., Li, S.: Logo detection with extendibility and discrimination. Multimedia Tools Appl. $72(2)$, 1285–1310 (2014)
11. Liu, W., Anguelov, D., Erhan, D., Szegedy, C., Reed, S.: SSD: single shot multibox detector. arXiv preprint arXiv:1512.02325 (2015)
12. Liu, W., Anguelov, D., Erhan, D., Szegedy, C., Reed, S., Fu, C.-Y., Berg, A.C.: SSD: single shot MultiBox detector. In: Leibe, B., Matas, J., Sebe, N., Welling, M. (eds.) ECCV 2016. LNCS, vol. 9905, pp. 21–37. Springer, Cham (2016). https://doi.org/10.1007/978-3-319-46448-0_2
13. Lowe, D.G.: Distinctive image features from scale-invariant keypoints. Int. J. Comput. Vis. $60(2)$, 91–110 (2004)
14. Nesterov, Y.: A method of solving a convex programming problem with convergence rate O(1/k2) (1983)
15. Oliveira, G., Frazão, X., Pimentel, A., Ribeiro, B.: Automatic graphic logo detection via fast region-based convolutional networks. arXiv preprint arXiv:1604.06083 (2016)
16. Pan, S.J., Yang, Q.: A survey on transfer learning. IEEE Trans. knowl. Data Eng. $22(10)$, 1345–1359 (2010)
17. Revaud, J., Douze, M., Schmid, C.: Correlation-based burstiness for logo retrieval. In: Proceedings of the 20th ACM International Conference on Multimedia, pp. 965–968. ACM (2012)
18. Romberg, S., Lienhart, R.: Bundle min-hashing for logo recognition. In: Proceedings of the 3rd ACM International Conference on Multimedia Retrieval, pp. 113–120. ACM (2013)
19. Romberg, S., Pueyo, L.G., Lienhart, R., Van Zwol, R.: Scalable logo recognition in real-world images. In: Proceedings of the 1st ACM International Conference on Multimedia Retrieval, p. 25. ACM (2011)
20. Yang, F., Bansal, M.: Feature fusion by similarity regression for logo retrieval. In: 2015 IEEE Winter Conference on Applications of Computer Vision (WACV), p. 959. IEEE (2015)

Retinal Vessels Segmentation Based on a Convolutional Neural Network

Nadia Brancati[1], Maria Frucci[1], Diego Gragnaniello[1], and Daniel Riccio[1,2(✉)]

[1] Institute for High Performance Computing and Networking National Research Council of Italy (ICAR-CNR), Naples, Italy
{nadia.brancati,maria.frucci,diego.gragnaniello}@cnr.it
[2] University of Naples "Federico II", Naples, Italy
daniel.riccio@unina.it

Abstract. We present a supervised method for vessel segmentation in retinal images. The segmentation issue has been addressed as a pixel-level binary classification task, where the image is divided into patches and the classification (vessel or non-vessel) is performed on the central pixel of the patch. The input image is then segmented by classifying all of its pixels. A Convolutional Neural Network (CNN) has been used for the classification task, and the network has been trained on a large number of samples, in order to obtain an adequate generalization ability. Since blood vessels are characterized by a linear structure, we have introduced a further layer into the classic CNN including directional filters. The method has been tested on the DRIVE dataset producing satisfactory results, and its performance has been compared to that of other supervised and unsupervised methods.

Keywords: Retinal image · Vessel segmentation
Convolutional Neural Network · Directional filters

1 Introduction and Background

The automatic segmentation of blood vessels from retinal fundus images has gained interest in the image processing community, due to its applicability to several problems in different fields. Precisely, the automatic analysis of retinal blood vessels is a basic step both for the diagnosis of several retinal pathologies [2] (e.g. diabetic retinopathy, arteriosclerosis, hypertension, and various cardiovascular diseases), and for person verification in biometric systems, since the retinal vascular structure is different for each individual [15]. Several methods have been proposed in the literature for retinal image segmentation since it remains a challenging task due to the complex nature of vascular structures, illumination variations and the anatomical variability between subjects. Existing methods can generally be divided into two categories: supervised and unsupervised. In general, the performance of supervised methods is superior to that of the unsupervised ones, but with a lower speed and higher computational complexity compare to

© Springer International Publishing AG, part of Springer Nature 2018
M. Mendoza and S. Velastín (Eds.): CIARP 2017, LNCS 10657, pp. 119–126, 2018.
https://doi.org/10.1007/978-3-319-75193-1_15

other unsupervised methods. Supervised methods are based on machine learning techniques and require a manually annotated set of training images in order to classify a pixel as either a vessel or non-vessel. They involve a k-NN classifier, a Support Vector Machine, a Bayesian classifier in combination with features obtained through the multi-scale analysis of Gabor wavelets, AdaBoost or a CNN, etc. [6,8,11,17,19–21]. Unsupervised segmentation methods work without any prior knowledge and are based on matched filtering, centerline tracking, mathematical morphology, or other rule-based techniques [3–5,7,12,22–24].

In this paper a supervised method based on a CNN and on the use of directional filters, as proposed in [7], is presented. The method has been tested on the DRIVE [19] dataset and its performance has been compared to that of other methods in the literature. The experimental results confirm the goodness of the proposed approach.

The rest of the paper is organized as follows: in Sect. 2, the proposed CNN architecture is presented; Sect. 3 describes the experiments, together with the training strategy, and shows the obtained results; and finally in Sect. 4 some conclusions are drawn.

2   The Method

Our solution is based on a CNN used as a pixel classifier and on the introduction of directional filters. As performed by the majority of authors in the field, only the green channel of RGB color retinal image is considered, since the vessels are characterized by the highest contrast in this channel. In this work, the vessel segmentation issue is addressed as a pixel-level binary classification task. The network computes the probability of a pixel being a vessel, using as input a patch of the image, representing a square window centered on the pixel itself. Next, the input image is segmented by classifying all its pixels. The CNN is trained on a high number of patches, in which the central pixel is annotated by using the relative ground truth included in the dataset. Details about the training dataset will be given in the next section.

A CNN is organized in stacked trainable stages, called *layers*, each composed of processing units operating on the output of the previous layer. A CNN consists of a number of convolutional and sub-sampling layers optionally followed by fully connected layers. A convolutional layer will have k *filters* (or kernels) devoted to produce k *feature maps*. Each map is then sub-sampled typically by means of pooling layers, a process that progressively reduces the spatial size of the representation, the amount of parameters and the computation in the network. The pooling can be of different types: Max, Average, Sum, etc., but max pooling is usually applied where the largest element from the current feature map within a window is considered [16]. The output from the convolutional and pooling layers consists in high-level features of the input image. The purpose of the Fully Connected Layers tecnique is to use these features to classify the input image into various classes based on the training dataset.

The layers of the proposed CNN architecture are:

- **A first non-learnable convolutional layer**: to obtain a better performance of our method, we have introduced a new layer representing directional filters to guide the training of the network, also on the linear behavior which characterizes blood vessels. In fact, vessels are thin and elongated structures whose pixels are aligned along different directions. Thus, by fixing a number of different orientations and by taking into account the gray-levels in a suitable window centered on a pixel p, directional information can be computed for p. Next, directional information is combined by higher layers of network to obtain more complex features. Differently from the filters of other Convolutional layers, the directional filters are not learned during the training process. The directional filters consist in twelve windows, each having a size of 7×7, like the ones presented in [7] (see Fig. 1). Each window represents a direction such that an angle of $15°$ between two successive directions is obtained.
- **Five convolutional layers**: all the filters of these layers have a size of 3×3 and a stride of 1. The first layer of this block learns 32 filters, the second and the third learn 64 filters and finally the fourth and fifth learn 128 filters. All the filters of these layers are initialized with a Xavier initialization [9] and they are equipped with rectification non linearity (ReLU) [13], i.e. the output volume is $max(0, e)$, where e is the outcome of convolution.
- **Five max-pooling layers**: max-pooling is performed after each convolutional layer. It is computed on a window of size 3×3 and the stride is set to 2.
- **Three fully connected layers**: the first two fully connected layers learn 256 filters, and these are used to learn non-linear combinations of the features provided from the previous layers. Moreover, to try to avoid overfitting, these two layers implement dropout regularization [18], where the ratio is set to 0.5. Finally, the number of filters of the last layer is 2, since in our case the classification problem is binary. All the filters of these layers are initialized with values sampled from the $N(0, 0.01)$ Gaussian distribution.

Table 1 summarizes the parameters of the CNN layers, where n-C stands for the Convolutional layer with non-learnable filters, C + P stands for the Convolutional layer followed by Max-pooling layer and FC stands for the Fully Connected layer.

Fig. 1. The 12 Directional filters implemented in the first layer of the proposed CNN.

Table 1. Summary of the proposed CNN architecture

	Layers								
	1	2	3	4	5	6	7	8	9
Type	n − C	C + P	C + P	C + P	C + P	C + P	FC	FC	FC
Number of filters	12	32	64	64	128	128	256	256	2
Filter Size	7×7	3×3	3×3	3×3	3×3	3×3	-	-	-
Convolution Stride	1	1	1	1	1	1	-	-	-
Pooling Size	-	2×2	2×2	2×2	2×2	2×2	-	-	-
Pooling Stride	-	2	2	2	2	2	-	-	-

3 Experiments

3.1 Training Strategy and Parameters Setting

The training phase consists in an iterative presentation of the patches together with their associated labels. Patches are randomly extracted from the set of training images of the DRIVE dataset. In particular, DRIVE contains 20 training images and 20 testing images and each image is associated with both a mask delimiting the Field of View (FOV) of the retinal image and two ground truths generated by two ophthalmologists. However, only patches completely contained in the FOV of the retinal images and only the ground truths of the first expert are taken into account. Each patch of an image I has a 27×27 size and is labeled as a vessel or non-vessel depending on whether its central pixel belongs to foreground or background of the ground truth associated with I. The experiments have been performed considering two types of training sets: a non-balanced set in which most of the patches are labeled as non-vessels and a balanced set including a balanced percentage of patches differently labeled. Precisely, the non-balanced training set is composed of $480,000$ random patches, while the balanced set includes about $700,000$ patches.

For the testing phase, patches are extracted by DRIVE test images. In particular, for each pixel p of a test image I, the patch centered on p is obtained and it is involved in the testing phase only if it is completely contained in the FOV of I.

The number of epochs of the network is set to 15, while the batch size is equal to 256. The learning rate initially is set to 10^{-2}, and it is decreased every six epochs by a factor of 10. To train our network, we used the NVIDIA Deep Learning GPU Training System [1] within the Caffe framework [10].

3.2 Results

A qualitative evaluation of the method is possible with reference to Fig. 2, where in each line the input image, the ground truth of the first expert, and the result of our segmentation method are shown from left to right.

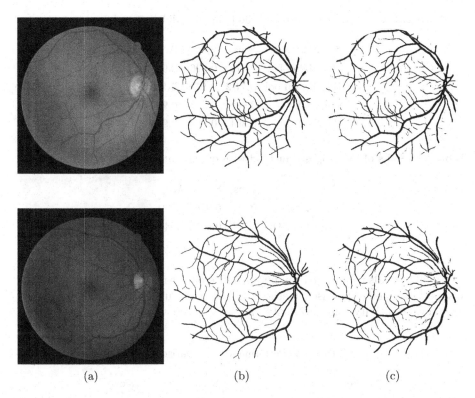

(a)   (b)   (c)

Fig. 2. Results of our method: (a) original image, (b) ground truth, (c) results of the proposed method.

We have quantitatively evaluated our method for both training sets, balanced and non-balanced, and in both cases with or without the max-pooling layers and also with or without directional filters. We have computed the Accuracy, Sensitivity and Specificity, as performed by the majority of researchers [14].

Quantitative results by applying different training strategies are given in Table 2. For the non-balanced training set, the best result is obtained without the max-pooling layers in respect of all the considered measures. For the balanced training set, we obtained a better performance as regards accurary and specifity without max-pooling layers, while a lower performance was obtained only as regards sensitivity. However, the non-balanced training set and the network without max-pooling layers provided the best performance in terms of accuracy and specificity and also gave a high sensitivity value.

To demonstrate how the introduction of directional filters in the network provides a better performance of our method, we computed the considered measures also with and without directional filters considering the $NoBal/NoPool$ strategy (see Table 3). We observed that the use of directional filters produces a high sensitivity value, while equivalent values are obtained for the remaining measures.

Table 2. Results of the proposed method applying different training strategies

Train. Strat	Sens	Spec	Acc
NoBal/Pool	0.773	0.976	0.948
NoBal/NoPool	**0.782**	**0.976**	**0.949**
Bal/Pool	0.909	0.921	0.919
Bal/NoPool	0.891	0.932	0.926

Table 3. Results of the proposed method with or without the directional filters (DF)

	Sens	Spec	Acc
With DF	0.782	0.976	0.949
Without DF	0.769	0.977	0.949

Finally, we also compared the performance of our method with that of other unsupervised and supervised methods in the literature. The average values of accuracy, sensitivity and specificity can be checked with reference to Table 4,

Table 4. Performance Comparisons

Methods	Sens	Spec	Acc
Supervised Methods			
2nd human observed	0.776	0.972	0.947
Staal et al. [19]	-	-	0.944
Soares et al. [17]	0.733	0.978	0.946
Marin et al. [11]	0.706	0.980	0.945
Strisciuglio et al. [20]	0.777	0.970	0.945
Vega et al. [21]	0.744	0.960	0.941
Fu et al. [8]	0.729	-	0.947
Fraz et al. [6]	0.741	0.981	0.948
Proposed Method	**0.782**	0.976	0.949
Unsupervised Methods			
Al-Rawi et al. [3]	-	-	0.953
Zhang et al. [23]	0.712	0.972	0.938
Yin et al. [22]	0.652	0.971	0.926
Mendonca and Campilho [12]	0.734	0.976	0.946
Azzopardi et al. [4]	0.765	0.970	0.944
Frucci et al. [7]	0.660	**0.985**	**0.956**
Zhao et al. [24]	0.742	0.982	0.954

where the highest values are in bold. Our method has a better performance as regards sensitivity with respect to all methods, a better performance as regards accuracy with respect to the supervised methods and, and a lower performance only as regards specificity with respect to some methods.

4 Conclusion

In this work, we have presented a supervised vessel segmentation method based on a Convolutional Neural Network. The adopted CNN architecture includes a specific layer to compute the directional features. The introduction of this layer allows an improved performance of the network in terms of sensitivity. The method provides results that are satisfactory both from a qualitative point of view and quantitatively. The performance of the method has been checked on the DRIVE dataset in terms of evaluation parameters such as accuracy, sensitivity and specificity. Comparisons have also been made with other unsupervised and supervised methods in the literature, showing that the suggested method has the highest performance in terms of sensitivity.

References

1. NVIDIA DIGITS. https://developer.nvidia.com/digits
2. Abràmoff, M.D., Garvin, M.K., Sonka, M.: Retinal imaging and image analysis. IEEE Rev. Biomed. Eng. **3**, 169–208 (2010)
3. Al-Rawi, M., Qutaishat, M., Arrar, M.: An improved matched filter for blood vessel detection of digital retinal images. Comput. Biol. Med. **37**(2), 262–267 (2007)
4. Azzopardi, G., Strisciuglio, N., Vento, M., Petkov, N.: Trainable cosfire filters for vessel delineation with application to retinal images. Med. Image Anal. **19**(1), 46–57 (2015)
5. Chutatape, O., Zheng, L., Krishnan, S.M.: Retinal blood vessel detection and tracking by matched Gaussian and Kalman filters. In: Proceedings of the 20th Annual International Conference of the IEEE Engineering in Medicine and Biology Society, 1998, vol. 6, pp. 3144–3149. IEEE (1998)
6. Fraz, M.M., Remagnino, P., Hoppe, A., Uyyanonvara, B., Rudnicka, A.R., Owen, C.G., Barman, S.A.: An ensemble classification-based approach applied to retinal blood vessel segmentation. IEEE Trans. Biomed. Eng. **59**(9), 2538–2548 (2012)
7. Frucci, M., Riccio, D., Sanniti di Baja, G., Serino, L.: Direction-based segmentation of retinal blood vessels. In: Beltrán-Castañón, C., Nyström, I., Famili, F. (eds.) CIARP 2016. LNCS, vol. 10125, pp. 1–9. Springer, Cham (2017). https://doi.org/10.1007/978-3-319-52277-7_1
8. Fu, H., Xu, Y., Wong, D.W.K., Liu, J.: Retinal vessel segmentation via deep learning network and fully-connected conditional random fields. In: 2016 IEEE 13th International Symposium on Biomedical Imaging (ISBI), pp. 698–701. IEEE (2016)
9. Glorot, X., Bengio, Y.: Understanding the difficulty of training deep feedforward neural networks. Proc. Int. Conf. Artif. Intell. Stat. **9**, 249–256 (2010)
10. Jia, Y., Shelhamer, E., Donahue, J., Karayev, S., Long, J., Girshick, R., Guadarrama, S., Darrell, T.: Caffe: convolutional architecture for fast feature embedding. In: Proceedings of the 22nd ACM International Conference on Multimedia. pp. 675–678. ACM (2014)

11. Marín, D., Aquino, A., Gegúndez-Arias, M.E., Bravo, J.M.: A new supervised method for blood vessel segmentation in retinal images by using gray-level and moment invariants-based features. IEEE Trans. Med. Imaging **30**(1), 146–158 (2011)

12. Mendonca, A.M., Campilho, A.: Segmentation of retinal blood vessels by combining the detection of centerlines and morphological reconstruction. IEEE Trans. Med. Imaging **25**(9), 1200–1213 (2006)

13. Nair, V., Hinton, G.E.: Rectified linear units improve restricted Boltzmann machines. In: Proceedings of the 27th International Conference on Machine Learning (ICML-10), pp. 807–814 (2010)

14. Powers, D.M.: Evaluation: from precision, recall and F-measure to ROC, informedness, markedness and correlation (2011)

15. Ramya, M., Sornalatha, M.: Personal identification based on retinal blood vessel segmentation. Int. J. Emerg. Technol. Comput. Sci. Electron. **7**(1), 164–168 (2014)

16. Scherer, D., Müller, A., Behnke, S.: Evaluation of pooling operations in convolutional architectures for object recognition. Artif. Neural Netw. ICANN **2010**, 92–101 (2010)

17. Soares, J.V., Leandro, J.J., Cesar, R.M., Jelinek, H.F., Cree, M.J.: Retinal vessel segmentation using the 2-D Gabor wavelet and supervised classification. IEEE Trans. Med. Imaging **25**(9), 1214–1222 (2006)

18. Srivastava, N., Hinton, G.E., Krizhevsky, A., Sutskever, I., Salakhutdinov, R.: Dropout: a simple way to prevent neural networks from overfitting. J. Mach. Learn. Res. **15**(1), 1929–1958 (2014)

19. Staal, J., Abràmoff, M.D., Niemeijer, M., Viergever, M.A., Van Ginneken, B.: Ridge-based vessel segmentation in color images of the retina. IEEE Trans. Med. Imaging **23**(4), 501–509 (2004)

20. Strisciuglio, N., Azzopardi, G., Vento, M., Petkov, N.: Supervised vessel delineation in retinal fundus images with the automatic selection of B-COSFIRE filters. Mach. Vis. Appl. **27**(8), 1137–1149 (2016)

21. Vega, R., Sanchez-Ante, G., Falcon-Morales, L.E., Sossa, H., Guevara, E.: Retinal vessel extraction using lattice neural networks with dendritic processing. Comput. Biol. Med. **58**, 20–30 (2015)

22. Yin, Y., Adel, M., Bourennane, S.: Automatic segmentation and measurement of vasculature in retinal fundus images using probabilistic formulation. Comput. Math. Methods Med. **2013** (2013)

23. Zhang, B., Zhang, L., Zhang, L., Karray, F.: Retinal vessel extraction by matched filter with first-order derivative of Gaussian. Comput. Biol. Med. **40**(4), 438–445 (2010)

24. Zhao, Y., Rada, L., Chen, K., Harding, S.P., Zheng, Y.: Automated vessel segmentation using infinite perimeter active contour model with hybrid region information with application to retinal images. IEEE Trans. Med. Imaging **34**(9), 1797–1807 (2015)

DNLM-IIFFT: An Implementation of the Deceived Non Local Means Filter Using Integral Images and the Fast Fourier Transform for a Reduced Computational Cost

Saúl Calderón Ramírez and Manuel Zumbado Corrales[✉]

Computing School, Costa Rican Institute of Technology,
PAttern Recognition and MAchine Learning Group (PARMA-Group),
159-7050 Cartago, Costa Rica
sacalderon@itcr.ac.cr, manu3193@gmail.com

Abstract. In this paper we propose an efficient implementation of the Deceived Non Local Means filter, using Integral Images and the Fast Fourier Transform, named DNLM-IFFT. The deceived non local means filter is part of the Deceived Weighted Averaging Filter Framework (DeWAFF), which defines an approach for image abstraction with a combination of unsharp masking for contrast and edges enhancement and weighted averaging filtering for noise reduction. The proposed optimization approach achieved a speedup factor up to 10.

1 Introduction

Image processing algorithms are commonly used nowadays in different fields, from medicine to sports (Aggarwal and Ryoo [1]; Ekin et al. [6]; Li et al. [8]). For example, in Sáenz et al. [12], the authors present applications of behavior analysis of cells in cancerous tissue, where an automatic cell segmentation and tracking system was proposed to assist researchers to achieve a faster and more accurate understanding of the behavior of a group of cells. The algorithms implemented in such systems for image enhancement, object segmentation and tracking, usually involve high computational cost, frequently making usage of these systems impractical. Hence, it becomes an important issue to optimize the execution time of cell tracking solutions in order to provide researchers in the field an efficient method, as researchers often need to process thousands of images from cell activity. The first stage in the proposed cell tracking pipeline in Sáenz et al. [12] corresponds to the preprocessing of the input image, designed to enhance edges and contrast and also to remove noise, for improving the cell segmentation and tracking stages performance.

This work presents a computational optimization of the Deceived Non Local Means filter (DNLM), a filter proposed in the Deceived Weighted Averaging Filter Framework (DeWAFF) (Calderón et al. [3]), for image enhancement and noise removal. The proposed optimization named as DNLM-IFFT, uses integral

© Springer International Publishing AG, part of Springer Nature 2018
M. Mendoza and S. Velastín (Eds.): CIARP 2017, LNCS 10657, pp. 127–134, 2018.
https://doi.org/10.1007/978-3-319-75193-1_16

images and the Fast Fourier Transform (FFT), as proposed for the original Non Local Means (NLM) filter in Wang et al. [16]. We present an evaluation of the proposed DNLM-IFFT approach for a speedup assessment. As demonstrated in Calderón et al. [3], the DNLM presented the best results in comparison to the deceived bilateral and the deceived scaled bilateral filters (part of the DeWAFF), as it is based in the NLM noise removal approach (Buades et al. [2]). However, the computational complexity of the DNLM brute force implementation often makes its usage impractical, as researchers frequently need to analyze thousands of images. For this reason, in this paper we focus on the optimization of the DNLM algorithm.

In Sect. 2, we describe important aspects of the DNLM filter. In Sect. 3, we present the proposed computational optimization of the DNLM. Section 5 presents the experiments and test results, comparing our strategy to the previous brute force implementation in order to demonstrate the viability of the proposed modification. Finally in Sect. 6 we present the conclusions and future work related.

2   Background

As presented in Calderón et al. [3], the DeWAFF consists in an image abstraction framework. Image abstraction is defined in Calderón et al. [3] as the enhancement of perceptually important characteristics of the image as contrast and edges, and the simplification of less relevant details for the application domain and noise. As demonstrated in Calderón and Siles [4]; Calderón et al. [3]; Sáenz et al. [12], image abstraction based preprocessing techniques are able to improve the performance of object segmentation, tracking and classification tasks.

The DeWAFF proposes the combination of an Adaptive Unsharp Masking (AUSM) based approach, as the one proposed in Ramponi and Polesel [11], with a weighted averaging filter, like the bilateral filter or the NLM filter. Such combination aims to achieve both contrast and edge enhancement, as also noise suppression. The combination proposed uses the AUSM image output F_{AUSM} as input of the weighted averaging filter algorithm, but uses the original input image U for weighting, *deceiving* the weighted averaging filter. Such combination diminishes the ringing effect, often present in USM filtered images. The following equation presents the general DeWAFF grasp for an input image U of $a \times b = c$ pixels, with $U^a = U$ and $U^b = F_{\mathrm{AUSM}}$ for ringing suppression purposes:

$$U'(p) = \left(\sum_{i \in \Omega_p} w_k\left(U^a, p, i\right) \right)^{-1} \left(\sum_{i \in \Omega_p} w_k\left(U^a, p, i\right) U^b(i) \right). \tag{1}$$

For notation simplicity, an image pixel is represented as $i = (x_i, y_i)$, and Ω_p stands for the $n \times n$ window centered in pixel p.

The weighting functions studied in Calderón et al. [3] were the Bilateral Filter (BF), the Scaled Bilateral Filter (SBF) and NLM filter:

$$w_1 = w_{\mathrm{BF}}, w_2 = w_{\mathrm{SBF}}, w_3 = w_{\mathrm{NLM}}. \tag{2}$$

The NLM kernel when used in Eq. 1 provides the best results under heavy and moderated noise conditions, given the similarity approach of weighting the neighborhood Φ_i of pixel i with $m \times m$ pixels, instead of using the pixels intensity similarity, as in the bilateral filter and the SBF approaches. The following equation presents the original non local means weighting grasp:

$$w_{\text{NLM}}(p, i) = \exp\left(-\frac{D(i, j)}{h^2}\right), \tag{3}$$

with the Euclidean distance between the neighborhoods given by:

$$D(i, j) = \| \Phi_i - \Phi_j \|^2, \tag{4}$$

where h is the kernel parameter which defines neighborhood weighting discrimination.

In the worst case scenario of using a window of $a \times b = c$ and a neighborhood of $a \times b = c$, pixels, the computational complexity of the NLM is $O(c^3)$, where c corresponds to the total number of pixels in the input image U.

Let us further examine the Euclidean distance between neighborhoods:

$$D(i, j) = \sum_u (\Phi_i(u) - \Phi_j(u))^2, \tag{5}$$

where $\sum_u$ stands for the summation of the pixels within the difference of the two neighborhoods Φ_i and Φ_j.

Developing the squared difference and rewritting the quadratic terms as $\Phi_i^2 = \sum_u \Phi_i(u)^2$, we get:

$$D(i, j) = \Phi_i^2 - 2\sum_u \Phi_j(u)\Phi_i(u) + \Phi_j^2 = \Phi_i^2 - 2\Phi_j \cdot \Phi_i + \Phi_j^2, \tag{6}$$

where $\Phi_j \cdot \Phi_i$ refers to the matrix dot product. The following example illustrates the calculation of the Euclidean distance between neighborhoods with an input image $U \in \mathbb{R}^{5\times5}$ defined in Table 1.

Table 1. Image example U.

	1	2	3	4	5
1	5	12	1	3	2
2	5	2	3	1	4
3	3	1	2	3	1
4	4	3	2	1	3
5	1	5	6	5	3

For such illustration purposes, we take a window $\Omega_{(3,3)} \in \mathbb{R}^{3\times3}$ and perform the calculation of the Euclidean distance between the neighborhoods $\Phi_{(3,3)}$ and $\Phi_{(2,2)}$:

$$\left\|\Phi_{(3,3)} - \Phi_{(2,2)}\right\|^2 = \left\|\begin{matrix} 2\,3\,1 & 5\,12\,1 \\ 1\,2\,3 - 5\,2\,3 \\ 3\,2\,1 & 3\,1\,2 \end{matrix}\right\| = 10.3923^2 = 108. \qquad (7)$$

Given the development of the quadratic difference between neighborhoods in the right part of Eq. 6, we can perform such calculation based in the element wise squared matrix of $U \in \mathbb{R}^{5\times5}$. This allows calculating the Euclidean distance done previously equivalent to perform:

$$\left\|\Phi_{(3,3)} - \Phi_{(2,2)}\right\| = \Phi^2_{(3,3)} - 2\Phi_{(2,2)} \cdot \Phi_{(3,3)} + \Phi^2_{(2,2)} \qquad (8)$$

$$= \sum \begin{pmatrix} 4\,9\,1 \\ 1\,4\,9 \\ 9\,4\,1 \end{pmatrix} - 2 \begin{pmatrix} 2\,3\,1 & 5\,12\,1 \\ 1\,2\,3 \cdot 5\,2\,3 \\ 3\,2\,1 & 3\,1\,2 \end{pmatrix} + \sum \begin{pmatrix} 25\,144\,1 \\ 25\,4\,9 \\ 9\,1\,4 \end{pmatrix} \qquad (9)$$

$$= 42 + -2 \cdot 78 + 222 = 42 - 156 + 222 = 108, \qquad (10)$$

consistent with the result obtained in Eq. 7.

3   Proposed Method: DNLM-IFFT

As seen in previous section, the computational cost of the brute force implementation of the DNLM is very high. To address such problem, there are different approaches which have been designed for the original NLM algorithm, and could be ported to the DNLM in order to lower computational complexity, as seen in Karnati et al. [7]; Liu et al. [9]; Wang et al. [16]. However, approximations or modifications in the computation of the NLM kernel must allow the decoupling of the weighted image U^a and the filtered image U^b, for a DNLM implementation, as stated in Eq. 1. Also, we prefer an optimization of the NLM over approximations as seen in Dauwe et al. [5]; Vignesh et al. [14]; Xue et al. [17].

In Wang et al. [16], an efficient computation of the weighting function $w_{\text{NLM}}(p, i)$ is proposed using integral images and the FFT. The approach achieves same numeric results compared to the brute force implementation. The optimization proposed in Wang et al. [16], modifies the computation of the weighting kernel, by performing the calculation of the Euclidean distance defined in Eq. 6 as follows:

- Use of integral images to calculate the terms Φ^2_i y Φ^2_j.
- Calculate the FFT to obtain the term $-2\Phi_j \cdot \Phi_i$ which corresponds to a sample of the auto-correlation with the signal Φ_j.

3.1   Integral Images

The integral image of I as defined in Viola and Jones [15], with a pair wise pixel notation $i = (x, y)$, is given by:

$$I_\Sigma(x, y) = \sum_{u \leq x, v \leq y} I(u, v), \qquad (11)$$

Table 2. Integral image for U^2.

	1	2	3	4	5
1	25(A)	169	170	179(B)	183
2	50	198	208	218	238
3	59	208	222	241	262
4	75(C)	233	251	271(D)	301
5	76	259	313	358	397

which is the summation of the pixels to the left and above of the parameter pixel $i = (x, y)$. Following the example exposed previously, the integral image of the pixel wise squared matrix of U is calculated as shown in Table 2. For a more efficient calculation of an integral image, it is worth to note that:

$$I_\Sigma(x, y) = I(x, y) - I_\Sigma(x - 1, y - 1) + I_\Sigma(x, y - 1) + I_\Sigma(x - 1, y), \quad (12)$$

which for instance, for the previous example means that:

$$I_\Sigma(3, 3) = 4 - 198 + 208 + 208 = 222. \quad (13)$$

To use the integral image to calculate the summation over a window limited by pixels $A = (x_0, y_0)$, $B = (x_1, y_0)$, $C = (x_0, y_1)$ and $D = (x_1, y_1)$, as illustrated in Table 2, we compute the following in general:

$$\sum_{x_0 < x \leq x_1, y_0 < y \leq y_1} I(x, y) = I(D) + I(A) - I(B) - I(C), \quad (14)$$

and for the example developed in this paper, for a window of dimensions 3×3 around the pixel $(3, 3)$ the corner pixels would be given as, $A = (1, 1)$, $B = (4, 1)$, $C = (1, 4)$ y $D = (4, 4)$, as specified in Table 2. Thus, the calculation of the summation of the pixels over such window is calculated as follows:

$$\sum_{x_0 < x \leq x_1, y_0 < y \leq y_1} I(3, 3) = 271 + 25 - 179 - 75 = 42, \quad (15)$$

which is consistent with the first term in the Eq. 10. Using the integral image to calculate the summation over a window of an input matrix with c pixels presents a computational cost of $O(c)$, and must be made only once for the NLM filtering.

4   Correlation and the Fast Fourier Transform

As stated previously, the term in the quadratic equation of the Euclidean distance between the neighborhoods of pixels i and j:

$$S(i, j) = N_i^2 - 2 \sum_{a=0}^{m-1} \sum_{b=0}^{m-1} N_j(a, b) N_i(a, b) + N_j^2 = N_i^2 - 2N_j \cdot N_i + N_j^2, \quad (16)$$

corresponding to the dot product of matrices $-2N_j \cdot N_i$, can be extracted by using the correlation of the neighborhood from pixel i, N_i. We can compute the dot product term using the FFT with a logarithmic computational complexity, lowering the overall DNLM computational cost.

5   Experiments and Results

We tested the computational execution time of the algorithm using images with resolutions of 480p, 720p and 1080p executed 10 times each, for a typical window size of 21×21 and a neighborhood size of 7×7, as recommended in Baudes et al. [2]. The results are shown in Table 4. The second experiment consists in evaluating the speedup by using different window and neighborhood sizes in an image of 1200×900 pixels. Its results are shown in Table 3.

Table 3. Time in seconds achieved with different image resolutions for the DNLM with a brute force implementation and the DNLM-IFFT.

Window size	Neighborhood size	DNLM	DNLM-IFFT	Speedup
21	7	566.21	51.90	10.91
21	15	652.97	72.60	8.99
21	21	707.81	184.44	3.84
43	7	2211.40	107.57	20.55
43	15	2644.30	388.02	6.81
43	21	2881.34	555.40	5.19
55	7	3548.80	328.17	10.81
55	15	4271.78	561.05	7.61
55	21	4504.16	298.80	15.07

All tests were executed in a desktop computer with an AMD Phenom FX-6300 processor at 4.6 GHz. Average values for these executions are displayed in Table 4.

Table 4. Time in seconds achieved with different image resolutions for the DNLM with a brute force implementation and the DNLM-IFFT.

Image size	DNLM	DNLM-IFFT	Speedup
480	156.3	16.3	9.6
720	480.3	47.9	10.0
1080	1093.6	110.3	9.9

6 Conclusions and Future Work

The proposed DNLM-IFFT approach achieves an effective speed-up up to an average factor of 10, as seen in Table 4. The results of the experiment according to Table 3, suggests that the DNLM-IFFT becomes more attractive with smaller neighborhood sizes, given a smaller speedup with bigger neighborhoods. For preprocessing low resolution video footages, execution time is reasonable, however, biomedical images and video analysis often require the analysis of thousands of images. As already mentioned, microbiology researchers from the University of Costa Rica need to analyze 170 000 images from glioblastoma tissue. This means that the preprocessing stage with the proposed DNLM-IFFT could take months, which is not appropriate.

As future work, we expect to explore alternative parallelization technologies, such as MPI and OpenMP for a cluster based on the Intel KNL architecture. In Cuomo [13], a GPU based parallelization approach for the Non Local Means filter was proposed, which suggests that a GPU based implementation of the DNLM might achieve greater performance gain when compared with a CPU based parallelization approach. Also, we can consider mixing the DNLM-IFFT proposed approach with other approximation grasps of the NLM, for instance using look-up tables, as seen in Mahmoudi and Sapiro [10].

The MATLAB code implemented for this paper can be downloaded from https://github.com/manu3193/DNLM-IFFTT.

References

1. Aggarwal, J.K., Ryoo, M.S.: Human activity analysis: a review. J. ACM Comput. Surv. CSUR **43**, 16 (2011)
2. Buades, A., Coll, B., Morel, J.M.: Neighborhood filters and PDEs. Numer. Math. **105**(1), 1–34 (2006)
3. Calderón, S., Sáenz, A., Mora, R., Siles, F., Orozco, I., Buemi, M.: Dewaff: a novel image abstraction approach to improve the performance of a cell tracking system. In: 2015 4th International Work Conference on Bioinspired Intelligence (IWOBI), pp. 81–88. IEEE (2015)
4. Calderón, S., Siles, F.: Deceived bilateral filter for improving the classification of football players from TV broadcast. In: IEEE 3rd International Conference and Workshop on Bioinspired Intelligence (2014)
5. Dauwe, A., Goossens, B., Luong, H.Q., Philips, W.: A fast non-local image denoising algorithm. In: Electronic Imaging 2008, p. 681210. International Society for Optics and Photonics (2008)
6. Ekin, A., Tekalp, A.M., Mehrotra, R.: Automatic soccer video analysis and summarization. IEEE Trans. Image Process. **12**(7), 796–807 (2003)
7. Karnati, V., Uliyar, M., Dey, S.: Fast non-local algorithm for image denoising. In: 2009 16th IEEE International Conference on Image Processing (ICIP), pp. 3873–3876. IEEE (2009)
8. Li, K., Miller, E.D., Chen, M., Kanade, T., Weiss, L.E., Campbell, P.G.: Computer vision tracking of stemness. In: 5th IEEE International Symposium on Biomedical Imaging: From Nano to Macro, ISBI 2008, pp. 847–850. IEEE (2008)

9. Liu, Y.L., Wang, J., Chen, X., Guo, Y.W., Peng, Q.S.: A robust and fast non-local means algorithm for image denoising. J. Comput. Sci. Technol. **23**(2), 270–279 (2008)
10. Mahmoudi, M., Sapiro, G.: Fast image and video denoising via nonlocal means of similar neighborhoods. IEEE Sig. Process. Lett. **12**(12), 839–842 (2005)
11. Ramponi, G., Polesel, A.: Rational unsharp masking technique. J. Electron. Imaging **7**(2), 333–338 (1998)
12. Sáenz, A., Calderón, S., Castro, J., Mora, R., Siles, F.: Deceived bilateral filter for improving the automatic cell segmentation and tracking in the NF-kB pathway without nuclear staining. In: Braidot, A., Hadad, A. (eds.) VI Latin American Congress on Biomedical Engineering CLAIB 2014. IFMBE, vol. 49, pp. 345–348. Springer, Heidelberg (2015). https://doi.org/10.1007/978-3-319-13117-7_89
13. Cuomo, S., De Michele, P., Piccialli, F.: 3D data denoising via nonlocal means filter by using parallel GPU strategies. Comput. Math. Methods Med. **2014**, Article no. 523862 (2014)
14. Vignesh, R., Oh, B.T., Kuo, C.C.J.: Fast non-local means (NLM) computation with probabilistic early termination. IEEE Sig. Process. Lett. **17**(3), 277–280 (2010)
15. Viola, P., Jones, M.: Robust real-time object detection. Int. J. Comput. Vis. **4**(34–47) (2001)
16. Wang, J., Guo, Y., Ying, Y., Liu, Y., Peng, Q.: Fast non-local algorithm for image denoising. In: 2006 IEEE International Conference on Image Processing, pp. 1429–1432. IEEE (2006)
17. Xue, B., Huang, Y., Yang, J., Shi, L., Zhan, Y., Cao, X.: Fast nonlocal remote sensing image denoising using cosine integral images. IEEE Geosci. Remote Sens. Lett. **10**(6), 1309–1313 (2013)

Utilizing Deep Learning and 3DLBP
for 3D Face Recognition

João Baptista Cardia Neto[1]([✉]) [iD] and Aparecido Nilceu Marana[2] [iD]

[1] São Carlos Federal University - UFSCAR, São Carlos, SP 13565-905, Brazil
joao.cardia@fatec.sp.gov.br
[2] UNESP - São Paulo State University, Bauru, SP 17033-360, Brazil
nilceu@fc.unesp.br

Abstract. Methods based on biometrics can help prevent frauds and do personal identification in day-to-day activities. Automated Face Recognition is one of the most popular research subjects since it has several important properties, such as universality, acceptability, low costs, and covert identification. In constrained environments methods based on 2D features can outperform the human capacity for face recognition but, once occlusion and other types of challenges are presented, the aforementioned methods do not perform so well. To deal with such problems 3D data and deep learning based methods can be a solution. In this paper we propose the utilization of Convolutional Neural Networks (CNN) with low-level 3D local features (3DLBP) for face recognition. The 3D local features are extracted from depth maps captured by a Kinect sensor. Experimental results on Eurecom database show that this proposal is promising, since, in average, almost 90% of the faces were correctly recognized.

Keywords: Biometrics · 3D face recognition
3D local features · Depth maps · Kinect · Deep learning
Convolutional Neural Networks

1 Introduction

In everyday life many situations require the assurance of a subject identity. Traditionally, this identification is made by utilizing something the subject knows (e.g. passwords, codes, personal data) or something he possesses (e.g. keys, id cards) [2]. The aforementioned approaches have severe drawbacks, since an imposter can steal or loan the id card of a genuine person or learn his/her knowledge or guess his/her password in order to circumvent the identification system.

Biometrics approaches have been proposed to solve such drawbacks by using physical (face, fingerprint, iris, etc.) or behavioral (gait, signature, voice, etc.) characteristics. The utilization of biometrics can help to reduce fraudulent operations [2].

Several automated facial recognition systems present better performance than the human visual system [14]. However, unfortunately, these results are

© Springer International Publishing AG, part of Springer Nature 2018
M. Mendoza and S. Velastín (Eds.): CIARP 2017, LNCS 10657, pp. 135–142, 2018.
https://doi.org/10.1007/978-3-319-75193-1_17

not sustained when faces are presented with partial occlusions or with variations in illumination, pose, and facial expressions. One solution to overcome this drawback is to use 3D information of the face [3,5,8]. Another solution is to use deep learning based methods, such as the Convolution Neural Networks (CNN) [12], for face recognition [17,20,22].

Convolution Neural Networks (CNN) tries to ensure robustness to shift and distortion combining three architectural ideas: local receptive fields, shared weights, and spatial (or temporal) subsampling [11]. Since face recognition suffers greatly from pose, illumination, and occlusion it becomes obvious that the application of CNN can help to create methods robust to the aforementioned drawbacks.

This work proposes utilizing CNN for 3D face recognition based on local features extracted from depth maps generated by Kinect devices[1]. The input to the Convolutional Neural Network is an image representation from the extracted characteristics.

2   Preprocessing and Local Features

In this section the steps for the preprocessing on the proposed method are presented. Each of these techniques have the responsibility to improve data quality and, consequently, increase the recognition results.

2.1   Symmetric Filling

The Symmetric Filling technique, proposed by [13], utilizes the left side of the face to increase point density by including the set of mirrored points from the right side of the face, and vice-versa. However, not all the mirrored points are useful because we only want to fill in the missing data (occluded regions, for instance).

Likewise in [13], the idea here is to add the mirrored point only if there is no neighboring point at that location. During this process, if the Euclidean distance from a mirrored point to its neighbors in the original point cloud is greater than a threshold value δ, then that point will be added to the original face data.

2.2   Iterative Closest Point

The Iterative Closes Point (ICP), proposed by [1], is a very well known solution for the problem of registration. It tries to find a rigid transformation that minimizes the least-square distance between two points.

Given two 3D point set (A and B), ICP performs the following three basic steps:

1. Pair each point of A to the closest point in B;
2. Compute the motion that gives the lowest Mean Square Error (MSE) between the points;
3. Apply the motion to the point set A and update the MSE.

---

[1] https://developer.microsoft.com/en-us/windows/kinect.

The three aforementioned steps are performed until the MSE is lower than a threshold τ. For a more complete description of the method the reader can refer to [1] or [4].

2.3 Savitzky-Golay Filter

The Savitzky-Golay filter is a lowpass filter based on least-squares polynomial approximation [21].

The Savitzky-Golay filter tries to find filter coefficients c_n that preserves the higher moments of the filtered data. It utilizes the idea of a window moving through the function data and approximate it by a polynomial of higher order, quadratic or quartic [18].

More details about this filter can be found in [19,21].

2.4 3DLBP Features

The LBP (Local Binary Pattern) operator, originally proposed by [16], only takes in consideration the sig of the comparison between a region and its kernel. However, one can observe that if two central points on different samples have higher (or lower) depth values than their neighbors, they will have the same operator value, even if they are from different subjects [6].

In order to deal with these cases the authors of [6] proposed the 3D Local Binary Patterns (3DLBP). This variation of the original operator considers not only the sig of the difference, but also the absolute depth difference.

In [6] it is stated that more than 93% of all depth differences (DD) with $R = 2$ are smaller than 7. Due to this property the absolute value of the DD is stored in three binary units (i_2, i_3, i_4). Therefore, it is possible to affirm:

$$|DD| = i_2 \cdot 2^2 + i_3 \cdot 2^1 + i_4 \cdot 2^0 \tag{1}$$

There is also i_1, a binary unit defined by:

$$i_1 = \begin{cases} 1 \text{ if DD} \geq 0; \\ 0 \text{ if DD} < 0. \end{cases} \tag{2}$$

Those four binary units are divided into four layers and, for each of those layers, four decimal numbers are obtained: P_1, P_2, P_3, and P_4. The value of the P_1 has the same value as the original LBP. For the present work the value at the center from each neighbourhood is replaced with the generated code. This way an image with the same height and width and four channels is generated.

3 Convolutional Neural Network

A Convolutional Neural Network consists of several sets of layers containing one or more planes. The input is made of images centered and normalized. Each small local region from this serves as an input to a unit in a plane of the previous layer [10]. Each of those layers has a fixed feature detector that is convolved utilizing a local window that scans the layers in the previous layers. These layers are the convolutional layers [10].

In the present work, for face recognition, the architecture defined in [9] is utilized. This network architecture is composed by eight layers with weights, being the first five convolutional and the others fully connected layers. The output from the last fully connected layer is the input of a softmax layer, with this the network generates a distribution among the class labels.

In our work, we use exactly the same CNN defined in [9] but with an input image of $100 \times 100 \times 4$ instead of $224 \times 224 \times 3$.

4 Proposed Method

In the present work a CNN is utilized for people identification. This is done by utilizing as input images generated by the 3DLBP method. The 3D Face Recognition method has four main stages: Data Preprocessing (Symmetric Filling, Iterative Closest Point, and Savitzky-Golay Filter), Generation of Depth Maps from the Point Cloud, Feature Extraction (3DLBP), and Classification (CNN). Figure 1 shows a block diagram of the proposed method.

4.1 Data Preprocessing

The first task in the preprocessing stage is the face segmentation. For this, all points that fall inside a sphere with radius $R = 70\,mm$, centered at the nose tip, are extracted. After, the face is translated to move the nose tip at the origin.

The second task in the preprocessing stage is the symmetric filling calculation, which is carried out as described in Subsect. 2.1.

After the Symmetric Filling process it is possible that the faces are not perfectly aligned, for correcting this the Iterative Closest Point (ICP) algorithm, described in Subsect. 2.2 is applied. This is, then, the third task in the preprocessing stage.

Finally, in the fourth task of the preprocessing stage, in order to smooth even further the input data, the Savitzky-Golay filter, described in Subsect. 2.3, is applied to the resulted point cloud.

4.2 Generation of Depth Maps from the Point Cloud

In our method, aiming to increase the robustness of the 3D face recognition from Kinect data, new depth maps are generated from the point cloud resulted from the preprocessing stage. Such point cloud is fitted to a smooth surface using an approximation approach based on grid fit[2]. The result of this process is a 100 x 100 matrix.

---

[2] http://mathworks.com/matlabcentral/fileexchange/8998-surface-fitting-using-gridfit.

4.3   Feature Extraction

In our method the low level local features, 3DLBP, used for 3D face recognition, are extracted from the new depth map generated in the previous stages. The 3DLBP method is presented in Subsect. 2.4.

In order to calculate the 3DLBP descriptor, the face image is divided in 64 micro-regions. For each region, a subimage is generate with the central value of the neighborhood being calculated by using the 3DLBP method. The final 3DLBP descriptor of the face is the concatenation of all subimages.

4.4   Matching

For the matching stage, we utilize the CNN defined in Sect. 3. The network outputs the probability of the current subject be in each of the 52 classes. The class with higher probability is the identity assigned to the subject. We utilized the Caffe Deep Learning Framework to implement the CNN [7].

To increase the volume of faces for each subject, after the Savitzky-Golay filter and before the generation of new depth maps, each point cloud is rotate from $-30°$ to $30°$ in steps of $10°$.

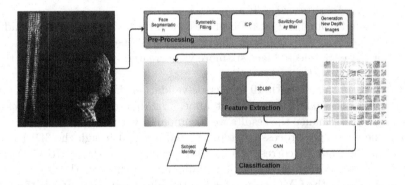

Fig. 1. A block diagram of the proposed method. The point cloud of a subjects face goes through a pre-processing step, after that the 3DLBP characteristics are extracted and, finally, the subject identity is found.

5   EURECOM Kinect Dataset

The proposed method was assessed on the database called EURECOM Kinect Dataset [15], captured using Kinect.

The EURECOM Dataset is composed of 52 subjects: 14 females and 38 males. There are two sets of images captured in the interval of 15 days and each set has nine categories: neutral, smiling, open mouth, lightning, occlusion of the eyes, occlusion of the mouth, occlusion of the right side of the face, left profile and right profile. For each face sample, there are three different data sources: depth map (in bitmap depth images and text files with all the values sensed by the Kinect), RGB image, and 3D .obj files. Each format has the annotation for: the left and right eyes, the tip of the nose, left and right side of the mouth and the chin. Figure 2 shows a set of images from the database.

Fig. 2. A set of images from a subject in the EURECOM database, in which it is possible to see different poses and facial expressions in the RGB (top row) and the depth map (bottom row) images.

6   Experimental Results

For validating the performance of the new method six different categories of faces were selected from the full database: neutral, smiling, open mouth, illumination differences, occlusion of the eyes and occlusion of the mouth faces. Each face from these six categories went through the pre-processing steps and, after applying the Savitz-Golay filter, it was rotated in six different degrees in the Y-axis ($-30°$, $-20°$, $-10°$, $10°$, $20°$, $30°$).

After going through the rotation process each rotated cloud point was added to the selected faces to partake in the generation of the new depth maps. This means that for each selected face six more were created. This increased six times the size of the dataset.

All tests were made with 10-fold cross-validation. Each fold was generated randomly and the same face did not appear at different folds. For assessing the method performance the results were compared with images that have been through the whole pre-processing steps but did not go through the 3DLBP feature extraction and the corresponding RGB images from the EURECOM face dataset. Table 1 shows the comparison results among those methods.

Figure 3 shows the CMC curves for the three methods that utilize the CNN for classification. It shows the curve for the average result for each experiment, it can be seen that there is a small variation but the overall system still performs well.

Table 1. Face recognition rates obtained by the three different methods: 3DLBP + CNN, depth maps + CNN, and RGB images + CNN.

Method	Lower	Higher	Average
3DLBP + CNN	71.33%	**98.62%**	**89.08%**
Depth Maps + CNN	6.65%	25.45%	14.76%
RGB + CNN	**77.75%**	93.34%	84.311%

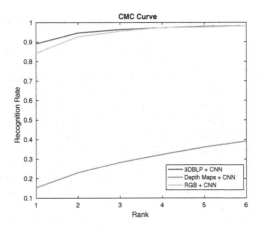

Fig. 3. CMC curves with the result for the methods utilizing the CNN for classification.

7   Conclusion

Looking at the results it is possible to see that utilizing CNN with 3DLBP performs much better than utilizing CNN directly with depth maps or CNN with RGB images. This highlights the importance of utilizing the 3DLBP feature extraction and representation.

Other important conclusion is that representing the features extracted as new images instead of concatenated histograms can be a solution to utilize 3D data in Convolutional Neural Networks.

The key to utilize 3D data with a CNN is to find the right way to represent the extracted features and use that representation as input.

References

1. Besl, P.J., McKay, N.D.: A method for registration of 3-D shapes. IEEE Trans. Pattern Anal. Mach. Intell. **14**(2), 239–256 (1992). https://doi.org/10.1109/34.121791
2. Bolle, R., Pankanti, S.: Biometrics. Personal Identification in Networked Society. Kluwer Academic Publishers, Norwell (1998)
3. Cardia Neto, J.B., Marana, A.N.: 3DLBP and HAOG fusion for face recognition utilizing kinect as a 3D scanner. In: Proceedings of the 30th Annual ACM Symposium on Applied Computing, pp. 66–73. ACM (2015)
4. Chetverikov, D., Stepanov, D., Krsek, P.: Robust Euclidean alignment of 3D point sets: the trimmed iterative closest point algorithm. Image Vis. Comput. **23**(3), 299–309 (2005). http://www.sciencedirect.com/science/article/pii/S0262885604001179
5. Drira, H., Amor, B.B., Srivastava, A., Daoudi, M., Slama, R.: 3D face recognition under expressions, occlusions, and pose variations. IEEE Trans. Pattern Anal. Mach. Intell. **35**(9), 2270–2283 (2013)

6. Huang, Y., Wang, Y., Tan, T.: Combining statistics of geometrical and correlative features for 3D face recognition. In: Proceedings of the British Machine Vision Conference, pp. 90.1–90.10. BMVA Press (2006). https://doi.org/10.5244/C.20.90
7. Jia, Y., Shelhamer, E., Donahue, J., Karayev, S., Long, J., Girshick, R., Guadarrama, S., Darrell, T.: Caffe: convolutional architecture for fast feature embedding. arXiv preprint arXiv:1408.5093 (2014)
8. Kakadiaris, I.A., Passalis, G., Toderici, G., Murtuza, M.N., Lu, Y., Karampatziakis, N., Theoharis, T.: Three-dimensional face recognition in the presence of facial expressions: an annotated deformable model approach. IEEE Trans. Pattern Anal. Mach. Intell. **29**(4), 640–649 (2007). https://doi.org/10.1109/TPAMI.2007.1017
9. Krizhevsky, A., Sutskever, I., Hinton, G.E.: Imagenet classification with deep convolutional neural networks. In: Advances in Neural Information Processing Systems, pp. 1097–1105 (2012)
10. Lawrence, S., Giles, C., Tsoi, A.C., Back, A.: Face recognition: a convolutional neural-network approach. IEEE Trans. Neural Netw. **8**(1), 98–113 (1997). http://ieeexplore.ieee.org/xpls/abs_all.jsp?arnumber=554195&tag=1
11. LeCun, Y., Bengio, Y.: The handbook of brain theory and neural networks. In: Convolutional Networks for Images, Speech, and Time Series, pp. 255–258. MIT Press, Cambridge, MA, USA (1998). http://dl.acm.org/citation.cfm?id=303568.303704
12. LeCun, Y., Kavukvuoglu, K., Farabet, C.: Convolutional networks and applications in vision. In: Proceedings of International Symposium on Circuits and Systems (ISCAS 2010). IEEE (2010)
13. Li, B., Mian, A., Liu, W., Krishna, A.: Using kinect for face recognition under varying poses, expressions, illumination and disguise. In: 2013 IEEE Workshop on Applications of Computer Vision (WACV), pp. 186–192 (2013)
14. Li, S.Z., Jain, A.K. (eds.): Handbook of Face Recognition, 2nd edn. Springer, London (2011). https://doi.org/10.1007/978-0-85729-932-1. http://dblp.uni-trier.de/db/books/daglib/0027896.html
15. Min, R., Kose, N., Dugelay, J.L.: Kinectfacedb: a kinect database for face recognition. IEEE Trans. Syst. Man Cybern. Syst. **44**(11), 1534–1548 (2014)
16. Ojala, T., Pietikäinen, M., Harwood, D.: A comparative study of texture measures with classification based on featured distributions. Pattern Recogn. **29**(1), 51–59 (1996)
17. Parkhi, O.M., Vedaldi, A., Zisserman, A.: Deep face recognition. In: BMVC. vol. 1, p. 6 (2015)
18. Press, W.H., Teukolsky, S.A., Vetterling, W.T., Flannery, B.P.: Numerical Recipes 3rd Edition: The Art of Scientific Computing, 3rd edn. Cambridge University Press, New York (2007)
19. Savitzky, A., Golay, M.J.E.: Smoothing and differentiation of data by simplified least squares procedures. Anal. Chem. **36**(8), 1627–1639 (1964). https://doi.org/10.1021/ac60214a047
20. Saxena, S., Verbeek, J.: Heterogeneous face recognition with CNNs. In: Hua, G., Jégou, H. (eds.) ECCV 2016. LNCS, vol. 9915, pp. 483–491. Springer, Cham (2016). https://doi.org/10.1007/978-3-319-49409-8_40
21. Schafer, R.W.: What is a Savitzky-Golay filter? [lecture notes]. IEEE Sig. Process. Mag. **28**(4), 111–117 (2011)
22. Schroff, F., Kalenichenko, D., Philbin, J.: Facenet: a unified embedding for face recognition and clustering. In: Proceedings of the IEEE Conference on Computer Vision and Pattern Recognition, pp. 815–823 (2015)

Efficient Hyperparameter Optimization in Convolutional Neural Networks by Learning Curves Prediction

Andrés F. Cardona-Escobar[1(✉)], Andrés F. Giraldo-Forero[1],
Andrés E. Castro-Ospina[1], and Jorge A. Jaramillo-Garzón[2]

[1] Instituto Tecnológico Metropolitano, Medellín, Colombia
andrescardona134713@correo.itm.edu.co
[2] Universidad de Caldas, Manizales, Colombia

Abstract. In this work, we present an automatic framework for hyper-parameter selection in Convolutional Neural Networks. In order to achieve fast evaluation of several hyperparameter combinations, prediction of learning curves using non-parametric regression models is applied. Considering that "trend" is the most important feature in any learning curve, our prediction method is focused on trend detection. Results show that our forecasting method is able to catch a complete behavior of future iterations in the learning process.

Keywords: Hyperparameter optimization · Deep learning
Learning curves · Forecasting · SVR · Singular spectrum analysis

1 Introduction

In several fields, the growing of available data is demanding more complex models for automatic processing. Besides, computational capabilities are also increasing allowing us to handle huge amounts of data [1]. During last years, pattern recognition problems have been solved using Deep Learning models, however, the training stage for these methods is highly time-consuming and thus, optimization of its free parameters can not be performed by metaheuristics based on collective behavior, such as Particle Swarm Optimization (PSO) or Cuckoo Search (CS), since those methods need an elevated number of evaluations. Hyperparameter selection on Deep Learning algorithms has been addressed by several techniques. The simplest method is Random Search (RS) applied in [2], where each free parameter is drawn from a probability distribution. The problem of (RS) is the low repeatability when few iterations are computed. On the other hand, sequential optimization methods like Bayesian Optimization (BO) [3], that only evaluate the objective function one time per iteration are more attractive. Despite these techniques are computationally tractable, validation process for one possible set of hyperparameters is still computationally expensive, therefore search spaces for hyperparameters should be carefully selected. To overcome

© Springer International Publishing AG, part of Springer Nature 2018
M. Mendoza and S. Velastín (Eds.): CIARP 2017, LNCS 10657, pp. 143–151, 2018.
https://doi.org/10.1007/978-3-319-75193-1_18

these facts, meta-models become a promising solution that allows to evaluate several solutions in less time than traditional approaches. In neural networks, learning curves present a monotonic trend that describes the behavior of performance which could be used for prediction. Learning curves prediction was used in [4], however, their authors employed undesirable prior assumptions over learning curves. In the present work we present an efficient approach for hyperparameter selection by forecasting future values of performance using a non-parametric method based on Support Vector Regression (SVR) [5]. Since some learning curves are not free of noise, Singular Spectrum Analysis (SSA) is used for smoothing.

2   Materials and Methods

Figure 1 depicts the proposed methodology which comprises the following stages:

(1) Construction of the Neural Network, (2) Obtaining the learning curves, (3) Learning curves smoothing, (4) Learning curves forecasting and (5) Hyperparameter optimization.

The next sections describe each stage in detail. An implementation of our methodology is freely available at:
https://labmirp@bitbucket.org/lab mirp/deep-hyperparameter-opt.git.

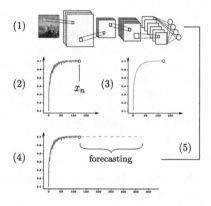

Fig. 1. General scheme of our methodology. Input image is taken the from preprocessed version of COREL [6]

2.1   Construction of the Neural Network

The most important step for hyperparameter optimization is the selection of the neural network topology, which depends on the type of data (e.g., Convolutional Neural Networks for image classification [7], Recurrent Neural Networks for speech recognition [8], Deep Belief Networks for secondary structure prediction in proteins [9] or combinations among them [10]). In this step, a fixed number of layers is carefully selected (for example; 4 convolutional layers, 1 flatten layer and 1 softmax layer). This stage is an important decision to get a high performance and is strongly dependent on the classification problem. After topology selection, the most important free parameters are set as hyperparameters and its search spaces are defined.

2.2   Obtaining the Learning Curves

The Neural Network generated in previous step is evaluated by predicting a learning curve created with accuracies for some mini-batches. All learning curves present a similar shape, their convergence depends on the problem, and thus, the number of mini-batches n employed for drawing the curve must be adapted. We employed a *variance* criterion for determining n as follows:

1. An initial evaluation on q points $\{x_i\}_{i=1}^q$ is performed and smoothed by the procedure described in the Sect. 2.3.
2. Some parameters are defined: the length of a sliding window p to calculate variance, the minimum variance allowed η and the minimum number of evaluations ψ to apply prediction.
3. New evaluations are incorporated and smoothed by Sect. 2.3, while the window defined in Item 2 is slipped to calculate the variance of k-th iteration $\sigma_k^2 = \frac{1}{p}\sum_{i=k}^{p+k}(x_k - \mu)^2$. Each variance σ_k^2 is compared with η. If the minimum value of variance allowed is higher than the calculated variance $(\eta > \sigma_k^2)$, then proceed with step 4, else step 3 is executed again.
4. The evaluations number at the end of step 3 is compared with ψ. If $\psi > q+k-1$, the number of evaluated points n is adjusted according to minimum number of defined evaluations $n = \psi$, else, the number of observations is set based on variance criterion $n = q + k - 1$.

2.3   Learning Curves Smoothing

The obtained learning curve from Sect. 2.2 presents high level of noise, making the forecast difficult. To overcome such problem, we apply smoothing over learning curves by Singular Spectrum Analysis (SSA) (see complete details in [11]). The idea is to perform time series decomposition according to the spectrum of the transition matrix. Summation of components arises in the original time series. Time series decomposition can be performed by following the next steps explained in [11]:

1. First step is to construct a Hankel matrix $\mathbf{H} \in \mathbb{R}^{L \times d}$ of the original time series $\{x_i\}_{i=1}^n$, where d is the length of the window and $L = n - d + 1$ is the number of rows for this matrix. $\mathbf{H}$ is given by:

$$\mathbf{H} = \begin{bmatrix} x_1 & x_2 & x_3 & \cdots & x_d \\ x_2 & x_3 & x_4 & \cdots & x_{d+1} \\ \vdots & \vdots & \vdots & \ddots & \vdots \\ x_L & x_{L+1} & x_{L+2} & \cdots & x_n \end{bmatrix} \tag{1}$$

2. Apply singular value decomposition (SVD) to $\mathbf{H}$, call U the eigenvectors of $\mathbf{H}\mathbf{H}^\top$ located in columns and organized according to singular values $\sqrt{\lambda_i}$.
3. Compute matrix $\mathbf{V}$ given by $\mathbf{H}^\top \mathbf{U}_i/\sqrt{\lambda_i}$ where $\mathbf{U}_i$ is the i-th column of matrix $\mathbf{U}$, then find $\{\widetilde{\mathbf{H}}_j\}_{j=1}^p$ where $p = rank(\mathbf{H})$ by using outer products, that is to say $\widetilde{\mathbf{H}}_j = \sqrt{\lambda_i}\,\mathbf{U}_i\,\mathbf{V}_i^\top, \ \forall j = 1,\ldots,p$
4. To select the adequate number of eigenvalues, we applied the approach of [12], where the amount of noise in the signal is unknown. This process is based

on singular value thresholding (SVT), the optimal threshold $\hat{\tau}$ depends on the learning curve $\mathbf{x}$ and Hankel matrix dimensions (L, d).

$$\hat{\tau}(\beta, \mathbf{x}) = \omega(\beta) \ median(diag(\mathbf{\Sigma})) \tag{2}$$

The equation above shows the approach to compute the threshold, in our case $\beta = \min(\frac{L}{d}, \frac{d}{L})$. Values for $\omega(\beta)$ are tabulated in [12].

5. Finally, summation of more relevant components is applied to get a group matrix $\mathbf{H}_I$. Then, anti-diagonal averaging over $\mathbf{H}_I$ is applied to get a denoised series $\{\tilde{x}_i\}_{i=1}^n$.

2.4   Learning Curves Forecasting

The learning curves forecasting procedure is carried out by mixing two known methods of prediction (iterative and direct forecasting) [13]. T models are trained for j-step ahead prediction, that is, the first model performs 1-step ahead prediction, the second model 2-step ahead, and so on. The j-th model employs as labels a vector $y_j = [x_{s+j}, ..., x_n]^T$. This method is an extension of the iterative approach, therefore, can be represented by the T recursive expressions given in Eq. (3) as is explained in [13].

$$\begin{aligned} \hat{x}_{1+n} &= f_1(x_1, ..., x_n) \\ \hat{x}_{2+n} &= f_2(x_1, ..., \hat{x}_{1+n}) \\ &\vdots \\ \hat{x}_{T+n} &= f_T(x_1, ..., \hat{x}_{T-1+n}) \end{aligned} \tag{3}$$

Each model is built by Support Vector Regression (SVR) implemented in [14]. The main purpose of this technique is to find an optimal hyperplane $f(\mathbf{x}) = \mathbf{w}^T\phi(\mathbf{x}) + b$, where data can be linearly adjusted. For this end, a mapping $\phi : \mathbb{R}^s \mapsto \mathbb{R}^N$ to a high-dimensional space is employed [5], as shown in figure of the Eq. (3). Hyperplane parameters $(\mathbf{w}, b)$ can be found by solving the optimization problem of the Eq. (4) proposed in [5].

$$\min_{\mathbf{w},b,\xi_i,\xi_i^*} \frac{1}{2}\|\mathbf{w}\|^2 + C\sum_{i=1}^m (\xi_i + \xi_i^*) \quad s.t \begin{cases} \mathbf{w}^T\phi(\mathbf{x}_i) + b - y_i \le \epsilon + \xi_i, \\ y_i - \mathbf{w}^T\phi(\mathbf{x}_i) - b \le \epsilon + \xi_i^*, \\ \xi_i, \xi_i^* \ge 0, \ i = 1, ..., m \end{cases} \tag{4}$$

2.5   Hyperparameters Optimization

In order to find the best hyperparameters, we employed Bayesian optimization due to its low computational cost and high exploration in black-box optimization problems. Bayesian optimization is an iterative procedure that involves the following steps described in [3]:

1. Performs regression with previous evaluations by a Gaussian process in order to find a surrogate model of the objective function that is used as prior.

2. The prior distribution computed before is employed to find a posterior distribution over the objective function.
3. An acquisition function is calculated based on the posterior and its maximum is found with the aim of establishing what point should be evaluated.
4. Evaluate and augment data to compute the prior again.

3 Experimental Setup

A big picture of performed experiments over each database is depicted in Fig. 2. Databases are divided in validation, train and test. Train and validation partition are used for find optimal hyperparameters following proposed methodology. At the end a neural network is trained and evaluated with the best hyperparameters.

3.1 Datasets

To validate our method we employed 3 known datasets; MNIST [15], Corel [16] and Devanagari [17]. The first one is MNIST, which contains 70,000 handwritting digits. This toy dataset has 10 classes (digits from 0 to 9), where each class is composed by the same number of images with a resolution of 28 × 28. For training and

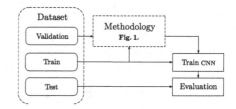

Fig. 2. Scheme for the experimental setup.

testing, this dataset was stratified and splitted in 3 folds; 45,000 images for training the model, 15,000 for validation (hyperparameter optimization) and 10,000 images for testing.

The second dataset is Corel, this dataset was created in [16], however, this work uses a preprocessed version employed in [18] formed by 10,800 RGB images of 120 × 80 pixels[1], there are 80 unbalanced classes (with different sizes), for solving the lack of images we applied data augmentation using a horizontal flip with a random rotation angle between 0–20, and the width/height shift range of 0–0.2.

Finally we used Devanagari, a handwriting recognition dataset formed by 92,000 images divided in 46 classes of 2,000 samples. For training and testing this dataset is sliced 85% and 15%, respectively. However, validation was carried out by taking the 10% of the training data using stratification.

---

[1] We thank to the professor Dacheng Tao for allowing us to use his preprocessed version of the Corel dataset available in https://sites.google.com/site/dctresearch/Home/content-based-image-retrieval.

3.2 Methodology

For the experiments, Keras [19] was employed as deep learning framework using Tensorflow [20] as backend. The employed topology for hyperparameter optimization in all datasets is depicted in Fig. 3. Since our classification problems are not challenging, our topologies does not have many layers, nevertheless, hyperparameter optimization is still computationally expensive.

To obtain the learning curve a length of the window is empirically ajusted over MNIST, COREL and Devanagari as $p = 5$. The parameters η and ψ are selected by the following rules of thumb: $\eta = \frac{1}{6}max\left\{\sigma^2\left(\{x_i\}_{i=1}^q\right)\right\}$, where the maximum variance allowed is calculated as one-third of the maximum variance on the first q points. The minimum number of evaluations is calculated as $\psi = T \times g \times k$, where g denote the number of points to adjust and k folds in cross-validation. For the smoothing process, the length of the window is half of the available points. Since our prediction method is based on lineal Support Vector Machines, we applied grid search to select optimal hyperparameters (C, ϵ) by 4 folds cross-validation. The number of machines used for each dataset was adjusted as follows: MNIST and Devaganari in $T = 3$, and Corel in $T = 8$. Bayesian optimization was carried out by the python package BayesianOptimization[2] with two initial points, 40 iterations and a $\kappa = 2.7$.

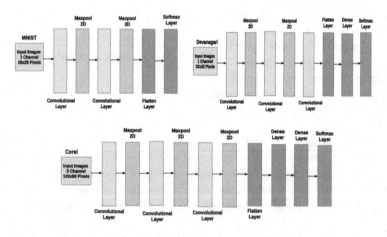

Fig. 3. Proposed topologies for hyperparameter optimization.

4 Results and Discussion

In order to evaluate the quality of the prediction model, we selected a learning curve for each dataset and computed Mean Squared Error (MSE), Mean Absolute Percentage Error (MAPE) and Mean Absolute Error (MAE) between actual

---

[2] This package is available in https://github.com/fmfn/BayesianOptimization.git.

Table 1. Evaluation of the forecasting method by 4 known measures in some parts of the obtained curves including the stopping criterion referred as SC. Bold results indicates the lowest error between two kernels for each prediction

Dataset	Stopping criterion	MSE ($\times 10^{-6}$)		MAPE (%)		MAE ($\times 10^{-3}$)	
		Lineal	RBF	Lineal	RBF	Lineal	RBF
MNIST	30%	7.40	**3.40**	0.252	**0.165**	2.50	**1.64**
	50%	112	**23.6**	0.899	**0.444**	8.92	**4.41**
	70%	**0.656**	1.47	**0.060**	0.0998	**0.596**	0.989
	SC	13.8	**0.599**	0.347	**0.0581**	3.44	**0.576**
Corel	30%	**1840**	58600	**6.39**	32.3	**39.1**	200
	50%	485	**344**	2.97	**2.39**	18.5	**14.6**
	70%	360	**265**	2.40	**2.05**	15.1	**12.8**
	SC	**1840**	58600	**6.39**	32.3	**39.1**	200
Devanagari	30%	43.1	**37.2**	0.617	**0.550**	6.10	**5.43**
	50%	22.6	**8.61**	0.405	**0.235**	4.00	**2.32**
	70%	18.6	**3.42**	0.391	**0.155**	3.86	**1.53**
	SC	150	**7.94**	1.18	**0.247**	11.6	**2.43**

and predicted values of the best set of hyperparameters obtained, on different percentages of the curve and with the stopping criterion (SC). Results are shown in Table 1.

As Table 1 shows, linear kernel performs better when learning curve is still increasing, therefore this kernel is better for trend detection. However, when the learning curve is getting flat RBF kernel outperforms linear kernel. Another important issue in the RBF kernel is the tuning process for γ values. Considering that noise affects the predictability of increasing learning curves, another important achievement of this work is the automatic selection of proper values in Singular Spectrum Analysis and the stopping criterion based on *variance*. As final prove we applied the proposed method in MNIST, Corel and Devanagari obtaining around 99%, 56% and 98% of accuracy, respectively. The traditional approach without using learning curves prediction got 99%, 60% and 98% of accuracy in the same datasets showing that learning curves prediction provides an efficient approach for hyperparameter optimization in Convolutional Neural Networks.

5   Conclusions and Future Work

In this work, we proposed an automatic framework for hyperparameter selection in Convolutional Neural Networks. This framework is based on prediction using kernel methods (SVMs). Despite RBF kernel is a common choice in machine learning problems, our results show that growth curves are well described using linear kernels, however, this kernel is sensitive when noise is considered, for this reason, noise is removed from original data using singular spectrum analysis. As future work, we are planning to use this method in another type of neural networks like LSTM, DBN, among others. In addition, it is important to find a robust early stopping criterion for starting the prediction of the learning curve.

References

1. Perera, C., Liu, C.H., Jayawardena, S., Chen, M.: A survey on internet of things from industrial market perspective. IEEE Access **2**, 1660–1679 (2014)
2. Bergstra, J., Bengio, Y.: Random search for hyper-parameter optimization. J. Mach. Learn. Res. **13**, 281–305 (2012)
3. Snoek, J., Larochelle, H., Adams, R.P.: Practical Bayesian optimization of machine learning algorithms. In: Advances in Neural Information Processing Systems (2012)
4. Domhan, T., Springenberg, J.T., Hutter, F.: Speeding up automatic hyperparameter optimization of deep neural networks by extrapolation of learning curves
5. Vapnik, V.: The Nature of Statistical Learning Theory. Springer Science & Business Media, Heidelberg (2013). https://doi.org/10.1007/978-1-4757-3264-1
6. Tao, D.: The COREL database for content based image retrieval. https://sites.google.com/site/dctresearch/Home/content-based-image-retrieval
7. Krizhevsky, A., Sutskever, I., Hinton, G.E.: ImageNet classification with deep convolutional neural networks. In: NIPS, pp. 1097–1105 (2012)
8. Graves, A., Mohamed, A., Hinton, G.: Speech recognition with deep recurrent neural networks. In: 2013 IEEE International Conference on Acoustics, Speech and Signal Processing (ICASSP), pp. 6645–6649. IEEE (2013)
9. Spencer, M., Eickholt, J., Cheng, J.: A deep learning network approach to ab initio protein secondary structure prediction. IEEE/ACM Trans. Comput. Biol. Bioinf. **12**(1), 103–112 (2015)
10. Busia, A., Collins, J., Jaitly, N.: Protein secondary structure prediction using deep multi-scale convolutional neural networks and next-step conditioning. arXiv preprint arXiv:1611.01503 (2016)
11. Golyandina, N., Nekrutkin, V., Zhigljavsky, A.: Analysis of time series structure: SSA and related techniques (2001)
12. Gavish, M., Donoho, D.L.: The optimal hard threshold for singular values is $4/sqrt(3)$. IEEE Trans. Inf. Theory **60**(8), 5040–5053 (2014)
13. Zhang, L., Zhou, W.D., Chang, P.C., et al.: Iterated time series prediction with multiple support vector regression models. Neurocomputing **99**, 411–422 (2013)
14. Pedregosa, F., Varoquaux, G., Gramfort, A., Michel, V., Thirion, B., Grisel, O., Blondel, M., Prettenhofer, P., Weiss, R., Dubourg, V., Vanderplas, J., Passos, A., Cournapeau, D., Brucher, M., Perrot, M., Duchesnay, E.: Scikit-learn: machine learning in Python. J. Mach. Learn. Res. **12**, 2825–2830 (2011)
15. LeCun, Y.: The MNIST database of handwritten digits (1998)
16. Wang, J.Z., Li, J., Wiederhold, G.: Simplicity: semantics-sensitive integrated matching for picture libraries. IEEE Trans. Pattern Anal. Mach. Intell. **23**(9), 947–963 (2001)
17. Acharya, S., Pant, A.K., Gyawali, P.K.: Deep learning based large scale handwritten Devanagari character recognition. In: 2015 9th International Conference on Software, Knowledge, Information Management and Applications (SKIMA), pp. 1–6. IEEE (2015)

18. Tao, D., Tang, X., Li, X., Wu, X.: Asymmetric bagging and random subspace for support vector machines-based relevance feedback in image retrieval. IEEE Trans. Pattern Anal. Mach. Intell. **28**(7), 1088–1099 (2006)
19. Chollet, F.: Keras (2015). https://github.com/fchollet/keras.git
20. Abadi, M., Agarwal, A., et al.: TensorFlow: large-scale machine learning on heterogeneous systems (2015). https://www.tensorflow.org/

Evaluation of Deep Feedforward Neural Networks for Classification of Diffuse Lung Diseases

Isadora Cardoso[1], Eliana Almeida[1], Héctor Allende-Cid[2]([⊠]),
Alejandro C. Frery[1], Rangaraj M. Rangayyan[3], Paulo M. Azevedo-Marques[4],
and Heitor S. Ramos[1]🄳

[1] Instituto de Computação, Universidade Federal de Alagoas, Maceió, AL, Brazil
{isadora.cardoso,eliana.almeida,acfrery,heitor}@laccan.ufal.br
[2] Escuela de Ingeniería Informática, Pontificia Universidad Católica de Valparaíso,
Valparaíso, Chile
hector.allende@pucv.cl
[3] Department of Electrical and Computer Engineering,
Schulich School of Engineering, University of Calgary, Calgary, AB, Canada
ranga@ucalgary.ca
[4] Department of Internal Medicine, Ribeirão Preto Medical School,
University of São Paulo, Ribeirão Preto, SP, Brazil
pmarques@fmrp.usp.br

Abstract. Diffuse Lung Diseases (DLDs) are a challenge for physicians
due their wide variety. Computer-Aided Diagnosis (CAD) are systems
able to help physicians in their diagnoses combining information provided by experts with Machine Learning (ML) methods. Among ML
techniques, Deep Learning has recently established itself as one of the
preferred methods with state-of-the-art performance in several fields. In
this paper, we analyze the discriminatory power of Deep Feedforward
Neural Networks (DFNN) when applied to DLDs. We classify six radiographic patterns related with DLDs: pulmonary consolidation, emphysematous areas, septal thickening, honeycomb, ground-glass opacities, and
normal lung tissues. We analyze DFNN and other ML methods to compare their performance. The obtained results show the high performance
obtained by DFNN method, with an overall accuracy of 99.60%, about
10% higher than the other studied ML methods.

Keywords: Computer-Aided Diagnosis
Deep Feedforward Neural Network · Deep Learning
Diffuse Lung Diseases · Machine Learning

1 Introduction

Diffuse Lung Diseases (DLDs) form a group of more than 150 pathological conditions, characterized by inflammation of the lung, which leads to breathing

© Springer International Publishing AG, part of Springer Nature 2018
M. Mendoza and S. Velastín (Eds.): CIARP 2017, LNCS 10657, pp. 152–159, 2018.
https://doi.org/10.1007/978-3-319-75193-1_19

dysfunctions [21]. Distinguishing among these diseases is a challenge for physicians due their wide variety. Computer-Aided Diagnosis (CAD) systems can help physicians in their diagnoses, combining information provided by experts with Machine Learning (ML) methods. Computers present several advantages over humans: they do not suffer from boredom, fatigue, and distractions; are consistent in their analyses; and they can combine the knowledge of several experts in a single analysis [18].

There are several studies using ML methods in diagnoses of several diseases, such as lung cancer [5] and breast cancer [26]. Among ML techniques, Deep Learning (DL) is a recent development that is providing state-of-the-art performance in several fields, such as speech recognition [19] and image recognition [12]. Litjens et al. [14] list over 300 contributions of DL in medical image analysis, most of which appeared in recent years.

In this paper, we analyze the discriminatory power of Deep Feedforward Neural Networks (DFNN) when applied to DLDs. DL methods present better stability, generalization ability, and scalability, when compared with simple neural networks [6]. We contrast this approach with other ML methods: k-Nearest Neighbors (kNN) [23], Support Vector Machines (SVM) [24], Logistic Regression (LR) [11], Gaussian Mixture Models (GMM) [15], and Artificial Neural Networks (ANN) [20], in order to compare their performance.

We evaluate the selected methods for the classification of the following six radiographic patterns related with DLDs: pulmonary consolidation (PC), emphysematous areas (EA), septal thickening (ST), honeycomb (HC), groundglass opacities (GG), and normal lung tissues (NL). For this purpose, we use the 28 texture and fractal features described in the work of Pereyra et al. [16]. In this work we only deal with the features, since the images from which they were obtained are not available due to practical issues. For this reason, we employed DFNN rather than the state-of-the-art Convolutional Neural Networks.

We applied dimensionality reduction methods prior to classification to remove irrelevant or redundant data, namely, Principal Component Analysis (PCA), Linear Discriminant Analysis (LDA), and Stepwise Selection – Forward (SSF), Backward (SSB), and Forward-Backward (SSFB).

The main contributions of this work are: **(I)** A comprehensive quantitative comparison of classification methods using dimensionality reduction for DLDs; and **(II)** The use of DFNN to improve the classification accuracy over state-of-the-art methods. This work may contribute to the analysis and selection of those features that can efficiently characterize the DLDs studied.

This article is organized as follows: Sect. 2 presents an overview of DFNN; Sect. 3 presents the material used and related work; Sect. 4 presents the methods used; Sect. 5 presents the main results; and, finally, Sect. 6 concludes this work.

2   Deep Feedforward Neural Networks

The goal of a DFNN is to approximate a desired function f^*, mapping an input x to a category y. A feedforward network defines a mapping $y = f(x; \theta)$ and learns the value of the parameters θ that result in the best function approximation.

The architecture of a DFNN, as in most neural networks, is usually described in groups of units called layers. The layers are arranged in a chain structure, with each layer being a function of the preceding layer. In a feedforward model, the information flows through the function being evaluated from x (the input layer), through the intermediate computations used to define a function f (the hidden layers), and finally to the output y (the output layer). The behavior of the hidden layers is not directly specified, leaving the learning algorithm to decide on how to use the layers to produce the desired output. The training data also do not specify what each layer should do. Instead, the learning algorithm must decide how to use the layers to best implement an approximation of f^* [9]. To avoid slow convergence, it is possible to use an adaptive learning method, such as AdaDelta [25].

Each unit that composes a layer is called a neuron. In a fully connected architecture, each neuron is connected to all the neurons in the next layer, transmitting a weighted combination of the input signal, as $\alpha = \sum_{i=1}^{n} w_i x_i + b$, where x_i and w_i denote the firing neuron's input values and their weights, respectively, and n is the number of input values. The bias b represents the neuron's activation threshold and is included in each layer of the network, except the output layer. The final output is fully determined by the weights linking neurons and biases [6].

Weighted values are aggregated and then the output signal is checked by a nonlinear activation function, activating the unit or not, and the value is transmitted by the connected neuron. There are several possibilities for the activation function. In this work, we used the hyperbolic tangent function: $f(\alpha) = (e^{\alpha} - e^{-\alpha})/(e^{\alpha} + e^{-\alpha})$, which is a rescaled and shifted logistic function. Its symmetry around 0 allows the training algorithm to converge faster [6], avoiding the saturation behavior of the top hidden layer observed with sigmoid networks [8].

Regularization methods can be used to reduce generalization error and hence prevent overfitting. Among the possibilities, we used Ridge and Lasso penalties: Ridge causes many weights to be small, and Lasso turns many weights into zero, which can add stability to the model [11].

3   Materials and Related Work

The material used is derived from the work of Pereyra et al. [16]: 247 high-resolution CT images from 108 different exams from the University Hospital at the Ribeirão Preto Medical School of the University of São Paulo, Brazil, previously selected and evaluated by thoracic radiologists. First, the DICOM images were normalized to 8 bits per pixel. Histogram equalization and centering at -400 and -800 Hounsfield Units (HU) were performed to highlight high- and low-attenuation patterns, respectively. Segmentation was made by using mathematical morphology (closing), binary thresholding (at -800 HU), and object selection. Then, a total of 3252 Regions of Interest (ROIs) (each of size 64×64 pixels) were grouped into categories, according to the radiologist report, that characterize each radiographic pattern: 529 ROIs for HC, 594 for GG, 589 for ST, 450 for PC, 501 for EA, and 589 for NL. After these procedures, the labels were reviewed and validated with radiologists for accuracy. From each ROI was extracted a set of 28 features: the first-order statistical (mean, median, standard deviation, skewness, and kurtosis); 14 texture features defined by Haralick et al. [10] (energy, difference moment, correlation, variance, inverse difference moment, sum of squares: average, sum of squares: variance, sum of squares: entropy, entropy, difference variance, difference entropy, information measures of correlation 1, information measures of correlation 2, maximal correlation coefficient) for the four possible neighborhood ($0°$, $45°$, $90°$, and $135°$), using one pixel as the reference distance; Laws' measures of texture energy [13] using the appropriate convolution masks [Laws' wave measures, Laws' wave measures (rotation-invariant), Laws' ripple measures, Laws' ripple measures (rotation-invariant), Laws' level measures]; statistical measures (standard deviation, mean, and median) of the power spectral density (PSD), obtained from the Discrete Fourier Transform (DFT); and fractal dimension (FD), using PSD of each ROI including the application of the tapered cosine window. Details about these features and analysis of texture in biomedical images are available in Rangayyan [18].

Using this data set of features, Pereyra et al. [16] achieved a correct classification rate of up to 82.62% using the kNN algorithm. Almeida et al. [2] performed a statistical analysis of the same features using the GMM with six components for each feature. GMM provided, for at least one class and 78.5% of the features, correct classification of at least 60%. Almeida et al. [1] extended their previous work by applying GMM to the five most significant features, generating fuzzy membership functions and obtaining 63% in average of correct classification.

In this work, we reproduce the main results of the aforementioned work and perform a comprehensive evaluation of different ML methods, including DFNN.

4   Methods

The dimensionality reduction methods used in the present study do not need setting of parameters. However, classification requires tuning the parameters empirically until the best results are found. We not only classify the full set of features but also the results from the dimensionality reduction methods, using the ML methods mentioned before. We validated the results by 10-fold-cross-validation.

To generate the results we used a machine with i5 processor and 4 GB of RAM memory, with no GPU. Our programming platform is the R language (version 3.3.2) [17] due to its high accuracy in computing statistical functions [3].

The same configuration used before by Pereyra et al. [16] for kNN was adopted, $k = 5$. However, for GMM, previously used by Almeida et al [2], we chose a different approach: an extension proposed by Fraley and Raftery [7], which uses the labels to choose the best parameterization of the covariance matrix of the data as well as the number of components in each class, turning GMM into a supervised algorithm.

For the other ML methods studied, we adopted the following parameters: **(I)** SVM: a kernel of inhomogeneous polynomials of degree 2, with the cost of constraint violation being 1; **(II)** LR: elastic net penalty [11], with elastic net mixing parameter set to 0.7; **(III)** ANN: *softmax* activation function, and training function BFGS quasi-Newton method, with a decay rate set to 10^{-5}. We used one hidden layer with different numbers of neurons for each method: full feature set (8 neurons), PCA (8), LDA (11), SSF (15), SSB (12), and SSFB (9).

With the DFNN we used 300 epochs; both Ridge and Lasso penalties used constraint values set to 10^{-5}, three hidden layers with 230 neurons, and the AdaDelta learning method with the hyperparameters $\rho = 0.99$ and $\epsilon = 10^{-8}$.

5   Results and Discussion

5.1   Results of Dimensionality Reduction Methods

The following subsets were achieved using dimensionality reduction methods: 28 features for the full set of features, 18 for SSB, 13 for PCA, 12 for SSFB, 11 for SSF, and 5 for LDA.

LDA produces the smallest vector, of dimension 5. For PCA, we could obtain an even greater reduction: only seven principal components retained 95% of the information (as measured by the variance); however, using 13 principal components (with 99% information retained) we achieved the best result. Nevertheless, since both LDA and PCA perform data transformations, it would still be necessary to extract all the features from the images. It is outside the scope of this work to analyze whether feature extraction is essential or not.

SSF obtained a vector with 11 features. In the case of SSF, SSB, and SSFB, it is not necessary to perform a transformation, i.e., they discard some features based on criteria, in our case, BIC [4] criterion.

Note that all the dimensionality reduction methods used resulted in substantial reduction: at least 10 features discarded by stepwise selection, and at least 13 dimensions represent 99% of the information. These results suggest the presence of insignificant information in the data set, which can happen due factors such as redundant features, noise caused by either the reconstruction of the image or feature acquisition, and irrelevant features.

5.2   Results of Classification

Table 1 shows the classification methods that provided the top few results along with the dimensionality reduction methods that were used to achieve them. We used macro-averaged measures, namely F1 score, specificity, sensitivity, and accuracy, to compare the results. These metrics are obtained from the confusion matrix. Macro-averaged measures are calculated using an unweighted average for each class separately, whereas micro-averaged measures are averages over instances, in a one-vs-all scheme. Therefore, while macro-averaged measures treats all the classes equally, micro-averaged measures favors larger classes [22].

Since our data set is almost uniformly distributed, values of the macro-averaged and micro-averaged measures are almost the same. Hence, we decided to show only the macro-averaged measures, and avoid redundant information.

We can describe the metrics used as follows: Specificity (Sp) is how effectively a classifier identifies negative labels (the ability of a test to correctly exclude those individuals who do not have a given disease); Sensitivity (Se) is how effectively a classifier identifies positive labels (the ability of a test to correctly identify those individuals who have a given disease); Precision (Pr) is when the amount of random variation in a test is small; and Accuracy (Ac) is when a test measures what it is supposed to measure. F1 score is the harmonic mean of precision and sensitivity, as follows: $F1 = 2(Pr \times Se)/(Pr + Se)$.

Table 1. Classification rates for the evaluated methods along with the corresponding dimensionality reduction algorithm

Classification	Dimensionality reduction	Macro specificity	Macro sensitivity	Macro F1	Overall accuracy
ANN	SSB	96.86%	84.48%	84.49%	84.40%
LR	Full	96.54%	82.84%	82.91%	82.81%
GMM	SSF	96.87%	84.61%	84.66%	84.46%
KNN	LDA	96.74%	83.87%	83.89%	83.82%
SVM	SSB	97.31%	86.73%	86.76%	86.64%
DFNN	Full	99.92%	99.62%	99.60%	99.60%

SVM presents the second best result among the evaluated methods, with an overall accuracy of 86.64%. DFNN is the winner, outperforming the others ML methods in all the measures analyzed. Note that DFNN presents an almost perfect score for all of the measures, especially at specificity, with 99.92%. It is worth mentioning that the overall accuracy using kNN (83.82%) and GMM (84.86%) surpass the accuracy of classification presented by Pereyra et al. [16] (82.62%) and Almeida et al. [1] (63.00%).

Table 2 shows the confusion matrix of DFNN. It can be seen that all the classes have a correct rate over 98%, with two of them, PC and HC, having a perfect score. The worst classification is presented by ST, with 98.88%. Pereyra et al. [16] describe this disease as having a consistently poor classification, but DFNN improves its classification by more than 20% over previous results.

The time and the respective confidence interval required for training each ML method are: 0.009 s ±0.3 ms for kNN with LDA; 8.83 s ±0.3 ms for ANN with SSB; 0.59 s ±8.8 ms for SVM with SSB; 4.64 s ±11.3 ms for LR using all features; 1.42 s ±80.3 ms for GMM with SSF; and 1375.45 s [±94.3 ms for DFNN using all features. It is possible to note that DFNN is the slowest method for training among the studied methods. However, the time required in predictions is negligible compared to the training time and is quite similar for all methods.

Table 2. Confusion matrix for DFNN

	PC	EA	ST	HC	NL	GG	Sum	Correct	Error
PC	**450**	0	0	0	0	0	450	100%	0.00%
EA	0	**500**	0	0	1	0	501	99.80%	0.19%
ST	0	0	**582**	6	0	1	589	98.81%	1.18%
HC	0	0	0	**529**	0	0	529	100%	0.00%
NL	0	1	0	0	**588**	0	589	99.83%	0.16%
GG	4	0	0	0	0	**590**	594	99.33%	0.67%
Correct	**450**	**500**	**582**	**529**	**588**	**590**	**3239**	99.60%	0.39%

6   Conclusion

The present work has performed an analysis of radiographic patterns related to DLDs, using data derived from the work of Pereyra et al. [16]. Dimensionality reduction methods were employed to reduce the dimensionality of the data. The best result, obtained using LDA, showed a reduction from 28 to 5 features, but LDA transforms the data into a new space, making it hard to know how each feature contributes to the data.

SVM, kNN, GMM, ANN, LR, and DFNN were used to classify the data. DFNN outperformed other methods in all comparison metrics used, with an overall accuracy of 99.60%, about 10% better than the second best ML method. The methods shown assist in the development of CAD systems for DLDs.

Acknowledgement. This work was partially funded by Fapeal, CNPq, and SEFAZ-AL. The work of Héctor Allende-Cid was supported by the project FONDECYT Initiation into Research 11150248.

References

1. Almeida, E., Rangayyan, R., Azevedo-Marques, P.: Fuzzy membership functions for analysis of high-resolution CT images of diffuse pulmonary diseases. In: 37th Annual International Conference of the IEEE Engineering in Medicine and Biology Society EMBC, pp. 719–722 (2015)
2. Almeida, E., Rangayyan, R., Azevedo-Marques, P.: Gaussian mixture modeling for statistical analysis of features of high-resolution CT images of diffuse pulmonary diseases. In: 2015 IEEE International Symposium on Medical Measurements and Applications (MeMeA), pp. 1–5 (2015)
3. Almiron, M., Almeida, E., Miranda, M.: The reliability of statistical functions in four software packages freely used in numerical computation. Braz. J. Probab. Stat. **23**(2), 107–119 (2009)
4. Bishop, C.M.: Pattern Recognition and Machine Learning. Information Science and Statistics. Springer, New York (2006)
5. Cai, Z., Xu, D., Zhang, Q., Zhang, J., Ngai, S., Shao, J.: Classification of lung cancer using ensemble-based feature selection and machine learning methods. Mol. BioSyst. **11**, 791–800 (2015)

6. Candel, A., Parmar, V., LeDell, E., Arora, A.: Deep Learning with H2O, 5th edn. H2O.ai Inc, Mountain View (2017)
7. Fraley, C., Raftery, A.: Model-based clustering, discriminant analysis, and density estimation. J. Am. Stat. Assoc. **97**, 611–631 (2000)
8. Glorot, X., Bengio, Y.: Understanding the difficulty of training deep feedforward neural networks. In: Proceedings of the International Conference on Artificial Intelligence and Statistics (AISTATS 2010). Society for Artificial Intelligence and Statistics (2010)
9. Goodfellow, I., Bengio, Y., Courville, A.: Deep Learning. MIT Press, Cambridge (2016)
10. Haralick, R.M., Shanmugam, K., Dinstein, I.: Textural features for image classification. IEEE Trans. Syst. Man Cybern. **6**, 610–621 (1973)
11. Hastie, T., Tibshirani, R., Friedman, J.: The Elements of Statistical Learning: Data Mining, Inference, and Prediction, 2nd edn. Springer, New York (2009). https://doi.org/10.1007/978-0-387-84858-7
12. He, K., Zhang, X., Ren, S., Sun, J.: Deep residual learning for image recognition. In: The IEEE Conference on Computer Vision and Pattern Recognition (CVPR), pp. 770–778 (2016)
13. Laws, K.: Rapid texture identification. In: Proceedings of SPIE Conference on Image Processing for Missile Guidance, vol. 238, pp. 376–380 (1980)
14. Litjens, G., Kooi, T., Bejnordi, B., Setio, A., Ciompi, F., Ghafoorian, M., Laak, J., Ginneken, B., Sánchez, C.: A survey on deep learning in medical image analysis. CoRR abs/1702.05747 (2017)
15. McLachlan, G., Peel, D.: Finite Mixture Models. Wiley, Hoboken (2004)
16. Pereyra, L., Rangayyan, R., Ponciano-Silva, M., Azevedo-Marques, P.: Fractal analysis for computer-aided diagnosis of diffuse pulmonary diseases in HRCT images. In: 2014 IEEE International Symposium on Medical Measurements and Applications (MeMeA), pp. 1–6 (2014)
17. R Core Team: R: A Language and Environment for Statistical Computing. R Foundation for Statistical Computing (2016). https://www.R-project.org/
18. Rangayyan, R.: Biomedical Image Analysis. CRC Press, Boca Raton (2004)
19. Ravanelli, M., Brakel, P., Omologo, M., Bengio, Y.: A network of deep neural networks for distant speech recognition. CoRR abs/1703.08002 (2017)
20. Ripley, B., Hjort, N.: Pattern Recognition and Neural Networks. Cambridge University Press, Cambridge (1995)
21. Rubin, A., Santana, A., Costa, A., Baldi, B., Pereira, C., et al.: Diretrizes de doenças pulmonares intersticiais da Sociedade Brasileira de Pneumologia e Tisiologia. Braz. J. Pulmonol. **38**(suppl. 2), S1–S133 (2012)
22. Sokolova, M., Lapalme, G.: A systematic analysis of performance measures for classification tasks. Inf. Process. Manag. **45**(4), 427–437 (2009)
23. Tan, P.N., Steinbach, M., Kumar, V.: Introduction to Data Mining. Addison-Wesley Longman Publishing Co., Inc., Boston (2005)
24. Zaki, M., Meira Jr., W.: Data Mining and Analysis: Fundamental Concepts and Algorithms. Cambridge University Press, Cambridge (2014)
25. Zeiler, M.D.: ADADELTA: An adaptive learning rate method. CoRR abs/1212.5701 (2012)
26. Zheng, B., Yoon, S., Lam, S.S.: Breast cancer diagnosis based on feature extraction using a hybrid of k-means and support vector machine algorithms. Expert Syst. Appl. **41**(4), 1476–1482 (2014)

Non-dermatoscopic Image Analysis
for the Recognition of Malignant Skin Diseases
with Convolutional Neural Network
and Autoencoders

Ricardo Coronado$^{(\boxtimes)}$ ⓘ, Alexander Ocsa ⓘ, and Oscar Quispe ⓘ

Universidad Nacional de San Agustín, Arequipa, Peru
{rcoronado,aocsa,oquispepo}@unsa.edu.pe

Abstract. Every year, people around the world are affected by different skin diseases or cancer. Nowadays, these can only be detected accurately by clinical analysis and skin biopsy. However, the diagnosis of this malignant disease does not ensure the survival of the patient, since many clinical cases are detected in the terminal phases. Only early diagnosis would increase the life expectancy of patients.

In this paper, we propose a method to recognition malignant skin diseases to identify malignant lesions in non-dermatoscopic images. For the method, we use Convolutional Neural Network and propose the use of autoencoders as another classification model that provides more information on the diagnosis. Experiments show that our proposal reaches up to 84.4% of accuracy in the well-known dataset of the ISIC-2016. In addition, we collect non-dermatoscopic images of skin lesions and developed a new dataset to demonstrate the advantage of our method.

Keywords: Skin cancer · Convolutional Neural Networks
Autoencoders · Image classification

1 Introduction

Malignant skin diseases take thousands of lives around of the world. For example, in 2016 in the United States, 83510 new cases of skin cancer have been diagnosed, from this, 13650 people have died [11]. The detection of this cancer is performed by clinical analysis, and the best clinical method used is ABCD [3,7]. This method analyzes the morphology of the lesion and its evolution. However, it requires a manual procedure and a high level of proficiency. As a solution for this problem, some researchers proposed computer-assisted methods, based on statistics, pattern recognition, machine learning, and deep learning, among others [2].

According to the state of the art, some researches achieves good results detecting malignant and benign lesions. However, this one could be insufficient in real scenarios due to correlations between diseases of different classes, it is common to find cases of benign lesions that become malignant over time. Find the subclass of a sample could provide more information for a specialist to make a successful diagnosis. In addition, the datasets analyzed are made up of dermatoscopic images, such data samples are inaccessible to people who don't have dermatoscopes. On the other hand, we have made up a dataset of non-dermatoscopic images, these one are samples of skin lesions taken with a conventional camera.

Many methods were used for the skin detection, but currently, the best results have been obtained with the use of Convolutional Neural Network (CNN), as demonstrated in [1]. In this work, we use the CNN architecture VGG-19 as [6], but we propose the use of Autoencoders (AEs) instead of fully-connected networks. Further, we have tested this one on dataset with 3 class and 11 sub-class. The main contribution of our work is the use of AEs as method of classification, to identify the kind of skin disease which a sample belongs. The result of this, will classify the samples as benign, premalignant or malignant diseases.

This paper is organized as follow: Sect. 2 presents the concepts for the development of the proposal, specifically about CNN and AEs, Sect. 3 describes the datasets used, Sect. 4 shows the proposed method, Sect. 5 shows the experimental results. Finally, in Sect. 6, presents the conclusions of the paper.

2 Background

The methods for detecting skin diseases are based on feature extraction. There are two approaches, a clinical analysis method [3] based on the specialist's experience and a method computer-aided that uses Machine Learning for processing samples [6,9].

2.1 Convolutional Neural Network (CNN)

Originally a CNN requires a lot of training to obtain good results, depending on the complexity of the training data. To reduce the time required and improve the accuracy results, some works such as [4], use transfer learning to initialize the filters of the network. This helps the process of feature extraction made on convolutional layers. On the other hand, fully connected layers are restarted to fine-tune the CNN and set the number of classes.

2.2 Classification by Reconstruction

An autoencoder (AE) can be seen as neural network that tries to reconstruct the input data, these are known as a class of unsupervised learning algorithms [5]. Unlike supervised algorithms, not need labels or class information, also AEs have been used as a method to pre-train a network and initialize its weights. According to [8], this research introduce the use of AEs as a classification method.

3 Datasets

In this section, we present a new dataset of non-dermatoscopic images, built using different sources[1]. This dataset consists of 2360 unsegmented images of

---

[1] http://rikardocorp.me:8000/.

medium and high quality, divided into three main classes (benign, pre-malignant and malignant), each class is divided into subclasses as shown in (Table 1).

Table 1. Skin diseases dataset

Category	Skin disease	Class	Sub-class	Train	Test	Total
Bening	Melanocytic Nevus	0	0	135	55	190
	Dermatofibroma	0	1	80	20	100
	Lentigo Solar	0	2	45	8	53
	Seborrheic Keratosis	0	3	249	75	324
Pre Malignant	Atypical Nevi	1	4	51	17	68
	Actinic Keratosis	1	5	60	20	80
	Keratoacanthoma	1	6	136	50	186
Malignant	Bowen	2	7	54	7	61
	Squamous Cell Carcinoma	2	8	49	11	60
	Melanoma	2	9	193	66	259
	Basal Cell Carcinoma	2	10	117	56	173

The sub-classes considered in this dataset were selected because they are the most common, lethal and easily confused by other lesions of less severity. This was done with the help of a specialist in dermatology and oncology, a professional at the National Institute of Neoplasm Diseases of Peru (INEN[2]). Finally, we pick 1554 samples for the final dataset, this was split in training (1169) and testing (385) samples.

In addition, we used 4 different datasets to validate our proposal; MNIST[3], CIFAR-10[4], SVHN[5] and the ISBI Challenge 2016 Dataset[6].

4 Proposed Method

Our proposed method is based on the use of Convolutional Neural Network (CNN) and Autoencoders (AE). For the evaluations, we measure the accuracy, precision, recall and fβ metric.

4.1 Data Preprocessing

Using a semi-automated process (Fig. 1), we segment the images using thresholding techniques [9], available in the Python library sklearn-image[7], we fine-tune

---

[2] http://www.inen.sld.pe/.
[3] http://yann.lecun.com/exdb/mnist/.
[4] https://www.cs.toronto.edu/~kriz/cifar.html.
[5] http://ufldl.stanford.edu/housenumbers/.
[6] https://isic-archive.com/.
[7] http://scikit-image.org/.

segmentation using a hand-craft tool available at github[8]. Then, we generate the images dataset at 224 × 224 pixels dimension.

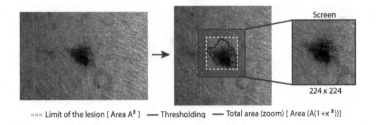

□□□ Limit of the lesion [Area A²]   —— Thresholding   —— Total area (zoom) [Area (A(1+x²))]

Fig. 1. Image segmentation process

Then, we generate synthetic data to increase the number of samples. For this, through the clinical analysis and specialist assistance (INEN), we pick the best samples of each subclass and performed rotations in 0, 90, 180 and 270 degrees to increase the training data by 33%. The testing data is immutable.

4.2   Features Extractor and Classifier

We use a VGG19 network architecture [6], which consists of 19 layers, as we can see in Fig. 2, this scheme conforms the feature extractor that we use. According to [10], to obtain the most general-purpose representation for learning is used the output of the last convolutional layer of the CNN. The original classifier is modified so that our network can classify three types of classes. Thus, the network weights are pre-trained on Imagenet[9] and the CNN was fine-tuned to the target dataset by transfer learning.

4.3   Clasification with Autoencoders

Before using AE as a classification method, we have to save the feature vectors of the first fully connected layer of the network, as shown in the Fig. 2, this one is the new representation of each image which consists of 4096 values. Then, we can train our first AE [5] which we will call global-AE. The global-AE is trained with the entire training dataset until the reconstruction error is minimized. Additionally, we generate n-autoencoders (n-AEs) which are cloned from global-AE, where n is the number of classes in the dataset (See Fig. 3).

Then, in the training phase, each training sample will only feed the AE associated with its class. What happens here is that each AE will be to specialize in reconstructing the data of its own class.

---

[8] https://github.com/rikardocorp/skin-diseases.

[9] http://www.image-net.org.

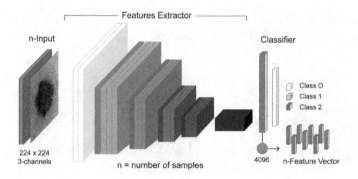

Fig. 2. Convolutional Neural Network VGG-19

For the test phase, each sample is tested by the n-AEs, to generate a reconstruction error vector by sample, as we see in the Fig. 3. Finally, we get the minimum reconstruction error for each vector to know the class to which the sample belongs.

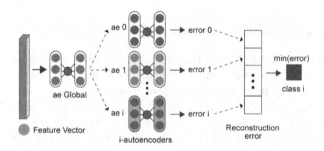

Fig. 3. Each AE consists of 5 layers (1 to feed, 2 for enconding and 2 for decoding)

5   Results

To validate the classification model with AEs, We train the CNN network with different datasets (MNIST, CIFAR-10, SVHN and ISBI) to get the accuracy classification and the feature vectors as described in Sect. 4.2. These feature vectors feed our method of classification with AE named CNN-AE described in Sect. 4.3. We setting our CNN with a minibatch of 30 samples, and learning rate between $<10^{-3}, 10^{-5}>$. In Table 2, we can see the results obtained by CNN and CNN-AE. Here we can observe that CNN-AE is comparable to CNN. If we focus only on the accuracy indicator, we get results that are in general slightly worse.

In addition, we compared our results with the results of the ISBI Contest Dataset. It is available on its ISBI-2016 webpage[10]. Table 3 shows these results.

---

[10] https://challenge.kitware.com/#phase/56fc2763cad3a54f8bb80e51.

Table 2. Comparison of CNN and CNN-AE classification models

	CIFAR-10		MNIST		SVHN		ISBI	
	CNN	CNN-AE	CNN	CNN-AE	CNN	CNN-AE	CNN	CNN-AE
Accuracy	0.910	0.896	0.990	0.978	0.909	0.860	0.846	0.844
Precision	0.911	0.895	0.990	0,977	0.903	0.846	0.759	0.762
Recall	0.910	0.896	0.990	0.977	0.904	0.855	0.759	0.702
$F\beta$	0.910	0.895	0.990	0.977	0.903	0.847	0.759	0.744

The winner of the contest has an Accuracy 1% higher than our model, while our Average Precision is slightly higher. However, according to the sensitivity metric, our model is better identifying True Positive, equivalent to cases of skin cancer.

Table 3. Proposed method (CNN, CNN-AE) vs the winner of ISBI contest

	Accuracy	Area under Roc	Avg. precision	Sensitivity	Specificity
1er Place	0.855	0.783	0.624	0.547	0.931
*CNN	0.846	0.758	0.651	0.613	0.904
*CNN-AE	0.844	0.702	0.610	0.466	0.937

Finally, we perform a comprehensive evaluation for the dataset we presented in this work, which is conformed by 3 classes and 11 sub-classes. First, we train the VGG19 CNN network to classify (VGG19 with 3 classes) which we will call CNN-3. In Fig. 4 shows confusion matrix for CNN-3 network.

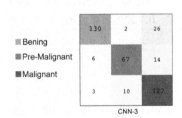

Fig. 4. Confusion Matrix for CNN-3 network for our own dataset

Know only the skin lesion class is not enough for an adequate diagnosis. It is important to know the sub-class (kind of disease) that is being detected. To achieve this, we use the CNN-3 network and use it as a feature extractor Sect. 4.2. Moreover, we performed the training VGG19 CNN network with AE to classify (11 sub-classes), which we will call CNN-AE-11 described in Sect. 4.3.

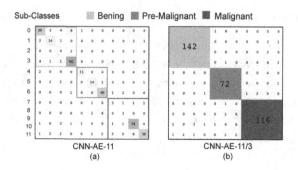

Fig. 5. Confusion Matrix for CNN-AE-11 and CNN-AE-11/3 for our own dataset

In Fig. 5(a), we see the confusion matrix of CNN-AE-11 with a accuracy (0.722) lowest that CNN-3 (0.841). However, if we analyze the results of CNN-AE-11, we can see that some samples were wrongly classified as sub-class, but were correctly classified as class, according to Table 1. Therefore, we group the results of CNN-AE-11 by class hits (benign, pre-malignant and malignant) and we will call CNN-AE-11/3, as we can see in the Fig. 5(b). Now, we can deduce that the accuracy of CNN-AE-11/3 is 85.71%, improving the accuracy of CNN-3 (84.15%). The results obtained in this test are shown in the Table 4.

Table 4. Evaluation for the own dataset, CNN-3, CNN-AE-11, and CNN-AE-11/3.

	Own dataset			
	Accuracy	Precision	Recall	FBetal
CNN-3	0.8415	0.8479	0.8333	0.8425
CNN-AE-11/3	**0.8571**	*0.9141*	*0.9217*	*0.9179*
CNN-11	0.1948	0.0177	0.0909	0.0211
CNN-AE-11	**0.7220**	*0.6424*	*0.6380*	*0.6365*

6   Conclusions

- This work has reached the first place of the ISBI-2016 Contest 3. Moreover, according to the sensitivity metric, our model is better identifying True Positive, equivalent to cases of skin cancer. So, our model is better due to the fact that there is a greater risk for sick people who are classified as healthy.
- Classification using autoencoders is a novel method for the malignant diseases diagnosis. It has shown comparable results as demonstrated in Table 2, even with unbalanced datasets. This feature is important for this kind of research, since there is no availability of large datasets with images, to build datasets.

– Finally, the detection of malignant diseases requires the analysis of all the information that we can obtain from a diagnostic method; even the errors provide information, patterns, and behavior, which resemble the clinical diagnosis that is performed by specialists.

Acknowledgements. This work has been partially funded by the Master Scholarship at the Universidad Nacional de San Agustín, which is an initiative of CITEC through a fund FONDECYT (Perú). We would like to thank research department of Instituto Nacional de Enfermedades Neoplásicas from Peru, for gently providing us his advice on the direction of this article.

References

1. Fornaciali, M., Carvalho, M., Bittencourt, F.V., de Avila, S.E.F., Valle, E.: Towards automated melanoma screening: Proper computer vision & reliable results. CoRR, abs/1604.04024 (2016)
2. Jafari, M.H., Nasr-Esfahani, E., Karimi, N., Reza Soroushmehr, S.M., Samavi, S., Najarian, K.: Extraction of skin lesions from non-dermoscopic images using deep learning. ArXiv e-prints, September 2016
3. Whited, J.D., Grichnik, J.M.: Does this patient have a mole or a melanoma? JAMA **279**(9), 696–701 (1998)
4. Kawahara, J., BenTaieb, A., Hamarneh, G.: Deep features to classify skin lesions. In: 2016 IEEE 13th International Symposium on Biomedical Imaging (ISBI), pp. 1397–1400, April 2016
5. Le, Q.V., Google Brain and Google Inc.: A tutorial on deep learning part 2: Autoencoders, convolutional neural networks and recurrent neural networks (2015)
6. Menegola, A., Fornaciali, M., Pires, R., Bittencourt, F.V., de Avila, S.E.F., Valle, E.: Knowledge transfer for melanoma screening with deep learning. CoRR, abs/1703.07479 (2017)
7. Nachbar, F., Stolz, W., Merkle, T., Cognetta, A.B., Vogt, T., Landthaler, M., Bilek, P., Braun-Falco, O., Plewig, G.: The ABCD rule of dermatoscopy. J. Am. Acad. Dermatol. **30**(4), 551–559 (1994)
8. Othman, E., Bazi, Y., Alajlan, N., Alhichri, H., Melgani, F.: Using convolutional features and a sparse autoencoder for land-use scene classification. Int. J. Remote Sens. **37**(10), 2149–2167 (2016)
9. Premaladha, J., Ravichandran, K.S.: Novel approaches for diagnosing melanoma skin lesions through supervised and deep learning algorithms. J. Med. Syst. **40**(4), 1–12 (2016)
10. Sablayrolles, A., Douze, M., Jégou, H., Usunier, N.: How should we evaluate supervised hashing? CoRR, abs/1609.06753 (2016)
11. Siegel, R., Miller, K., Jemal, A.: Cancer statistics. CA Cancer J. Clin. **66**(1), 7–30 (2016)

Non-stationary Multi-output Gaussian Processes for Enhancing Resolution over Diffusion Tensor Fields

Jhon F. Cuellar-Fierro[1]($\boxtimes$), Hernán Darío Vargas-Cardona[1],
Mauricio A. Álvarez[2], Andrés M. Álvarez[1], and Álvaro A. Orozco[1]

[1] Automatic Researh Group, Universidad Tecnológica de Pereira, Pereira, Colombia
{jfcuellar,hernan.vargas,andres.alvarez1,aaog}@utp.edu.co
[2] University of Sheffield, Sheffield, UK
mauricio.alvarez@sheffield.ac.uk

Abstract. Diffusion magnetic resonance imaging (dMRI) is an advanced technique derived from magnetic resonance imaging (MRI) that allows the study of internal structures in biological tissue. Due to acquisition protocols and hardware limitations of the equipment employed to obtain the data, the spatial resolution of the images is often low. This inherent lack in dMRI is a considerable difficulty because clinical applications are affected. The scientific community has proposed several methodologies for enhancing the spatial resolution of dMRI data, based on interpolation of diffusion tensors fields. However, most of the methods have considerable drawbacks when they interpolate strong transitions, such as crossing fibers. Also, relevant clinical information from tensor fields is modified when interpolation is performed. In this work, we propose a probabilistic methodology for interpolation of diffusion tensors fields using multi-output Gaussian processes with non-stationary kernel function. First, each tensor is decomposed in shape and orientation features. Then, the model interpolates the features jointly. Results show that proposed approach outperforms state-of-the-art methods regarding resolution enhancement accuracy on synthetic and real data, when we evaluate interpolation quality with Frobenius and Riemann metrics. Also, the proposed method demonstrates an adequate characterization of both stationary and non-stationary fields, contrary to previous approaches where performance is seriously reduced when complex fields are interpolated.

1 Introduction

Since the appearance of Magnetic Resonance Imaging (MRI) several years ago, many clinical applications have been developed. As well as a variety of methodologies for improving diagnosis of various neurological diseases, e.g., Parkinson, Epilepsy, [1]. Recently, the diffusion MRI (dMRI) has emerged as a non-invasive method that allows exploring, visualizing, and evaluating qualitatively and quantitatively biological structures, such as white matter tracts, cortical gray substance, cardiac fibers, among others [2]. dMRI describes the diffusion of water

© Springer International Publishing AG, part of Springer Nature 2018
M. Mendoza and S. Velastín (Eds.): CIARP 2017, LNCS 10657, pp. 168–176, 2018.
https://doi.org/10.1007/978-3-319-75193-1_21

particles through a 2nd order tensor $\mathbf{D} \in \mathbb{R}^{3 \times 3}$, which fully represents the particle mobility along each spatial direction, yielding anisotropic diffusions [3]. Therefore, the diffusion tensor of each voxel in a given dMRI is represented by a 3×3 symmetric and positive definite matrix, being necessary at least six independent measurements along different directions.

However, dMRI has a considerable difficult with the spatial resolution of acquired images. This problem is due to clinical acquisition protocols, and technological limitations of MRI scanners [4]. Spatial resolution of dMRI is commonly in a range of 1 to $2\,mm^3$ for each voxel. Nonetheless, in medical imaging it is often desired to obtain information from smaller structures, leading to a lack of precision in some clinical applications [5]. Different methodologies for interpolation of diffusion tensor fields have been proposed for enhancing dMRI resolution. Diffusion tensors have to satisfy some restrictions. For example, the determinant of matrices must change monotonously to avoid the swelling effect [5], and they must be positive definite (PD). The first attempt for improving resolution of dMRI is based on direct interpolation, where each component of the tensor is interpolated independently in a Euclidean space [6]. However, this technique generates swelling effect [6]. To preserve some constraints of diffusion tensors (i.e., PD tensors), parametric approaches have been developed using Cholesky factorization [7]. Nevertheless, the fractional anisotropy (FA) is modified. Other proposed approaches [8] simultaneously seek to eliminate the swelling effect and preserve the symmetry and PD property of diffusion tensors by interpolation in Riemannian spaces. However, their computational cost is high. To overcome computational cost issue and PD constraint, the Log-Euclidean interpolation [6] applies a logarithmic transformation to the matrices before operating on them, but its performance is reduced in heterogeneous fields. An alternative study called Feature-based interpolation (FbI), proposed by [5], developed an approach for interpolation of diffusion tensors by decomposing the tensors into eigenvalues (direction) and Euler angles (orientation). Then, each feature is interpolated linearly. The previously mentioned methods [5–8], have difficulties to describe non-stationary fields: strong transitions and abrupt changes among tensors such as crossing fibers.

In this work, we present a methodology for interpolation of diffusion tensor fields based on multi-output Gaussian processes with a non-stationary kernel function (NmoGp). First, we decompose each tensor of the field in three eigenvalues and three Euler angles, following the method proposed in [5]. Thus, each tensor is represented by six features indexed in an independent variable that represents spatial coordinates (x, y, z). The main goal is to describe non-stationary diffusion tensor fields allowing an accurate interpolation of complex and noisy dMRI data. To do this, we use an expressive kernel to construct the covariance matrix of the multi-output Gaussian process. The expressive kernel is a convex combination of several kernels or the same kernel with different hyperparameters. The methodology introduced is an extension of [9] that we call moGp. In this case, we are interested in a more robust description of complex fields (i.e. crossing fibers) using a non-stationary kernel function. We compare our approach (NmoGp) against the stationary moGp [9] and the FbI proposed by [5], evaluating Frobenius and Riemann distances.

2   Materials and Methods

2.1   dMRI and Tensor Fields

The dMRI is a measurement of diffusion of water particles inside biological tissues. This diffusion is fully described using a 2nd order tensor $\mathbf{D} \in \mathbb{R}^{3 \times 3}$, which defines shape, orientation and direction [3]. The diffusion tensor is expressed as a 3×3 symmetric positive definite matrix whose elements $D_{ij} = D_{ji}$ with $i, j \in \{x, y, z\}$. It is necessary six independent gradient diffusion measures ($S_k \in \mathbb{R}^+, k = \{1, \ldots, 6\}$) and a baseline image without gradient field ($S_0 \in \mathbb{R}^+$) for estimating the coefficients of a tensor (D_{ij}) from Stejskal-Tanner equation [3]: $S_k(\mathbf{x}) = S_0(\mathbf{x})e^{-b\hat{\mathbf{g}}_k^\top D(\mathbf{x})\hat{\mathbf{g}}_k}$, where, S_k is k^{th} dMRI, $\hat{\mathbf{g}}_k \in \mathbb{R}^3$ is a factor that controls strength and timing of gradients.

2.2   Feature-Based Interpolation Framework

A scheme for tensorial interpolation that keeps restrictions of a diffusion tensor was proposed by [5]. It consists of decomposing a diffusion tensor in six features: three eigenvalues (direction) and three Euler Angles (orientation). Each tensor of the field is decomposed as $\mathbf{D} = \mathbf{E}\varLambda\mathbf{E}^\top$, where $\varLambda = \mathrm{diag}(\lambda_1, \lambda_2, \lambda_3) \in \mathbb{R}^+$ is a diagonal matrix of eigenvalues and $\mathbf{E}$ is a matrix whose columns holds the eigenvector of the tensor $\mathbf{D}$ as follows:

$$\mathbf{E} = \begin{bmatrix} v_{11} & v_{12} & v_{13} \\ v_{21} & v_{22} & v_{23} \\ v_{31} & v_{32} & v_{33} \end{bmatrix}. \tag{1}$$

For preserving the PD property, we apply a logarithmic transformation to eigenvalues as suggested in [5]. A tensor can be represented as a feature vector $\mathbf{y} \in \mathbb{R}^6$ indexed in positions coordinates, $\mathbf{x} = [x, y, z]^\top$, yielding: $\mathbf{y}(\mathbf{x}) = [\ln\lambda_1(\mathbf{x}), \ln\lambda_2(\mathbf{x}), \ln\lambda_3(\mathbf{x}), \alpha(\mathbf{x}), \beta(\mathbf{x}), \gamma(\mathbf{x})]^\top$, where Euler angles are given by: $\alpha = \mathrm{artan2}(v_{12}, v_{11}), \beta = \mathrm{artan2}(-v_{13}, \sqrt{v_{11}^2 + v_{12}^2}), \gamma = \mathrm{artan2}(v_{23}, v_{33})$ and $\mathrm{artan2}(a, b)$ is the four-quadrant arctangent of the real arguments a and b.

2.3   Multi-output Gaussian Processes and Non-stationary Kernel

The model for multi-output is defined as a collection of random variables. Such that any finite of them follows a joint Gaussian distribution. The random variables are associated with a set of different processes $\{f_d\}_{d=1}^{M=6}$, evaluated at different values of $\mathbf{x}$ [10]. Therefore, it is assumed the vector-value function $\mathbf{f}$ as a Gaussian process. Mathematically is $\mathbf{f} \sim \mathcal{GP}(\mathbf{m}(\mathbf{X}), \mathbf{K}(\mathbf{X}, \mathbf{X}))$ where $\mathbf{m}(\mathbf{X})$ is a M-dimensional vector whose components are the mean functions $\{m_d(\mathbf{x})\}_{d=1}^{M=6}$ of each output, and $\mathbf{K}(\mathbf{X}, \mathbf{X})$ is a $NM \times NM$ positive defined matrix, with entries,

$(\mathbf{K}(x_i, x_j))_{d,d'}$, for $i, j = 1, \ldots, N$ and $d, d' = 1, \ldots, M$, (being N and M the number of training samples and outputs). The predictive distribution for a new input vector x_* is a Gaussian distribution [11] and it is given by (2):

$$p(\mathbf{f}(x_*)|\mathbf{S}, \mathbf{f}, x_*, \phi) = \mathcal{N}(\mathbf{f}_*(x_*), \mathbf{K}_*(x_*, x_*)) \tag{2}$$

with, $\mathbf{f}_*(x_*) = \mathbf{K}_{x_*}^\top (\mathbf{K}(\mathbf{X}, \mathbf{X}) + \boldsymbol{\Sigma})^{-1} \bar{\mathbf{y}}$, and $\mathbf{K}_*(x_*, x_*) = \mathbf{K}(x_*, x_*) - \mathbf{K}_{x_*}^\top (\mathbf{K}(\mathbf{X}, \mathbf{X}) + \boldsymbol{\Sigma})^{-1} \mathbf{K}_{x_*}$, where $\mathbf{K}_{x_*} \in \mathbf{R}^{M \times NM}$ has entries $(\mathbf{K}(x_*, x_j))_{d,d'}$ $i = 1, \ldots, N$ and $d, d' = 1, \ldots, M$, $\boldsymbol{\Sigma}$ is a diagonal matrix whose elements are noise of the observation, $\bar{\mathbf{y}}$ is an NM vector obtained concatenating the output vectors, $\mathbf{S}$ is the training data, and ϕ represents the set of hyperparameters of the covariance function and the variances of noise for each output $\{\sigma_d^2\}_{d=1}^{M=6}$. There are several possible choices of covariance for multi-output problem, in this work we consider the linear model of coregionalization (LMC) [10]. Usually, the kernel employed in moGp is one of the common stationary functions available in the literature (RBF, Mattern, rational quadratic, among others). On the other hand, we define a non-stationary kernel as an expressive mixture according to the model proposed by [12]:

$$k(x, x') = \sum_{i=1}^{r} \sigma(w_i(x)) k_i(x, x') \sigma(w_i(x')), \tag{3}$$

where $w_i(x) : \mathbb{R}^P \to \mathbb{R}^1$ is the weighting function, with P the dimensional input. The expressiveness of this function determines how many changes can occur in the data. $w_i(x) = \sum_{j=1}^{v} a_j \cos(\omega_j x + b_j)$. $\sigma(z) : \mathbb{R}^1 \to [0, 1]$, is the warping function, that is computed as a convex combination over the weighting function $\sigma(w_i(x)) = \exp(w_i(x))/\sum_{i=1}^{r} \exp(w_i(x))$, $\sum_{i=1}^{r} \sigma(w_i(x)) = 1$ inducing a partial discretization over latent functions. This function produces non-stationarity, since it depends of the input variable x, and $k_i(x, x')$ can be any stationary kernel. It is expected that latent functions have different kernel structures or a same form with different hyperparameters.

2.4   Experimental Setup and Datasets

First, we test the proposed model in a simulation of crossing fibers. This dataset is obtained from FanDTasia Toolbox [13]. Also, we evaluated real dMRI data of the head from a healthy subject. We employ 25 gradient directions with a value b equal to $1000S/mm^2$. Both datasets are splitted in a 50% for training and 50% for validation. Original datasets are the gold standard fields. We compare our model with methodologies proposed by [5] and [9]. Finally, to quantitatively evaluate the models, we use two distance metrics defined over matrices: the Frobenius norm (Frob) and Riemann distance (Riem) [6], given by (4) and (5).

$$\text{Frob}(\mathbf{D}_1, \mathbf{D}_2) = \sqrt{\text{trace}\left[(\mathbf{D}_1 - \mathbf{D}_2)^\top (\mathbf{D}_1 - \mathbf{D}_2)\right]}, \tag{4}$$

$$\text{Riem}(\mathbf{D}_1, \mathbf{D}_2) = \sqrt{\text{trace}\left[\log(\mathbf{D}_1^{-1/2}\mathbf{D}_2\mathbf{D}_1^{-1/2})^\top \log(\mathbf{D}_1^{-1/2}\mathbf{D}_2\mathbf{D}_1^{-1/2})\right]}, \tag{5}$$

where $\mathbf{D}_1$ and $\mathbf{D}_2$ are the interpolated and the gold standard tensor, respectively. The error metrics are computed in all the test data, and we report the mean, standard deviation and confidence interval for each model.

3 Experimental Results and Discussion

3.1 Crossing Fibers

Crossing fibers data is one of most difficult field to interpolate, because transition among tensors is very abrupt. We test the NmoGp over a 2D crossing fibers field of 22×22. We use five latent functions ($r = 5$) and we employ a rational quadratic kernel. The Fig. 1(a) shows the crossing fibers field, Fig. 1(b) is the downsampled tensor field (used for training), and Fig. 1(c) illustrates the validation tensors. Also, we show Riemann error maps in Figs. 1(d), (e) and (f). If we observe in detail, error values are very low over smooth regions of the tensor field for all methods (blue zones in error maps of Fig. 1). However, when we evaluate strong transition regions, interpolation is not straightforward, and the error is higher in this areas. Specifically, over this critic region of the field, our model achieves more precision than comparison methods. We can explain this because the non-stationary kernel can capture the dynamic of the tensor field. While it is true that there are some errors when we interpolate new tensors in the crossing fibers region, we can say that NmoGp outperforms to moGp and the FbI method, when interpolation is challenging due to strong changes in neighboring tensors.

Table 1. Frob and Riem distances for crossing fibers of NmoGp, moGp, and FbI methods

Model	Riem	Confidence interval	Frob (10e-4)	Confidence interval
NmoGp	0.191 ± 0.277	$[0.156, 0.227]$	17.342 ± 25.191	$[1.410, 2.058]$
moGp	0.213 ± 0.283	$[0.176, 0.249]$	19.589 ± 26.136	$[1.622, 2.294]$
FbI	0.266 ± 0.362	$[0.220, 0.313]$	24.228 ± 33.375	$[1.993, 2.851]$

Also, according to results in Table 1, the FbI model obtained the higher error (Frobenius and Riemann). FbI model has a considerable lack because it does not consider correlation among the six features extracted from tensor decomposition. This approach only interpolates linearly and separately each feature. On the other hand, moGp and NmoGp interpolate the six features simultaneously. The idea is to share a correlation among features. For this reason, there is an additional information allowing a better estimation of new data. The main difference between moGp and NmoGP is the type of kernel for constructing the covariance matrix inside the model. Thereby, moGp works with a single kernel (stationary) while our NmoGP is based on a non-stationary kernel. We see in Table 1, that global error of the NmoGp approach is lower than the moGp and FbI method.

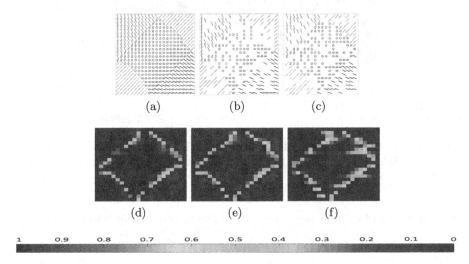

Fig. 1. Interpolation over crossing fiber fields, (a) gold standard, (b) training field, (c) test field. (d), (e) and (f) Riemann norm error maps, for the NmoGp, moGp and FbI model, respectively.

3.2   Real DT Field

We test the methods over a 2D diffusion tensor field obtained from real dMRI. The data corresponds to an axial slice with 40×40 tensors of the head. Figures 2(a), (b) and (c) correspond to the gold standard, training and test data respectively. Figures 2(d), (e) and (f) show the Riemann error maps. Again, Table 2 show the error distance for all tested methods and the confidence interval. A real dMRI tensor field is noisy and very heterogeneous. Therefore, simple interpolation methods fail to achieve a good accuracy. Also, probabilistic methods such as moGp and NmoGp have the robustness property. Again, NmoGp improves to the comparison methods according to outcomes of Table 2. This result is very relevant, because it confirms that proposed method can interpolate tensorial data with good accuracy, no matter the type of dataset. Finally, we can establish that insertion of a non-stationary kernel in multi-output Gaussian process increases its performance significantly.

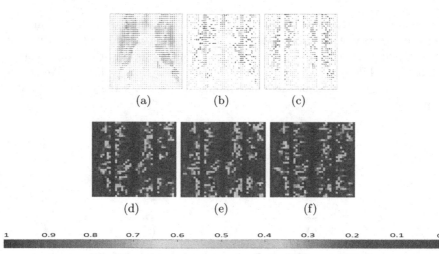

Fig. 2. Interpolation over a 2D real crossing fiber field, (a) gold standard, (b) training field, (c) test field. (d), (e) and (f) Riem norm error maps for the NmoGp, moGp and FbI model, respectively.

Table 2. Frob and Riem distances for real data of NmoGp, moGp, and FbI method

Model	Riem	Confidence interval	Frob	Confidence interval
NmoGp	11.222 ± 1.041	[1.051, 1.193]	0.725 ± 0.553	[0.686, 0.764]
moGp	11.458 ± 0.958	[1.078, 1.213]	0.745 ± 0.497	[0.709, 0.780]
FbI	11.569 ± 1.044	[1.085, 1.227]	0.756 ± 0.504	[0.720, 0.791]

4   Conclusions and Future Work

In this work, we presented a probabilistic methodology for interpolation of diffusion tensor fields. The model decomposes the tensors in six features describing the main properties of a diffusion tensor (direction and orientation). We index the extracted features in spatial coordinates to interpolate them using a moGp with a non-stationary kernel function. The structure of the kernel function employed in this method combines several kernels with different properties. The purpose is to characterize complex fields such as crossing fibers. In this context, the non-stationary kernel allows differentiating between strong and uniform transitions. In this way, the interpolation of new tensors is more accurate in comparison to ordinary approaches. We tested NmoGp against state-of-the-art methods in two different datasets: simulation of crossing fibers, and a dMRI segment. Outcomes proved that proposed model outperforms to the comparison methods evaluating accuracy with Frobenius and Riemann distances. Although, our proposed approach is accurate and robust for interpolation of complex tensor fields, there is an important issue: initialization of parameters and hyperparameters is not straightforward. Currently, we use a cross-validation procedure. As future work, we would like to extend non-stationary kernel functions to more complex models such as generalized Wishart processes and tractography procedures.

Acknowledgments. This research is developed under the project "Desarrollo de un sistema de soporte clínico basado en el procesamiento estócasitco para mejorar la resolución espacial de la resonancia magnética estructural y de difusión con aplicación al procedimiento de la ablación de tumores" financed by COLCIENCIAS with code 111074455860 under the program: *744 Convocatoria para proyectos de ciencia, tecnología e innovación en salud 2016.* H.D. Vargas is funded by Colciencias under the program: *Formación de alto nivel para la ciencia, la tecnología y la innovación - Convocatoria 617 de 2013* with code 111065740687. We thank to the program of master in electrical engineering of the Universidad Tecnológica de Pereira.

References

1. Sąsiadek, M.J., Szewczyk, P., Bladowska, J.: Application of diffusion tensor imaging (DTI) in pathological changes of the spinal cord. Med. Sci. Monit. **18**(6), RA73–RA79 (2012)
2. Del Cura, J.L., Pedraza, S.: Radiologia Esencial. Toma 2 (2010)
3. Le Bihan, D., Mangin, J.-F., Poupon, C., Clark, C.A., Pappata, S., Molko, N., Chabriat, H.: Diffusion tensor imaging: concepts and applications. J. Magn. Reson. Imaging **13**(4), 534–546 (2001)
4. Gupta, V., Ayache, N., Pennec, X.: Improving DTI resolution from a single clinical acquisition: a statistical approach using spatial prior. In: Mori, K., Sakuma, I., Sato, Y., Barillot, C., Navab, N. (eds.) MICCAI 2013. LNCS, vol. 8151, pp. 477–484. Springer, Heidelberg (2013). https://doi.org/10.1007/978-3-642-40760-4_60
5. Yang, F., Zhu, Y.-M., Magnin, I.E., Luo, J.-H., Croisille, P., Kingsley, P.B.: Feature-based interpolation of diffusion tensor fields and application to human cardiac DT-MRI. Med. Image Anal. **16**(2), 459–481 (2012)
6. Arsigny, V., Fillard, P., Pennec, X., Ayache, N.: Log-Euclidean metrics for fast and simple calculus on diffusion tensors. Magn. Reson. Med. **56**(2), 411–421 (2006)
7. Wang, Z., Vemuri, B.C., Chen, Y., Mareci, T.H.: A constrained variational principle for direct estimation and smoothing of the diffusion tensor field from complex DWI. IEEE Trans Med. Imag. **23**(8), 930–939 (2004)
8. Batchelor, P.G., Moakher, M., Atkinson, D., Calamante, F., Connelly, A.: A rigorous framework for diffusion tensor calculus. Magn. Reson. Med. **53**(1), 221–225 (2005)
9. Cardona, H.D.V., Orozco, Á.A., Álvarez, M.A.: Multi-output Gaussian processes for enhancing resolution of diffusion tensor fields. In: 2016 IEEE 38th Annual International Conference of the Engineering in Medicine and Biology Society (EMBC), pp. 1111–1114. IEEE (2016)
10. Alvarez, M.A., Rosasco, L., Lawrence, N.D., et al.: Kernels for vector-valued functions: a review. Found. Trends® Mach. Learn. **4**(3), 195–266 (2012)
11. Rasmussen, C.E.: Gaussian Processes for Machine Learning. MIT Press, Cambridge (2006)

12. Herlands, W., Wilson, A., Nickisch, H., Flaxman, S., Neill, D., Van Panhuis, W., Xing, E.: Scalable Gaussian processes for characterizing multidimensional change surfaces. arXiv preprint arXiv:1511.04408 (2015)
13. Barmpoutis, A., Vemuri, B.C.: A unified framework for estimating diffusion tensors of any order with symmetric positive-definite constraints. In: Biomedical Imaging From Nano to Macro, pp. 1385–1388. IEEE (2010)

Improving the Classification of Volcanic Seismic Events Extracting New Seismic and Speech Features

Millaray Curilem[1(⊠)], Camilo Soto[1], Fernando Huenupan[1],
Cesar San Martin[1], Gustavo Fuentealba[1], Carlos Cardona[2],
and Luis Franco[2]

[1] Facultad de Ingeniería y Ciencias, Universidad de La Frontera,
Francisco Salazar, 01145 Temuco, Chile
{millaray.curilem, fernando.huenupan, cesar.sanmartin,
gustavo.fuentealba}@ufrontera.cl
[2] Observatorio Vulcanológico de los Andes Sur,
Rudecindo Ortega, 03850 Temuco, Chile
{carlos.cardona, luis.franco}@sernageomin.cl

Abstract. This paper presents a study on features extracted from the seismic and speech domains that were used to classify four groups of seismic events of the Llaima volcano, located in the Araucanía Region of Chile. 63 features were extracted from 769 events that were labeled and segmented by experts. A feature selection process based on a genetic algorithm was implemented to select the best descriptors for the classifying structure formed by one SVM for each class. The process identified a few features for each class, and a performance that overcame the results of previous similar works, reaching over that 95% of exactitude and showing the importance of the feature selection process to improve classification. These are the newest results obtained from a technology transfer project in which advanced signal processing tools are being applied, in collaboration with the Southern Andes Volcano Observatory (OVDAS), to develop a support system for the monitoring of the Llaima volcano.

Keywords: Volcanic seismicity · Pattern recognition · SVM
Feature selection

1 Volcanic Seismology and Pattern Recognition

Because of its geographical location in the pacific ring of fire, Chile has an intense volcanic activity, throughout its territory. The "Observatorio Vulcanológico de los Andes Sur" (OVDAS) is the state agency responsible for establishing systems to continuously monitor and record forty three of the most active Chilean volcanoes. This monitoring is mainly of seismological type, as seismicity is highly related to volcanic activity [1]. The most common seismic events are Tremor (TR) and Long period (LP) events, which are related to the sustained and transitory fluid flow through the volcanic conduits, respectively. Volcano Tectonic (VT) events are related to rock failure inside the volcanic structure [2]. In recent years, automatic analysis of these

© Springer International Publishing AG, part of Springer Nature 2018
M. Mendoza and S. Velastín (Eds.): CIARP 2017, LNCS 10657, pp. 177–185, 2018.
https://doi.org/10.1007/978-3-319-75193-1_22

events has grown rapidly. This task is not trivial because each volcano has a particular behavior. In many studies, the problem is addressed in two stages: extraction of signal parameters and classification.

Several approaches have been used to address the classification problem of volcanic seismicity, including Support Vector Machines [3], Self Organizing Maps [4], Hidden Markov Models (HMM) [5], Multi-Layer Perceptron (MLP) [6], Gaussian Mixture Models (GMM) [7], among others. These studies show the diversity of techniques for implementing the classification stage.

Although these works achieved good results, it is necessary to improve them to reach acceptable performances in the real time and online applications. One approach is to improve the quality of the events' descriptors (features). In this sense it is important to underline the specificities of each volcano. The feature extraction process defines which information is extracted from the signals to facilitate their discrimination. Literature on volcanoes presents many features: Linear predictor coding and a signal parameterization to extract information from the waveform can be found in [8]. In [9], autocorrelation functions, obtained using Fast Fourier Transform, represent the spectral contents, and the sum and ratios of the amplitude were used. In other works, wavelet transform gives the possibility to use multiband analysis of the frequencies [3, 10]. Cepstral coefficients were extracted in [11]. The phase information of the events was included to the analysis in [12] and the amplitude statistics (mean, standard deviation, skewness and kurtosis), together with the power of the events were studied in [13].

In addition, many methods have been applied in the literature to perform feature selection in volcanoes [7], like the discriminant method [14], principal component analysis [15] and genetic algorithms [6].

The main motivation of this study is to improve the discrimination capacity of the classifiers by improving the quality of the features. This is why a high number of features were extracted from the events and a selection process, based on a genetic algorithm was performed to evaluate which is the best feature combination for each class.

2   Materials and Methods

Figure 1 depicts the block diagram of the proposed method. The signal database contains events already classified and segmented by experts. The automatic classification stage performs the discrimination between events, according to a subset of

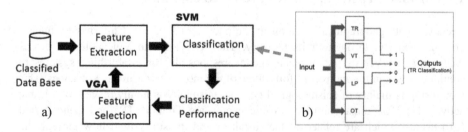

Fig. 1. (a) General working structure of the proposed method and (b) "one versus all" structure of the classification step.

features extracted from the data. The optimal subset of features is found using the validation error of the classifier as the unique fitness criterion. This error is used by the genetic algorithm to select new features, searching for the best subset.

2.1 Classified Database

The Llaima volcano is located in the Araucanía Region (38° 41'S–71° 44'W). The signals were obtained from the LAVE station, located 6 km from the crater. Only the Z component was considered. The records belong to the period 2010 to 2013. The events considered were of type LP, TR and VT. A contrast group was created, named "other" (OT) that contained events that did not belong to the first groups (like tectonic earthquakes, noise, ice breaking, etc.). The database had 769 events, 296 LP, 173 TR, 134 VT and 166 OT. The events were stored in files with variable length, according to their duration. The features were extracted from these files, thus, in this work the features were calculated from variable length windows that covered the entire events.

2.2 Feature Extraction

Most of the extracted features were obtained from our research on volcanic seismicity references, however, some of them belong to other processing domains, like speech. The features marked in gray in Table 1 were introduced as new features for this study.

All the extracted features were linearly normalized between [− 1, 1] and stored in a matrix of 769 rows (all the events) and 63 columns, each one is a specific feature, according to the order presented in Table 1. Each feature is explained next.

Energy: measures the size of the event and is calculated as the sum of the square of the samples [3].

LTA/STA: It is the ration of a long time window average (LTA) and a short time window average (STA) of the event [9]. This ratio is calculated along the signal. When it exceeds a threshold, it may define the beginning of a seismic event. Here it was calculated at the beginning of the event.

Table 1. Overview of the features extracted from each volcano's event

1	Energy	25	WAV Entropy
2	LTA/STA	26-30	LPC1-5
3-9	STA WAV1-7	31	Maximum Amplitude
10-16	WAV1-7 Energy	32	Median Amplitude
17	Variance	33	Mean Amplitude
18	Skewness	34-46	DCT1-13
19	Kurtosis	47-59	DCTLOG1-13
20	WAV4 Variance	60	Duration
21	WAV4 Skewness	61	Mean 5 Picks
22	WAV4 Kurtosis	62	Skewness Envelope
23	Pitch	63	Kurtosis Envelope
24	1st. Circular Moment		

STA WAV1-7: The STA indexes, calculated at the beginning of the event, were also calculated in the spectrum of 7 frequency bands, thus obtaining 7 features: STA WAV1 to STA WAV7. The considered frequency bands are presented in Table 2.

WAV1-7 Energy: This is the relative energy per wavelet band [3]. It is obtained decomposing the event in the 7 bands of Table 2, using a Daubechies mother wavelet type five. The percentage of energy is calculated as the ratio of the energy of the band over the energy of the event.

Table 2. Frequency bands for the feature extractions.

Band	1	2	3	4	5	6	7
Frequency (Hz)	25–50	12,5–25	6,25–12,5	3,13–6,25	1,56–3,13	0,78–1,56	0,39–0,78

Variance, Skewness and Kurtosis: The variance, the skewness and the kurtosis are extracted from the events in the time domain [6]. The variance is the expected value of the squared deviation from the mean of the event. The skewness is a statistical parameter that measures the asymmetry of the distribution of the seismic event. The kurtosis is a statistical parameter that measures the sharpness of the distribution of the seismic event. These features measure the shape of the event.

WAV4 Variance, WAV4 Skewness and WAV4 Kurtosis: These features are calculated for the 5th frequency band (from 1.56 to 3.13 Hz).

Pitch: This feature is mainly used in speech processing but it has been applied here to the events. It is related to the lowest frequency at which an object vibrates.

First Circular Moment: reflects the behavior of the phase part of the volcanic signals [12]. The phase is obtained using the Hilbert transform in the range of [0 2π). This feature is mainly used in biomedical signal processing.

Entropy: is a measure of the uncertainty of a distribution. It is also a measure of the "quantity of information" contained in a signal.

LPC 1-5: Linear predictive coding (LPC) is a tool used mostly in audio and speech signal processing, for representing the spectral envelope of a signal in a compressed form [8]. It uses a model based on the fact that a signal can be modeled as the linear combination of N previous samples multiplied by coefficients plus a prediction error. This work considered only 5 of all the possible LPC coefficients.

Maximum Amplitude, Mean and Median: The maximum amplitude is the maximum value of the segmented event [3, 6]. The mean of the event is obtained from its absolute value. The median calculates the median of the samples of the event. All are extracted in the time domain.

DCT1-13 and LOGDCT1-13: The discrete cosine transform (DCT) is mainly used in audio and image processing. Like the discrete Fourier transform (DFT), the DCT expresses a signal as the sum of sinusoidal signals, with different frequencies and amplitudes. However, unlike DFT that works with complex exponentials, the DCT only works with cosines. The DCT has a good ability to compact the energy to the transformed domain, that is, it concentrates most of the information in few coefficients. 13 DCT and 13 LOG-DCT were extracted from the events.

Duration: The duration is extracted from the segmented event in the time domain, counting the number of samples from the start to the end of the event. The automatic extraction of this feature is not trivial, as it requires a start and end detection system. This system is being developed in a parallel work, so for the present study, this characteristic was manually defined by an OVDAS analyst.

Mean of the 5 Frequency Peaks: The events are transformed using the FFT [3, 6]. The 5 highest peaks are detected and their mean is calculated to obtain this feature.

Skewness Envelope and Kurtosis Envelope: The envelope of a signal is related to its external form, that is, the slow variations of the amplitude in the time domain. The envelope is obtained using the Hilbert transform of the signal and calculating its absolute value. In this work, the kurtosis and the skewness of the envelope were used as shape features.

2.3 Classification

The classification step was performed using support vector machines (SVM). SVM tackles classification problems by nonlinearly mapping input data into high-dimensional feature spaces, wherein a linear decision hyperplane separates two classes [16]. To do so, SVM transforms the input space where the data is not linearly separable into a higher-dimensional space called a feature space through functions called kernels. One of the most generally used is the RBF Kernel that was applied here.

To train SVM it is necessary to adjust two hyperparameters: σ defines the width of the Gaussian used in the RBF and c determines the trade-off between the complexity of the model and the error tolerated.

Since SVM is a two-class classifier, a "one vs. all" structure was implemented using four classifiers (one per class). The training process of the classifiers applied a 2-fold cross validation strategy (bilateral cross-validation) due to the high computational cost of the feature selection strategy. The validation error of each trained classifier is obtained as the mean of the validation error of the 2-folds.

To evaluate the performance of each classifier, a contingency table was built. Four statistical indices were calculated for each classifier: sensitivity (Se), specificity (Sp), exactitude (Ex) and error (Er), obtained from the following Eqs. 1 to 4.

$$Se = \frac{TP}{TP + FN}; \tag{1}$$

$$Sp = \frac{TN}{TN + FP}; \tag{2}$$

$$Ex = \frac{TP + TN}{n};$$ (3)

$$Er = \frac{FP + FN}{n}$$ (4)

where TP (true positives) is the number of events correctly classified for each class; TN (true negatives) is the number of events correctly classified as negatives, for each class; FP (false positives) and FN (false negatives) is the number of events classified erroneously and n is the total number of events.

2.4 Feature Selection

A genetic algorithm (GA) [17] performed the search of the best feature subset. The chromosome of each individual was defined as a string of 63 bits, where a "1" represented the presence of the corresponding feature, according to the number of Table 1, and the "0" represented its absence. Each generation had 64 individuals and the maximum number of generations was 200. The first 63 individuals in the initial population had one of the 63 features (diagonal matrix). The last individual (number 64) had all the features set to "1". The percentage of elitism was set to 20% and mutation to 10%.

The Vasconcelos GA [18] was used, since in previous works it reached better solutions than the traditional GA approach [6]. The main characteristics of this algorithm is that it maintains a highly diverse population along the generations, because the selection and crossover operators combine the genetic information of good and bad solutions. Good individuals survive from one generation to another, through elitism.

The performance of each individual was measured by the validation error of the classifier, trained with the features retrieved by its chromosome. In addition to the validation error, chromosomes with a high number of features were penalized. Therefore, the performance of each individual is given by Eq. 5:

$$Performance = Validation\ Error + \frac{Number\ of\ Features}{63}*25\%$$ (5)

3 Results

The simulations were carried out with Matlab of Mathworks, in an Intel Core™ i5 CPU with 3 GHz and 8 GB RAM. The results of the best classifiers, after the 200 generations, are presented in Table 3. This table presents the contingency table and the performance of the best classifiers for each class. It also shows the values of the hyper parameters, for each classifier and the features selected for the best models.

Table 3. Results of the best classifiers after 200 generations.

LAVE Station	LP	OT	TR	VT
LP	274	1	0	9
OT	0	153	0	0
TR	0	0	170	0
VT	0	0	0	113
Not Assigned	22	12	3	12
TOTAL	296	166	173	134
Sensitivity	92,57%	92,17%	98,27%	84,33%
Specificity	97,89%	100,00%	100,00%	100,00%
Error	4,16%	1,69%	0,39%	2,73%
Exactitude	95,84%	98,31%	99,61%	97,27%
Sigma	2^2	2^{-2}	2^{-6}	2^2
C	2^9	2^1	2^1	2^5
Best Feature Set	WAV4 Energy, WAV6 Energy, Duration, Mean Amplitude.	Duration DCT1.	Energy WAV5 Energy	WAV6 Energy WAV7 Energy Kurtosis WAV5 Variance WAV5 Skewness

The "Not Assigned" events correspond to events that were classified positive by more than one classifier or negative by all the classifiers. Thus, the results presented in Table 3 may improve if a confidence step is added to the classifiers. This confidence step has to be able to give an output, even when the classifiers do not agree.

4 Discussion and Conclusions

In a previous work [3], the authors demonstrated that the features selected for the classification of the seismic events of the Villarrica volcano [6], also performed well for the Llaima volcano. However, the global performance decreased from 90 to 80%. This paper presents a study on new characteristics extracted from the seismic and speech processing areas. This study evaluates their impact in the discrimination of four groups of events from the Llaima volcano: the LP, TR, VT and a contrast group, OT.

A feature selection was necessary to define which of the 63 extracted features led to a better performance of the classifiers. It is essential that the set of features that represent the signals is appropriate to the classification problem, so that it divides the space into decision regions associated to the classes to be distinguished. In addition, the number of features has to be reduced to simplify the complexity of the classifier. This is why a feature selection process is needed. Different methods applied to the feature selection process demonstrate the complexity of this task. It is also important to observe that features that are not individually relevant may become so when used in combination with others, and features that are individually relevant may not be useful due to possible redundancies when combined with other. This is why we applied a strategy that evaluates subsets of features. The classification performance was the objective function and the GA performed the search of the best subset. The resulting classifiers reached performances superior to 95%, improving the results of previous works.

LP and VT are more difficult to classify than TR, which needed only two features to reach a high sensitivity. The energy of the low frequencies, from 0.78 to 6.25 Hz are the most important discriminators for LP and VT, while the energy of the 1.56 to 3.13 Hz is the best discriminator for the TR. It is interesting to note the importance of the duration feature, as a good descriptor for LP and OT events. It is also interesting to see that DCT, a feature from the speech domain, was considered important to discriminate the OT group.

The results obtained in this work were very promising, as the addition of new features improved the classification performance. Future works need to go-on in the study of new features and other selection methods. The performances reached in the off-line experiments were considered satisfactory by the OVDAS experts.

Acknowledgements. We would like to thank FONDEF as this study is being supported by the project FONDEF IDeA IT15I10027.

References

1. Chouet, B.: A seismic model for the source of long-period events and harmonic tremor. In: Gasparini, P., Scarpa, R., Aki, K. (eds.) Volcanic Seismology, pp. 133–156. Springer, Heidelberg/New York (1992). https://doi.org/10.1007/978-3-642-77008-1_11
2. Lahr, J.C., Chouet, B.A., Stephens, C.D., Power, J.A., Page, R.A.: Earthquake classification, location and error analysis in a volcanic environment: implications for the magmatic system of the 1989–1990 eruptions at Redoubt Volcano, Alaska. J. Volcanol. Geoth. Res. **62**(1–4), 137–151 (1994)
3. Curilem, M., Vergara, J., San Martin, C., Fuentealba, G., Cardona, C., Huenupan, F., Chacón, M., Khan, S., Hussein, W., Becerra, N.: Pattern recognition applied to seismic signals of the Llaima volcano (Chile): an analysis of the events' features. J. Volcanol. Geoth. Res. **282**, 134–177 (2014)
4. Langer, H., Falsaperla, S., Messina, A., Spampinato, S., Behncke, B.: Detecting imminent eruptive activity at Mt Etna, Italy, in 2007–2008 through pattern classification of volcanic tremor data. J. Volcanol. Geoth. Res. **200**(1–2), 1–17 (2011)
5. Bhatti, S.M., Khan, M.S., Wuth, J., Huenupan, F., Curilem, M., Franco, L., Becerra-Yoma, N.: Automatic detection of volcano-seismic events by modeling state and event duration in hidden Markov models. J. Volcanol. Geoth. Res. **324**, 134–143 (2016)
6. Curilem, G., Vergara, J., Fuentealba, G., Acuña, G., Chacón, M.: Classification of seismic signals at Villarrica volcano (Chile) using neural networks and genetic algorithms. J. Volcanol. Geoth. Res. **180**, 1–8 (2009)
7. Cortés, G., García, L., Álvarez, I., Benítez, C., de la Torre, T., Ibáñez, J.: Parallel System Architecture (PSA): an efficient approach for automatic recognition of volcano-seismic events. J. Volcanol. Geoth. Res. **271**, 1–10 (2014)
8. Scarpetta, S., Giudicepietro, F., Ezin, E.C., Petrosino, S., Del Pezzo, E., Martíni, M., Marinaro, M.: Automatic classification of seismic signals at Mt Vesuvius Volcano, Italy, using neural networks. Bull. Seismol. Soc. Am. **95**(1), 185–196 (2005)
9. Langer, H., Falsaperla, S., Powell, T., Thompson, G.: Automatic classification and a-posteriori analysis of seismic event identification at Soufriere Hills volcano, Montserrat. J. Volcanol. Geoth. Res. **153**, 1–10 (2006)

10. Erlebacher, G., Yuen, D.A.: A wavelet toolkit for visualization and analysis of large data sets in earthquake research. Pure. Appl. Geophys. **161**, 2215–2229 (2004)
11. Ibáñez, J., Benítez, C., Gutiérrez, L., Cortés, G., García-Yeguas, A., Alguacil, G.: The classification of seismo-volcanic signals using Hidden Markov Models as applied to the Stromboli and Etna volcanoes. J. Volcanol. Geoth. Res. **187**, 218–226 (2009)
12. San-Martin, C., Melgarejo, C., Gallegos, C., Soto, G., Curilem, M., Fuentealba, G.: Feature extraction using circular statistics applied to volcano monitoring. In: Bloch, I., Cesar, Roberto M. (eds.) CIARP 2010. LNCS, vol. 6419, pp. 458–466. Springer, Heidelberg (2010). https://doi.org/10.1007/978-3-642-16687-7_61
13. Joevivek, V., Chandrasekar, N., Srinivas, Y.: Improving seismic monitoring system for small to intermediate earthquake detection. Int. J. Comput. Sci. Secur. **4**(3), 308–315 (2010)
14. Álvarez, I., García, L., Cortés, G., Benítez, C., de La Torre, A.: Discriminative feature selection for automatic classification of volcano-seismic signals. IEEE Geosci. Remote Sens. Lett. **9**(2), 151–155 (2012)
15. Unglert, K., Radić, V., Jellinek, A.M.: Principal component analysis vs. self-organizing maps combined with hierarchical clustering for pattern recognition in volcano seismic spectra. J. Volcanol. Geoth. Res. **320**, 58–74 (2016)
16. Vapnik, V.: The Nature of Statistical Learning Theory. Springer, New York (1995). https://doi.org/10.1007/978-1-4757-3264-1
17. Holland, J.H.: Adaptation in Natural and Artificial Systems: An Introductory Analysis with Applications to Biology, Control, and Artificial Intelligence. The MIT Press, Cambridge (1992). 228 p.
18. Kuri-Morales, A.: Pattern recognition via Vasconcelos' Genetic Algorithm. In: Sanfeliu, A., Martínez Trinidad, J.F., Carrasco Ochoa, J.A. (eds.) CIARP 2004. LNCS, vol. 3287, pp. 328–335. Springer, Heidelberg (2004). https://doi.org/10.1007/978-3-540-30463-0_40

An Automatic System for Computing Malaria Parasite Density in Thin Blood Films

Allisson Dantas Oliveira[1] , Bruno M. Carvalho[1(✉)] , Clara Prats[2] ,
Mateu Espasa[3] , Jordi Gomez i Prat[3] , Daniel Lopez Codina[2] ,
and Jones Albuquerque[4]

[1] Department of Informatics and Applied Mathematics, UFRN, Natal, Brazil
allissondantas@gmail.com, bruno@dimap.ufrn.br
[2] BarcelonaTech, Universitat Politècnica de Catalunya, Barcelona, Spain
{clara.prats,daniel.lopez-codina}@upc.edu
[3] Microbiology Department, Vall dHebron University Hospital, Barcelona, Spain
{mespasa,j.gomez}@vhebron.net
[4] Department of Statistics and Informatics, UFRPE, Recife, Brazil
jones.albuquerque@gmail.com

Abstract. Malaria is a major worldwide health problem, specially in countries with tropical climates and remote areas. In this paper, we present an automatic system for estimating malaria parasite density in thin blood smears. The proposed approach is based on simple image processing methods that can be implemented efficiently even on low budget devices. The method has been tested on images acquired under different illumination and acquisition setups and has produced encouraging results, achieving a sensitivity of 89.3%.

Keywords: Malaria · Parasite density · Medical image processing

1 Introduction

Malaria is a major worldwide public health problem. In 2014, 97 countries have been reported with continuous transmission of Malaria, and about 3.3 billion people are at risk of being contaminated [9]. In recent decades, researchers have looked for cost-effective solutions to assist health professionals in public control of epidemics and diseases, including, but not limited to, real-time diagnostic systems and epidemiological events simulations [5,8]. Specifically, Image Processing techniques have been used been used successfully in the diagnosis of many diseases. In this article, we propose the usage of an automatic system for detecting Malaria parasites in thin blood films and computing parasite density, a fundamental information needed for a successfull treatment.

Malaria is caused by a protozoan the genus *Plasmodium*, where the transmission vector is a female of mosquitoes *Anopheles*. There are 5 species of Plasmodium can infect humans: *P. falciparum*, *P. vivax*, *P. malariae*, *P. ovale* and *P. knowlesi* [9].

© Springer International Publishing AG, part of Springer Nature 2018
M. Mendoza and S. Velastín (Eds.): CIARP 2017, LNCS 10657, pp. 186–193, 2018.
https://doi.org/10.1007/978-3-319-75193-1_23

The gold standard for malaria diagnosis, widely used around the world, is a microscopic examination of blood that uses blood films stained with Giemsa dye [7,9]. Two techniques have been used for slide staining to red blood cells (RBC): thick blood and thin blood. The microscopy technique used in the thick blood method slide stainer is the most common and inexpensive to diagnose Malaria. The thin blood technique is used more specifically for identifying the morphology of the parasite *Plasmodium* spp. and in correctly identifying the species [12], and it is preferred for routine estimation because the parasites are easy to see and count in it. Accurate estimates of the number of parasites infecting, as well as the species, are very important in determining an appropriate course of treatment. If done properly, there will be an improvement in the treatment of patients and response to the drugs, especially with the *P. falciparum* species [7].

Here, we propose and evaluate a segmentation approach to count and identify RBCs that are healthy and infected with *P. falciparum* species in *ring-stage*. This restriction to one specie was used because of a broader access to images of thin blood films containing *P. falciparum*, even though the proposed method can be directly applied to identify the other species. However, a new stage of classification would be necessary to correctly identify the species.

1.1 Related Work

Recently, several papers have shown the feasibility of a detection system using image processing and artificial intelligence for medical aplications. Linder et al. [6] employed computer vision techniques to identify candidate regions based on color and size of objects to extract features using Local Binary Patterns (LBPs), local contrast and Scale-Invariant Feature Transform (SIFT) descriptors. The features extracted on the previous step are the input on linear SVM (*Support Vector Machines*) for classification of *P. falciparum* ring-stage trophozoites.

Ross et al. [13] proposed using feedforward neural networks for classification of 4 species of malaria. The neural networks are fed with image features that are extracted based on colour, texture and geometry of the objects. Another contribution is the morphological and threshold selection to identify erythrocytes (red blood cells). Pirnstill et al. [11] proposed the use of a polarized microscope, thus, exploring the properties of polarized light to identify and describe structures and properties of materials. This type of microscopy has a higher cost and requires a specific training. However, Pirnstill et al. [11] used a high fidelity and high optical resolution cell-phone based polarized light microscopy system, a much less expensive system.

Kaewkamnerd et al. [4] proposed a method divided into five phases: image acquisition; preprocessing; image segmentation; feature extraction and classification, to be used on an automatic device equiped with a motor adapted to a microscope for both detection and classification of the malaria parasite species into *P. falciparum* and *P. vivax*. In another study, Chakrabortya et al. [2] proposes an algorithm that uses morphological operations and pixel discrimination based on color to identify malaria parasites from thick smear images of *P. vivax*.

2   Proposed Method

The general workflow for the proposed method is given in Fig. 1. In the remainder of this section, we will describe its major steps and talk about the reasons that impacted their development.

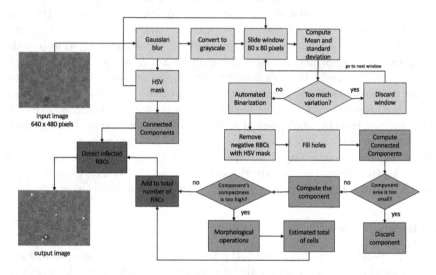

Fig. 1. Workflow of the proposed method. The yellow, blue and magenta boxes correspond to steps described in Sects. 2.2, 2.3 and 2.4, respectively. (Color figure online)

2.1   Image Acquisition

In this article, we deal with thin blood films, due to the fact that they allow us a better visualization of the RBCs and the morphological characteristics of the malaria parasites [12]. The data set of images was acquired and labelled by experienced parasitologists of the Microbiology Department (Drassanes Unit) of Vall d'Hebron Hospital, Barcelona, Spain, and was captured using light microscopes under an oil-immersion objective (100X) and a Nikon E5400 camera. It can be seen in Sect. 3 that, since the acquisition is performed by different specialists using different microscopes, the overall appeareance of the thin blood films change considerably over the set of images. Our method has to deal with those differences, since we want it to be of general use.

2.2   Image Preprocessing

The first steps performed in the workflow shown in Fig. 1 are part of what we call pre-processing, starting with a Gaussian blurring [3]. This reduces the noise present in the image at the cost of a small loss of signal, as we use a 9×9 kernel to perform this operation. The resulting image is used in two other steps, a gray level conversion, whose result will be used for segmenting the RBCs, and an HSV conversion, whose result will be used to produce an HSV mask or binary image. The segmentation of the RBCs can be performed on the intensity channel alone, since the color components do not play a major role in discriminating between RBCs and the background. The Hue component of the HSV converted image will be used to detect potential plasmodia, platelets and white blood cells (WBCs), since these biological structures acquire more intense purplish colors that RBCs when stained by Giemsa [9].

Then, an adaptive binarization is performed by using a sliding window with 80 × 80 pixels. This is done because of the variations observed on the corners of some of the acquired images, where we can see different lightning and excess blurring due to change of focus. For each 80 × 80 window, we compute the mean and standard deviation of the intensities and discard the window if the standard deviation is smaller that 0.03. That means that there aren't RBCs on this window, only background, and so, the window can be discarded. The binarization of each window is performed by using the well-known nonparametric approach proposed by Otsu [10]. However, in this process, we do not take into account the intensities of pixels with high intensity stained by Giemsa, detected in the HSV mask. Figure 2 shows some results of the adaptive Otsu binarization.

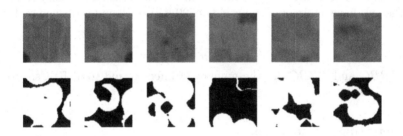

Fig. 2. Otsu without Giemsa objects

We observed that some RBCs have lighter nucleus, causing the segmentation to produce cells with holes in their middle, which can impact the RBC counting step. In order to deal with this problem, we detect the outer borders of RBCs and perform a hole filling. After this step, the RBCs are sent to the connected component detection step.

2.3 Connected Component Analysis

After the holes are filled in the previous step, we analize the binary image and detect the connected components, i.e., the connected regions of pixels with the same label [14], that in this case are the contiguous regions segmented as RBCs. For each connected component we can obtain geometrical features such as area, perimeter, compactness, that can be used to estimate the total number of RBCs in an image. We empirically determined values for some of these features that allow us to estimate the number of RBCs in a single connected component. For example, we defined the minimum and maximum area thresholds for a single RBC are 500 and 1,500, respectively, while we decide if we should decompose the connected components in more than one RBC by using the value of compactness $C = 1$, which is defined by Eq. 1:

$$C = \frac{P^2}{4\pi A},$$ (1)

where P is the perimeter and A is the area of the connected component.

Thus, if a connected component does not meet the area and compactness criteria defined, we try to decompose it by performing morphologic erosion [3]. An example of a connected component image can be seen on Fig. 3(f)[1]. When the RBCs are conneted through a thin bridge, this process breaks the component into more than one. If the remainder of the connected component still does not meet the criteria defined for a single RBC, we then estimate the number of RBCs in the cluster of cells by dividing the area of the cluster by 1500. After the RBC cluster decomposition, the eroded RBCs are grown again by using morphologic dilation. An accurate estimation of the number of healthy RBCs is important because this is a number used to determine the parasite density, which has an impact on the treatment. The parasite density or parasitemia is considered valid only when we have a minimum counting of 500 RBCs [1], and is defined by Eq. 2:

$$P = \frac{IRBC}{TRBC} \cdot 100,$$ (2)

where $IRBC$ and $TRBC$ are the number of infected and total RBCs, respectively.

2.4   Detecting Infected Cells

After detecting the RBCs, we use the result of the HSV mask to look for infected RBCs, as well as platelets and WBCs, since they have higher concentrations of Giemsa stain. The HSV mask is produced by looking only to the Hue channel of the HSV image and selecting the pixels with high H values. An example of such mask can be seen on Fig. 3(b)[2]. According to the World Health Organization [12], on thin blood films, the parasites are located inside the RBCs. Thus, we compute the centroids of the structures detected on the HSV mask and determine on which RBC they are located by using the binary mask with the connected components. If the size of the strucure detected on the HSV mask is not within defined minimum and maximum sizes for Plasmodium trophozoites and the centroid of the structure is not inside a RBC, then it is labelled as other element type, which include WBCs, platelets and small defects. Otherwise, me mark the RBC as an infected cell.

---

[1] We used different colors to show different connected components on this image.

[2] We used red circles to label RBCs, yellow circles to label trophozoites and white circles to label other objects.

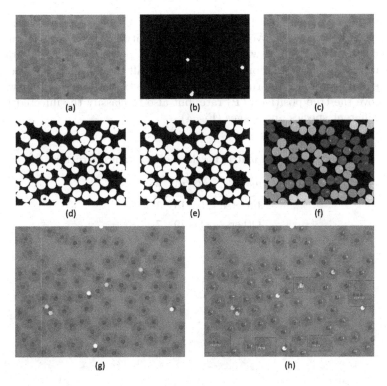

Fig. 3. The proposed method workflow for a given image: (a) input image, (b) HSV mask, (c) blurred grayscale, (d) adaptive Otsu, (e) hole filling, (f) connected components, (g) ground truth image labelled by an specialist and (h) detected and counted cells. (Color figure online)

3   Experimental Results

In order to test the efficiency of our method, we performed some experiments on an Intel Core i7-6500U (2.60 GHz 16 GB memory) computer. We used a data set containing 50 images of thin blood films infected with *P. falciparum* trophozoites. This data set was divided into a 20 image training data set and a 30 image testing set, and the images were labelled by an specialist. Table 1 shows the total number of cells and other structures detected by the specialist and the automated method.

Table 1. Total number of RBCs and other structures detected on test data set

Samples	RBCs	Infected	Others
Ground truth	2664	30	48
Segmented	2577	42	50

Table 2 shows the results of the classification of Giemsa detected structures based on location and size. We achieved a sensitivity of 89.3% for the infected RBCs, i.e., the trophozoites, without using any shape feature. The addition of shape features for discrimination will allow us not onlt to reduce the false positive (FP) rate and even relax the rules for detecting the trophozoites so that we can also improve the true positive (TP) rate, but also to classify the infected RBCs into the different species and stages (Fig. 4).

Table 2. Confusion matrix for the Giemsa stained structures

Type	TP	FP	FN
Infected	25	16	3
Others	41	9	7

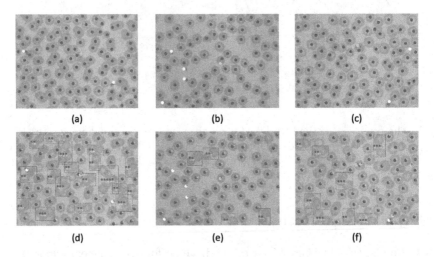

Fig. 4. Three examples of thin blood films, labelled by an specialist (top row) and automatically labelled by the proposed method (bottom row).

4   Conclusion

In this paper, we proposed an automatic system for estimating malaria parasite density in thin blood smears. The proposed approach is a combination of well known image processing methods that allows us to detect and classify healthy and infected RBCs on images acquired under different illumination conditions and acquisition setups.

This is a first step towards the development of a full-fledged system that will be capable of performing classification using shape features and of identifying different stages of at least 4 species types of the Plasmodium. Moreover, we also plan to identify and classify the WBCs into its subtypes. The idea is that the system should have a small footprint, i.e., low computation and memory requirements, thus, allowing its usage on low budget devices.

Acknowledgments. This work was supported by the Graduate Program in Systems and Computer Science (PPgSC/UFRN) and funded by CAPES-Brazil.

References

1. CDC: Diagnostic procedures (2016). https://www.cdc.gov/dpdx/diagnostic procedures. Accessed 06 Sept 2017
2. Chakrabortya, K., Chattopadhyayb, A., Chakrabarti, A., Acharyad, T., Dasguptae, A.: A combined algorithm for malaria detection from thick smear blood slides. Health Med. Inform. **6**(1), 179 (2015)
3. Gonzalez, R., Woods, R.: Digital Image Processing. Pearson/Prentice Hall, Upper Saddle River (2008)
4. Kaewkamnerd, S., Uthaipibull, C., Intarapanich, A., Pannarut, M., Chaotheing, S., Tongsima, S.: An automatic device for detection and classification of malaria parasite species in thick blood film. BMC Bioinform. **13**(Suppl 17), S18 (2012)
5. Leal Neto, O.B., Cesar, A., Jones, A., Constanca, B.: The schisto track: a system for gathering and monitoring epidemiological surveys by connecting geographical information systems in real time. JMIR mHealth and uHealth **2**, e10 (2014)
6. Linder, N., Turkki, R., Walliander, M., Mårtensson, A., Diwan, V., Rahtu, E., Pietikainen, M., Lundin, M., Lundin, J.: A malaria diagnostic tool based on computer vision screening and visualization of *Plasmodium falciparum* candidate areas in digitized blood smears. PLoS ONE **9**(8), e104855 (2014)
7. Moody, A.: Rapid diagnostic tests for malaria parasites. Clin. Microbiol. Rev. **15**(1), 66–78 (2002)
8. Oliveira, D.A., Prats, C., Espasa, M., Zarzuela Serrat, F., Montañola Sales, C., Silgado, A., Codina, L.D., Arruda, E.M., Gomez i Prat, J., Albuquerque, J.: The malaria system microapp: a new, mobile device-based tool for malaria diagnosis. JMIR Res. Protoc. **6**(4), e70 (2017)
9. World Health Organization: World malaria report 2014. Technical report, World Health Organization (2014)
10. Otsu, N.: A threshold selection method from gray-level histograms. IEEE Trans. Syst. Man Cybern. **9**(1), 62–66 (1979)
11. Pirnstill, C.W., Cote, G.L.: Malaria diagnosis using a mobile phone polarized microscope. Sci. Rep. **5**, 13368 (2015)
12. WHO Press: Basic malaria microscopy. World Health Organization, Geneva, Switzerland (1991)
13. Ross, N., Pritchard, C., Rubin, D., Dusao, A.: Automated image processing method for the diagnosis and classification of malaria on thin blood smears. Med. Biol. Eng. Comput. **44**(5), 427–436 (2006)
14. Szeliski, R.: Computer Vision: Algorithms and Applications. Texts in Computer Science. Springer, London (2010). https://doi.org/10.1007/978-1-84882-935-0

A Local Branching Heuristic for the Graph Edit Distance Problem

Mostafa Darwiche[1,2], Romain Raveaux[1(✉)],
Donatello Conte[1], and Vincent T'Kindt[2]

[1] Laboratoire d'Informatique (LI), Université François Rabelais Tours, Tours, France
{mostafa.darwiche,romain.raveaux,donatello.conte}@univ-tours.fr
[2] Laboratoire d'Informatique (LI), ERL-CNRS 6305,
Université François Rabelais Tours, Tours, France
tkindt@univ-tours.fr

Abstract. In graph matching, Graph Edit Distance (GED) is a well-known distance measure for graphs and it is a NP-Hard minimization problem. Many heuristics are defined in the literature to give approximated solutions in a reasonable time. Some other work have used mathematical programming tools to come up with Mixed Integer Linear Program (MILP) models. In this work, a heuristic from Operational Research domain, is proposed and adapted to handle GED problem. It is called Local Branching and operates over a MILP model, where it defines neighborhoods in the solution space by adding the local branching constraint. A black-box MILP solver is then used to intensify the search in a neighborhood. This makes the solution search very fast, and allow exploring different sub-regions. Also, it includes a diversification mechanism to escape local solutions and in this work this mechanism is modified and improved. Finally, it is evaluated against other heuristics in order to show its efficiency and precision.

Keywords: Graph Edit Distance · Graph matching
Local branching heuristic

1 Introduction

A powerful and well-known tool to represent patterns and objects is offered by the graph-based representation. Graphs are able to depict the components of a pattern by the mean of vertices, and the relational properties between them using edges. Moreover, through the attributes (labels) that are assigned to vertices and edges, a graph can carry more information and characteristics about the pattern. Finding the (dis)similarities between two graphs requires the computation and the evaluation of the best matching between them. Since exact isomorphism rarely occurs in pattern analysis applications, the matching process must be error-tolerant, i.e., it must tolerates differences in the topology and/or its labeling. Finally, graphs can be used in tasks such as classification and clustering, by studying and comparing them. Error-tolerant graph matching problem is known

© Springer International Publishing AG, part of Springer Nature 2018
M. Mendoza and S. Velastín (Eds.): CIARP 2017, LNCS 10657, pp. 194–202, 2018.
https://doi.org/10.1007/978-3-319-75193-1_24

to be difficult, mainly due to its computational complexity, especially for large graphs. Graphs have become popular in the context of *Structure-Activity Relationships* (SAR), since they provide a natural representation of the atom-bond structure of molecules. Each vertex of the graph then represents an atom, while an edge represents a molecular bond [9]. SAR is an important application field for graph matching.

Graph Edit Distance (GED) problem is an error-tolerant graph matching problem. It provides a dissimilarity measure between two graphs, by computing the cost of editing one graph to transform it into another. The set of edit operations are substitution, insertion and deletion, and can be applied on both vertices and edges. Solving the GED problem consists of finding the set of edit operations that minimizes the total cost. GED, by concept, is known to be flexible because it has been shown that changing the edit cost properties can result in solving other matching problems like maximum common subgraph, graph and subgraph isomorphism [3]. Also, GED problem has many applications such as Image Analysis, Handwritten Document Analysis, Malware Detection, Bio- and Chemoinformatics and many others [12]. GED is a NP-Hard problem [13], so the optimal solution cannot be obtained in polynomial time. Nevertheless, many heuristics have been proposed to compute good solutions in reasonable amount of time. The works in [10,11] presented fast algorithms that mainly solve the linear sum assignment problem for vertices, and then deduce the edges assignment. The vertices cost matrix includes information about the edges, through estimating the edges assignment cost implied by assigning two vertices. However, one drawback in this approach is that it takes into account only local structures, rather than the global one. Other algorithms based on beam search are presented in [4,8]. The first builds the search tree for all vertices assignment, then only the beam-size nodes are processed. While the second solves the vertices assignment problem, and then using beam search, it tries to improve the initial assignment by switching vertices, and re-computing the total cost. On the other hand, GED problem is addressed by the means of mathematical programming tools. Two types of mathematical formulations can be found in the literature: linear models as in [6,7] and quadratic as in [2].

This work proposes the use of *Local Branching* (LocBra) heuristic, which is well-known in *Operational Research* domain. It is presented originally in [5] as a general metaheuristic for MILP models. It makes use of a MILP solver in order to explore the solution space, through a defined branching scheme. As well, it involves techniques, such as intensification and diversification during the exploration. To benefit from the efficiency of these techniques, a LocBra version is designed and adapted to handle the GED problem. Also, the original diversification is modified to consider information about the problem that will improve the exploration and return better solutions. Since LocBra depends on a MILP model, $MILP^{JH}$ is chosen in the implementation of LocBra. Because to the best of our knowledge, it is one of the most efficient for GED problem [7]. Henceforth, the heuristic is referred to as *LocBra_GED*. Subsequently, it is evaluated and compared with existing competitive heuristic algorithms. The remainder is organized as follows: Sect. 2 presents the definition of GED problem, followed with a review of $MILP^{JH}$ model. Then, Sect. 3 details the proposed heuristic. And Sect. 4 shows the results of the computational experiments. Finally, Sect. 5 highlights some concluding remarks.

2 GED Definition and $MILP^{JH}$ Model

An attributed graph is a 4-tuple $G = (V, E, \mu, \xi)$ where, V is the set of vertices, E is the set of edges, such that $E \subseteq V \times V$, $\mu : V \to L_V$ (resp. $\xi : E \to L_E$) is the function that assigns attributes to a vertex (resp. an edge), and L_V (resp. L_E) is the label space for vertices (resp. edges).

Next, given two graphs $G = (V, E, \mu, \xi)$ and $G' = (V', E', \mu', \xi')$, GED is the task of transforming one graph source into another graph target. To accomplish this, GED introduces the vertices and edges edit operations: $(u \to v)$ is the substitution of two nodes, $(u \to \epsilon)$ is the deletion of a node, and $(\epsilon \to v)$ is the insertion of a node, with $u \in V, v \in V'$ and ϵ refers to the empty node. The same logic goes for the edges. The set of operations that reflects a valid transformation of G into G' is called a complete edit path, defined as $\lambda(G, G') = \{e_1, \ldots, e_k\}$ where e_i is an elementary vertex (or edge) edit operation and k is the number of operations. GED is then

$$d_{min}(G, G') = \min_{\lambda \in \Gamma(G,G')} \sum_{e_i \in \lambda} c(e_i) \tag{1}$$

where $\Gamma(G, G')$ is the set of all complete edit paths, d_{min} represents the minimal cost obtained by a complete edit path $\lambda(G, G')$, and c is the cost function that assigns the costs to elementary edit operations.

$MILP^{JH}$ is a model proposed in [6], that solves the GED problem. The main idea consists in determining the permutation matrix minimizing the L_1 norm of the difference between adjacency matrix of the input graph and the permuted adjacency matrix of the target one. The details about the construction of the model can be found in [6]. The model is as follows:

$$\min_{P,S,T \in \{0,1\}^{N \times N}} \sum_{i=1}^{N} \sum_{j=1}^{N} c\left(\mu(u_i), \mu'(v_j)\right) P^{ij} + \left(\frac{1}{2} \times const \times (S + T)^{ij}\right) \tag{2}$$

such that

$$(AP - PA' + S - T)^{ij} = 0 \ \forall i, j \in \{1, N\} \tag{3}$$

$$\sum_{i=1}^{N} P^{ik} = \sum_{j=1}^{N} P^{kj} = 1 \ \forall k \in \{1, N\} \tag{4}$$

where A and A' are the adjacency matrices of graphs G and G' respectively, $c : (\mu(u_i), \mu'(v_j)) \to \mathbb{R}^+$ is the cost function that measures the distance between two vertices attributes. As for P, S and T, they are the permutation matrices of size $N \times N$, and of boolean type, with $N = |V| + |V'|$. P represents the

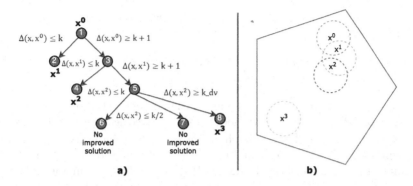

Fig. 1. Local branching flow. (a) Depicts the left and right branching. (b) Shows the neighborhoods in the solution space

vertices matching e.g. $P^{ij} = 1$ means a vertex $i \in V \cup \{\epsilon\}$ is matched with vertex $j \in V' \cup \{\epsilon\}$. While S and T are for edges matching. Hence, the objective function (Eq. 2) minimizes both, the cost of vertices and edges matching. As for constraint 3, it is to make sure that when matching two couples of vertices, the edges between each couple have to be mapped. Constraint 4 guarantees the integrity of P. This model has a limitation that it does not consider the attributes on edges, so edge substitution costs 0 while deletion and insertion have a $const \in \mathbb{R}^+$ cost.

3   Local Branching Heuristic for *GED*

The general MILP formulation is of the form:

$$\min_{x} \; c^T x \tag{5}$$

$$Ax \geq b \tag{6}$$

$$x_j \in \{0, 1\}, \forall j \in B \neq 0 \tag{7}$$

$$x_j \in \mathbb{N}, \forall j \in I \tag{8}$$

$$x_j \in \mathbb{R}, \forall j \in C \tag{9}$$

where the variable index set is split into three sets (B, I, C), respectively stands for binary, integer and continuous. $MILP^{JH}$ can be written in the same form, where the sets $I = C = \{\phi\}$ since all the variables are binary.

As presented in [5], LocBra heuristic is a local search approach that makes use of MILP solver to explore the neighborhoods of solutions through a branching scheme. In addition, it involves mechanisms such as intensification and diversification. Starting from an initial solution x^0, it defines the *k-opt neighborhood* $N(x^0, k)$, with k a given integer. In other words, the neighborhood set contains the solutions that are within a distance no more than k from x^0 (in the sense of

Hamming distance). This implies adding the following *local branching constraint* to the base $MILP^{JH}$ model:

$$\Delta(x, x^0) = \sum_{j \in S^o} (1 - x_j) + \sum_{j \in B \setminus S^o} x_j \leq k \qquad (10)$$

such that, B is the index set of binary variables defined in the model, and $S^0 = \{j \in B : x_j^0 = 1\}$. This new model is then solved leading to the search of the best solution in $N(x^0, k)$. This phase corresponds to intensifying the search in a neighborhood e.g. node 2 in Fig. 1a. If a new solution x^1 is found, the constraint (Eq. 10) is replaced by $\Delta(x, x^0) \geq k + 1$ (node 3 in Fig. 1a). This constraint makes sure that visited solutions (e.g. x^0) will not be selected twice. Next, a new constraint Eq. 10 is added but now with x^1 to explore its neighborhood. The process is repeated until a stopping criterion is met e.g. a *total time limit* is reached. Moreover, it cannot be generalized that an improved solution could be found, due to reasons such as node time limit is reached or the problem may become infeasible. For instance, assuming that at node 6 (Fig. 1a) the solution of model $MILP^{JH}$ plus equation $\Delta(x, x^2) \leq k$ does not lead to a feasible solution in the given time limit. It might be interesting to apply a complementary intensification phase, by adding constraint $\Delta(x, x^2) \leq k/2$ and solving the new model. If again, no feasible solution is found (e.g. node 7 of Fig. 1a), then a diversification phase is applied by adding the constraint $\Delta(x, x^2) \geq k\_dv$ (with k_dv a positive integer), to jump to another point in the solution space (e.g. node 8). The diversification constraint simply forces the algorithm to change the region in the search space by choosing any solution that has k_dv differences from the current solution (the authors in [5] uses another diversification constraint). Figure 1b shows the evolution of the solution search and the neighborhoods.

The key point of this heuristic is the selection of the variables while branching. To adapt the method to GED, $\Delta(x, x^i)$ is replaced by $\Delta'(x, x^i)$ where x contains only the set of binary variables that represent the vertices assignment (edges assignment are excluded). The reason behind this relies on the fact that edges assignment are driven by the vertices assignment, i.e. deleting one vertex implies deleting all edges that are connected to it, this is based on the definition of the *GED* problem. Another improvement is proposed for the diversification mechanism, where also not all binary variables are included but a smaller set of *important* variables is used instead. The diversification constraint is then $\Delta''(x, x^i) = \sum_{j \in S_{imp}^i} (1 - x_j) + \sum_{j \in B_{imp} \setminus S_{imp}^i} x_j \geq k\_dv$ with B_{imp} is the index set of binary important variables and $S_{imp}^i = \{j \in B_{imp} : x_j^i = 1\}$. The selection of these variables is based on the assumption that one variable is considered important if changing its value from $1 \rightarrow 0$ (or the opposite) highly impacts the objective function's value. This, in turn, helps skipping local solutions and change the matching. Accordingly, B_{imp} is defined by computing a special cost matrix $[C_{ij}]$ for each possible assignment of a vertex $i \in V \cup \{\epsilon\}$, to a vertex $j \in V' \cup \{\epsilon\}$. Each value $C_{ij} = c_{ij} + \theta_{ij}$, where c_{ij} is the vertex operation cost induced by assigning vertex i to vertex j, and θ_{ij} is the cost of assigning the set of edges $E_i = \{(i, v) \in E\}$ to $E_j = \{(j, v') \in E'\}$. This assignment problem, of size

$max(|E_i|, |E_j|) \times max(|E_i|, |E_j|)$, is solved by the Hungarian algorithm which requires $(O(max(|E_i|, |E_j|)^3))$ time. Next, the standard deviation is computed at each row of the matrix $[C_{ij}]$, resulting in a vector $[\sigma_i]$. The values in $[\sigma_i]$ are split into two clusters min and max. Finally, for every σ_i belonging to max cluster, the indexes of all binary variables that represent a vertex assignment for i are added to B_{imp}. Consequently, the local structure of a vertex is considered to assess its influence on the objective function value. Preliminary experiments, not reported here, have shown that such diversification helps improving the local branching heuristic better than the original diversification defined in [5].

4   Computational Experiment

Database: There are many graph databases that represent chemical molecules e.g. MUTA, PAH, MAO [1] in the literature. In the current experiment, MUTA database is chosen because it contains different subsets of small and large graphs and known to be difficult. There are 7 subsets, each of which has 10 graphs of same size (10 to 70 vertices) and a subset of also 10 graphs with mixed graph sizes. Each pair of graphs is considered as an instance. Therefore, a total of 800 instances (100 per subset) are considered in this experiment.

Comparative heuristics and experiment settings: To solve the $MILP^{JH}$, the solver CPLEX 12.6.0 is used in mixed-integer programming mode. $LocBra\_GED$ algorithm is implemented in C language. The tests were executed on a machine with the following configuration: Windows 7 (64-bit), Intel Xeon E5 4 cores and 8 GB RAM. $LocBra\_GED$ is applied on all instances with the following parameter values: $k = 20$, $k\_dv = 30$, $total\_time\_limit = 900s$, $node\_time\_limit = 180s$. Since $LocBra\_GED$ uses CPLEX with time limit, it is interesting to study the performance of the solver itself with the same time limit in order to see whether the proposed heuristic actually improves the solution of the problem, the method is called *CPLEX-900s*. Other competitive heuristics are included in the experiment such as *BeamSearch* and *SBPBeam*. Both are based on beam-search method, so their performance varies based on the choice of the beam size, therefore two versions of each are considered: *BeamSearch 5*, *BeamSearch 15000*, *SBPBeam 5* and *SBPBeam 400*. All parameter values are set based on preliminary experiments that are not shown here. For instance, beam sizes 15000 and 400 increases the execution time of *BeamSearch* and *SBPBeam* to be around $900s$. For each heuristic, the following values are computed for each subset of graphs: t_{avg} is the average CPU time in seconds for all instances. Then, $d_{min}, d_{avg}, d_{max}$ are the deviation percentages between the solutions obtained by one heuristic, and the optimal solutions. Given an instance I and a heuristic H, deviation percentage is equal to $\frac{solution_I^H - optimal_I}{optimal_I} \times 100$, with $optimal_I$ is the optimal solution for I. Lastly, η_I represents the number of solutions obtained by a heuristic that are equal to the optimal ones.

Results and analysis: Table 1 shows the results of the experiment. Looking at η_I, $LocBra\_GED$ and *CPLEX 900s* heuristics seem to have the highest values among the others. Moreover, $LocBra\_GED$ has higher values than *CPLEX*

Table 1. *LocBra_GED* vs. literature heuristics on MUTA instances.

	S	10	20	30	40	50	60	70	Mixed
	opt_I	100	100	100	99	92	71	35	91
LocBra_GED	t_{avg}	0.17	1.12	212.36	359.45	552.21	693.54	474.97	276.79
	d_{min}	0.00	0.00	0.00	0.00	0.00	0.00	0.00	0.00
	d_{avg}	**0.00**	**0.00**	**0.00**	0.06	**0.00**	**0.33**	**0.76**	**0.22**
	d_{max}	0.00	0.00	0.00	3.90	0.00	3.57	8.57	3.43
	η_I	**100**	**100**	**100**	97	**92**	**63**	**28**	84
CPLEX 900s	t_{avg}	0.13	1.02	141.07	241.21	412.36	651.43	458.20	246.90
	d_{min}	0.00	0.00	0.00	0.00	0.00	0.00	0.00	0.00
	d_{avg}	**0.00**	**0.00**	**0.00**	**0.00**	0.29	0.57	1.80	0.29
	d_{max}	0.00	0.00	0.00	0.00	6.42	6.86	8.61	6.86
	η_I	**100**	**100**	**100**	**99**	84	58	22	**87**
BeamSearch 5	t_{avg}	**0.00**	**0.00**	**0.01**	**0.03**	**0.07**	**0.11**	**0.16**	**0.08**
	d_{min}	0.00	0.00	0.00	0.00	0.00	0.00	0.00	0.00
	d_{avg}	15.17	36.60	47.21	58.73	73.73	66.47	57.76	27.02
	d_{max}	110.00	124.59	147.37	186.67	200.00	146.37	210.71	120.00
	η_I	35	10	10	10	10	10	10	11
BeamSearch 15000	t_{avg}	8.57	80.65	167.48	278.70	436.20	610.82	777.43	782.78
	d_{min}	0.00	0.00	0.00	0.00	0.00	0.00	0.00	-
	d_{avg}	1.35	26.66	47.45	52.53	65.77	66.34	78.36	-
	d_{max}	30.00	142.31	165.52	180.00	150.00	157.63	226.79	-
	η_I	88	12	10	10	10	10	10	-
SBPBeam 5	t_{avg}	0.01	0.10	0.45	1.37	3.18	5.54	10.91	2.90
	d_{min}	0.00	0.00	0.00	0.00	0.00	0.00	0.00	0.00
	d_{avg}	20.43	44.90	76.45	82.74	101.63	101.39	107.02	30.31
	d_{max}	90.00	127.87	206.90	204.71	314.29	198.50	280.36	133.71
	η_I	15	10	10	10	10	10	10	10
SBPBeam 400	t_{avg}	0.84	10.02	47.65	139.78	321.21	587.96	1144	279.17
	d_{min}	0.00	0.00	0.00	0.00	0.00	0.00	0.00	0.00
	d_{avg}	20.43	44.90	76.45	82.74	101.63	101.39	107.02	30.00
	d_{max}	90.00	127.87	206.90	204.71	314.29	198.50	280.36	133.71
	η_I	15	10	10	10	10	10	10	10

900s for all subsets except Mixed with 87 against 84 solutions equal to the optimal. Now considering the d_{avg}, again *LocBra_GED* and *CPLEX 900s* has the smallest deviations w.r.t. the optimal solutions (always less than 2%). For easy instances (subsets 10 to 40), both heuristics have scored 0%, except for subset 40 where *LocBra_GED* has 0.06% average deviation, but still very close to *CPLEX 900s*. For hard instances (subsets 50 to 70), *LocBra_GED* has the lowest deviations (always less than 1%). The same conclusion can be seen when checking the deviations for Mixed subset. The d_{avg} values for beam-search based methods are very high comparing to the first two heuristics, where on hard instances they reach over 100% (e.g. *SBPBeam* for subset 70 has 107%). Regarding the execution time of the heuristics, *BeamSearch 5* is the fastest with t_{avg} always less than $0.2s$, but in terms of solution quality it is very far from the optimal with d_{avg} up to 73%. And even after increasing the beam size for both *BeamSearch* and *SBPBeam* both are not able to provide better solutions and the average deviations remain very high. *BeamSearch 15000* was not applicable on subset Mixed because it did not return feasible solutions for all the instances, so the deviations and η_I are not computed. In summary, *LocBra_GED* is the most accurate and has provided near optimal solutions.

5   Conclusion

In this work, a local branching heuristic is proposed for the *GED* problem. As a conclusion, *LocBra_GED* significantly improves the literature heuristics and provides near optimal solutions. This is basically due, to the analysis and the branching scheme combined with the efficiency of CPLEX when solving $MILP^{JH}$ model. A second important factor is the diversification procedure that is problem dependent and really helps escaping local optima solutions. Next, more heuristic algorithms will be tested against *LocBra_GED*, and more techniques will be investigated in order to boost the solution of the method and also to allow dealing with graphs that have attributes on their edges.

References

1. Abu-Aisheh, Z., Raveaux, R., Ramel, J.-Y.: A graph database repository and performance evaluation metrics for graph edit distance. In: Liu, C.-L., Luo, B., Kropatsch, W.G., Cheng, J. (eds.) GbRPR 2015. LNCS, vol. 9069, pp. 138–147. Springer, Cham (2015). https://doi.org/10.1007/978-3-319-18224-7_14
2. Bougleux, S., Brun, L., Carletti, V., Foggia, P., Gaüzère, B., Vento, M.: Graph edit distance as a quadratic assignment problem. Pattern Recogn. Lett. **87**, 38–46 (2016)
3. Bunke, H.: On a relation between graph edit distance and maximum common subgraph. Pattern Recogn. Lett. **18**(8), 689–694 (1997)
4. Ferrer, M., Serratosa, F., Riesen, K.: Improving bipartite graph matching by assessing the assignment confidence. Pattern Recogn. Lett. **65**, 29–36 (2015)
5. Fischetti, M., Lodi, A.: Local branching. Math. Program. **98**(1–3), 23–47 (2003)
6. Justice, D., Hero, A.: A binary linear programming formulation of the graph edit distance. IEEE Trans. Pattern Anal. Mach. Intell. **28**(8), 1200–1214 (2006)
7. Lerouge, J., Abu-Aisheh, Z., Raveaux, R., Héroux, P., Adam, S.: Exact graph edit distance computation using a binary linear program. In: Robles-Kelly, A., Loog, M., Biggio, B., Escolano, F., Wilson, R. (eds.) S+SSPR 2016. LNCS, vol. 10029, pp. 485–495. Springer, Cham (2016). https://doi.org/10.1007/978-3-319-49055-7_43
8. Neuhaus, M., Riesen, K., Bunke, H.: Fast suboptimal algorithms for the computation of graph edit distance. In: Yeung, D.-Y., Kwok, J.T., Fred, A., Roli, F., de Ridder, D. (eds.) SSPR /SPR 2006. LNCS, vol. 4109, pp. 163–172. Springer, Heidelberg (2006). https://doi.org/10.1007/11815921_17

9. Raymond, J.W., Willett, P.: Maximum common subgraph isomorphism algorithms for the matching of chemical structures. J. Comput. Aided Mol. Des. **16**(7), 521–533 (2002)
10. Riesen, K., Neuhaus, M., Bunke, H.: Bipartite graph matching for computing the edit distance of graphs. In: Escolano, F., Vento, M. (eds.) GbRPR 2007. LNCS, vol. 4538, pp. 1–12. Springer, Heidelberg (2007). https://doi.org/10.1007/978-3-540-72903-7_1
11. Serratosa, F.: Fast computation of bipartite graph matching. Pattern Recogn. Lett. **45**, 244–250 (2014)
12. Stauffer, M., Tschachtli, T., Fischer, A., Riesen, K.: A survey on applications of bipartite graph edit distance. In: Foggia, P., Liu, C.-L., Vento, M. (eds.) GbRPR 2017. LNCS, vol. 10310, pp. 242–252. Springer, Cham (2017). https://doi.org/10.1007/978-3-319-58961-9_22
13. Zeng, Z., Tung, A.K., Wang, J., Feng, J., Zhou, L.: Comparing stars: on approximating graph edit distance. Proc. VLDB Endow. **2**(1), 25–36 (2009)

Emotion Assessment by Variability-Based Ranking of Coherence Features from EEG

Iván De La Pava[(✉)], Andres Álvarez-Meza, and Alvaro-Angel Orozco

Automatic Research Group, Faculty of Engineering,
Universidad Tecnológica de Pereira, Pereira, Colombia
ide@utp.edu.co

Abstract. The automatic assessment of emotional states has important applications in human-computer interfaces and marketing. Several approaches use a dimensional characterization of emotional states along with features extracted from physiological signals to classify emotions elicited from complex audiovisual stimuli; however, the classification accuracy remains low. Here, we develop an emotion assessment approach using a variability-based ranking scheme to reveal relevant coherence features from electroencephalography (EEG) signals. Our method achieves higher classification accuracies than comparable state-of-the-art methods and almost matches the performance of multimodal strategies that require information from several physiological signals.

Keywords: Emotion assessment · Electroencephalography
Coherence features · Relevance analysis

1 Introduction

The automatic recognition of emotional states is an active area of research due to its wide range of possible applications, including human-computer interfaces (HCI), targeted publicity, automatic video tagging, among others [2,8]. Usually, emotional states are studied using either a discrete or a dimensional representation. The discrete representation associates them to a series of specific emotions, e.g. joy, anger, fear, sadness, etc. On the other hand, the dimensional approach uses a set of latent dimensions that describe each emotional state regarding the physiological responses that emotions induce. The dimensional representation has the advantage of allowing a broad range of emotions to be described as a combination of the basic dimensions. Commonly, the arousal/valence dimensional spaces are considered, where each emotion is represented in terms of the active or passive response to a stimulus (arousal dimension), and the positive or negative response to the same stimulus (valence dimension) [5].

Once the emotional state representation has been established, the type of data from which those states can be assessed must be selected. Both audiovisual (facial expressions and speech patterns) and physiological data have been used for this purpose [2]. The latter have the advantage of not being regulated by

© Springer International Publishing AG, part of Springer Nature 2018
M. Mendoza and S. Velastín (Eds.): CIARP 2017, LNCS 10657, pp. 203–211, 2018.
https://doi.org/10.1007/978-3-319-75193-1_25

the subject. In particular, the information provided by electroencephalography (EEG) signals has well-established connections to cognitive processes and emotional states [3,4]. Nonetheless, the use of EEG signals for emotion recognition is far from being straightforward, since complex spatiotemporal relationships between EEG channels have to be characterized to capture patient and stimuli specific variations associated with each emotional state. In [10] power spectral density features extracted from EEG are used for emotional state classification in the arousal/valence dimensional space. Others have proposed more complex feature extraction schemes based on wavelet decomposition [2], brain connectivity measures and graph theory [4,6], or a characterization by spatially compact regions of interest [8]. In [11] a multimodal approach using EEG and other six physiological signals, along with an ensemble deep learning classifier, was used to recognize binary arousal or valence states. However, the accuracies achieved by these approaches when complex stimuli, such as music videos, are used to elicit the emotional response remain relatively low, and there is still uncertainty about which features of the EEG signal are the most relevant.

Here, we introduce a patient dependent approach for emotional state classification in the arousal/valence dimensional space, using a functional brain connectivity based characterization of EEG signals. The functional brain connectivity measure, termed coherence, is used to extract frequency and spatiotemporal dependent features from EEG. In turn, these features are ranked according to a variability-based relevance analysis, and input to a K-nearest neighbor classifier. Obtained results show that our method outperforms the state-of-the-art techniques concerning the classification accuracy in a public database while using a simple classifier and features extracted only from EEG signals. Also, it almost matches the performance of multimodal strategies that require information from several physiological signals. The remainder of the paper is organized as follows: Sect. 2 introduces the theoretical foundations, Sect. 3 describes the experiments and obtained results, and Sect. 4 presents the conclusions.

2   Variability-Based Ranking of the Coherence Features

Let $\Psi = \{\boldsymbol{X}_n \in \mathbb{R}^{C \times M}\}_{n=1}^{N}$ be an EEG set holding N trials of an emotion elicitation experiment with C channels and M samples. Besides, let $\boldsymbol{Y} \in \{-1, 1\}^N$ be a label set for a particular experiment; where the n-th element y_n corresponds to the emotion dimension class obtained for trial $\boldsymbol{X}_n$. Our goal is to infer the class label y_n from a set of features extracted from $\boldsymbol{X}_n$. Thus, we use a functional brain connectivity analysis, termed magnitude square coherence or coherence, to code spatiotemporal and frequency dependencies among EEG channels as follows: let $\boldsymbol{x}_c, \boldsymbol{x}_{c'} \in \mathbb{R}^M$ be a pair of simultaneously recorded EEG signals belonging to $\boldsymbol{X}_n$, their coherence $\gamma_{cc'}(f) \in [0, 1]$ at frequency f is given by:

$$\gamma_{cc'}(f) = \frac{|S_{cc'}(f)|^2}{S_{cc}(f)S_{c'c'}(f)}, \tag{1}$$

where $S_{cc'}(f) \in \mathbb{C}$ is the cross-spectrum between $\boldsymbol{x}_c$ and $\boldsymbol{x}_{c'}$, and $S_{cc}(f)$, $S_{c'c'}(f) \in \mathbb{C}$ are the auto-spectral densities of $\boldsymbol{x}_c$ and $\boldsymbol{x}_{c'}$, respectively. Namely, the cross-spectrum is the Fourier transform of the cross-correlation function between the two signals. Analogous relationships hold for the auto-spectral densities. $\gamma_{cc'}(f)$ is a linear measure of the relationship stability between $\boldsymbol{x}_c$ and $\boldsymbol{x}_{c'}$ with respect to power asymmetry and phase behavior [9]. Thus, a matrix $\boldsymbol{\Gamma}_n \in \mathbb{R}^{P \times C(\frac{C-1}{2})}$ holding the coherence values for a range of frequencies $\boldsymbol{f} \in \mathbb{R}^P$ between all pairwise, non-repeating EEG channel combinations characterizes the trial $\boldsymbol{X}_n$ regarding linear functional relationships in the frequency and spatial domains. The latter because of the spatial location of the EEG electrode associated with each channel. Temporal information can also be included in the coherence based characterization by splinting each channel $\boldsymbol{x}_c$ into Q segments through a windowing procedure, using a window of length $L < M$, so that for each channel there is a set $\{\boldsymbol{z}_c^j \in \mathbb{R}^L\}_{j=1}^Q$. The coherence analysis is then performed over each element of a set of matrices $\{\boldsymbol{Z}_n^j \in \mathbb{R}^{C \times L}\}_{j=1}^Q$, holding row vectors $\boldsymbol{z}_c^j$. The result is a set of coherence matrices $\boldsymbol{\zeta}_n = \{\boldsymbol{\Gamma}_n^j \in \mathbb{R}^{P \times C(\frac{C-1}{2})}\}_{j=1}^Q$ which could be used directly to infer y_n. However, the high dimensionality of $\boldsymbol{\zeta}_n$ poses a discrimination hurdle, and the need to rank the features contained in $\boldsymbol{\zeta}_n$ arises, so that only the most relevant ones are used in further classification stages.

In this sense, a variability dependent relevance analysis, based on Principal Component Analysis (PCA), is performed as follows: vector concatenation is applied to $\boldsymbol{\zeta}_n$ to yield a vector $\boldsymbol{\lambda}_n \in \mathbb{R}^D$, with $D = P \times C(\frac{C-1}{2}) \times Q$. Then, the N coherence vectors $\boldsymbol{\lambda}_n$, corresponding to each trial, are stacked to form a matrix $\boldsymbol{\Lambda} \in \mathbb{R}^{N \times D}$ with the coherence information of the entire EEG dataset $\boldsymbol{\Psi}$. Afterwards, a relevance index vector $\boldsymbol{\rho} \in \mathbb{R}^D$ is computed as follows:

$$\boldsymbol{\rho} = \mathbb{E}\left(|\rho_d \boldsymbol{\alpha}_d| \ : \ \forall d \ \in \ D' \leq D\right), \tag{2}$$

where $\rho_d \in \mathbb{R}^+$ and $\boldsymbol{\alpha}_d \in \mathbb{R}^D$ are the eigenvalues and eigenvectors of the covariance matrix $\boldsymbol{\Lambda}^\top \boldsymbol{\Lambda}/D$, and D' depends on the percentage of variance retained from the input data [1]. Under the assumption that emotions are better assessed in the frequency domain [4,10], the relevance vector $\boldsymbol{\rho}$ is averaged over the frequency using knowledge about the structure of $\boldsymbol{\lambda}_n$. The average vector $\bar{\boldsymbol{\rho}} \in \mathbb{R}^P$ can then be used to rank $\boldsymbol{\lambda}_n$ according to the most discriminant features, that is, the frequencies that present the highest variability in the coherence values. Therefore, only the $p < P$ features $\boldsymbol{\lambda}_n^p \in \mathbb{R}^H$, with $H = p \times C\left(\frac{C-1}{2}\right) \times Q$, corresponding to the most relevant frequencies as indicated by the p largest values of $\bar{\boldsymbol{\rho}}$, are selected to discriminate emotional states.

3 Experiments and Results

To test the proposed method for emotional state classification, we used the pre-processed version of the Database for Emotion Analysis Using Physiological Signals (DEAP) [5]. This database consists of EEG records taken from 32 subjects while performing 40 trials of an emotion elicitation experiment. In each experiment the subjects were exposed to a music video, then they rated their response to the each video in two scales, from 1 to 9, in the valence and arousal emotional dimensions. The valence scale ranged from unpleasant to pleasant, while the arousal scale ranged from inactive or calm to active or excited. The EEG data were acquired using a 32 channel BioSemi ActiveTwo system. The preprocessed dataset underwent artifact removal, a frequency down-sampling to 128 Hz, and a bandpass filtering from 4–45 Hz. Besides, the data was averaged to the common reference and segmented into 60 s trials. Since our goal is to use EEG data of a particular subject to infer his or her emotional response to a stimulus, we recast the emotion state assessment task as two binary classification problems. The rating scales for both variance and arousal were divided into low (from 1 to 5) and high (from 5 to 9) levels, and given class labels −1 and 1, respectively. To characterize the EEG data corresponding to each trial, the coherence was computed as explained in section Sect. 2, using 12 s long Hamming window with a 50% overlap to segment the signals and add temporal variation to the coherence data [2]. Only the coherence values for frequencies between 5 Hz and 45 Hz were considered, because of abnormally high variance present beyond the cutoff frequencies of the bandpass filter applied to the data. Then, to rank the coherence features in the frequency domain, the variability-based analysis described in section Sect. 2 was performed, setting the percentage of retained variance to 98%. Next, the features ranked according to the average relevance vector $\bar{\rho} \in \mathbb{R}^P$, $P = 161$, were used as inputs, in a progressive and accumulative fashion, to a classification algorithm. The classification procedure was carried out using a Gaussian similarity-based K-nearest neighbor classifier. The number of nearest neighbors of the classifier was selected for each subject from the range $K = \{1, 3, 5, 7, 9, 11\}$ as the one producing the highest classification accuracy. We employed a nested cross-validation scheme of 10 repetitions, where 70% of trials were used as the training set and the remaining 30% as the test set.

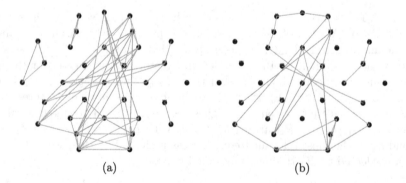

(a)   (b)

Fig. 1. Graphical representation of the coherence between EEG channels for a DEAP subject. The plots show the coherence for the trials rated as eliciting (a) minimum valence, and (b) maximum valence in the original rating scale ranging from 1 to 9. (Coherence plots obtained with the Matlab toolbox HERMES [7]).

Figure 1 shows a graphical representation of the coherence between EEG channels for the trials rated as eliciting minimum valence (Fig. 1(a)), and maximum valence (Fig. 1(b)) in one of the subjects of the database. The nodes of the plots represent the spatial distribution of the 32 EEG channels associated with the positions of the corresponding electrodes on the scalp (the uppermost nodes correspond to the electrodes located closer to the anterior part of the head). The lines joining the nodes represent coherence values, at a given frequency, between the joined nodes or channel pairs larger than a threshold level. Our method aims to exploit the difference in brain functional connectivity generated by different emotional states, as measured by the coherence, to identify those states from EEG data.

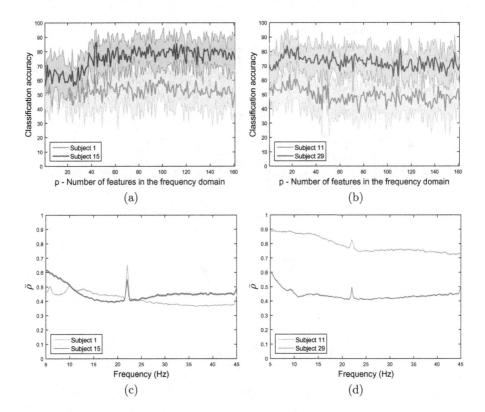

Fig. 2. Top row: average classification accuracy for two different subjects as a function of the number of relevant coherence features p for (a) valence and (b) arousal. Bottom row: average relevance vectors $\bar{\rho}$ used to rank the coherence features, the vectors in (c) were used to rank the features in (a), and analogously for (d) and (b).

To take advantage of the wealth of information provided by the coherence, while avoiding the problems posed by the high dimensionality of these data, we introduced a variability based relevance analysis of the coherence features. This analysis allows ranking the features so that they can be input to the classifier in a progressive and accumulative way. Figure 2(a) and (b) show the effects on the classification accuracy, for the valence and arousal emotional dimensions, of using

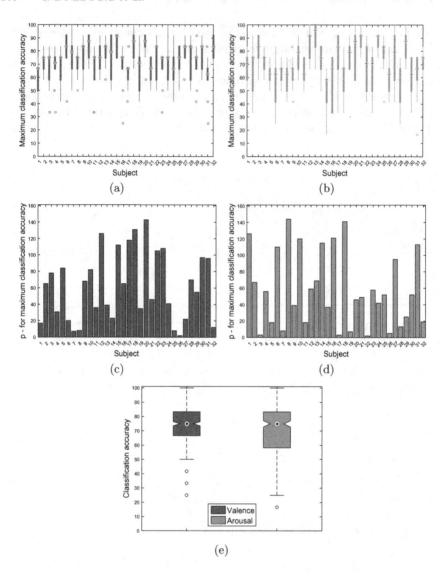

Fig. 3. Maximum classification accuracy distributions for (a) valence, and (b) arousal, holding the number of coherence features in (c) and (d), respectively. (e) shows the maximum valence and arousal classification accuracy distributions for all subjects.

an increasing number p of features as inputs to the classifier. The first feature used corresponds to the most relevant one in the frequency domain according to the relevance vector $\bar{\rho}$. Figure 2(c) and (d) show the average values of $\bar{\rho}$ used to rank the features employed to obtain Fig. 2(a) and (b), respectively. The behavior of $\bar{\rho}$ for each subject is remarkably stable, with small variations for the different folds of the cross-validation scheme. This behavior points to the feasibility of using a variability index obtained from a training set to rank the coherence features of a new input from the same subject. Larger variations in $\bar{\rho}$ are evident in an inter-subjects comparison. Such comparison also shows that, in average, the most relevant features correspond to the theta (θ: 4–8 Hz) and alpha (α: 8–13 Hz) frequency bands. These frequency bands, and especially the alpha band, have been widely used as indicators of emotional states [4]. The peak midway between the cutoff frequencies of the bandpass filter applied to the data is more likely be the result of variation introduced in the coherence by the filter itself, rather than being a reflection of relevant neurophysiological activity. Another significant result that can be observed in Fig. 2 is that the method achieves its best performance, with respect to the classification accuracy, for a p value that can be lower than the total number of coherence features P (Fig. 2(a) and (b)). Besides, the value of p that produces the maximum classification accuracy varies among subjects and between the emotion dimensions. These results are presented more clearly in Fig. 3.

Figure 3(a) and (b) show the maximum classification accuracies for each subject for the level of valence and arousal, respectively. Figure 3(c) and (d) show the number of coherence features p at which the system reached the maximum accuracies for each subject. Figure 3(c) and (d) highlight the fact that the number of input features required to attain the best performance of the system varies both among subjects and within subjects when either valence or arousal are assessed. When the adequate p is selected, the system achieves median classification accuracies for every subject in the database between 60% and 91% for the valence level, and between 55% and 96% for the arousal level. The overall performance of the system under the same condition stated above is shown in Fig. 3(e). The maximum classification accuracy distributions for all subjects show a median accuracy of 75% for both the valence and arousal levels. Those results put the performance of the proposed method at the top of similar approaches in the state-of-the-art that tackle the emotion assessment problem and test their methods in the DEAP database, and only slightly below the multimodal approach proposed in [11] that uses not only EEG but also other six physiological signals. A comparison with such methods is presented in Table 1.

Table 1. Classification accuracy [%] for all subjects in DEAP.

Approach	Arousal	Valence
Koelstra et al. [5]	62.00	57.60
Soleymani et al. [10]	62.10	50.50
Gupta et al. [4]	65.00	60.00
Padilla-Buritica et al. [8]	52.80	58.60
Daimi et al. [2]	66.90	65.30
Yin et al. (multimodal strategy) [11]	**77.19**	**76.17**
Our method	**75.00**	**75.00**

4   Conclusions

In this study, we developed a patient-specific method to assess from EEG signals the emotional state, in the arousal/valence dimensional space, elicited by audio-visual stimuli. Our method first characterizes the EEG signals using coherence, a functional brain connectivity measure. Next, the frequency and spatiotemporal temporal information contained in the coherence features is ranked according to a variability analysis based on PCA, under the assumption that those features with the highest degree of variability will be the most discriminant, then the classification is carried out using a K-nearest neighbor classifier. Our approach outperforms most of the state-of-the-art methods that use EEG for emotional state assessment regarding the classification accuracy and obtains results close to those of a multimodal approach that uses data from several physiological signals. As future work, our method can likely be further improved by the utilization of a more robust classifier, e.g., support vector machine, and by the inclusion of directionality in the brain connectivity analysis., e.g., partial directed coherence.

Acknowledgments. This work was supported by the project "Desarrollo de un sistema de apoyo al diagnóstico no invasivo de pacientes con epilepsia fármaco-resistente asociada a displasias corticales cerebrales: método costo-efectivo basado en procesamiento de imágenes de resonancia magnética" with code 1110-744-55778, and author I. De La Pava was supported by the program "Doctorado Nacional en Empresa - Convoctoria 758 de 2016", both funded by Colciencias. The authors would also like to thank Cristian Alejandro Torres Valencia for his valuable contribution to this work.

References

1. Álvarez-Meza, A.M., et al.: Time-series discrimination using feature relevance analysis in motor imagery classification. Neurocomputing **151**, 122–129 (2015)
2. Daimi, S.N., et al.: Classification of emotions induced by music videos and correlation with participants rating. Expert Syst. Appl. **41**(13), 6057–6065 (2014)
3. Gonuguntla, V., et al.: Identification of emotion associated brain functional network with phase locking value. In: 2016 IEEE 38th Annual International Conference of the EMBC, pp. 4515–4518. IEEE (2016)
4. Gupta, R., et al.: Relevance vector classifier decision fusion and EEG graph-theoretic features for automatic affective state characterization. Neurocomputing **174**, 875–884 (2016)
5. Koelstra, S., et al.: Deap: a database for emotion analysis; using physiological signals. IEEE Trans. Affect. Comput. **3**(1), 18–31 (2012)
6. Lee, Y.Y., Hsieh, S.: Classifying different emotional states by means of EEG-based functional connectivity patterns. PLoS One **9**(4), e95415 (2014)
7. Niso, G., et al.: HERMES: towards an integrated toolbox to characterize functional and effective brain connectivity. Neuroinformatics **11**(4), 405–434 (2013)
8. Padilla-Buritica, J.I., Martinez-Vargas, J.D., Castellanos-Dominguez, G.: Emotion discrimination using spatially compact regions of interest extracted from imaging EEG activity. Front. Comput. Neurosci. **10**, 1–11 (2016)

9. Sakkalis, V.: Review of advanced techniques for the estimation of brain connectivity measured with EEG/MEG. Comput. Biol. Med. **41**(12), 1110–1117 (2011)
10. Soleymani, M., et al.: Multimodal emotion recognition in response to videos. IEEE Trans. Affect. Comput. **3**(2), 211–223 (2012)
11. Yin, Z., et al.: Recognition of emotions using multimodal physiological signals and an ensemble deep learning model. Comput. Methods Programs Biomed. **140**, 93–110 (2016)

Fusion of Deep Learning Descriptors for Gesture Recognition

Edwin Escobedo Cardenas$^{(\boxtimes)}$ and Guillermo Camara-Chavez

Federal University of Ouro Preto, Ouro Preto, MG, Brazil
edu.escobedo88@gmail.com, gcamarac@gmail.com

Abstract. In this paper, we propose an approach for dynamic hand gesture recognition, which exploits depth and skeleton joint data captured by Kinect^{TM} sensor. Also, we select the most relevant points in the hand trajectory with our proposed method to extract keyframes, reducing the processing time in a video. In addition, this approach combines pose and motion information of a dynamic hand gesture, taking advantage of the transfer learning property of CNNs. First, we use the optical flow method to generate a flow image for each keyframe, next we extract the pose and motion information using two pre-trained CNNs: a CNN-flow for flow-images and a CNN-pose for depth-images. Finally, we analyze different schemes to fusion both informations in order to achieve the best method. The proposed approach was evaluated in different datasets, achieving promising results compared to other methods, outperforming state-of-the-art methods.

Keywords: Keyframe extraction · Hand gesture recognition
Pose and motion information · Convolutional neuronal networks
Fusion methods

1 Introduction

Actually, with greatest technological advance, more emphasis is being placed on non-verbal communication due to its speed and expressiveness in interaction, it is performed through gestures involving the gesture recognition area, which is one of the main component in the recent research field of human computer interaction (*HCI*) and recognized as a valuable technology for several applications due to its potential in areas such as video surveillance, robotics, multimedia video retrieval, etc. [9,13]. Hand gestures are one of the most common categories of gesture recognition used for communication and interaction. Furthermore, hand gesture recognition is seen as a first step towards sign language recognition, where each little difference in motion or hand configuration can change completely the meaning of a sign. So, the recognition of fine-grained hand movements represent a major research challenge.

Hand-crafted spatio-temporal features were widely used in gesture recognition [19]. Many vision-based algorithms were introduced to recognize dynamics

© Springer International Publishing AG, part of Springer Nature 2018
M. Mendoza and S. Velastín (Eds.): CIARP 2017, LNCS 10657, pp. 212–219, 2018.
https://doi.org/10.1007/978-3-319-75193-1_26

hand gestures [9,14]. Others well-known feature detection methods used were HOG/HOF [11], HOG3D [10], SIFT [12], etc. In the same way, to exploit the trajectory information, Shin et al. [17] proposed a geometric method using Bezier curves. Escobedo *et al.* [3,4] proposed convert the trajectory to spherical coordinates to describe the spatial and temporal information of the movements and to avoid problems with the user position changes.

Recent studies have demonstrated the power of deep convolutional neural networks(CNNs), it become an effective approach for extracting high-level features from data [16]. Wu *et al.* [21] proposed a novel method called Deep Dynamic Neural Networks for multimodal gesture recognition using deep neural nets to automatically extract relevant information from the data, they integrated two distinct feature learning methods, one for processing skeleton features and the other for RGB-D data. Besides, they used a feature learning model with a HMM to incorporate temporal dependencies. An interesting property of the CNNs, is the transfer of pre-trained network parameters to problems with limited training data, this has show success in different computer vision areas, achieving equal or better results compared to the state-of-the-art methods [15].

Based on the previous study, we propose a dynamic hand gesture recognition approach combining motion and pose information computed from depth and skeleton data captured by a Kinect^{TM}device. In contrast to previous works, we use the method proposed in [3] to extract keyframes, this method exploits the spatial information of both arms, detecting the dominant hand. This method analyses the *3D* trajectory skeleton to detect points which represent frames with more differentiated pose. As we have a fixed keyframe number, the proposed method becomes independent of the repeated use of time series techniques as Hidden Markov Model (HMM) or Dynamic Time Warping (DTW), reducing the processing time for a gesture. Finally, in this paper we investigate different fusion methods for human pose and motion features computed from two pre-trained CNNs, the hand user posture together with the motion information are decisive to determine a good classification. We report experimental results for different datasets composed of a set of fine-grained gestures.

The remainder of this paper is organized as follows. In Sect. 2, we describe and detail our proposed hand gesture recognition system. Experiments and Results are presented in Sect. 3. The Conclusions and future works are presented in Sect. 4.

2   Method Overview

Our approach consists of four main stages, as shown in Fig. 1. In the first stage, hand gesture information is captured by Kinect^{TM}device. In the second stage, we preprocess the trajectory information to obtain the keyframes. In the third stage, we compute motion and pose features using two CNNs and finally, these features are fused by different methods to generate a unique feature vector, which is used as an input into our classifier.

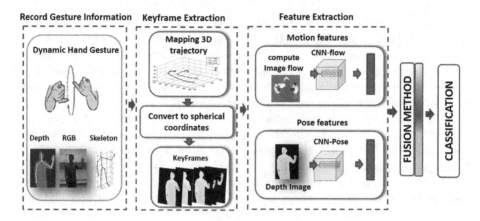

Fig. 1. Hand Gesture Recognition proposed model.

2.1 Record Gesture Information

The Kinect^{TM} sensor *V1* was used to capture hand gestures. The key of gesture recognition success is the depth camera device, which consists of an infrared laser projector and an infrared video camera. Furthermore, this device provides intensity and depth data, and Cartesian coordinates of 20 human body skeleton joints.

2.2 Keyframe Extraction Method

One of the principal challenges in hand gesture video analysis is the time variability that arises when every user makes a gesture with different speeds. Work with all frames is inefficient and take a long time, so it is necessary to choose some keyframes. Therefore in this approach, we used the method proposed in [3], which is an improve of [4] to extract these keyframes. These make our hand gesture recognition system invariant to temporal variations and avoid the repeated use of time series techniques.

2.3 Feature Extraction

A dynamic hand gesture has two essential parts: the body pose information and its motion. According to this, we further borrow inspiration from [2,5] to represent our dynamic hand gesture approach. Unlike [2], we do not create body regions since this step is unnecessary and only increases the processing time. Another difference is that we use depth image keyframes instead of the RGB, avoiding illumination changes and complex background interference. To generate our final feature vector, we first compute optical flow. For that, we apply the method used in [6] for each consecutive pair of keyframes. According to [2], the values of the motion field v_x, v_y are transformed to the interval $[0; 255]$ by $\tilde{v}_{x|y} = av_{x|y} + b$ where $a = 16$ and $b = 128$. The values below 0 and above

255 are truncated. We save the transformed flow maps as images with three channels corresponding to motion $\tilde{v}_x, \tilde{v}_y$ and the flow magnitude. So far, we have N keyframes and $N - 1$ flow images. Figure 2 shows an example from a dynamic gesture with its N keyframes and its corresponding flow images.

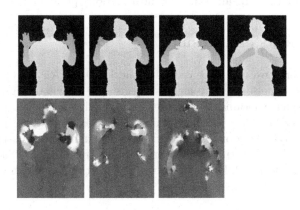

Fig. 2. Example from a dynamic gesture with its $N = 4$ keyframes and its corresponding $N - 1$ flow images.

Now, given a keyframe and its corresponding flow image, we use two distinct CNNs to compute pose and motion features. Both networks contain 5 convolutional and 3 fully-connected layers. The output of the second fully-connected layer with $k = 4096$ values is used as a keyframe descriptor.

For depth images, we use the publicly available *imagenet-vgg-f* network from [1] (our CNN-pose). For flow images, we use the motion network provided by [6] (CNN-flow). Computing at the end, two vector of $N \times 4096$ and $(N - 1) \times 4096$ to be fused.

2.4 Fusion Methods

Finally, we follow the ideas proposed in [2,5]. We considered and analyzed different schemes for fusing both informations. We distinguished two different fusion methods: direct fusion and fusion by aggregation.

Fusion by Aggregation. Here, the two vectors do not need to have the same dimensions, since individual information is extracted separately through a function (max, mean, max-min) per column. At the end, both are concatenated in a single vector.

$$y = cat(f(x^a), f(x^b)) \tag{1}$$

where f can be a max, mean or max-min operator and *cat* is the concatenation operator.

Direct Fusion. In this case, both vector maps need to have the same size, because the function is applied in both vectors at the same time, following next rules:

$$y_{ij} = f(x_{ij}^a, x_{ij}^b) \tag{2}$$

where f can be a max, mean, sum or concatenation operator.

3   Experiments and Results

In our experiments we use three datasets: the UTD-MHAD [7] which contains 27 human actions performed by eight subjects, the Brazilian Sign Language - LIBRAS [4] which contains 20 gestures performed by two subjects and the recently SHREC-2017 [18] which contains sequences of 14 hand gestures performed in two ways: using one finger and the whole hand. Each gesture was performed between 1 and 10 times by 28 participants, resulting in 2800 sequences. Each dataset provides a different protocol to make the classification stage. Our experiments follow these specifications.

3.1   Fusion Methods Evaluation

To demonstrated the utility of the keyframe extraction process, we conducted two previous experiments. We randomly select 10 signals from the LIBRAS test dataset, we measure its processing time by using the keyframe extraction method and without it, *i.e.*, working with all frames. The results are shown in Fig. 3, where we can observe the processing time is almost constant in the 10 random gestures. In contrast, when we work with all frames, the processing time varies according to the frame number, thus we show that the process of keyframe extraction accelerates the execution time, remaining almost constant.

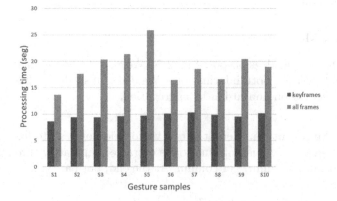

Fig. 3. Processing time recording to compare ten random S_i gestures from the LIBRAS dataset. Using the keyframe extraction algorithm we can wee that the processing time is almost constant.

We also make a comparison by measuring the overall performance obtained from the test dataset. The results are shown in Table 1, we observe that the results are almost similar for both cases, presenting a low standard deviation (SD). Thus, the performance is not significantly affected, we demonstrate the robustness of the keyframe extraction method used in our approach.

Table 1. Performance comparison between our approach to use the keyframe extraction algorithm and using all frames. We compared using different methods of fusion by aggregation. Results show that the difference is minimum.

Fusion method	Using keyframes	Using all frames	SD
Aggregation min-max	98.25	98.50	**0.18**
Aggregation max	95.99	96.25	**0.18**
Aggregation mean	96.74	97.50	**0.54**
SD	**1.15**	**1.13**	

After demonstrating the robustness of our approach, we conducted a set of experiments to find the best fusion method of flow and pose features, thereby we divided the experiments into two parts: first, we conducted experiments using fusion by aggregation, where we used the max, min and max-min operators. The second experiment was conducted by direct fusion. It is worth highlighting that these experiments can be performed only on keyframes, since this fusion method requires to have vectors with fixed size. The experiments were performed using the max, min, sum, mean and concatenation operators. All experiments were performed on the previously mentioned datasets. The results obtained are shown in Table 2 and we observe that in some cases the fusion by aggregation presents better results on UTD-MHAD and SHERC-2017 datasets, using the max-min operator. In the case of the LIBRAS dataset, the direct fusion presents better results when applying the max operator. This can be due to the LIBRAS dataset is formed from a set of sign language gestures with structured movements.

Table 2. Summary of our experiments using different schemes of fusion. The table shows the results in three different datasets.

Dataset	Fusion methods						
	Aggregation			Direct			
	Max	Mean	Max-min	Sum	Max	Mean	Concat
UTD-MHAD	93.54	93.73	**94.65**	91.30	94.63	92.49	93.75
SHREC-14	75.68	75.79	**75.97**	74.82	75.69	73.63	74.10
LIBRAS	96.74	95.99	98.25	97.74	**98.50**	96.24	98.00

Finally, we compared our best results with other methods that used the LIBRAS and UTD-MHAD datasets. Table 3 shows the results obtained and it is deduced that the fusion of motion and pose features represent better a gesture. Here, emerge the importance of the fusion of features by searching a method that best suits the problem.

Table 3. Comparison of the our approach with other method in the literature

Method	UTD -MHAD	LIBRAS
Escobedo and Camara [3]	84.89	98.54
Chéron et al. [2] (Only Kinect data)	66.10	-
Wang et al. [20]	85.81	-
Hou et al. [8]	86.97	-
Aggregation (max-min)	**94.65**	**98.25**
Direct fusion (max)	**94.63**	**98.50**

4   Conclusion

In this paper, we propose a new approach that fuses pose and motion features of a dynamic hand gesture. We exploited the transfer learning property from CNNs by extracting information from two pre-trained models. We conclude that the fusion stage is very important and decisive to the correct performance of our method. Another important issue is the definition of a method to extract keyframes, the processing time is considerably decrease when the keyframe extraction process is used, without affecting the performance of our approach. This is a very important characteristic when applied in real time applications.

Finally, the robustness of our method was shown in the experiments when compared with other methods of the literature. As future works, we propose to exploit new fusion methods, research on new $3D$ CNN architectures to extract best features from depth images and apply it on more hand gesture datasets.

Acknowledgements. The authors thank UFOP and Pro-Rectory of Research and Post-Graduation (PROPP).

References

1. Chatfield, K., Simonyan, K., Vedaldi, A., Zisserman, A.: Return of the devil in the details: delving deep into convolutional nets. arXiv preprint arXiv:1405.3531 (2014)
2. Chéron, G., Laptev, I., Schmid, C.: P-CNN: pose-based CNN features for action recognition. In: Proceedings of the IEEE International Conference on Computer Vision, pp. 3218–3226 (2015)
3. Escobedo, E., Camara, G.: A new approach for dynamic gesture recognition using skeleton trajectory representation and histograms of cumulative magnitudes. In: 2016 29th SIBGRAPI Conference on Graphics, Patterns and Images (SIBGRAPI), pp. 209–216. IEEE (2016)

4. Escobedo-Cardenas, E., Camara-Chavez, G.: A robust gesture recognition using hand local data and skeleton trajectory. In: 2015 IEEE International Conference on Image Processing (ICIP), pp. 1240–1244. IEEE (2015)

5. Feichtenhofer, C., Pinz, A., Zisserman, A.: Convolutional two-stream network fusion for video action recognition. In: Proceedings of the IEEE Conference on Computer Vision and Pattern Recognition, pp. 1933–1941 (2016)

6. Gkioxari, G., Malik, J.: Finding action tubes. In: Proceedings of the IEEE Conference on Computer Vision and Pattern Recognition, pp. 759–768 (2015)

7. Han, F., Reily, B., Hoff, W., Zhang, H.: Space-time representation of people based on 3D skeletal data: a review. Comput. Vis. Image Underst. **158**, 85–105 (2017)

8. Hou, Y., Li, Z., Wang, P., Li, W.: Skeleton optical spectra based action recognition using convolutional neural networks. IEEE Trans. Circuits Syst. Video Technol. (2016)

9. Kim, D., Hilliges, O.D., Izadi, S., Olivier, P.L., Shotton, J.D.J., Kohli, P., Molyneaux, D.G., Hodges, S.E., Fitzgibbon, A.W.: Gesture recognition techniques. US Patent 9,372,544, 21 June 2016

10. Klaser, A., Marszałek, M., Schmid, C.: A spatio-temporal descriptor based on 3D-gradients. In: 2008–19th British Machine Vision Conference on BMVC, p. 275:1. British Machine Vision Association (2008)

11. Laptev, I.: On space-time interest points. Int. J. Comput. Vis. **64**(2/3), 107–123 (2005)

12. Lowe, D.G.: Object recognition from local scale-invariant features. In: The Proceedings of the Seventh IEEE International Conference on Computer Vision, vol. 2, pp. 1150–1157. IEEE (1999)

13. Pisharady, P.K., Saerbeck, M.: Recent methods and databases in vision-based hand gesture recognition: a review. Comput. Vis. Image Underst. **141**, 152–165 (2015)

14. Rautaray, S.S., Agrawal, A.: Vision based hand gesture recognition for human computer interaction: a survey. Artif. Intell. Rev. **43**(1), 1–54 (2015)

15. Sharif Razavian, A., Azizpour, H., Sullivan, J., Carlsson, S.: CNN features off-the-shelf: an astounding baseline for recognition. In: Proceedings of the IEEE Conference on Computer Vision and Pattern Recognition Workshops, pp. 806–813 (2014)

16. Shin, H.C., Roth, H.R., Gao, M., Lu, L., Xu, Z., Nogues, I., Yao, J., Mollura, D., Summers, R.M.: Deep convolutional neural networks for computer-aided detection: CNN architectures, dataset characteristics and transfer learning. IEEE Trans. Med. Imaging **35**(5), 1285–1298 (2016)

17. Shin, M.C., Tsap, L.V., Goldgof, D.B.: Gesture recognition using Bezier curves for visualization navigation from registered 3-D data. Pattern Recogn. **37**(5), 1011–1024 (2004)

18. SHREC-2017: 3D hand gesture recognition using a depth and skeletal dataset (2017). http://www-rech.telecom-lille.fr/shrec2017-hand/

19. Trindade, P., Lobo, J., Barreto, J.P.: Hand gesture recognition using color and depth images enhanced with hand angular pose data. In: 2012 IEEE Conference on Multisensor Fusion and Integration for Intelligent Systems (MFI), pp. 71–76. IEEE (2012)

20. Wang, P., Li, Z., Hou, Y., Li, W.: Action recognition based on joint trajectory maps using convolutional neural networks. In: Proceedings of the 2016 ACM on Multimedia Conference, pp. 102–106. ACM (2016)

21. Wu, D., Pigou, L., Kindermans, P.J., Nam, L., Shao, L., Dambre, J., Odobez, J.M.: Deep dynamic neural networks for multimodal gesture segmentation and recognition (2016)

Ultimate Leveling Based on Mumford-Shah Energy Functional Applied to Plant Detection

Charles Gobber[1], Wonder A. L. Alves[1,2(✉)], and Ronaldo F. Hashimoto[2]

[1] Informatics and Knowledge Management Graduate Program,
Universidade Nove de Julho, São Paulo, Brazil
wonder@uni9.pro.br
[2] Department of Computer Science, Institute of Mathematics and Statistics,
Universidade de São Paulo, São Paulo, Brazil

Abstract. This paper presents a filter based on energy functions applied to the ultimate levelings which are powerful image operators based on numerical residues. Within a multi-scale framework, these operators analyze a given image under a series of levelings. Thus, contrasted objects can be detected if a relevant residue is generated when they are filtered out by one of these levelings. During the residual extraction process, it is very common that undesirable regions of the input image contain residual information that should be filtered out. These undesirable residual regions often include desirable residual regions due to the design of the ultimate levelings which consider maximum residues. In this paper, we improve the residual information by filtering out residues extracted from undesirable regions. In order to test our approach, some experiments were conducted in plant dataset and the results show the robustness of our approach.

Keywords: Ultimate levelings · Morphological trees
Mumford-Shah energy functional · Plant detection

1 Introduction

In Mathematical Morphology several image operators are defined by the difference between two operators, such as morphological gradient which is defined as the difference between dilation and erosion. Such operators are called *residual operators*. In more general approach, some residual operators are defined by extraction of residual values from of an indexed family of operators, for example, the skeleton by maximal discs and the ultimate attribute opening. It is worth noting that ultimate attribute openings belong to a wider class of residual operators called *ultimate levelings* [1]. This class of residual operators analyzes the evolution of the residual values between two consecutive operators on a scale-space of levelings. Residual values of these operators can reveal important contrast information in images. In a addition to residues, other information associated

© Springer International Publishing AG, part of Springer Nature 2018
M. Mendoza and S. Velastín (Eds.): CIARP 2017, LNCS 10657, pp. 220–228, 2018.
https://doi.org/10.1007/978-3-319-75193-1_27

with them can be obtained at the moment of the residual extraction, such as operator properties that produced the maximum residual value.

Examples of ultimate levelings include the maximum difference of openings (resp., closings) by reconstruction [2], differential morphological profiles [3], ultimate attribute openings (resp., closings) [4], differential attribute profiles [5], shape ultimate attribute openings (resp., closings) [6] and ultimate grain filters [7]. They have successfully been used as a preprocessing step in various applications such as texture features extraction, segmentation of high-resolution satellite imagery, text location, segmentation of building façades and restoration of historical documents.

Given the above considerations, the main contribution of this paper is the proposal of an energy filter to be a criterion to select tree nodes in order to perform the *ultimate levelings* [1]. This filter is based on the seminal work of Mumford and Shah [8] and the energy attribute introduced by Xu et al. [9]. In addition, plant image detection and segmentation tasks were selected to test the proposed energy filter.

2   Theoretical Background

Image representations through trees have been proposed in recent years to carry out tasks of image processing and analysis such as filtering, segmentation, pattern recognition, contrast extraction, registration, compression and others. In this scenario, the first step consists of constructing a representation of the input image by means of a tree; then the task of image processing or analysis is performed through information extraction from or modifications in the tree itself, and finally an image is reconstructed from the modified tree.

In order to build the trees considered in this paper, we need the following definitions. First, we consider images as mappings from a Cartesian grid $\mathcal{D} \subset \mathbb{Z}^2$ to a discrete set of $k \geq 1$ integers $\mathbb{K} = \{0, 1, \ldots, k-1\}$. (More precisely, $f : \mathcal{D} \to \mathbb{K}$). These mappings can be decomposed into *lower (strict)* and *upper (large)* level sets, i.e., for any $\lambda \in \mathbb{K}$, $\mathcal{X}_{\downarrow}^{\lambda}(f) = \{p \in \mathcal{D} : f(p) < \lambda\}$ and $\mathcal{X}_{\lambda}^{\uparrow}(f) = \{p \in \mathcal{D} : f(p) \geq \lambda\}$. One important property of these level sets is that they are nested, i.e., $\mathcal{X}_{\downarrow}^{0}(f) \subseteq \mathcal{X}_{\downarrow}^{1}(f) \subseteq \ldots \subseteq \mathcal{X}_{\downarrow}^{k}(f)$ and $\mathcal{X}_{k}^{\uparrow}(f) \subseteq \mathcal{X}_{k-1}^{\uparrow}(f) \subseteq \ldots \subseteq \mathcal{X}_{0}^{\uparrow}(f)$. It is possible to show that any image f can be reconstructed using either the family of lower or upper sets, i.e.,

$$\forall p \in \mathcal{D}, f(p) = \sup\{\lambda : p \in \mathcal{X}_{\lambda}^{\uparrow}(f)\} = \inf\{\lambda - 1 : p \in \mathcal{X}_{\downarrow}^{\lambda}(f)\}.$$

From these sets, we define two other sets $\mathcal{L}(f)$ and $\mathcal{U}(f)$ composed by the connected components (CCs) of the lower and upper level sets of f, i.e., $\mathcal{L}(f) = \{\tau \in CC(\mathcal{X}_{\downarrow}^{\lambda}(f)) : \lambda \in \mathbb{K}\}$ and $\mathcal{U}(f) = \{\tau \in CC(\mathcal{X}_{\lambda}^{\uparrow}(f)) : \lambda \in \mathbb{K}\}$, where $CC(\mathcal{X})$ denotes the sets of either 4 or 8-CCs of $\mathcal{X}$, respectively. The ordered pairs consisting of the CCs of the lower and upper level sets and the usual inclusion set relation, i.e., $(\mathcal{L}(f), \subseteq)$ and $(\mathcal{U}(f), \subseteq)$, induce two dual trees [10,11] called *component trees*. It is possible to combine them into a single tree in order to obtain the so-called *tree of shapes*. In fact, note that, if $\tau \in \mathcal{L}(f)$ then the CCs of $\mathcal{D}\backslash\tau$ are in $\mathcal{U}(f)$ (or vice versa, $\tau \in \mathcal{U}(f) \Rightarrow CC(\mathcal{D}\backslash\tau) \subseteq \mathcal{L}(f)$). Then, let $\mathcal{P}(\mathcal{D})$ denote the *powerset* of $\mathcal{D}$ and let $sat : \mathcal{P}(\mathcal{D}) \rightarrow \mathcal{P}(\mathcal{D})$ be the operator of saturation [11] (or filling holes). Then, $\mathcal{SAT}(f) = \{sat(\tau) : \tau \in \mathcal{L}(f) \cup \mathcal{U}(f)\}$ be the family of CCs of the upper and lower level sets with holes filled. The elements of $\mathcal{SAT}(f)$, called *shapes*, are nested by the inclusion relation and thus the pair $(\mathcal{SAT}(f), \subseteq)$ induces a tree which is called *tree of shapes* [11].

Tree of shapes, and also component trees, are complete representations of images and they can be stored using compact and non-redundant data structures. In the case of component trees $(\mathcal{L}(f), \subseteq)$ and $(\mathcal{U}(f), \subseteq)$, these structures are known, respectively, as *min-trees* and *max-trees*. In these structures, each pixel $p \in \mathcal{D}$ is associated only to the smallest CC of the tree containing it; and through parenthood relationship, it is also associated to all its ancestor nodes. Then, we denote by $SC(\mathcal{T}, p)$ the smallest CC containing p in tree $\mathcal{T}$. Similarly, we say $p \in \mathcal{D}$ is a *compact node pixel* (CNP) of a given CC $\tau \in \mathcal{T}$ if and only if τ is the smallest CC containing p, i.e., $\tau = SC(\mathcal{T}, p)$. Moreover, we denote by $\hat{\tau}$ the set composed by only CNPs of τ, i.e., $\hat{\tau} = \{p \in \mathcal{D} : \tau = SC(\mathcal{T}, p)\}$. Figure 1 shows examples of min-tree, max-tree, and tree of shapes, where CNPs are highlighted in red.

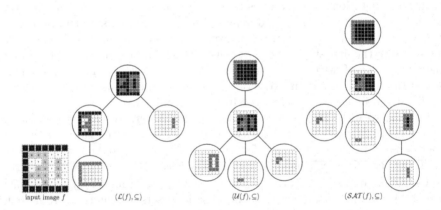

Fig. 1. Min-tree, max-tree and tree of shapes as compact representations of $(\mathcal{L}(f), \subseteq)$ and $(\mathcal{U}(f), \subseteq)$, and $(\mathcal{SAT}(f), \subseteq)$, respectively, of the input image f. Only CNPs are stored and they are highlighted in red. (Color figure online)

3   Ultimate Levelings Based on Energy Functions

Ultimate levelings constitute a wider class of residual operators defined from a scale-space of levelings $\{\psi_i : i \in \mathcal{I}\}$ [1,7,12]. An ultimate leveling analyzes the evolution of residual values from a family of consecutive primitives, i.e. $r_i^+(f) = [\psi_i(f) - \psi_{i+1}(f) \vee 0]$ and $r_i^-(f) = [\psi_i(f) - \psi_{i+1}(f) \vee 0]$, keeping the maximum positive and negative residues for each pixel. Thus, contrasted objects can be detected if a relevant residue is generated when they are filtered out by one of these levelings. More precisely, the ultimate leveling $\mathcal{R}$ is defined for any image f as follows:

$$\mathcal{R}(f) = \mathcal{R}^+(f) \vee \mathcal{R}^-(f), \tag{1}$$

where $\mathcal{R}^+(f) = \sup_{i \in \mathcal{I}}\{r_i^+(f)\}$ and $\mathcal{R}^-(f) = \sup_{i \in \mathcal{I}}\{r_i^-(f)\}$.

Residual values of these operators can reveal important contrasted structures in the image. In addition to these residues, other associated information can be obtained such as properties of the operators that produced the maximum residual value. For example, Beucher [13] introduced a function $q_{\mathcal{I}_{\max}} : \mathcal{D} \to \mathcal{I}$ that associates to each pixel the major index that produces the maximum non-null residue, i.e.,

$$p \in \mathcal{D}, q_{\mathcal{I}_{\max}}(p) = \begin{cases} q_{\mathcal{I}_{\max}}^+(p), \text{ if } [\mathcal{R}_\theta(f)](p) > [\mathcal{R}_\theta^+(f)](p), \\ q_{\mathcal{I}_{\max}}^-(p), \text{ otherwise.} \end{cases} \tag{2}$$

where $q_{\mathcal{I}_{\max}}^+(p) = \max\{i + 1 : [r_i^+(f)](p) = [\mathcal{R}_\theta^+(f)](p) > 0\}$ and $q_{\mathcal{I}_{\max}}^-(p) = \max\{i + 1 : [r_i^-(f)](p) = [\mathcal{R}_\theta^-(f)](p) > 0\}$.

The ultimate levelings can be efficiently implemented thanks to the theorem proposed in Alves et. al [10], which shows that an increasing family of levelings $\{\psi_i : i \in \mathcal{I}\}$ can be obtained through a sequence of pruned trees $(T_f^0, T_f^1, \ldots, T_f^{\mathcal{I}MAX})$ from the structure of the max-tree, min-tree or tree of shapes T_f constructed from the image f. Then, the i-th positive (resp. negative) residue $r_i^+(f)$ can be obtained from the nodes $\mathcal{N}r(i) = T_f^i \backslash T_f^{i+1}$, i.e., $\forall \tau \in \mathcal{N}r(i)$,

$$r_{T_f^i}^+(\tau) = \begin{cases} \mathtt{level}(\tau) - \mathtt{level}(\mathtt{Parent}(\tau)), \text{ if } \mathtt{Parent}(\tau) \notin \mathcal{N}r(i), \\ \mathtt{level}(\tau) - \mathtt{level}(\mathtt{Parent}(\tau)) \\ \quad + \ r_{T_f^i}^+(\mathtt{Parent}(\tau)), \qquad \text{otherwise,} \end{cases} \tag{3}$$

where $\mathtt{level}$ and $\mathtt{Parent}$ are functions that represent the gray level and the parent node of τ in T_f, respectively. Thus, the i-th positive (resp. negative) residue $r_i^+(f)$ is given as follows:

$$\forall p \in \mathcal{D}, [r_i^+(f)](p) = \begin{cases} r_{T_f^i}^+(\mathcal{SC}(T_f^i, p)), \text{ if } \mathcal{SC}(T_f^i, p) \in \mathcal{N}r(i), \\ 0, \qquad \text{otherwise.} \end{cases} \tag{4}$$

Those facts lead to efficient algorithms for computing ultimate levelings and its variations [7,14].

Ultimate levelings are operators that extract residual information from primitive families. During the residual extraction process, it is very common that undesirable regions of the input image contain residual information that should be filtered out. These undesirable residual regions often include desirable residual regions due to the design of the ultimate levelings which consider maximum residues. Thus, we improve the residual information by filtering out residues extracted from undesirable regions. To decide whether a residue $r_i^+(f)$ (resp., $r_i^-(f)$) is filtered out or not, just checking nodes $\tau \in \mathcal{N}r(i)$ that satisfy a given filtering criterion $\Omega : \mathcal{P}(\mathcal{D}) \to \{\text{desirable}, \text{undesirable}\}$. Thus, we only calculate the ultimate leveling $\mathcal{R}$ for residues r_i^+ (resp., r_i^-) such that satisfy the criterion Ω. So, positive (resp. negative) residues are redefined as follows:

$$\forall \tau \in \mathcal{N}r(i), r_{T_f^i}^{\Omega+}(\tau) = \begin{cases} r_{T_f^i}^+(\tau), \text{ if } \exists C \in \mathcal{N}r(i) \\ \qquad \text{such that } \Omega(C) \text{ is desirable;} \\ 0, \qquad \text{otherwise,} \end{cases} \tag{5}$$

and thus redefined the ultimate levelings with strategy for filtering from undesirable residues as follows: $\mathcal{R}_\Omega(f) = \mathcal{R}_\Omega^+(f) \vee \mathcal{R}_\Omega^-(f)$, where, $\mathcal{R}_\Omega^+(f) = \sup\{r_i^{\Omega+}(f) : i \in \mathcal{I}\}$ and $\mathcal{R}_\Omega^-(f) = \sup\{r_i^{\Omega-}(f) : i \in \mathcal{I}\}$.

3.1 Filtering Based on Mumford-Shah Energy

In this section, we present the main contribution of this paper, that is, the construction of a criterion Ω to filter undesirable residues from the ultimate levelings based on energy functions. This means that some desirable residual regions are selected according to the energy functional attribute $\kappa : \mathcal{P}(\mathcal{D}) \to \mathbb{R}^+$ and a parameter threshold $\varepsilon \in \mathbb{R}^+$. Thus, to decide whether a residue $r_i^+(f)$ (resp., $r_i^-(f)$) is desirable, simply check if there exists at least one node $\tau \in \mathcal{N}r(i)$ that satisfies the filtering criterion:

$$\Omega(\tau) = \begin{cases} \text{desirable,} & \text{if } \kappa(\tau) > \varepsilon, \\ \text{undesirable, otherwise.} \end{cases} \tag{6}$$

Thus, we present a brief review of how we obtained the attribute κ from a morphological tree $\mathcal{T}$. First, we remember that a large and growing body of literature has investigated the use of energy functions to segment regions of images [8,9,15]. One class of interested energy-based segmentation problem is the one that involves the minimization of a two-term energy based function $E_\nu = D + \nu C$, where D is called goodness-of-fit term, C denotes the regularization term and ν is a parameter. For example, one seminal instance of two-term energy based function is the piecewise-constant Mumford-Shah functional [8] that is given for any image f as follows:

$$E_\nu(f, P) = \sum_{R \in P} (D(R) + \nu \cdot C(R)), \tag{7}$$

where $P = \{R_1, R_2, \ldots, R_n\}$ is the set of regions of a given partition on $\mathcal{D}$, $C(R) = |\partial R|$ is the cardinality of the set of contour pixels of $R \in P$ and $D(R) = \sum_{p \in R}(f(p) - \mu_R)^2$ is the sum of squared errors and μ_R is the mean of f within the region R.

In fact, a tree $\mathcal{T}$ provides an associated partition of the image f through the CNPs, i.e., $P_\mathcal{T} = \{\hat{\tau} : \tau \in \mathcal{T}\}$. Thus, considering Eq. 7 as a minimization problem, we are interested in finding the simplified version $\mathcal{T}^*$ of $\mathcal{T}$ such that has minimum energy, i.e., $\mathcal{T}^* = \arg\min\{E_\nu(f, P_{\mathcal{T}'}) : \mathcal{T}' \text{ is a simplified version of } \mathcal{T}\}$.

On one hand, previous studies have primarily concentrated on minimizing the Mumford-Shah energy function using curve evolution methods. On the other hand, based on Ballester et al. [15], Xu et al. [9] proposed an efficient greedy algorithm to minimize energy E_ν subordinated to the *tree of shapes* and it is based on a measure called *energy variation*. This measure is defined as the difference between the energy of the tree $\mathcal{T}$ with and without a node $\tau \in \mathcal{T}$, and it is given by:

$$\Delta E_\nu(\mathcal{T}, \tau) = E_\nu(f, P_\mathcal{T}) - E_\nu(f, P_{\mathcal{T} \setminus \{\tau\}}). \qquad (8)$$

Thus, since only the regions (denoted by $\hat{\tau}$, $\hat{\tau}'$ and $\hat{\tau}''$) obtained by the CNPs of $\tau \in \mathcal{T}$, $\texttt{Parent}(\tau) \in \mathcal{T}$ and $\tau \cup \texttt{Parent}(\tau) \in \mathcal{T} \setminus \{\tau\}$ are, respectively, in the symmetric difference between $P_\mathcal{T}$ and $P_{\mathcal{T} \setminus \{\tau\}}$, we can rewrite Eq. 8 by:

$$\begin{aligned}
\Delta E_\nu(\mathcal{T}, \tau) &= E_\nu(f, P_\mathcal{T}) - E_\nu(f, P_{\mathcal{T} \setminus \{\tau\}}) \\
&= [D(\hat{\tau}) + \nu \cdot C(\hat{\tau})] + [D(\hat{\tau}') + \nu \cdot C(\hat{\tau}')] - [D(\hat{\tau}'') + \nu \cdot C(\hat{\tau}'')] \quad (9) \\
&= D(\hat{\tau}) + D(\hat{\tau}') - D(\hat{\tau}'') + \nu \cdot (C(\hat{\tau}) + C(\hat{\tau}') - C(\hat{\tau}'')).
\end{aligned}$$

Recently, Xu et al. [9] provided a new approach to minimize the piecewise-constant Mumford-Shah energy functional. Thus, based on the energy variation ΔE_ν, the *energy functional attribute* κ is defined:

$$\begin{aligned}
\kappa(\tau) &= \frac{-\Delta E_\nu(\mathcal{T}, \tau)}{C(\hat{\tau}) + C(\hat{\tau}') - C(\hat{\tau}'')} + \nu \\
&= \frac{-[D(\hat{\tau}) + D(\hat{\tau}') - D(\hat{\tau}'')]}{C(\hat{\tau}) + C(\hat{\tau}') - C(\hat{\tau}'')} - \frac{\nu \cdot (C(\hat{\tau}) + C(\hat{\tau}') - C(\hat{\tau}''))}{C(\hat{\tau}) + C(\hat{\tau}') - C(\hat{\tau}'')} + \nu \\
&= -[D(\hat{\tau}) + D(\hat{\tau}') - D(\hat{\tau}'')] \cdot \frac{1}{C(\hat{\tau}) + C(\hat{\tau}') - C(\hat{\tau}'')} \\
&= \frac{[S(f, \hat{\tau})]^2}{|\hat{\tau}|} + \frac{[S(f, \hat{\tau}')]^2}{|\hat{\tau}'|} - \frac{[S(f, \hat{\tau}'')]^2}{|\hat{\tau}''|} \cdot \frac{1}{|\partial\hat{\tau}| + |\partial\hat{\tau}'| - |\partial\hat{\tau}''|},
\end{aligned} \qquad (10)$$

where $S(f, R)$ is the sum of graylevels of all pixels of R and $|\cdot|$ denotes the cardinality. The energy functional attribute κ is obtained by the greedy algorithm proposed by Xu et al. [9]. As already mentioned before, in this paper, this measure is used in Eq. 6 just for the filtering criterion.

4   Results and Experiments

One interesting vision task problem is the plant bounding box detection. Actually an important image plant dataset is provided by Minervine et al. [16]. In respect of plant image bounding box detection task, the dataset is composed by 70 images divided in three subsets: Ara2012; Ara2013-Canon and Ara2013-RPi. In addition, all images have size of 3108×2324 and 2592×1944 pixels.

The representation of our approach is showed in Fig. 2. First, the max-tree of the green channel of the input image is constructed. After, the ultimate leveling based on Mumford-Shah energy functional is applied. As some images present incomplete regions, then a post-processing to merge those regions is computed, and, finally, we can obtain the bounding box coordinates.

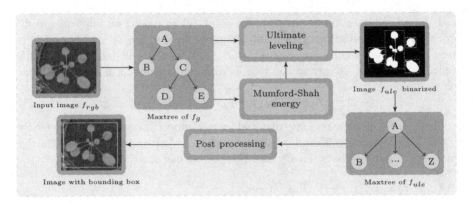

Fig. 2. Representation of our approach. The image f_g is the green channel of the input image f_{rgb} and f_{ule} is the output image generated by the energy-based ultimate leveling. (Color figure online)

In order to evaluate our approach, we chose the measurements presented by Minervine et al. [16]. Those measurements are: SBD that provides a measure about accuracy; DIC that is the difference between the plant detection using the bounding box annotated and the bounding box predicted and $|DIC|$ is the absolute value of DIC. The results obtained using our approach are shown in Table 1 and they reveal its robustness. In addition, to our knowledge, these results are the unique ones in bounding box plant detection task obtained from Minervine et al. [16] plant dataset presented in the literature. The value choice for the parameter ε for the filtering attribute was determined by cross-validation tuning parameter. The best result showed in Table 1 was obtained with $\varepsilon = 190$. In addition, Fig. 3 shows the result of the cross-validation tuning parameter for

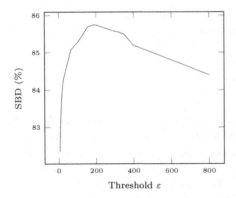

Fig. 3. Accuracy values obtained with the cross-validation tuning. It shows the importance of the filter to increase the accuracy.

Table 1. Results obtained with $\varepsilon = 190$. Reported values are mean and standard deviation (in parenthesis).

	SBD [%]	DiC	$|DiC|$
Ara2012	92.0	0.20	0.20
	(±2.8)	(±0.4)	(±0.4)
Ara2013-Canon	87.5	−0.10	0.30
	(±2.7)	(±0.5)	(±0.4)
Ara2013-RPi	80.3	0.10	0.40
	(±3.9)	(±0.8)	(±0.6)
All	85.8	0.00	0.30
	(±5.7)	(±0.6)	(±0.5)

different values of ε. It reveals the importance of the energy filter to increase the accuracy in our approach.

5 Conclusion

This paper presented a filter based on the Mumford-Shah energy functional applied to ultimate levelings operators. As it is known, during the residual extraction process, it is very common that undesirable regions of the input image contain residual information that should be filtered out. In this sense, we design a filter based on the Mumford-Shah energy functional to filter out residues extracted from undesirable regions. The results obtained applying the filter in plant detection reveal the robustness of our approach. In future works, we intend to explore other energy functions as Snakes or active contour.

References

1. Alves, W.A., Hashimoto, R.F., Marcotegui, B.: Ultimate levelings, computer vision and image understanding
2. Li, W., Haese-Coat, V., Ronsin, J.: Residues of morphological filtering by reconstruction for texture classification. Pattern Recogn. **30**(7), 1081–1093 (1997)
3. Pesaresi, M., Benediktsson, J.: A new approach for the morphological segmentation of high-resolution satellite imagery. IEEE Trans. Geosci. Remote Sens. **39**(2), 309–320 (2001)
4. Retornaz, T., Marcotegui, B.: Scene text localization based on the ultimate opening. In: Proceedings of the 8th International Symposium on Mathematical Morphology and its Applications to Image and Signal Processing, vol. 1, pp. 177–188 (2007)
5. Dalla Mura, M., Benediktsson, J.A., Waske, B., Bruzzone, L.: Morphological attribute profiles for the analysis of very high resolution images. IEEE Trans. Geosci. Remote Sens. **48**(10), 3747–3762 (2010)
6. Hernández, J., Marcotegui, B.: Shape ultimate attribute opening. Image Vis. Comput. **29**(8), 533–545 (2011)
7. Alves, W.A.L., Hashimoto, R.F.: Ultimate grain filter. In: IEEE International Conference on Image Processing, Paris, France, pp. 2953–2957 (2014)
8. Mumford, D., Shah, J.: Optimal approximations by piecewise smooth functions and associated variational problems. Commun. Pure Appl. Math. **42**(5), 577–685 (1989)
9. Xu, Y., Géraud, T., Najman, L.: Hierarchical image simplification and segmentation based on Mumford-Shah-salient level line selection. Pattern Recogn. Lett. Part 3 **83**, 278–286 (2016)
10. Alves, W.A.L., Morimitsu, A., Hashimoto, R.F.: Scale-space representation based on levelings through hierarchies of level sets. In: Benediktsson, J.A., Chanussot, J., Najman, L., Talbot, H. (eds.) ISMM 2015. LNCS, vol. 9082, pp. 265–276. Springer, Cham (2015). https://doi.org/10.1007/978-3-319-18720-4_23
11. Caselles, V., Monasse, P.: Geometric Description of Images as Topographic Maps, Berlin. Springer, Heidelberg (2010). https://doi.org/10.1007/978-3-642-04611-7

12. Alves, W., Morimitsu, A., Castro, J., Hashimoto, R.: Extraction of numerical residues in families of levelings. In: 2013 26th SIBGRAPI - Conference on Graphics, Patterns and Images (SIBGRAPI), pp. 349–356 (2013)
13. Beucher, S.: Numerical residues. Image Vis. Comput. **25**(4), 405–415 (2007)
14. Fabrizio, J., Marcotegui, B.: Fast implementation of the ultimate opening. In: Proceedings of the 9th International Symposium on Mathematical Morphology, pp. 272–281 (2009)
15. Ballester, C., Caselles, V., Igual, L., Garrido, L.: Level lines selection with variational models for segmentation and encoding. J. Math. Imaging Vis. **27**(1), 5–27 (2007)
16. Minervini, M., Fischbach, A., Scharr, H., Tsaftaris, S.A.: Finely-grained annotated datasets for image-based plant phenotyping. Pattern Recogn. Lett. **81**, 80–89 (2016)

Sclera Segmentation in Face Images Using Image Foresting Transform

Jullyana Fialho Pinheiro, João Dallyson Sousa de Almeida[✉],
Geraldo Braz Junior, Anselmo Cardoso de Paiva, and Aristófanes Corrêa Silva

Universidade Federal do Maranhão, Av. dos Portugueses, 1966 - Vila Bacanga,
São Luís, MA 65065-545, Brazil
jullyanafialho92@gmail.com, jdallyson@gmail.com, ge.braz@gmail.com,
anselmo.c.paiva@gmail.com, aricsilva@gmail.com

Abstract. The sclera is the part of the eye surround the iris, it is white and presents blood vessels that can be used for biometric recognition. In this paper, we propose a new method for sclera segmentation in face images. The method is divided into two steps: (1) the eye location and the (2) sclera segmentation. Eyes are located using Color Distance Map (CDM), Histogram of Oriented Gradients (HOG) descriptor and Random Forest (RF). The sclera is segmented by Image Foresting Transform (IFT). The first step has an accuracy of 95.95%.

Keywords: Sclera segmentation · Histogram of oriented gradients
Random Forest · Image Foresting Transform

1 Introduction

The location of the eye and the extraction of its features, as iris, corners, and sclera, are an important area for computer vision and machine learning. It can be used in applications such as facial recognition, safety control, and driver behavior analysis [1].

Several works have been produced in eye location field. A great number of these researches are based on the physical properties of eyes, as in [2,3]. The work [4] uses template matching for eye detection by founding the correlation of a template eye T with various overlapping regions of the face image.

The segmentation of sclera has been studied mainly in the biometrics systems field. The approach implemented in [5], uses Fuzzy C-means clustering, a clustering method which divides one cluster of data into two or more related clusters. Abhijit [6] uses CDM, to segment the skin around the sclera, and saturation level at HSV color space, to applied a threshold based on the intensity of its pixels.

This paper proposes a new process for human eye detection in face images and sclera segmentation using HOG descriptor (Histogram of oriented gradients), Random Forest and IFT (Image Foresting Transform).

M. Mendoza and S. Velastín (Eds.): CIARP 2017, LNCS 10657, pp. 229–236, 2018.
https://doi.org/10.1007/978-3-319-75193-1_28

The organization of the paper is as follows: in Sect. 2, the methodology proposed in this article is presented. In Sect. 3, we introduce results and, lastly, we present the conclusions in Sect. 4.

2   Methodology

The approach proposed in this paper is shown in Fig. 1. The method starts with the preprocessing to reduce lighting problems present in the images. The eye candidates are located based on the Color Distance Map (CDM) [7], and its features are extracted and selected by Histogram of Oriented Gradients (HOG) [9], and Best First (BF) [11], respectively. The classification is performed by Random Forest. The found eyes have its sclera segmented with Image Foresting Transform (IFT) [13]. The classifier was selected by Auto-Weka tool [16] based on the performance obtained.

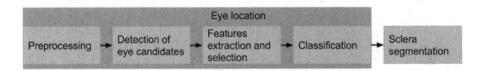

Fig. 1. Methodology

2.1   Eye Location

In this section, we present the approach used to locate eyes in images of faces. The method starts with the preprocessing to reduce lighting problems, next the skin is segmented, followed by the detection of eye candidate and finally by the classification.

Preprocessing. To reduce or eliminate lighting problem in the images, the Color Badge [8] is applied. Color Badge is a novel tone mapping operator based on the Light Random Sprays Retinex algorithm. It converts high dynamic range images in low dynamic range images.

Skin Segmentation and Detection of Eye Candidates. The second step in the eye localization is the skin segmentation. In this stage, a mask (Fig. 2a) is created by classifying each pixel into skin or not-skin labels using the Color Distance Map (CDM) [7]. The mask is composed of two maps where natural lighting and flash lighting conditions are extracted. The maps are defined by Eqs. 1 and 2.

$$
MAP1 = \begin{cases} 1, \text{if } ((R > 95, G > 40, B > 20) \text{ and} \\ \quad (\max (R, G, B) - \min (R, G, B) > 15) \text{ and} \\ \quad (\ |R - G| > 15, R > G, R > B)) \\ 0, \text{otherwise} \end{cases} \tag{1}
$$

$$MAP2 = \begin{cases} 1, \text{ if } ((R > 220, G > 210, B > 170) \text{ and} \\ \quad (\ |R - G| \leq 15, B < R, B > G)) \\ 0, \text{ otherwise} \end{cases} \quad (2)$$

where R, G, and B are the red, green, and blue component values of a RGB image. Next, noise removal (Fig. 2b), hole filling with successive closing and opening operations, and the largest skin-color region extraction (Fig. 2d) are applied. The following step is to apply arithmetic operations methods to extract the eye candidates. Using Fig. 2c and the mask (Fig. 2d) obtained before, the operation described in Eq. 3 is implemented.

$$IMAGE = \begin{cases} 1, \text{ if } (MAP1 \text{ and } MAP2) == 0 \text{ and } MASK == 255 \\ 0, \text{ otherwise} \end{cases} \quad (3)$$

The remaining sets of pixels will have their mass center located. These centers will be considered the Eye candidates(Fig. 2f) and will have their features extracted with the HOG descriptor.

Fig. 2. The skin segmentation process (a) image, (b) segmented skin, (c) noise removal, (d) largest skin region, (e) Eq. 3 result and (f) ROI based on centers in original image. The black labels were used to preserve the identity of the individuals.

Feature Extraction and Selection. The HOG descriptor was introduced by Dalal and Triggs [9] as features for pedestrian recognition, although, it has demonstrated that it is capable of describing other objects [10].

The HOG algorithm result is a discrete group of features that describe the image. The number of cells and orientation bins defines the number of features. The configuration used on this work, defined empirically, was: window size of 16×16 and cell size of 8×8. Generating 144 attributes. Every eye candidate has its features extracted with HOG.

With HOG features extracted, we perform a feature selection to retain just the best features, the algorithm used for feature selection was the Best First (BF) [11]. BF algorithm searches the space of attribute subsets by exploring the most promising set with a backtracking facility [20]. The parameters used on BF were defined by the Auto-Weka.

Classification. The candidates classification were made using Random Forest (RF). RF is an ensemble learning algorithm for classification. It works by building a set of predictors trees where each tree is dependent on the values of a random vector sampled [19]. The RF implemented on WEKA [20] was used to generate the model for the classification. The dataset used was [12]. The Auto-Weka tool were used [16] to estimate the parameters.

2.2 Sclera Segmentation

Using the ROI (Region of interest) of the eye region obtained from the previous step, the sclera is segmented with IFT (Image Foresting Transform). IFT is a tool created to transform an image processing problem into a minimum-cost path forest problem using a graph derived from the image [13]. It works by creating a minimum-cost path that connects each node of the graph to a seed based on the similarity between neighbors. Building a forest that covers the whole image. The pixel of the image has 3 attributes: the cost of the path between seed and pixel, its predecessor pixel on the path, and the label of its seed. That way if a new minimum path is located, the label, cost, and predecessor are changed, [13]. With the set of seeds defined the IFT can separate foreground pixels from background pixels. To place the seeds in the sclera and the rest of the eye, the iris of the eye must be located.

The algorithm is executed in the saturation channel of the color model HSV, Fig. 3.

Fig. 3. Example a image in of eye in the S channel of the HSV model

Iris Location. The location of the iris is obtaining using Hough Transform to find circumferences on the gray scale image. The circle found by the Hough Transform has its center and radius refined by two additional steps, this procedure is based on [17].

With the approximate location of iris center obtained, the first step is the circumference shift. It consists in modifying the center of the found circumference to its nearest neighbors in order to obtain a periphery with the lowest intensity of their pixels. An example of this is step is show in Fig. 4a, where the circumference in yellow is the first location of the iris, the red circle is the location after the circumference shift application.

The second step is to increase and decrease the radius of the circle looking for the radius where the intensity of the pixels are smaller. An example of this is step is show in Fig. 4b, where the circumference in yellow is the first radius of the iris, the red circle is the radius after the radius shift application.

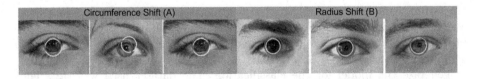

Fig. 4. Circumference (A) and Radius (B) shift examples. (Color figure online)

Seeds Placement. The background seeds were placed on the border of the image. The sclera seeds were obtain based on an adaptation of [14], that uses the eye geometry to find the location of the sclera. The 2 seeds corresponding to the sclera are placed on the right and left side of the iris at a distance of 1.1 and at 25 horizontal degrees of the found radius, this is elucidated in Fig. 5, where the black point indicates the 2 seeds to be used. With the background and foreground pixels defined, the regions will grow until the image is complete segmented.

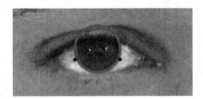

Fig. 5. Seeds used for the sclera on the IFT

3   Results and Discussion

A number of experiments were performed to measure the reliability of all steps of the proposed method. The following sections explain the image databases used as well as the tests applied using the proposed method.

3.1   Image Databases

Two image databases were utilized. The dataset presented in [12] was used for the location of the eyes, as well as for the sclera segmentation. Images have a dimension 2048×1536 pixels and contains faces of 45 individuals in 5 different poses (225 Images). This dataset was used in a strabismus research and is composed of individuals who have ocular deviations. The UBIRIS.v2 [15] database was used only on the sclera segmentation because it contains exclusively eye images. This dataset simulates un-constrained conditions with realistic noise factors (200 Images).

3.2   Experiments on the Eye Localization Method

The detection of the eye candidates were measured based on the percent of eyes on the dataset that were not considered a candidate. The database contains 450 eyes in its 225 images, of those, 97.7% were found by the method. Figure 6a show an example of an eye lost by the algorithm, where the blue rectangle indicates correctly found candidates.

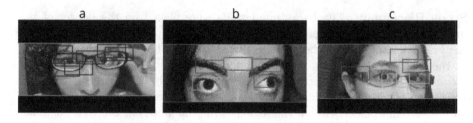

Fig. 6. Example of eye detection method (Color figure online)

The eye location had an accuracy of 98.13%, a precision of 98.1% and a recall of 98.1%. Incorrect classifications are shown in Fig. 6b–c. The false-negative and false-positive were obtained mainly at candidates where the format of the eye is not totally comprehended inside the ROI. Figure 6b shows false-negative samples, while Fig. 6c shows false-positive samples, where the red rectangles correspond to candidates classified as non-eye, while the green rectangles correspond to candidates classified as eye.

3.3 Experiments on the Sclera Segmentation Method

To evaluate the method for sclera segmentation we compare the area present on the scleras segmented manually and automatically. The segmentation was measure based on [18] that defines 2 Equations to calculate the precision and recall, Eq. 4. Example of segmented scleras are shown in Fig. 7.

$$precision = NPAM/NPRS$$
$$recall = NPAM/NRMS$$

(4)

where NPAM = Number of pixels retrieved in the sclera region by the automatically segmented mask, NPRS = Number of pixels retrieved in the automatically segmented mask, and NRMS = Number of pixels in the sclera region in the manually segmented mask.

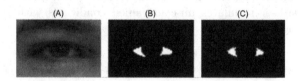

Fig. 7. Examples of sclera segmentation on the UBIRISv2 database. (A) Original image, (B) Manual segmented mask and (C) Automatically generated mask

The method had 86.02% of precision and 84.15% of recall on the UBIRISv2 database, and 80.39% of precision and 79.95% of recall on the database presented in [12]. Analyzing the results obtained in conjunction with the results present in [18], an improvement can be observed in relation to the best segmentation presented, which was 85.21% for precision and 80.21% for recall. Several papers also use the UBIRISv2 database to perform sclera segmentation, but the purpose of those articles consists mostly of biometric recognition and do not present measures focused on the evaluation of the quality of sclera segmentation, as [15].

4   Conclusions

In this paper, a new eye location and sclera segmentation method have been proposed. The method performs the detection of eyes, iris and realizes a segmentation of the sclera in face images.

The results for the eye location method using the [12] showed that the location method has a high accuracy and it is robust in the image of faces without a straight point for gaze reference. The sclera segmentation method was tested in both the [12] and the UBIRIS.v2 database. The segmented scleras show that the method tends to segment the most white parts of the sclera and tends to lose scleras where the white appearance is not present or is less evident.

The results shown that the variation of the illumination in the images compromised the segmentation in the two databases.

For future work, it is important to present more robust preprocessing techniques, in order to reduce the influence of lighting on segmentation. The number of seeds used for the sclera segmentation can also be evaluated, as that is no limit number for seeds.

Our research group acknowledges financial support from FAPEMA (Grant-Number: UNIVERSAL-01082/16), CNPQ (GrantNumber: 423493/2016-7) and FAPEMA/CAPES.

References

1. Khosravia, M.H., Safabakhsh, R.: Human eye sclera detection and tracking using a modified time-adaptive self-organizing map. Pattern Recognit. **41**, 2571–2593 (2008)
2. Zhu, Z., Ji, Q., Fujimura, K., Lee, K.: Combining Kalman filtering and mean shift for real time eye tracking under active IR illumination. In: International Conference on Pattern Recognition (2002)
3. Haro, A., Flickner, M., Essa, I.: Detecting and tracking eyes by using their physiological properties, dynamics, and appearance. In: IEEE International Conference on Computer Vision and Pattern Recognition (2000)
4. Bhoi, N., Mohanty, M.N.: Template matching based eye detection in facial image. Int. J. Comput. Appl. (0975–8887) (2010)
5. Das, A., Pal, U., Ballester, M.A.F., Blumenstein, M.: A new efficient and adaptive sclera recognition system. In: Computational Intelligence in Biometrics and Identity Management (CIBIM) (2014)
6. Alkassar, S., Woo, W.L., Dlay, S.S., Chambers, J.A.: A novel method for sclera recognition with images captured on-the-move and at-a-distance. In: Biometrics and Forensics (IWBF) (2016)

7. Abdullah-Al-Wadud, M., Chae, O.: Skin segmentation using color distance map and water-flow property. In: Information Assurance and Security, ISIAS 2008. Kyung Hee University, Seoul (2008)
8. Banić, N., Lončarić, S.: Color badger: a novel retinex-based local tone mapping operator. In: Elmoataz, A., Lezoray, O., Nouboud, F., Mammass, D. (eds.) ICISP 2014. LNCS, vol. 8509, pp. 400–408. Springer, Cham (2014). https://doi.org/10.1007/978-3-319-07998-1_46
9. Dalal, N., Triggs, B.: Histograms of oriented gradients for human detection for image classification. In: IEEE Computer Society Conference on Computer Vision and Pattern Recognition (CVPR 2005), San Diego, CA, USA, vol. 1, pp. 886–893 (2005)
10. Fleyeh, H., Roch, J.: Benchmark evaluation of HOG descriptors as features for classification of traffic signs. J. Digit. Imag. (2013)
11. Liu, H., Yu, L.: Toward integrating feature selection algorithms for classification and clustering. IEEE Trans. Knowl. Data Eng. **17**(4), 491–502 (2005)
12. De Almeida, J.D.S., Silva, A.C., Teixeira, J.A.M., Paiva, A.C., Gattass, M.: Computer-aided methodology for syndromic strabismus diagnosis (2015)
13. de Miranda, P.A.V.: Image segmentation by the image foresting transform (2013)
14. Das, A., Pal, U., Ballester, M.A.F., Blumenstein, M.: Sclera recognition using dense-SIFT. In: Intelligent Systems Design and Applications (ISDA) (2013)
15. Proenca, H., Filipe, S., Santos, R., Oliveira, J., Alexandre, L.: The UBIRIS.v2: a database of visible wavelength images captured on-the-move and at-a-distance. IEEE Trans. PAMI **32**, 1529–1535 (2010)
16. Thornton, C., Hutter, F., Hoos, H., Leyton-Brown, K.: The auto-WEKA: combined selection and hyperparameter optimization of classification algorithms. In: Proceedings of KDD (2013)
17. Zheng, Z., Yang, J., Yang, L.: A robust method for eye features extraction on color image. Pattern Recognit. Lett. **26**, 2252–2261 (2005)
18. Das, A., Pal, U., Ferrer, M.A., Blumenstein, M.: SSRBC 2016: sclera segmentation and recognition benchmarking competition (2016)
19. Gislason, P.O., Benediktsson, J.A., Sveinsson, J.R.: Random forests for land cover classification. Pattern Recognit. Lett. **27**, 294–300 (2006)
20. Weka 3: Data Mining Software in Java. http://www.cs.waikato.ac.nz/ml/weka/

Use of Machine Learning to Improve the Robustness of Spatial Estimation of Evapotranspiration

David Fonseca-Luengo[1]([✉]) [iD], Mario Lillo-Saavedra[2], L. O. Lagos[2],
Angel García-Pedrero[3], and Consuelo Gonzalo-Martín[3]

[1] Faculty of Natural Resources, Universidad Católica de Temuco, Temuco, Chile
dfonseca@uct.cl
[2] Faculty of Agricultural Engineering, Universidad de Concepción,
Vicente Méndez Avenue 595, Chillán, Chile
[3] Facultad de Informática, Universidad Politécnica de Madrid,
Campus Montegancedo, Boadilla del Monte, 28660 Madrid, Spain

Abstract. Estimation of the crop water requirement is critical in the optimization of the agricultural production process, due to that yield and costs are directly affected by this estimation. Nowadays, remote sensing is a useful tool for estimating Evapotranspiration (ET), since it is possible to map their spatial and temporal variability. ET models using satellite images have been developed in the last decades, using in most cases the surface energy balance which has generated good ET representation in different study sites. One of these models is METRIC (Mapping Evapo-Transpiration at high Resolution using Internalized Calibration), which estimates ET using mainly data from Landsat 7 and 5 images, and a physical-empirical basis to solve the surface energy balance. The main drawback of the METRIC model is the low robustness in the selection of two parameters called anchor pixels. Even though the rules to select anchor pixels are standardized, the procedure requires a user to choose the area where these pixels will be selected. In this sense, ET estimation is highly sensible to this selection, producing important differences when different anchor pixels are selected. In this study, a machine learning method is implemented through the GEOBIA (Geographic Object Based Image Analysis) approach for the identification of anchor objects, changing the focus from the pixels to the objects. Image segmentation and classification processes are used for an adequate selection of anchor objects, considering spectral and contextual information. The main contribution of this work proves that it is not necessary to choose an area to select the anchor parameters, improving the numerical stability of the model METRIC and increasing the robustness of the ET estimation. Results were validated by comparing the original selection of anchor pixels, as well as in-situ ET estimation using data obtained from Surface Renewal Stations, in sugar beet crops.

Keywords: Random Forest · METRIC model · GEOBIA approach

© Springer International Publishing AG, part of Springer Nature 2018
M. Mendoza and S. Velastín (Eds.): CIARP 2017, LNCS 10657, pp. 237–245, 2018.
https://doi.org/10.1007/978-3-319-75193-1_29

1 Introduction

Estimation of the crop water requirement is critical in the optimization of agricultural production process, due to the yield and costs are directly affected by this estimation. Water requirement estimation is performed through the quantification of ET, using local estimation or quantification models. Beside, in agriculture, the local ET estimation not necessarily represent the real amount of crop ET, due to the spatial and temporal variability caused by the soil heterogeneity, vegetation types, climate variability and phenological stage, among others.

Remote Sensing allows ET quantification in both local and regional scales, monitoring the crops considering spatial and temporal variability. Thus, remote sensing is a useful tool to ET estimation. In [8], a description of the main methods used in ET models based on remote sensing was presented.

METRIC model estimates crop ET as a residual of the surface energy balance. It is based on the SEBAL model (Surface Energy Balance Algorithm for Land) developed by [4]. In the SEBAL model, to estimate sensible heat fluxes, a relationship between the difference of temperature (dT) (at different heights) and surface temperature (T_s) for two extreme conditions pixels, named as *anchor pixels* (hot and cold pixels), was proposed.

In [5] a sensitivity analysis of the METRIC model was performed in relation to several input data, as well as the quality of the information from satellite imagery and *anchor pixels*. The results showed that the METRIC model is particularly sensitive to selection of *anchor pixels*, especially to the hot pixel.

Considering above mentioned, one of the main drawback of METRIC model is the low robustness of the ET_{inst} respect to the selection of the *anchor pixels*. In the original version of METRIC, these pixels were selected by an operator. With the aim to standardize this selection and avoid the effect of different operator criteria an automation was proposed [2]. Although that this automation standardizes the strategy of the pixels selection, requires an identification of an area where to find these anchor pixels, being a relevant step than can affect the robustness of the model.

Considering that GEOBIA (Geographic Object Based Image Analysis) approach allows dividing the image into segments that are useful to generate more features than from pixels [9], and to avoid problems related to high variability of pixels [3,6]. GEOBIA approach comprises a segmentation and classification steps. In this sense, the Random Forest (RF) method can be considered as a good option to be carried out in this approach, due RF keeps a low bias and generates a robust classifier considering the noise in the training data set [10].

Following this idea, the main objective of this work is to use a machine learning method for automatic selection of *anchor objects* (*AO*), allowing to increase the robustness in the ET estimation using the METRIC model. This is carried out considering a GEOBIA methodology that use segmentation and classification steps.

2   Materials and Methods

2.1   Ground Data and Images

The study site is located in the coordinates 36° 32′ 11″ S and 72° 8′ 25″ W. It corresponds to a rotation sugar beet crop of Sandrina, Magnolia and Donella varieties. The plantation frame is 0.5 m between rows and 0.07 m between plants, with a center pivot irrigation system with I-Wob emitters. The total area of the farm is around 450 ha, and the area of crop covered for the center pivot (study site) is around 17 ha. The climate is warm temperate, the annual mean temperature is 14 °C, with short dry season and annual rainfall that varies from 1,000 to 1,300 mm. The soil is a well drained deep sandy loamy.

The in-situ instrumentation corresponds to the Surface Renewal (SR) system, consisting of a net radiometer sensor mounted at 1.5 m above the canopy; a pyrometer; two plates for soil heat flux measurement; thermocouples for soil temperature, a sonic anemometer, measurements of temperature and air humidity; rain gauge, and measurements of soil moisture.

Images were captured by the ETM+ sensor on board of Landsat-7 satellite. Agricultural seasons 2011–2012 and 2012–2013, corresponding to the path 233, and row 85, were used. Year and DOY (day of year) of used images are: 2011-345, 2011-361, 2012-12, 2012-28, 2012-92, 2012-316, 2012-364, 2013-30, 2013-46, 2013-62, 2013-78 and 2013-126. In addition, can be mentioned that the gap filling method detailed in [12] was implemented.

2.2   Objects Anchor Selection

The proposed methodology consists in a classifier that learns taking into account different attributes from all segments belonging manually delineated regions of interest (ROI), similar to [6], where was concluded that RF classifier trained using image segments is most accurate than pixel and windowing approaches.

In order to accomplish an accurate classification, different land cover conditions had to be considered. In this sense, following same criteria of [11], annual season was divided into triplets aiming to consider different climatic and vegetative characteristics, in addition two humidity conditions for each season were considered, i.e. wet and dry conditions, based on the previous precipitation for each training scene.

Training data set. To avoid that the training data were imbalanced, data set was defined in order to generate an uniform distribution between all classes, and then they were equally represented [7].

Training data set was generated by supervised ROI delineation, using different false color combinations as a reference to identify the different classes. These ROI were delineated considering small areas to achieve pure information of each class. Classes were defined considering that the main objective was the identification of objects that have similar conditions to anchor pixels, i.e., objects with full vegetation coverage and high vigor (cold pixel), and objects of bare agricultural soil (hot pixel).

To accomplish this, initial classes were identified. Then, the classes for cold and hot objects were separated using contextual rules. The classes are described below, where in brackets the abbreviation and the percentage respect to the total training area are detailed: Vegetation with high vigor (VC_{HV}, 16%), Vegetation with low vigor (VC_{LV}, 13%), Forest and shrub (20%), Urban (10%), Water bodies (7%), Stubble (11%), Mixture between soil and vegetation ($Mixture_{S-V}$, 12%), and Bare soil (11%).

Next, the classes of interest for AO are: homogeneous VC_{HV} and Bare soil, both located away from Water bodies (to avoid Bare soil that be part of banks of a river) and Urban (to avoid non agricultural Bare soil and VC_{HV}) classes.

Using the same steps as training data set, validation data sets were generated over each scene, but considering to delineate entire covers of the same classes used in training aiming to evaluate classification process.

Segmentation. This process was performed by the *simple linear iterative clustering* (SLIC) algorithm categorized as a *superpixel* method. Due the high correlation between ET and NDVI (Normalized Difference Vegetation Index), the NDVI was selected as the input to the SLIC algorithm. For the training step, an over segmentation was carried out, allowing that the classifier can extract more segments from the training ROI, increasing the samples for the training data set.

Image classification. In this study, RF algorithm extracts information from a set of training ROI, considering all segments that are completely and partially inside them. Using RF, were generated classification models for specific conditions, considering the season (Spring, Autumn and Summer) and the humidity condition (wet or dry) based on a water balance model carried out in the scene using soil information and data from a weather station.

The features used in RF are: surface reflectance for blue, green, red, near infrared (NIR), shortwave infrared-1 (SIR1) and shortwave infrared-2 (SIR2) bands; surface temperature in Kelvin degrees, surface albedo, NDVI, Soil-adjusted vegetation index (SAVI), 3 components of Tasseled Cap transformation (Brightness, Greenness and Wetness), Water index (NDWI) and Entropy. Due to differences in the ranges of values for each map, these were scaled to a specific range for each map. The metrics extracted from the segments (for training and implementation of classification models) are mean, median, standard deviation and mode.

Contextual analysis. This analysis allowed to avoid AO that can be erroneously selected, for example, from an urban or from banks of river covers. In this sense, are rejected all candidates to AO that have an Urban or Water class neighbor.

Filtering. In order to select the final AO, statistical filtering was done considering the extreme temperature in the selection of both $AO's$. Then, the coolest and hottest objects were selected as final candidates to cold and hot AO, respectively, similar to [2]. Finally, the $AO's$ that have the least standard deviation in the temperature are selected as final AO.

2.3   METRIC Model with GEOBIA Approach

Details about algorithms of METRIC model are detailed in [1]. In the present research, two METRIC implementations are carried out, the first one was performed over original pixels maps of Landsat images (METRIC$_{Geobia-pixels}$), while the second implementation was performed over segments (METRIC$_{Geobia-segments}$). In both implementations, AO selected are used in METRIC model, making it possible to compare both results and to find differences attributable to segmentation process.

2.4   Validation Process

Validation was performed considering to prove that homogeneity of the ET segments, and robustness of the proposed methodology are improved from the use of GEOBIA approach. Thereby, validation of ET segments homogeneity was done considering ET pixels generated by METRIC$_{Geobia-pixels}$, that belong to segments generated by METRIC$_{Geobia-segments}$. This analysis provides an indicator of how homogeneous the segments are considering ET values generated with same parameters of METRIC model (AO).

Robustness validation was performed comparing the response of the fraction of reference evapotranspiration (ET$_r$F) in function to different candidates of AO. This validation method allowed to compare both implementations of METRIC model: selecting anchor pixels with classical rules, and METRIC model with GEOBIA approach.

In addition, a validation of ET$_r$F magnitudes was done comparing estimated (METRIC$_{Geobia-pixels}$) versus measured ET$_r$F (henceforth called ET$_r$F$_{measured}$). In this regard, segments that are inside of measurement station fetch are averaged to be able to obtain a correct comparison between estimated and measured values. ET$_r$F$_{measured}$ is estimated considering the ET measured by SR and ET$_r$ estimated using meteorological data, i.e., ET$_r$F$_{measured}$ = ET$_{SR}$/ET$_r$.

Validation of estimated ET from images was carried out using the average of 3×3 window of pixels centered in the location of SR.

3   Results and Discussion

3.1   Training and Classification

In the RF method, a portion of the data set used in the training are available to carried out an internal evaluation of the classification model, through the out-of-bag (OOB). The OOB errors in the classification process showed that the stabilization is accomplished around 300 trees in all scenes. Meanwhile, according to the features importance estimated using the OOB error, in all scenes the principal features were blue, near infrared, surface temperature, NDVI, Water index, greenness, and entropy bands, using the *mean* metric. Almost of other metrics had low importance, specially the standard deviation.

Confusion matrix was used to evaluate and visualize the performance of classification process. Only for purposes of saving space in writing, only Confusion matrix for 2011-345 training scene is shown in Table 1. Analyzing all the matrices of confusion, it can be observed that the ensembles tend to misclassifying between Forest and shrub, VC_{HV} and VC_{LV}; and between Bare soil with Urban and $Mixture_{S-V}$. In addition, $Mixture_{S-V}$ is misclassified by all rest of classes. Also is observed that VC_{HV} is detected in some covers of Water class because similar low temperatures are found in both classes. In addition, this misclassification can be explained since all validation ROI for water class are very small, then these small ROI of water can be integrated in VC_{HV} class that are located around a river or reservoirs. About the classes that will be candidates to AO in the post classification process, VC_{HV} have main errors when is erroneously classified by VC_{LV}, but using simple conditions this error can be solved in a post classification process (e.g., considering temperature). Bare soil have a good accuracy, having small misclassification with Urban and $Mixture_{S-V}$ classes. Even though all Urban ROI were not detected, the advantage is that this class was not misclassified with other classes. Through the comparison of generated classes with validation ROI delineated manually by a user, overall accuracies for all scenes ranged between 83% and 90%.

Table 1. Confusion matrix for scene 2011-345

	Forest and shrub	Water	Stubble	Urban	VC_{HV}	VC_{LV}	$Mixture_{S-V}$	Bare soil	%
Forest and shrub	**94.72**	0.00	0.00	0.00	4.80	0.47	0.00	0.00	100.00
Water	0.03	**89.17**	0.18	2.62	1.15	5.37	1.49	0.00	100.00
Stubble	0.00	0.00	**77.41**	0.00	0.90	1.51	0.00	20.18	100.00
Urban	0.00	0.00	0.00	**86.91**	0.00	13.09	0.00	0.00	100.00
VC_{HV}	21.00	0.00	0.59	0.00	**68.53**	9.87	0.00	0.00	100.00
VC_{LV}	0.00	0.00	0.00	2.03	5.08	**92.55**	0.34	0.00	100.00
$Mixture_{S-V}$	0.00	0.00	0.10	1.91	29.73	6.21	**62.05**	0.00	100.00
Bare soil	8.44	0.00	14.66	0.00	2.93	26.85	0.00	**47.13**	100.00
%	124.19	89.17	92.93	93.47	113.13	155.92	63.87	67.31	**82.78**

3.2   METRIC Model with GEOBIA Approach

Robustness validation. Figure 1 shows the variation in the ET_rF generated through $METRIC_{Geobia-pixels}$ using different AO identified in the image. With the goal of to prove the improvement of the model sensitivity respect to the anchor parameters, a sensitive analysis was done considering traditional pixel approach over the same scenes, resulting that for 2012-092 and 2012-364 scenes the ET_rF ranged from 0.6 to 1.15, and from 0.6 to 0.95, respectively. On the other hand, results in Fig. 1 shows that ET_rF ranged from 0.85 to 1.1 for scene 2012-092, and from 0.7 to 0.9 for scene 2012-364, considering the new AO approach, reducing the variation of ET_rF. Moreover, 0.05 was the average of standard deviation for ET_rF in all scenes showed in Fig. 1, with a maximum of 0.075, indicating that this minimal variation is consistent in all scenes.

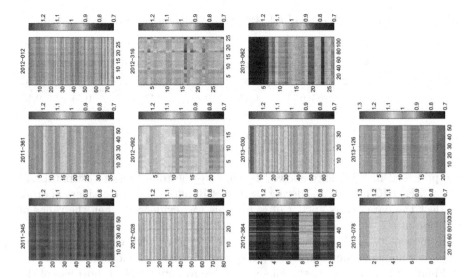

Fig. 1. Variation in ET_rF generated by all AO identified by the proposed methodology in all scenes. Variations in columns represent hot AO, while variations in rows represent cold AO. (Color figure online)

Validation with ground measurements. Results show that the RMSE is slightly greater in METRIC$_{Geobia-pixels}$ than METRIC$_{Original}$ (0.17 and 0.12, respectively). But, in contrast, the MSE (mean signed error) for METRIC$_{Geobia-pixels}$ is less than in METRIC$_{Original}$ (0.02 and -0.11, respectively). This indicates that while METRIC$_{Geobia-pixels}$ have estimated points that overestimate and underestimate ET_rF, all together are closer than estimated using METRIC$_{Original}$.

Homogeneity validation. Table 2 shows the analysis for *Vegetation with high vigor* and *Vegetation with low vigor* classes, including the standard deviation of ET pixels generated by METRIC$_{Geobia-pixels}$ belonging to objects generated with METRIC$_{Geobia-segments}$ (STD$_{intra}$), and the absolute difference of ET between each objects and their neighbors ($|Diff_{neighbors}|$). STD$_{intra}$ averaged for all objects was 0.35 mm day^{-1} for all scenes, which is a good result considering that all objects of these classes were included in this analysis. In addition, $|Diff_{neighbors}|$ averaged for all scenes was 0.6 mm day^{-1} and the magnitude had a relation with the season analyzed, i.e., in summer $|Diff_{neighbors}|$ tends to be greater than other seasons due that in summer there are many irrigated covers with high ET unlike their neighbors of natural landscape with water restriction.

Table 2. Homogeneity analysis of ET segments for all scenes. Units are in mm day^{-1}

	2011		2012								2013				
	345	361	012	028	044	092	108	316	348	364	030	046	062	078	126
STD$_{intra}$	0.52	0.28	0.45	0.37		0.37		0.19		0.39	0.43		0.37	0.34	0.14
$|Diff_{neighbors}|$	0.92	0.57	0.85	0.71		0.54		0.25		0.57	0.79		0.65	0.56	0.16

4   Conclusions

The proposed methodology allowed to improve the robustness in estimation of ET by METRIC model considering the use of Random Forest classifiers in order to select the parameters anchor objects. This was evidenced when the range of variation of ET$_r$F was decreased.

Homogeneity analysis showed that in ET objects with classes VC$_{HV}$ and VC$_{LV}$ are homogeneous and correctly separated from one another.

Finally, the proposed methodology was able to improve the robustness of the METRIC model decreasing the variation in the ET estimation considering all candidates of anchor objects, but showed no improvements in the accuracy of estimated ET.

References

1. Allen, R., Tasumi, M., Trezza, R.: Satellite-based energy balance for mapping evapotranspiration with internalized calibration (METRIC). Model. J. Irrig. Drain. Eng. **133**(4), 380–394 (2007)
2. Allen, R.G., Burnett, B., Kramber, W., Huntington, J., Kjaersgaard, J., Kilic, A., Kelly, C., Trezza, R.: Automated calibration of the metric-landsat evapotranspiration process. JAWRA J. Am. Water Resour. Assoc. **49**(3), 563–576 (2013)
3. Peña Barragan, J., Ngugi, M., Plant, R., Six, J.: Object-based crop identification using multiple vegetation indices, textural features and crop phenology. Remote Sens. Environ. **115**, 1301–1316 (2011)
4. Bastiaanssen, W., Meneti, M., Feddes, R., Holtslag, A.: A remote sensing surface energy balance algorithm for land (SEBAL). Formulation. J. Hydrol. **212–213**, 198–212 (1998)
5. Choragudi, V.N.R.K.: Sensitivity analysis on mapping evapotranspiration at high resolution using internal calibration (METRIC). Civil Engineering Theses, Dissertations, and Student Research. Paper 35 (2011)
6. Corcoran, J., Knight, J., Pelletier, K., Rampi, L., Wang, Y.: The effects of point or polygon based training data on randomforest classification accuracy of wetlands. Remote Sens. **7**(4), 4002–4025 (2015)
7. Galar, M., Fernandez, A., Barrenechea, E., Bustince, H., Herrera, F.: A review on ensembles for the class imbalance problem: bagging-, boosting-, and hybrid-based approaches. IEEE Trans. Syst. Man Cybern. Part C Appl. Rev. **42**(4), 463–484 (2012)
8. Gowda, P., Chavez, J., Colaizzi, P., Evett, S., Howell, T., Tolk, J.: ET mapping for agricultural water management: present status and challenges. Irrig. Sci. **26**(3), 223–237 (2008)

9. Hay, G., Castilla, G.: Geographic object-based image analysis (GEOBIA): a new name for a new discipline. Object-Based Image Anal. 75–89 (2008)
10. Prasad, A.M., Iverson, L.R., Liaw, A.: Newer classification and regression tree techniques: bagging and random forests for ecological prediction. Ecosystems **9**(2), 181–199 (2006)
11. Sexton, J.O., Urban, D.L., Donohue, M.J., Song, C.: Long-term land cover dynamics by multi-temporal classification across the Landsat-5 record. Remote Sens. Environ. **128**, 246–258 (2013)
12. Storey, J., Scaramuzza, P., Schmidt, G., Barsi, J.: Landsat 7 scan line corrector-off gap filled product development. In: Pecora, vol. 16, pp. 23–27 (2005)

Automatic Peripheral Nerve Segmentation in Presence of Multiple Annotators

Julián Gil González[(✉)], Andrés M. Álvarez, Andrés F. Valencia,
and Álvaro A. Orozco

Faculty of Engineering, Universidad Tecnológica de Pereira,
Pereira 660003, Colombia
{jugil,andres.alvarez1,andres.valencia,aaog}@utp.edu.co

Abstract. Peripheral Nerve Blocking (PNB) is a technique commonly used to perform regional anesthesia. The success of PNB procedures lies of the accurate location of the target nerve. The ultrasound images (UI) have frequently been used aiming to locate nerve structures in the context of PNB procedures. This type of images allows a direct visualization of the target nerve, and the anatomical structures around it. Notwithstanding, the nerve segmentation in UI by an anesthesiologist is not straightforward since these images are affected by several artifacts; hence, the accuracy of nerve segmentation depends on the anesthesiologist expertise. In this sense, we face a scenario where we have manual multiple nerve segmentations performed by several anesthesiologists with different levels of expertise. In this paper, we propose a nerve segmentation approach based on supervised learning. For the classification step, we compare two schemes based on the concepts "Learning from crowds" aiming to code the information of multiple manual segmentations. Attained results show that our approach finds a suitable UI approximation by ensuring the identification of discriminative nerve patterns according to the opinions given by multiple specialists.

1 Introduction

Recently, regional anesthesia has become an attractive alternative for general anesthesia in the context of medical surgeries, mainly because it improves postoperative mobility and reduces morbidity and mortality [1]. Regional anesthesia comprises the administration of an anesthetic substance in the area surrounding a nerve structure to block the transmission of nociceptive information (this procedure is known as Peripheral Nerve Blocking, PNB) [2]. In this regard, the success of regional anesthesia depends on the accurate location of the target nerve [3]. The use of ultrasound images (UI) has gained considerable interest to locate nerve structures in PNB procedures [1]. This method allows a direct visualization of the target nerve and the anatomical structures around it [4]. Notwithstanding, the localization of nerve structures in ultrasound images is a challenging task for the specialists (in this case anesthesiologists) due to these kind of images are affected by several artifacts such as attenuation, acoustic

© Springer International Publishing AG, part of Springer Nature 2018
M. Mendoza and S. Velastín (Eds.): CIARP 2017, LNCS 10657, pp. 246–254, 2018.
https://doi.org/10.1007/978-3-319-75193-1_30

shadows, and speckle noise [5]. Thus, the accurate delimitation of a given target depends on the operator (anesthesiologist) experience [6].

The above problem can be minimized using automatic nerve-segmentation systems, which are intended to assist the anesthesiologist with the aim of locating nerve structures in PNB procedures. Nevertheless, to build a system with these specifications, it is necessary to access to the actual label, that is, we need ultrasound images indicating which regions correspond to a nerve (usually, this process is performed by an anesthesiologist) [1,6]. In practice, the above is considered as a problem since it is not possible for the specialists to accurately identify nerve structures in an ultrasound image, considering that the speckle noise and the artifacts difficult the delimitation of anatomical structures [7]. In this sense, the obtained labels do not correspond to the Ground Truth but a subjective interpretation (possibly noisy) given by the specialist based on his experience and his training. For the automatic segmentation of nerve structures, without the knowledge of the Ground Truth, a set of noisy annotations (manual segmentation) from various specialists could be used. In this case, it is necessary to use the manual segmentation provided by multiple experts with the aim of building a segmentation model that allows to measure the performance of the annotators based on the parametrization of the ultrasound images to deal with the subjectivity presented in the labeled regions.

In the presence of multiple annotators, the labels from several experts have been used in different ways with the aim of building automatic systems for nerve segmentation. For example, in [6], the authors consider as the gold standard the annotations from one specialist. On the other hand, in [1] the authors use the Majority voting from the annotations as the ground truth. However, these approaches have some problems, for instance, if we only use the labels from one of the annotators, the segmentations results would be biased by the expertise of the annotators. Similarly, in the Majority Voting approach, it is considered that all the annotators are equally reliable, which is not common in real scenarios [8]. Another way to deal with the problem of not having the gold standard is to use a recent trend in machine learning named "Learning from multiple annotators". The area of learning with multiple annotators is relatively new; its aim is to perform supervised learning task when the gold standard is not available, and we just have access to multiple annotations provided by several experts or annotators. This area has been applied to problems such as, regression [9], classification [10], and sequence labeling [11].

In this paper, we present a method for the automatic segmentation of nerve structures depicted in ultrasound images considering the scenario where the ground truth is not available. In particular, we use two classification schemes with multiple annotators aiming to combine the manual segmentation from different experts to reveal discriminant patterns associated with the nerve structures. One of them is based on Logistic Regression, where the annotator performance depends only on the true label and is measured in terms of sensitivity and specificity [12]. The second scheme is a model based on Logistic Regression, where it is assumed that the annotator performance depends on both, the true label and the instance that the annotator is labeling [13]. We hypothesize that by using classifications schemes considering that consider the non-availability of the ground truth, it is possible to reduce the subjectivity present in the labeled regions. There are a few works that consider the information of several annotators to build nerve-segmentation systems [1,6]. However, these works use basic schemes to deal with this information (majority voting or an approach where only the information from one annotator is considered), these approaches are not suitable since they consider that the experts have the same level of expertise. In this sense, the main contribution of our work lies in to develop an automatic nerve-segmentation system, which captures the segmentation expertise from different specialist considering a non-homogeneity in the performance of experts. The obtained results show that our approach finds a suitable UI approximation by ensuring the identification of discriminative nerve patterns according to the opinions given by multiple specialists. Indeed, our proposal outperforms state-of-the-art approaches that carry out nerve segmentation in terms of a Dice coefficient assessment.

2   Materials and Methods

2.1   Multi-annotator Classifications Schemes

For the training of a typical classification problem (i.e. a classification scheme without considering multiple annotators) we dispose a training set $\mathscr{D} = \{(\mathbf{x}_i, t_i)\}_{i=1}^{N}$, with N samples, where $\mathbf{x}_i$ is an instance known as the D−dimensional feature vector and t_i is the label associated to x_i, which is assumed as the "ground truth". However, in this work, we take into account the case where the ground-truth is not available for the training, and in contrast, we only have access to an amount of labels (possibly noisy) provided by R experts o annotators [12]. In this regard, the training set in the context of multiple annotators is $\mathscr{D} = \{(\mathbf{x}_i, \mathbf{y}_i)\}_{i=1}^{N}$, where $\mathbf{y}_i = y_i^1, \ldots, y_i^R$ are the annotations for the i-th sample given by the R annotators. In this work, we use two classification schemes with multiple annotators to deal with the problem of automatic segmentation of nerve structures. The following is a brief description of these methods, where they establish a random variable $\mathbf{z} = [z_1, \ldots, z_N]$, which represents the unknown ground-truth for the i-th sample.

Logistic regression with multiple annotators (LFC). We follow the multi-annotator classification model proposed in [12]. The annotator performance is measured in terms of sensitivity α^r and specificity β^r, where $\alpha^r = p(y^r = 1|z = 1)$, $\beta^r = p(y^r = 0|z = 0)$. Hence, we use the training dataset to construct a multiple-annotator classification based on logistic regression [14]. In this sense, given the samples and the annotations, we need to estimate the parameters associated with the performance of each annotator $\boldsymbol{\alpha} = [\alpha^1, \ldots, \alpha^R]$,

$\boldsymbol{\beta} = [\beta^1, \ldots, \beta^R]$, and the parameters associated with the classifier $\boldsymbol{w}$. For estimating these parameters, we employ an Expectation-Maximization (EM) algorithm. The likelihood function is given as

$$p(\mathscr{D}|\boldsymbol{\theta}) = \prod_{i=1}^{N} \left[p_i \prod_{r=1}^{R} (\alpha^r)^{(y_i^r)} (1-\alpha^r)^{(1-y_i^r)} + (1-p_i) \prod_{r=1}^{R} (\beta^r)^{(1-y_i^r)} (1-\beta^r)^{(y_i^r)} \right],$$

where $\boldsymbol{\theta} = \{\boldsymbol{\alpha}, \boldsymbol{\beta}, \boldsymbol{w}\}$, and p_i is computed by means of a "Logistic Regression" function [14]. The EM algorithm is performed from the following steps:

E-step: The conditional expectation of the log-likelihood yields

$$\mathbb{E}[\ln(p(\mathscr{D}, \mathbf{z}|\boldsymbol{\theta}))] = \sum_{i=1}^{N} \mathbb{E}[z_i] \ln(a_i p_i) + (1 - \mathbb{E}[z_i]) \ln(b_i(1-p_i)),$$

where $a_i = \prod_{r=1}^{R} (\alpha^r)^{(y_i^r)} (1-\alpha^r)^{(1-y_i^r)}$, $b_i = \prod_{r=1}^{R} (\beta^r)^{(1-y_i^r)} (1-\beta^r)^{(y_i^r)}$, and $\mathbb{E}[z_i]$ is the estimated ground truth which follows $\mathbb{E}[z_i] = \mu_i = \dfrac{a_i p_i}{a_i p_i + b_i (1-p_i)}$.

M-step: Given the estimated gold standard μ_i and the training data, we estimate the parameters $\boldsymbol{\theta}$ by maximizing the conditional expectation of the log-likelihood computed in the E-step. The annotators performance parameters are updated using

$$\alpha^r = \frac{\sum_{i=1}^{N} \mu_i y_i^r}{\sum_{i=1}^{N} \mu_i}, \quad \beta^r = \frac{\sum_{i=1}^{N} (1-\mu_i)(1-y_i^r)}{\sum_{i=1}^{N} (1-\mu_i)}.$$

Finally, the parameters related with the logistic regression classifier, can be calculated by using similar equations to the single annotator context, where the true labels are changed for soft labels given by μ_i. See [14].

Modeling annotator expertise: Learning when everybody knows about something (MAE). We follow the multi-annotator classification schemes proposed in [13]. This model is an extension of the proposed model in [12]. Unlike the model **LFC**, the model **MAE** consider that the label given by the annotator r depends on the unknown true label z_i and the instance $\mathbf{x}_i$ that he is labeling, in this sense

$$p(y_i^r|\mathbf{x}_i, z_i) = (1 - \eta_r(\mathbf{x}_i))^{|y_i^r - z_i|} \eta_r(\mathbf{x}_i)^{1 - |y_i^r - z_i|},$$

where $\eta_r(\mathbf{x}_i)$ follows a Logistic regression model $\eta_r(\mathbf{x}_i) = (1 + \exp(-\boldsymbol{\lambda}_r^\top \mathbf{x}_i))^{-1}$. Given the dataset, we need to estimate the parameters associated with the performance of each annotator $\boldsymbol{\Lambda} = [\boldsymbol{\lambda}_1, \ldots, \boldsymbol{\lambda}_R]$ and the parameters associated with the classifier $\boldsymbol{w}$ based on "Logistic Regression" [14]. For estimating these parameters, we employ an Expectation-Maximization (EM) algorithm. The likelihood function is given as

$$p(\mathscr{D}|\boldsymbol{\phi}) = \prod_{i=1}^{N} \left[p_i \prod_{r=1}^{R} (1 - \eta_r(\mathbf{x}_i))^{1-y_i^r} \eta_r(\mathbf{x}_i)^{y_i^r} + (1-p_i) \prod_{r=1}^{R} (1 - \eta_r(\mathbf{x}_i))^{y_i^r} \eta_r(\mathbf{x}_i)^{1-y_i^r} \right],$$

where $\phi = \{\Lambda, w\}$, and p_i is computed by means of a "Logistic Regression" function [14]. The EM algorithm is performed from the following steps:

E-step: The conditional expectation of the log-likelihood is defined as

$$\mathbb{E}[\ln(p(\mathscr{D}, \mathbf{z}|\phi))] = \sum_{i=1}^{N} \mathbb{E}[z_i] \ln(c_i p_i) + (1 - \mathbb{E}[z_i]) \ln(d_i(1 - p_i)),$$

where $c_i = \prod_{r=1}^{R} (1 - \eta_r(\mathbf{x}_i))^{1-y_i^r} \eta_r(\mathbf{x}_i)^{y_i^r}$, and $d_i = \prod_{r=1}^{R} (1 - \eta_r(\mathbf{x}_i))^{y_i^r} \eta_r(\mathbf{x}_i)^{1-y_i^r}$, and $\mathbb{E}[z_i]$ is the estimated ground truth which follows $\mathbb{E}[z_i] = \mu_i = \dfrac{c_i p_i}{c_i p_i + d_i(1 - p_i)}$.

M-step: Given the estimated gold standard μ_i and the training data, we estimate the parameters ϕ by maximizing The conditional expectation of the log-likelihood computed in the E-step. To compute the parameters λ related to the model, we use gradient-based methods. Next we provided the first order derivate w.r.t. λ

$$\frac{\partial \mathbb{E}[\ln(p(\mathscr{D}, \mathbf{Z}|\theta))]}{\partial \lambda_r} = \sum_{i=1}^{N} (-1)^{y_i^r} (1 - 2\mu_i) \eta_r(\mathbf{x}_i) (1 - \eta_r(\mathbf{x}_i)) \mathbf{x}_i$$

Finally, the parameters related with the logistic regression classifier can be calculated by using similar equations to the single annotator context, where the true labels are changed for soft labels given by μ_i. See [14].

3   Results and Discussions

Ultrasound imaging dataset: To validate the nerve segmentation approach based on classification with multiple annotators, we use a dataset named *UI-UTP*, which consists of recordings of ultrasound images from patients who underwent regional anesthesia using the Peripheral nerve blocking procedure. This dataset is composed of 48 ultrasound images from the ulnar nerve (21 images) and median nerve (27 images). Each ultrasound image was collected using a Sonosite Nano-Maxx device (the resolution of each image is 640×480 pixels). Each image in the dataset was labeled by three specialists in anesthesiology to indicate the location of the nerve structures.

Segmentation Scheme training and testing: A leave-one-out validation scheme is employed to compute the system performance regarding the nerve segmentation in the context of multiple annotators. The nerve segmentation considering multiple experts comprises the following stages: First, we use Graph Cuts Segmentation [15] to define a region of interest (ROI) in which the nerve region is probably located. Then, a median filter is applied over the ROIs to reduce the speckle noise effect while enhancing the UI quality. Then, each filtered image is divided into different regions by using SLIC-superpixel [16], and each of these superpixel is parametrized from the non-linear Wavelet transform (for details, see [6]). Now, in this work, the segmentation problem is considered as a binary classification, where each parametrized superpixel is classified as nerve or background. However, as we have previously pointed out, it is not possible to obtain the ground truth (i.e. the labels indicating which superpixel is a nerve region and what it is not) since in this case, the labels correspond to a subjective assessment given by an anesthesiologist. According to the above, we use two schemes for binary classification in the context of multiple annotators, one of them is a model based on Logistic Regression, where the annotator performance depends on the true label and is measured in terms of sensitivity and specificity [12] (**LFC**). The second scheme is a model based on Logistic Regression, where it is assumed that the annotator performance depends on both, the true label and the instance that the annotator is labeling [13] (**MAE**). Finally, we use a methodology based on morphological operators aiming to improve the segmentations results (see [6]). The system performance is measured in terms of the Dice coefficient (DC) that quantifies the overlap between the UI segmented based on the proposed segmentation approach and the label given by the specialists. In addition, we consider two common approaches to deal classification problems with multiple annotators, the first approach is to use and scheme based on Logistic Regression, where the labels from each one of the annotators are considered as gold standard (**LR-EX1, LR-EX2, LR-EX3**); the second methodology is based on Logistic Regression, where we consider as true labels the majority voting from the annotations (**LR-MV**).

Obtained results: To visually compare the attained nerve identification in the context of multiple experts, Fig. 1 exposes some segmentation results regarding each multi-annotator classification approach. Overall, all classification methods allow to identify relevant patterns to segment nerves from the UI. Nevertheless, due to the absence of the true label, typical classifications methods (**LR-EX1, LR-EX2, LR-EX3**, and **LR-MV**) fail in the complete identification of nerves by generating false negatives (i.e., classify as background nerve regions). The above is a significant issue since the anesthesiologist needs an accurate delimitation of the nerve structure aiming to define the point where the anesthetic should be spread. Unlike these classification methods, our approach based on multiple annotators (specifically **MAE**) reduces the number of false negatives considerably in the segmented image offering a better identification of the nerve structure. Table 1 shows the results of the morphological validation in terms of the DC for the leave-one-out validation scheme. We perform a statistical significance analysis based on the equal mean test for the segmentation approaches considered in this work. This test allows to determine which method provides a higher performance in terms of the Dice coefficient. From the results exposed, the segmentation scheme based on multiple annotators (specifically the multiple annotators model proposed in [13]) outperforms state-of-the-art approaches which are based on typical supervised learning schemes (i.e. supervised learning approaches without considering multiple annotators). These results can be explained in the sense that the multiple annotators schemes are based on the

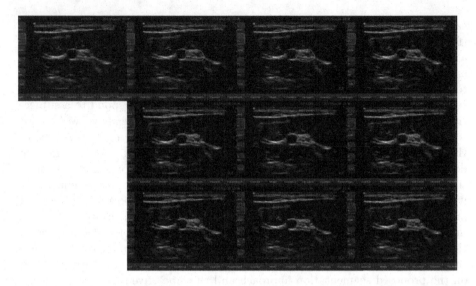

Fig. 1. Segmentation results for an ulnar nerve. On the top from left to right, we show the original image and the labels provided by the three experts. On the second row from left to right, we expose the segmentation results provided by **LR-EX1**, **LR-EX2**, and **LR-EX3**. Finally, on the bottom from left to right we show the segmentation results for **LR-MV**, **LFC**, and **MAE**.

Table 1. Nerve segmentation validation in terms of the Dice coefficient.

	Dice coefficient $\mu \pm \sigma$
LR-Ex1	0.6380 ± 0.0035
LR-Ex2	0.6536 ± 0.0043
LR-Ex3	0.6304 ± 0.0042
LR-MV	0.6414 ± 0.0052
LFC [12]	0.6446 ± 0.0011
MAE [13]	$\mathbf{0.6557 \pm 0.0015}$

fact that the gold standard is not available for the training stage, where the true label is estimated from the manual segmentations provided by different experts in anesthesiology. In contrast, segmentation approaches based on typical supervised learning assume as a gold standard the manual segmentations from one of the annotators, which implies that the classifier predictions will be biased according to annotator expertise.

4   Conclusion

In this paper, we discuss a first attempt for the design of nerve-segmentation systems based on classification with multiple annotators. In this sense, we perform the nerve identification by considering the case where the ground truth is not available. In fact, this consideration is not far from reality since the nerve identification in UI depends on the specialist expertise. So, we use multi-annotator classification schemes to estimate the unavailable true label and the classifier parameters jointly. We tested our strategy in a real-world nerve segmentation dataset captured by "The Automatics Research Group-Universidad Tecnológica de Pereira," which holds UI images of ulnar and median nerves. The experimental results showed that the segmentation methodology based on the information from different experts outperforms state-of-the-art alternatives for nerve segmentation in terms of the Dice coefficients. Hence, the proposed method have a better interpretation of the patterns associated with the nerves by combining the manual segmentation given by multiple anesthesiologists. As future work, authors plan to use more robust multi-annotators classification schemes (for example approaches based on deep-learning) to improve further the quality of the nerve segmentation.

Acknowledgment. This work was funded by Colciencias under the project: "Desarrollo de un sistema de identificación de estructuras nerviosas en imágenes de ultrasonido para la asistencia de bloqueo de nervios periféricos. Aplicación al tratamiento de dolor agudo traumático y prevención del dolor neuropático crónico" (code: 1110-744-55958).

References

1. Hadjerci, O., Hafiane, A., Conte, D., Makris, P., Vieyres, P., Delbos, A.: Computer-aided detection system for nerve identification using ultrasound images: a comparative study. Inform. Med. Unlocked **3**, 29–43 (2016)
2. Hadjerci, O., Hafiane, A., Makris, P., Conte, D., Vieyres, P., Delbos, A.: Nerve detection in ultrasound images using median Gabor binary pattern. In: Campilho, A., Kamel, M. (eds.) ICIAR 2014. LNCS, vol. 8815, pp. 132–140. Springer, Cham (2014). https://doi.org/10.1007/978-3-319-11755-3_15
3. Denny, N.M., Harrop-Griffiths, W.: Editorial I: location, location, location! ultrasound imaging in regional anaesthesia. Br. J. Anaesth. **94**(1), 1–3 (2005)
4. Shi, J., Schwaiger, J., Lueth, T.C.: Nerve block using a navigation system and ultrasound imaging for regional anesthesia. In: 2011 Annual International Conference of the IEEE Engineering in Medicine and Biology Society, EMBC, pp. 1153–1156. IEEE (2011)
5. Noble, J.A., Boukerroui, D.: Ultrasound image segmentation: a survey. IEEE Trans. Med. Imaging **25**(8), 987–1010 (2006)
6. González, J.G., Álvarez, M.A., Orozco, Á.A.: A probabilistic framework based on SLIC-superpixel and Gaussian processes for segmenting nerves in ultrasound imagess. In: 2016 38th Annual International Conference of the IEEE Engineering in Medicine and Biology Society (EMBC), pp. 4133–4136. IEEE (2016)
7. Chan, V.W.S., Nova, H., Abbas, S., McCartney, C.J.L., Perlas, A., et al.: Ultrasound examination and localization of the sciatic nerve: a volunteer study. J. Am. Soc. Anesth. **104**(2), 309–314 (2006)
8. Rodrigues, F., Pereira, F., Ribeiro, B.: Learning from multiple annotators: distinguishing good from random labelers. Pattern Recogn. Lett. **34**(12), 1428–1436 (2013)

9. Groot, P., Birlutiu, A., Heskes, T.: Learning from multiple annotators with Gaussian processes. In: Honkela, T., Duch, W., Girolami, M., Kaski, S. (eds.) ICANN 2011. LNCS, vol. 6792, pp. 159–164. Springer, Heidelberg (2011). https://doi.org/10.1007/978-3-642-21738-8_21
10. Rodrigues, F., Pereira, F.C., Ribeiro, B.: Gaussian process classification and active learning with multiple annotators. In: ICML, pp. 433–441 (2014)
11. Rodrigues, F., Pereira, F., Ribeiro, B.: Sequence labeling with multiple annotators. Mach. Learn. **95**(2), 165–181 (2014)
12. Raykar, V.C., Yu, S., Zhao, L.H., Valadez, G.H., Florin, C., Bogoni, L., Moy, L.: Learning from crowds. J. Mach. Learn. Res. **11**(Apr), 1297–1322 (2010)
13. Yan, Y., Rosales, R., Fung, G., Schmidt, M.W., Valadez, G.H., Bogoni, L., Moy, L., Dy, J.G.: Modeling annotator expertise: learning when everybody knows a bit of something. In: AISTATS, pp. 932–939 (2010)
14. Bishop, C.M.: Pattern Recogn. Mach. Learn. **128**, 1–58 (2006)
15. Boykov, Y., Funka-Lea, G.: Graph cuts and efficient ND image segmentation. Int. J. Comput. Vis. **70**(2), 109–131 (2006)
16. Achanta, R., Shaji, A., Smith, K., Lucchi, A., Fua, P., Susstrunk, S.: SLIC superpixels. Department and School of Computer and Communication Sciences, EPFL, Lausanne, Switzerland, Technical report, 149300 (2010)

Sparse Hilbert Embedding-Based Statistical Inference of Stochastic Ecological Systems

Wilson González-Vanegas[(✉)],
Andrés Alvarez-Meza, and Álvaro Orozco-Gutierrez

Automatic Research Group, Faculty of Engineerings,
Universidad Tecnológica de Pereira, Pereira, Colombia
wilgonzalez@utp.edu.co

Abstract. The growth rate of a population has been an important aspect in ecological and biologic applications. Since a non-linear stochastic behavior is linked to this type of systems, the inference of the model parameters is a challenging task. Approximate Bayesian Computation (ABC) can be used for leading the intractability of the likelihood function caused by the model characteristics. Recently, some methods based on Hilbert Space Embedding (HSE) have been proposed in the context of ABC; nevertheless, the relevance of the observations and simulations are not contemplated. Here, we develop a Sparse HSE-based distance, termed SHSED, to compare distributions associated with two random variables through sparse estimations of the densities in a Reproducing Kernel Hilbert Space (RKHS). Namely, SHSED highlights relevant information using a sparse weighted representation of data within an ABC-based inference. Our method improves the inference accuracy of a Ricker map-based population model in comparison with other state-of-the-art ABC-based approaches.

Keywords: Hilbert space embeddings · Ricker model
Sparse methods · Statistical inference

1 Introduction

In population dynamics, the growth rate of a particular population is a widely known field of interest, e.g., ecology and biology applications. Several discrete-time methods have been proposed to model the population growth for single-species with different characteristics. However, the Ricker map is one of the most important ones, since it stands for a general approach to discretize almost any continuous model of population dynamics while taking into account the stochastic behavior proper to nature. Besides, it is capable not only of stability but also of limits cycles and chaos [12]. Given an observed data from a population, the goal is to find the parameters of the growth population model, i.e., Ricker model. Since this model is stochastic, traditional techniques, such as least squares, are

© Springer International Publishing AG, part of Springer Nature 2018
M. Mendoza and S. Velastín (Eds.): CIARP 2017, LNCS 10657, pp. 255–262, 2018.
https://doi.org/10.1007/978-3-319-75193-1_31

not appropriate for leading the uncertainty, demanding for a Bayesian approach. It is well-known that in all Bayesian-based statistical inference tasks, the likelihood function performs an important role as long as it states the probability of the observed data under a particular given statistical model. Then, using the Bayes theorem, we can update a priori knowledge about the set of needed parameters into a posterior distribution. For straightforward models, it is possible to derive an analytic expression for the likelihood function, and the evidence can be computed with relative ease; nonetheless, for complex models, to find an exact formula for the likelihood function is often intractable [9].

To deal with this intractability, free-likelihood techniques like Approximate Bayesian Computation (ABC) have emerged. The idea behind ABC is to assess an auxiliary model with different parameter values drawn from a prior distribution to obtain simulations that are compared to the observed data [10]. The most basic ABC technique uses a rejection of samples using a comparison between summary statistics of the observed and simulated data. In particular, ABC methods based on the rejection of samples present a high dependence on the choice of sufficient summary statistics [7]. A poor selection of summary statistics can lead to an additional bias that can be difficult to quantify, thus, unless the selected summary statistics are sufficient, inference results in a partial posterior rather than the desired full one [5].

To avoid the selection of summary statistics, ABC methods based on kernel embeddings have been proposed, where probability density functions are set over the observed and simulated data to be compared in a RKHS. Authors in [5] proposed to use empirical approximations for the probability measures embedded in the RKHS, standing for a comparison based uniquely on the characteristic kernel, leading to an approach that considers one free parameter. On the other hand, authors in [13] proposed to use Parzen-window density estimation to approximate the probability measures, producing a robust estimator regarding the number of samples required in ABC. However, the relevance of the samples is not considered, which is crucial for highly dynamic models, i.e., the Ricker model.

In this work, we introduce a Sparse Hilbert Space Embedding-based distance, named SHSED, to compare distributions based on sparse estimations of the densities in RKHSs. In this sense, SHSED highlights relevant information employing a Sparse Kernel Density Estimation of probability measures associated with the observed and simulated data in the context of ABC. Obtained results show that our proposed technique improves the accuracy in terms of the relative error when inferring highly dynamic models as the Ricker map, in comparison to state-of-the-art ABC-based methods.

The remainder of this paper is organized as follows: Sect. 2 describes the mathematical background of SHSED. Section 3 describes the experimental setup and the obtained result. Finally, the conclusions are outlined in Sect. 4.

2 Sparse Hilbert Space Embedding-Based Distance for Approximate Bayesian Computation

Let us consider a situation where a generative model can be simulated through a conditional probability sampling from $p(x|\theta)$, where $x \in \mathcal{X}$ is a random variable with domain $\mathcal{X}$ and $\theta \in \Theta$ holds the parameters. Besides, the computation of the likelihood $p(y|\theta)$ for the observed data $y \in \mathcal{X}$ is intractable. Since the posterior $p(\theta|y) \propto p(y|\theta)\zeta(\theta)$ cannot be calculated, being $\zeta(\theta)$ a prior over θ, neither exact nor sampled posterior are feasible. Therefore, Approximate Bayesian Computation (ABC)-based techniques aim to solve such an inference issue by computing the likelihood from simulation [5]. In particular, a straightforward ABC-based approach comprises the rejection of simulated data x according to the distance function $d_{\mathcal{X}} : \mathcal{X} \times \mathcal{X} \to \mathbb{R}^+$. So, an approximate posterior can be estimated as follows: $\hat{p}(\theta|y;\epsilon) \propto \hat{p}(y|\theta;\epsilon)\zeta(\theta)$, where $\hat{p}(y|\theta;\epsilon) = \int_{\Omega(y;\epsilon)} p(x|\theta)dx$, $\Omega(y;\epsilon) = \{x : d_{\mathcal{X}}(x,y) < \epsilon\}$, and $\epsilon \in \mathbb{R}^+$. However, applying a distance directly on $\mathcal{X}$ is often difficult because of the large amount of features and/or observations. Moreover, fixing the ϵ value is crucial to achieve an accurate ABC estimation. Alternative strategies includes a feature mapping $s = \vartheta(x)$ before calculating the distance, where $s \in \mathcal{S}$ is a summary statistics space and $\vartheta : \mathcal{X} \to \mathcal{S}$ [4]; nonetheless, such summary statistics often leak information for complex models. Then, to code complex relations among samples, the approximate likelihood $\hat{p}(y|\theta;\epsilon)$ can be computed as the convolution of the true likelihood $p(y|\theta)$ and the kernel function $\kappa : \mathcal{X} \times \mathcal{X} \to \mathbb{R}$, which imposes a constraint to the rejection of samples as the inner product $\kappa(x,y) = \langle \phi(x), \phi(y) \rangle_{\mathcal{H}}$ in the RKHS, $\mathcal{H}$, where $\phi : \mathcal{X} \to \mathcal{H}$. Although various kernel functions can be tested, the Gaussian function is preferred since it aims at finding $\mathcal{H}$ with universal approximating ability, not to mention its mathematical tractability. Thus, the Gaussian kernel in the ABC context yields: $\kappa_G(d_{\mathcal{X}}(x,y);\sigma) = \exp(-d_{\mathcal{X}}^2(x,y)/2\sigma^2)$, where $\sigma \in \mathbb{R}^+$ stands for the kernel bandwidth.

Here, we introduce a Hilbert Space Embedding-based distance (HSED) for enhancing the Gaussian-based similarity computation in nonlinear stochastic systems exhibiting dynamic observations. Thus, let us assume that both the observed and a simulated data can be modeled as a pair of random variables $Y \sim P_Y$ and $X \sim P_X$, respectively, given the marginal probability distributions P_Y and P_X. Besides, let $\{x^{(i)}\}_{i=1}^{Q_x} \sim P_X$ and $\{y^{(i)}\}_{i=1}^{Q_y} \sim P_Y$ be a pair of independent and identically distributed sample sets of size Q_x and Q_y. Our HSED is defined as follows [8]: $d_{\mathcal{H}}^2(P_X, P_Y) = \|\hat{\mu}[P_X] - \hat{\mu}[P_Y]\|_{\mathcal{H}}^2$, where $\hat{\mu}[P_X], \hat{\mu}[P_Y] \in \mathcal{H}$ are the embedding approximations of $\mu[P_X] = \boldsymbol{E}_X\{\phi(X)\}$ and $\mu[P_Y] = \boldsymbol{E}_Y\{\phi(Y)\}$, through weighted averaging as: $\hat{\mu}[P_X] = \sum_{i=1}^{Q_x} \hat{f}(x^{(i)})\kappa_G(d_e(x^{(i)},\cdot);\sigma_K)$, with $\hat{f}(x^{(i)}) = \boldsymbol{\alpha}^\top \boldsymbol{a}_i^{(\sigma_X)}$ and $\hat{\mu}[P_Y] = \sum_{i=1}^{Q_y} \hat{g}(y^{(i)})\kappa_G(d_e(y^{(i)},\cdot);\sigma_K)$, with $\hat{g}(y^{(i)}) = \boldsymbol{\beta}^\top \boldsymbol{b}_i^{(\sigma_Y)}$; d_e stands for the Euclidean distance, $\boldsymbol{\alpha} \in [0,1]^{Q_x}$, $\boldsymbol{\beta} \in [0,1]^{Q_y}$, $\|\boldsymbol{\alpha}\|_1 = 1$, $\|\boldsymbol{\beta}\|_1 = 1$ and $\hat{f}$ and $\hat{g}$ are the approximated densities associated with P_X and P_Y, respectively. Moreover, $\boldsymbol{a}_i^{(\sigma_X)}$ and $\boldsymbol{b}_i^{(\sigma_Y)}$ are the i-th column vectors of the matrices $\boldsymbol{A}^{(\sigma_X)} \in \mathbb{R}^{Q_x \times Q_x}$, $\boldsymbol{B}^{(\sigma_Y)} \in \mathbb{R}^{Q_y \times Q_y}$ holding elements

$a_{ij}^{\sigma_X} = \kappa_G(d_e(x^{(i)}, x^{(j)}); \sigma_X)$, and $b_{ij}^{(\sigma_Y)} = \kappa_G(d_e(y^{(i)}, y^{(j)}); \sigma_Y)$, respectively, and $\sigma_K, \sigma_X, \sigma_Y \in \mathbb{R}^+$. Bearing in mind the convolution properties for Gaussian functions [6], the HSED is rewritten as:

$$\hat{d}_{\mathcal{H}}^2(P_X, P_Y) = \boldsymbol{\alpha}^\top \boldsymbol{A}^{(\sigma_{KX})} \boldsymbol{\alpha} + \boldsymbol{\beta}^\top \boldsymbol{B}^{(\sigma_{KY})} \boldsymbol{\beta} - 2\boldsymbol{\alpha}^\top \boldsymbol{L}^{(\sigma_L)} \boldsymbol{\beta}, \tag{1}$$

where $\sigma_{KX} = \sqrt{\sigma_K^2 + 2\sigma_X^2}$, $\sigma_{KY} = \sqrt{\sigma_K^2 + 2\sigma_Y^2}$, $\sigma_L = \sqrt{\sigma_K^2 + \sigma_X^2 + \sigma_Y^2}$, and the matrix $\boldsymbol{L}^{(\sigma_L)} \in \mathbb{R}^{Q_x \times Q_y}$ holds elements $l_{i,j}^{(\sigma_L)} = \kappa_G(d_e(x^{(i)}, y^{(j)}); \sigma_L)$. The HSED in Eq. (1) highlights relevant samples concerning the weight values $\boldsymbol{\alpha}, \boldsymbol{\beta}$ to reject or accept a given sample in ABC. In turn, the probability density functions are estimated using informative samples through sparseness in the weighting coefficients. In this regard, a constrained quadratic optimization problem is formulated based on a Integrated Mean Squared Error, $\varepsilon(\boldsymbol{\alpha}) = \int_X (f(x^{(i)}) - \hat{f}(x^{(i)}))^2 dx$, as follows [3]:

$$\underset{\boldsymbol{\alpha}}{\operatorname{argmin}} \, \varepsilon(\boldsymbol{\alpha}) = \boldsymbol{\alpha}^\top \boldsymbol{A}^{(\sqrt{2}\sigma_X)} \boldsymbol{\alpha} - \frac{2}{Q_x} \boldsymbol{\alpha}^\top \boldsymbol{A}^{(\sqrt{2}\sigma_X)} \boldsymbol{1}; \quad \text{s.t.} \quad ||\boldsymbol{\alpha}||_1 = 1, \quad \alpha_i \geq 0. \tag{2}$$

The optimization problem in Eq. (2) can be solved using a Sequential Minimal Optimization (SMO) algorithm. Likewise, the $\boldsymbol{\beta}$ weights values are fixed through the minimization of $\varepsilon(\boldsymbol{\beta})$. Therefore, our Sparse HSED (SHSED) focuses relevant observations in $\hat{d}_{\mathcal{H}}^2(P_X, P_Y)$ based on the sparse weights $\boldsymbol{\alpha}, \boldsymbol{\beta}$. In practice, given N samples $\{x_n^{(i)} \backsim P_{X_n}\}_{n=1}^N$ from $p(x|\theta_n)$, with $\theta_n \backsim \zeta(\theta)$, and the observation $y \backsim P_Y$, a weighted sample set $\Psi = \{(\theta_n, w_n)\}_{n=1}^N$ is calculated by fixing:

$$w_n = \kappa_G(d_{\mathcal{H}}(P_{X_n}, P_Y); \epsilon) / \sum_{n=1}^N \kappa_G(d_{\mathcal{H}}(P_{X_n}, P_Y); \epsilon), \tag{3}$$

which can be employed to approximate $p(\theta|y)$ using posterior expectation as follows: $\hat{p}(\theta|y) = \sum_{n=1}^N w_n \kappa_G(d_e(\theta, \theta_n); \sigma_\theta)$, with $\sigma_\theta \in \mathbb{R}^+$. Algorithm 1 summarizes an ABC approach based on the introduced SHSED.

Algorithm 1. ABC based on SHSED

Input: Observed data: $\{y^{(j)} \backsim P_Y\}_{j=1}^{Q_y}$, prior: $\zeta(\theta)$, bandwidths: $\epsilon, \sigma_K, \sigma_X, \sigma_Y, \sigma_\theta \in \mathbb{R}^+$.
Output: Posterior estimation: $\hat{p}(\theta|y)$.
1: **for** $n = 1, \cdots, N$ **do**
2:     $\theta_n \backsim \zeta(\theta)$ ▷ Draw a solution θ_n from the prior.
3:     $x_n \backsim p(x|\theta_n)$; $\{x_n^{(i)} \backsim P_{X_n}\}_{i=1}^{Q_x}$ ▷ Draw a sample from the model.
4:     $\tilde{w}_n = \kappa_G(d_{\mathcal{H}}(P_{X_n}, P_Y); \epsilon)$ ▷ Compute the n-th weight value using SHSED.
5: **end for**
6: $w_n = \tilde{w}_n / \sum_{n=1}^N \tilde{w}_n$ ▷ Normalize the weights
7: $\hat{p}(\theta|y) = \sum_{n=1}^N w_n \kappa_G(d_e(\theta, \theta_n); \sigma_\theta)$,   ▷ Compute the posterior.

---

3   Experiments and results

To evaluate the SHSED performance, we consider two statistical inference task: a toy problem that comprises a Poisson mixture model and a nonlinear ecological dynamic system termed the Ricker model [12]. As benchmark, the straightforward ABC Rejection and two state-of-the-art ABC methods based on HSE are studied. The former, Maximum Mean Discrepancy (MMD)-based ABC [5], uses empirical density approximations: $\hat{f}(x^{(i)}) = 1/Q_x \sum_{j=1}^{Q_x} \delta(x^{(i)} - x^{(j)})$, $\hat{g}(y^{(i)}) = 1/Q_y \sum_{j=1}^{Q_y} \delta(y^{(i)} - y^{(j)})$, being $\delta(\cdot)$ the Dirac delta function. The latter, Parzen-based ABC [13], uses traditional Parzen window estimators as follows: $\hat{f}(x^{(i)}) = 1/Q_x \sum_{j=1}^{Q_x} \kappa_G(d_e(x^{(i)}, x^{(j)}); \sigma_X)$, $\hat{g}(y^{(i)}) = 1/Q_y \sum_{j=1}^{Q_y} \kappa_G(d_e(y^{(i)}, y^{(j)}); \sigma_Y)$. Besides, as quantitative assessment, the following relative error is used: $\mathcal{E}_{\theta^{(z)}} = 100 \times ||\theta^{(z)} - \sum_{n=1}^{N} w_n \hat{\theta}_n^{(z)}||/||\theta^{(z)}||$, where $\theta^{(z)}$ is the value of the z-th target parameter and $\hat{\theta}_n^{(z)}$ is the n-th ABC-based approximation with weight w_n.

Poisson mixture model results. We consider a finite Poisson mixture model of the form: $p(x|\boldsymbol{\lambda}, \boldsymbol{\pi}) = \sum_{c=1}^{C} \pi_c \exp(-\lambda_c)\lambda_c^x/x!$, where $\boldsymbol{\pi} = \{\pi_c\}_{c=1}^{C}$ are the mixing coefficients holding the condition $\sum_{c=1}^{C} \pi_c = 1$, $\boldsymbol{\lambda} = \{\lambda_c\}_{c=1}^{C}$ are the number of average events for each Poisson density ($\lambda_1 < \lambda_2 < \cdots < \lambda_C$), and C is the number of components [11]. For concrete testing, we aim to estimate the posterior $p(\boldsymbol{\pi}|\boldsymbol{\lambda}, x)$, with $C = 2$ and $\boldsymbol{\lambda} \in \{1, 8\}$. Moreover, a Dirichlet distribution is imposed over $\boldsymbol{\pi}$ as prior [11], that is, $\boldsymbol{\pi} \sim \text{Dirichlet}(1, 1)$. Initially, we generated ten samples for $\boldsymbol{\pi}$ from the prior, then, 50 observations are computed for each of them using the model. Figure 1 shows the obtained inference results by fixing $\sigma_X = \sigma_Y = 0.33$ and $\sigma_K = 2.154$. As seen, the proposed SHSED extracts relevant information from simulations when suitable values for σ_X and σ_Y are selected. In fact, Fig. 1(b) shows how sample π_6 is the farthest from the true values of $\boldsymbol{\pi}$ but its associated simulation x_6 in Fig. 1(a) is the closest from the observed data; nevertheless, Fig. 1(c) shows that our sparse approach assigns a high value for the distance, leading to a low weight. On the other hand, the proposed SHSED approach assigned high weights to the samples π_3, π_4, and π_5, which exihibit close distance values to the true parameters in $\boldsymbol{\pi}$ but distant dependencies with simulations x_3, x_4, and x_5 concerning the observed data.

Now, to test the robustness of the ABC-based estimations regarding the number of simulations, we build an observed dataset holding 50 samples drawn from the Poisson mixture model, with $\boldsymbol{\pi} \in \{0.3, 0.7\}$. Afterward, to generate the simulated data, we calculated candidates from the prior and drew 50 samples from the model to compute the posterior. This procedure is repeated 100 times. The number of simulations (N) is increased from 10 to 300 using a step size of 10. For the ABC rejection, we use the Euclidean distance to compare the observed and simulated data with $\epsilon = 10$; this threshold was defined empirically. For the MMD, the Parzen, and the SHSED-based ABC methods, we use a soft threshold value of $\epsilon = 0.158$. Figures 1(d) and (e) show the error bar curves of the posterior distribution for the mixing coefficients. As seen, the SHSED improves the posterior estimation leading to the lowest errors and deviations for $\sigma_X = \sigma_Y = 0.406$. Our sparse-based weighting favors the identification of relevant structures in an RKHS before computing the dependencies between observed and simulated data within an ABC framework. Additionally, note that for low kernel bandwidth values in Eq. (2), the SHSED approach converges to the MMD and Parzen-based ones, since the weights $\boldsymbol{\alpha}$ and $\boldsymbol{\beta}$ get closer to $1/Q_x$ and $1/Q_y$, respectively.

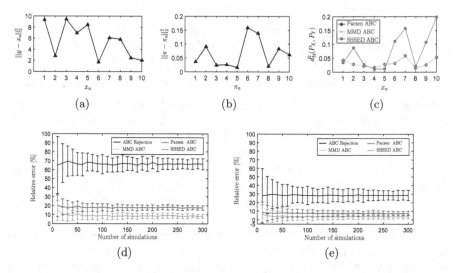

Fig. 1. Poisson mixture model results. (a) Euclidean distance between the observed and simulated data. (b) Euclidean distance between candidates sampled from the prior and the true values of the parameters. (c) Different distances based on HSE between distributions associated with the observed and simulated data. (d) and (e) Relative error bar curves for the posterior of the mixing coefficients: π_1 and π_2 respectively.

Ricker model results. This nonlinear ecological dynamic system can be modeled as follows [1]: $\ln(M^{(t)}) = \ln(r) + \ln(M^{(t-1)}) - M^{(t-1)} + e^{(t)}$, where $M^{(t)} \in \mathbb{R}$ is the size of some animal population at time t, $e^{(t)} \backsim \mathcal{N}(0, \sigma_e^2)$, being $\sigma_e \in \mathbb{R}^+$ the standard deviation of the innovations, and $\ln(r)$ is related to the growth rate parameter of the model ($r \in \mathbb{R}^+$). Additionally, the observation y is a time series drawn from a Poisson distribution: $y \backsim \text{Poisson}(\phi M^{(t)})$, with $y \in \mathbb{N}$, where ϕ is a scale parameter [2]. Thus, the Ricker model is completely parametrized by $\boldsymbol{\theta} = [\ln(r), \phi, \sigma_e]$. In this experiment, we draw 50 samples from the model with $\boldsymbol{\theta} = [3.8, 10, 0.3]$ by fixing the following priors [1]: $\ln(r) \backsim \mathcal{N}(4, 0.5)$; $\phi \backsim \mathcal{X}^2(10)$; $\sigma_e \backsim \text{invgamma}(3, 1.3)$. The kernel bandwidth for the ABC rejection approach is fixed empirically as $\epsilon = 75$, meanwhile for the MMD, the Parzen and, the SHSED, we set $\epsilon = 0.158$. Besides, we use $\sigma_X = \sigma_Y = 0.0158$ to compute the sparse weights and $\sigma_K = 2.5$. According to the results in Figs. 2(a) to (c), the MMD-based posteriors are almost identical to the ones obtained by the Parzen-based approach. Furthermore, notice how the posterior distribution for $\ln(r)$ and σ_e parameters obtained by our approach has their maximum closer to the true value in comparison to the benchmarks, and the rejection-based ABC has the worst performance. In the case of ϕ, it can be seen that the results obtained by the SHSED are close enough to the other ABC techniques based on HSED.

Lastly, to test the stability of the inference, we repeated 100 times the procedure to approximate the posterior of the model parameters concerning the relative error assessment. Obtained results in Figs. 2(d) to (f) reveals how the proposed SHSED reaches the lowest uncertainty for the $\ln(r)$ and ϕ inferences. In the case of σ_e, since the true value is low ($\sigma_e = 0.3$), a small change in the form of the posterior produces a substantial change in the expected value; hence, relevant changes in the index error are gathered. Remarkably, the lowest uncertainty for σ_e is related to the ABC rejection method due to the smoother form of the posterior in comparison to other HSED methods; nonetheless, SHSED achieves the best overall performance (see Fig. 2(g)).

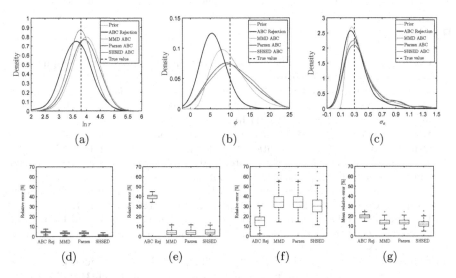

Fig. 2. Ricker model results. (a)–(c) Different ABC-based estimated posteriors for $\{\ln r, \phi, \sigma_e\}$ with $\sigma_\theta = \{0.253, 2.190, 0.082\}$, respectively. (d)–(f) Relative error boxplots for the posterior of the Ricker model parameters: $\ln r$, ϕ, and σ_e, respectively. (g) Overall performance of the different ABC methods in terms of the mean relative error.

4   Conclusions

We introduce a novel Sparse Hilbert Space Embedding-based distance, termed SHSED, to compute data dependencies within Approximate Bayesian Computation (ABC) inference frameworks. Remarkably, our SHSED compares distributions associated with two random variables through sparse estimations of the densities in RKHSs. To test the introduced approach, we consider two statistical inference tasks: a Poisson mixture model and a nonlinear ecological dynamic system, termed the Ricker model. Attained results demonstrated how our proposal outperforms state-of-the-art ABC-based inference approaches, including the well-known ABC rejection and ABC alternatives based on HSE. Indeed,

SHSED highlights relevant information from dynamic observations and simulations before their quantitive comparison in an RKHS. As future work, authors plan to develop an automatic selection of the RKIIS regarding the input data dynamics. Moreover, applying ABC approaches using HSE for multivariate data is a research line of interest.

Acknowledgments. Research under grants provided by the project "Desarrollo de una plataforma para el cálculo de confiabilidad en la operación interdependiente de los sistemas de gas natural y sector eléctrico de Colombia que permita evaluar alternativas de inversión y regulación para optimizar los costos de operación" with code: 1110-745-58696, funded by Colciencias. Moreover, author W. González-Vanegas was supported by Colciencias under the 706 agreement, "Jóvenes Investigadores e Innovadores-2015".

References

1. Golchi, S., Campbell, D.A.: Sequentially constrained Monte Carlo. Comput. Stat. Data Anal. **97**, 98–113 (2016)
2. Gutmann, M.U., Corander, J.: Bayesian optimization for likelihood-free inference of simulator-based statistical models. J. Mach. Learn. Res. **17**(125), 1–47 (2016). http://jmlr.org/papers/v17/15-017.html
3. Hong, X., Chen, S.: A fast algorithm for sparse probability density function construction. In: 2013 18th International Conference on DSP, pp. 1–6. IEEE (2013)
4. Joyce, P., Marjoram, P., et al.: Approximately sufficient statistics and Bayesian computation. Stat. Appl. Genet. Mol. Biol. **7**(1), 26 (2008)
5. Park, M., Jitkrittum, W., Sejdinovic, D.: K2-ABC: approximate Bayesian computation with kernel embeddings. arXiv preprint arXiv:1502.02558 (2015)
6. Principe, J.C.: Information Theoretic Learning: Renyi's Entropy and Kernel Perspectives. Springer Science & Business Media, Heidelberg (2010). https://doi.org/10.1007/978-1-4419-1570-2
7. Robert, C.P., Cornuet, J.M., Marin, J.M., Pillai, N.S.: Lack of confidence in approximate bayesian computation model choice. Proc. Nat. Acad. Sci. **108**(37), 15112–15117 (2011)
8. Smola, A., Gretton, A., Song, L., Schölkopf, B.: A hilbert space embedding for distributions. In: Hutter, M., Servedio, R.A., Takimoto, E. (eds.) ALT 2007. LNCS (LNAI), vol. 4754, pp. 13–31. Springer, Heidelberg (2007). https://doi.org/10.1007/978-3-540-75225-7_5
9. Toni, T., Welch, D., Strelkowa, N., Ipsen, A., Stumpf, M.P.: Approximate Bayesian computation scheme for parameter inference and model selection in dynamical systems. J. R. Soc. Interface **6**(31), 187–202 (2009)
10. Turner, B.M., Van Zandt, T.: A tutorial on approximate Bayesian computation. J. Math. Psychol. **56**(2), 69–85 (2012)
11. Viallefont, V., Richardson, S., Green, P.J.: Bayesian analysis of poisson mixtures. J. Nonparametric Stat. **14**(1–2), 181–202 (2002)
12. Wood, S.N.: Statistical inference for noisy nonlinear ecological dynamic systems. Nature **466**(7310), 1102–1104 (2010)
13. Zuluaga, C.D., Valencia, E.A., Álvarez, M.A., Orozco, Á.A.: A parzen-based distance between probability measures as an alternative of summary statistics in approximate Bayesian computation. In: Murino, V., Puppo, E. (eds.) ICIAP 2015. LNCS, vol. 9279, pp. 50–61. Springer, Cham (2015). https://doi.org/10.1007/978-3-319-23231-7_5

Fingerprint Presentation Attack Detection Method Based on a Bag-of-Words Approach

Lázaro Janier González-Soler[1](✉), Leonardo Chang[3],
José Hernández-Palancar[1], Airel Pérez-Suárez[1], and Marta Gomez-Barrero[2]iD

[1] Advanced Technologies Application Center (CENATAV),
7a # 21406 e/ 214 y 216, Rpto. Siboney, Playa, 12200, Havana, Cuba
{jsoler,jhernandez,asuarez}@cenatav.co.cu
[2] da/sec - Biometrics and Internet Security Research Group,
Hochschule Darmstadt, Darmstadt, Germany
marta.gomez-barrero@h-da.de
[3] Tecnologico de Monterrey, Campus Estado de México, Carretera al Lago de
Guadalupe Km. 3.5, 52926 Atizapán de Zaragoza, Estado de México, Mexico
leonardochang36@gmail.com

Abstract. Fingerprint-based biometric systems are not entirely secure
due to their vulnerability to presentation attacks. In this paper, we propose a new presentation attack method based on a Bag-of-Words approach, which by combining local and global information of fingerprint
can correctly identify bona fine presentations from presentation attacks.
The experimental evaluation of our proposal, over the well-known LivDet
2011 dataset, showed an Average Classification Error of 4.73%, outperforming the state of the art.

1 Introduction

Fingerprint-based biometric systems are commonly used in several contexts since
they provide higher accuracy and in many cases, they increase user convenience
with respect to traditional credential-based systems. Despite being very popular, fingerprint-based biometric systems are vulnerable to presentation attacks,
in which synthetic artifacts are presented to the biometric sensor. The task of
determining if a fingerprint comes from a live subject (i.e., it is a bona fide presentation) or it comes from an artificial fingertip (i.e., it is a presentation attack)
is a critical issue that has received a considerable attention in the recent years.

In this context, several software-based methods have been reported in literature for Presentation Attack Detection (PAD) [1,2]. These methods focus on
skin properties, which can be classified as dynamic (e.g., skin distortion and
perspiration) or static (e.g., texture, ridge frequency and ridge-valley structure).
In contrast to hardware-based methods, these approaches are a less expensive
alternative since they do not require a special hardware. In addition, most of
the software-based methods use a single image and do not employ additional
invasive biometric measurements. However, they can be more challenging, as

© Springer International Publishing AG, part of Springer Nature 2018
M. Mendoza and S. Velastín (Eds.): CIARP 2017, LNCS 10657, pp. 263–271, 2018.
https://doi.org/10.1007/978-3-319-75193-1_32

they require the extraction of discriminative features to distinguish a bona fide presentation from a presentation attack.

Taking into account that real and fake fingerprint images exhibit different appearance characteristics, many texture descriptors have been used for discriminating presentation attacks from bona fide presentations (e.g., LPQ [3], MBLTP [4], WLD [5], BSIF [6]), and combinations of individual algorithms (e.g., SURF-PHOG-Gabor [7]). The main drawback of the texture-based methods is that they depend both on the material used for building the artificial fingerprint and the sensor used for acquiring the fingerprint images. More specifically, different materials utilized for building artificial fingerprint replicas exhibit different ridge structures, which can be identified based on discriminative multi-scale appearance description methods.

An appearance-based technique that has shown remarkable results in object classification tasks is the Bag of Words (BoW) approach [8]. In this paper, we propose a new PAD method based on the BoW algorithm, which computes local descriptors at fixed points on a regular grid. In our methodology, several scaling factors are studied in order to represent both local and global information of a fingerprint, thereby allowing to distinguish a bona fide presentation from a presentation attack.

In order to evaluate the performance of the proposal and to allow the reproducibility of the presented results, several experiments were conducted on the well-known LivDet 2011 dataset. The proposed method achieved an Average Classification Error (ACE) of 4.73%, significantly outperforming the top state-of-the-art results. The performance of the proposal was also evaluated using the recent ISO standard evaluation metrics specifically tailored for PAD [9].

The remainder of this paper is organized as follows: in Sect. 2, we describe the proposed method. The experimental evaluation, comparing the performance of our proposal against top state-of-the-art techniques, is presented in Sect. 3. Finally, conclusions and future work directions are presented in Sect. 4.

2   Proposed Method

Bag-of-words approaches have been largely used for image appearance description and they have obtained remarkable results in object categorization tasks. In this paper, we propose a method for detecting fingerprint presentation attacks which is based on a BoW approach. Figure 1 shows an overview of the proposed method, which pipeline is composed of the following three steps: (i) extraction of Scale Invariant Feature Transform (SIFT) descriptors, (ii) encoding of the SIFT features with a spatial histogram of visual words, and (iii) classification of the image descriptor by a Support Vector Machine (SVM) through a feature map.

For local features extraction, we follow the Pyramid Histogram Of visual Words (PHOW) approach of [10], in which SIFT descriptors are computed densely at points on a regular grid with a fixed spacing (e.g., 5 pixels). In order to allow scale variation between fingerprints, descriptors are computed over four circular patches with different radii σ. Therefore, each point in the grid is represented by four SIFT descriptors. To efficiently compute the descriptors, we use the implementation provided in [11], which delivers a speed-up of up to 60x by

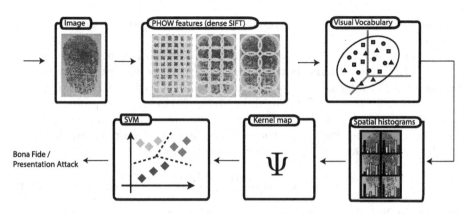

Fig. 1. Method overview. First, dense SIFT descriptors are computed at different scales. Then this features are encoded using a previously learned visual vocabulary. Finally, the fingerprint descriptor is classified using SVM through a feature map.

exploiting the uniform sampling and overlapping between descriptors, and by implementing linear interpolation with integral image convolution.

For image encoding, spatial histograms are used in order to incorporate spatial relationship between patches. In this context, the fingerprint image is partitioned into increasingly fine sub-regions and the dense extracted features inside each sub-region are vector quantized into K visual words using a k-means clustering. Once the BoW has been built, histograms inside each sub-region are computed and they are then stacked for describing the fingerprint image.

Finally, for bona fide or presentation attack classification, a homogeneous kernel map is used to transform a χ^2 SVM into a linear one [12], providing fast computation with high accuracy.

3 Experimental Evaluation

In this section, the results of several experiments for evaluating the performance of the proposed method are presented. The objectives of the experimental evaluation are threefold, namely: (i) analyse the impact of the different parameters on the method performance, (ii) compare the performance of our proposal against the top state-of-the-art approaches and (iii) analyse the computational time of the proposed algorithm for different parameter configurations.

3.1 Experimental Protocol

The experiments were conducted on the publicly available database provided by LivDet 2011 [13]. The database contains 16,000 images acquired with four different sensors, namely, Biometrika FX2000, Digital 4000B, Italdata ET10 and Sagem MSO300. For each sensor, 2,000 images from presentation attacks and 2,000 images from bona fide presentations were captured. Presentation attacks were built using five different materials: gelatin, eco flex, latex, silgum and wood glue [14].

The proposed method was implemented in Matlab using the VLFeat library [11] for SIFT computation, k-means and SVM. All the experiments were performed on a PC with an Intel i5-3470 processor at 3.2 GHz, 8 GB RAM.

In all the experiments we followed the LivDet 2011 evaluation protocol, where half of the database is used for training and the other half for testing. In order to allow a fair comparison with the available literature, we reported the ACE, which is the standard metric in LivDet competitions

$$ACE = \frac{FSAR + FLRR}{2}, \tag{1}$$

where False Spoofing Acceptance Rate (FSAR) is the percentage of misclassified spoof fingerprints (i.e., presentation attacks) and False Live Rejection Rate (FLRR) is the percentage of misclassified live fingerprints (i.e., bona fide presentations).

In addition, we also reported the accuracy our proposal attains in terms of Attack Presentation Classification Error Rate (APCER) and Bona Fide Presentation Classification Error Rate (BPCER), which are recent ISO standard metrics used for evaluating PAD schemes [9].

3.2   Experimental Results

In the first set of experiments, we study the impact of the different key parameters; the visual vocabulary sizes (K) and the scale values (σ), on the performance and efficiency of our proposal. In particular, we have chosen the following ranges: $K = \{256, 512, 1024, 2048, 4096, 8192\}$ and $\sigma = \{1, 1.40, 1.96, 2.74, 3.84\}$. For determining the best performance of the PAD method, all combinations of K and σ are evaluated, hence resulting in 30 different experiments.

Figure 2 shows the impact on the ACE of scale variation for the different visual vocabulary sizes, for each sensor in the dataset. As it can be observed, most of the curves have a stationary behaviour (Biometrika, Digital and Sagem sensors) with the exception of the ItalData sensor, which shows a more random trend for different scales. Both trends hence indicate a lack of a significant impact of σ on the overall accuracy of the PAD scheme. Therefore, for applications where computational cost is a critical issue, any σ value can be used without considerably affecting the system accuracy.

In Fig. 3, we show the impact on the classification accuracy of the different vocabulary sizes (K) at different scale values (σ) for each sensor in the LivDet 2011 dataset. It can be appreciated that the size of the vocabulary has a significant impact on the classification accuracy. Figure 3 shows that again the Biometrika, Digital and Sagem sensors exhibit a similar behaviour, where the best results are obtained for $K = 1024$. On the other hand, for the ItalData sensor, the best results are obtained for the biggest vocabularies, $K = 8192$.

In the second experiment, we computed the ACE achieved by our proposal and we compared it with those yielded by the top state-of-the-art methods; these results are showed in Table 1. As it can be observed, several configurations of our proposal achieve state-of-the-art results for the LivDet 2011. In particular, the best reported result presented in [14], based on deep learning, is outperformed by 14%. Compared to other appearance-based methods, our proposal outperforms the best two results by 19% [7] and 21% [15] of ACE, respectively.

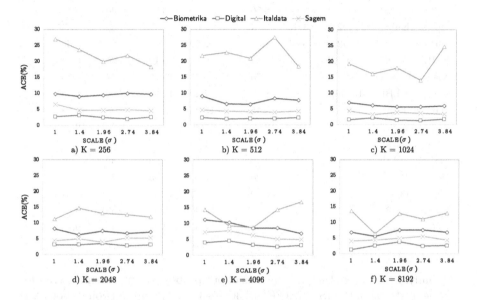

Fig. 2. ACE per sensor on several scales (σ) for different visual vocabulary sizes (K).

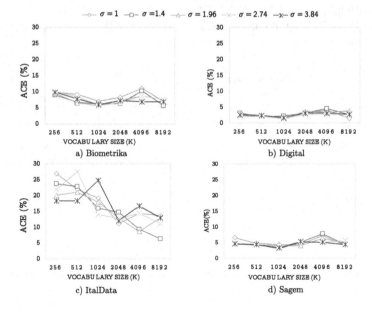

Fig. 3. ACE per scale for each sensor using different sizes of visual vocabulary.

Table 1. Comparison of ACE. For each sensor, the lowest value is highlighted in bold.

Methods	Biometrika	Digital	ItalData	Sagem	Avg. ACE
Our method (K = 1024, σ = 1.40)	6.05	2.15	16.05	**3.15**	6.85
Our method (K = 1024, σ = 2.74)	5.65	**1.25**	13.95	3.65	6.13
Our method (K = 2048, σ = 1.00)	8.15	3.15	11.15	4.35	6.70
Our method (K = 4096, σ = 1.96)	8.50	3.30	8.65	6.30	6.69
Our method (K = 8192, σ = 1.4)	**5.60**	2.60	6.35	4.35	**4.73**
Our method (K = 8192, σ = 3.84)	7.00	2.50	11.40	4.40	6.33
SURF+PHOG+Gabor 2016 [7]	7.89	6.25	8.10	5.36	6.90
Xia et al. method 2016 [15]	6.70	4.75	11.75	3.34	6.64
Wavelet + LBP 2016 [16]	8.20	7.20	11.85	7.86	8.78
Jiang and Liu 2015 [17]	8.45	15.35	14.85	5.26	10.98
Johnson and Schuckers 2014 [18]	18.4	7.80	15.20	6.70	12.03
Nogueira et al. 2014 [14]	9.90	1.90	**5.09**	7.86	6.19
MSLBP 2014 [19]	7.30	2.50	14.80	5.30	7.48
WLD 2013 [5]	7.20	8.00	12.65	3.66	7.87
BSIF 2013 [6]	6.80	3.55	13.65	4.86	7.22
MBLTP 2013 [4]	10.00	6.90	16.30	5.90	9.77
LPQ 2012 [3]	10.40	8.00	13.20	5.30	9.23

Finally, we evaluate the performance of our proposal using recent metrics for PAD defined within the ISO/IEC FDIS 30107-3 [9]. The proportion of bona fide presentations incorrectly classified as presentation attacks for a fixed APCER of 10% (BPCER10) and 5% (BPCER20) are also reported. Figure 4 shows the Detection Error Trade-Off Curve (DET) as well as the BPCER10 and BPCER20 values for the four LivDet 2011 sensors.

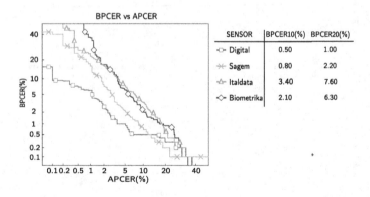

Fig. 4. DET curves, BPCER10 and BPCER20 values for the four sensors.

As it can be observed from Fig. 4, for smaller values of APCER (higher security thresholds), Digital sensor reports smallest values of BPCER. Specifically, it shows a BPCER value (BPCER10 = 0.5%), which is 7 times smaller than the one reported by Italdata sensor (BPCER10 = 3.4%). These results show that our proposed method is suitable for applications demanding very high security, in which presentation attacks may be frequently employed.

In the third experiment, we study the computational efficiency of the proposed algorithm for different parameters configurations. Table 2 shows the average performance of the proposal over different vocabulary sizes (K) at different scale values (σ). As it can be observed, different K and σ values have a slight impact in the average computational efficiency of the proposed method. Moreover, these efficiency results indicate that higher vocabulary sizes (K) and higher scale values (σ) worsen the computational efficiency of the proposal, at the same time, its accuracy is improved. On the other hand, an opposite effect occurs when K is close to 256. It should be noted that, in all cases, the efficiency value reported for each parameter combination is always below 810 ms; therefore, we can conclude that our proposal can be utilized in any real-time PAD system.

Table 2. Performance in milliseconds of our method for different configuration.

	$\sigma = 1$	$\sigma = 1.40$	$\sigma = 1.96$	$\sigma = 2.74$	$\sigma = 3.84$
K = 256	630.4	690.9	687.7	695.5	734.3
K = 512	642.3	657.9	672.9	697.7	732.1
K = 1024	673.3	680.7	700.3	722.1	753.3
K = 2048	677.5	706.2	721.4	729.0	758.2
K = 4096	692.3	707.7	713.8	753.5	805.9
K = 8192	700.6	742.6	752.6	774.3	807.8

4   Conclusions

In this paper, we proposed a new PAD method based on a BoW approach. The experimental evaluation conducted over the publicly available database LivDet 2011 assessed the performance of our proposal with respect to the top state-of-the-art methods. Experiments focused on parameter optimization showed that the scaling factor σ does not have a significant impact on the overall accuracy of the PAD, whilst the size of vocabulary (K) does. The proposal achieved the best result at $K = 8192$ and $\sigma = 1.4$, decrease the sate-of-the-art by 14%. This result shows the high capacity of the representation based on BoW in the task of PAD. Finally, the computational efficiency evaluation showed that our proposal is suitable for real-time applications, where the overall process took between 600 and 800 ms.

As future work, in order to improve the distinctiveness of the proposed descriptor, we plan to tackle some of the limitations of the histogram-based representation by using VLAD and Fisher Vector representations which are known to be much more powerful, in terms of representation, than the plain histogram.

Acknowledgement. This work was partly supported by the German Federal Ministry of Education and Research (BMBF) as well as by the Hessen State Ministry for Higher Education, Research and the Arts (HMWK) within the Center for Research in Security and Privacy (CRISP).

References

1. Marasco, E., Ross, A.: A survey on antispoofing schemes for fingerprint recognition systems. ACM Comput. Surv. (CSUR) **47**(2), 28 (2015)
2. Sousedik, C., Busch, C.: Presentation attack detection methods for fingerprint recognition systems: a survey. Iet Biometrics **3**(4), 219–233 (2014)
3. Ghiani, L., Marcialis, G.L., Roli, F.: Fingerprint liveness detection by local phase quantization. In: ICPR 2012, pp. 537–540. IEEE (2012)
4. Jia, X., Yang, X., Zang, Y., Zhang, N., Dai, R., Tian, J., Zhao, J.: Multi-scale block local ternary patterns for fingerprints vitality detection. In: ICB 2013, pp. 1–6. IEEE (2013)
5. Gragnaniello, D., Poggi, G., Sansone, C., Verdoliva, L.: Fingerprint liveness detection based on Weber local image descriptor. In: 2013 IEEE Workshop on BIOMS, pp. 46–50. IEEE (2013)
6. Ghiani, L., Hadid, A., Marcialis, G.L., Roli, F.: Fingerprint liveness detection using binarized statistical image features. In: BTAS, pp. 1–6. IEEE (2013)
7. Dubey, R.K., Goh, J., Thing, V.L.: Fingerprint liveness detection from single image using low-level features and shape analysis. IEEE Trans. Inf. Forensics Secur. **11**(7), 1461–1475 (2016)
8. Csurka, G., Dance, C.R., Fan, L., Willamowski, J., Bray, C.: Visual categorization with bags of keypoints. In: Workshop ECCV, pp. 1–22 (2004)
9. ISO: Information technology biometric presentation attack detection part 3: Testing and reporting, JTC 1/SC 37, Geneva, Switzerland ISO/IEC FDIS 30107–3:2017 (2017)
10. Bosch, A., Zisserman, A., Munoz, X.: Image classification using random forests and ferns. In: IEEE International Conference on Computer Vision (2007)
11. Vedaldi, A., Fulkerson, B.: VLFeat: an open and portable library of computer vision algorithms (2008). http://www.vlfeat.org/
12. Vedaldi, A., Zisserman, A.: Efficient additive kernels via explicit feature maps. Pattern Anal. Mach. Intellingence **34**(3), 480–492 (2011)
13. Yambay, D., Ghiani, L., Denti, P., Marcialis, G.L., Roli, F., Schuckers, S.: Livdet 2011-fingerprint liveness detection competition 2011. In: 2012 5th IAPR International Conference on Biometrics (ICB), pp. 208–215. IEEE (2012)
14. Nogueira, R.F., de Alencar Lotufo, R., Machado, R.C.: Evaluating software-based fingerprint liveness detection using convolutional networks and local binary patterns. In: 2014 IEEE Workshop on BIOMS Proceedings, pp. 22–29. IEEE (2014)
15. Xia, Z., Lv, R., Zhu, Y., Ji, P., Sun, H., Shi, Y.-Q.: Fingerprint liveness detection using gradient-based texture features. Signal Image Video Process. **11**(2), 381–388 (2017)

16. Xia, Z., Yuan, C., Sun, X., Sun, D., Lv, R.: Combining wavelet transform and LBP related features for fingerprint liveness detection. IAENG Int. J. Comput. Sci. **43**(3), 290–298 (2016)
17. Jiang, Y., Liu, X.: Spoof fingerprint detection based on co-occurrence matrix. Int. J. SERSC **8**(8), 373–384 (2015)
18. Schuckers, S., Johnson, P.: Fingerprint pore analysis for liveness detection. US Patent App. 14/243,420, 2 April 2014
19. Jia, X., Yang, X., Cao, K., Zang, Y., Zhang, N., Dai, R., Zhu, X., Tian, J.: Multi-scale local binary pattern with filters for spoof fingerprint detection. Inf. Sci. **268**, 91–102 (2014)

Multi-objective Overlapping Community Detection by Global and Local Approaches

Darian H. Grass-Boada[1]([✉]), Airel Pérez-Suárez[1], Andrés Gago-Alonso[1],
Rafael Bello[2], and Alejandro Rosete[3]

[1] Advanced Technologies Application Center (CENATAV), Havana, Cuba
{dgrass,asuarez,agago}@cenatav.co.cu
[2] Department of Computer Science,
Universidad Central "Marta Abreu" de Las Villas, Santa Clara, Cuba
rbellop@uclv.edu.cu
[3] Facultad de Ingeniería Informática, Universidad Tecnológica de la Habana
"José Antonio Echeverría" (CUJAE), Havana, Cuba
rosete@ceis.cujae.edu.cu

Abstract. Overlapping community detection on social networks has
received a lot of attention nowadays and it has been recently addressed
as Multi-objective Optimization Evolutionary Algorithms. In this paper,
we introduce a new algorithm, named MOGLAOC, which is based on
the Pareto-dominance based MOEAs and combines global and local
approaches for discovering overlapping communities. The experimental
evaluation over four classical real-life networks showed that our proposal
is promising and effective for overlapping community detection in social
networks.

Keywords: Social network analysis
Overlapping community detection
Multi-objective Optimization Evolutionary Algorithm

1 Introduction

The analysis of complex networks has been received a lot of attention nowadays
due to its applications on several contexts which include bioinformatics, sociology
and security, among others. Several data mining techniques have been applied
in order to extract knowledge from social networks, specifically, the detection of
communities plays an important role in the analysis of these networks [1]. This
technique aims to organise the nodes of a network in groups or communities
such that nodes belonging to the same community are densely interconnected
but spare connected with the remaining nodes in the network [2].

Taking into account the NP-hard nature of the community detection problem, most reported approaches use heuristics in order to search for a set of nodes
that optimises an objective function which captures the intuition of community
[2]. Consequently, most of community detection algorithms focused on solving

© Springer International Publishing AG, part of Springer Nature 2018
M. Mendoza and S. Velastín (Eds.): CIARP 2017, LNCS 10657, pp. 272–280, 2018.
https://doi.org/10.1007/978-3-319-75193-1_33

a single-objective optimization problem; however, these approaches face some difficulties: (1) the optimization of only one function confines the solution to a particular community structure, (2) many of them may fail due to the *resolution limit problem* [3], and (3) returning one single partition may not be suitable when the network has many potential structures. To overcome the aforementioned problems, many community detection algorithms model the problem as a Multi-objective optimization problem, and specifically, they used Multi-objective Optimization Evolutionary Algorithms (MOEAs) to solve it.

Unfortunately, most of reported MOEAs focused on discovering disjoint communities, although according to Palla et al. in [4], most real-world networks have overlapping community structure. To the best of our knowledge, only the MOEAs proposed in [1,5–8] addressed the overlapping community detection problem.

The MEA_CDPs algorithm [1] uses an undirected representation of the solution and the classical NSGA-II framework with the reverse operator, in order to search for the solutions optimising three objective functions. On the other hand, iMEA_CDPs [6] uses the same representation as MEA_CDPs but it uses other objective functions. Besides, iMEA_CDPs proposes to employ the NSGA-II or the MOEA/D as the optimization framework, together with the PMX and simple mutation operators. IMOQPSO [7] uses a center-based representation of the solution together with a combination of QPSO and HSA optimization frameworks, in order to find a set of nodes that optimises two previously defined objective functions. OMO [5] employs a representation based on adjacencies between edges of the network together with two objective functions and the NSGA-II framework. Finally, MCMOEA [8] detects first the set of maximal cliques of the network and then, it builds the maximal-clique graph. Starting from this transformation, MCMOEA uses a representation based on labels and the MOEA/D framework in order to detect the communities optimizing two objective functions.

In this paper, we propose a new algorithm based on Pareto-dominance based MOEAs [12] which combines global and local approaches for discovering overlapping communities (MOGLAOC). Our proposal starts by detecting a set of seeds which are then used to build the overlapping communities. With this aim we introduced in the classical Pareto-dominance based MOEAs framework an *expansion, improving* and *merging* steps which allow our proposal to detect overlapping zones in the network, to improve the overlapping quality of these zones, and to merge communities having a high overlapping. The experimental evaluation of our proposal over four classical real-world social networks showed that it is promising and effective for overlapping community detection.

The remainder of this paper is organized as follow: in Sect. 2, we introduce the MOGLAOC algorithm. The experimental evaluation, showing the performance of our proposed algorithm, over four real-life networks is presented in Sect. 3. Finally, conclusions and future work are presented in Sect. 4.

2   The MOGLAOC Algorithm

Let $G = \langle V, E \rangle$ be a given network, where V is the set of vertices and E the set of edges among the vertices. A multi-objective community detection problem aims to search for a partition P^* of G such that:

$$F(P^*) = min_{P \in \Omega} \left(f_1(P), f_2(P), \ldots, f_r(P) \right), \tag{1}$$

where P is a partition of G, Ω is the set of feasible partitions, r is the number of objective functions, f_i is the ith objective function and $min(\cdot)$ is the minimum value. With the introduction of the multiple objective functions there is usually no absolute optimal solution, thus, the goal is to find a set of *Pareto* optimal solutions [2].

A commonly used way to solve a multi-objective community detection problem is by using MOEAs. The general Pareto-dominance based MOEAs framework consists on the following four steps: (a) to generate an initial population of chromosomes (i.e., solutions to the problem at hand), taking into account a predefined representation, (b) to apply the evolutionary operators over the current population in order to build the next generation and to move through the solution space, (c) to evaluate the current and new populations by using a predefined set of objective functions, and (d) to apply a predefined heuristic for keeping and improving the best solutions found so far. Usually, steps b, c and d are repeated a predefined number of times or until a specific stop criterium is fulfilled.

The contributions our proposal introduces to the Pareto-dominance based MOEAs framework for overlapping community detection are focused on the inclusion of an *expansion*, *improving* and *merging* steps, which in turn are inserted between the above mentioned steps c and d. Following, in Sect. 2.1 we briefly describe how our proposal addresses the general steps of the MOEAs framework and then, Sect. 2.2 describes in details the *expansion*, *improving* and *merging* steps our proposal introduces.

2.1   General Steps of MOGLAOC

The main idea of our proposal is to use the steps of classic Pareto-dominance based MOEAs framework in order to detect a set of disjoint seed clusters, where each seed cluster represents the set of objects that a community should not share with any other community.

In order to detect these seed clusters we use the PESA-II [9] as the optimization mechanism. Taking this into account, we use the locus-based adjacency graph encoding [10] for representing each chromosome of the initial population generated at step a; the decoding of a chromosome requires the identification of all connected components, which in turn will be our seed clusters. For generating a chromosome the ith genotype composing the chromosome is built by randomly selecting a neighbour of node i. The uniform two-point crossover operator [10] is selected for crossover and for mutation, some genes are randomly selected and substituted by other randomly selected adjacent nodes [11].

For evaluating the quality of a set of seeds clusters S we employ the *intra* and *inter* objective functions proposed in [2], which measure the intra-link and inter-link strength of S, respectively. These functions are defined as follows:

$$Intra(S) = 1 - \sum_{S_i \in S} \frac{|E(S_i)|}{m} \qquad Inter(S) = \sum_{S_i \in S} \left(\frac{\sum_{v \in S_i} |N(v)|}{2 \cdot m} \right)^2, \tag{2}$$

where $E(S_i)$ is the number of edges inside seed S_i, m is the total number of edges in the network, and $N(v)$ is the set of adjacent vertices of vertex v. In order to address the step (d) we use the mechanism PESA-II includes for keeping and improving the best seed clusters found so far.

2.2   Expansion, Improving and Merging Steps

Let $S = \{S_1, S_2, \ldots, S_k\}$ be the set of disjoint seed clusters represented by a chromosome of the current population. Overlapping vertices are supposed to be those vertices that belong to more than one community and in order to be correctly located inside a community they need to have edges with vertices in those communities. The *expansion* step aims to detect these vertices through a greedy randomise local search procedure (GRASP) over each S_i.

For detecting the zones containing overlapping vertices we soften the initial criterium used for building the seeds. With this aim each seed cluster S_i is processed for determining which vertices outside the community share a significant number of their adjacent vertices with the community; that is, the *potential* overlapping vertices.

Let $S_i \in S$ be a seed cluster and $\partial S_i \subseteq S_i$ the set of vertices of S_i having neighbours outside S_i. The strength of ∂S_i is denoted as $Str(\partial S_i)$ and it is computed as the ratio between the number of edges the vertices of ∂S_i have with vertices inside S_i, and the number of edges the vertices of ∂S_i have with vertices inside and outside S_i. The greater the value of $Str(\partial S_i)$ the greater the number of inner edges ∂S_i has and consequently, the better ∂S_i is. A vertex $u \notin S_i$ is considered a *candidate* to be included in S_i iff u is adjacent to at least one vertex in ∂S_i and $Str(\partial S_i') - Str(\partial S_i) > 0$, where $S_i' = S_i \cup \{v\}$.

In order to iteratively expand a seed cluster S_i, the following steps are performed: (1) determining the set L of vertices which are *candidate* to be included in S_i, (2) applying the roulette wheel selection method over the set L, where the probability of being selected that a vertex $v \in L$ has is computed by using the increase v produces in $Str(\partial S_i)$, and (3) repeating steps 1 and 2 while $L \neq \emptyset$.

Once the *expansion* step finished, the *improving* step is performed in order to locally improve each overlapping zone detected. From our point of view, any overlapping vertex is expected to have adjacent vertices belonging to different communities and possibly, belonging to different overlapping zones. In this paper, we propose to measure the overlapping quality of a vertex belonging to an overlapping zone by using two properties we call *uniformity* and *simple betweenness*.

Let Z be an overlapping zone detected and $C_Z = \{C_1, C_2, \ldots, C_m\}$ the set of communities that set up Z. Let $v \in Z$ be an overlapping vertex. Let $N_{C_Z}(v)$ be the set of adjacent vertices of v that belong to at least one community in C_Z. Let $G_v = \{G_v^1, G_v^2, \ldots, G_v^l\}$ be the set of communities or overlapping zones containing the vertices in $N_{C_Z}(v)$. A property we will expect v satisfies is to have the vertices in $N_{C_Z}(v)$ equally distributed over the groups of G_v. The *uniformity* of v, denoted as $U(v)$, measures how much the distribution of vertices in $N_{C_Z}(v)$ deviates from the expected distribution of $N_{C_Z}(v)$ and it is computed as follows:

$$U(v) = 1 - \sum_{G_v^i \in G_v} abs \left(\frac{\left| N_{C_Z}(v) \cap G_v^i \right|}{|N_{C_Z}(v)|} - \frac{1}{|G_v|} \right), \tag{3}$$

where $abs(\cdot)$ is the absolute value. $U(v)$ takes values in $[0, 1]$ and the higher its value the better well-balanced v is.

Let $N'_{C_Z}(v)$ be the set of adjacent vertices of v that belong to at most one community in C_Z. Another property we would expect an overlapping vertex $v \in Z$ to have is to be an *intermediary* between any pair of its adjacent vertices in $N'_{C_Z}(v)$; that is, the shortest path connecting any pair of vertices $u, w \in N'_{C_Z}(v)$ should be the path made of the undirected edges (u, v) and (v, w). The *simple betweenness* of v, denoted as $SB(v)$, measures how much intermediary v is and it is computed as follows:

$$SB(v) = \frac{2 \cdot \sum_{i=1}^{|C_Z|-1} \sum_{j>i}^{|C_Z|} \left(1 - \frac{|E(C_i, C_j)|}{\left| N'_{C_Z}(v) \cap C_i \right| \cdot \left| N'_{C_Z}(v) \cap C_j \right|} \right)}{|C_Z| \cdot (|C_Z| - 1)} \tag{4}$$

where $E(C_i, C_j) = \left\{ (u, w) \in E \mid u, w \in N'_{C_Z}(v) \wedge u \in C_i \wedge w \in C_j \right\}$ is the set of edges between vertices in $N'_{C_Z}(v)$, with one vertex in C_i and the other one in C_j. $SB(v)$ takes values in $[0, 1]$ and the higher its value the best intermediary v is.

We would like to highlight that both the uniformity and simple betweenness concepts can be straightforward generalised in order to be applied to an overlapping zone. Let $U_{ave}(Z)$ be the initial average uniformity of the vertices belonging to an overlapping zone Z. A vertex $v \in Z$ is a candidate to be removed from Z iff $U(v) < U_{ave}(Z)$. On the other hand, a vertex $u \in N(v|C_Z), v \in Z$, is a candidate to be added to Z iff $U(u) > U_{ave}(Z)$. Any addition or removal of a candidate vertex from Z that transforms Z into Z' is considered as *viable* iff $(U(Z') + SB(Z')) - (U(Z) + SB(Z)) > 0$.

Let $O = \{Z_1, Z_2, \ldots, Z_j\}$ be the set of all the overlapping zones detected after the expansion step. The heuristic proposed for improving these zones is as follows: (1) computing the initial average uniformity of each zone $Z_i \in O$, (2) detecting the set T of viable transformations to apply over O, (3) selecting and performing the transformation $t \in T$ which produces the higher improvement in its overlapping zone, and (4) to repeat steps 2 and 3 while $T \neq \emptyset$.

Let $C = \{C_1, C_2, \ldots, C_k\}$ be the set of communities detected after the improving step. Although it is allowable for communities to overlap, what is most important for each community is to have a subset of vertices that makes the community different from the remaining ones. The *merging* step aims to reduce the redundancy in the detected communities, by iteratively merging those communities having a high overlapping.

Let $C_i \in C$ a community. The *distinctiveness* of C_i, denoted as D_{C_i}, is computed as the difference between the number of edges of C_i composed of vertices belonging only to C_i, and the number of edges of C_i composed of at least one vertex C_i shares with another community. Let C_i and C_j be two communities which overlap each other. C_i and C_j are candidates to be merged iff $D_{C_i} \leq 0$ or $D_{C_j} \leq 0$.

The heuristics for merging communities having a high overlapping degree is as follows: (1) detecting the set PC of pairs of communities which are candidate to be merged, (2) applying the roulette wheel selection method over the set PC, where the probability of selection of each pair is computed by using the highest absolute value of the distinctiveness of the two communities forming the pair, and (3) repeating steps 1 and 2 while $PC \neq \emptyset$. The set of overlapping communities remaining after this step is evaluated by using the intra and inter objective functions described in Sect. 2.1, in order to keep a set of non-dominated solutions.

3 Experimental Results

In this section, the results of several experiments testing the MOGLAOC algorithm are presented. The experiments were focused on: (1) to compare the accuracy attained by MOGLAOC against the one attained by MEA_CDP [1], IMO-QPSO [7], iMEA_CDP [6] and OMO [5] algorithms, and (2) to evaluate the number of communities as well as the overlapping degree of these communities, for the best solutions found by our proposal for each network.

In our experiments we use four real-life networks: the American College Football network, the Zachary's Karate Club network, the Bottlenose Dolphins network, and the Krebs' books on American politics network; these networks can be downloaded from http://konect.uni-koblenz.de/networks. Table 1 shows the characteristics of these networks.

Table 1. Overview of the networks used in our experiments

Networks	# of nodes	# of edges	Ave. degree	# communities
American Cool. Football	115	613	10.66	12
Zachary's Karate Club	34	78	4.58	2
Bottlenose Dolphins	62	159	5.129	2
Krebs' books	105	441	8.4	3

In the first experiment we used the NMI external evaluation measure [6] for computing the accuracy of each algorithm. NMI takes values in $[0, 1]$ and it evaluates a set of communities based on how much these communities resemble a set of communities manually labelled by experts, where 1 means identical results and 0 completely different results. For each network, we executed MOGLAOC thirty times and computing the NMI value attained by the best solution of the Pareto front. Table 2 showed the average NMI value attained by each algorithm over the networks; the average values for MEA_CDP, IMOQPSO, iMEA_CDP and OMO algorithms were taken from their original articles. The "X" in Table 2 means that IMOQPSO does not report any results on the Krebs' books network.

Table 2. Best NMI average values attained by each algorithm. Highest values appears bold-faced

Networks	MEA_CDP	IMOQPSO	iMEA_CDP	OMO	MOGLAOC
American Cool. Football	0.495	0.462	0.593	0.33	**0.65**
Zachary's Karate Club	0.52	**0.818**	0.629	0.375	0.75
Bottlenose Dolphins	0.549	**0.886**	0.595	0.41	0.69
Krebs' books	0.469	X	**0.549**	0.39	0.449
Ave. ranking position	3.25	2.75	2.25	4.75	2.0

As it can be seen from Table 2, MOGLAOC attains comparable results with those of the related algorithms. Our proposal clearly outperforms the other algorithms in the American Cool. Football network which is a difficult network, while it is second in Zachary's Karate Club and Bottlenose Dolphins networks, outperformed only by the IMOQPSO, which in turn it includes some operations that make it a highly computational expensive algorithm. In the last row of Table 2 we also showed the average ranking position attained by each algorithm and as it can be observed, our proposal attains the best results. From the above experiments on real-world networks, we can say that MOGLAOC is promising and effective for overlapping community detection in complex networks.

In the second experiment, we compute the average number of communities MOGLAOC detects when it attains its highest NMI value, as well as the overlapping among these communities. The results of this experiment are showed in Table 3.

Table 3. Average number of communities and overlapping degree for the best solutions found by MOGLAOC

Networks	Ave # communities	Ave. overlapping degree
American Cool. Football	9.7	1.1
Zachary's Karate Club	2	1.108
Bottlenose Dolphins	2	1.2
Krebs' books	3.8	1.33

As it can be seen from Table 3, our proposal detects a number of communities which, in the average case, is close to the real number of communities existing in the networks. Moreover, our proposal do not produce solutions having high overlapping degree which means it is able to detect overlapping zones but it does not profit from this overlapping for boosting its accuracy.

4   Conclusions

In this paper, we proposed a new algorithm, named MOGLAOC, for discovering overlapping communities through a combination of global and local optimization approaches. MOGLAOC introduced three steps to the classical Pareto-dominance based MOEAs framework which allow it to detect overlapping zones, to optimise the quality of these zones and to reduce redundancy in the solutions. Unlike previously reported algorithms, our proposal defined two properties that should satisfy any vertex belonging to an overlapping zone.

The MOGLAOC algorithm was evaluated over four real-life networks in terms of its accuracy and it was compared against four algorithms of the related work. The experimental evaluation showed our proposal attains comparable results to that of the evaluated state-of-the-art algorithms. Moreover, this evaluation showed that MOGLAOC is promising and effective for overlapping community detection in complex networks. Another conclusion is that MOGLAOC detects a number of communities which is close to that existing in the networks. Besides, our proposal produces communities having low overlapping degree.

As future work, we would like to further evaluate the MOGLAOC algorithm over synthetical networks in order to have a better insight about its behaviour under different conditions.

References

1. Liu, J., Zhong, W., Abbass, H., Green, D.G.: Separated and overlapping community detection in complex networks using multiobjective evolutionary algorithms. In: IEEE Congress on Evolutionary Computation (CEC) (2010)
2. Shi, C., Yan, Z., Cai, Y., Wu, B.: Multi-objective community detection in complex networks. Appl. Soft Comput. **12**(2), 850–859 (2012)
3. Fortunato, S., Barthelemy, M.: Resolution limit in community detection. Proc. Nat. Acad. Sci. **104**(1), 36–41 (2007)
4. Palla, G., Derényi, I., Farkas, I., Vicsek, T.: Uncovering the overlapping community structure of complex networks in nature and society. Nature **435**, 814–818 (2005)
5. Liu, B., Wang, C., Wang, C., Yuan, Y.: A new algorithm for overlapping community. In: Proceeding of the 2015 IEEE International Conference on Information and Automation Detection, pp. 813–816 (2015)
6. Liu, C., Liu, J., Jiang, Z.: An improved multi-objective evolutionary algorithm for simultaneously detecting separated and overlapping communities. Int. J. Nat. Comput. **15**(4), 635–651 (2016)
7. Li, Y., Wang, Y., Chen, J., Jiao, L., Shang, R.: Overlapping community detection through an improved multi-objective quantum-behaved particle swarm optimization. J. Heuristics **21**(4), 549–575 (2015)
8. Wen, X., Chen, W.N., Lin, Y., Gu, T., Zhang, H., Li, Y., Yin, Y., Zhang, J.: A maximal clique based multiobjective evolutionary algorithm for overlapping community detection. IEEE Trans. Evol. Comput. **21**(3), 363–377 (2016)
9. Corne, D., Jerram, N., Knowles, J., Oates, M.: PESA-II: region-based selection in evolutionary multi-objective optimization. In: Proceedings of the 3rd Annual Conference on Genetic and Evolutionary Computation (GECCO 2001), San Francisco, CA, pp. 283–290 (2001)

10. Mukhopadhyay, A., Maulik, U., Bandyopadhyay, S.: A survey of multiobjective evolutionary clustering. ACM Comput. Surv. **47**(4), 61:1–61:46 (2015)
11. Pizzuti, C.: A multiobjective genetic algorithm to find communities in complex networks. IEEE Trans. Evol. Comput. **16**(3), 418–430 (2012)
12. Zhou, A., Qu, B.Y., Li, H., Zhao, S.Z., Suganthan, P.N., Zhang, Q.: Multiobjective evolutionary algorithms: a survey of the state of the art. Swarm Evol. Comput. **1**(1), 32–49 (2011)

Impulse Response Estimation of Linear Time-Invariant Systems Using Convolved Gaussian Processes and Laguerre Functions

Cristian Guarnizo[1(✉)] and Mauricio A. Álvarez[2]

[1] Engineering PhD Program, Universidad Tecnológica de Pereira, Pereira, Colombia
cdguarnizo@utp.edu.co
[2] Department of Computer Science, The University of Sheffield, Sheffield, UK
mauricio.alvarez@sheffield.ac.uk

Abstract. This paper presents a novel method to estimate the impulse response function of Linear Time-Invariant systems from input-output data by means of Laguerre functions and Convolved Gaussian Processes. We define a new non-stationary covariance function that encodes the convolution between the Laguerre functions and the input. The input (excitation) is modelled by a Gaussian Process prior. Thus, we are able to estimate the system's impulse response by performing maximum likelihood estimation over the model hyperparameters. Besides, the proposed model performs well in missing and noisy data scenarios.

Keywords: Convolved Gaussian Process · Impulse response function
Laguerre function

1 Introduction

Ordinary differential equations (ODEs) are used to describe Linear Time-Invariant (LTI) systems in many fields, such as, Biology, Engineering, Economics, among others [3]. Generally, an ODE can be represented as

$$b_p \frac{\mathrm{d}^p f(t)}{\mathrm{d}t^p} + \ldots + b_1 \frac{\mathrm{d}f(t)}{\mathrm{d}t} + b_0 f(t) = u(t), \tag{1}$$

where p is the order of the differential equation, b_i (with $i = \{0, 1, \ldots, p\}$) is a coefficient that weights the i-th derivative of $f(t)$ w.r.t. t, $u(t)$ is the forcing function and $f(t)$ is the solution function. Equation (1) can be rewritten as the following convolution,

$$f(t) = \int_0^t h(\tau) u(t - \tau) \mathrm{d}\tau, \tag{2}$$

where $h(t)$ is function of all the b_i's and is also known as the impulse response function (IRF).

In [2] is introduced a Gaussian Process (GP) model where the covariance function of the output is build from the convolution of a smoothing kernel and the

© Springer International Publishing AG, part of Springer Nature 2018
M. Mendoza and S. Velastín (Eds.): CIARP 2017, LNCS 10657, pp. 281–288, 2018.
https://doi.org/10.1007/978-3-319-75193-1_34

covariance of a white noise process. Later, this model is extended in Latent Force models (LFMs) [1] by replacing the smoothing kernel with the system's IRF. Thus, the IRF is encoded within the covariance function in the LFM framework. Besides, LFMs are an interesting tool given that we are able to make system identification by learning the model from the available output data.

In many system identification applications, the order the ODE and its parametrization is unknown. This problem have been addressed by approximating the IRF using Orthogonal Basis Functions (OBFs) [8,10]. Laguerre functions are a type of OBFs that are characterized by having only one pole or parameter that controls the waveform of the basis functions. Furthermore, Laguerre functions have been used in system identification tasks, as in [5,12].

In this paper, we propose to estimate the IRF using the Laguerre functions encoded in the covariance function of a Convolved Gaussian Process (CGP). Laguerre parameters are learned during the maximization of the CGP marginal likelihood function.

This paper is organized as follows. In Sect. 2, a brief overview of CGP theory is given, with the mathematical construction of all covariance functions required to perform system identification. Then, in Sect. 3 the Laguerre functions are introduced. Using the theory developed in previous section, the proposed model is presented in Sect. 4. Some results showing the ability of the model to estimate the IRF under different scenarios are explored in Sect. 5. Finally, a discussion about similar approaches is given in Sect. 6.

2   Convolved Gaussian Processes

In this section an overview of the mathematical foundation for CGPs is given. We start by considering a multi-output multi-input (MIMO) dynamical system with D outputs and Q inputs. In this system, the d-th output is described by

$$y_d(t) = f_d(t) + \epsilon_d(t), \tag{3}$$

with $\epsilon_d(t) \sim \mathcal{N}(0, \sigma_d^2)$ and

$$f_d(t) = \int_0^t h_d(t - \tau) \sum_{q=1}^Q u_q(\tau) \mathrm{d}\tau, \tag{4}$$

where $h_d(t)$ is a smoothing kernel which corresponds to the system's IRF [1]. The set of inputs $\{u_q(z)\}_{q=1}^Q$ are independent GPs, and each one is defined as

$$u_q \sim \mathcal{GP}(0, k_{u_q, u_q}(z, z')). \tag{5}$$

where $k_{u_q, u_q}(t, t')$ is the covariance function used to model the input functions. As in [1] we adopted the square exponential covariance function defined as follows

$$k_{u_q, u_q}(z, z') = \exp\left(-\frac{(z - z')^2}{l_q^2}\right),$$

where l_q is the lengthscale associated to the q-th input. From Eqs. (3), (4) and (5), we are able to model the set of outputs $\{y_d(t)\}_{d=1}^{D}$ as a GP with zero mean and covariance function given by

$$k_{y_d,y_{d'}}(t,t') = k_{f_d,f_{d'}}(t,t') + \sigma_d^2 \delta_{d,d'} \delta_{t,t'},$$

with

$$k_{f_d,f_{d'}}(t,t') = \sum_{q=1}^{Q} k_{f_d,f_{d'}}^{(q)}(t,t'),$$

where $k_{f_d,f_{d'}}^{(q)}(t,t')$ is defined as

$$\int_0^t h_d(t-\tau) \int_0^{t'} h_{d'}(t'-\tau') k_{u_q,u_q}(\tau,\tau') d\tau' d\tau.$$

Additionally, we require to define the cross-covariance between outputs and inputs which is defined as

$$k_{f_d,u_q}(t,t') = \int_0^t h_d(t-\tau) k_{u_q,u_q}(\tau,t') d\tau.$$

Covariance function $k_{f_d,u_q}(t,t')$ allows us to make predictions from the output domain to the input domain, and vice versa. This property is exploited in the proposed model defined in Sect. 4.

3   Laguerre Functions

Laguerre functions form a complete and orthonormal set of basis, where each function is defined as [5]

$$l_m(t) = \mathcal{L}_m(t) \exp(-\gamma t),$$

where γ is a free parameter known as Laguerre scale [4], and

$$\mathcal{L}_m(t) = \sqrt{2\gamma} \sum_{k=0}^{m} \frac{(-1)^k m! 2^{m-k}}{k![(mk)!]^2} (2\gamma t)^{m-k}.$$

The m-th degree polynomial $\mathcal{L}_m(t)$ is called the m-th Laguerre polynomial. It is also interesting to note that the Laguerre function $l_m(t)$ has m zero crossings defined by the zeros of $\mathcal{L}_m(t)$. Additionally, each Laguerre function represents a dynamical system characterized by one pole with multiplicity, e.g. the Laplace transform of the m-th Laguerre function is given by

$$L_m(s) = \sqrt{2\gamma} \frac{(\gamma-s)^m}{(\gamma+s)^{m+1}},$$

where the Laguerre scale parameter γ controls the position of the pole and the zeros of each Laguerre function.

4   Proposed Method

In LFMs, the IRF is assumed to be known before hand, this allows to the CGP to predict the input values solely from the output data. Here, we are interested in approximating the IRF by means of Laguerre functions. In order to do so, we require to make use of input-output data. Thus, the IRF of the d-th output is approximated by the Laguerre functions as

$$\hat{h}_d(t) = \sum_{m=0}^{M} c_{d,m} l_{d,m}(t), \tag{6}$$

where $c_{d,m}$ weights the m-th Laguerre function of the d-th output. Then (4) becomes

$$f_d(t) = \sum_{m=0}^{M} c_{d,m} \int_0^t l_m(t-\tau) \sum_{q=1}^{Q} u_q(\tau) d\tau. \tag{7}$$

Next, we proceed to define the covariance functions for the new model described in (7). Thus, the covariance function $k_{f_d,f_{d'}}^{(q)}(t,t')$ becomes

$$\int_0^t \hat{h}_d(t-\tau) \int_0^{t'} \hat{h}_{d'}(t'-\tau') k_{u_q,u_q}(\tau,\tau') d\tau' d\tau,$$

and $k_{f_d,u_q}(t,t')$ is rewritten as

$$\sum_{m=0}^{M} c_{d,m} \int_0^t l_{d,m}(t-\tau) k_{u_q,u_q}(\tau,t') d\tau.$$

Note that, if $m > 0$, then the convolutions for the above covariance functions have no closed form. Consequently, we approximate the convolutions by using discrete sums as in [6]. This increases the computation time for the evaluation of the covariance function, but it also allows the model to use any covariance function for the inputs.

4.1   Hyperparameter Learning

Let us assume we are given observations of Q-inputs and D-outputs in vectors $\{u_q\}_{q=1}^{Q}$ and $\{y_d\}_{d=1}^{D}$, respectively. Thus we condition the proposed GP model on this finite set of observations, the model becomes into the following multivariate normal distribution

$$\mathbf{g} = \begin{bmatrix} \mathbf{y} \\ \mathbf{u} \end{bmatrix} \sim \mathcal{N}(\mathbf{0}, \mathbf{K_g}),$$

where $\mathbf{g}$ is obtained by stacking in one column the vectors $\mathbf{y}$ and $\mathbf{u}$. Furthermore, the covariance function for $\mathbf{g}$ is defined as follows,

$$\mathbf{K_g} = \begin{bmatrix} \mathbf{K}_{y,y} & \mathbf{K}_{f,u} \\ \mathbf{K}_{u,f} & \mathbf{K}_{u,u} \end{bmatrix}, \tag{8}$$

where each $\mathbf{K}_{i,j}$ (i or j can be either y, f or u) is obtained by evaluating the covariance function $k_{i,j}(t, t')$ at the time inputs $\mathbf{t}$ associated to the observations. From the above definitions, the set of parameters $\theta = \{l_q, \sigma_d^2, \alpha_d, c_{d,m}\}$ (where m indexes the Laguerre functions as described in (6)) are learned by maximizing the logarithm of the marginal likelihood, which is given by

$$\log p(\mathbf{g}|\mathbf{t}) = -\frac{1}{2}\mathbf{g}^\top \mathbf{K}_\mathbf{g}^{-1}\mathbf{g} - \frac{1}{2}\log|\mathbf{K}_\mathbf{g}| - \frac{N}{2}\log(2\pi)$$

where N is the total number of data points [7].

5   Results

In this section, two different numerical problems are given to illustrate the properties of the proposed model.

5.1   Approximation of the Impulse Response

In this experiment we show the ability of the proposed model to recover the impulse response function by means of the Laguerre functions that are encoded within the covariance function. We first generate 50 data points equally spaced along the range $[0, 4]$ s, by sampling $\mathbf{u}$ from the GP defined in (5) with $lq = 0.8$. Then, the output data $\mathbf{f}$ is obtained by applying $\mathbf{u}$ through a second order dynamical system characterized by the following IRF

$$h(t) = \frac{1}{\omega}\exp\left(-\frac{b_1 t}{2}\right)\sinh(\omega t),$$

with $b_0 = 1$, $b_1 = 4$ and $\omega = \sqrt{b_1^2 - 4b_0}/2$. By stacking data vectors $\mathbf{f}$ and $\mathbf{u}$ we proceed to learn a model with $M = 10$ Laguerre basis by using the procedure described in Sect. 4.1. The model is learned 10 times with different initializations for the parameters θ. In Fig. 1 is shown the mean and two standard deviations calculated from the 10 IRFs learned by the proposed model. The mean function fits well the true IRF, but the standard deviation indicates that some of the obtained approximations tend to be less accurate at the end of the available data. This effect is produced when using high order Laguerre functions to fit smooth IRFs. As discussed in [5], this problem can be addressed by forcing the higher order weights $c_{d,m}$ to be zero. In other words, overfitting might be reduced by setting to zero the coefficients of Laguerre functions related to high frequency components.

5.2   Prediction of Missing Input/Output Values

For this experiment, we are interested in predicting missing values of inputs and outputs. First, we configured a MIMO system with 2 inputs and 2 outputs described by the following equations

$$\frac{d^4 f_1(t)}{dt^4} + 4\frac{d^3 f_1(t)}{dt^3} + 9\frac{d^2 f_1(t)}{dt^2} + 14\frac{df_1(t)}{dt} + 8f_1(t) = \bar{u}(t),$$

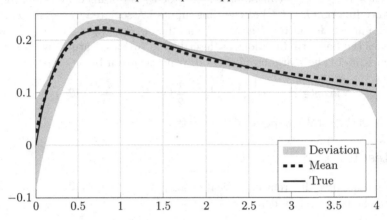

Fig. 1. Mean and two standard deviations estimated from the impulse responses learned at example Sect. 5.1.

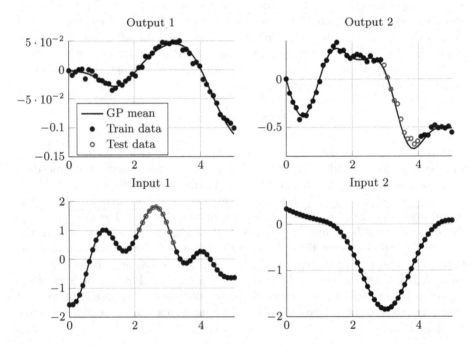

Fig. 2. Prediction of missing data (Test data) for the MIMO system described in Sect. 5.2

$$\frac{\mathrm{d}f_2(t)}{\mathrm{d}t} + f_2(t) = \bar{u}(t),$$

with $\bar{u}(t) = \sum_{q=1}^{2} u_q(t)$. Then, 50 data points are generated for inputs and outputs variables (we follow similar steps as the ones described in Sect. 5.1).

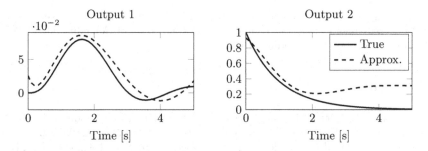

Fig. 3. Impulse response approximation for the MIMO system described in Sect. 5.2

Additionally, output data is corrupted by an additive white noise process as shown in (3). In order to show the ability of the proposed model to deal with missing data, the dataset is divided into 2 subsets (training and testing) as shown in Fig. 2.

Output 2 missing data is predicted by the data from output 1 and the inputs, while input 1 missing data is predicted only from the outputs (regarding the covariance function defined in (8)). In Fig. 3 is shown the IRFs estimated for both outputs. The estimated IRF for output 1 is close enough, but the estimation of the IRF for output 2 is clearly affected by the missing data. Nevertheless, predictions for the testing data are well fitted.

6   Discussion and Future Work

Our experiments demonstrated that the proposed model is able to estimate the continuous-time IRF of LTI systems in the presence of noise corruption and missing input and output data. The IRF can be approximated in a non-parametric manner by placing a GP prior over this function. In [11] the Gaussian Process Convolution Model (GPCM) is introduced. The GPCM is described as a continuous-time non-parametric window moving average process. One advantage of this model is that it can be considered in the frequency domain as well. Another non-parametric approximation is proposed in [9], where a specific covariance matrix is built by using a stable spline kernel.

The above mentioned methods require to approximate the posterior distribution of the IRF by using special algorithms, i.e. the GPCM requires a variational inference approach, meanwhile the work in [9] an Expectation-Maximization algorithm is adopted. In contrast, for the model proposed here, the IRF is estimated using a set of orthogonal basis (Laguerre functions) which included a dynamical representation. Additionally, the proposed model is learned using the standard GP training method [7].

This model can be easily extended by using another set of orthonormal basis. For example, selecting a set of basis which allows to find a closed form for the convolutions is recommended.

Acknowledgements. Authors would like to thank Convocatoria 567 from the Administrative Department of Science, Technology and Innovation of Colombia (COLCIENCIAS) for the support and funding of this work.

References

1. Álvarez, M.A., Luengo, D., Lawrence, N.D.: Linear latent force models using Gaussian processes. IEEE Trans. Pattern Anal. Mach. Intell. **35**(11), 2693–2705 (2013)
2. Boyle, P., Frean, M.: Dependent Gaussian processes. In: Saul, L.K., Weiss, Y., Bottou, L. (eds.) Advances in Neural Information Processing Systems, vol. 17, pp. 217–224. MIT Press (2005). http://papers.nips.cc/paper/2561-dependent-gaussian-processes.pdf
3. Dass, S.C., Lee, J., Lee, K., Park, J.: Laplace based approximate posterior inference for differential equation models. Stat. Comput. **27**(3), 679–698 (2017). https://doi.org/10.1007/s11222-016-9647-0
4. Haber, R., Keviczky, L.: Nonlinear System Identification - Input-Output Modeling Approach. Kluwer Academic, Dordrecht (1999)
5. Israelsen, B.W., Smith, D.A.: Generalized laguerre reduction of the volterra kernel for practical identification of nonlinear dynamic systems. CoRR abs/1410.0741 (2014). http://arxiv.org/abs/1410.0741
6. Lawrence, N.D., Sanguinetti, G., Rattray, M.: Modelling transcriptional regulation using Gaussian processes. In: Schölkopf, B., Platt, J.C., Hoffman, T. (eds.) Advances in Neural Information Processing Systems, vol. 19, pp. 785–792. MIT Press (2007). http://papers.nips.cc/paper/3119-modelling-transcriptional-regulation-using-gaussian-processes.pdf
7. Rasmussen, C., Williams, C.: Gaussian Processes for Machine Learning. Adaptive Computation and Machine Learning. MIT Press, Cambridge (2006)
8. Reginato, B.C., Oliveira, G.H.C.: On selecting the MIMO generalized orthonormal basis functions poles by using particle swarm optimization. In: 2007 European Control Conference (ECC), pp. 5182–5188, July 2007
9. Risuleo, R.S., Bottegal, G., Hjalmarsson, H.: Kernel-based system identification from noisy and incomplete input-output data. In: 2016 IEEE 55th Conference on Decision and Control (CDC), pp. 2061–2066, December 2016
10. Stanislawski, R., Hunek, W.P., Latawiec, K.J.: Modeling of nonlinear block-oriented systems using orthonormal basis and radial basis functions. In: International Conference on Systems Engineering, pp. 55–58 (2008)
11. Tobar, F., Bui, T.D., Turner, R.E.: Learning stationary time series using Gaussian processes with nonparametric kernels. In: Cortes, C., Lawrence, N.D., Lee, D.D., Sugiyama, M., Garnett, R. (eds.) Advances in Neural Information Processing Systems, vol. 28, pp. 3501–3509. Curran Associates, Inc. (2015)
12. Wahlberg, B.: System identification using Laguerre models. IEEE Trans. Autom. Control **36**(5), 551–562 (1991)

A Novel Hybrid Data Reduction Strategy and Its Application to Intrusion Detection

Vitali Herrera-Semenets$^{(\boxtimes)}$, Osvaldo Andrés Pérez-García,
Andrés Gago-Alonso, and Raudel Hernández-León

Advanced Technologies Application Center (CENATAV),
7a # 21406, Rpto. Siboney, Playa, 12200, Havana, Cuba
{vherrera,osvaldo.perez,agago,rhernandez}@cenatav.co.cu

Abstract. The presence of useless information and the huge amount of data generated by telecommunication services can affect the efficiency of traditional Intrusion Detection Systems (IDSs). This fact encourage the development of data preprocessing strategies for improving the efficiency of IDSs. On the other hand, improving such efficiency relying on the data reduction strategies, without affecting the quality of the reduced dataset (i.e. keeping the accuracy during the classification process), represents a challenge. Also, the runtime of commonly used strategies is usually high. In this paper, a novel hybrid data reduction strategy is presented. The proposed strategy reduces the number of features and instances in the training collection without greatly affecting the quality of the reduced dataset. In addition, it improves the efficiency of the classification process. Finally, our proposal is favorably compared with other hybrid data reduction strategies.

Keywords: Data mining · Data reduction · Instance selection
Feature selection

1 Introduction

Nowadays, the volume of data generated from using telecommunication services is considerably large, which causes Big Data challenges in the network traffic [1]. For example, AT&T long distance customers alone generate over 300 million records (instances) per day [2]. The presence of non-relevant information or noise in the data could affect the performance of the learning methods used to detect events representing attacks. Additionally, such events are executed by malicious users known as intruders, causing millions in losses and damaging the prestige of the affected companies. In order to analyze such volume of data and quickly detect which events are associated to an attack, it is necessary to apply intrusion detection techniques.

In literature, there are two main approaches based on data mining for the intrusion detection problem: (1) supervised approach, which requires a labeled training set T [3], and (2) unsupervised approach, where previous knowledge

© Springer International Publishing AG, part of Springer Nature 2018
M. Mendoza and S. Velastín (Eds.): CIARP 2017, LNCS 10657, pp. 289–297, 2018.
https://doi.org/10.1007/978-3-319-75193-1_35

is not required [4]. In this paper, we propose a novel hybrid data reduction strategy (following the supervised approach), called HDR, that combines both feature selection and instance selection to obtain a reduced training set $S \subset T$, providing high efficiency without affecting the quality of the reduced dataset too much. We evaluate the quality of the data during the classification process. For this, we use three standard measures: Accuracy, Recall and False Positive Rate.

2   Related Work

Data can be reduced in terms of the number of rows (instances) or in terms of the number of columns (features) [5]. In general, three main approaches have been proposed: (1) feature selection based, (2) instance selection based and hybrid, where feature selection and instance selection are combined [6].

The feature selection algorithms look up the most relevant features of the dataset. In this way, only a subset of features from the underlying data is used in the analytical process. On the other hand, the instance selection algorithms obtain a reduced subset of instances S from the original training set T, so that S does not contain superfluous instances. Instance selection approach can either start with $S = \emptyset$ (incremental methods) or $S = T$ (decremental methods) [7]. Incremental methods obtain S by selecting instances from T [8], while decremental ones obtain S by deleting instances from T [9].

According to the strategy used for selecting instances, the algorithms can be divided into two groups: Wrapper and Filter [7]. In Wrapper algorithms, the classifier is used in the selection process, the instances which do not affect the classification accuracy are removed from T. On the other hand, Filter algorithms are independent from the classifiers and the selection criterion is based on different heuristics.

In [6], a hybrid method for intrusion detection is proposed. In this case, the feature selection process is performed by the OneR [10] algorithm. Then, an expensive clustering algorithm, called Affinity Propagation [11], is used for instance selection process. Also, in order to improve the performance and scalability of the method, they implemented a distributed solution using MapReduce.

However, the main problem of the instance selection methods is the high runtime when large datasets are processed, which makes unfeasible their application in some cases, and directly affects the performance of the hybrid approaches.

3   Our Proposal

In this section, we introduce the Hybrid Data Reduction (HDR) strategy. The HDR strategy has three main phases: (1) feature selection phase, (2) relabeling phase and (3) instance reduction phase.

Feature Selection Phase
Most of the reported works in these scenarios use only one feature selection metric, and do not take advantage of the possibilities that the combination

of different metrics can offer; since different metrics could measure different information in the features. Therefore, as final result, different features with the same level of data representativeness could be selected. In this sense, our hypothesis is that a better management of these metrics could lead to a better selection of the final set of features. After a study, we determined that the most commonly used measures can be grouped into three categories: entropy based (Information Gain, Gain Ratio, and Symmetric Uncertainty), statistical based (Chi-square), and instance based (Relief and ReliefF) [12].

In our proposal, we use three different algorithms (one representative from each category): ReliefF, Chi-squared Ranking Filter and Information Gain Ranking Filter. The ReliefF algorithm estimates how well a feature can differentiate instances from different classes by searching for the nearest neighbors of the instances from the same and different classes. Chi-squared is a nonparametric statistical measure that estimates the correlation between the distribution of an attribute and the distribution of the class. In the case of Information Gain, it measures the amount of information that a feature can provide about whether an instance belongs to one class or another.

In Algorithms 1 (lines 4–8) and 2, the proposed feature selection strategy is described. Notice that for each feature selection algorithm, the score mean is computed, and the features whose values exceed the score mean are selected. Finally, the union of the three resulting sets is returned.

Algorithm 1. HDR

Input: T: *training set*
Output: S: *reduced training set*

1 $S \leftarrow T$
2 $F \leftarrow \emptyset$ // *selected features set*
3 $L \leftarrow \emptyset$ // *generated labels set*

/*Feature selection phase*/
4 $F_{RF} \leftarrow$ Selector (ReliefF_FS (S))
5 $F_{CHI} \leftarrow$ Selector (Chi_Square_FS (S))
6 $F_{IG} \leftarrow$ Selector (InfoGain_FS (S))
7 $F \leftarrow F_{UFS} \cup F_{CHI} \cup F_{IG}$
8 $S \leftarrow$ Dimensionality-Reduction (F, S)

/*Relabeling phase*/
9 $L \leftarrow$ Label-Generation (S)
10 $S \leftarrow$ Relabeling (S, L)

/*Instance reduction phase*/
11 $S \leftarrow$ Duplicated-Removing (S)

12 **return** S

Algorithm 2. Selector

Input: P_A: *Set of scores assigned by A*
Output: F_A: *Set of features selected from A*

1 $F_A \leftarrow \emptyset$ //
2 $\overline{p}_A \leftarrow$ Mean-Score(P_A)
3 **foreach** $p_f \in P_A$ **do**
4   **if** $p_f > \overline{p}_A$ **then**
5 | $F_A \leftarrow F_A \cup \{f\}$
6   **end**
7 **end**
8 **return** F_A

Relabeling Phase

After reducing the number of features in S, the relabeling process is carried out to generate new labels for the selected features values. In order to gain efficiency, we use the k-means algorithm to generate the labels during the *Label − Generation* function (see Algorithm 1, line 9). The purpose of applying a clustering method as part of the relabeling process is to search for similar values and group them into clusters.

For each selected numerical feature f_i, the k-means algorithm is executed over the set of values taken by f_i in S, denoted by V_i. Each of the obtained clusters contains a range of numerical values, which are represented by a unique numerical label (see Algorithm 1, line 10). The use of these clusters allow us to cover feature values that do not exist in S and are included in the classification stage.

For example, suppose that in the training phase a feature f_1 takes values in the set $V_1 = \{0, 0.2, 0.3, 0.6, 0.7, 1.0\}$, and during the relabeling process a cluster $c_1 = [0, 0.2, 0.3, 0.6]$ and $c_2 = [0.7, 1.0]$ are obtained. Then, in the classification stage, it is required to classify a new transaction, in which the feature f_1 takes the value $0.9 \notin V_1$, but 0.9 falls within the range of c_2, therefore it can be classified.

Instance Reduction Phase

Finally, the instance reduction phase is performed over the relabeled training collection. Here, $Duplicated - Removing$ function, as its name suggests, removes duplicate instances from S (see Algorithm 1, line 11). Notice that an instance is a duplicate instance if there is at least another instance with the same feature values and class. The result is a reduced training collection. This collection is used by a classifier to build a classification model.

4   Experimental Results

In this section, we aim to evaluate the novel hybrid data reduction strategy introduced in this paper; comparing its efficiency and the quality of the reduced dataset against the hybrid algorithm proposed by Chen et al. [6].

The experiments were conducted using two different datasets: KDD'99 [13] and CDMC 2013 [14]. The KDD'99 dataset has been widely used for testing network intrusion detection approaches, and it is considered a benchmark dataset. This dataset provides instances, each one containing 41 features out of which 9 are discrete and 32 are continuous. The training set consists of 494021 instances, while the testing set contains 311029 instances. In our experiments, we classify all the instances into two types, "normal" or "attack". In the case of CDMC 2013, it is an intrusion detection dataset collected from a real intrusion detection system. Each instance of this dataset contains 7 numerical features and a label indicating whether an instance is related with an attack or not. CDMC 2013 consists of 71758 "normal" labeled instances, and 6201 "attack" labeled instances. Without loss of generality, the dataset was splitted into two subsets: one for training (40000 instances), and another for testing (37959 instances).

Our experiments were performed on a PC equipped with 2.5 GHz Intel Core 2 Quad CPU and 4.00 GB DDR2 of RAM, running Windows 8. In Chen et al. [6], the KDD'99 dataset was evaluated using a specific configuration, which was the same used in our experiments. On the other hand, the CDMC 2013 dataset was not evaluated in [6], therefore, to establish a fair comparison, we adjusted the configuration of its proposal in order to obtain a reduced dataset with a similar size to the one obtained by HDR. We decided to test our strategy using different k values for the relabeling process.

In Fig. 1 we show the runtime of Chen *et al.* [6] method and HDR for different volumes of KDD'99 dataset. Notice that the execution time of HDR and the k value are directly proportional. This is because the k intervals in which a feature value is divided may have different sizes, and we have to check all of them (in the worst case) to assign the new labels. According to the performance shown in Fig. 1, we can conclude that regardless of the k value used, the HDR strategy is more efficient than the proposal of Chen *et al.* [6], and the HDR algorithm scales linearly regarding to k. Because the size of KDD'99 dataset is more than five times the size of CDMC 2013 dataset, and due to space reasons, we consider that it is enough to show this experiment using only the KDD'99 dataset.

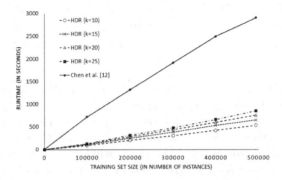

Fig. 1. Performances of Chen *et al.* [6] and HDR for different volumes of KDD'99 training dataset.

After the reduced training set S is obtained, two traditional classifiers are used in our experiments to evaluate the quality of our data reduction strategy, such classifiers are KNN and SVM. Both classifiers are commonly accepted for classification in intrusion detection scenario [15,16]. Similar to Chen *et al.* [6], for KNN, we set the parameter $k = 1$. To evaluate the quality of the reduction process, we used three standard measures: Accuracy (see Eq. 1), Recall (see Eq. 2) and False Positive Rate (FPR) (see Eq. 3).

$$Acc = \frac{TP + TN}{TP + TN + FP + FN} \cdot 100, \tag{1}$$

$$Recall = \frac{TP}{TP + FN} \cdot 100, \tag{2}$$

$$FPR = \frac{FP}{FP + TN} \cdot 100, \tag{3}$$

where TP, TN, FP and FN represents the true positive, true negative, false positive and false negative, respectively.

Table 1. Results achieved using KDD'99 dataset.

Classifier	Data reduction strategy	Acc	Recall	FPR	Instances	Features
SVM	Baseline	92.30	90.8	1.7	494021	41
SVM	Chen et al. [6]	90.05	87.9	2.1	63234	17
SVM	HDR (k = 10)	87.03	86.1	2.7	22822	17
SVM	HDR (k = 15)	88.31	86.9	2.4	38679	17
SVM	HDR (k = 20)	88.83	87.3	2.1	55999	17
SVM	HDR (k = 25)	90.26	88.5	2.0	60540	17
KNN	Baseline	92.81	91.2	0.7	494021	41
KNN	Chen et al. [6]	89.51	88.8	1.1	63234	17
KNN	HDR (k = 10)	87.12	86.7	2.2	22822	17
KNN	HDR (k = 15)	87.93	87.6	1.6	38679	17
KNN	HDR (k = 20)	89.10	88.4	1.2	55999	17
KNN	HDR (k = 25)	89.97	89.1	1.0	60540	17

In Tables 1 and 2, we show the values of the three quality measures mentioned above together with the number of instances and features, for HDR and Chen et al. [6] proposal, over the two datasets evaluated. Taking into account that in [6] it is necessary to predefine the number of features to be selected, in order to make a fair comparison, we decided that it will be the same as the amount of features selected by our proposal. Additionally, we include the results using the original training set without any reduction (baseline).

From Tables 1 and 2, it can be seen that the performance of HDR improves as the k values increase. In Table 1, we can see that for $k = 25$ the HDR strategy reaches better accuracy with less instances than Chen et al. [6] proposal. Additionally, HDR strategy obtains better results with respect to the recall and the FPR quality measures. Similar results are shown in Table 2 for $k = 20$ in the CDMC 2013 dataset.

Note that despite the considerable reduction of the dataset, the quality measures obtained using HDR with $k = 25$ do not vary greatly regarding to those obtained using the baseline. This shows that the quality of the reduced data was not affected too much. In addition, according to HDR behavior during the experiments, if a higher k value is defined, the dataset is reduced to a lesser extent, but the quality of the reduced data is improved, which is clearly shown by the improvement in the quality measures.

Our last experiment is to demonstrate how our proposal contributes to the classifiers efficiency during the classification process. For space reasons, and because KNN is much slower than SVM, we decided to show only the results achieved by KNN. Such results are shown in Fig. 2 for both KDD'99 and CDMC 2013 datasets respectively.

As it can be seen in Fig. 2(a), by using KNN with HDR it was possible to reduce the time consumed by 36 % regarding to the baseline, which means that KNN using HDR with $k = 25$ is 15807 s faster than KNN using the original

Table 2. Results achieved using CDMC 2013 dataset.

Classifier	Data reduction strategy	Acc	Recall	FPR	Instances	Features
SVM	Baseline	96.42	95.9	0.2	40000	7
SVM	Chen et al. [6]	94.21	93.0	1.2	984	4
SVM	HDR (k = 10)	91.47	91.3	2.1	239	4
SVM	HDR (k = 15)	94.16	92.7	1.5	500	4
SVM	HDR (k = 20)	94.76	93.1	0.9	825	4
SVM	HDR (k = 25)	95.41	94.0	0.4	1086	4
KNN	Baseline	96.57	96.1	0.1	40000	7
KNN	Chen et al. [6]	94.78	93.9	1.0	984	4
KNN	HDR (k = 10)	92.25	91.9	1.8	239	4
KNN	HDR (k = 15)	94.72	93.3	1.1	500	4
KNN	HDR (k = 20)	95.11	94.2	0.6	825	4
KNN	HDR (k = 25)	95.87	94.9	0.3	1086	4

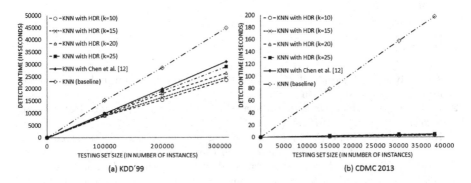

Fig. 2. Comparative results of classification time using KNN for KDD'99 and CDMC 2013 testing sets.

dataset (baseline). In addition, a better efficiency than the Chen et al. [6] proposal was achieved during the classification process using HDR.

On the other hand, for CDMC 2013 dataset, we also achieved improvements in efficiency during the classification process (see Fig. 2(b)). This time, the KNN classifier using HDR with $k = 20$ achieved better efficiency than KNN using Chen et al. [6] proposal. In this sense, KNN using HDR with $k = 20$, it was able to reduce the detection time by 98 % regarding to the baseline, being in turn 0.8 % faster than KNN using Chen et al. [6] proposal.

As we saw in this experiment, HDR improves the performance of KNN classifier by decreasing the time spent during the classification process. This can be very useful in scenarios such as intrusion detection, where real-time data analysis may be necessary. Our proposal is capable of achieving a better efficiency than the baseline, without greatly affect the quality of the data during the reduction process.

5   Conclusions

In this paper, we present a novel hybrid data reduction strategy composed of three main phases: (1) feature selection phase, (2) relabeling phase and (3) instance reduction phase. The experimental results show that HDR greatly reduces the original dataset, without affecting the quality of the reduced dataset too much. Also, it was able to further reduce the dataset than the other hybrid approach evaluated, and even so, the classifiers using HDR achieved better results. On the other hand, our proposal does not require large computational resources. HDR is able to process large volumes of data with good performance on a standard PC. Improving the performance of the classifiers in terms of the time elapsed during the classification process is another contribution of our proposal, which can be very useful in scenarios where it is necessary to process data in real time. In addition, HDR did not show any significant increase of its running time, which makes it feasible to process large volumes of data, where data reduction is essential. However, an interesting direction for future work is to evaluate our proposal in other scenarios with higher dimensional spaces, in terms of features.

References

1. Zuech, R., Khoshgoftaar, T.M., Wald, R.: Intrusion detection and big heterogeneous data: a survey. J. Big Data **2**(1), 1–41 (2015)
2. Cortes, C., Pregibon, D.: Signature-based methods for data streams. Data Mining Knowl. Discov. **5**(3), 167–182 (2001)
3. Kotsiantis, S.B.: Supervised machine learning: a review of classification techniques. In: Proceedings of the 2007 Conference on Emerging Artificial Intelligence Applications in Computer Engineering, Amsterdam, The Netherlands, pp. 3–24 (2007)
4. Ghahramani, Z.: Unsupervised learning. In: Bousquet, O., von Luxburg, U., Rätsch, G. (eds.) ML -2003. LNCS (LNAI), vol. 3176, pp. 72–112. Springer, Heidelberg (2004). https://doi.org/10.1007/978-3-540-28650-9_5
5. Aggarwal, C.C.: Data Mining: The Textbook. Springer, Heidelberg (2015)
6. Chen, T., Zhang, X., Jin, S., Kim, O.: Efficient classification using parallel and scalable compressed model and its application on intrusion detection. Expert Syst. Appl. **41**(13), 5972–5983 (2014)
7. Olvera-Lopez, J.A., Carrasco-Ochoa, J.A., Martínez-Trinidad, J.F., Kittler, J.: A review of instance selection methods. Artif. Intell. Rev. **34**(2), 133–143 (2010)
8. Chou, C.H., Kuo, B.H. and Chang, F.: The generalized condensed nearest neighbor rule as a data reduction method. In: 18th International Conference on Pattern Recognition (ICPR 2006), vol. 2, pp. 556–559 (2006)
9. Wilson, D.R., Martínez, T.R.: Reduction techniques for instance-based learning algorithms. Mach. Learn. **38**(3), 257–286 (2000)
10. Holte, R.C.: Very simple classification rules perform well on most commonly used datasets. Mach. Learn. **11**(1), 63–90 (1993)
11. Frey, B.J., Dueck, D.: Clustering by passing messages between data points. Science **315**(5814), 972–976 (2007)

12. Liu, W., Liu, S., Gu, Q., Chen, J., Chen, X., Chen, D.: Empirical studies of a two-stage data preprocessing approach for software fault prediction. IEEE Trans. Reliab. **65**(1), 38–53 (2016)
13. KDDCup 1999: Computer network intrusion detection. http://kdd.ics.uci.edu/databases/kddcup99/kddcup99.html. Accessed 25 Feb 2017
14. Song, J.: CDMC2013 intrusion detection dataset. Department of Science & Technology Security, Korea Institute of Science and Technology Information (KISTI) (2013)
15. Horng, S.J., Su, M.Y., Chen, Y.H., Kao, T.W., Chen, R.J., Lai, J.L., Perkasa, C.D.: A novel intrusion detection system based on hierarchical clustering and support vector machines. Expert Syst. Appl. **38**(1), 306–313 (2011)
16. Aburomman, A.A., Reaz, M.B.I.: A novel SVM-kNN-PSO ensemble method for intrusion detection system. Appl. Soft Comput. **38**, 360–372 (2016)

Clustering-Based Undersampling to Support Automatic Detection of Focal Cortical Dysplasias

Keider Hoyos-Osorio[1]([✉]), Andrés M. Álvarez[1], Álvaro A. Orozco[1],
Jorge I. Rios[1], and Genaro Daza-Santacoloma[2]

[1] Automatic Researh Group, Universidad Tecnológica de Pereira, Pereira, Colombia
{jkhoyos,andres.alvarez1,aaog,jirios}@utp.edu.co
[2] Instituto de Epilepsia y Parkinson del Eje Cafetero, Pereira, Colombia
research@neurocentro.co

Abstract. Focal Cortical Dysplasias (FCDs) are cerebral cortex abnormalities that cause epileptic seizures. Recently, machine learning techniques have been developed to detect FCDs automatically. However, dysplasias datasets contain substantially fewer lesional samples than healthy ones, causing high order imbalance between classes that affect the performance of machine learning algorithms. Here, we propose a novel FCD automatic detection strategy that addresses the class imbalance using relevant sampling by a clustering strategy approach in cooperation with a bagging-based neural network classifier. We assess our methodology on a public FCDs database, using a cross-validation scheme to quantify classifier sensitivity, specificity, and geometric mean. Obtained results show that our proposal achieves both high sensitivity and specificity, improving the classification performance in FCD detection in comparison to the state-of-the-art methods.

Keywords: Imbalance learning · Clustering · Bagging

1 Introduction

Patients with pharmacoresistant epilepsy can be treated surgically, but first, it is necessary to identify with high precision the anatomical region of the cerebral cortex where the anomaly that triggers epileptogenic crises is located. Among the most common abnormalities are Focal Cortical Dysplasias (FCDs). FCDs produce a disorder in the electrical functioning of the brain and they can be caused by genetic factors, lack of oxygen during brain development, parasites, among others [1]. Currently, in order to locate these abnormalities, invasive electrophysiological tests are performed, but they generate discomfort in patients. In this sense, it is necessary to develop a non-invasive FCD detection tools.

Recently, machine learning methods that recognize morphological/intensity patterns associated with dysplasias from magnetic resonance images (MRIs)

© Springer International Publishing AG, part of Springer Nature 2018
M. Mendoza and S. Velastín (Eds.): CIARP 2017, LNCS 10657, pp. 298–305, 2018.
https://doi.org/10.1007/978-3-319-75193-1_36

[2–4] have been proposed to support the non-invasive diagnosis of patients with pharmacoresistant epilepsy. However, the automatic detection of FCDs by machine learning algorithms is hindered by the class imbalance problem in their input data. Such imbalance is due to the fact that there are substantially fewer lesional vertex than healthy ones (points of brain surface). The fundamental issue with the imbalanced learning problem is the ability of imbalanced data to significantly compromise the performance of most standard learning algorithms, which assume or expect balanced class distributions or equal misclassification costs. Therefore, when they are presented with complex imbalanced data sets, the result can be biased classifiers towards the majority class [5].

To address this issue, in [3] a random bagging approach is used and a set of "base-level " classifiers is constructed, each trained using logistic regression, by an iterative-reweighted least squares (IRLS) algorithm, identifying 14 out of 24 FCD lesions (58%), nevertheless this methodology is computationally expensive because the classifier is trained many times [5]. Hong et al. in [4] propose randomly select non-lesional vertex, balancing their number in both classes, implementing a linear discriminant analysis and providing a detection rate of 74% in adult cohorts. On the contrary, Adler et al. in [2] do not consider the class imbalance problem, they implement a neural network using novel surface-based features obtaining a sensitivity of 73%.

In this paper we propose a novel methodology for automatic detection of FCD. We use in our experiments a public dataset that contains cerebral surface morphological data of 22 subjects whit histological confirmed FCD. In first place, we use a cluster-based under-sampling method to select the most representative samples to train several neural networks, building a majority vote classifier and achieving both high sensitivity (95.11%) and specificity (93.2%) results.

Our paper contains: a brief description of several morphological and intensity features of brain cortex (Sect. 2.1), description of cluster-based under-sampling approach (Sect. 2.2) and bagging approach (Sect. 2.3), the experimental setup (Sect. 3), results and discussion (Sect. 4) and finally, conclusions (Sect. 5).

2   Methods

2.1   MRI Feature Estimation for Automatic Detection of Dysplasias

Let $\mathbf{V} \in \mathbb{R}^{W \times H \times L}$ be an MRI over which is generated a model of the GM-WM (gray matter - white matter) and GM-CSF (gray matter - cerebrospinal fluid) surfaces using the algorithm proposed in [6]. Once these surfaces are reconstructed, it can be measured several cortical features such as cortical thickness, gray-white matter intensity contrast, curvature, and sulcal depth and intensity at each vertex of the 3D cortical reconstruction. Thickness cortex is calculated as the average minimum distance between each vertex on the pial and white matter surfaces. Gray-white matter intensity contrast is calculated as the ratio of the gray matter to the white matter signal intensity. The gray matter signal intensity is sampled at a distance of 30% of the cortical thickness above the gray-white matter boundary. The white matter signal intensity is sampled 1 mm below the gray-white matter boundary. FLAIR intensity is sampled at the gray-white matter boundary as well as at 25%, 50% and 75% depths of the cortical thickness and at $-0.5\,\text{mm}$ and $-1\,\text{mm}$ below the gray-white matter boundary. Mean curvature is measured at the gray-white matter boundary as $\frac{1}{r}$, where r is the radius of an inscribed circle and is equal to the mean of the principal curvatures k1 and k2 [7]. Other novel features have been recently introduced in [2] to identify local changes in thickness/intensity on MRI surface reconstructions which are insensitivity to motion artifact.

Finally, we can obtain an array $\mathbf{X} \in \mathbb{R}^{n \times p}$, where n is the number of samples in the dataset and each vertex is represented as $\mathbf{x}_i \in \mathbb{R}^p$ characterized by p features, which can be labeled as healthy vertex (0) or lesional vertex (1), creating a target vector $\mathbf{t} = \{t_i \in [0, 1] : i = 1, \ldots, n\}$.

2.2 Relevant Sampling for Imbalance Problem in Dysplasias Recognition

Many datasets in real applications involve imbalanced class distribution problem. This is the case of detection of focal cortical dysplasias since there are substantially fewer vertex labeled as lesional than vertex labeled as non-lesional.

In supervised classification, we are given an observed training set $\mathbf{X} \in \mathbb{R}^{n \times p}$ over which a predictive model $\mathcal{C}$ is to be induced. In imbalanced data distributions, the majority class $\mathbf{X}_{maj}$, outnumber the minority class $\mathbf{X}_{min}$. Representing $|\mathbf{X}_{min}|$ as the number of samples in the minority class, and $|\mathbf{X}_{maj}|$ as the number of samples in the majority class, where $|\mathbf{X}_{min}| < |\mathbf{X}_{maj}|$, the use of under-sampling methods in imbalanced learning applications consists of the modification of an imbalanced dataset by some mechanisms in order to provide a balanced distribution [5], which means that $|\mathbf{X}_{min}| = |\mathbf{X}^*_{maj}|$ where $\mathbf{X}^*_{maj}$ is the under-sampled majority class set. Hence, it is important to select the suitable training data for classification in the imbalanced class distribution problem due to the training data significantly influence the classification accuracy [8].

We address this issue by using a Cluster-Based Under-Sampling technique (CBUS) to select the most representative data to train the classifier [8]. To find the most relevant samples, the first step is to partition the dataset in several clusters by using K-means algorithm. The main objective is to find an assignment of data points to clusters, as well as a set of means $\{\boldsymbol{\mu}_k \in \mathbb{R}^p : k = 1, \ldots, K\}$, such that the sum of the squares of the euclidean distances of each data point to its closest vector $\boldsymbol{\mu}_k$, is a minimum. This can be achieved by minimizing an objective function given by

$$J = \sum_{i=1}^{n} \sum_{k=1}^{K} r_{nk} \|\mathbf{x}_i - \boldsymbol{\mu}_k\|^2, \tag{1}$$

where

$$r_{nk} = \begin{cases} 1 & \text{if } k = \underset{j}{\operatorname{argmin}} \|\mathbf{x}_i - \boldsymbol{\mu}_j\|^2 \\ 0 & \text{otherwise} \end{cases} \tag{2}$$

Once the data have been clustered, a suitable number of majority class samples is randomly selected from the kth cluster, by considering the ratio of the number of majority class samples to the number of minority class samples in each cluster as follows:

$$|\mathbf{X}^{*(k)}_{maj}| = (r \times |\mathbf{X}_{min}|) \times \frac{|\mathbf{X}^{(k)}_{maj}|/|\mathbf{X}^{(k)}_{min}|}{\sum_{i=1}^{K}|\mathbf{X}^{(k)}_{maj}|/|\mathbf{X}^{(k)}_{min}|}, \tag{3}$$

where r is the expected ratio of $|\mathbf{X}^*_{maj}|$ to $|\mathbf{X}_{min}|$ (generally set to 1), building an under-sampled matrix $\mathbf{X}^* = \bigcup_{k=1}^{K} \mathbf{X}^{*(k)}_{maj} \cup \mathbf{X}_{min}$.

2.3 Bagging Approach

To enhance the performance of classifiers in the imbalanced data distribution problem, a technique known as bagging is often used. The bagging methods consist in construct a set of N(odd) classifiers $\mathcal{C}_1, \mathcal{C}_2, \ldots, \mathcal{C}_N$, each trained on all the minority class instances and N equal-sized samples of random majority class instances. When a new instance $\mathbf{x}_{new}$ is to be classified, each trained classifier makes a prediction, and the final prediction $\hat{t}$ is taken as the majority vote [9].

Here we propose a novel strategy to implement a bagging approach in order to train N models, but using the most relevant majority class samples: First, the database is cluster-based under-sampled according to Eq. 3, by establishing the expected imbalance ratio r as the number N of models to train, in order to obtain r equal sized classes with the most representative majority class instances, then training N classifiers. This new approach can be summarized in Fig. 1.

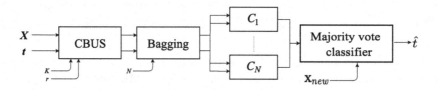

Fig. 1. Novel strategy to address the class imbalance, oriented to automatic detection of FCDs.

3 Experimental Setup

Our purpose is to assess the predictive performance of the CBUS technique combined with an ensemble method known as bagging, as an alternative to address the class imbalance for the automatic recognition of FCDs.

The proposed method is tested in a public FCDs dataset used in [2].[1] Due to bioethical issues the original MRIs are not shared publicly, even though a matrix which contains cerebral surface morphological data of 22 patients with FCD is published, which is represented as $\Psi = \left\{ \mathbf{X}^{(m)} \in \mathbb{R}^{n^{(m)} \times p} : m = 1, \ldots, 22 \right\}$, where the number of features is $p = 28$. The number of vertex in each brain surface reconstruction varies from patient to patient, but the total number of vertex (samples) is 3307529.

FreeSurfer software v5.3 [6] was used to generate cortical reconstructions and to coregister FLAIR scans to T1-weighted images. Then it was employed to compute several morphological/intensity features, as reported in Sect. 2.1. Manual lesion masks were created for the 22 participants, on axial slices of the volumetric scan and they are then registered onto the cortical surface reconstructions (see Fig. 2), in order to assign the labels to each sample.

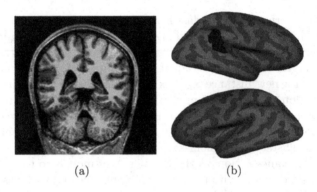

(a)   (b)

Fig. 2. Lesions are identified combining information from T1 images, previous radiological reports and reports from multi-disciplinary team meetings, after, the lesion mask are created on (a) MRI and on (b) surface cortex reconstruction. (Images taken from [2]).

The features included in the dataset are cortical thickness, gray-white matter intensity contrast, sulcal depth, mean curvature, FLAIR intensity samples at six cortical depths, local cortical deformation, "dough-nut" thickness, "dough-nut" intensity contrast, "dough-nut" FLAIR intensity at different cortical depths. Besides, a set of normalized inter-hemispheric asymmetry measures for cortical thickness, gray-white matter intensity contrast, the FLAIR intensity samples and local cortical deformation. This dataset exhibit a significant imbalance on the order of 42:1.

With regard to class imbalance, we propose two techniques: CBUS, and a combination of CBUS and bagging as mentioned in Sect. 2.3. To implement CBUS the data is clustered into $K = 4$ groups. When CBUS is implemented alone the expected imbalance ratio is set to 1 ($r = 1$), balancing at order of

---

[1] Available online in https://doi.org/10.17863/CAM.6923.

1:1. Then, CBUS is implemented for each patient, and the under-sampled data is concatenated resulting in a sub-sampled input matrix $\mathbf{X}^*$ with $n = 151340$ samples, $p = 28$ features, and a target vector $\mathbf{t} = \{t_i \in [0,1] : i = 1 \ldots, 151340\}$, both with the same number of samples of the majority and minority classes. To combine CBUS and bagging, five classifiers ($r = N = 5$) are trained on all the lesional vertex and on five equal-sized samples of the most relevant non-lesional vertex.

As benchmark, we compare our results with the methodologies to address the class imbalance oriented to automatically detection of FCD proposed in [2–4]: without under-sampling (WUS), random under-sampling (RUS) and bagging approach on five random samples of majority class, each one of equal size of the minority class. For all schemes, Neural Network classifiers (NN) are trained using surface based measures from samples (vertex) selected by each method. A single hidden layer neural network is chosen as the classifier because it can be rapidly trained on large datasets [2]. The number of hidden neurons in the network is determined through a principal component analysis applied to the input features, using the number of components that explained over 99% of the variance. In our case, 11 principal components are required. Finally, we employ a 10-fold cross-validation analysis to test the performance of our methods. G-mean, sensitivity and specificity of the classifiers outputs are computed because these measures give us a class-by-class performance estimate. ROC curves are also employed to assess the performance of each approach. Finally, the overall results of each method are compared.

4   Results and Discussion

Table 1 shows the results obtained for proposed methods; WUS, RUS, random bagging, CBUS and the combination of CBUS and bagging. Without under-sampling the classifier achieves a high specificity (99.7%), nevertheless the sensitivity is low (49.8%). This happens because the classifier is biased towards the healthy class due to the high class imbalance. When input data is randomly under-sampled the sensitivity results improve, but the specificity decreases. However, the G-mean value, which encompasses sensitivity and specificity information, increases significantly. The bagging approach improves the G-mean value, with respect to the two previews methods. For CBUS this figure is even higher, this is explained because this technique selects the most relevant samples from clusters where data is concentrated. The combination of CBUS and bagging achieves further improvement in the G-mean and sensitivity results. Figure 3, presents the ROC curves for the five proposed methods. The ROC curves show that our approach performed better than the remaining methods.

The study developed in [2] does not consider the class imbalance, however, they post-process the output probability maps from the classifier and cluster into neighbor-connected vertex, additionally, they discard the smallest clusters. Their automated lesion detection method is considered successful if this cluster overlapped the lesion mask. Under this criterion, their method was able to detect

Table 1. Methods results comparison

Method	Imbalance order	G-mean (%)	Sensitivity (%)	Specificity (%)
WUS	42:1	70.46 ± 1.34	49.8 ± 1.22	99.7 ± 1.28
RUS	1:1	86.19 ± 2.28	84.1 ± 2.36	88.4 ± 1.82
Bagging	5:1	89.46 ± 0.71	88.95 ± 0.78	89.98 ± 0.85
CBUS	1:1	92.19 ± 1.18	92.6 ± 1.92	91.8 ± 2.17
CBUS-Bagging	**5:1**	**94.15 ± 0.21**	**95.11 ± 0.49**	**93.2 ± 0.33**

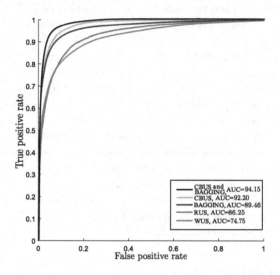

Fig. 3. ROC curves obtained for different methods studied

16 out of 22 FCD (73%). Unfortunately, we do not have the original MRIs, so we can not validate our method using an analogous strategy. However, the results at the vertex level show that our methods outperform the state-of-the-art approaches that address the class imbalance for the automatic detection of FCDs. It is important to note, that although our method has a satisfactory performance, it presents limitations regarding its computational cost, due to the necessity of training 5 neural networks.

5   Conclusions

In this study, we propose a novel strategy to address the class imbalance oriented to the automatic detection of focal cortical dysplasias. Our method is based on measures of cerebral surface morphological/intensity features and uses Cluster-Based Under-Sampling combined with a bagging process. Then, the relevant samples are employed to train neural networks classifiers. The method was tested on a matrix of FCDs data from an online repository. The results at the vertex level show G-mean (94.15%), sensitivity (95.11%), and specificity (93.2%)

values. So, our approach outperforms comparable methods that address the class imbalance in the detection of FCDs.

As future work, authors plan to test the introduced method based on a cluster-wise validation from MRIs. Besides, support vector machines classifiers will be coupled with our clustering-based and bagging strategy to reveal relevant samples using reproducing kernel Hilbert spaces.

Acknowledgments. This research is developed under the project *Desarrollo de un sistema de apoyo al diagnóstico no invasivo de pacientes con epilepsia farmacoresistente asociada a displasias corticales cerebrales: Método costo-efectivo basado en procesamiento de imágenes de resonancia magnética,* financed by COLCIENCIAS with code 111074455778. Thanks for the support to the master in electrical engineering program of the Universidad Tecnológica de Pereira.

References

1. Najm, I.M., Tassi, L., Sarnat, H.B., Holthausen, H., Russo, G.L.: Epilepsies associated with focal cortical dysplasias (FCDS). Acta neuropathologica **128**(1), 5–19 (2014)
2. Adler, S., Wagstyl, K., Gunny, R., Ronan, L., Carmichael, D., Cross, J.H., Fletcher, P.C., Baldeweg, T.: Novel surface features for automated detection of focal cortical dysplasias in paediatric epilepsy. NeuroImage Clin. **14**, 18–27 (2017)
3. Ahmed, B., Brodley, C.E., Blackmon, K.E., Kuzniecky, R., Barash, G., Carlson, C., Quinn, B.T., Doyle, W., French, J., Devinsky, O., et al.: Cortical feature analysis and machine learning improves detection of "MRI-negative" focal cortical dysplasia. Epilepsy Behav. **48**, 21–28 (2015)
4. Hong, S.-J., Kim, H., Schrader, D., Bernasconi, N., Bernhardt, B.C., Bernasconi, A.: Automated detection of cortical dysplasia type II in MRI-negative epilepsy. Neurology **83**(1), 48–55 (2014)
5. He, H., Garcia, E.A.: Learning from imbalanced data. IEEE Trans. Knowl. Data Eng. **21**(9), 1263–1284 (2009)
6. Dale, A.M., Fischl, B., Sereno, M.I.: Cortical surface-based analysis: I. segmentation and surface reconstruction. Neuroimage **9**(2), 179–194 (1999)
7. Pienaar, R., Fischl, B., Caviness, V., Makris, N., Grant, P.E.: A methodology for analyzing curvature in the developing brain from preterm to adult. Int. J. Imaging Syst. Technol. **18**(1), 42–68 (2008)
8. Yen, S.-J., Lee, Y.-S.: Cluster-based under-sampling approaches for imbalanced data distributions. Expert Syst. Appl. **36**(3), 5718–5727 (2009)
9. Wallace, B.C., Small, K., Brodley, C.E., Trikalinos, T.A.: Class imbalance, redux. In: 2011 IEEE 11th International Conference on Data Mining (ICDM), pp. 754–763. IEEE (2011)

Ultrasonic Assessment of Platelet-Rich Plasma by Digital Signal Processing Techniques

Julián A. Villamarín[1], Yady M. Jiménez[1], Tatiana Molano[1],
Edgar W. Gutiérrez[1(✉)], Luis F. Londoño[2], David Gutiérrez[1],
and Daniela Montilla[3]

[1] Antonio Nariño University, Bogotá, Colombia
{jvilla22,yadjimenez,lemolano,edgargu,
davidguierrez}@uan.edu.co
[2] Polytechnic Colombian Jaime Isaza Cadavid, Bello, Colombia
lflondono@elpoli.edu.co
[3] Valparaíso University, Valparaíso, Chile
daniela.montilla@postgrado.uv.cl

Abstract. This paper presents the implementation of an ultrasonic non-invasive and non-destructive system for the acoustic characterization of bovine Platelet-Rich Plasma based on advanced digital signal processing techniques. The system comprises the development of computational procedures that allow spectral estimation of parameters such as the angular coefficients with linear frequency dependence and the measurement of the speed of sound of regions of concern in sample studies. The results show that the relationship of acoustic parameters obtained from backscattered ultrasonic signals contributes to the hematological prediction of platelet concentration based on linear regression model.

Keywords: Attenuation · Backscattering · Blood

1 Introduction

Platelets rich plasma (PRP) is an important bioactive substance in processes that stimulate tissue regeneration and healing, acting as an autologous source of growth factors and in turn stimulating the secretion of proteins involved in tissue regeneration [1]. Currently, the characterization of plasma for therapeutic applications in animals [2] as well as in humans causes great interest, since the quality of the platelet concentration present in the PRP should be sufficient to guarantee the regenerative potential [3]. Considering the different methods of platelet quantification, based on the direct observational analysis performed under a conventional microscope (susceptible to errors by the observer's appreciation), to automated methods with hematology equipment able to identify different cell types from a single whole blood sample [4], it is well known that the methods mentioned above present variability in sensitivity and specificity, as well as disadvantages in relation to the costs of operation, maintenance and negative impacts generated to the environment, by the use of chemical reagents [5], which suggests the need of looking for alternative methodologies or techniques.

© Springer International Publishing AG, part of Springer Nature 2018
M. Mendoza and S. Velastín (Eds.): CIARP 2017, LNCS 10657, pp. 306–313, 2018.
https://doi.org/10.1007/978-3-319-75193-1_37

On the other hand, ultrasound-based technologies are shown to be advantageous in the non-invasive characterization of tissues, fluids and materials [6], in addition to having an optimal cost-benefit ratio. Characterization of biological fluids has been proposed recently, particularly at the hematological level, in which blood coagulation [7], erythrocyte aggregation [8, 9] and estimation of glucose concentration parameters [10] have been proposed, among others performed in humans and animal models [11], reflecting the scope of ultrasonic characterization techniques.

Taking into account the potential of high frequency mechanical wave characterization techniques and their validation for the estimation of concentrations of substances, the present study presents the implementation of a non-invasive non-destructive system of quantitative characterization of animal blood plasma, based on spectral energy loss estimation algorithms and time of flight measurements, which provide information associated with platelet concentration indices.

2   Materials and Method

2.1   Hematological Samples

Twelve samples of bovine blood each one with 5 ml were obtained from the jugular vein in the faculty of veterinary medicine of the Antonio Nariño University (UAN). All experimental tests were carried out with the consent of the research group on welfare, health and animal production (UAN). The protocol for obtaining samples of platelet-rich plasma consisted first of the provision of hematological samples in EDTA tubes at rate of centrifugation of 2400 rpm during 10 min. The samples were processed for histological analysis, mounted on 75 mm × 25 mm slides, and then treated with Masson's trichrome staining. They were analyzed under an optical microscope at 100x, followed by the corresponding quantification of platelets in Neubauer's chamber (see Fig. 1).

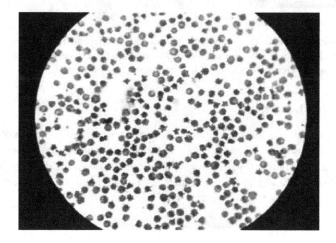

Fig. 1. Morphological analysis and platelet count. 100× (Masson's trichrome staining).

2.2   Characterization System

The acoustic non-invasive and non-destructive characterization system comprises: (a) an ultrasonic field emission system (*Olympus 5072PR*), (b) a positioning machine and (c) a data acquisition and processing system. The ultrasonic field emission system is configured in pulse-echo mode to provide an excitation energy of 26 µJ for an ultrasound transducer (*Panametrics A019s*) with a central frequency of 5 MHz at bandwidth of 6 MHz at −6 dB. It is stimulated with a pulse repetition frequency of 100 Hz. The positioning machine uses a ROHS stepper motor (28BYJ-48) driven by an *Arduino* microcontroller, which allows a control step size of 1 mm by coupling of an endless screw. This system also includes a mechanical support to hold the ultrasound transducer and an acoustic glass tank on which a liquid is attached aiming the coupling between the acoustic wave transfer and the hematological sample. At last, the data acquisition system includes a *Gw Instek GDS-1102 A-U* oscilloscope, which stores the ultrasonic backscatter signals in the interest region (RoI) with a sampling rate of 50 m/s. The stored signal files are subsequently processed by computational procedures implemented using *Matlab* (*R2014b*). Figure 2 describes the experimental set up components.

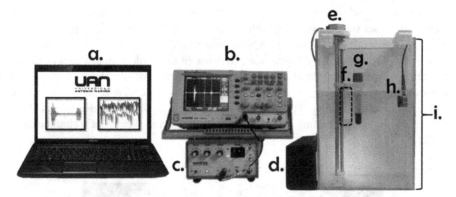

Fig. 2. Ultrasonic characterization system design and implemented system. a. Processing system. b. Oscilloscope. c. Pulse generator. d. Microcontroller. e. Stepper motor and endless screw. f. Backing. g. Hematological tube. h. Transducer. i. Acoustic tank.

2.3   Ultrasonic Inspection

The bovine platelet-rich plasma samples contained in the clinical hematology tubes were subjected to a high frequency and low power acoustic field at a focal length of 6 cm, where a more uniform wave front with minimization of the diffraction effects is guaranteed. For each sample, twenty signals were obtained during the vertical sweep

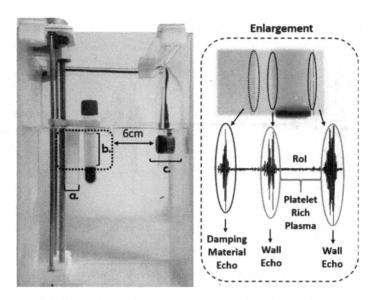

Fig. 3. Backscattering signal in RoI of platelet-rich plasma.

from the basal – line of blood plasma aiming the digital signal processing. An example of the ultrasonic backscattering signal collected in the RoI is shown in Fig. 3.

2.4 Computational Procedures

The characterization of the study samples was performed using digital signal processing algorithms, which estimate three acoustic parameters: speed of sound, acoustic attenuation and the backscatter integrated coefficient, being the last two based on spectral analysis techniques. The calculation of the speed of sound is important taking as it contributes to the understanding about the relationship between the acoustic impedance and the density of the study medium, as well as provides information that might be associated with the compressibility characteristics and particles concentration of biological fluids studied, which allows its characterization. Experimentally, the speed of sound (c) was determined by measuring the time of flight (ToF) of a wave travelling along the propagation distance (x) in the RoI, detecting echoes of high amplitude produced by the high reflection coefficient of the polypropylene walls of the hematological tube, associated with acoustic impedance change between the tube wall and the blood plasma. According to the *ToF*, the speed of sound was estimated based on the expression:

$$c = \frac{2x}{ToF} \tag{1}$$

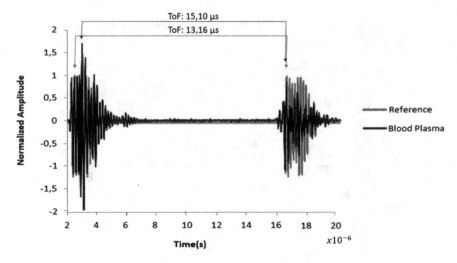

Fig. 4. Ultrasonic backscattering signals from platelet-rich plasma and reference medium.

Figure 4 shows an example of the echoes selected on the walls of the blood test tube aiming the calculation of the speed of sound. The distance of propagation in the colloidal system is twice the diameter of the hematological tube (considering that the acoustic inspection was carried out in pulse echo mode). The speed of sound algorithm was adjusted from experimental measurements in water which was used as reference liquid.

The calculation of the acoustic attenuation coefficient is a parameter that describes the loss of acoustic energy caused by the propagation medium. Its value depends on the medium in question. In soft tissue characterization procedures its calculation is given by the following equation:

$$\alpha = 10log\left(\frac{Vx}{Vi}\right)/2 \tag{2}$$

Considering the power law $\alpha = \beta f^n$, the frequency-dependent attenuation coefficient (β) can be estimated assuming that experimentally, the frequency dependence index n for soft tissues can be approximated linearly.

Computationally, the acoustic attenuation is measured based on the estimation of the angular coefficient given by the linear adjustment between the normalized power spectral subtraction obtained from ultrasound echoes in platelet-rich plasma and water. The spectral subtraction guarantees the elimination of energy loss effects caused by the hematological tube walls and alignment errors in the incident ultrasonic field (Fig. 5).

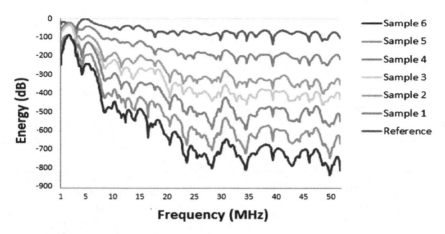

Fig. 5. Spectral differences for different platelet-rich plasma samples and its respective comparison with reference spectrum in water.

3 Results

Acoustic parameters values were estimated for blood plasma samples. These values are displayed in Table 1.

A linear relationship between platelet concentration and experimental estimation of speed of sound and loss of spectral energy (attenuation) was found in Fig. 6. The speed of sound can be adjusted linearly showing the increase of plasma platelet concentration. For the speed of sound, the linear regression is described by the mathematical model seen in Fig. 6, $y = 0.9061x + 1218$, indicating a growth rate of 1 m/s for each 1.103×10^3 platelet per mm^3. The results obtained were compared with existing data reported in the literature [12], for measurements of speed of sound for blood cell concentration [13, 14], which also indicates a linear proportionality. On the other hand, attenuation estimation presented a linear relation according to the increase of platelets, evidencing an angular coefficient of 2×10^{-6} that describes the resolution for predicting concentration values.

It is clearly observed that these acoustic descriptors present a proportional relation to the cell concentration and allow a better understanding of the interaction process wave - colloidal system, from which it is possible to predetermine and quantify platelet concentration.

Table 1. Acoustic parameters estimated by signal processing techniques

Sample	Speed of sound average (m/s)	Attenuation average (dB/cm-MHz)	Platelets (10^3/mm^3)
1	1.398	0.000308	287
2	1.590	0.000281	283
3	1.433	0.000163	280
4	1.438	0.000134	225
5	1.431	0.000141	225
6	1.385	0.00011	209

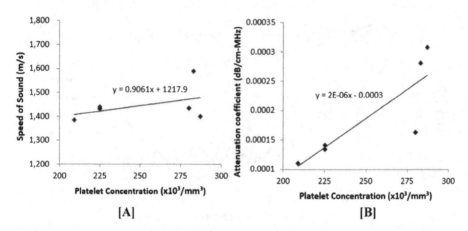

Fig. 6. A. Average speed of sound profile for different platelet concentration. B. Profile of the acoustic attenuation average for different platelet concentration.

4 Conclusions

The computational procedures for the ultrasonic characterization of Platelet-Rich Plasma from bovine blood samples showed the potential of acoustic parameters such as speed of sound and attenuation for the estimation of platelet concentration. The calculations made possible the estimation of a linear regression model that allowed to infer platelet concentration values, evidencing that for speed of sound variations of approximately 53 m/s there is an approximate variation of 55.18×10^3 platelets per mm^3. Finally, the experimental results showed a linear proportionality between the concentration cells and the estimated parameters demonstrating the relevance of the ultrasonic quantitative characterization technique to infer in measurement of substance concentrations in other applications.

References

1. Amable, P.R., Carias, R.B., Teixeira, M.V., da Cruz Pacheco, I., do Amaral, R.J.C., Granjeiro, J.M., Borojevic, R.: Platelet-rich plasma preparation for regenerative medicine: optimization and quantification of cytokines and growth factors. Stem Cell Res. Ther. **4**(3), 67 (2013)
2. Hessel, L.N., Bosch, G., van Weeren, P.R., Ionita, J.C.: Equine autologous platelet concentrates: a comparative study between different available systems. Equine Vet. J. **47**(3), 319–325 (2015)
3. Ehrenfest, D.M.D., Andia, I., Zumstein, M.A., Zhang, C.-Q., Pinto, N.R., Bielecki, T.: Classification of platelet concentrates (Platelet-Rich Plasma-PRP, Platelet-Rich Fibrin-PRF) for topical and infiltrative use in orthopedic and sports medicine: current consensus, clinical implications and perspectives. Muscles Ligaments Tendons J. **4**(1), 3–9 (2014)
4. O'Shea, C.M., Werre, S.R., Dahlgren, L.A.: Comparison of platelet counting technologies in equine platelet concentrates. Vet. Surg. **44**(3), 304–313 (2015)

5. Briggs, C., Harrison, P., Machin, S.J.: Continuing developments with the automated platelet count. Int. J. Lab. Hematol. **29**(2), 77–91 (2007)
6. Vargas, A., Amescua-Guerra, L., Bernal, A., Pineda, C.: Principios físicos básicos del ultrasonido, sonoanatomía del sistema musculoesquelético y artefactos ecogeográficos. Acta Ortopédica Mexicana **22**(6), 361–373 (2008)
7. Franceschini, E., Yu, F.T., Destrempes, F., Cloutier, G.: Ultrasound characterization of red blood cell aggregation with intervening attenuating tissue-mimicking phantoms. J. Acoust. Soc. Am. **127**(2), 1104–1115 (2010)
8. Yu, F.T., Cloutier, G.: Experimental ultrasound characterization of red blood cell aggregation using the structure factor size estimator. J. Acoust. Soc. Am. **122**(1), 645–656 (2007)
9. Ma, X., Huang, B., Wang, G., Fu, X., Qiu, S.: Numerical simulation of the red blood cell aggregation and deformation behaviors in ultrasonic field. Ultrason. Sonochem. **38**, 604–613 (2016)
10. Schwerthoeffer, U., Winter, M., Weigel, R., Kissinger, D.: Concentration detection in water-glucose mixtures for medical applications using ultrasonic velocity measurements. In: 2013 IEEE International Symposium on Medical Measurements and Applications (MeMeA), pp. 175–178. Conference, Gatineau (2013)
11. Huang, C.C.: High-frequency attenuation and backscatter measurements of rat blood between 30 and 60 MHz. Phys. Med. Biol. **55**(19), 5801–5815 (2010)
12. Nam, K., Yeom, E., Ha, H., Lee, J.: Simultaneous measurement of red blood cell aggregation and whole blood coagulation using high-frequency ultrasound. Ultrasound Med. Biol. **38**(3), 468–475 (2012)
13. Callé, R., Ossant, F., Gruel, Y., Lermusiaux, P., Patat, F.: High frequency ultrasound device to investigate the acoustic properties of whole blood during coagulation. Ultrasound Med. Biol. **34**(2), 252–264 (2008)
14. Treeby, B.E., Zhang, E.Z., Thomas, A.S., Cox, B.T.: Measurement of the ultrasound attenuation and dispersion in whole human blood and its components from 0–70 MHz. Ultrasound Med. Biol. **37**(2), 289–300 (2011)

3D Probabilistic Morphable Models for Brain Tumor Segmentation

David A. Jimenez[(⊠)], Hernán F. García,
Andres M. Álvarez, and Álvaro A. Orozco

Grupo de Investigación en Automática,
Universidad Tecnológica de Pereira, Pereira, Colombia
{david.jimenez,hernan.garcia,andres.alvarez1,aaog}@utp.edu.co

Abstract. Segmenting abnormal areas in brain volumes is a difficult task, due to the shape variability that the brain tumors exhibit between patients. The main problem in these processes is that the common segmentation techniques used in these tasks, lack of the property of modeling the shape structure that the tumor presents, which leads to an inaccurate segmentation. In this paper, we propose a probabilistic framework in order to model the shape variations related to abnormal tissues relevant in brain tumor segmentation procedures. For this purpose the database of the Brain Tumor Image Segmentation Challenge (Brats) 2015 is used. We use a Probabilistic extension of the 3D morphable model to learn those tumor variations between patients. Then from the trained model, we perform a non-rigid matching to fit the deformed modeled tumor in the medical image. The experimental results show that by using Probabilistic morphable models, the non-rigid properties of the abnormal tissues can be learned and hence improve the segmentation task.

Keywords: Brain tumor segmentation · 3D brain models
Shape fitting · Probabilistic morphable models

1 Introduction

Brain tumor segmentation is a challenging research topic, due to the need to locate accurately the abnormalities in a given brain tissue [1]. Since these abnormalities varies among patients, the clinical information derived from the medical diagnostic can lead to a patient-specific surgery plan which is a demanding task for the medical experts [2]. Besides, the analysis of the texture information between healthy and abnormal tissues in multidimensional data, makes the appearance modeling of the brain tissue an important factor in magnetic resonance imaging analysis [3,4].

In order to extract the tumor area, most of the common segmentation techniques make use of the texture information to find a given contour that wraps

© Springer International Publishing AG, part of Springer Nature 2018
M. Mendoza and S. Velastín (Eds.): CIARP 2017, LNCS 10657, pp. 314–322, 2018.
https://doi.org/10.1007/978-3-319-75193-1_38

the abnormal tissue [3]. To this end, most of the state-of-the-art methods employ active contours [5], texture descriptors [6], and fuzzy approaches [7], to locate the abnormal regions and then establish the area to extract. The main problems of these approaches are based on the inabilities of modeling the anatomical structure of the abnormal tissues, such as non-rigid properties of the shapes, greatly variability in size and different tumor location between patients [1,8].

Morphable models (MM) are well known for make use of the prior information related to a given object [9]. Hence, by using this prior knowledge we can model the shape contour related to a given tumor [1,10]. However, classic approaches of these models lack the ability of modeling shape uncertainties (shape deformations among objects) which in the brain tumor segmentation context leads to an inaccurate segmentation [3,11]. Probabilistic principal component analysis (PPCA), has the property of modeling the density of complex data structures [12], which combined with morphable models can improve the generalizability of complex shape structures related to different brain tumors [8].

In this paper, we propose a probabilistic morphable model for brain tumor segmentation. We use a probabilistic approach to the classic point distribution model (PDM) in order to learn the shape variations related to those structures that present abnormal tissues (brain tumors). From the trained probabilistic morphable model, we perform a fitting process based on non-rigid iterative closest point (ICP) [13], to match the modeled shape of the tumor to the brain volume (MRI). Our contribution, is based on the probabilistic representation of the point distribution model that allows us to model complex shape structures related to large deformations over different patients (i.e. different brain tumors). The rest of the paper proceeds as follows. Section 2 provides a detailed discussion of materials and methods. Section 3 presents the experimental results and some discussions about the proposed method. The paper concludes in Sect. 4, with a summary and some ideas for future research.

2   Materials and Methods

2.1   Database

In this work we used the Brain Tumor Image Segmentation Challenge[1] (Brats) 2015 [14]. This Database contains high-grade tumors, Low-grade tumors and labels maps made by experts based on landmarks. The tumors of this database are located in different brain regions. The label map showed in Fig. 1, have four different labels (1) for Necrosis (Green), (2) for the Edema (Yellow), (3) for Non-enhancing tumor (Red) and (4) for Enhancing tumor (Blue). We used the MRI T1 images with resolution of 240×240 pixels and $1\,mm \times 1\,mm \times 1\,mm$ voxel size.

---

[1] https://www.smir.ch/BRATS/Start2015.

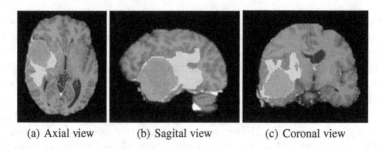

(a) Axial view   (b) Sagital view   (c) Coronal view

Fig. 1. Sample of a given abnormal tissue for the Brast2015 database. (Color figure online)

2.2 3D Brain Tumor Model

3D Model: In order to build a morphable model, we use the volume label information for each patient. From these volumes, we extract the cloud of points that represent the 3D shape of the tumor. We use marching cubes algorithm[2] to extract the real world coordinates $(g(h) \rightarrow \Omega)$ [9].

Model Training: Let us define a given $3D$ shape as $\boldsymbol{\Gamma}_i = [\mathbf{x_i}, \mathbf{y_i}, \mathbf{z_i}] \in \mathbb{R}^{Mx3}$ where $\mathbf{x_i}$, $\mathbf{y_i}$, $\mathbf{z_i} \in \mathbb{R}^{Mx1}$ contain the x, y and z locations of M vertex in the point cloud data related to the tumor. Hence, the training set is composed of N-$3D$ tumor shapes as $\mathcal{S} = \{\mathbf{s_i}\}_{i=1}^{N}$, where $\mathbf{s_i} = [\mathbf{x_i}^\top, \mathbf{y_i}^\top, \mathbf{z_i}^\top]^\top$ is the vectorized form of $\boldsymbol{\Gamma}_i$. This vectorized representation allows the assumption that the shape variation can be modeled using a normal distribution $\hat{\mathbf{s}} \sim \mathcal{N}(\bar{\mathbf{s}}, \boldsymbol{\Sigma})$, where the mean ($\bar{\mathbf{s}}$) and covariance matrix $\boldsymbol{\Sigma}$, are estimated as follows:

$$\bar{\mathbf{s}} = \frac{1}{N}\sum_{i=1}^{N}\mathbf{s_i}, \quad \boldsymbol{\Sigma} = \frac{1}{N-1}\sum_{i=1}^{N}(\mathbf{s_i} - \bar{\mathbf{s}})(\mathbf{s_i} - \bar{\mathbf{s}})^T. \tag{1}$$

We use a probabilistic variation of PCA that adds isotropic Gaussian noise to the model. This variation is called Probabilistic Principal Component Analysis (PPCA) [12]. The benefits of adding a probabilistic extension to the model are the ability to model missing data (non-rigid properties of the brain tumors). The PPCA approximation seeks to related given shape $\mathbf{s}$ (($M \times 3$)-dimensional observation vector) to a corresponding q-dimensional latent vector $\mathbf{z}$ (with less dimensionality). Let us set the linear relationships as,

$$\mathbf{s} = \mathbf{W}\mathbf{z} + \bar{\mathbf{s}} + \xi, \tag{2}$$

where $\mathbf{W} \in \mathbb{R}^{(M \times 3) \times q}$ relates the two sets of variables, ξ is the isotropic Gaussian noise ($\xi \sim \mathcal{N}(0, \sigma^2 I)$). We assume as prior, a standard Gaussian distribution

---

[2] We use the implementation available at http://iso2mesh.sourceforge.net/cgi-bin/index.cgi.

($\mathbf{z} \sim \mathcal{N}(0, I)$). Hence, the data follows the Gaussian distribution $\mathbf{s} \sim \mathcal{N}(\bar{\mathbf{s}}, \mathbf{C})$, where covariance of the model is $\mathbf{C} = \mathbf{W}\mathbf{W}^\top + \sigma^2 I$. Hence, the logarithmic likelihood is given by,

$$\mathcal{L} = -\frac{N}{2}\left((M \times 3)\log(2\pi) + \log|\mathbf{C}| + \mathrm{tr}(\mathbf{C}^{-1}\boldsymbol{\Sigma})\right). \tag{3}$$

By maximizing the log-likelihood for $\mathbf{W}$ and σ^2, we obtain:

$$\mathbf{W}_{\mathbf{ML}} = \boldsymbol{\Lambda}_q(\mathbf{b_q} - \sigma^2\mathbf{I})^{1/2}\mathbf{R}, \tag{4}$$

$$\sigma_{ML}^2 = \frac{1}{N-q}\sum_{j=q+1}^{d}\lambda_j, \tag{5}$$

where $\boldsymbol{\Lambda_i}$ are the eigenvectors of matrix $\boldsymbol{\Sigma}$, $\mathbf{b_q}$ is a diagonal matrix with the eigenvalues $(\lambda_1, \lambda_2, \ldots, \lambda_q)$ and $\mathbf{R}$ is an arbitrary $q \times q$ orthogonal rotation matrix [12].

From this approximation of PPCA, the eigenvalues and eigenvectors of the covariance matrix are used to create new shape instances from the mean shape as:

$$\hat{\mathbf{s}} \approx \bar{\mathbf{s}} + \boldsymbol{\Lambda}\mathbf{b}, \tag{6}$$

where $\boldsymbol{\Lambda} = (\boldsymbol{\Lambda}_1|\boldsymbol{\Lambda}_2|\ldots|\boldsymbol{\Lambda}_t)$ and $\mathbf{b}$ is a t-dimensional vector representing the shape parameters of the model (those who control the shape variability). The value of t is chosen such that the model represents the 98% of the variance, this is because there are eigenvalues that do not contribute to the reconstruction of $\mathbf{s}$ and can be considered as noise of the data, so this value of t allows to filter some noise in data. Finally, Fig. 2 shows the scheme of the proposed model. First, we model the shape variations using two morphable models, one based on PPCA and the other one using the traditional PCA approach. Secondly, from the trained models, we select the most relevant eigenvalues (λ) to represent the Volume of

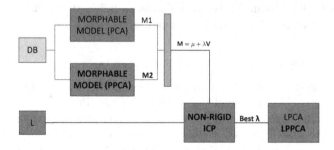

Fig. 2. DB stands for database, L stands for the prior (region-growing). Scheme of the proposed method. We compare our approach by building the classic morphable model based on PCA framework. Both models are matched to a given shape using non-rigid ICP approach.

a given tumor. Finally, region growing is used as shape prior to initialize the model and then Non-rigid Iterative Closest Point (ICP) algorithm is performed as the fitting strategy by using a local affine regularization [13].

3   Results and Discussions

In the following section we show the results for our 3D morphable model for brain tumor segmentation. On Sect. 3.1, we show the results of the 3D shape model for the brain tumor using a our approach. Section 3.2 shows the results for the fitting process of the morphable model using non-rigid ICP.

3.1   Training Results for 3D-MM

Figure 3, shows the mean shape derived from the training set. The figure reflects a surface not too smooth, because of the manual segmentation that includes some error and noise to the data. From these shape we can establish a given fitting process by deforming the modeled shape structure.

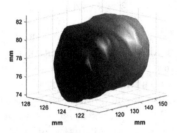

Fig. 3. Structure of the mean shape for the brain tumor morphable model.

Figure 4 shows the variability of the training set using our approach. The figure shows that by inducing isotropic Gaussian noise, the model captures relevant shape variations depicted by the model eigenvalues.[3]

3.2   Fitting Results

In order to match the morphable model to the brain volume a fitting strategy is needed. In this case, non-rigid ICP (Iterative Close Point) is performed. The model used as shape prior the region growing segmentation, then we deform the prior contour to match the real tumor volume as shown in Fig. 2.

Figure 5 shows a problem with the classic morphable model based on PCA. Here, Fig. 5(d) shows that the modeled tumor is inaccurate due to the abnormal representation of some corners of the shape contour. Nevertheless, Fig. 5(c)

---

[3] We set the shape of the training tumors Γ_i to have the same size, that means all vertices were adjusted to the smallest one.

Fig. 4. Effects of varying the first 2 shape parameters of the PDM for the analyzed brain tumor. Each shape parameter ranges from $-3\sqrt{\lambda_i}$ starting with the left column until $3\sqrt{\lambda_i}$ for the right column

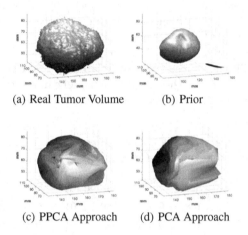

(a) Real Tumor Volume   (b) Prior

(c) PPCA Approach   (d) PCA Approach

Fig. 5. 3D-MM fitting.

shows that by using the probabilistic approach, the modeled tumor using PPCA exhibit better representation of the shape structure for a given tumor. In addition, another subjective evaluation made for the model was the Laplacian curvature showed in Fig. 6. In the right top corner the real volume has some high curvatures (red colors), which in our approach these curvatures are closest in the location to the real ones. However, they has lower curvatures, in some regions where the shape structure performs the deformation using the fitting strategy. Nevertheless, there are some dissimilarities between the curvature of the labeled shape and the modeled one. The main reason for this behavior is the reduction of vertexes in all of the shapes of the training set and the non-smoothed information that the ground-truth labels present.

Finally, Table 1 shows a quantitative evaluation of the segmentation accuracy of our approach. We compare our approach with the classic morphable model based on PCA and a common strategy for shape segmentation based on Fuzzy

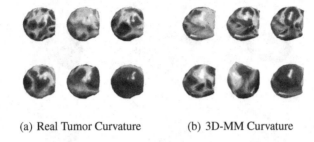

(a) Real Tumor Curvature   (b) 3D-MM Curvature

Fig. 6. Curvature of the real volume and the approximated by the probabilistic 3D-MM. (Color figure online)

C-means (FCM)[4]. The results show that our approach is more accurate that the classic morphable model based on PCA. For instance the Volume Ratio shows better results for our approach in comparison with PCA approach and Fuzzy C-means. However Fuzzy C-means shows a lower Dice coefficient (0.88 for FCM and 0.94 for our approach) since semantic rules for shape segmentation can lack of generalizability for those contours related to brain tumors. Besides, by using a probabilistic approach the morphable model can represent more complex structures related given shape (contour of the brain tumor).

Table 1. Model performance of the fitting strategy in terms of structural similarity index (SSIM), Volume Ratio and Dice Coefficient.

Tumor	SSIM	Volume Ratio	Dice
Classic 3D-MM	0.87 ± 0.026	0.89 ± 0.014	0.89 ± 0.0129
FCM	0.86 ± 0.021	0.89 ± 0.011	0.88 ± 0.0013
Ours	$\mathbf{0.91 \pm 0.021}$	$\mathbf{0.94 \pm 0.015}$	$\mathbf{0.94 \pm 0.0126}$

4 Conclusions and Future Works

In this paper, we propose a probabilistic morphable model for brain tumor segmentation using a probabilistic approach to the classic point distribution model (PDM) in order to learn the shape variations related to those structures that present abnormal tissues (brain tumors). From the trained PDM, we perform a fitting process based on non-rigid ICP, to match the modeled shape of the tumor to a given brain volume (MRI image). Our method deforms accurately a trained PDM in order to match a given brain tumor. Besides, by inducing isotropic Gaussian noise to the model, we can model those nonrigid properties that brain tumors present. The results show that by placing a Region growing method as

---

[4] We use the implementation available at https://github.com/ab93/SIFCM.

shape prior of the morphable model, the proposed approach can improve the accuracy of the segmented tumor. As future works, we want to analyze a morphable model framework using Gaussian Processes in order to obtain a more informative representation of the shape structure related to a set of abnormal tissues.

Acknowledgments. This research is developed under the project "Desarrollo de un sistema de soporte clínico basado en el procesamiento estócasitco para mejorar la resolución espacial de la resonancia magnética estructural y de difusión con aplicación al procedimiento de la ablación de tumores" financed by COLCIENCIAS with code 111074455860 under the program: *744 Convocatoria para proyectos de ciencia, tecnología e innovación en salud 2016*. H.F. García is funded by Colciencias under the program: *Formación de alto nivel para la ciencia, la tecnología y la innovación - Convocatoria 617 de 2013* with code 111065740687. Thanks to the master program of electrical engineer of the UTP for the support.

References

1. Gordillo, N., Montseny, E., Sobrevilla, P.: State of the art survey on MRI brain tumor segmentation. Magn. Reson. Imaging **31**(8), 1426–1438 (2013)
2. Nabizadeh, N., Kubat, M.: Automatic tumor segmentation in single-spectral MRI using a texture-based and contour-based algorithm. Expert Syst. Appl. **77**, 1–10 (2017)
3. Işin, A., Direkoğlu, C., Şah, M.: Review of MRI-based brain tumor image segmentation using deep learning methods. Procedia Comput. Sci. **102**, 317–324 (2016)
4. Menze, B., Jakab, A., Bauer, S., Kalpathy-Cramer, J., Farahani, K., et al.: The multimodal brain tumor image segmentation benchmark (BRATS). IEEE Trans. Med. Imaging 33 (2014)
5. Namías, R., D'Amato, J.P., del Fresno, M., Vénere, M., Pirró, N., Bellemare, M.E.: Multi-object segmentation framework using deformable models for medical imaging analysis. Med. Biol. Eng. Comput. **54**(8), 1181–1192 (2016)
6. Nabizadeh, N., Kubat, M.: Brain tumors detection and segmentation in MR images. Comput. Electr. Eng. **45**(C), 286–301 (2015)
7. Lladó, X., Oliver, A., Cabezas, M., Freixenet, J., Vilanova, J.C., Quiles, A., Valls, L., Ramió-Torrentí, L., Rovira, À.: Segmentation of multiple sclerosis lesions in brain MRI: a review of automated approaches. Inf. Sci. **186**(1), 164–185 (2012)
8. Devi, C.N., Chandrasekharan, A., Sundararaman, V., Alex, Z.C.: Neonatal brain MRI segmentation: a review. Comput. Biol. Med. **64**, 163–178 (2015)
9. Nair, P., Cavallaro, A.: 3-D face detection, landmark localization, and registration using a point distribution model. IEEE Trans. Multimed. **11**(4), 611–623 (2009)
10. Sjöberg, C., Johansson, S., Ahnesjö, A.: How much will linked deformable registrations decrease the quality of multi-atlas segmentation fusions? Radiat. Oncol. **9**(1), 1–8 (2014)
11. Sjöberg, C., Ahnesjö, A.: Multi-atlas based segmentation using probabilistic label fusion with adaptive weighting of image similarity measures. Comput. Methods Programs Biomed. **110**(3), 308–319 (2013)
12. Tipping, M.E., Bishop, C.M.: Probabilistic principal component analysis. J. Roy. Stat. Soc. Ser. B **61**, 611–622 (1999)

13. Amberg, B., Romdhani, S., Vetter, T.: Optimal step nonrigid ICP algorithms for surface registration. In: CVPR. IEEE Computer Society (2007)
14. Kistler, M., Bonaretti, S., Pfahrer, M., Niklaus, R., Büchler, P.: The virtual skeleton database: an open access repository for biomedical research and collaboration. J. Med. Internet Res. **15**(11), e245 (2013)

Multi Target Tracking Using Determinantal Point Processes

Felipe Jorquera, Sergio Hernández[✉], and Diego Vergara

Laboratorio Geoespacial, Universidad Católica del Maule, Talca, Chile
f.jorquera.uribe@gmail.com, shernandez@ucm.cl,
diego.vergara.letelier@gmail.com
http://www.ucm.cl

Abstract. Multi Target Tracking has many applications such as video surveillance and event recognition among others. In this paper, we present a multi object tracking (MOT) method based on point processes and random finite sets theory. The Probability Hypothesis Density (PHD) filter is a MOT algorithm that deals with missed, false and redundant detections. However, the PHD filter, as well as other conventional tracking-by-detection approaches, requires some sort of pre-processing technique such as non-maximum suppression (NMS) to eliminate redundant detections. In this paper, we show that using NMS is sub-optimal and therefore propose Determinantal Point Processes (DPP) to select the final set of detections based on quality and similarity terms. We conclude that PHD filter-DPP method outperforms PHD filter-NMS.

Keywords: Multi object tracking · Tracking-by-detection
Determinantal Point Processes

1 Introduction

Multi Object Tracking (MOT) is a challenging computer vision task with applications in video surveillance, event recognition and crowd monitoring among others. Conventional MOT methods based on tracking-by-detection, require an object detection step which locates target detections within an image and associate those detections with trajectories.

The general framework for the object detection step is to test image patches using a sliding window and a trained classifier. Then, redundant responses are usually suppressed by non-maximum suppression (NMS) [10]. Otherwise, object proposals entail a large number of raw detection responses around a true object. However, it is hard to detect occluded objects by this way. The detector is trained to distinguish between target classes and not to differentiate between intra-class variations. Therefore, when multiple objects are occluded, the detector assigns high confidence scores to foreground detections and low confidence scores to other objects. Therefore, the NMS scheme generates false negatives when multiple detections occur closely. On the other hand, if object detection is performed using strong appearance cues, this issue can be alleviated to some extent.

ⓒ Springer International Publishing AG, part of Springer Nature 2018
M. Mendoza and S. Velastín (Eds.): CIARP 2017, LNCS 10657, pp. 323–330, 2018.
https://doi.org/10.1007/978-3-319-75193-1_39

Currently, some of the top-performing trackers rely on strong affinity models [7]. Sparse appearance models are used in LINF1 [3], online appearance updates in MHT_DAM [6], integral channel feature appearance models in oICF [5] and aggregated local flow of long-term interest point trajectories in NOMT [1]. Recently, Deep Learning approaches have been proposed for tracking applications, e.g., MDPNN16 [13] uses Recurrent Neural Networks to encode appearance, motion, and interaction, and JMC [15] uses deep matching to improve the affinity measure.

Otherwise, whenever strong or weak detections are to be used, the data association step must handle multiple or missed detections. Current MOT algorithms tackle the data association step into an optimization problem, where the cost function is built upon pairwise similarity costs. While still being able to deliver consistent trajectories, most MOT algorithms based on optimization techniques underestimate the true number of tracks. The Probability Hypothesis Density (PHD) filter is a MOT algorithm that propagates the first-order moment of a multi-object density, using a moment-matching approximation of the true posterior [9]. Therefore, the method has the appealing property of being able to recursively estimate the number of tracks and their locations [11].

In this paper, we propose a tracking-by-detection approach based on the Probability (PHD) filter. Due to the strong dependency of the estimated number of tracks and the number of detections, Determinantal Point Processes (DPP) are used as an alternative to NMS.

2   Determinantal Point Processes

In order to solve the aforementioned problems of NMS, DPPS have been proposed to pedestrian detections [8]. DPPs can select a subset of detections by merging their objectness (detection scores) and individualness (similarity between detections candidates) into a single objective function.

Let N be the number of items and $\mathcal{Z} = \{1, 2, \ldots, N\}$ be the corresponding index set and $Z \subset \mathcal{Z}$ be a subset of indices for the selected items. Each item i is represented by its quality q_i and similarity S_{ij} to another item j. Using their qualities and similarities, we can compute a positive definite kernel matrix $L_Z = [q_i q_j S_{ij}]_{i,j \in Z}$. In this way, the DPP likelihood is a measure for the joint probability $\mathcal{P}_L(Z)$ of the selected indices. Then, we can find the most probable subset by solving the following optimization problem:

$$Z^* = \arg\max_{Z \subset \mathcal{Z}} (det(L_Z)) = \arg\max_{Z \subset \mathcal{Z}} (\prod_{i \in Z} q_i^2) \det([S_{ij}]_{i,j \in Z}). \tag{1}$$

Quality Term. The quality term is defined based on the assumption that wrong detections with high confidence scores degrade the accuracy detection and that a bounding box with a small number of raw detections inside it is more likely to contain ground truth detections [8]. Then, let $s = \{s_1, s_2, \ldots, s_N\}$ be set of scores for N raw detections, s_i^o the detection score, n_i the number of raw detections inside the current bounding box and λ a penalty constant, the detection score can be computed as $s_i = s_i^o \exp(-\lambda n_i)$. Thus, the quality score can be represented as follows:

$$q = \alpha s + \beta, \tag{2}$$

where α and β are weights for the quality term needed to balance the detection scores of different detectors.

Similarity Term. The similarity term combines appearance and spatial information of detections. Then, appearance is determined by the correlation of feature descriptors of bounding boxes, i.e. if we denote y_i as detection features vector for the i-th detection, the appearance term S^c can be computed as $S_{ij}^c = y_i^T y_j$. Also, the spatial information term is designed to give high correlation to multiple boxes around a single detection. Let S^s be the spatial individualness term, it is defined as $S_{ij}^s = \frac{|\sigma_i \cap \sigma_j|}{\sqrt{|\sigma_i||\sigma_j|}} \in [0,1]$, where σ_i denote a set of pixel indices belonging to the i-th detection box. Thus, the similarity term is defined merging S^c and S^s into a single diversity feature as follows:

$$S = \mu S^c + (1 - \mu)S^s, \tag{3}$$

where $0 \leq \mu \leq 1$ determines the relative importance of each feature term (Fig. 1).

2.1 Optimization Step

At the eliminating redundant detections step, the DPP mode finding (i.e. optimization for Eq. (1)) can be tackled using the following greedy algorithm [8]:

Algorithm 1. Greedy algorithm for solving (1)

    Input : $q, S, \mathcal{Z}, \epsilon$
    Output: Z^*
1 $Z^* = \emptyset$
2 **while** $\mathcal{Z} \neq \emptyset$ **do**
3     $j^* = \arg\max_{j \in \mathcal{Z}} (\prod_{i \in Z^* \cup \{j\}} q^2) \det(S_{Z^* \cup \{j\}})$
4     $Z = Z^* \cup \{j^*\}$
5   **if** $P_L(Z)/P_L(Z^*) > 1 + \epsilon$ **then**
6         $Z^* \leftarrow Z$
7   delete j^* from $\mathcal{Z}$
8   **else**
9   break
10   **end**
11 **end**

---

3 PHD Filter - DPP Algorithm Description

The PHD filter is an MOT algorithm that is based on random finite sets and point processes theory [12]. Although we introduce DPPs as a novel preprocessing technique, our implementation follows the standard PHD filter algorithm [16]. In order to compute the quality and similarity terms, we use a Support Vector Machine classifier with Histograms of Oriented Gradients features [2]. Therefore, our MOT algorithm can be summarized as follows:

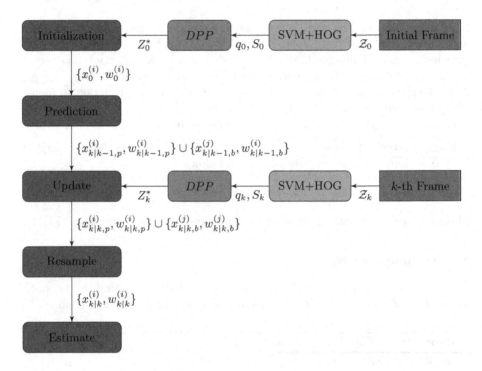

Fig. 1. PHD filter-DPP description.

- **Initialization.** At the initial time $k = 0$, the PHD filter states set $X_0 = \{x_0^{(1)} \dots, x_0^{(N_0)}\}$ is initialized from the observation set $Z_0^* = \{z_0^{(1)}, \dots, z_0^{(M_0)}\}$, i.e., DPP results. Also, a weight $w_0^{(i)} = 1/N_0$ is assigned to each one of the states.
- **Prediction.** From each one of the states $x_{k-1}^{(i)}$, we can draw $w_{k|k-1,p}^{(i)} \sim \pi(\cdot|x_{k-1}^{(i)})$ and $w_{k|k-1,p}^{(i)} = p_s w_{k-1}^{(i)}$, where p_s is the probability of survival. We use a measurement driven birth intensity approach. From the observation set Z_k^* (DPP pre-processed detections) a set of new particles is sampled from a birth density $x_{k|k-1}^{(j)} \sim b_k(\cdot|z_k^{(j)})$. Then, let v_k be the expected number of newborn objects at time k, weights $w_{k|k-1,b}^{(j)} = v_k/N_k^b$ are assigned to each new born state.

– **Update.** At this step, each state is updated using the DPP observations Z_k^*, according to:

$$w_{k|k,p}^{(i)} = (1 - p_D)w_{k|k-1,p}^{(i)} + \sum_{z \in Z_k^*} \frac{p_D g_k(z|x_{k|k-1,p}^{(i)})w_{k|k-1,p}^{(i)}}{\mathscr{L}} \quad (4)$$

and

$$\mathscr{L} = \kappa + \sum_{i=1}^{N_k^b} w_{k|k-1,b}^{(i)} + \sum_{i=1}^{N_k} p_D g_k(z|x_{k|k-1,p}^{(i)})w_{k|k-1,p}^{(i)}, \quad (5)$$

where κ is a constant probability of clutter and p_D is also a constant probability of detection.

– **Resample.** Given a threshold T, the states with weights $w_k^{(i)} < T$ are pruned. Then, the weights for the surviving particles are normalized to get the expected number targets $N_{k|k} = \sum_i w_k^{(i)}$.

– **Target State Estimation.** Cluster particles $x_{k|k}^{(i)}$ using the E-M algorithm to get $N_{k|k}$ states. Connect each one of the states at time k to only one of the tracks collected until time $k-1$.

4 Experiments

Firstly, we discuss the experimental settings for evaluating the proposed and existing methods, and then present the empirical results.

4.1 Experimental Settings

In order to demonstrate the advantages of the proposed method, we use the OSPA distance proposed by Schuhmacher et al. [14], which can be interpreted as a combination of two components referred to "localization" and "cardinality" errors.

Also, the proposed and existing methods were evaluated on sequence S1L1 (see Fig. 2) from PETS2009 [4] dataset, which contains 190 frames and exhibit regular crowd movement in a relatively dense queue in two directions.

    (a)   (b)   (c)

Fig. 2. Sequence S1L1 (PETS2009). (a) Ground-truth, (b) DPP, (c) NMS

4.2   Evaluation Results

In Table 1 the DPP parameters are shown, where α and β are the weight constants needed to balance detection scores of different detectors, ϵ is the acceptance parameter for Algorithm 1, λ is the penalty constant for overlap between detection and μ determine the importance of each detection feature descriptor. Also, Table 2 shows parameters of the SVM+HOG pedestrian detector, where G is the coefficient to regulate the NMS threshold and H is the threshold distance between detection features and SVM classifying plane. Furthermore, in Table 3 PHD Filter parameters are shown, where N_0 is the initial number of states, v_k is the expected number of newborn particles, κ is the probability of clutter, p_S is the survival probability and T is the resample threshold.

Table 1. DPP

Parameter	Values
Alpha α	0.9
Beta β	1.1
Epsilon ϵ	0.1
Lambda λ	−0.1
Mu μ	0.2, 0.4, 0.6, 0.8, 1.0

Table 2. SVM-HOG

Parameter	Values
Group	1
Threshold G	
Hit Threshold H	0.0, 0.2, 0.4, 0.6, 0.8, 1.0

Table 3. PHD Filter

Parameter	Values
Initial particles number N_0	100
Newborn objects number v_k	100
Probability of survival p_S	0.9
Probability of clutter κ	10^{-3}
Probability of detection p_D	0.7
Resample threshold T	1000

We note the correlation between the performance of the PHD filter-DPP method and the μ parameter (see Fig. 3a). Best performance of the proposed approach is achieved when $\mu = 0.7$, then we can conclude that even implementing strong features (HOG), solely using appearance individualness is not robust

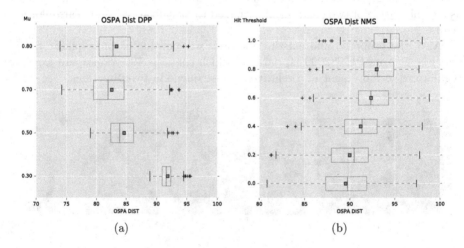

(a)   (b)

Fig. 3. OSPA distance comparison

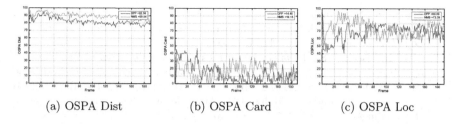

(a) OSPA Dist   (b) OSPA Card   (c) OSPA Loc

Fig. 4. Performance comparison between DPP v/s NMS

enough. Also, from Fig. 3b we can see a high correlation between H parameter and NMS performance. Best performance is achieved when $H = 0.0$.

Therefore, we set parameters specified by Tables 1, 2 and 3 and optimal values for μ and H ($\mu = 0.7$, $H = 0.0$, see Fig. 3). Thus, we note that proposed approach outperforms to the existing method (see Fig. 4).

Both implementations were developed by using C++ with OpenCV 3.1 and can be found in the author repository[1].

5   Conclusion

In this paper, we present a novel multi-target tracking method based on tracking-by-detection approach using DPP to introduce individualness and similarity between detections. The results show that the suppression of redundant detections using the proposed method outperforms NMS.

Acknowledgments. This work was supported by CONICYT/FONDECYT grant, project **Robust Multi-Target Tracking using Discrete Visual Features**, code 11140598.

References

1. Choi, W.: Near-online multi-target tracking with aggregated local flow descriptor. In: Proceedings of the IEEE International Conference on Computer Vision, pp. 3029–3037 (2015)
2. Dalal, N., Triggs, B.: Histograms of oriented gradients for human detection. In: IEEE Computer Society Conference on Computer Vision and Pattern Recognition, CVPR 2005, vol. 1, pp. 886–893. IEEE (2005)
3. Fagot-Bouquet, L., Audigier, R., Dhome, Y., Lerasle, F.: Improving multi-frame data association with sparse representations for robust near-online multi-object tracking. In: Leibe, B., Matas, J., Sebe, N., Welling, M. (eds.) ECCV 2016. LNCS, vol. 9912, pp. 774–790. Springer, Cham (2016). https://doi.org/10.1007/978-3-319-46484-8_47

---

[1] http://github.com/fjorquerauribe/multitarget-tracking.

4. Ferryman, J., Shahrokni, A.: Pets 2009: dataset and challenge. In: 2009 Twelfth IEEE International Workshop on Performance Evaluation of Tracking and Surveillance (PETS-Winter), pp. 1–6. IEEE (2009)

5. Kieritz, H., Becker, S., Hübner, W., Arens, M.: Online multi-person tracking using integral channel features. In: 2016 13th IEEE International Conference on Advanced Video and Signal Based Surveillance (AVSS), pp. 122–130. IEEE (2016)

6. Kim, C., Li, F., Ciptadi, A., Rehg, J.M.: Multiple hypothesis tracking revisited. In: Proceedings of the IEEE International Conference on Computer Vision, pp. 4696–4704 (2015)

7. Leal-Taixé, L., Milan, A., Schindler, K., Cremers, D., Reid, I., Roth, S.: Tracking the trackers: an analysis of the state of the art in multiple object tracking. arXiv preprint arXiv:1704.02781 (2017)

8. Lee, D., Cha, G., Yang, M.-H., Oh, S.: Individualness and determinantal point processes for pedestrian detection. In: Leibe, B., Matas, J., Sebe, N., Welling, M. (eds.) ECCV 2016. LNCS, vol. 9910, pp. 330–346. Springer, Cham (2016). https://doi.org/10.1007/978-3-319-46466-4_20

9. Mahler, R.P.: Multitarget Bayes filtering via first-order multitarget moments. IEEE Trans. Aerosp. Electron. Syst. **39**(4), 1152–1178 (2003)

10. Pirsiavash, H., Ramanan, D., Fowlkes, C.C.: Globally-optimal greedy algorithms for tracking a variable number of objects. In: 2011 IEEE Conference on Computer Vision and Pattern Recognition (CVPR), pp. 1201–1208. IEEE (2011)

11. Ristic, B., Clark, D., Vo, B.N., Vo, B.T.: Adaptive target birth intensity for PHD and CPHD filters. IEEE Trans. Aerosp. Electron. Syst. **48**(2), 1656–1668 (2012)

12. Ristic, B.: Particle Filters for Random Set Models. Springer, Heidelberg (2013). https://doi.org/10.1007/978-1-4614-6316-0

13. Sadeghian, A., Alahi, A., Savarese, S.: Tracking the untrackable: learning to track multiple cues with long-term dependencies. arXiv preprint arXiv:1701.01909 (2017)

14. Schuhmacher, D., Vo, B.T., Vo, B.N.: A consistent metric for performance evaluation of multi-object filters. IEEE Trans. Signal Process. **56**(8), 3447–3457 (2008)

15. Tang, S., Andres, B., Andriluka, M., Schiele, B.: Multi-person tracking by multicut and deep matching. In: Hua, G., Jégou, H. (eds.) ECCV 2016. LNCS, vol. 9914, pp. 100–111. Springer, Cham (2016). https://doi.org/10.1007/978-3-319-48881-3_8

16. Vo, B.N., Singh, S., Doucet, A.: Sequential Monte Carlo methods for multitarget filtering with random finite sets. IEEE Trans. Aerosp. Electron. Syst. **41**(4), 1224–1245 (2005)

Boosted Projection: An Ensemble of Transformation Models

Ricardo Barbosa Kloss[(✉)], Artur Jordão, and William Robson Schwartz

Smart Surveillance Interest Group, Department of Computer Science,
Universidade Federal de Minas Gerais, Belo Horizonte, Brazil
{rbk,arturjordao,william}@dcc.ufmg.br

Abstract. Computer vision problems usually suffer from a very high dimensionality, which can make it hard to learn classifiers. A way to overcome this problem is to reduce the dimensionality of the input. This work presents a novel method for tackling this problem, referred to as Boosted Projection. It relies on the use of several projection models based on Principal Component Analysis or Partial Least Squares to build a more compact and richer data representation. We conducted experiments in two important computer vision tasks: pedestrian detection and image classification. Our experimental results demonstrate that the proposed approach outperforms many baselines and provides better results when compared to the original dimensionality reduction techniques of partial least squares.

Keywords: Dimensionality reduction · Machine learning · Ensemble
Partial Least Squares · Computer vision

1 Introduction

Problems such as pedestrian detection and image classification are active research topics in computer vision due to their applications in surveillance and robotics. However, these problems usually suffer from very high dimensional data. Therefore, the employment of dimensionality reduction techniques is common to achieve reduction in training and testing cost and, most importantly, as a workaround to the curse of dimensionality, as can be seen in the work of Guyon and Elisseeff [10].

It has been shown that performing dimensionality reduction on the raw features improves data representation while also reduces computational cost [11,14, 16,17]. The idea is to use a projection on the raw data to find a new space that uses a smaller number of features, improving the classification performance. For instance, the Partial Least Squares (PLS) technique [15] was employed to project to a low dimensionality space, achieving better results than the state-of-the-art on the task of hyperspectral face recognition with spatiospectral information by Uzair et al. [20]. In another case, Akata et al. [1] employed Principal Component Analysis (PCA) to reduce the dimensionality of their SIFT [13] descriptors to circumvent the curse of dimensionality. It is important to note that such approaches

© Springer International Publishing AG, part of Springer Nature 2018
M. Mendoza and S. Velastín (Eds.): CIARP 2017, LNCS 10657, pp. 331–338, 2018.
https://doi.org/10.1007/978-3-319-75193-1_40

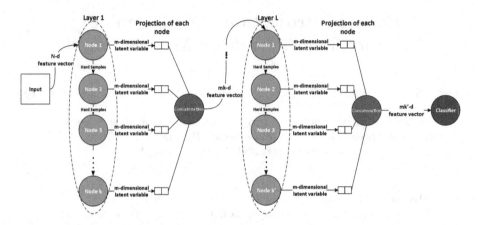

Fig. 1. General structure of the proposed framework. A node learns a projection using the hard samples found by the previous node in the same layer. Then, new layers can be created using the concatenation of the projections of each node in the previous layer as input.

could also be used together with deep learning techniques to potentially compact and improve deep features.

Besides the use of projection methods, another way of improving the classification is the employment of ensemble classifiers. These classifiers have been widely employed in several machine learning tasks [2,6,11,19] due to their simplicity and improved results. In general, ensemble classifiers are composed of weak classifiers (e.g., decision stumps) that are able to achieve a powerful classification when combined.

Inspired by the aforementioned studies, this work proposes a novel method that combines dimensionality reduction and ensemble techniques to find richer variables than those estimated by a single projection model. As illustrated in Fig. 1, the units of our ensemble are modeled as nodes built iteratively. Each node is composed of a projection model and a weak classifier learned with a distinct subset of samples hard to classify (referred to as *hard samples*[1]) by the previous node. The idea behind this procedure is to create projection models yielding rich and discriminative features for different categories of samples (from samples easier to harder to classify). It is also possible to cascade the projection obtained by a set of nodes to a new set of nodes, this is represented by multiple layers, as seen in Fig. 1. The goal of using multiple layers is to reduce possible redundancy among nodes.

Our proposed approach is similar to the Adaptive Boosting [9] in the sense that it reweighs samples according to their classification score. However, while the adaptive boosting focuses on grouping classifiers, we use the classifier as a tool to find hard samples. This is also similar to the hard mining employed by Dalal and Triggs [7], except that we do not search for these samples on a vali-

---

[1] Hard samples are samples that were misclassified when presented to a classifier.

dation set but on the training set itself. Another difference is that we iteratively search for hard samples to construct multiple models, whereas Dalal and Triggs employ this operation to construct a single model. Since we use the boosting-based idea of reweighing hard samples applied to a projection context, we refer to our method as *Boosted Projection*. The proposed method takes advantage of the characteristics of the underlying projection method, for instance reducing redundancy, and it might also present richer features due to the employment of the ensemble scheme to find hard samples.

The main contribution of this paper is the proposal of a novel method for feature projection, which demonstrates the possibility to enhance data representation exploiting the advantages provided by dimensionality reduction techniques and ensemble methods. In addition, the main advantage compared to other dimensionality reduction methods is that the nodes in our method explore different subsets of the data, where the deeper the node, the harder it is to classify its subset of samples, which can better convey discriminative information.

Our experimental results analyze the impact of the proposed Boosted Projection coupled with PLS [22], a method for dimensionality reduction that has been extensively explored in many tasks across the literature. To evaluate the proposed method, we test it in the pedestrian detection, using the INRIA pedestrian detection benchmark [7], and in the image classification task, using the CIFAR-10 [12] benchmark. On the former task, our method achieves a log-average miss-rate of 33.92, an improvement of 3.34 percentage points (p.p.) over the work of Jordão and Schwartz [11], a PLS based random forest, which to the best of our knowledge is the most recent PLS-based approach for pedestrian detection. On the latter task, we achieve an average accuracy of 76.25%, which is 1.01 p.p. better than the single PLS approach with the same number of dimensions.

2 Proposed Approach

In this section, we explain the proposed boosted projection method. We aim at creating an ensemble of projection methods that can yield richer and more discriminative features than a single projection method. To achieve such goal, we propose a framework for dimensionality reduction that explores discriminative information by employing a classifier to find samples that are hard to classify *(hard samples)* and using this information to learn new projections that prioritize the fitting of such samples.

The first unit of our framework is the *node*. It receives features of samples as input and learns a projection model that generates latent variables, used as new features. This projection is performed using a traditional projection method such as PCA [23] or PLS [22] (while the former estimates projections that maximize the variance of the data, the latter finds projection that maximizes the covariance between the data and class labels). The latent variables are used by a classifier to learn a discriminative model used to find which samples have been misclassified according to a threshold applied to the classifier responses. The threshold is initialized with an arbitrary value and is decreased for each new node to make samples easier to classify and guarantee that the nodes have more diversity. This is important since in ensemble methods, a high diversity is desirable [4,24].

After the hard samples have been identified for a given node, a subsequent node is created by learning a projection and a new classifier considering such samples, which will lead to a new subset of hard samples. This procedure continues until convergence, which can be met either by reaching a maximum number of nodes or by having no more misclassified samples. Thus, each node learns a model based on different subsets of samples, where the deeper the node, the harder it is to classify its subset of samples.

We refer to the set of learned nodes built until the convergence as a *layer*, as illustrated in Fig. 1. The output of a layer is the concatenation of the latent variables of each node of that layer. After all the nodes of a layer have been built, their output can be used as the input to create a new layer. The benefit of using additional layers is that if there is redundancy in some nodes, probably due to of a lack of diversity, the nodes of the next layer would be able to filter the redundancy out.

The output of the last layer is used to learn a final classifier, which might be different than the one used to find hard samples. It is important to emphasize that the output of the last layer is composed of projections learned for different subsets of samples and based on different features, causing each node's output to be an expert on that subset. This is an advantage of the proposed approach over a single projection method, such as PLS. On the other hand, a single projection-based approach would find the projection that is best for the average of the whole data, which would fail to model multi-modal distributions, for instance.

3   Experimental Results

In this section, we present our experiments and results achieved. First, we evaluate our method on the INRIA pedestrian detection benchmark, using log-average miss rate and compare it with other baseline approaches. Then, we evaluate our method on the task of image classification with the CIFAR-10 benchmark, using the average accuracy to measure the performance of the proposed approach against a single projection method.

3.1   Pedestrian Detection

The studies of Schwartz et al. [17] and Jordão and Schwartz [11] are the main works concerning PLS based dimensionality reduction applied to pedestrian detection. Therefore, to show that our method achieves better results than traditional projection-based approaches, we employed it to the pedestrian detection problem and compare with other methods that use comparable features (HOG and LBP).

To validate our results, we adopt the evaluation protocol used by state-of-the-art works, which is called *reasonable set* [8] (a detailed discussion regarding this protocol can be found in [3,8]), where the results are reported on the log-average miss rate. In addition, similarly to Jordão et al. [11], we used the TUD pedestrian dataset as validation set to calibrate some parameters of our method.

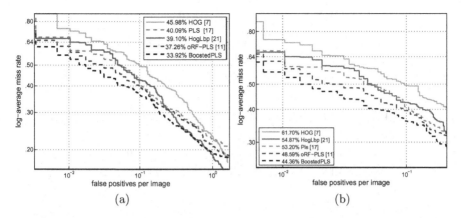

Fig. 2. Comparison of our boosted PLS projection with other approaches found in the literature. (a) results using the log-average miss-rate (lower values are better) of 10^{-2} to 10^{0} (standard protocol); (b) results using the area from 10^{-2} to 10^{-1}.

Our first experiment evaluates the impact of the number of nodes and layers to compose the boosted projection. On the validation set, the best values for these parameters were 20 and 3, respectively. The remaining parameter to be defined in our method is the number of components (latent variables), for each PLS that will compose a node of a layer (see Sect. 2). Similarly to Jordao and Schwartz [11] and according to the results achieved on the validation set, we note that this parameter is the most important parameter for PLS-based methods since varying this parameter from 3 to 6, the log-average miss rate decreases from 53.20 to 50.07, respectively. Based on the results obtained in the validation, our final PLS based boosted projection was set to 3 layers, where each layer is composed of 20 nodes that project the data to a 6-dimensional feature space using the PLS.

In our next experiment, we compare the results of the proposed Boosted Projection with other baseline methods. To provide a fair comparison, we considered the results reported by the authors in their works. Figure 2(a) shows the log-average miss rate (on the standard protocol) achieved by the methods on the INRIA person dataset. Our method achieved a log-average miss rate of 33.92, outperforming the works of Schwartz et al. [17] and Jordão and Schwartz [11] in 6.17 and 3.34 percentage points, respectively. On the area on the curve from 10^{-2} to 10^{-1} (Fig. 2(b)), which represents a very low false positive rate, our Boosted PLS still outperforms the other methods, a evidence of being robust to hard detections. Moreover, our proposed method uses around 30 times fewer features, compared to the work of Schwartz et al. [17], and achieves more accurate detections. It is also important to reinforce that our method is able to outperform detectors that employ more complex features, for instance, HOG-LBP [21], while using simple HOG-features (the same setup as proposed by Dalal and Triggs [7]).

Our last experiment regarding pedestrian detection evaluates the computational cost compared to the work of Jordão and Schwartz [11], which, to the best of our knowledge, is the most recent PLS-based method. We execute, on the same

Table 1. Results on the CIFAR-10 test dataset.

Method	Average accuracy	Dimensionality
SVM	71.62%	4096
PLS + SVM	69.10%	10
PLS + SVM	74.52%	30
PLS + SVM	75.24%	200
PLS + SVM	75.39%	500
Boosted PLS (1 layer)	75.94%	$20 \times 10 = 200$
Boosted PLS (2 layers)	**76.25%**	$20 \times 10 = 200$

hardware configuration of [11], the detection on a 640×480 image for 10 times, as is done by Jordão et al. [11], and we compute its confidence interval using 95% of confidence. Our method obtained a confidence interval of [67.02 s, 67.77 s] against [270.20 s, 272.44 s] (results reported by the authors) of [11], showing that the methods present statistical differences at the execution time. Therefore, according to the results, our method is four times faster than [11]. This reduction in computational cost is possible due to the use of fewer PLS projections than [11] (even if we consider the optimal case, where each oblique decision tree consists of only one PLS projection, see [11] for details).

3.2   Image Classification

To evaluate our method on the image classification task, we used the CIFAR-10 [12], a dataset with 32×32 pixels RGB colored images of ten classes. We extract deep features without data augmentation using the Keras Framework [5] with a VGG16 [18] based network. We tune the C parameter of the Support Vector Machine (SVM), with linear kernel, by employing stratified cross validation on the training set. The parameter was found to be optimal as $C = 0.1$, and we used this parameter for all models. Finally, we use the average of the accuracy of each class as the evaluation metric, which is the number of true positives over the sum of the number of true positives and false negatives (higher values are better).

The first column of Table 1 specifies the method used, which is either a SVM, a PLS transformation followed by a SVM, or the boosted version of PLS. The boosted version also uses a SVM to mine for hard samples and as the final classifier. The second and third columns show the results and the number of dimension for each method, respectively. Regarding the dimensionality (third column of Table 1), in the case of the SVM, the dimensionality is the original dimensionality of the features, 4096, however, with the dimensionality reduction method, PLS, the reported dimensionality is the number of components used. Furthermore, with the Boosted methods, we report the dimensionality as the product of the number of nodes and the number of components of the underlying projection method.

According to Table 1, we find the impact of the number of components in Table 1, the more components the models have, the better its representation. We can also see a correlation between the accuracy and the number of components. Employing the Boosted PLS transformation, with 10 components, and classifying with SVM (Boosted PLS) also achieved better results than with just the PLS transformation of the same model (PLS + SVM). In this case, the improvement is likely due to the fact that while PLS transformations minimize an average error, but thanks to the specialization of each node, our framework can preserve information of small distributions in the data that falls far from the average.

Regarding the PLS transformation (rows 2 to 5 in Table 1), we can see that from 10 components to 30 components there is an improvement of 5.42 percentage points, between 200 components and 500 there is only an improvement of a 0.15, illustrating the saturation of the components learned. Our method, however, with 200 dimensions, was able to improve over the PLS model with 500 components by 0.86 percentage points.

4   Conclusions

We presented a novel method for dimensionality reduction, the *Boosted Projection*. The method focuses on the idea of using several projection models (e.g., Partial Least Squares), to build an ensemble with more compact and richer representation of the raw features. We conducted experiments on two important computer vision tasks: pedestrian detection and image classification. In the first, we demonstrate that the proposed method outperforms many pedestrian detectors, using simple features. In addition, our method is more accurate and faster than [11], one of the most recent detector based on PLS. In the second task, we demonstrate that the proposed method is able to compute features richer than a single dimensionality reduction method.

Acknowledgments. The authors would like to thank the Brazilian National Research Council – CNPq, the Minas Gerais Research Foundation – FAPEMIG (Grants APQ-00567-14 and PPM-00540-17) and the Coordination for the Improvement of Higher Education Personnel – CAPES (DeepEyes Project). The authors gratefully acknowledge the support of NVIDIA Corporation with the donation of the GeForce Titan X GPU used for this research.

References

1. Akata, Z., Perronnin, F., Harchaoui, Z., Schmid, C.: Good practice in large-scale learning for image classification. PAMI **36**(3), 507–520 (2014)
2. Avidan, S.: Ensemble tracking. PAMI **29**(2), 261–271 (2007)
3. Benenson, R., Omran, M., Hosang, J., Schiele, B.: Ten years of pedestrian detection, what have we learned? In: Agapito, L., Bronstein, M.M., Rother, C. (eds.) ECCV 2014. LNCS, vol. 8926, pp. 613–627. Springer, Cham (2015). https://doi.org/10.1007/978-3-319-16181-5_47
4. Brown, G., Kuncheva, L.I.: "Good" and "Bad" diversity in majority vote ensembles. In: El Gayar, N., Kittler, J., Roli, F. (eds.) MCS 2010. LNCS, vol. 5997, pp. 124–133. Springer, Heidelberg (2010). https://doi.org/10.1007/978-3-642-12127-2_13

5. Chollet, F.: Keras (2015). https://github.com/fchollet/keras
6. Cogranne, R., Fridrich, J.: Modeling and extending the ensemble classifier for steganalysis of digital images using hypothesis testing theory. IEEE Trans. Inf. Forensics Secur. **10**(12), 2627–2642 (2015)
7. Dalal, N., Triggs, B.: Histograms of oriented gradients for human detection. In: CVPR (2005)
8. Dollár, P., Wojek, C., Schiele, B., Perona, P.: Pedestrian detection: an evaluation of the state of the art. PAMI **34**, 743–761 (2012)
9. Freund, Y., Schapire, R.E.: A desicion-theoretic generalization of on-line learning and an application to boosting. In: Vitányi, P. (ed.) EuroCOLT 1995. LNCS, vol. 904, pp. 23–37. Springer, Heidelberg (1995). https://doi.org/10.1007/3-540-59119-2_166
10. Guyon, I., Elisseeff, A.: An introduction to variable and feature selection. J. Mach. Learn. Res. **3**(Mar), 1157–1182 (2003)
11. Jordão, A., Schwartz, W.R.: Oblique random forest based on partial least squares applied to pedestrian detection. In: ICIP (2016)
12. Krizhevsky, A., Hinton, G.: Learning multiple layers of features from tiny images (2009)
13. Lowe, D.G.: Distinctive image features from scale-invariant keypoints. Int. J. Comput. Vis. **60**(2), 91–110 (2004)
14. Martis, R.J., Acharya, U.R., Min, L.C.: ECG beat classification using PCA, LDA, ICA and discrete wavelet transform. Biomed. Sig. Process. Control **8**(5), 437–448 (2013)
15. Rosipal, R., Krämer, N.: Overview and recent advances in partial least squares. In: Saunders, C., Grobelnik, M., Gunn, S., Shawe-Taylor, J. (eds.) SLSFS 2005. LNCS, vol. 3940, pp. 34–51. Springer, Heidelberg (2006). https://doi.org/10.1007/11752790_2
16. Sánchez, J., Perronnin, F., Mensink, T., Verbeek, J.: Image classification with the fisher vector: theory and practice. Int. J. Comput. Vis. **105**(3), 222–245 (2013)
17. Schwartz, W., Kembhavi, A., Harwood, D., Davis, L.: Human detection using partial least squares analysis. In: ICCV (2009)
18. Simonyan, K., Zisserman, A.: Very deep convolutional networks for large-scale image recognition. arXiv preprint arXiv:1409.1556 (2014)
19. Takemura, A., Shimizu, A., Hamamoto, K.: Discrimination of breast tumors in ultrasonic images using an ensemble classifier based on the adaboost algorithm with feature selection. IEEE Trans. Med. Imaging **29**(3), 598–609 (2010)
20. Uzair, M., Mahmood, A., Mian, A.: Hyperspectral face recognition with spatiospectral information fusion and PLS regression. TIP **24**(3), 1127–1137 (2015)
21. Wang, X., Han, T.X., Yan, S.: An HOG-LBP human detector with partial occlusion handling. In: ICCV (2009)
22. Wold, H.: Partial least squares. In: Encyclopedia of Statistical Sciences (1985)
23. Wold, S., Esbensen, K., Geladi, P.: Principal component analysis. Chemometr. Intell. Lab. Syst. **2**(1–3), 37–52 (1987)
24. Woźniak, M., Graña, M., Corchado, E.: A survey of multiple classifier systems as hybrid systems. Inf. Fusion **16**, 3–17 (2014)

Full-Quaternion Color Correction in Images for Person Re-identification

Reynolds León Guerra$^{(\boxtimes)}$, Edel B. García Reyes, and Francisco J. Silva Mata

Advanced Technologies Application Center (CENATAV), Havana, Cuba
`rleon@cenatav.co.cu`

Abstract. Nowadays, video surveillance systems are very used to safe-guard airport areas, train stations, public places, among others. Using these systems, the person re-identification is an automated task. How-ever, there are many problems that affect the good performance of per-son re-identification algorithms. For example, illumination changes in the scenes is one of the essential problems. It increases the false colors on the person appearance. Moreover, in the extraction of low level fea-tures (color) there is a need to obtain reliable colors of the image or person appearance. To this end, we propose a new algorithm using full-quaternions for color image representation. The quaternionic trigonom-etry and the Quaternion Fast Fourier Transform are used to improve person image in the frequency domain. Finally, to see the transformed image, an adaptive gamma function is developed. Experiment results on two datasets (VIPeR and GRID) show consistent improvements over the state-of-the-art approaches.

Keywords: Full-quaternion · Person image · Color
Fourier transforms

1 Introduction

Automation of tasks linked with video surveillance have received a lot of atten-tion recently by computer vision researchers [1]. This is because cameras located in avenue or public places allow person tracking, anomaly detection or person re-identification (Re-Id). The aim of person re-identification is identifying the same person in different cameras with non-overlapping camera views [2]. The task of Re-Id is a challenging problem caused by illumination changes, occlu-sion, pedestrian poses, background clutters, viewpoints and so on [3].

All illumination changes generate variance in the person appearance [4], for instance, false colors in person's clothes. It is clear because it is necessary to get true colors to decrease the error using descriptors as color histograms. Moreover, a stage in Re-Id is the feature extraction [5] that has a lot of influence on the image comparison (in a second stage). As explained above, color is an important feature for person re-identification.

The problem of looking for robust colors to illumination change is called color constancy (also color invariant, photometric normalization or color correction).

© Springer International Publishing AG, part of Springer Nature 2018
M. Mendoza and S. Velastín (Eds.): CIARP 2017, LNCS 10657, pp. 339–346, 2018.
https://doi.org/10.1007/978-3-319-75193-1_41

To address this problem, several alternative have been proposed for Re-Id algorithms. Ma et al. [6] propose using the Gray World algorithm to improve the parts of the person body. Kviatkovsky et al. [7] make groups of colors in a log-chromaticity space of the upper/lower parts of human body assuming that color clouds are invariant under some conditions. Jung et al. [8] use Shades of Gray algorithm to improve the person images. Eisenbach et al. [9] improve the image, taking as reference the person's clothes and the variations of illumination when the person passes through the scene in a learning process. Allouch [10] applies Retinex algorithm to improve images. Monari [11] developed an algorithm using shadow information of the person to estimate local illumination. Liao et al. [12] developed a version the multiscale Retinex algorithm to get vivid colors in person images. However, previous works for color correction in Re-Id do not take into consideration the connections among each color channel in the person image.

Person images have edges and isolated colors that arise from the self shadow or some anomaly (these isolated colors and edge are considered as false colors, because they have no relation to the pixel colors nearby), and are ignored by the algorithms, for instance, Gray World, which was designed for generic images (as landscapes). We propose a new algorithm for color correction in person images using full-quaternion (to preserve the relationships among colors) and the Quaternion Fast Fourier Transforms (QFFT) for the removal of the false colors previously mentioned.

Contributions of this paper are summarized as follows: First, a full-quaternion is designed using local and neighborhood information of each pixel. Second, in the frequency domain the person image is modified through its spectral module. Third, the image is displayed with an adaptive gamma function.

This paper is organized as follows. In Sect. 2, the algebra of the quaternion is explained to understand the following sections. The proposed approach, the full-quaternion, the QFFT and gamma function are explained in Sect. 3. Experimental results are presented and analyzed in Sect. 4. Finally, the conclusions are set out in Sect. 5.

2   Algebra of the Quaternions

In 1843, Hamilton presents the quaternions [13], denoted with letter $\mathcal{H}$. If $q \in \mathcal{H}$, then it is represented as follows:

$$\{q = t + xi + yj + zk | (t, x, y, z) \in \mathcal{R}\} \tag{1}$$

Where the complex operators $\mathbf{i}, \mathbf{j}, \mathbf{k}$ have the next rules $\{i^2 = j^2 = k^2 = ijk = -1, ij = k = -ji, ki = j = -ik, jk = i = -kj\}$. This shows that the multiplication of quaternions is not commutative. Other form to represent the quaternion is:

$$q = Sq + Vq \tag{2}$$

Where $(Sq = t)$ denotes the real or scalar part and Vq denotes the vector or imaginary part.

$$Vq = xi + yj + zk \in \mathcal{H}(Vq) \tag{3}$$

If $Sq = 0$, the quaternion is a pure quaternion and $q = Vq$, if $Sq \neq 0$ the quaternion is a full-quaternion. Other properties and operations are:

$$|q| = (t^2 + x^2 + y^2 + z^2)^{1/2} \tag{4}$$

$$q = Sq - Vq \tag{5}$$

Where the expression (4) and (5) are the module and conjugate respectively. If $|q| = 1$ and $q \in \mathcal{H}(Vq)$, then it is called unit pure quaternion.

Its representation in polar form is:

$$q = |q|e^{\tau\phi} = |q|(\cos\phi + \tau\sin\phi) \tag{6}$$

Where:

$$\phi = \tan^{-1}(\frac{|Vq|}{Sq}) \tag{7}$$

$$\tau = \frac{Vq}{|Vq|} \tag{8}$$

are respectively eigenangle and eigenaxis. The sine, hyperbolic cosine and inverse hyperbolic cosine [14] are represented as follows:

$$sin(q) = ((\sin(Sq)\cosh(|Vq|)) + (\cos(Sq)\sinh(|Vq|)\tau)) \tag{9}$$

$$cosh(q) = (exp(q) + 1/exp(q))/2 \tag{10}$$

$$acosh(q) = log(q + \sqrt{q^2 - 1}) \tag{11}$$

If $\hat{q} = a + bi + cj + dk$ and $\tilde{q} = w + xi + yj + zk$ and $(\hat{q}, \tilde{q}) \in \mathcal{H}$, then add, subtraction and multiplication are as follows:

$$\hat{q} \pm \tilde{q} = (a \pm w) + (b \pm x)i + (c \pm y)j + (d \pm z)k \tag{12}$$

$$\hat{q}\tilde{q} = (S\hat{q}S\tilde{q} - V\hat{q} \cdot V\tilde{q}) + (S\hat{q}V\tilde{q} + S\tilde{q}V\hat{q} + V\hat{q} \times V\tilde{q}) \tag{13}$$

Where, $< \cdot >$ is the scalar product and $\times$ is the vector cross product.

3 The Proposed Approach

Person image is represented to the domain $q \in \mathcal{H}$ (full-quaternion) where the real part is a value that represents the clear-dark contrast effect in the image. After, the image is transformed using hyperbolic cosine and with the Quaternion Fast Fourier Transform are modified the coefficients (module). Finally the person image is retrieved through the inverse QIFFT, the inverse hyperbolic cosine and the gamma function.

3.1   Full-Quaternion

An image has three color channels (Red, Green, Blue) and this can be represented by $f_{(x,y)}$. Nevertheless, the information among color channels is not associated and this is a constraint to the image analysis (the algorithms work only with the intensities of each channel). Also, all the colors depend of their neighbor colors to be analyzed. For example, in a patch with size 3×3 the neighbor pixels are gray and the central pixel is black, then the black color (noise) is generally omitted and the patch is classified as a gray homogeneous region. To solve this difference among colors, we start from the expression (9), as $f_{(x,y)} \in \mathcal{H}(Vq)$ then $Sq = 0$ implying that vector part can only be used in the expression (9) . The expression (8) is to normalize a quaternion, getting a unit pure quaternion. It is applied to all pixels, where take the value $|q| = 1$, and each pixel is located in the surface of a unitary sphere. It is deductible that $\sinh(|Vq|)$ acts as a scale factor on τ increasing or decreasing the size of the quaternion. We called clear-dark contrast effect (CDCE) to the increment or decrement of τ, because the dark colors are close to zero and light colors are close to higher values (taking the RGB cube as reference), for example the pink or white.

The transformation $f_{(x,y)} \longrightarrow f_{(x,y)}^H \in \mathcal{H}$, is performed as follows. A sliding window is built and the value of the function $(\sinh(|Vq|))$ of the pixels neighboring the central pixel is obtained in each position. The value of the real part of the full-quaternion is the average value of the neighborhood, as explained above.

$$\xi = \{\frac{1}{\eta} \sum_{\nu=1}^{\eta} log(\sinh(|Vq|) + 1)_\nu\} + ri + gj + bk \tag{14}$$

Where, $\xi \in f_{(x,y)}^H$, $\eta = 8$, $\{r, g, b\}$ are color channels.

3.2   Hyperbolic Cosine and Fourier Transforms

Differently from the other works, we consider that a modification of the coefficients of the spectral module in the person image allows obtaining more reliable colors (considering the false colors as noise in the signal). The quaternion hyperbolic cosine (see expression 10) is used to integrate the value of the real part with the vector part of the quaternion where a new full quaternion is obtained before applying the QFFT [15]. The QFFT allows transforming the frequency coefficients using filters in the spectral module where is possible to eliminate noises present in the spatial domain.

$$F(p,s) = S \sum_{m=0}^{m-1} \sum_{n=0}^{n-1} e^{-\mu 2\pi(\frac{pm}{M}) + (\frac{sn}{N})} f(m,n) \tag{15}$$

$$f(m,n) = S \sum_{p=0}^{p-1} \sum_{s=0}^{s-1} e^{\mu 2\pi(\frac{pm}{M}) + (\frac{sn}{N})} F(p,s) \tag{16}$$

Where, $S = \sqrt{\frac{1}{MN}}$, the expression (15) and (16) are the direct and inverse Quaternion Fourier Transforms respectively, μ is unit pure quaternion, p and s are frequency coefficients, m and n are spatial coordinates of the image. The filter used is a Gaussian filter applying a convolution on Υ (Module):

$$\zeta = \Upsilon * \Gamma \tag{17}$$

Where, $\{*\}$ is convolution operator and Γ is the Gaussian filter. Returning the image to the spatial domain is done as follows:

$$f(m,n) = exp(S \sum_{p=0}^{p-1} \sum_{s=0}^{s-1} e^{\mu 2\pi(\frac{pm}{M})+(\frac{sn}{N})}((\zeta + F(p,s))\beta)) \tag{18}$$

Where, β is a scale factor. After the image obtained in the spatial domain $f(m,n)$ is transformed by the expression (11).

3.3   Visualization

A problem with QIFFT is that $f(m,n) \in \mathcal{H}$, to see the image in $\mathcal{H}(Vq)$ is developed a gamma function as follows:

$$f_{(x,y)} = |C_i|_{(x,y)}^{1/|t|} \tag{19}$$

Where, C is color channel, $i \in \{r,g,b\}$, (x,y) are spatial coordinates of the image. The gamma function is adaptive because each channel is affected by the real part t of full-quaternion.

4   Experimental Results

Our experiments are to evaluate the best performance of the color correction algorithms in person images used for person re-identification. However, to the best of our knowledge there are no datasets with ground truth for true colors in person image (for person re-identification). Therefore, to evaluate the algorithm, we use as metric the Area Under Curve (AUC) for Cumulated Matching Characteristics (CMC) curve [16].

The datasets used are VIPeR and GRID. The VIPeR [17] has the image of the same person when passing in front of two different cameras; it contains 1264 images of people divided into 632 images for camera A and camera B. It has strong variations in views and changes in lighting. The characteristics in GRID [18,19] are: it contains 250 probe images (camera A) captured in one view and 250 images of the same person (in camera B), also there are 775 additional images that are not in the camera A, that is why we reject them and only work with 250 images. The images have strong variations in color and pose.

The parameters used in the experiment for our algorithm called **FqCC** (Full-quaternion Color Correction) are the following: $\mu = ((i + j + k)/\sqrt{3})$, $\beta = 1.5$.

The window and standard deviation of the Gaussian filter are respectively 3 and 0.8. The person image is segmented with a parsing person algorithm [20] to get the trunk and legs (see Fig. 1(i) and (j)). The feature vector is composed of the histograms of the color channels -red, green and blue- of the trunk and legs (because there are pixels that belong to clothing), also the hue and contrast histograms of the whole image using the $\sinh(|Vq|)$ for contrast and the hue channel of the HSV color space. The Bathacharyya distance [21] is used three times for the person marching for 16, 32 and 64 bins in the feature vector histograms (see Table 1). The algorithms used in the comparison are applied in the pre-processing step for person re-identification such as: multiscale retinex (MSR) adapted in the work of Liao [12], Shades of Grey (SofG) and Grey-World (GW) the framework [22]. Others algorithms as maxRGB (Mrgb), Grey-Edge (GE), Weighted Grey-Edge (WGE) [23] are used and original images (OI) without transformations, too.

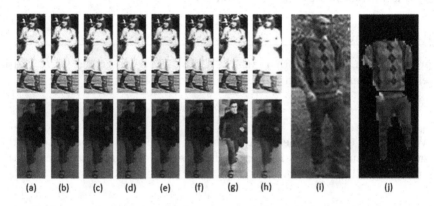

(a)   (b)   (c)   (d)   (e)   (f)   (g)   (h)   (i)   (j)

Fig. 1. Shows different results of the algorithms used for color correction (row above are VIPeR images ans row below are GRID images), (a) OI, (b) SofG, (c) GW, (d) Mrgb, (e) GE, (f) WGE (g) MSR, (h) **FqCC**. The image (j) is the trunk and legs of the image (i). (Color figure online)

In Fig. 1 (h - above) it can be seen that our algorithm removes the noise (black small regions in the dress) by white color in correspondence with true color of the dress and the arm edges are smoothed unlike the other algorithms. Also in Fig. 1 (h - below) image contrast is improved and yellow color decreases (considered as noise). The results obtained are shown in Table 1 where our algorithm gets the best results for a value of 0.7308 (to 16 bins) in the VIPeR dataset and 0.5876 (to 32 bins) in the GRID dataset. The values obtained in the GRID dataset are lower than those of VIPeR dataset due to the high noise level of the person images in the GRID dataset. The results show that not always the algorithms used to improve the colors in the images achieve their goals, as it is the case of SofG, GW, GE and WGE (with VIPeR) that obtained values smaller than the experiment with the image without transforming it and MSR, SofG, GW,

Table 1. Experimets with Bathacharyya distance.

Methods	VIPeR - 632			GRID - 250		
	16 bins	32 bins	64 bins	16 bins	32 bins	64 bins
FqCC	**0.7308**	0.7281	0.7251	0.5812	**0.5876**	0.5775
Original image	0.7076	0.7035	0.7021	0.5674	0.5726	0.5779
Multiscale retinex	0.7153	0.7116	0.7107	0.5526	0.5400	0.5339
Shades of grey	0.6834	0.6787	0.6766	0.5435	0.5404	0.5468
Grey-world	0.6759	0.6725	0.6713	0.5578	0.5642	0.5708
MaxRGB	0.7140	0.7091	0.7065	0.5607	0.5668	0.5686
Grey-edge	0.6906	0.6854	0.6828	0.5731	0.5785	0.5859
Weighted grey-edge	0.6931	0.6882	0.6857	0.5665	0.5731	0.5806

Mrgb and WGE in the GRID dataset. However, the algorithm herein proposed obtains the best results by contributing to the elimination of false colors (edges and isolated colors). This results are also, by the transformations of the image in the frequency domain and the incorporation of (CDCE) as an element of the full-quaternion.

5 Conclusions

Obtaining a value (real part), based on the clear-dark contrast effect, allows the construction of a full-quaternion and applying transformations in the frequency domain make it possible to correct the color and to improve the image. It was possible to establish an adaptive gamma function to display the image without losing information of the real part. For future works, we are going to design a local approach to color correction and design a metric of image quality (without ground truth).

References

1. Wu, Z.: Human Re-Identification. Springer, Heidelberg (2016). https://doi.org/10.1007/978-3-319-40991-7
2. Deng, X., Ma, B., Chang, H., Shan, S., Chen, X.: Deep second-order Siamese network for pedestrian re-identification. In: Lai, S.-H., Lepetit, V., Nishino, K., Sato, Y. (eds.) ACCV 2016. LNCS, vol. 10112, pp. 321–337. Springer, Cham (2017). https://doi.org/10.1007/978-3-319-54184-6_20
3. Wang, H., Gong, S., Xiang, T.: Highly efficient regression for scalable person re-identification. arXiv preprint arXiv:1612.01341 (2016)
4. Schumann, A., Gong, S., Schuchert, T.: Deep learning prototype domains for person re-identification. arXiv preprint arXiv:1610.05047 (2016)
5. de Carvalho Prates, R.F., Schwartz, W.R.: Appearance-based person re-identification by intra-camera discriminative models and rank aggregation. In: 2015 International Conference on Biometrics (ICB), pp. 65–72. IEEE (2015)

6. Ma, B., Li, Q., Chang, H.: Gaussian descriptor based on local features for person re-identification. In: Jawahar, C.V., Shan, S. (eds.) ACCV 2014. LNCS, vol. 9010, pp. 505–518. Springer, Cham (2015). https://doi.org/10.1007/978-3-319-16634-6_37

7. Kviatkovsky, I., Adam, A., Rivlin, E.: Color invariants for person reidentification. IEEE Trans. Pattern Anal. Mach. Intell. **35**(7), 1622–1634 (2013)

8. Jung, J., et al.: Normalized metadata generation for human retrieval using multiple video surveillance cameras. Sensors **16**(7), 963 (2016)

9. Eisenbach, M., et al.: Learning illumination maps for color constancy in person reidentification (2013)

10. Allouch, A.: People re-identification across a camera network. Master of Science Thesis, KTH Royal Institute of Technology, Stockholm, Sweden (2010)

11. Monari, E.: Color constancy using shadow-based illumination maps for appearance-based person re-identification. In: 2012 IEEE Ninth International Conference on Advanced Video and Signal-Based Surveillance (AVSS), pp. 197–202. IEEE (2012)

12. Liao, S., et al.: Person re-identification by local maximal occurrence representation and metric learning. In: Proceedings of the IEEE Conference on Computer Vision and Pattern Recognition, pp. 2197–2206 (2015)

13. Morais, J.P., Georgiev, S., Sprig, W.: Real Quaternionic Calculus Handbook. Springer, Basel (2014). https://doi.org/10.1007/978-3-0348-0622-0

14. Turner, J.D.: Quaternion analysis tools for engineering and scientific applications. J. Guidance Control Dyn. **32**(2), 686–693 (2009)

15. Ell, T.A., Sangwine, S.J.: Hypercomplex fourier transforms of color images. IEEE Trans. Image Process. **16**(1), 22–35 (2007)

16. Bolle, R.M., et al.: The relation between the ROC curve and the CMC. In: Fourth IEEE Workshop on Automatic Identification Advanced Technologies, vol. 2005, pp. 15–20. IEEE (2005)

17. Gray, D., Brennan, S., Tao, H.: Evaluating appearance models for recognition, reacquisition, and tracking. In: Proceedings of the IEEE International Workshop on Performance Evaluation for Tracking and Surveillance (PETS) (2007)

18. Liu, C., Gong, S., Loy, C.C., Lin, X.: Person re-identification: what features are important? In: Fusiello, A., Murino, V., Cucchiara, R. (eds.) ECCV 2012. LNCS, vol. 7583, pp. 391–401. Springer, Heidelberg (2012). https://doi.org/10.1007/978-3-642-33863-2_39

19. Loy, C.C., Xiang, T., Gong, S.: Time-delayed correlation analysis for multi-camera activity understanding. Int. J. Comput. Vis. **90**(1), 106–129 (2010)

20. Luo, P., Wang, X., Tang, X.: Pedestrian parsing via deep decompositional network. In: Proceedings of the IEEE International Conference on Computer Vision, pp. 2648–2655 (2013)

21. Naik, N., Patil, S., Joshi, M.: A scale adaptive tracker using hybrid color histogram matching scheme. In: 2009 2nd International Conference on Emerging Trends in Engineering and Technology (ICETET), pp. 279–284. IEEE (2009)

22. Van De Weijer, J., Gevers, T., Gijsenij, A.: Edge-based color constancy. IEEE Trans. Image Process. **16**(9), 2207–2214 (2007)

23. Gijsenij, A., Gevers, T., Van De Weijer, J.: Improving color constancy by photometric edge weighting. IEEE Trans. Pattern Anal. Mach. Intell. **34**(5), 918–929 (2012)

Long Short-Term Memory Networks Based in Echo State Networks for Wind Speed Forecasting

Erick López[1(✉)], Carlos Valle[1], Héctor Allende[1], and Esteban Gil[2]

[1] Departamento de Informática, Universidad Técnica Federico Santa María,
Valparaíso, Chile
{elopez,cvalle,hallende}@inf.utfsm.cl
[2] Departamento de Ingeniería Eléctrica, Universidad Técnica Federico Santa María,
Valparaíso, Chile
esteban.gil@usm.cl

Abstract. Integrating increasing amounts of wind generation require power system operators to improve their wind forecasting tools. Echo State Networks (ESN) are a good option for wind speed forecasting because of their capacity to process sequential data, having achieved good performance in different forecasting tasks. However, the simplicity of not training its hidden layer may restrict reaching a better performance. This paper proposes to use an ESN architecture, but replacing its hidden units by LSTM blocks and to train the whole network with some restrictions. We tested the proposal by forecasting wind speeds from 1 to 24 h ahead. Results demonstrate that our proposal outperforms the ESNs performance in terms of different error metrics such as MSE, MAE and MAPE.

Keywords: Wind speed forecasting · Echo State Network
Long Short-Term Memory Network · Multivariate time series

1 Introduction

In order to integrate wind into an electric grid, it is necessary to estimate how much energy will be generated in the next few hours. Nevertheless, this task is highly complex because wind power depends on wind speed, which has a random nature. It is worth to mention that the prediction errors may increase the operating costs of the electric system, as system operators would need to use peaking generators to compensate for an unexpected interruption of the resource, as well as reducing the reliability of the system [14].

In this paper we focus on wind speed forecasting. Among the different alternatives [6], we chose recurrent neural networks (RNN) because they propose a network architecture that can process temporal sequences naturally, relating events from different time periods. Between different RNN proposals stand out

© Springer International Publishing AG, part of Springer Nature 2018
M. Mendoza and S. Velastín (Eds.): CIARP 2017, LNCS 10657, pp. 347–355, 2018.
https://doi.org/10.1007/978-3-319-75193-1_42

the Echo State Networks [5] and Long Short-Term Memory Networks [4]. Both networks have reported good performance in different time series forecasting tasks [7,12,13,16].

However, the simplicity of ESN may restrict improving its performance. On the other hand, the complexity of LSTM demands high computation time, making difficult to reach optimum performance. Given the above, in this paper we propose improving the performance of ESN by means LSTM blocks within of its hidden layer (reservoir), and test it to forecast a multivariate time series with multi-ahead steps adding 1 lag. First, we create a networks of 3 layers (input-hidden-output). The hidden layer will be formed by a high number of LSTM blocks, and it will have a sparse connection between blocks. Then the network is trained similarly as proposed in [8], but keeping the sparse connections and the spectral ratio. To finalize, we re-adjust the output layer weights with a ridge regression. Although we are forecasting a random vector (with 4 features), the proposal will be evaluated using standard time series metrics [10] only for wind speed forecasting. We consider wind speed, wind direction, ambient temperature, and relative humidity as input features of the Multivariate time series from three geographical points of Chile.

The rest of the paper is organized as follows. In Sect. 2 we describe briefly the ESN and LSTM models. Section 3 describes our proposal. Next, we describe the experimental setting on which we tested the method for different data sources in Sect. 4. Finally, the last section is devoted to conclusions and future work.

2   Echo State Network and Long Short-Term Memory Network

The Echo State Network (ESN) proposed by Jaeger [5] is a special kind of RNN, classified inside of the paradigm Reservoir Computing [9]. The model consist of 3 layers (input, hidden and output), where the hidden layer (called reservoir) is formed for high number of neurons and with a low connectivity rate (allowing self-connections). Initially all weights are initialize random (usually using a normal o uniform centred on zero). Next, only the output layer weights are fixed using a ridge-regression.

Let $\{U(t)\}_{t=1,...,T}$ a time series. Let r the number of neurons from reservoir; W is an $r \times r$ matrix with connections weights inside of reservoir; W_{out} is an $(r + n) \times k$ matrix where k is the number of neurons from output layer and n is the number of neurons from input layer. A restriction additional is than the spectral radius of W (maximum eigenvalue in absolute value) must be less or equal to 1. Then, the connection matrix of the reservoir is randomly initialized, W_{start}. Next, it updates the matrix by $W = W_{start} \cdot \alpha / \rho(W_{start})$, where $\rho(W_{start})$ is the spectral radius of W_{start}, and $\alpha \approx 1$ is established as the influence of an input $U(t)$ vanishes in the reservoir through time. The output signal of hidden neurons is calculated by expression $s_j(t) = (1 - a) \cdot s_j(t - 1) + a \cdot f(net_j(t))$, where $s_j(t)$ is output of jth hidden neurons at time t, $a \in [0,1]$ is a leaking rate that regulates the speed update of the reservoir dynamics, $f(.)$ is an activation function, and $net_j(t) = \sum_{i=1}^{r+n} x_i^{[j]} \cdot w_{ij}$ such that $x_i^{[j]}$ is the ith input signal to the jth hidden neuron and w_{ij} is the weight that connects the ith input to jth neuron.

Furthermore, it is necessary to define a parameter θ corresponding to the number of steps that the network will use to update the states of the reservoir without being considered in the process of adjusting the output layer. The above is because the states are initialized to zero, $s(0) = 0$.

Let S a matrix $(T - \theta) \times (r + n)$ with states of all output layer neurons for $t \in [(\theta + 1), T]$. Then, the output layer weights are calculate as:

$$W_{out} = (S' \cdot S + \lambda \cdot I)^{-1} \cdot S' \cdot Y,$$

where S' is transpose of S, λ is a parameter of regularization, I is identity matrix, Y is a $(T - \theta) \times k$ matrix with the target outputs.

Another class of RNN are the Long Short-Term Memory (LSTM) networks [4], which replaces the traditional neuron in the hidden layer (perceptron) by a memory block. This block is composed of one or more memory cells and three gates for controlling the information flow passing through the blocks, by using sigmoid activation functions with range $[0, 1]$. Each memory cell is a self-connected unit called "Constant Error Carousel" (CEC), whose activation is the state of the cell. Currently there are different variants to the original model [3], however, in this paper we are considering the original proposal of Gers [2], using only one memory cell per block. It should be noted that the Gers model truncates the gradients for the learning process.

3   Proposal

Although the ESN has reported a good performance [7,13], we can improve the way that it is trained by adjusting the weights of the layers that have been fixed (sacrificing its simplicity). For example, in [11] can be found a proposal to learn all the weights of the network, tested over a classification problem, showing a improve over a traditional ESN. In contrast, the LSTM network is trained adjusting all its weights, it usually achieves a good performance in forecasting tasks [12,16], however, depending on the optimization method utilized, it may converge slow. Inspired on both networks, LSTM and ESN, we propose an ESN-based network, that uses LSTM memory blocks. Initially all weights are randomly initialized, but connection matrix among blocks is sparse (similar to ESN) with 10 connected units on average. The input weights matrix (W_{input}) also is sparse. For speeding up the training process, based in [8], we start adjusting only the weights of output layer by ridge regression. Next, we train the whole network by a gradient method in an online form, but keeping the sparse connections and fixing the output weights (no update). Finally, we rescaled the recurrent weights matrix by the spectral ratio and we use a ridge regression to re-adjust the weights of the output layer. Algorithm 1 describes our proposal. It considers a network of three layers (input-hidden-output), where the hidden layer is composed by memory blocks and the output layer by simple perceptron

---

Algorithm 1. Proposal LSTM based ESN

Input: A set of instances $\{U(t), Y(t)\}_{t=1,...,T+H}$.
1: Randomly initialize all the weights from a uniform distribution $[-0.1; 0.1]$.
2: The recurrent weights matrix is scaled for getting a spectral ratio equal to α.
3: Fix the output layer weights same as ESN using $\{U(t), Y(t)\}_{t=1,...,T}$.
4: $Error_{best} \leftarrow$ CalculateForecastError$(\{Y(t), \hat{Y}(t)\}_{t=T+1,...,T+H})$
5: Save(network)
6: Define $i = 0$, attempts=1, MaxEpoch $= 1$, $\zeta = 10$, MaxAttemtps=10.
7: **while** $i <$ MaxEpoch **do**
8:      $i \leftarrow i + 1$
9:      **for** $t = 1$ **to** T **do**
10:        $\hat{Y}_{tmp}(t) \leftarrow$ ForwardPropagate$(U(t),$ network$)$
11:        UpdateWeights$(Y(t), \hat{Y}_{tmp}(t),$ network$)$
12:        Keep the sparse connectivity, making zero the originally inactive weights.
13:        Cleaned all weights that exceed ζ (except them from output layer).
14:      **end for**
15:     Repeat the step 2 and 3.
16:     $Error_{validation} \leftarrow$ CalculateForecastError$(\{Y(t), \hat{Y}(t)\}_{t=T+1,...,T+H})$
17:     **if** $Error_{validation} < Error_{best}$ then update $Error_{best}$ and save the **network**; else we increase **attempts** in 1 unit.
18:     **if** attempts $>$ MaxAttempts then stop while loop.
19: **end while**
Output: Last network saved

---

units. Moreover, T is the length of the training series, H is the length of ahead step in forecasting. In the first stage of the algorithm (steps 1 to 3) we initialize the network and we fix the output layer weights by proposed method in [8]. Next (steps 4 to 5), with the network trained in the previous stage, we calculate a validation error to use as reference and save the network. In step 6, some necessary parameters are set for following steps (MaxEpoch, ζ and MaxAttemtps must be set before starting the while loop). In steps 9 to 14, are typical steps for online learning, with the additional step 12 and 13. From step 16 to 17, we calculate a validation error and compare if it is lower than previous error, with the aim of saving the network with the lowest validation error found until the moment. If the current error is not lower, we increase the counter attempts. Finally, if we did not find a lower validation error after 10 failed attempts (not necessarily consecutive), we stop the search even though we have not reached the MaxEpoch.

4   Experiments and Results

In order to assess our proposal, we use three data sets from different geographic points of Chile: Dataset 1, code b08, (22.54°S, 69.08°W); Dataset 2, code b21, (22.92°S, 69.04°W); and Dataset 3, code d02, (25.1°S, 69.96°W). These data are provided by the Department of Geophysics of the Faculty of Physical and

Mathematical Sciences of the University of Chile, commissioned by the Ministry of Energy of the Republic of Chile[1].

We worked with the hourly time series, with no missing values. The attributes considered for the study are: wind speed at 10 m height (m/s), wind direction at 10 m height (degrees), temperature at 5 m (°C), and relative humidity at 5 m height. The series starting at 00:00 on December 1, 2013 to 23:00 on March 31, 2015. Each feature is scaled to $[-1, 1]$ using the min-max function.

To evaluate the accuracy of model, each available series is divided in $R = 10$ subsets using a sliding window approach of approximately 3 months with a shift of 1000 points (40 days approximately), as depicted in Fig. 1.

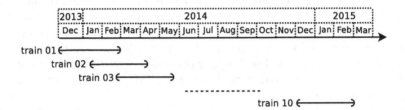

Fig. 1. Sliding window approach to evaluate model accuracy.

To check the performance over multi-step ahead forecasting, the above configuration is repeated 10 times with different seeds to initialize the random number generator. Then for each seed we measure the accuracy of the model considering the average error over the 24 ahead step, and also we compute the average over the R subsets, getting three standard metrics [10] based on $e_r(T + h|T) = y_r(T + h) - \hat{y}_r(T + h|T)$: MSE, MAE and MAPE.

$$\text{MSE} = \frac{1}{R} \sum_{r=1}^{R} \left(\frac{1}{24} \sum_{h=1}^{24} (e_r(T + h|T))^2 \right), \quad \text{MAE} = \frac{1}{R} \sum_{r=1}^{R} \left(\frac{1}{24} \sum_{h=1}^{24} |e_r(T + h|T)| \right),$$

$$\text{MAPE} = \frac{1}{R} \sum_{r=1}^{R} \left(\frac{1}{24} \sum_{h=1}^{24} \left| \frac{e_r(T + h|T)}{y_r(T + h)} \right| \right).$$

Here $y_r(T + h)$ is the unnormalized target at time $T + h$. T is the index of the last point, r is the index for each subset, and h is the number of steps ahead; $\hat{y}_r(T + h|T)$ is the target estimation of the model at the time $T + h$. The forecasting of several steps ahead was made with the multi-stage approach prediction [1]. In this work, we use 100 neurons in the hidden layer and we add one lag (since this network theoretically is able to model the sequential dynamics naturally), therefore as the time series has 4 features, the input layer has 4 neurons. First, we tuned the parameters of ESN: the leaking rate a, the regularization coefficient λ, and the spectral ratio α. The parameters a and α are

[1] http://walker.dgf.uchile.cl/Mediciones/.

obtained from uniform distribution $[0, 1]$ following the idea of a random search. While the parameter $\lambda \in \{10^{-10}, 10^{-9}, ..., 10^{-4}\}$. The previous tuning is run using W_{input} sparse or non-sparse. The parameters of the model are selected with seed 29. After tuning parameters, we train the model 10 times (as we explained above) with seeds 1, 2, 3, 5, 7, 11, 13, 17, 19, 23. For Dataset 1 the best parameters values are $a = 0.2640$, $\lambda = 1e{-}07$ and $\alpha = 0.9987$. For Dataset 2 are $a = 0.7559$, $\lambda = 1e{-}06$ and $\alpha = 0.9504$. Finally for Dataset 3 are $a = 0.7943$, $\lambda = 1e{-}07$ and $\alpha = 0.8294$. In case of Dataset 1 and Dataset 2 the best performance was obtained when W_{input} is sparse. However, for Dataset 3 not-sparse W_{input} outperforms. Our proposal run different values for `MaxEpoch`: 0 (E1), 1 (E2) and 20 (E3), using the best λ and α (we do not use the parameter a). While `MaxAttempts` and ζ are set as shown in Algorithm 1. The optimization algorithm used for learning is AdaDelta [15] with the parameters $\rho_a = 0.95$ and $\epsilon_a = 1e{-}8$.

The results show that the proposed method achieves a better performance in the three used metrics. Figure 2 shows the boxplot of MSE generated using the 10 seeds mentioned above. It is clear that our proposal gets better performance and also the error decreases while `MaxEpoch` increases. Table 1 presents a summary of all the results obtained for each metric in each dataset, but for space reasons we only show ESN and E3, and the results are rounded to 2 decimals. Is important to say that ESN, E1 and E2 present outliers, but in Fig. 2 is not shown for that the difference among the width of boxes is easy to see.

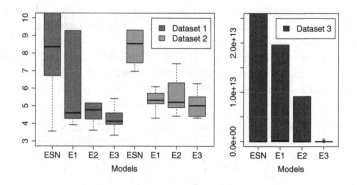

Fig. 2. Boxplot with MSE for each data set. (Color figure online)

A particular case occurs in Dataset 3, since the scale of the errors is very large. Reviewing in detail, the performance of the algorithm significantly changes depending on the seed that is used. However, our proposal despite to get high values, it keeps the errors lower than ESN (see blue boxplots in Fig. 2).

Given than the best tuning of ESN for Dataset 3 was achieved with non-sparse W_{input}, but our initial proposal considers to use a sparse connection, we repeat the experiments. Then, the best tuning for ESN using sparse W_{input} are: $a = 0.4065$, $\lambda = 1e{-}06$ and $\alpha = 0.9454$. The Table 2 shows a summary of the results got with the new parameters. It is interesting to observe that now

ESN reaches a **Mean**, **Q3** and **Max.** lower than those obtained with the non-sparse version, but the median is increased when we use MSE and MAE. While our proposal also improves your results, overcoming to ESN in the **Min.**, **Q1** and **Q2**.

Table 1. Summary of experiments results.

	Dataset 1		Dataset 2		Dataset 3	
	ESN	E3	ESN	E3	ESN	E3
MSE						
Min.	3.56	3.34	6.98	4.31	8.57	7.83
Q1	6.80	3.96	7.50	4.49	9.56	9.42
Q2	8.36	4.13	8.54	5.01	10.90	6326.38
Mean	1.78e+05	4.26	3.56e+20	5.09	9.95e+24	3.07e+10
Q3	2.11e+04	4.55	9.22	5.47	2.03e+21	2.05e+07
Max.	1.08e+06	5.42	3.56e+21	6.27	5.76e+25	3.07e+11
MAE						
Min.	1.46	1.41	2.11	1.60	2.33	2.22
Q1	1.87	1.47	2.23	1.65	2.47	2.41
Q2	2.09	1.53	2.32	1.76	2.58	8.99
Mean	21.72	1.55	3.86e+08	1.75	1.05e+11	5056.97
Q3	18.20	1.58	2.42	1.82	2.57e+09	332.91
Max.	134.99	1.75	3.86e+09	1.97	5.71e+11	4.82e+4
MAPE						
Min.	0.38	0.37	0.39	0.30	0.43	0.39
Q1	0.47	0.39	0.43	0.32	0.50	0.41
Q2	0.51	0.39	0.45	0.33	0.54	1.13
Mean	5.55	0.40	1.62e+08	0.33	1.05e+10	5.52e+02
Q3	4.71	0.40	0.48	0.34	2.71e+08	32.04
Max.	34.72	0.45	1.62e+09	0.37	6.04e+10	5303.02

Table 2. Results for Dataset 3 using non-sparse W_{input}.

	MSE		MAE		MAPE	
	ESN	E3	ESN	E3	ESN	E3
Min.	6.83	5.70	2.22	1.84	0.40	0.35
Q1	11.93	6.64	2.70	2.07	0.46	0.38
Q2	15.03	8.90	2.93	2.30	0.48	0.42
Mean	3.61e+10	1.58e+17	5250.69	1.04e+07	493.69	8.74e+05
Q3	35.23	2858.66	3.51	8.31	0.54	1.00
Max.	3.61e+11	1.58e+18	52480.43	1.04e+08	4932.27	8.74e+06

5   Conclusions and Future Work

This work presents an alternative to ESN using LSTM blocks in the hidden layer, with the additional advantage that the input and hidden layers can be trained. We observed that our proposal outperforms ESN in most cases tested. Depending on the data set used, the results may exhibit a high variability, as in it Dataset 3. In the last case, despite we do not outperform ESN in all indicators, is possible to improve the results since the experiments show that it improve as `MaxEpoch` increases, but in that case the proposed algorithm would lose more efficiency. In future work we plan to research its performance as we increase the number of lags on the time series. It would also be of interest to explore the performance as the number of layers increases.

Acknowledgments. This work was supported in part by Fondecyt 1170123 and in part by Basal Project FB0821.

References

1. Cheng, H., Tan, P.-N., Gao, J., Scripps, J.: Multistep-ahead time series prediction. In: Ng, W.-K., Kitsuregawa, M., Li, J., Chang, K. (eds.) PAKDD 2006. LNCS (LNAI), vol. 3918, pp. 765–774. Springer, Heidelberg (2006). https://doi.org/10.1007/11731139_89
2. Gers, F.: Long Short-Term Memory in Recurrent Neural Networks. Ph.D. thesis, École Polytechnique Fédérale de Laussanne (2001)
3. Greff, K., Srivastava, R.K., Koutník, J., Steunebrink, B.R., Schmidhuber, J.: LSTM: a search space odyssey. IEEE Trans. Neural Netw. Learn. Syst. **PP**(99), 1–11 (2016)
4. Hochreiter, S., Schmidhuber, J.: Long short-term memory. Neural Comput. **9**(8), 1735–1780 (1997)
5. Jaeger, H.: The "echo state" approach to analysing and training recurrent neural networks. GMD report 148, GMD - German National Research Institute for Computer Science (2001)
6. Jung, J., Broadwater, R.P.: Current status and future advances for wind speed and power forecasting. Renew. Sustain. Energy Rev. **31**, 762–777 (2014)
7. Liu, D., Wang, J., Wang, H.: Short-term wind speed forecasting based on spectral clustering and optimised echo state networks. Renew. Energy **78**(C), 599–608 (2015)
8. López, E., Valle, C., Allende, H., Gil, E.: Efficient training over long short-term memory networks for wind speed forecasting. In: Beltrán-Castañón, C., Nyström, I., Famili, F. (eds.) CIARP 2016. LNCS, vol. 10125, pp. 409–416. Springer, Cham (2017). https://doi.org/10.1007/978-3-319-52277-7_50
9. Lukoeviius, M., Jaeger, H.: Reservoir computing approaches to recurrent neural network training. Comput. Sci. Rev. **3**(3), 127–149 (2009)
10. Madsen, H., Pinson, P., Kariniotakis, G., Nielsen, H.A., Nielsen, T.S.: Standardizing the performance evaluation of short-term wind prediction models. Wind Eng. **29**(6), 475–489 (2005)
11. Palangi, H., Deng, L., Ward, R.K.: Learning input and recurrent weight matrices in echo state networks. CoRR, abs/1311.2987 (2013)

12. Schmidhuber, J., Gagliolo, M., Gomez, F.: Training recurrent networks by evolino. Neural Comput. **19**(3), 757–779 (2007)
13. Sheng, C., Zhao, J., Liu, Y., Wang, W.: Prediction for noisy nonlinear time series by echo state network based on dual estimation. Neurocomputing **82**, 186–195 (2012)
14. Sideratos, G., Hatziargyriou, N.D.: An advanced statistical method for wind power forecasting. IEEE Trans. Power Syst. **22**(1), 258–265 (2007)
15. Zeiler, M.D.: ADADELTA: an adaptive learning rate method. CoRR, abs/1212.5701 (2012)
16. Zhao, Z., Chen, W., Wu, X., Chen, P.C.Y., Liu, J.: Lstm network: a deep learning approach for short-term traffic forecast. IET Intell. Transport Syst. **11**(2), 68–75 (2017)

On the Use of Pre-trained Neural Networks for Different Face Recognition Tasks

Leyanis López-Avila[✉], Yenisel Plasencia-Calaña, Yoanna Martínez-Díaz,
and Heydi Méndez-Vázquez

Advanced Technologies Application Center, 7ma A # 21406, Playa, Havana, Cuba
{lelopez,yplasencia,ymartinez,hmendez}@cenatav.co.cu

Abstract. Deep Convolutional Neural Networks (DCNN) are the state-of-the-art in face recognition. In this paper, we study different representations obtained from a pre-trained DCNN, in order to determine the best way in which they can be used in different tasks. In particular, we evaluate the use of intermediate representations independently or combined with a Fisher Vector approach, or with a Bilinear model. From our study, we found that convolutional features may be more suitable than the features obtained from the last fully connected layers for different applications.

Keywords: Convolutional neural networks · Deep learning
Face recognition · Transfer learning

1 Introduction

Face recognition is one of the core problems in computer vision and it has been an active research topic in the last decades [14]. Most of existing methods have shown to work well on images or videos that are collected in controlled scenarios, but their performance often degrades significantly when there are large variations in pose, illumination, expression, aging, cosmetics, and occlusion, among others. To deal with this problem, different works have focused on learning invariant and discriminative representations from face images and videos [12].

In the last years, deep Convolutional Neural Networks (CNN) have demonstrated impressive performances on face recognition and are the state-of-the-art on this an other computer vision tasks [11,13]. CNNs are deep networks designed from the beginning to work with the large amounts of parameters that are fed into a network by an image, which exploit the structure and peculiarities of the input in order to learn discriminative features. To achieve this, they adopt a structure of alternating layers that achieves a reduction in characteristics (weights, parameters and features), and also decreases training times.

CNNs are hard to train because they have many hyperparameters (*e.g.* learning rate, momentum, different types of regularization, activation functions) and the layers of the networks can vary in type, number and width. The best network for a given problem is some combination of all those hyperparameters, so

© Springer International Publishing AG, part of Springer Nature 2018
M. Mendoza and S. Velastín (Eds.): CIARP 2017, LNCS 10657, pp. 356–364, 2018.
https://doi.org/10.1007/978-3-319-75193-1_43

researchers have to go through a vast space of possible combinations to find the best one. To perform all these combinations, great computational capacity is needed. However, once the model is learned, it can be used for other tasks without any further network training. This is beneficial for users who do not have the computer or data resources needed for training.

When using a pre-trained network, the standard choice is to compute the image descriptors at the end of the CNN. In this work, we aim at analyzing the behaviour of different representations obtained from a pre-trained network, in order to determine the most effective way in which the discriminative power of deep learning can be achieved without any training effort, and in different tasks. The adaptation of a trained method for a similar problem is called transfer learning, and the standard approach for CNN features is to retrain a new network starting from the first layers and weights learned by the trained network. Here we avoid retraining and analyze some of the different posibilities that exist for using an already trained network.

2 Using a Pre-trained Network for Face Recognition

There are different CNNs models publicly available in the literature. In this work, we choose the VGG-Face [11], because it is one of the most widely used for face recognition with outstanding results on several datasets, even very near to the best performing ones from commercial systems.

The VGG-Face network [11] has 11 blocks. The first eight blocks contain convolutional layers, while the last three blocks fully connected layers. The input to the network is a face image of size 224×224 with the average face image (computed from the training set) subtracted; and the output of the last convolutional layer, before the linear class predictor and the softmax layers is a 4096 vector descriptor. For every test image, ten 224×224 patches are obtained, by using the center and four corners, and applying horizontal flip to them. To enable multi-scale testing, the face image is first scaled to three different sizes of 256, 384 and 512 pixels, and the cropping procedure is repeated for each of them (see Fig. 1). In total 30 feature vectors are obtained which are averaged for obtaining the final face descriptor. This kind of net descriptor has been used in previous face recognition studies [8]. Besides, we propose to use other representations by using features extracted from not-fully connected blocks. They can be used like a tensor or a vector if we calculate the average from each descriptor in the tensor (see Fig. 1). We are going to refer to these descriptors as cube descriptor and average pooling descriptors respectively. A tensor can be read in different ways, but we use it as in [2], then a tensor is composed by several descriptors with size: $1 \times number\ of\ filters$ in the convolutional layer.

Apart from using a net descriptor directly, they can be used in combination with other methods. Here we will explore the use of Fisher Vector encoded convolutional features (FV-DCNN) [2] and Bilinear Models [3].

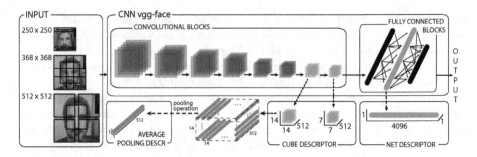

Fig. 1. Schematic view of the VGG-Face representations. The scales and crops of the face images used as input for the network are shown in the left and the convolutional blocks that were used for obtaining the intermediate descriptors are enlarged showing their dimensions.

2.1   FV-DCNN

The Fisher Vector (FV) method [12] has shown competitive performance on unconstrained face recognition. The original FV representation is obtained by densely computing local descriptors (e.g., SIFT), at different scales, which are aggregated into a high-dimensional vector by assuming a parametric generative model for the data and stacking the derivatives of its log-likelihood with respect to all its parameters. Usually, a Gaussian Mixture Model (GMM) with parameters $\theta = \{\pi_1, \mu_1, \sigma_1, ..., \pi_k, \mu_k, \sigma_k\}$ denoting the weight, mean vector and diagonal covariances of the K mixture components, is assumed as generative model. Let $S = \{\mathbf{s}_1, \mathbf{s}_2, ..., \mathbf{s}_N\} \in \mathbb{R}^Q$ be a set of local descriptors. The FV representation of S based on the GMM is given by $F = \left(\phi_1^{(1)}, \phi_1^{(2)}, ..., \phi_K^{(1)}, \phi_K^{(2)}\right)$ where the entries $\phi_{ik}^{(1)}$ and $\phi_{ik}^{(2)}$, $i = 1, ..., Q$, of the vectors $\phi_k^{(1)}$, and $\phi_k^{(2)}$ are defined as follows:

$$\phi_{ik}^{(1)} = \frac{1}{N\sqrt{\pi_k}} \sum_{n=1}^{N} \alpha_k(\mathbf{s}_n) \left(\frac{s_{in} - \mu_{ik}}{\sigma_{ik}}\right), \tag{1}$$

$$\phi_{ik}^{(2)} = \frac{1}{N\sqrt{2\pi_k}} \sum_{n=1}^{N} \alpha_k(\mathbf{s}_n) \left[\left(\frac{s_{in} - \mu_{ik}}{\sigma_{ik}}\right)^2 - 1\right], \tag{2}$$

where $\alpha_k(\mathbf{s}_n)$ is the soft assignment of $\mathbf{s}_n$ to component k. Finally, the FV is further improved by applying power normalization followed by l_2 normalization.

Recently, it was proposed in [2] to encode deep convolutional features with FV approach for unconstrained face verification. Based on this work, we extract the CNN descriptors from the last convolutional block of the VGG-Face network as described previously and encode them by applying FV. In order to meet the assumption of diagonal covariances for GMM, all descriptors are decorrelated by using PCA before feeding into FV encoding.

2.2   Bilinear Model

Another proposal in the literature for the use of trained networks is the bilinear model, initially proposed for image classification [7] and later used for face recognition [3]. Bilinear combination of two or more feature extractors allows considering extra information from pairwise interactions between initial features, which can contribute to increase discriminative ability in classification. This kind of mixing model combines the advantages of previous approaches for fine-grained recognition tasks, such as part-based and texture based models [7].

A bilinear model is defined in [7] as a quadruple $B = (f_A, f_B, P, C)$. Here f_A and f_B are feature functions, P is a pooling function and C is a classification function. A feature function is a mapping $f : L \times I \rightarrow R^{c \times D}$ that takes an image I and a location l and outputs a feature of size $c \times D$. A bilinear combination, for I and l is given by $bilinear(l, I, f_A, f_B) = f_A(l, I)^T * f_B(l, I)$. In this work we are going to combine three feature functions: A pre-trained CNN VGG-Face truncated at the end of convolutional blocks (as is proposed in [3] for face recognition), and the same descriptors but projected with PCA and FV methods.

3   Using a Pre-trained Network for Heterogeneous Face Recognition

An interesting application where the use of pre-trained models might be beneficial is for heterogeneous face recognition. It encompasses several modalities, and also in this case, state-of-the-art performance is achieved by deep learning approaches [6,10] with complex networks and training strategies. Our intuition is that the information contained in the pre-trained model can be useful to tackle these problems without having to retrain it or use a transfer learning approach.

In our research we take face-sketch recognition as a case of study, but we believe the same can be applied for different heterogeneous problems. Recently, a transfer learning approach was proposed in [9] for this task, in which a deep believe network is re-trained with sketch-image pairs, using as the initial weights, the ones provided by the trained model. Instead, we may use less computationally expensive approach on top of the deep features, by carefully selecting the appropriate information to be extracted from the trained model. We start from the convolutional features obtained from the network as discussed in the previous section, and improve them by metric learning which is very fast to train.

Metric Learning: The main goal of metric learning is to learn the weights related to a Mahalanobis distance of the form: $(x - y)^t M(x - y)$, where the matrix M is positive-semidefinite, while x and y are objects from a dataset. This is learned in such a way that the discriminative information for the problem is emphasized. The main idea is to converge to weights that bring objects belonging to the same class closer while pulling objects from different classes apart. These weights can be incorporated directly into the vectorial representations and the Euclidean distance can be computed on top of them. Here we use Linear Discriminant Analysis (LDA) [1], which can be seen as a type of metric learning method which learns a projection such that it maximizes the between or inter-class scatter over the within or intra-class scatter. This is solved using a closed-form expression based on a generalized eigenproblem.

4   Experimental Analysis

4.1   Performance Evaluation for Face Recognition

In order to evaluate and compare the different descriptors, we conducted experiments on the Labeled Face in the Wild (LFW) database [5]. It contains 13233 face images from 5749 people taken from Yahoo! News, with different variations in pose, scale, clothing, expression, focus, resolution among others. Since we aim at not training the network, we use the closed set protocol, where the dataset is divided in ten splits, with 300 genuines and 300 impostor comparisons for each of them. We used the aligned version of the images (LFW-a). The Euclidean distance and Cosine similarity were used as distance measures. We test the different descriptors that were described in Sect. 2. In particular we evaluate the original net descriptor from the 11th block of the network (b11) and the two different descriptors obtained from the third convolutional layer of the 8th block (b8): the concatenated cube descriptor (conc) and the average pooling descriptor (avg). We also test the FV-DCNN representation and the different bilinear combinations. In all cases we evaluate also the benefits of applying Principal Components Analysis (PCA) and L2 vector normalization.

Table 1 presents the obtained Equal Error Rates (EER) and the False Rejection Rate for a fixed False Acceptance Rate of 0.1% (FRR@0.1). The best results for every configuration is highlighted in bold. It can be seen that the original net descriptor (b11-norm) using Euclidean distance achieves an EER of 4.70% and all the different configurations evaluated outperform this result. This suggests that when using a pre-trained network for a given task, intermediate representations might achieve better results. It is also interesting to note that usually the feature vectors obtained from CNN networks are normalized and using intermediate representations we obtained better results without normalization. Besides, by applying PCA and using for classification vectors of only 20 or even 10 dimensions, results are equal to or better than the results found when using representations of higher dimensionalities. By using more complex representations such as FV-DCNN and Bilinear models good results are obtained but not better than simple intermediate representations from the network. As it was suggested in [7], for this combined configurations new training is needed.

4.2   Performance Evaluation for Face Sketch Recognition

From the analysis presented in the previous section, we consider that the average pooling descriptor from the third convolutional layer in block 8 (b8-avg) provides the best compromise between accuracy and efficiency; therefore we will use it

Table 1. EER (%) and FRR@0.1 (%) for different descriptors in LFW database.

Representation	Dim	Euclidean		Cosine	
		EER	FRR	EER	FRR
VGG-FACE					
b11	4096	15.3	54.2	5.10	27.7
b11-norm	**4096**	**4.70**	**25.7**	**4.60**	**25.7**
b11-PCA20	20	9.40	53.7	6.70	44.5
b11-PCA20-norm	20	6.30	43.1	59.3	44.2
b8-conc	$7 \times 7 \times 512$	0.80	8.00	4.30	43.4
b8-conc-norm	$7 \times 7 \times 512$	4.40	35.5	4.70	64.5
b8-PCA30	30×49	0.80	7.80	1.30	19.3
b8-PCA30-norm	30×49	4.10	34.7	4.10	36.7
(b8-PCA30)-conc-PCA20	**20**	**0.80**	**7.80**	**4.00**	**30.0**
(b8-PCA30-norm)-conc-PCA20	20	3.80	33.8	2.80	29.3
b8-avg	**512**	**0.80**	**7.80**	**4.30**	**46.5**
b8-avg-norm	512	4.20	35.1	4.20	56.4
b8-avg-PCA10	10	0.90	7.40	4.30	35.0
b8-avg-PCA10-norm	10	0.70	9.30	3.70	34.5
FV-DCNN					
fv-dcnn	3840	2.20	30.4	14.3	20.0
fv-dcnn-norm	3840	3.60	34.7	41.7	46.2
fv-dcnn-PCA30	**30**	**1.20**	**18.9**	**1.30**	**19.3**
fv-dcnn-norm-PCA30	30	63.8	100	64.8	100
BILINEAR					
bl-(b8)+(b8)	**512×512**	**3.30**	**61.2**	**4.30**	**39.9**
bl-(b8-norm)+(b8-norm)	512×512	4.20	35.1	4.20	70.5
bl-(b8-PCA20)+(b8-PCA20)	20×20	1.60	26.8	4.00	28.7
bl-(b8-PCA20-norm)+(b8-PCA20-norm)	20×20	5.70	39.4	2.80	29.2
bl-(fv-dcnn-PCA30)+(b8-PCA20)	**30×20**	**1.80**	**23.7**	**1.20**	**19.6**
bl-(fv-dcnn-norm-PCA30)+(b8-PCA20-norm)	30×20	4.30	34.3	2.90	30.7

for the experiments on face sketch recognition. Our goal here is to study the performance of these descriptors in a different problem for which the network was not trained for.

For the face sketch recognition evaluation we used the PRIP Viewed Software-Generated Composite (PRIP-VSGC) [4] dataset (see Fig. 2), which was created from photographs from 123 subjects from the AR database and three composites created for each subject using FACES (American and Asian users) and Identi-Kit softwares. Both mug-shots and sketches were normalized to the size required by

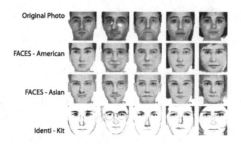

Fig. 2. Images from the PRIP-VSGC database.

Table 2. Recognition rate (%) at Rank-10 in PRIP-VSGC for sketch vs. image

Representation	Asian	American	Identi-Kit
Mittal [9]	48.10	56.00	**52.08**
b8-avg	43.90	61.78	17.88
b8-avg+LDA	46.34	**67.47**	17.88
b8-c2-avg	27.64	51.21	13.00
b8-c2-avg+LDA	51.21	64.22	26.82
b7-c2-avg	21.95	33.33	16.26
b7-c2-avg+LDA	**52.84**	63.41	42.27

the VGG-Face network. With the aim of comparison, we used the experimental protocol in [9], which takes as gallery the 123 subjects mugshots and as probe the sketches. Since in this case the image modality is very different from the original training of the network, we also evaluate previous layers representations, i.e.: the second layer from the same 8th block (b8-c2-avg) and the second layer from the 7th block (b7-c2-avg). The obtained results for each descriptor with and without metric learning are depicted in Table 2. It can be seen in Table 2 that in general the performance is improved by using lower layers, except for the American users, in which the sketches are more similar to the original images. Due to this higher resemblance to real photos, this can be expected, since for higher layers the network is more specialized. When using lower layers, the improvement is particularly larger for the Identi-Kit software, where the sketches are more different from real face images, they seem more like a drawing. Therefore, the lower layers which are less specialized for face images than the upper layers, are able to represent better the different face modalities. In general it is corroborated that the use of a metric learning is of particular importance since it is able of emphasize the discriminative information from the descriptors. The results are comparable with the ones obtained in [9], where a more specialized approach specifically trained for face sketch recognition was used. Besides, it is highly convenient that the intrinsic dimensionality achieved by the selected convolutional features is very low, and therefore the final representations can be very compact with a low memory footprint.

5   Conclusions

In this paper we study different approaches to exploit the information provided by a pre-trained convolutional network as well as its application for other domains with some similarities to the one for which the network was trained for. From our analysis, we found that the best performing features from the network were obtained with an average pooling descriptor from the last convolutional block. This representation is more compact and by applying PCA a very low dimensional representation can be obtained maintaining a high discriminative power. This can be useful for specific applications such as large scale face recognition. Besides, we found that by using intermediate representations vector normalization does not provide good results. In the case of heterogeneous face recognition we found that the pre-trained network performs similar to state-of-the-art approaches, after learning a metric for the specific problem with the provided features. We also found that lower blocks are better for problems where the imaging modality is more different from the one used for training the network. The explanation for this is that in top blocks the network is more specialized for the problem that it was trained for.

References

1. Belhumeur, P.N., Hespanha, J.P., Kriegman, D.J.: Eigenfaces vs. fisherfaces: recognition using class specific linear projection. IEEE Trans. Pattern Anal. Mach. Intell. **19**(7), 711–720 (1997)
2. Chen, J.C., Zheng, J., Patel, V.M., Chellappa, R.: Fisher vector encoded deep convolutional features for unconstrained face verification. In: ICIP, pp. 2981–2985. IEEE (2016)
3. Chowdhury, A.R., Lin, T.Y., Maji, S., Learned-Miller, E.G.: One-to-many face recognition with bilinear CNNs. In: WACV, pp. 1–9. IEEE Computer Society (2016)
4. Han, H., Klare, B., Bonnen, K., Jain, A.K.: Matching composite sketches to face photos: a component-based approach. IEEE Trans. Inf. Forensics and Secur. **8**(1), 191–204 (2013)
5. Huang, G.B., Ramesh, M., Berg, T., Learned-Miller, E.: Labeled faces in the wild: a database for studying face recognition in unconstrained environments. Technical report, Technical Report 07–49, University of Massachusetts, Amherst (2007)
6. Jin, Y., Lu, J., Ruan, Q.: Coupled discriminative feature learning for heterogeneous face recognition. IEEE Trans. Inf. Forensics Secur. **10**(3), 640–652 (2015)
7. Lin, T.Y., RoyChowdhury, A., Maji, S.: Bilinear CNN models for fine-grained visual recognition. CoRR abs/1504.07889 (2015)
8. Mehdipour Ghazi, M., Kemal Ekenel, H.: A comprehensive analysis of deep learning based representation for face recognition. In: Proceedings of the IEEE Conference on Computer Vision and Pattern Recognition Workshops, pp. 34–41 (2016)
9. Mittal, P., Vatsa, M., Singh, R.: Composite sketch recognition via deep network - a transfer learning approach. In: International Conference on Biometrics, ICB 2015, Phuket, Thailand, 19–22 May 2015, pp. 251–256 (2015)
10. Ouyang, S., Hospedales, T., Song, Y.Z., Li, X., Loy, C.C., Wang, X.: A survey on heterogeneous face recognition. Image Vis. Comput. **56**(C), 28–48 (2016)

11. Parkhi, O.M., Vedaldi, A., Zisserman, A.: Deep face recognition. In: BMVC, vol. 1, p. 6 (2015)
12. Simonyan, K., Parkhi, O.M., Vedaldi, A., Zisserman, A.: Fisher vector faces in the wild. In: BMVC, vol. 2, p. 4 (2013)
13. Taigman, Y., Yang, M., Ranzato, M., Wolf, L.: Deepface: closing the gap to human-level performance in face verification. In: Proceedings of the IEEE Conference on Computer Vision and Pattern Recognition, pp. 1701–1708 (2014)
14. Zhao, W., Chellappa, R., Phillips, P.J., Rosenfeld, A.: Face recognition: a literature survey. ACM Comput. Surv. **35**(4), 399–458 (2003)

Many-Objective Ensemble-Based Multilabel Classification

Marcos M. Raimundo and Fernando J. Von Zuben[⊠]

LBiC/DCA/FEEC - University of Campinas, Campinas, SP, Brazil
{marcos,vonzuben}@dca.fee.unicamp.br

Abstract. This paper proposes a many-objective ensemble-based algorithm to explore the relations among the labels on multilabel classification problems. This proposal consists in two phases. In the first one, a many-objective optimization method generates a set of candidate components exploring the relations among the labels, and the second one uses a stacking method to aggregate the components for each label. By balancing or not the relevance of each label, two versions were conceived for the proposal. The balanced one presented a good performance for recall and F1 metrics, and the unbalanced one for 1-Hamming loss and precision metrics.

Keywords: Multilabel classification · Many-objective optimization
Multiobjective optimization · Ensemble of classifiers · Stacking

1 Introduction

Multilabel classification is a generalization of the conventional classification problem in machine learning when, instead of assigning a unique, relevant label for each object, it is possible to assign more than one label per object. A straightforward approach, called Binary Relevance (BR), ignores any possible relationship among the labels and learns one classifier per label, for example, using kNN with Bayesian inference [23]. BR is computationally efficient, but it is not capable of exploring the relations among the labels to improve generalization. The main proposals devoted to promoting task relationship rely on Label Powerset, Classification Chains, and Multitask Learning.

Label powerset consists in transforming the multilabel problem into a multiclass one by creating a class for each combination of original labels. Despite exploring the relationship of labels, this proposal promotes an exponential growth of classes in the multiclass equivalent problem. Some solutions for this issue were proposed: converting the powerset process in random subsets of labels which are aggregated by simple voting [21]; excluding the labels on the multiclass equivalent problem characterized by few objects [12]; heuristically subsampling to overcome unbalanced data [2].

Considering an ordered sequence of labels, **Classification Chains** create a sequence of classifiers, each one taking the predicted relevance of the labels provided by classifiers previously trained. The considered sequence can be nominal

© Springer International Publishing AG, part of Springer Nature 2018
M. Mendoza and S. Velastín (Eds.): CIARP 2017, LNCS 10657, pp. 365–373, 2018.
https://doi.org/10.1007/978-3-319-75193-1_44

or random [13], and the architecture can be a tree instead of a sequence [10], so that the prediction depends on the parents of the label. Also, the classification can be based on the relevance probability [5].

Multitask learning creates binary relevance classifiers jointly exploring the relation of labels by structure learning [1]. This can be done by modeling the dependence among the labels using Ising-Markov Random Fields, further applied to constrain the flexibility of the task parameters adjustment [6], or using a multi-target regression proposal that explores multiple output relations in data streams [7].

Other methods were considered to extend these main proposals. **Ensembles** were proposed to increase robustness by resampling [12]; generating ensemble components using powersets in random sets of labels [21], and filtering then using genetic algorithms and rank-based proposals [4]; generating multiple classifiers by changing the label order in classification chains [13], and using many state-of-the-art multilabel classifiers to compose ensembles with different aggregation methods [19]. **Meta-learning** methods, instead of predicting the relevant labels found by binary relevance, predict the labels with higher membership degree, such that the number of predicted labels are estimated by a previously trained cardinality classifier [15, 20], or by a fixed optimal number of labels [11]. **Multi-objective optimization** was used to: create ensembles by optimizing a novel accuracy metric that takes into account the correlation of the labels and a diversity ensemble metric using evolutionary multiobjective optimization [16]; train an RBF network considering different sets of performance metrics as conflicting objectives [17,18]; and make feature selection in ML-kNN classifiers [22].

In this paper, we propose a novel ensemble method that uses a many-objective optimization approach to generate components exploring the relations among the labels (by weighted averaging the loss on each label), followed by a stacking method to aggregate the components for each label.

2 Multiobjective Optimization

Based on the Noninferior Set Estimation (NISE) method [3], the multiobjective optimization used in this work is an adaptive process that iteratively calculates a parameter vector **w** of the weighted method (Definition 1) using the previously found solutions and its related parameter vectors. The applied method, called MONISE, differs from NISE by being capable of dealing with a high number of objectives (many-objective problems) [9].

Definition 1. *The definition of the weighted method is as follows:*

$$
\begin{aligned}
&\underset{\mathbf{x}}{minimize}\ \ \mathbf{w}^{\top}\mathbf{f}(\mathbf{x})\\
&subject\ to\ \ \mathbf{x} \in \Omega,\\
&\qquad\qquad \mathbf{f}(\mathbf{x}) : \Omega \to \Psi, \Omega \subset \mathbb{R}^{n}, \Psi \subset \mathbb{R}^{m}\\
&\qquad\qquad \mathbf{w} \in \mathbb{R}^{m}, \mathbf{w}_{i} \geq 0\ \forall i \in \{1, 2, \ldots, m\}.
\end{aligned}
\tag{1}
$$

Parameter finding. Let us consider a utopian solution $\mathbf{z}^{utopian}$, composed of the infimum of all objectives, as well as a set $P, |P| \geq 1$ of solutions described by the objective vector $\mathbf{f}(\mathbf{x}^i)$, $i \in P$, and the weights $\mathbf{w}^i$, $i \in P$, used to find the solution based on the weighted method (Definition 1). We can then iteratively determine the approximation $\bar{\mathbf{r}}$ and the relaxation $\underline{\mathbf{r}}$ of the current frontier.

The **frontier relaxation** is given by the optimal solution of the weighted method. So, it is possible to conclude that $\mathbf{w}^{i\top}\underline{\mathbf{r}} \geq \mathbf{w}^{i\top}\mathbf{f}(\mathbf{x}^i)$, $\forall \underline{\mathbf{r}} \in \Psi$. The **frontier approximation** is also given by the optimality of the weighted method. Given a weight vector $\mathbf{w}$, the efficient solution must satisfy $\mathbf{w}^\top\bar{\mathbf{r}} \leq \mathbf{w}^\top\mathbf{f}(\mathbf{x}^i)$, $\forall i \in P$.

These concepts act as constraints to the best and worst possible solutions for the weighted method. The parameter selection procedure must find a new parameter vector $\mathbf{w}$ that leads to the maximum gap between $\underline{\mathbf{r}}$ and $\bar{\mathbf{r}}$. The problem formalized in Definition 2 is used to achieve this goal [9].

Definition 2

$$
\begin{aligned}
\underset{\mathbf{w},\bar{\mathbf{r}},\underline{\mathbf{r}}}{minimize} \quad & -\mu = \mathbf{w}^\top\underline{\mathbf{r}} - \mathbf{w}^\top\bar{\mathbf{r}} \\
subject\ to \quad & \mathbf{w}^i\underline{\mathbf{r}} \geq \mathbf{w}^i\mathbf{f}(\mathbf{x}^i) \ \forall i \in P \\
& \mathbf{w}^\top\bar{\mathbf{r}} \leq \mathbf{w}^\top\mathbf{f}(\mathbf{x}^i) \ \forall i \in P \\
& \underline{\mathbf{r}} \geq \mathbf{z}^{utopian}, \underline{\mathbf{r}} \leq \bar{\mathbf{r}}, \mathbf{w} \geq 0, \mathbf{w}^\top\mathbf{1} = 1.
\end{aligned}
\tag{2}
$$

The **initialization** of the algorithm consists of finding $\mathbf{z}^{utopian}$, by optimizing each objective separately (finding the infimum of each objective), and finding a solution with an arbitrary $\mathbf{w}$. Each **iteration** is composed of: 1. Finding a new parameter $\mathbf{w}^k$ using Definition 2; 2. Using this parameter to find a new solution $\mathbf{f}(\mathbf{x}^k)$; 3. Queue up the new solution k in the set P. The **stop criterion** is achieved when the number of solutions found achieves a limit R.

3   Statistical Model

The multilabel classification problem involves a set of N samples, also called objects, where $\mathbf{x}_i \in \mathbb{R}^d : i \in \{1,\ldots,N\}$ corresponds to the input feature vector and $\mathbf{y}_i \in \{0,1\}^L : i \in \{1,\ldots,N\}$ is the output feature vector we are willing to predict. In multilabel classification any output vector value $\mathbf{y}_i^l$ indicates the membership degree of label l to sample i, and any sample can have 0 to L labels assigned.

A binary relevance approach using logistic regression creates a classifier for each label l choosing the Bernoulli distribution as the predictive distribution $p(y|z) = z^y(1-z)^{1-y}$ and the softmax function $f(\mathbf{x}, \theta) = \frac{e^{\theta_1^\top \mathbf{x}}}{e^{\theta_0^\top \mathbf{x}} + e^{\theta_1^\top \mathbf{x}}} \in [0,1]$ as the input-output model. The optimization problem applied to find the optimal parameter vector θ^l is given by:

$$\min_{\theta^l} l(\theta^l, \mathbf{x}, \mathbf{y}^l) = -\sum_{i=1}^{N} \left[\frac{1}{n_1^l} \mathbf{y}_i^l \ln \left(f(\mathbf{x}_i, \theta^l) \right) + \frac{1}{n_0^l} (1 - \mathbf{y}_i^l) \ln \left(1 - f(\mathbf{x}_i, \theta^l) \right) \right]$$
(3)

where n_1^l is the number of 1s in label l and n_0^l is the number of 0s in label l.

Notice that, in this approach, the aim is to find the optimal parameters θ^l for specific label l using the general input vectors $\mathbf{x}_i \in \mathbb{R}^d : i \in \{1, \ldots, N\}$ and specific output values $\mathbf{y}_i \in \{0, 1\} : i \in \{1, \ldots, N\}$. In our approach, instead of finding a parameter vector (and a classifier) for each label, given a weighting factor $\mathbf{v}^l$ for each label, we use a joint optimization to find a single classifier, thus guiding to:

$$\min_{\theta} \sum_{l=1}^{L} \mathbf{v}_l l(\mathbf{x}, \mathbf{y}^l, \theta) + \lambda ||\theta||^2 \equiv \sum_{l=1}^{L} \mathbf{w}_l l(\mathbf{x}, \mathbf{y}^l, \theta) + \mathbf{w}_{L+1} ||\theta||^2.$$
(4)

where $||\theta||^2$ is the regularization component, λ is the regularization parameter, $\mathbf{w}_i = \frac{\mathbf{v}_i}{\sum_{k=1}^{L} \mathbf{v}_k + \lambda} \forall i \in \{1, \ldots, L\}$, and $\mathbf{w}_{L+1} = \frac{\lambda}{\sum_{k=1}^{L} \mathbf{v}_k + \lambda}$.

4   Stacking and Proposed Methodology

Stacking is an ensemble methodology that uses the outcome of the ensemble components (learning machines trained by an ensemble generation methodology, represented in Step 1 of Fig. 1a) to train another classifier (Step 2 of Fig. 1a) which will be responsible for making the prediction. In our proposal, the many-objective optimization method is responsible for generating R classifiers to compose the ensemble, and we create a stack classifier to predict each label l using logistic regression as the classification model:

$$\min_{\widehat{\theta}^l} - \sum_{i=1}^{N} \left[\frac{1}{k_1^l} \mathbf{y}_i^l \ln \left(f(\mathbf{z}_i, \widehat{\theta}^l) \right) + \frac{1}{k_0^l} (1 - \mathbf{y}_i^l) \ln \left(1 - f(\mathbf{z}_i, \widehat{\theta}^l) \right) \right]$$
(5)

where $\mathbf{z}_i^j$ is the degree of membership predicted by the j-th ensemble component with relation to the i-th sample.

In **Step 1**, ensemble components $(\{\theta_1, \theta_2, \ldots, \theta_l\})$ are generated by finding a set of efficient solutions, to the formulation of Eq. 4 using the methodology described in Sect. 2 (Fig. 1b). This generator of many-objective ensemble components can be seen as a **feature vector mapping** $(fm(\theta, x))$, so that each mapped feature is a classifier (Fig. 1c) associated with a distinct weight vector. From Eq. 4 we can realize that each classifier will give a distinct weight to the loss at each label and also to the regularization term. In **Step 2** (represented in Fig. 1d), the output of all efficient solutions are aggregated using a stacking approach [14], which can be seen as a cross-validation procedure in the mapped feature space $\widehat{x}$ using the model of Eq. 5. The training procedure in the second step is done for each label l employing the same feature vector $\widehat{x}$, but adopting the output $\widehat{y}^l$ of the worked label.

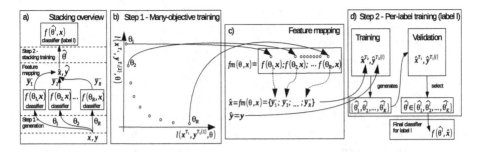

Fig. 1. Many-objective training followed by a stacking aggregation representation.

The set of classifiers were generated by a weighted average of the label losses. For this reason, not all classifiers will have a good performance for a specific label, thus requiring a more flexible aggregation, such as stacking, to create a final classifier.

5 Experimental Setting

5.1 Definition of Datasets and Evaluation Metrics

To evaluate the potential of the proposed multiobjective ensemble-based methodology we consider 6 datasets[1]. Table 1 provides the main aspects of these datasets. Aiming at obtaining better statistical results, we used a 10-fold split to create 10 independent test sets with 10% of the samples, and the remaining samples are randomly divided into 75% for training and 25% for validation for the baseline algorithms and 50% for T_1 set and 50% for T_2 set for the proposed method. T_1 set was used to create the ensemble components (Fig. 1b), and T_2 set to train the stacked classifiers (Fig. 1d). T_1 set was used again to select the model in the stacking training procedure (Fig. 1d).

The used evaluation metrics were: 1-Hamming Loss (1-hl), precision, accuracy, recall, F1 and Macro-F1 [6,8], all of those metrics associated with a quality measure in the interval $[0,1]$ so that higher values indicate a better method.

5.2 Contenders for a Comparative Analisys and Versions of the Proposed Method

To create a solid baseline we compared our method with 5 other approaches: Binary Relevance, Classification Chains [13], RakelD [21], Label Powerset[2]. All of those methods were implemented using Logistic Regression[3] as the base classifier

---

[1] Available at mulan.sourceforge.net/datasets-mlc.html.

[2] Implementations available at http://scikit.ml/.

[3] Available at http://scikit-learn.org.

Table 1. Description of the benchmark datasets.

Name	Instances	Attributes	Labels	Cardinality	Density of 1s
Emotions	593	72	6	1.869	0.311
Scene	2407	294	6	1.074	0.179
Flags	194	19	7	3.392	0.485
Yeast	2417	103	14	4.237	0.303
Birds	645	260	19	1.014	0.053
Genbase	662	1186	27	1.252	0.046

and had its parameters selected using hyperopt[4] with 50 evaluations to tune regularization strength and 50 more evaluations if the method involves another parameter (RakelD). Since the proposed and baseline algorithms use logistic regression as the base classifier, the attributes are considered as a vector of real numerical values.

In our proposal, we generate $10 * (L + 1)$ ensemble components, and the parameter selection on the stacking phase was implemented by cross-validation with 50 evaluations. We developed two versions of our proposal. These versions were created by balancing or not the importance of a label according to the stacking by changing the constants k_1^l and k_0^l on Eq. 5. In the unbalanced approach (described as MOn) k_1^l and k_0^l were set to 1, and in the balanced approach (described as MOb) k_1^l was set to the number of 1 s for this labels and k_0^l for the specific number of 0s.

6   Results

To promote an extensive comparison we presented the results from two perspectives. Figure 2 presents the average performance, calculated over the 10-folds, for all evaluated metrics for each dataset. And to make a more incisive evaluation, we used a Friedman paired test with $p = 0.01$ comparing all folders for all datasets, followed by a Finner post-hoc test with the same p, if Friedman test were rejected. Table 2 contains the evaluated method in the rows, and, for each performance metric, the Friedman ranking in the first column, how many methods are worse than the evaluated metric in the second column, and how many methods are better in the third column. This ordering relation (better and worse) is accounted only if there is a statistical significance according to the Finner post-hoc test. Looking to RakelD for the precision score, it is statistically significantly better than the worst ranked method: MOb (4.92), and statistically significantly worse than the three better-ranked methods: MOn (1.99), BR (2.74) and CC (2.94).

---

[4] Available at https://github.com/hyperopt/hyperopt.

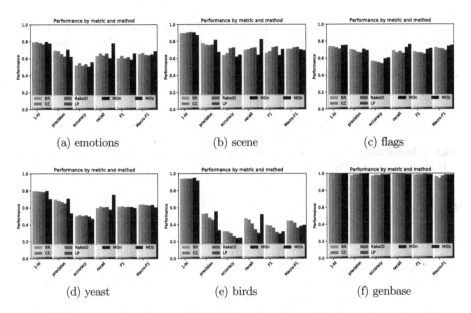

(a) emotions   (b) scene   (c) flags

(d) yeast   (e) birds   (f) genbase

Fig. 2. Average performance of the evaluated methods for each metric in each dataset.

Table 2. Average ranking and statistical comparisons for each metric.

Method	1-hl			Precision			Accuracy			Recall			F1			Macro-F1		
	Rank	>	<	Rank	>	<	Rank	>	<	Rank	>	<	Rank	>	<	Rank	>	<
BR	3.29	1	1	2.74	3	0	3.79	0	1	3.73	1	1	3.68	0	0	3.22	0	0
CC	3.29	1	1	2.94	3	1	2.64	3	0	3.16	1	1	2.9	1	0	2.87	0	0
RakelD	3.64	1	1	3.95	1	3	3.35	0	0	3.59	1	1	3.3	1	0	3.45	0	0
LP	3.89	0	1	4.43	0	3	3.12	1	0	3.85	1	1	3.54	0	0	4.01	0	0
MOn	2.23	5	0	1.99	4	0	4.31	0	2	4.96	0	5	4.37	0	3	3.75	0	0
MOb	4.63	0	4	4.92	0	4	3.75	0	1	1.69	5	0	3.17	1	0	3.66	0	0

7   Concluding Remarks

In this work, we successfully proposed a many-objective ensemble-based classifier to multilabel classification. Analyzing both Fig. 2 and Table 2, it is possible to see that MOn is the best-ranked classifier on 1-hl and precision but falling away on recall, accuracy, and F1. MOb is one of the best-ranked classifiers on recall and F1, but has difficulties on 1-hl and precision. These findings indicate that these two classifiers are biased for some metrics, exhibiting complementary performance. This behavior is due to the low density of the datasets, and to the fact that the non-balanced stacked model focuses the prediction on the 0s, thus producing high precision, as long as the balanced approach is predicting 1s more frequently, explaining the high recall score.

This scenario where an approach has a good performance on specific metrics at the expense of performance loss for other metrics can be useful in some applications. Given the absence of a dominant method for all metrics, our proposals can be seen as valuable choices in metric-driven scenarios. Also, since the complementary behavior was generated changing parameters, further exploration using ensembles of many-objective trained classifiers can promote good classifiers with different performance profiles.

Acknowledgments. This research was supported by grants from FAPESP, process #14/13533-0, and CNPq, process #309115/2014-0.

References

1. Caruana, R.: Multitask learning. Mach. Learn. **75**(1), 41–75 (1997)
2. Charte, F., Rivera, A.J., del Jesus, M.J., Herrera, F.: MLeNN: a first approach to heuristic multilabel undersampling. In: Corchado, E., Lozano, J.A., Quintián, H., Yin, H. (eds.) IDEAL 2014. LNCS, vol. 8669, pp. 1–9. Springer, Cham (2014). https://doi.org/10.1007/978-3-319-10840-7_1
3. Cohon, J.L., Church, R.L., Sheer, D.P.: Generating multiobjective trade-offs: an algorithm for bicriterion problems. Water Resour. Res. **15**(5), 1001–1010 (1979)
4. Costa, N., Coelho, A.L.V.: Genetic and ranking-based selection of components for multilabel classifier ensembles. In: Proceedings of the 2011 11th International Conference on Hybrid Intelligent Systems, HIS 2011, pp. 311–317 (2011)
5. Dembczy, K.: Bayes optimal multilabel classification via probabilistic classifier chains. In: Proceedings of the 27th International Conference on Machine Learning, pp. 279–286 (2010)
6. Gonçalves, A.R., Von Zuben, F.J., Banerjee, A.: Multi-label structure learning with Ising model selection. In: Proceedings of 24th International Joint Conference on Artificial Intelligence, pp. 3525–3531 (2015)
7. Osojnik, A., Panov, P., Džeroski, S.: Multi-label classification via multi-target regression on data streams. In: Japkowicz, N., Matwin, S. (eds.) DS 2015. LNCS (LNAI), vol. 9356, pp. 170–185. Springer, Cham (2015). https://doi.org/10.1007/978-3-319-24282-8_15
8. Madjarov, G., Kocev, D., Gjorgjevikj, D., Džeroski, S.: An extensive experimental comparison of methods for multi-label learning. Pattern Recogn. **45**(9), 3084–3104 (2012)
9. Raimundo, M.M., Von Zuben, F.J.: MONISE - many objective non-inferior set estimation, pp. 1–39 (2017). arXiv:1709.00797
10. Ramírez-Corona, M., Sucar, L.E., Morales, E.F.: Hierarchical multilabel classification based on path evaluation. Int. J. Approx. Reason. **68**, 179–193 (2016)
11. Ramón Quevedo, J., Luaces, O., Bahamonde, A.: Multilabel classifiers with a probabilistic thresholding strategy. Pattern Recogn. **45**(2), 876–883 (2012)
12. Read, J., Pfahringer, B., Holmes, G.: Multi-label classification using ensembles of pruned sets. In: Proceedings - IEEE International Conference on Data Mining, ICDM, pp. 995–1000 (2008)
13. Read, J., Pfahringer, B., Holmes, G., Frank, E.: Classifier chains for multi-label classification. Mach. Learn. **85**(3), 333–359 (2011)
14. Rokach, L.: Ensemble-based classifiers. Artif. Intell. Rev. **33**(1–2), 1–39 (2010)

15. Satapathy, S.C., Govardhan, A., Raju, K.S., Mandal, J.K.: Emerging ICT for Bridging the Future - Proceedings of the 49th Annual Convention of the Computer Society of India (CSI) Volume 2. Advances in Intelligent Systems and Computing, vol. 338, pp. 1–4. Springer, Heidelberg (2015). https://doi.org/10.1007/978-3-319-13731-5

16. Shi, C., Kong, X., Yu, P.S., Wang, B.: Multi-label ensemble learning. In: Gunopulos, D., Hofmann, T., Malerba, D., Vazirgiannis, M. (eds.) ECML PKDD 2011. LNCS (LNAI), vol. 6913, pp. 223–239. Springer, Heidelberg (2011). https://doi.org/10.1007/978-3-642-23808-6_15

17. Shi, C., Kong, X., Fu, D., Yu, P.S., Wu, B.: Multi-label classification based on multi-objective optimization. ACM Trans. Intell. Syst. Technol. 5(2), 1–22 (2014)

18. Shi, C., Kong, X., Yu, P.S., Wang, B.: Multi-objective multi-label classification. In: Proceedings of the 12th SIAM International Conference on Data Mining, SDM 2012, pp. 355–366 (2012)

19. Tahir, M.A., Kittler, J., Bouridane, A.: Multilabel classification using heterogeneous ensemble of multi-label classifiers. Pattern Recogn. Lett. 33(5), 513–523 (2012)

20. Tang, L., Rajan, S., Narayanan, V.K.: Large scale multi-label classification via metalabeler. In: Proceedings of the 18th International Conference on World Wide Web, p. 211 (2009)

21. Tsoumakas, G., Katakis, I., Vlahavas, I.: Random k-labelsets for multilabel classification. IEEE Trans. Knowl. Data Eng. 23(7), 1079–1089 (2011)

22. Yin, J., Tao, T., Xu, J.: A multi-label feature selection algorithm based on multi-objective optimization (2015)

23. Zhang, M.L., Zhou, Z.H.: ML-KNN: a lazy learning approach to multi-label learning. Pattern Recogn. 40(7), 2038–2048 (2007)

Bio-Chemical Data Classification by Dissimilarity Representation and Template Selection

Victor Mendiola-Lau[1], Francisco José Silva Mata[1(✉)],
Yenisel Plasencia Calaña[1], Isneri Talavera Bustamante[1],
and Maria de Marsico[2]

[1] Advanced Technologies Application Center, Havana, Cuba
{vmendiola,fjsilva,yplasencia,italavera}@cenatav.co.cu
[2] Universita degli Studi de Roma, Rome, Italy
demarsico@di.uniroma1.it

Abstract. The identification and classification of bio-chemical sub-stances are very important tasks in chemical, biological and forensic analysis. In this work we present a new strategy to improve the accuracy of the supervised classification of this type of data obtained from different analytical techniques that combine two processes: first, a dissimilarity representation of data and second, the selection of templates for the refinement of the representative samples in each class set.

In order to evaluate the performance of our proposal, a comparative study between three approaches is presented. As a baseline, entropy template selection (ETS) is performed in the original feature space and selected templates are used for training. The underlying concept of the other two alternatives, is the combination of Dissimilarity Representations and ETS. The first alternative performs ETS in the original feature space and uses the selected templates as prototypes for the generation of the dissimilarity space and as training set. The second one represents the data in the dissimilarity space, and next ETS is performed.

The experimental results showed that an adequate combination of the representation in the dissimilarity the space and the selection of templates based on entropy, outperformed the baseline in accuracy and/or efficiency for the majority of the problems studied.

Keywords: Dissimilarity representation · Entropy
Template selection · Classification · Bio-chemical data

1 Introduction

Classification is one of the most common tasks of pattern recognition. It is based on learning the structure of groups of objects with similar patterns. In this learning process, it is assumed that there are training examples that are representative and contain sufficient information to find a good (generalizable) model for predefined groups/classes of those examples. Therefore, new unseen

M. Mendoza and S. Velastín (Eds.): CIARP 2017, LNCS 10657, pp. 374–381, 2018.
https://doi.org/10.1007/978-3-319-75193-1_45

objects can be assigned into these groups reliably. The formalized representation of objects is an important aspect in the determination of a good class description.

The identification and classification of bio-chemical substances are very important tasks in chemical, biological and forensic analysis. Nowadays, bio-chemical data are analyzed as vectors of discretized data where the variables have no connection, and other aspects of their functional nature shape differences, are also ignored. The goal is to find new data representations with the capability to extract more information from the data taking into account its specific characteristics, which should improve classification results.

Dissimilarity Representation (DR) is rather a new approach which can take the functional information into account, and has shown to be advantageous in problems where the number of objects is small, and also when they are represented in high dimensional spaces, which are both common characteristics of bio-chemical spectral data sets.

Profound studies of the dissimilarity on chemical and bio-chemical data sets have not been carried out. There are already some results on spectral data in general [5,14], demonstrating its advantages for classification but a generalization of this behavior for the diversity of spectral and non-spectral analysis techniques and the selection of dissimilarity measures more appropriate for the typicity of the data has not yet been sufficiently researched. In consequence, there is still no consensus on the real scope of the improvements that can be obtained with the use of dissimilarity-based representations in a broader set of analytical techniques, underpinned by thorough experimentation and analysis of the results.

Given the wide physical and chemical nature of the data studied, the notion dissimilarities/proximities between samples play a key factor. Another factor to take into account is the representation chosen for building the dissimilarity space [12]. Recent works [4,10] have addressed this problem by selecting templates based on an entropy criterion. A suitable combination of these two factors is key for the improvement of the classification process. In this work, some approaches for combining dissimilarity representations with an entropy-based template selection are proposed (see Fig. 1):

- *(i) ENTR+CLAS* (a) perform ETS in the original feature space and the selected templates used as a training set (baseline).
- *(ii) ENTR+DR+CLAS* (b) perform ETS as in *(i)*, the selected templates are used as prototypes to build a dissimilarity space in which they are also used as a training set for classifiers.
- *(iii) DR+ENTR+CLAS* (c) build a dissimilarity space and in which to perform ETS and classifier validation.

2   Materials and Methods

The data sets used in this study come from different analytical techniques and sources, thus allowing for representativeness. Some data sets such as *IR Fuel* (Infrared spectroscopy of fuel) [13], *GC Cannabis* (Gas Chromatography of cannabis) [10], and *UV Cannabis* (Ultraviolet spectroscopy of cannabis) [10]

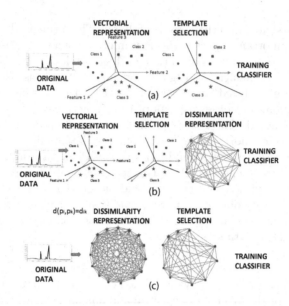

Fig. 1. Approaches for combining dissimilarity representations with entropy.

were obtained from our laboratory. *MVDA Alcohol* (Alcohols in medical applications) [8], *NIR Tablets* (NIR spectroscopy of tablets) [1], *NIR Tecator* (NIR spectroscopy of meat) [16], *Raman Porkfat* (Raman spectroscopy of porcine tissue) [2] and *Raman Tablets* (Raman spectroscopy of tablets) [1]; come from known sources. Due to the variety of the different analytical techniques, traditional data preprocessing methods like Multiplicative Scatter Correction (MSC) [7] and Savitzky-Golay (SG) derivatisation [15] were used.

3   Dissimilarity Representations

In the traditional framework of Pattern Recognition, feature representations are usually employed for objects. The Dissimilarity Representation [6], consists of an alternative representation for objects, where proximities and/or dissimilarities among them play a key role. Let X a set of objects in some space (which might not be a vector space), and $T = \{x_1, x_2, \ldots, x_N\}$ a finite sample of such space ($T \subset X$) consisting of N samples. Let $d : X \times X \rightarrow \mathbb{R}^+$ be a dissimilarity measure denoting the notion of resemblance between objects in T. A dissimilarity representation of an object x in T, denoted as $D(x,T) = [d(x,p_1), d(x,p_2), \cdots, d(x,p_N)]$, is a vector containing the dissimilarities between x and the objects in T. While building a dissimilarity representation, a representation set $R = \{r_1, r_2, \ldots, r_k\}$ ($R \in X$), is often employed. Using this representation set, an object x can be represented as $D(x,R) = [d(x,r_1), d(x,r_2), \cdots, d(x,r_k)]$, a vector containing the dissimilarities between x and each *prototype* in R. Each new *feature* corresponds to the dissimilarity with some prototype and the dimension of the space is determined by the amount of prototypes used.

The Dissimilarity Representation requires a suitable measure that expresses the relation among objects, in this case between the spectra. For substances, the concentration of its components and their presence is a key factor for a proper comparison. Therefore, the dissimilarity measures employed should take this relation into account, which different for every kind of spectra or data content [9]. In such manner, a more powerful representation can be obtained, allowing for a better discrimination.

4   Entropy-Based Template Selection

In previous works [4,10], the information gain has been used to select the best templates for improving classification accuracy. In analytic chemistry, the analysis performed on a substance is usually replicated, yielding almost identical replicated samples. The entropy-based template selection strategy proposed in [4] allows to identify the most representative templates in a given set. Thus, it is possible to reduce the size of the training set for classifiers by removing less significant samples. This procedures is outlined next.

Let G be a set of templates that can be expressed as the union of several subsets (groups), i.e., $G = \cup_{k=1}^{K} G_k$ and $G_k \cap G_h = \varnothing, \forall k \neq h$. Let $s : G \times G \to \mathbb{R}^+$ be a similarity measure defined over templates in G and the comparison of templates v and $g_{i,k} \in G_k$ be denoted as $s_{i,v} = s(v, g_{i,k})$, which is normalized to be a value in the interval $[0, 1]$ [4]. Considering a distance/dissimilarity measure $d_{i,v} = d(v, g_{i,k})$, it is possible to obtain a similarity measure $s_{i,v} = \frac{1}{1+d(v,g_{i,k})}$, which has the advantages of being normalized in $[0, 1]$ and avoid numeric errors for very small distance/dissimilarity value. Assuming that template v was correctly assigned to the class k, each $s_{i,v}$ value can be interpreted as the probability that template v conforms to $g_{i,k}$, that is $s_{i,v} = p(v \approx g_{i,k})$.

The entropy of the probability distribution obtained by applying (4) to the whole G_k with respect to a probe v is defined as follows:

$$H(G_k, v) = -\frac{1}{\log_2 |G_k|} \sum_{i=1}^{|G_k|} s_{i,v} \log_2 (s_{i,v}), \tag{1}$$

where $\frac{1}{\log_2(|G_k|)}$ is a normalization factor. Finally, the entropy value for G_k is computed by considering each template $g_{j,t} \in G_k$ as a probe v:

$$H(G_k) = -\frac{1}{\log_2 |Q|} \sum_{q_{i,j} \in Q} s_{i,j} \log_2 (s_{i,j}), \tag{2}$$

where Q represents the set of pairs $q_{i,j} = (q_{i,k}, q_{j,k})$ of elements in G_k such that $s_{i,j} > 0$. $H(G_k)$ represents a measure of heterogeneity for G_k, and it can be used to select a subset of representative samples out of it. The procedure described in [4] is based on an ordering of the gallery templates according to a representativeness criterion.

After performing sample ordering according, a *top-percentage* criterion [10] was used to select templates that guarantee a suitable representativeness of the gallery. In this work, the classification behavior for different training sets yielded by the *top-percentage* criterion was analyzed.

5   Experimental Results and Discussion

As previously mentioned in Sect. 2, a representative experimental set of 8 spectral and non-spectral data sets were employed for evaluating the proposed strategies. Although tests were carried out in all data sets, the ones showing the general behavior for each analytical technique are discussed. A total of 14 dissimilarity measures were used for data representation in dissimilarity spaces and those achieving the highest classification accuracy for each data set are presented. The dissimilarity measures that achieved the best results were *DShape* (Shape) [11], *Correlation distance* (Corr) [3] and *Bray Curtis* (Bray) [3]. The validation protocol consisted of averaging the results of 5 independent runs of a 10-fold cross validation procedure for each approach: *ENTR+CLAS* (baseline), *ENTR+DR+CLAS* and *DR+ENTR+CLAS*. Traditional classifiers such as LDA and 1-NN were employed for validation.

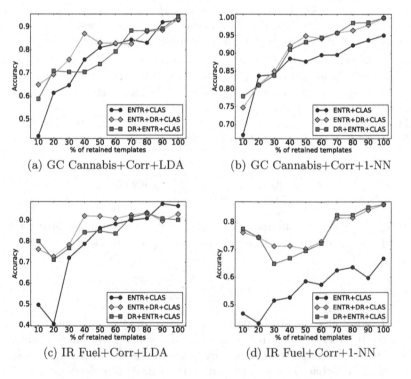

Fig. 2. Classification accuracies for different data sets.

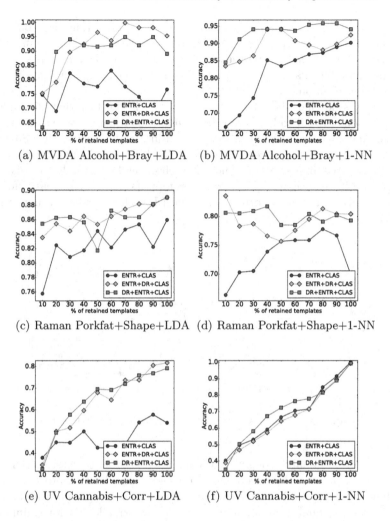

(a) MVDA Alcohol+Bray+LDA   (b) MVDA Alcohol+Bray+1-NN

(c) Raman Porkfat+Shape+LDA (d) Raman Porkfat+Shape+1-NN

(e) UV Cannabis+Corr+LDA   (f) UV Cannabis+Corr+1-NN

Fig. 3. Classification accuracies for different data sets.

Figures 2 and 3 show a substantial improvement in the classification accuracy of our proposals for the majority of data sets. Moreover, the $ENTR+DR+CLAS$ strategy managed to obtain a very good accuracy while maintaining a reduced training set size and a lower number of features, thus reducing the time and spatial complexity of the classification process. The dissimilarity representation proved to be advantageous for both classifiers studied. First, the 1-NN in feature spaces is very sensitive to outliers since its decision is only based on the nearest object from the training set to the test object. If this nearest object is an outlier or is noisy, the classification decision will be wrong. However, in the dissimilarity space, the dissimilarity to an object or prototype accounts only for one feature; therefore outlier objects are compensated by the other prototypes which are not noisy. By compensating the influence of outliers, the wrong decision can be corrected. In addition, the LDA in a high dimensional feature space suffers from the curse of dimensionality and the small sample size problem and usually must be regularized. The dissimilarity space offers an intrinsic regularization for LDA by prototype selection.

Another issue that should not be overlooked is the fact that dissimilarity representations proved to be more robust than the orginal feature space for un-processed data sets (see Fig. 4) (also observed in *Raman Tablets* with Savitzky-Golay smoothing). In preprocessed data sets, the 1-NN classifier showed improve-ments for the baseline strategy. The 1-NN classifier heavily relies in distances computation in the original space, and therefore, its accuracy increased for pre-processed data sets. On the other hand, the LDA classifier had a more robust behavior for unprocessed data sets.

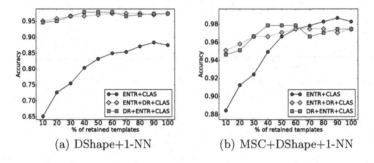

(a) DShape+1-NN   (b) MSC+DShape+1-NN

Fig. 4. MSC preprocessing effect on *NIR Tecator* data set.

An exploratory analysis ought to be conducted in order to determine the best number of templates to be selected for an optimal classification process. On the other hand, as observed in the majority of data sets, a compromise can be made among a near-optimal accuracy and a considerably smaller set for training the classifiers.

6   Conclusions and Future Work

In this work, new approaches for combining dissimilarity representations with an entropy-based template selection strategy were proposed. Our proposals showed an improvement over operating in the original feature space for classification tasks. It is also important to stress that our proposals were also able to deal with the curse of dimensionality present in some of the data sets. Moreover, the pro-posed strategies can be used for discovering more compact training sets by means of an exploratory analysis, while allowing to preserve, and sometimes improve, the classification accuracy. As could be observed, performing an entropy-based template selection in the feature space and using the selected instances as pro-totypes for a dissimilarity representation not only achieved very good results, but also yielded a more compact representation and increased the classification performance. As future work, a more extensive and detailed analysis ought to be conducted concerning the suitability of the existing dissimilarity measures for the different types of spectral data studied. Also, further tests should be carried out on a wider range of classification strategies.

References

1. Quality and Technology website (2017). http://www.models.life.ku.dk/tablets
2. Quality and Technology website (2017). http://www.models.life.ku.dk/ RAMANporkfat
3. Cha, S.H.: Comprehensive survey on distance/similarity measures between probability density functions. City **1**(2), 1 (2007)
4. De Marsico, M., Nappi, M., Riccio, D., Tortora, G.: Entropy-based template analysis in face biometric identification systems. Sig. Image Video Process. **7**(3), 493–505 (2013)
5. Duin, R.P., Pekalska, E.: The dissimilarity space bridging structural and statistical pattern recognition. Pattern Recogn. Lett. **33**(7), 826–832 (2012)
6. Duin, R.P., et al.: The dissimilarity representation for pattern recognition: foundations and applications, vol. 64. World Scientific (2005)
7. Helland, I.S., Næs, T., Isaksson, T.: Related versions of the multiplicative scatter correction method for preprocessing spectroscopic data. Chemometr. Intell. Lab. Syst. **29**(2), 233–241 (1995)
8. Infometrix: Infometrix website (2017). https://infometrix.com/pirouette/
9. Kumar, V., Chhabra, J.K., Kumar, D.: Performance evaluation of distance metrics in the clustering algorithms. J. Comput. Sci. **13**(1), 38–52 (2014)
10. Mendiola-Lau, V., Mata, F.J.S., Martínez-Díaz, Y., Bustamante, I.T., de Marsico, M.: Automatic classification of herbal substances enhanced with an entropy criterion. In: Beltrán-Castañón, C., Nyström, I., Famili, F. (eds.) CIARP 2016. LNCS, vol. 10125, pp. 233–240. Springer, Cham (2017). https://doi.org/10.1007/978-3-319-52277-7_29
11. Paclik, P., Duin, R.: Classifying spectral data using relational representation. NA (2003)
12. Plasencia Calaña, Y.: Prototype selection for classification in standard and generalized dissimilarity spaces. Ph.D. thesis, TU Delft (2015)
13. Porro Munoz, D.: Classification of continuous multi-way data via dissimilarity representation (2013)
14. Porro-Muñoz, D., Talavera, I., Duin, R.P., Hernández, N., Orozco-Alzate, M.: Dissimilarity representation on functional spectral data for classification. J. Chemom. **25**(9), 476–486 (2011)
15. Press, W.H., Teukolsky, S.A.: Savitzky-Golay smoothing filters. Comput. Phys. **4**(6), 669–672 (1990)
16. Thodberg, H.H.: StatLib–Datasets Archive website (2017). http://lib.stat.cmu. edu/datasets/tecator

Abnormal Event Detection in Video Using Motion and Appearance Information

Neptalí Menejes Palomino[1](✉) and Guillermo Cámara Chávez[2]

[1] Universidad Católica San Pablo, Arequipa, Peru
neptalimenejes@gmail.com
[2] Computer Science Department, Federal University of Ouro Preto,
Ouro Preto, MG, Brazil
gcamarac@gmail.com

Abstract. This paper presents an approach for the detection and localization of abnormal events in pedestrian areas. The goal is to design a model to detect abnormal events in video sequences using motion and appearance information. Motion information is represented through the use of the velocity and acceleration of optical flow and the appearance information is represented by texture and optical flow gradient. Unlike literature methods, our proposed approach provides a general solution to detect both global and local abnormal events. Furthermore, in the detection stage, we propose a classification by local regions. Experimental results on UMN and UCSD datasets confirm that the detection accuracy of our method is comparable to state-of-the-art methods.

Keywords: Abnormal event detection · Video analysis
Spatiotemporal feature extraction · Video surveillance
Computer vision

1 Introduction

In recent years, abnormal event detection in video has attracted more attention in the computer vision research community due to the increased focus on automated surveillance systems to improve security in public places. Traditionally, video surveillance systems are monitored by human operators, who alert if there are any suspicious events on the scene. However, human eye is susceptible to distraction or tiredness due to long hours of monitoring. In addition, the relationship between human operators and the large number of cameras is disproportionate, making it more difficult to detect and respond to all anomalous events that occur on the scene. This motivates the need for an automated abnormal event detection framework using computer vision technologies.

An abnormal event has no consistent definition, as it varies according to the context. In general, it is defined as an event which stands out from the normal behavior within a particular context. In this work, we consider the context of pedestrian walkways, where the normal behavior is performed by people walking.

© Springer International Publishing AG, part of Springer Nature 2018
M. Mendoza and S. Velastín (Eds.): CIARP 2017, LNCS 10657, pp. 382–390, 2018.
https://doi.org/10.1007/978-3-319-75193-1_46

Events which involve speed violations and the presence of abnormal objects are considered to be abnormal. We present an approach for the detection and localization of abnormal events in video using motion and appearance information. Motion information is represented through the use of the velocity and acceleration of optical flow and appearance information is represented by the textures and optical flow gradient. To represent these features we use non-overlapping spatio-temporal patches. Unlike literature methods, our model provides a general solution to detect both, global and local abnormal events. Furthermore, the detection stage presents problems of perspective distortion, this occur due to the objects near the camera appear to be large, while objects away from the camera appear to be small. To address these problems, we propose a classification by local regions.

The remainder of the paper is organized as follows. Section 2 reviews previous work on abnormal event detection. An overview on the proposal is provided in Sect. 3. Section 4 presents the results of our method on several public datasets. Finally, conclusions are provided in Sect. 5.

2   Related Works

Depending on the context and the specific application, abnormal event detection generally falls into two categories: trajectory analysis and motion analysis. The first is based on object tracking and typically requires an uncrowded environment to operate. There are some trajectory-based approaches for object tracking [13–15], these methods can obtain satisfactory results on traffic monitoring, however, fail in crowded scenes, since they cannot get good object trajectories due to the number of objects or events occurring simultaneously. Motion analysis is better suited for crowded scenes by analyzing patterns of movement rather than attempting to distinguish objects. Most of the state-of-the-art methods use local features for abnormal event representation, by considering the spatio-temporal information. For crowded scenes, according to the scale of interest, abnormal event detection can be classified into two [1]: Global Abnormal Event (GAE) detection and Local Abnormal Event (LAE) detection. The former focuses on determining whether the scene contains an abnormal event or not, while the latter determines the location where the event is occurring.

To detect GAE, some methods have been proposed. Mehran *et al.* [2] propose a method to detect abnormal crowd behavior by adopting the Social Force Model (SFM), then use Latent Dirichlet Allocation to detect abnormality. Wang and Snoussi [7] extract Histograms of Oriented Optical Flow (HOOF) for the detection of GAE, and a Bayesian framework for crowd escape behavior detection [9]. To detect LAE, most approaches extract motion and appearance features from local 2D patches or local 3D cubes. Kratz and Nishino [10] extract spatio-temporal gradient to fit a Gaussian model, and then use HMM to detect abnormal events. Mahadevan *et al.* [11] proposed a detection framework which uses mixture of dynamic textures to represent both motion and appearance features. Raghavendra *et al.* [12] introduce the Particle Swarm Optimization method to optimize the force interaction calculated using SFM. Furthermore, some methods based on optical flow estimation have been proposed. Ryan *et al.* [6] used optical flow vectors to model the motion information and proposed the textures of optical flow features to capture the smoothness of the optical flow field across a region. Cong *et al.* [8] presented a novel algorithm for abnormal event detection based on the sparse reconstruction cost for multilevel histogram of optical flows.

Recently, deep learning technique has been also used to abnormal event detection. Xu *et al.* [18] proposed a model that trains a Convolutional Neural Network (CNN) using spatio-temporal patches from optical flow images as input. Revathi and Kumar [20] use the deep learning classifier to detect abnormal event. Ravanbakhsh *et al.* [19] introduce a novel Binary Quantization Layer and then propose a temporal CNN pattern measure to represent motion in crowd.

3   Proposed Method

In this section, we present our approach for abnormal event detection, illustrated in Fig. 1. We use different sets of features to model the normal events and to detect different anomalies related to speed violations, access violations to restricted areas and the presence of abnormal objects in the scene.

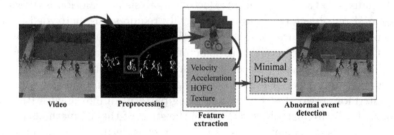

Fig. 1. Workflow of the proposed approach.

3.1   Feature Extraction

In order to restrict the analysis to regions of interest and to filter out distractions (*eg.* waving trees, illumination changes, etc.), we perform foreground extraction on each incoming frame. In this work, we calculate the foreground by the binarization of the subtraction of two consecutive frame, generating a foreground mask. In addition, to estimate the optical flow field of each frame, our approach considers the Lucas-Kanade pyramidal optical flow algorithm. It is important to note that only the optical flow of the foreground pixels is calculated.

Our approach divides the video sequence into non-overlapping spatio-temporal local patches of $m \times n \times t$. For each spatio-temporal patch P, we extract features based on motion, such as velocity and acceleration of optical flow, and features based on appearance, such as textures and gradients of optical flow. The extraction process of each of the features is shown below.

Velocity: The optical flow velocity feature is the summation of optical flow vectors inside a spatio-temporal patch [4].

Acceleration: The optical flow acceleration feature extracts information about the temporal variation of optical flow [5]. To model this feature, we calculate the optical flow of each incoming frame and then compute the optical flow magnitude of each foreground pixel, the magnitude is normalized between $[0, 255]$ to create a magnitude image. Again, we compute the optical flow of the magnitude image to calculate the acceleration. Finally, for each spatio-temporal patch the acceleration information is calculated by a summation of the acceleration vectors.

Textures of Optical Flow: The textures of optical flow measure the uniformity of the motion [6]. This feature is computed from the dot product of flow vectors at different offsets p and $p' = p + \delta$, where δ denotes the displacement of the pixel p.

Histogram of Optical Flow Gradient (HOFG): This feature extracts information about the spatial variation of optical flow [5]. To model this feature, we first treat both horizontal and vertical optical flow components as a separate image and then compute the gradients of each image using Sobel operators. For each spatio-temporal patch, we generate a four-bins histogram in each image using a soft binning-based approach and finally concatenate both histograms into a single eight-dimensional feature vector.

3.2 Abnormal Event Detection

In real world, usually normal event samples are predominant, and abnormal event samples have little presence. Because of this, the detection of abnormal events becomes even more difficult. In this paper, to address this problem, we use the classification method proposed in [3].

In [3], the classification stage uses the minimum distance. The main idea of this classification method is to search trained pattern that are similar to the incoming pattern. That is, if the incoming pattern is similar enough to some of the known patterns, then it is considered as a normal pattern. Otherwise, it will be considered as an abnormal event. However, the classification based on the minimum distance presents problems of detection due to some videos present the perspective distortion in the scene. That is, objects near to the camera appear to be large while distant objects appear to be small. This can significantly affect the feature extraction methods as the extracted features will vary according to their depth in the scene. In this paper, to address the problem of perspective distortion and to obtain better results, we propose a classification based on local regions. This classification method divides the scene into local spatial regions according to the depth of the scene, where normal patterns are modeled in each local region.

4 Experimental Results

We test our proposed approach by combining the proposed features (see Sect. 3) on UMN dataset [16] and UCSD dataset [11]. Our experiments are divided in two parts. The first part shows a comparison of our proposed classification method with 1 and 4 classification regions. Then, the second part compares our results with other methods published in the literature.

Experimental Setup and Datasets: The final feature vector contains two features describing the horizontal and vertical optical flow fields (velocity), two optical acceleration features (horizontal and vertical directions), three features of optical flow textures and eight-dimensional histogram of optical flow gradient components, four components for each vertical and horizontal directions. Therefore, we have a fifteen-dimensional feature vector for each spatio-temporal patch. In addition, we experimentally set a fixed spatio-temporal patch size of $20 \times 20 \times 7$. The criterion used to evaluate abnormal events detection accuracy was based on frame-level and to measure the performance of the proposed method, we have calculated the ROC curve, the area under the curve (AUC) and the equal error rate (EER).

UMN dataset consists of three different scenes of crowded escape events with a 320×240 resolution. The normal events are pedestrians walking randomly and the abnormal events are human spread running at the same time. There are a total of 11 video clips in the dataset. In the training phase the first 300 frames of each video were used to model normal events and the remaining frames were used in the test phase. UCSD dataset includes two sub-datasets, *Ped1* and *Ped2*. The crowd density varies from sparse to very crowded. The training sets are all normal events and the testing set includes abnormal events like cars, bikes, motorcycles and skaters. Ped1 contains 34 video clips for training and 36 video clips for testing with a 158×238 resolution, and Ped2 contains 16 video clips for training and 12 video clips for testing with a 360×240 resolution.

Global Abnormal Event Detection: We use UMN dataset [16] to detect GAE. The results are shown in Fig. 2, where columns (a), (b) and (c) represent the results on UMN dataset (scenes 1, 2 and 3, respectively). Furthermore, the first row shows the results for 1 region of classification, while the second for 4 regions. The ROC curves of the proposed method are shown in Fig. 3(a).

In Table 1, we show the quantitative results of our experiment on the UMN dataset, using two ways of classification (1 and 4 regions). In scenarios 1 and 3, our method using the classification with a single region achieves an AUC of 0.9985 and 0.9954, respectively. These results overcome the classification with 4 regions. Meanwhile, in scene 2, the result using 4 regions of classification achieves an AUC of 0.9486, which overcomes the result of the classification using a single region.

This happens due to the videos in scene 2 have problems of perspective distortion. Table 2 provides the quantitative comparisons to the state-of-the-art

(a) UMN scene1   (b) UMN scene2   (c) UMN scene3   (d) UCSD ped1   (e) UCSD ped2

Fig. 2. Results of the proposed method on the UMN and UCSD dataset.

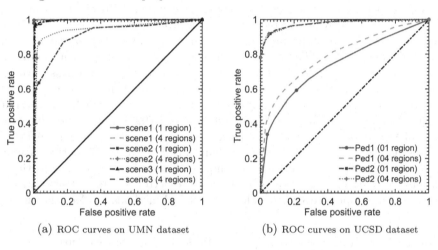

(a) ROC curves on UMN dataset   (b) ROC curves on UCSD dataset

Fig. 3. The ROC curves for the detection of GAE on the UMN dataset.

Table 1. Quantitative results of the proposed method on the UMN dataset.

Our method	AUC (1 region)	AUC (4 regions)
Scene 1	**0.9985**	0.9962
Scene 2	0.9182	**0.9486**
Scene 3	**0.9954**	0.9951

Table 2. Quantitative comparison of the proposed method with the state-of-the-art methods on the UMN dataset.

Methods	AUC
Sparse scene 1 [8]	0.995
Sparse scene 2 [8]	**0.975**
Sparse scene 3 [8]	0.964
Scene 1 [17]	0.936
Scene 2 [17]	0.775
Scene 3 [17]	0.966
Our proposed scene 1	**0.998**
Our proposed scene 2	0.948
Our proposed scene 3	**0.995**

methods. The AUC of our method in scenes 1 and 3 overcomes the state-of-the-art methods, which are 0.998 and 0.995, respectively. However, the AUC in scene 2 is comparable to the results of literature.

Local Abnormal Event Detection: We use the UCSD dataset to detect EAL. The results are shown in Fig. 2, where columns (d) and (e) represent the results on Ped1 and Ped2 datasets, respectively. Furthermore, we show the classification of 1 and 4 regions in the first and second row, respectively. The ROC curves of the our method on the UCSD dataset are shown in Fig. 3(b).

Table 3 shows the quantitative results of our experiment on the UCSD dataset, using two ways of classification (1 and 4 regions). The results on the Ped1 dataset using 4 regions of classification achieves an EER of 29.28% and an AUC of 0.7923, this result overcomes the classification using a single region due to perspective distorsion problem presented in this dataset. On the other hand, the results on the Peds2 dataset using a single region of classification achieves a EER of 07.24% and an AUC of 0.9778, this result overcomes the classification using 4 regions. Table 4 shows quantitative comparison of the proposed method with the state-of-the-art methods on the UCSD dataset. In the Ped1 dataset, our method achieves an EER of 29.2% and an AUC of 0.792, being competitive with most of the reported methods in the literature. On the other hand, on Ped2, our method achieves an EER of 07.2% and an AUC of 0.977, outperforming all reported results. It is important to emphasize, that state-of-the-art methods that use deep learning techniques [18,19] achieve better results on Ped1 dataset. However, our proposal overcomes the results of all methods on Ped2 dataset.

Table 3. Quantitative results of the proposed method on the UCSD dataset.

Our method	01 region		04 regions	
	EER	AUC	EER	AUC
Ped1	32.33%	0.737	**29.28%**	**0.792**
Ped2	**07.24%**	**0.977**	07.81%	0.976

Table 4. Quantitative comparison of the proposed method with the state-of-the-art methods on the UCSD dataset.

Methods	Ped1		Ped2	
	EER	AUC	EER	AUC
Social force [2]	31.0%	–	42.0%	–
MDT [11]	25.0%	–	25.0%	–
Reddy [4]	22.5%	–	20.0%	–
Textures [6]	23.1%	0.838	12.7%	0.939
HOFM [3]	33.3%	0.715	19.0%	0.899
CNN [18]	16.0%	0.921	17.0%	0.908
TCNN [19]	**08.0%**	**0.957**	18.0%	0.884
Our proposed	29.2%	0.792	**07.2%**	**0.977**

In our experiments we observed that the results on video sequences that don't have problems of perspective distortion, the performance of our method overcomes all state-of-the-art methods including techniques that use deep learning. This is because our proposed extracts and combines information of motion and appearance. However, our results fall down when videos have problems of perspective distortion. To address this problem we propose a classification by local regions, which improves the performance of our method, as can be observed in Fig. 3(a), (b).

5 Conclusions

In this paper, we propose a new approach based in the motion and appearance information for abnormal event detection (GAE and LAE) in crowded scenes. Motion features are suitable for detecting abnormal events with high speed motion. However, some abnormal events have a normal speed, to address this type of events, we introduce the use of appearance features. Furthermore, we propose a new classification method based on local regions to address the problems of perspective distortion in videos.

We evaluated the performance of our proposed method on the UCSD and UMN datasets. On the UCSD Ped1 dataset and scene 2 of UMN dataset, our results are comparable with the literature methods. On the other hand, our results on the UCSD Ped2 dataset achieves an EER of 07.2% and an AUC of 0.977, and on the UMN dataset achieves an AUC Of 0.998 and 0.995 in the scenes 1 and 3, respectively. According to these results, our method achieved better results and overcome the state-of-the-art results.

Acknowledgment. This work was supported by grant 011-2013-FONDECYT (Master Program) from the National Council for Science, Technology and Technological Innovation (CONCYTEC-PERU).

References

1. Li, T., Chang, H., Wang, M., Ni, B., Hong, R., Yan, S.: Crowded scene analysis: a survey. In: TCSVT 2015, pp. 367–386. IEEE (2015)
2. Mehran, R., Oyama, A., Shah, M.: Abnormal crowd behavior detection using social force model. In: CVPR 2009, pp. 935–942. IEEE (2009)
3. Colque, R., Júnior, C., Schwartz, W.: Histograms of optical flow orientation and magnitude to detect anomalous events in videos. In: SIBGRAPI 2015 (2015)
4. Reddy, V., Sanderson, C.: Improved anomaly detection in crowded scenes via cell-based analysis of foreground speed, size and texture. In: CVPRW 2011 (2011)
5. Nallaivarothayan, H., Fookes, C., Denman, S.: An MRF based abnormal event detection approach using motion and appearance features. In: AVSS 2014 (2014)
6. Ryan, D., Denman, S., Fookes, C., Sridharan, S.: Textures of optical flow for real-time anomaly detection in crowds. In: AVSS 2011, pp. 230–235 (2011)
7. Wang, T., Snoussi, H.: Histograms of optical flow orientation for abnormal events detection. In: PETS 2013, pp. 45–52. IEEE (2013)
8. Cong, Y., Yuan, J., Liu, J.: Abnormal event detection in crowded scenes using sparse representation. Pattern Recogn. **46**(7), 1851–1864 (2013)
9. Wu, S., Wong, H.-S., et al.: A Bayesian model for crowd escape behavior detection. In: CSVT 2014, vol. 24, no. 1, 85–98. IEEE (2014)

10. Kratz, L., Nishino, K.: Anomaly detection in extremely crowded scenes using spatio-temporal motion pattern models. In: CVPR, pp. 1446–1453 (2009)
11. Mahadevan, V., Li, W., et al.: Anomaly detection in crowded scenes. In: CVPR 2010, pp. 1975–1981. IEEE (2010)
12. Raghavendra, R., Del Bue, A., Cristani, M., Murino, V.: Optimizing interaction force for global anomaly detection in crowded scenes. In: ICCV 2011 (2011)
13. Guler, S., Farrow, M.K.: Abandoned object detection in crowded places. In: Proceedings of the PETS, pp. 18–23. Citeseer (2006)
14. Gul, S., Meyer, J.T., Hellge, C., Schierl, T., Samek, W.: Hybrid video object tracking in H. 265/HEVC video streams. In: MMSP, pp. 1–5. IEEE (2016)
15. Ngo, D.V., Do, N.T., Nguyen, L.A.T.: Anomaly detection in video surveillance: a novel approach based on sub-trajectory. In: ICEIC, pp. 1–4. IEEE (2016)
16. Unusual crowd activity dataset of University of Minnesota. http://mha.cs.umn.edu/movies/crowdactivity-all.avi
17. Shi, Y., Gao, Y., Wang, R.: Real-time abnormal event detection in complicated scenes. In: ICPR 2010, pp. 3653–3656. IEEE (2010)
18. Xu, D., Ricci, E., Yan, Y., Song, J., Sebe, N.: Learning deep representations of appearance and motion for anomalous event detection (2015)
19. Ravanbakhsh, M., Nabi, M., Mousavi, H., Sangineto, E.: Plug-and-play CNN for crowd motion analysis: an application in abnormal event detection (2016)
20. Revathi, A.R., Kumar, D.: An efficient system for anomaly detection using deep learning classifier. Signal Image Video Process. 11, 1–9 (2016)

Efficient and Effective Face Frontalization for Face Recognition in the Wild

Nelson Méndez$^{(\boxtimes)}$, Luis A. Bouza, Leonardo Chang, and Heydi Méndez-Vázquez

Advanced Technologies Application Center, 7ma A ♯ 21406, Playa, Havana, Cuba
{nllanes,lbouza,lchang,hmendez}@cenatav.co.cu

Abstract. Face image alignment is one of the most important steps in a face recognition system, being directly linked to its accuracy. In this work we propose a method for face frontalization based on the use of 3D models obtained from 2D images. We first extend the 3D Generic Elastic Model method in order to make it suitable for real applications, and once we have the 3D dense model of a face image, we obtain its frontal projection, introducing a new method for the synthesis of occluded regions. We evaluate the proposal by frontalizing the face images on LFW database and compare it with other frontalization techniques using different face recognition methods. We show that the proposed method allows to effectively align the images in an efficient way.

Keywords: Face frontalization · 3D face modeling · Face recognition

1 Introduction

Face recognition is one of the most used biometric techniques because of its potential applications. Despite the advances in this area, the accuracy of face recognition systems usually degrades under non-controlled scenarios. Among other factors, pose variation is considered one of the most challenging problems and different approaches have been developed for facing it. Pose-invariant face recognition methods can be roughly divided into four groups [4]: pose-robust feature extraction, multi-view subspace learning, face synthesis based on 2D methods and face synthesis based on 3D methods. Among them, frontal face image synthesis based on 3D methods, has the advantage of effectively reducing the differences on pose between 2D images and the real 3D domain, by using only one image, and without needing a large amount of multi-pose training data.

Existing 3D-based face synthesis approaches are still far from perfect, and new methods are still needed [4]. Two specific problems that are in the center of research is to develop more efficient approaches and to deal with self-occluded regions when frontalizing.

In this paper a new method for face image frontalization is presented. The proposal is based on the 3D Generic Elastic Model (3DGEM) approach, aiming

© Springer International Publishing AG, part of Springer Nature 2018
M. Mendoza and S. Velastín (Eds.): CIARP 2017, LNCS 10657, pp. 391–398, 2018.
https://doi.org/10.1007/978-3-319-75193-1_47

at synthesize a 3D dense face model more efficiently. Besides, we introduce a new method for effectively recover the information of occluded regions. In order to provide a logical and coherent exposition, this paper is divided as follow: Sect. 2 reviews related work; Sect. 3 makes a general description of 3DGEM approach and describes the proposal; later, in Sect. 4, several experimental results are given, in order to prove the accuracy and efficiency of the method; and finally, conclusions and future work ideas are presented in Sect. 5.

2   Related Work

The process of synthesizing a frontal image by using a 3D model has three main steps: face landmarks detection and correspondence, 3D face modeling and 2D frontal face image rendering. In this work we focus on the last two problems.

2.1   3D Face Modeling from a Single 2D Image

Existing methods for face frontalization using a single 2D image can be classified as 2D [10,18] or 3D based [3,6]. Very good results have been obtained in both categories, but since head rotations occur in the 3D space, 3D methods seems to be more effective [4]. In particular, 3D Morphable Models (3DMM) [2] has reported the synthesis of high quality 3D face models using a single input image. In the last years, several works have been proposed in order to improve this technique. Among them, the 3D Generic Elastic Models (3DGEM) [7], is one of the methods that achieves acceptable visual quality with a high performance. Different works have been then developed in order to improve 3DGEM's synthesis quality [9] and its expression-robust property [15]. The improvement of its performing time is another research topic for this technique. A synthesis process between one or two seconds is the best synthesis time reported in the literature for 3DGEM [9], which is still inappropriate for real time applications.

Recently, other approaches different from 3DGEM have been proposed, but in general they are also time consuming. For example, in [14] a person-specific method is proposed by combining the simplified 3DMM and the Structure-from-Motion methods to improve reconstruction quality. However, the proposal incurs a high computational cost and requires large training data. Yin et al. [21] combine elements of 3DMM with convolutional neural networks, but as any deep learning approach, requires a large amount of training images from different persons in different poses.

2.2   Filling of Self-occluded Regions

Different strategies have been proposed for filling of self-occluded parts when frontalizing. The most straightforward approach is to interpolate by using neighboring regions. One of the most popular strategy is to use the face symmetry and used mirrored pixels [5], but this can produce incoherent face texture when there are some differences on both sides of the face. Aiming at reducing the effect caused by occlusions, Hassner et al. [6] design a "soft symmetry" method that takes into account an occlusion degree estimation. But there are other affecting factors like illumination that should be taken into account. The lighting-normalized face frontalization (LNFF) method [3] estimates the illumination invariant quotient image from the visible parts, and uses it for rendering the estimated lighting on the self-occluded part. The method reports very good results but the process of rendering based on the Quotient Image is time consuming and can not model specular lights or cast shadows in an effective way [3].

3   Proposal

Our method is based on the 3D Generic Elastic Models (3DGEM) [17] approach. An overview of the synthesis process of 3DGEM can be observed in Fig. 1. In the offline or training stage the mean landmark points of the training images are computed. Then these landmarks are transformed in such way to be aligned with their corresponding depth-map. Later, a Delaunay triangulation is performed in order to create a sparse 2D mesh, which finally is refined by some subdivision algorithm, resulting in a dense 2D mesh. Heo [8] proposed the Loop subdivision algorithm to be used. The overall result of the offline stage is a mesh in which, every vertex has its corresponding depth value at the depth-map as a result of to the alignment step. On the other hand, the online or synthesis stage starts with face and landmarks detection on the input image. Then, facial landmarks are spatially transformed to be in the same scale and position of the trained 2D mesh. Afterwards, triangulation and subdivision are performed in the same way as they were executed in the training phase, so trained and subject 2D dense meshes have a 1-on-1 relationship between their vertices. This relationship is used to provide z-coordinates to subject's mesh, obtaining a 3D dense mesh to which, finally, subject's texture is applied.

In this work we aim at maintaining the visual quality of the 3D face models obtained by the 3DGEM approach, but improving its facial synthesis processing time. We propose to improve the efficiency of the dense 3D mesh construction process in the synthesis stage and also to enhance the quality of the filling of self-occluded regions.

3.1   Efficient Dense 3D Mesh Construction

One of the most time consuming steps in the original 3DGEM approach is the mesh subdivision process that is followed by the computation of the correspondences between the dense 2D mesh of a given image and the generic depth map. These correspondences must be computed for every image during training and testing, given that the original used Delaunay triangulation cannot guarantee a mesh with the same configuration.

In order to gain in efficiency, we propose to remove the mesh subdivision and correspondence process in the online stage. To do this, it is first necessary to modify the offline stage. The Delaunay triangulation is replaced by a fixed one,

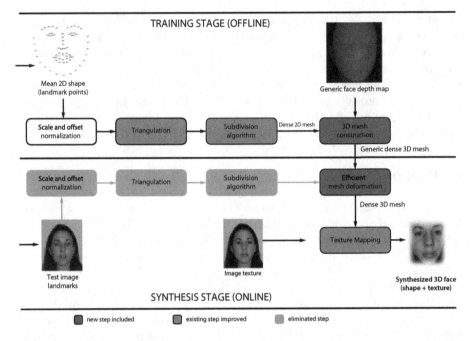

Fig. 1. 3DGEM synthesis process diagram. In blue the new steps included in our proposal are depicted, i.e., generic 3D mesh construction in the offline stage and an efficient mesh deformation algorithm in the syntesis stage. These new steps replace the steps filled in gray in order to speed up the synthesis process. In orange, the improved steps of 3DGEM are also illustrated. This figure is based on a diagram exposed by Prabhu et al. in [17]. (Color figure online)

and instead of using a raw subdivision algorithm, an adaptive mesh subdivision technique is applied, which iteratively obtains six triangles on every step, leading to a higher quality and a less complex 3D mesh. Then, the 3D dense mesh is generated in the training stage by using the aligned 2D dense mesh (resulting from the subdivision step) and its corresponding depth-map. Finally, in order to obtain the dense 3D mesh on the online stage, we introduce an efficient 3D mesh deformation algorithm described in [13]. This method uses bounded biharmonic deformations to efficiently and accurately deform a 3D mesh by using a few reference 2D points. In our case the dense 3D mesh learned in the offline stage is deformed taking as input 14 points belonging to the eyes, mouth and nose from the automatic landmarks detected on a given image.

These changes to the 3DGEM pipeline are the key-components of our proposal that allow a faster synthesis process. The differences between 3DGEM and our proposal are highlighted in Fig. 1.

3.2 Symmetric Interpolation for Self-occlusion

Once we have the dense 3D mesh for the test image and their corresponding 2D points, every triangular region in the mesh is filled with the texture values

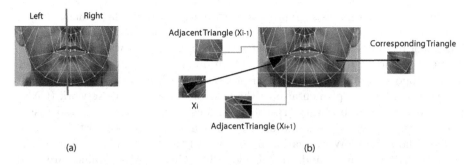

Fig. 2. Use of symmetric interpolation for self-occluded regions. In (a) the symmetric regions of the face and a sample triangulation are shown. In (b) the adjacent and corresponding triangles according to face symmetry are shown.

according the corresponding region in the image. However, because of pose variations, there are some regions in which no texture information is available. In 3DGEM, this information is completed by interpolating the texture in the corresponding region based on the symmetry of the face. In order to avoid interpolation artifacts in the resulting texture, in this paper we propose to include the adjacent triangles of the occluded regions in the interpolation process.

For better understanding we divided the face into two regions, left X and right X', as is depicted in Fig. 2(a). As can be seen, in these regions every triangle formed by the points of the face has its corresponding triangle on the opposite side of the face. A given triangle in X is denoted as X_i and its corresponding triangle in X' as X_i'.

Let X_{i+1} and X_{i-1} be the triangles adjacent to triangle X_i, respectively. Let $x_{int} = \Upsilon(x_1, x_2)$ be a linear interpolation function between two given pixel values. For every pixel value $x_i \in X_i$ exist two pixels values $x_{i+1} \in X_{i+1}$ and $x_{i-1} \in X_{i-1}$ given by an assignment function and its corresponding pixel $x_i' \in X_i'$. Then, our interpolation function is defined as:

$$x_i = \Upsilon \left(\frac{x_{i-1} + x_{i+1}}{2}, x_i' \right). \tag{1}$$

This process is illustrated on Fig. 2(b), and allows us to obtain the estimation of the values of the occluded regions in a simple but effective way.

4   Experimental Evaluation

In order of evaluating our approach we conducted experiments on the popular Labeled Faces in the Wild (LFW) database [12]. It contains 13233 images of 5749 people downloaded from the Web. For verification evaluation, the data is divided into 10 disjoint splits, which contain different identities and come with a list of 600 pre-defined image pairs for evaluation: 300 genuine comparisons and 300 impostors.

We use our method as a preprocessing step applied to the face images before the verification process and compare it with the provide images with different alignment techniques: LFW-a, aligned with a commercial software [20]; with funneling [11]; and with deep funneling [10]. In order to evaluate the performance of our proposal for different face descriptors, we use three state-of-the-art methods with available implementations: the Convolutional Neural Network (CNN) provided by the DLib library [1]; the VGG-face deep network [16]; and Fisher Vector method [19]. DLib is also used for face and landmarks detection.

We follow the LFW evaluation protocol, and the Receiver Operating Characteristic (ROC) curves and Area Under ROC Curve (AUC) values are used as evaluation metrics. In Table 1, the AUC values obtained for the three face recognition methods with the different alignment techniques are provided. The obtained ROC curves are shown in Fig. 3.

As can be seen from the Fig. 3 and Table 1, in general the proposed frontalization technique achieves the best results for the three different tested methods, showing its capabilities and benefits in the face recognition process.

Figure 4 shows some example results of our frontalization method applied to several images from the LFW database. It can be seen how the method is able to correct different levels of pose variations and effectively reconstruct the self-occluded regions.

Table 1. AUC values in LFW for different alignment methods.

Alignment method	DLib	Fisher vector	vgg-face
Deep funneling	**0.9991**	0.8636	0.8061
Funneling	0.9924	0.8736	0.8507
LFW-a	0.9960	0.8743	0.9048
Our proposal	**0.9991**	**0.9024**	**0.9125**

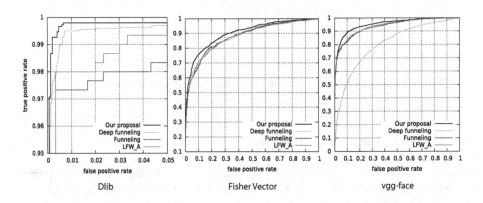

Fig. 3. ROC curves for the three face recognition methods with different alignments.

Fig. 4. Example results of our proposed frontalization method in some images of the LFW dataset. First row: original images. Second row: images frontalized with the proposed method.

4.1   Computational Efficiency Experiments

The computational efficiency of the proposed method in the synthesis stage has been also evaluated and compared to other methods. We recorded the computational processing time over 200 randomly selected images from the LFW dataset and reported the average processing time for an image. Compared to the 75 ms exhibited by LFW3D [6] and the 1276 ms exhibited by the original 3DGEM [17] for frontalizing an image, our proposal took 33 ms (i.e., 2.3x and 38x faster, respectively). Both LFW3D and 3DGEM were evaluated using the publicly available codes provided by their respective authors. Our proposed method was implemented in C++ using OpenCV. All the experiments were conducted on a single thread of an Intel i7-4770 CPU @3.40 GHz, 8 GB RAM.

5   Conclusions

In this paper we present a new method for face image frontalization. The proposal is based on the 3DGEM approach but introducing several changes that allows to perform efficiently the online stage. Our proposal exhibited a speed-up of 38x with respect to the original 3DGEM method, making it more feasible for real applications. Besides, we designed a simple method for filling the occluded regions using adjacent and opposite-side regions, that shows to recover the missing facial information in an effective way. The proposed method was compared in the challenging LFW dataset with three different alignment techniques, with three state-of-the-art recognition methods, and achieved the best results in all the experiments.

References

1. DLib C++ library kernel description. http://dlib.net/. Accessed 02 May 2017
2. Blanz, V., Vetter, T.: A morphable model for the synthesis of 3D faces. Techical Report SIGGRAPH (1999)

3. Deng, W., Hu, J., Wu, Z., Guo, J.: Lighting-aware face frontalization for unconstrained face recognition. Pattern Recogn. **68**, 260–271 (2017)
4. Ding, C., Tao, D.: A comprehensive survey on pose-invariant face recognition. ACM Trans. Intell. Syst. Technol. (TIST) **7**(3), 37 (2016)
5. Ding, L., Ding, X., Fang, C.: Continuous pose normalization for pose-robust face recognition. IEEE Sig. Process. Lett. **19**(11), 721–724 (2012)
6. Hassner, T., Harel, S., Paz, E., Enbar, R.: Effective face frontalization in unconstrained images. In: Proceedings of the IEEE Conference on Computer Vision and Pattern Recognition, pp. 4295–4304 (2015)
7. Heo, J.: 3D generic elastic models for 2D pose synthesis and face recognition. Technical Report (2009)
8. Heo, J., Savvides, M.: 3-D generic elastic models for fast and texture preserving 2-D novel pose synthesis. Trans. Inf. Forensics Secur. **7**(2), 563–576 (2012)
9. Heo, J., Savvides, M.: Gender and ethnicity specific generic elastic models from a single 2D image for novel 2D pose face synthesis and recognition. IEEE Trans. Pattern Anal. Mach. Intell. **34**(12), 2341–2350 (2012)
10. Huang, G., Mattar, M., Lee, H., Learned-Miller, E.G.: Learning to align from scratch. In: NIPS, pp. 764–772 (2012)
11. Huang, G.B., Jain, V., Learned-Miller, E.: Unsupervised joint alignment of complex images. In: ICCV, pp. 1–8. IEEE (2007)
12. Huang, G.B., Ramesh, M., Berg, T., Learned-Miller, E.: Labeled faces in the wild: a database for studying face recognition in unconstrained environments. Technical Report 07–49, University of Massachusetts, Amherst (2007)
13. Jacobson, A.: Algorithms and interfaces for real-time deformation of 2D and 3D shapes. Ph.D. Thesis, ETH Zurich (2013)
14. Jo, J., Choi, H., Kim, I.J., Kim, J.: Single-view-based 3D facial reconstruction method robust against pose variations. Pattern Recogn. **48**(1), 73–85 (2015)
15. Moeini, A., Moeini, H., Faez, K.: Pose-invariant facial expression recognition based on 3D face reconstruction and synthesis from a single 2D image. In: ICPR, pp. 1746–1751 (2014)
16. Parkhi, O.M., Vedaldi, A., Zisserman, A.: Deep face recognition. In: BMVC, vol. 1, p. 6 (2015)
17. Prabhu, U., Heo, J., Savvides, M.: Unconstrained pose-invariant face recognition using 3D generic elastic models. Trans. Pattern Anal. Mach. Intell. **33**(10), 1952–1961 (2011)
18. Sagonas, C., Panagakis, Y., Zafeiriou, S., Pantic, M.: Robust statistical face frontalization. In: Proceedings of the IEEE International Conference on Computer Vision, pp. 3871–3879 (2015)
19. Simonyan, K., Parkhi, O.M., Vedaldi, A., Zisserman, A.: Fisher vector faces in the wild. In: BMVC, vol. 2, p. 4 (2013)
20. Wolf, L., Hassner, T., Taigman, Y.: Effective unconstrained face recognition by combining multiple descriptors and learned background statistics. IEEE Trans. Pattern Anal. Mach. Intell. **33**(10), 1978–1990 (2011)
21. Yin, X., Yu, X., Sohn, K., Liu, X., Chandraker, M.: Towards large-pose face frontalization in the wild. arXiv preprint arXiv:1704.06244 (2017)

Path-Gradient – A Theory of Computing Full Intensity-Transition Between Two Points

Syed Ahmed Nadeem[1]([✉]), Eric A. Hoffman[2], and Punam K. Saha[1,2]

[1] Department of Electrical and Computer Engineering,
University of Iowa, Iowa City, IA, USA
syedahmed-nadeem@uiowa.edu
[2] Department of Radiology, University of Iowa, Iowa City, IA, USA

Abstract. A major challenge for path-based segmentation methods is to select the optimum scale capturing the total intensity variation across object interfaces without losing small-scale structures. Minimum barrier distance (MBD) attempts to alleviate this issue using a unique path-cost function that computes the maximum intensity variation on the path. Two major concerns of MBD are related to high computational complexity and convoluted trajectory of the optimum path between two points on either side of an object interface limiting benefits of MBD. Here, we introduce the notion of path-gradient (PG) that exhibits similar behavior as MBD for object segmentation with significantly reduced computation. The formulation of PG allows the addition of a regularization term in path cost, which improves segmentation still at considerably reduced computation cost than regular MBD. Efficient algorithms for computing PG and regularized PG are presented and their segmentation performances are compared with that of MBD.

Keywords: Digital topology · Image segmentation · Minimum barrier distance
Path cost · Dynamic programming

1 Introduction

Large data volumes acquired daily in many imaging application areas including medical imaging, remote sensing, industrial applications, character recognition, etc. have created the need for automated and quantitative assessment of targeted knowledge. It is imperative to define target objects in an image before extracting quantitative or descriptive information. This process of locating and delineating a target object in an image is referred to as segmentation, which has a rich history, and is a fundamental task in many computerized imaging applications [1]. There are three major segmentation paradigms based on: shape [2], boundary [3], and region-continuity [4]. Shape-based approaches [2] take advantage of the inherent correlation among structures in a shape to formulate an analytic model representing the mean and variance of shape instances in a family. Boundary-based segmentation methods locate and delineate an object boundary under a pre-defined optimization criterion. Region-based approaches attempt to grow object regions in an image from a set of seeds under a pre-defined criterion of region continuity [5, 6].

© Springer International Publishing AG, part of Springer Nature 2018
M. Mendoza and S. Velastín (Eds.): CIARP 2017, LNCS 10657, pp. 399–407, 2018.
https://doi.org/10.1007/978-3-319-75193-1_48

Continuity of an object in a region-based segmentation method is imposed using prior definitions of path-strength or path-cost [7]. Path-strength and path-cost are formulated using pre-defined local affinity or step-cost functions, respectively. Intensity gradient makes major contributions in the computation of such local functions. A major challenge in formulating such functions is the dichotomy between capturing the total intensity variation across object interfaces and the preservation of small-scale structures. It defines a tradeoff for selection of the optimum scale satisfying the fine balance between the object scale and the gradient (transition) width, and a fixed choice of scale is bound to fail due to variability among images and even at different locations within an image. Recently, there have been attempts, e.g., the minimum barrier distance (MBD) [8, 9], to define the path-cost in a manner that computes the maximum intensity variation along the entire path without depending upon accurate gradient computation within the local function. Two major concerns with MBD are related to high computational complexity and convoluted trajectory of the optimum path between two points on either side of an object interface limiting benefits of MBD.

In this paper, we introduce a new formulation of path-cost that captures the entire geodesic gradient over the path instead of total intensity variation used in MBD and demonstrate its applications in image segmentation. Specifically, we introduce the formulation of path-gradient (PG) based region growing and present its theoretical properties and an efficient algorithm for its computation. The cost-function of PG is modified to add regularization term that stops convoluted trajectories of an optimum path while allowing an optimum path to follow the continuity within an object structure. An efficient algorithm for computation of regularized path-gradient (RPG) is presented. Finally, the computational complexity and segmentation results of MBD, PG, and RPG are presented and discussed.

2 Theory and Algorithms

2.1 Definition of Path-Gradient and Regularized Path-Gradient

In this section, we introduce the notion of "path-gradient". An n-dimensional rectilinear grid is represented by $\mathbb{Z}_+^n$, where $\mathbb{Z}_+$ is the set of positive integers. An element of $\mathbb{Z}_+^n$ will be abbreviated as a *spatial element* or *spel*, e.g., a pixel for $n = 2$ and a voxel for $n = 3$. In a digital space, adjacencies are used to define path-continuity. In two-dimension (2-D), standard 4- and 8-adjacencies are used, while 6-, 18-, or 26-adjacencies are used in 3-D [7]. We will use α to denote an adjacency relation. The pair $\left(\mathbb{Z}_+^n, \alpha\right)$ denotes a *digital space*. A *digital image* on $\left(\mathbb{Z}_+^n, \alpha\right)$ is defined by a triple $\left(\mathbb{Z}_+^n, \alpha, f\right)$, where $f : \mathbb{Z}_+^n \to \mathbb{Z}|$ $\mathbb{Z}$ is the set of all integers, is the image *intensity function*; note that some imaging techniques, e.g., CT imaging, generate digital images with negative intensity values. A *digital path*, or simply a *path* $\pi_{p,q}$ is a sequence of spels $\langle p_0 = p, p_1, \cdots, p_{l-1} = q \rangle$, where every two successive spels are α-adjacent. The gradient-strength of a path $\pi_{p,q}$ is defined as follows:

$$\tau\left(\pi_{p,q}\right) \;=\; \max_{r \in \pi_{p,q}} |f(p) - f(r)|. \tag{1}$$

Using the above formulation of gradient-strength of a path, the path-gradient or PG in a digital setup from p to q, where $p,\ q \in \mathbb{Z}_+^n$, is defined as follows:

$$\Delta(p,q) \;=\; \min_{\pi_{p,q} \in \Pi_{p,q}} \tau\left(\pi_{p,q}\right). \tag{2}$$

It is obvious from the above formulation that the PG function Δ fails to satisfy the symmetry property, i.e., $\Delta(p,\ q) \neq \Delta(q,\ p)$. On the other hand, it satisfies the triangular rule, i.e., $\forall p,\ q,\ r \in \mathbb{Z}_+^n$,

$$\Delta(p,\ q) \leq \Delta(p,\ r) + \Delta(r,\ q). \tag{3}$$

Here, it may be worthy to examine its relation to MBD [8, 9]. Both PG and MBD are formulated using intensity differences along paths. The major difference in their formulation is that, in MBD, the difference is computed between any two variable points on a path, while, for PG, the difference is computed from the starting point on a path, which is fixed and may be interpreted as the "seed" in imaging applications. This subtle change in the formulation of gradient-strength of a path significantly simplifies its computational algorithm allowing for the addition of a regularization term, which enhances its performance in image segmentation. The gradient-strength of a path for regularized path-gradient or RPG is defined as follows:

$$\ddot{\tau}\left(\pi_{p,q}\right) \;=\; (1 - K) \max_{r \in \pi_{p,q}} |f(p) - f(r)| + KL\left(\pi_{p,q}\right), \tag{4}$$

where, K is the regularization constant, and L is the path-length, i.e., the number of spels on the path $\pi_{p,q}$. Finally, the RPG is computed as follows:

$$\ddot{\Delta}(p,q) \;=\; \min_{\pi_{p,q} \in \Pi_{p,q}} \ddot{\tau}\left(\pi_{p,q}\right). \tag{5}$$

Like PG, RPG is not symmetric but satisfies the triangular rule.

2.2   Algorithms

In this section, we present efficient algorithms for computing PG and RPG using wave propagation with dynamic programming. It can be shown that an efficient wave propagation algorithm for accurate computation of a path-cost function requires the existence of a geodesic from any spel to another spel under the specific cost function. A path $\pi_{p,q}$ between two spels p, q is a geodesic if for any r on $\pi_{p,q}$, the sub-paths $\pi_{p,r}$ and $\pi_{r,q}$ on $\pi_{p,q}$ are minimum cost paths between p, r and r, q, respectively. The existence of geodesic ensures that, during wave propagation, the function value at a spel can be determined from the function-value and additional information (if needed) at the specific neighbor preceding on the geodesic. It is interesting to note that the triangular property does not guarantee the validity of geodesic from one spel to another. For example, the minimum barrier distance (MBD) satisfies the triangular property, while it fails to guarantee the existence of a geodesic between every two spels. It can be shown that PG guarantees the existence of a geodesic from a spel to another. Note that, due to non-symmetry of the PG function, the geodesic from p to q may be different from that from q to p. In the following, a wave-propagation algorithm is presented to compute PG from a seed spel s to any spel in the image domain. Theorems 1 and 2 state that the algorithm terminates in a finite number of steps, and it accurately computes PG from s at termination. In this paper, proofs of the theorems are omitted due to the space limitation.

Algorithm 1. Computation of Path-Gradient on a Digital Image.

begin algorithm

Input: An input image $(\mathbb{Z}_+^n, \alpha, f)$ and a seed spel s

Output: Computed PG image $(\mathbb{Z}_+^n, \alpha, f_{PG})$

Auxiliary Data Structures: A queue Q of spels

  1. *for* all $p \in \mathbb{Z}_+^n$ within the image domain

  2.   set $f_{PG}(p) = \infty$

  3. set $f_{PG}(s) = 0$

  4. initiate a queue Q containing s

  5. *while* Q is not empty

  6.   pop a spel p from Q

  7.   *for* $q \in N_\alpha^*(p)$ /* $N_\alpha^*(p)$: excluded α-neighborhood of p */

  8.   *if* $f_{PG}(q) > \max(|f(q) - f(s)|, f_{PG}(p))$

  9. $f_{PG}(q) = \max(|f(q) - f(s)|, f_{PG}(p))$

  10.   push q

end algorithm

Theorem 1.   For any digital image $(\mathbb{Z}_+^n, \alpha, f)$ with a bounded image domain, for any seed spel s, Algorithm 1 for computing PG terminates in a finite number of iterations.

Theorem 2.   For any digital image $(\mathbb{Z}_+^n, \alpha, f)$ with a bounded image domain, for any seed spel s, at the termination of Algorithm 1, the output function f_{PG} accurately computes the PG value $\Delta(s, \cdot)$ at individual spels following Eq. (2).

The algorithm for RPG is relatively more complex than Algorithm 1 for computation of PG. The fundamental difficulty in computing RPG is that the existence of a geodesic from one spel to another is not guaranteed for RPG. To overcome this challenge, we develop a dynamic programming algorithm for RPG computation where the queue of spels is maintained in the ascending order of path length from the seed spel. It can be shown that the algorithm terminates in a finite number of steps, and the above subtle change allows accurate computation of RPG.

Algorithm 2. Computation of Regularized Path-Gradient on a Digital Image.

begin algorithm

Input: An input image $(\mathbb{Z}_+^n, \alpha, f)$ and a seed spel s

Output: Computed RPG image $(\mathbb{Z}_+^n, \alpha, f_{\text{RPG}})$

Auxiliary Data Structures: (1) An image $(\mathbb{Z}_+^n, \alpha, L)$ where $L(p)$ represents the length of the current optimum path to p from the seed spel s; (2) an image $(\mathbb{Z}_+^n, \alpha, P)$ where $P(p)$ gives the path-gradient term of the current optimum path to p from s; (3) an image $(\mathbb{Z}_+^n, \alpha, L_{\text{pre}})$ where $L_{\text{pre}}(p)$ gives the length of the immediately previous optimum path to p from the seed spel s; (4) An image $(\mathbb{Z}_+^n, \alpha, P_{\text{pre}})$ where $P_{\text{pre}}(p)$ gives the path-gradient term of the immediately previous optimum path to p from s; and (5) an queue Q of spels ordered by the value of $L(\cdot)$

1. *for* all $p \in \mathbb{Z}_+^n$ within the image domain
2.   set $f_{\text{RPG}}(p) = P(p) = P_{\text{pre}}(p) = L(p) = L_{\text{pre}}(p) = \infty$
3. set $P(s) = f_{\text{RPG}}(s) = 0, L(s) = 1, P_{\text{pre}}(s) = L_{\text{pre}}(s) = \infty$
4. initiate a queue Q containing s
5. *while* Q is not empty
6.   pop a spel p from Q
7.   *for* $q \in N_\alpha^*(p)$ /* $N_\alpha^*(p)$: excluded α-neighborhood of p */
8.   *if* $f_{\text{RPG}}(q) > (1 - K)\max(|f(q) - f(s)|, P_{\text{pre}}(p)) + K(L_{\text{pre}}(p) + 1)$
9.   *if* $L(q) < L_{\text{pre}}(p) + 1$
10. $L_{\text{pre}}(q) = L(q)$
11. $P_{\text{pre}}(q) = P(q)$
12. $P(q) = \max(|f(q) - f(s)|, P_{\text{pre}}(p))$
13. $L(q) = L_{\text{pre}}(p) + 1$
14. $f_{\text{RPG}}(q) = (1 - K)P(q) + KL(q)$
15.   push q into the ordered queue Q based on $L(q)$
16.   *else if* $f_{\text{RPG}}(q) > (1 - K)\max(|f(q) - f(s)|, P(p)) + K(L(p) + 1)$
17.   *if* $L(q) < L(p) + 1$
18. $L_{\text{pre}}(q) = L(q)$
19. $P_{\text{pre}}(q) = P(q)$
20. $P(q) = \max(|f(q) - f(s)|, P(p))$
21. $L(q) = L(p) + 1$
22. $f_{\text{RPG}}(q) = (1 - K)P(q) + KL(q)$
23.   push q into the ordered queue Q based on $L(q)$

end algorithm

Theorem 3. For any digital image $(\mathbb{Z}_+^n, \alpha, f)$ with a bounded image domain, for any seed spel s, and the regularization constant $K \in [0, 1]$, Algorithm 2 for computing RPG terminates in a finite number of iteration.

Theorem 4. For any digital image $(\mathbb{Z}_+^n, \alpha, f)$ with a bounded image domain, for any seed spel s, at the termination of Algorithm 2, the output function f_{RPG} accurately computes the RPG value $\ddot{\Delta}(s, \cdot)$ at individual spels following Eq. (5).

3   Experimental Results

We present preliminary experimental results to—(1) demonstrate fundamental differences and similarities between three path-cost based segmentation region-growing segmentation methods, namely MBD, PG, and RPG and (2) examine the effectiveness of RPG in segmenting tree-like objects and compare it with MBD and PG.

For the first experiment, an image (size: 640×360), shown in Fig. 1, was generated and downgraded with blur (7×7 window averaging), white Gaussian noise (contrast-to-noise ratio: 8), and background intensity inhomogeneity (100% of the object-background contrast uniformly distributed across image-width). Optimum paths using the three methods MBD, PG and RPG between several pairs of pixels were examined. Two sets of pixel-pairs were used to examine the optimum paths between two points across object-background interface and those connecting two distant pixels within the same objects. In the top row of Fig. 1, pixels pairs were selected across the object-background interface. Different pixel pairs were selected at regions with different inhomogeneity to examine their behavior under the combined field of a slow varying inhomogeneity and rapid intensity variation across object-background interface. In the bottom row of the figure, both pixels in a pair was chosen inside the object region, while separated by a long geodesic distance to examine whether the regularization terms influence the optimum path to move out and in of the object region. In the top row, MBD and PG generated convoluted trajectories of optimum paths for two pixels across the object-background interface artificially lowering the minimum barrier distance or path-gradient value between them. A small regularization term in RPG reduces the convolutedness of the optimum path while ensuring that the path remains confined inside the object for a pair of distant pixels within the same object.

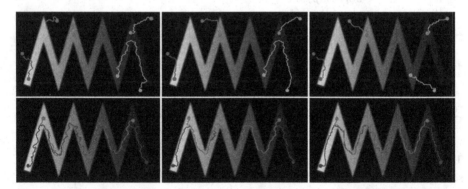

Fig. 1. Comparison of optimum cost-paths computed using different methods. Top row from left to right: optimum cost-paths using MBD, PG, and RPG generated between four different pairs of nearby pixel pairs one inside the object, while the other in the background (inter-object pixel pairs). See text for details.

To examine the effectiveness of RPG in segmenting tree-like objects, a silhouette image of a tree (size: 1770 × 2400) was downloaded from an internet site,[1] which was downgraded with blur (17 × 17 window averaging), white Gaussian noise (contrast-to-noise ratio: 8), and background intensity inhomogeneity (75% of the object-background contrast uniformly distributed across image-height). Individual methods were applied to the degraded test image using a single seed point and resultant distance transform maps were thresholded [10] to get binary segmentation results. The degraded test image and segmentation results using a single seed are presented in the top row of Fig. 2. It is noticeable from these segmentation results that MBD and PG produce similar segmentation results while RPG performs better than both methods in terms of the extents of tree branches. Segmentation results using the three methods on the same image with two seeds are presented in the bottom row of the figure. As observed from the segmentation results of the bottom row, the segmentation performances of MBD and PG are similar, while RPG significantly improves the segmentation. More interestingly, the improvement in the segmentation result using the second seed is greater for RPG as compared to that of MBD or PG. These experimental results confirm the advantages of the regularization term in the path cost function.

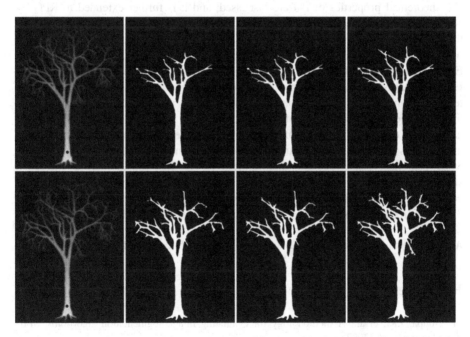

Fig. 2. Comparison of segmentation results using different methods. Top row from left to right: An image of tree with the seed marked in magenta; segmentation results using MBD, PG, and RPG. Bottom row: Same as top row but using two seeds. Segmentation results using MBD and PG are visually similar, while the improvements using RPG are apparent.

---

[1] http://www.clipartkid.com/tree-silhouette-clip-art-silhouettes-clipart-best-clipart-best-vx39H9-clipart/.

To assess the computational performance of the three methods, the 2-D image of Fig. 2 was used, while a 3-D image (size: $256 \times 256 \times 256$) was generated from the CT-based segmentation of pulmonary airway mask after degrading with blur ($3 \times 3 \times 3$ window averaging), white Gaussian noise (contrast-to-noise ratio: 8), and background intensity inhomogeneity (75% of the object-background contrast uniformly distributed in the axial direction). For the 2-D image, a single seed pixel was placed at the middle of the main tree-trunk, while a seed voxel was selected inside the trachea for the 3-D airway phantom image. Execution time was calculated as the mean over 100 executions of each method on a given image. The execution times for MBD, PG and RPG were 18.89 s, 1.03 s, and 3.54 s, respectively for the 2-D image and 43.15 s, 3.87 s, and 17.7 s, respectively for the 3-D image.

4   Conclusions

In this paper, we have introduced a new notion of PG as means to capture the full intensity transition between two points, which overcomes the fundamental challenge related to scale in region growing methods using conventional gradient computation. The theoretical properties of PG are discussed, and it is further extended to RPG by adding a regularization term in the path-cost function to stop convoluted trajectories of the optimum path between two points across object-interface, while allowing an optimum path to follow a curved path satisfying the continuity of an object region. Efficient algorithms for computation of both PG and RPG have been presented. Experimental results have demonstrated similarity in performance in MBD and PG in terms of consistency of behavior of optimum paths as well as segmentation of a tree-like object in the presence of image degradation, while PG is computed significantly more efficiently than MBD. Improvements in the consistency of optimum paths as well as segmentation results using RPG have been experimentally demonstrated. More elaborately, the regularization term in RPG added the ability to enforce a straightness on the convoluted trajectories of an optimum path while maintaining continuity within an object structure. Finally, it has been shown that, both in 2-D and 3-D, RPG requires more computation time than PG, while less than MBD.

Acknowledgements. This work was supported by the NIH grant R01 HL112986.

References

1. Udupa, J.K., et al.: A methodology for evaluating image segmentation algorithms. SPIE Med. Imaging (2002)
2. Cootes, T.F., et al.: Active shape models - their training and application. Comput. Vis. Image Underst. **61**, 38–59 (1995)
3. Kass, M., Witkin, A., Terzopoulos, D.: Snakes: active contour models. Int. J. Comput. Vis. **1**, 321–331 (1988)
4. Udupa, J.K., Saha, P.K.: Fuzzy connectedness and image segmentation. Proc. IEEE **91**, 1649–1669 (2003)

5. Saha, P.K., Udupa, J.K., Odhner, D.: Scale-based fuzzy connected image segmentation: theory, algorithms, and validation. Comput. Vis. Image Underst. **77**, 145–174 (2000)
6. Falcão, A.X., Stolfi, J., de Alencar Lotufo, R.: The image foresting transform: theory, algorithms, and applications. IEEE Trans. Pattern Anal. Mach. Intell. **26**(1), 19–29 (2004)
7. Saha, P.K., Strand, R., Borgefors, G.: Digital topology and geometry in medical imaging: a survey. IEEE Trans. Med. Imaging **34**(9), 1940–1964 (2015)
8. Ciesielski, K.C., et al.: Efficient algorithm for finding the exact minimum barrier distance. Comput. Vis. Image Underst. **123**, 53–64 (2014)
9. Strand, R., et al.: The minimum barrier distance. Comput. Vis. Image Underst. **117**(4), 429–437 (2013)
10. Saha, P.K., Udupa, J.K.: Optimum threshold selection using class uncertainty and region homogeneity. IEEE Trans. Pattern Anal. Mach. Intell. **23**, 689–706 (2001)

A Robust Indoor Scene Recognition Method Based on Sparse Representation

Guilherme Nascimento[(✉)] [ID], Camila Laranjeira [ID], Vinicius Braz,
Anisio Lacerda [ID], and Erickson R. Nascimento [ID]

Universidade Federal de Minas Gerais (UFMG), Belo Horizonte, Brazil
{gsn2009,camilalaranjeira,viniciusbraz30}@ufmg.br,
anisio@decom.cefetmg.br, erickson@dcc.ufmg.br

Abstract. In this paper, we present a robust method for scene recognition, which leverages Convolutional Neural Networks (CNNs) features and Sparse Coding setting by creating a new representation of indoor scenes. Although CNNs highly benefited the fields of computer vision and pattern recognition, convolutional layers adjust weights on a global-approach, which might lead to losing important local details such as objects and small structures. Our proposed scene representation relies on both: global features that mostly refers to environment's structure, and local features that are sparsely combined to capture characteristics of common objects of a given scene. This new representation is based on fragments of the scene and leverages features extracted by CNNs. The experimental evaluation shows that the resulting representation outperforms previous scene recognition methods on Scene15 and MIT67 datasets, and performs competitively on SUN397, while being highly robust to perturbations in the input image such as noise and occlusion.

Keywords: Indoor scene recognition · Sparse coding
Convolutional Neural Networks

1 Introduction

Scene recognition is one of the most challenging tasks on image classification, since the characterization of a scene is performed by using visual features of the objects that compose it and its spatial layout. In other words, a scene is the result of objects compositions. A given environment is more likely to be classified as a bathroom when equipped with a bath and a shower. This is especially true when considering indoor images, which are the focus of this work. The rationale

---

G. Nascimento—This work is supported by grants from Vale Institute of Technology, CNPq, CAPES, FAPEMIG and CEFETMG; CNPq under Proc. 456166/2014-9 and 431458/2016-2; and FAPEMIG under Procs. APQ-00783-14 and APQ-03445-16, and FAPEMIG-PRONEX-MASWeb, Models, Algorithms and Systems for the Web APQ-01400-14.

M. Mendoza and S. Velastín (Eds.): CIARP 2017, LNCS 10657, pp. 408–415, 2018.
https://doi.org/10.1007/978-3-319-75193-1_49

is that besides the constituent objects, the image of a room (i.e., an indoor scene) is similar to every scene and it is hard to distinguish among them.

In this paper, we propose a novel method, which leverages features extracted by Convolutional Neural Networks (CNNs) using a sparse coding technique to represent discriminative regions of the scene. Specifically, we propose to combine global features from CNNs and local sparse feature vectors mixed with max spatial pooling. We built an over-complete dictionary whose base vectors are feature vectors extracted from fragments of a scene. This approach makes our image representation less sensitive to noise, occlusion and local spatial translations of regions and provides discriminative vectors with global and local features.

The main contributions of this paper are: (i) a robust method that simultaneously leverages global and local features to the scene recognition task, and (ii) a thorough set of experiments to validate our requirements and the proposed scene recognition method. We evaluate our method on three different datasets (Scene15 [6], MIT67 [12] and SUN397 [18]) comparing it to previously proposed methods in the literature, and perform a robustness test against the work of Herranz et al. [3]. The experimental results show that our method outperforms the current state of the art on Scene15 and MIT67, while performing competitively on SUN397. Additionally, when subjected to image perturbations (i.e., noise and occlusion), our proposal outperforms Herranz et al. [3] on all three datasets, surpassing it by a large margin on the most challenging one, SUN397.

Related Work. Most solutions proposed in the past decade for scene recognition exploited handcrafted local [4] and global descriptors [10]. Methods to combine these descriptors and to build an image representation varied from Fisher Vectors [13] to Sparse Representation [11]. Sparse representation methods reached a great performance on image recognition, becoming popular in the last few years. Yang et al. [19] and Gao et al. [2] encoded SIFT [7] descriptors into a single sparse feature vector with the max-pooling method, achieving the best results on the Caltech-101 and the Scene-15 datasets.

In recent years, methods based on CNNs achieved state of the art on the task of scene recognition. In Zhou et al. [21], the authors introduce the Places dataset and present a CNN method to learn deep features for scene recognition. The authors reached an accuracy of 70.8% on MIT67. Another line of investigation is to combine features from CNNs. Dixit et al. [1] use a network to extract posterior probability vectors from local image patches and combine them using a Gaussian mixture model. Features extracted from networks trained on ImageNet [9] and Places [21] are combined, achieving 79% accuracy on MIT67.

Herranz et al. [3] analyzed the impact of object features in different scales in combination with global holistic features provided by scene features. According to the authors, the aggregation of local features brings robustness to semantic information in scene categorization. Herranz et al. [3] held the state of the art before Wang et al. [17] propose the architecture called PatchNet. This architecture models local representations from scene patches and aggregates them into a global representation using a newly semantic encoding method called VSAD.

Differently from previous works, our approach is flexible to being used with any feature representation, not being restricted to a single dataset or network architecture. This is a key advantage, since the use of different sources of features (e.g., ImageNet for object features and Places for structure) cannot be easily handled by training a single CNN or an autoencoder representation. Additionally, our method is attractive because it presents a small number of hyperparameters (the sparsity and dictionary size), which makes it much easier to train. Typical CNN or autoencoder approaches require selecting the batch size, learning rate and momentum, architecture, optimization algorithm, activation functions, the number of neurons in the hidden layer and dropout regularization.

2   Methodology

Our methodology is composed of four steps, namely: (i) feature extraction and dictionary building; (ii) feature sparse coding; (iii) pooling, and (iv) concatenation. The final feature vector feeds a Linear SVM used to classify the image. The three first steps are illustrated in Fig. 1.

The overall idea of our approach is to extract features from two semantic levels of the image. Firstly, a global representation is computed from the entire image, encoding the structural features of the scene. Then, we move a sliding window over the image with $\frac{I}{2}$ and $\frac{I}{4}$ window sizes, where I represent the image dimensions, computing a feature vector for each fragment of the scene. These fragments contain local features that are potentially from objects or object's parts. The stride was fixed to half of the window dimensions in both scales. The rationale is that features extracted from smaller regions will convey information regarding the constituent objects of the scenes. Thus, by combining these features we can represent an indoor scene as a composition of objects.

Feature Extraction and Dictionary Building. Although our method can be instantiated with any feature extractor, such as a bag-of-features model, we use

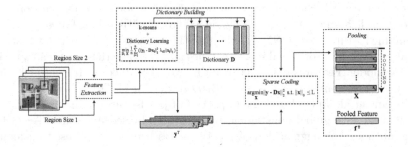

Fig. 1. Overview of the sparse representation pipeline. For each type of local feature, a single dictionary is built and optimized for both scales, and later used to encode the sparse vectors in **X**. The final feature $\mathbf{f}^T$ of a given scale is computed by pooling all sparse representations calculated on that scale.

Convolutional Neural Networks to extract the features. We extract semantic features from the fc7 layer of VGGNet-16 [14]. For global information, the CNN was trained on Places [21], while two types of local information were extracted: the same model trained on Places to acquire information regarding local structures, and a second model trained on ImageNet to encode object features.

The feature extraction step provided a feature vector $\mathbf{y} \in \mathbb{R}^d$ for each fragment of the image. This process is performed for all samples, grouping the $\mathbf{y}$ vectors into k clusters using the k-means algorithm. Hence, a dictionary for scale i is represented by $K_i = [\mathbf{v}_{1,i}, \ldots, \mathbf{v}_{k,i}] \in \mathbb{R}^{d \times k}$, where $\mathbf{v}_{j,i} \in \mathbb{R}^d$ represents the jth cluster in scale i. We define the matrix $\mathbf{D_0}$ as the concatenation of multiple c scales matrices $\{K_i | 1 \leq i \leq c\}$: $\mathbf{D_0} = [K_1, K_2, \ldots, K_c]$.

It is worth noting that, since we are considering a set of over-complete basis, we need $\cup_{i=0}^{c} K_i \gg d$, where K_i is the number of scale matrices in the dictionary and d is the dimension of each feature vector.

We concatenate the dictionaries built on different patch sizes into a single dictionary, respecting the nature of the feature. This leaves us with two dictionaries, one built from features extracted using the model trained on Places, and the other using the model trained on ImageNet. The idea is improving the probability of finding a match in different scales, which offers a better discrimination and repeatability power. Once we have built the initial dictionary $\mathbf{D_0}$, we adjust the dictionary to matrix $\mathbf{D}$ by solving

$$\min_{\mathbf{D}, \mathbf{X}} \frac{1}{n} \sum_{i=1}^{n} \left(\|\mathbf{y_i} - \mathbf{D x_i}\|_2^2 \quad \lambda_{dl} \|\mathbf{x_i}\|_1 \right), \tag{1}$$

where $\|\cdot\|_2$ is the $L2$-norm, $\|\cdot\|_1$ is the $L1$-norm, λ_{dl} is a regularization parameter for the dictionary learning and the vector $\mathbf{x_i}$ is the ith column of matrix $\mathbf{X}$. We applied a dictionary learning algorithm [8], which solves Eq. 1 alternating between $\mathbf{D}$ and $\mathbf{X}$ as variables, i.e., it minimizes over $\mathbf{D}$ while keeping $\mathbf{X}$ fixed. The dictionary $\mathbf{D_0}$ is used to start the optimization process.

Sparse Coding. Despite the large number of possible objects that can compose a scene, each class of indoor environments has a small number of characteristic types of objects. The dictionary $\mathbf{D}$ provides a linear representation of each image fragment using a small number of basis functions. Thus, we find the sparse vector $\mathbf{x}$ by modeling the composition of a scene as a sparse coding problem.

Considering the new domain of representation defined by the dictionary $\mathbf{D}$, the representation of a feature vector $\mathbf{y}_i$ extracted from a sample of indoor class i, can be rewritten as $\mathbf{y}_i = \mathbf{D x}_i$, where $\mathbf{x}_i$ is a vector whose entries are zero except those associated with the fragments of class i.

We use as a basis for the vector representation the columns $\mathbf{D}_i$ of a dictionary that is mixed with weights $\mathbf{x}$ to infer the vector $\mathbf{y}$, which is the vector that reconstructs the input $\mathbf{y}$ with the smallest error. Each column $\mathbf{D}_i$ is a descriptor representing an image fragment of the scene. We use a sparsity regularization term to select a small number of columns in the dictionary $\mathbf{D}$.

Let $\mathbf{y} \in \mathbb{R}^d$ be a descriptor extracted from an image patch and $\mathbf{D} \in \mathbb{R}^{d \times kc}$ ($d \ll kc$). The coding can be modeled as the following optimization problem:

$$\mathbf{x}^* = \underset{\mathbf{x}}{\operatorname{argmin}} \|\mathbf{y} - \mathbf{D}\mathbf{x}\|_2^2 \quad s.t. \quad \|\mathbf{x}\|_0 \leq L, \tag{2}$$

where $\|\cdot\|_0$ is the $L0$-norm, indicating the number of nonzero values, and L controls the sparsity of vector $\mathbf{x}$. The final vector $\mathbf{x}^* \in \mathbb{R}^{kc}$ is the set of basis weights that represents the descriptor $\mathbf{y}$ as a linear and sparse combination of fragments of scenes.

Although the minimization problem of Eq. 2 is NP-hard, greedy approaches, such as Orthogonal Matching Pursuit (OMP) [16] or $L1$ norm relaxation, also know as LASSO [15], can be used to effectively solve it. In this paper, we use OMP because it achieved the best results on our tests.

The pooling process refers to the final step of the diagram presented in Fig. 1. To create the vector encoding the scene features, we compute the final feature vector $\mathbf{f} \in \mathbb{R}^{kc}$ by a pooling function $\mathbf{f} = \mathcal{F}(\mathbf{X})$, where $\mathcal{F}$ is defined on each column of $\mathbf{X} \in \mathbb{R}^{m \times kc}$. The rows of matrix $\mathbf{X}$ represent the sparse vectors of each feature vector extracted from m sliding-windows. We create kc-dimensional vectors according to the maximum pooling function, since according to Yang et al. [19], it gives the best results for sparse coded local features.

These steps are executed for both scales, using both the model trained on Places and ImageNet. At the end, five features vectors are concatenated into one: the global features as originally output by the CNN, and four local features.

3   Experiments

We evaluated our approach on the standard datasets for scene recognition benchmark: Scene15, MIT67 and SUN397. The specific attributes of each dataset allowed us to analyze different properties of our method. We trained the models on Places for global and local descriptors, and ImageNet for local descriptors, with VGGNet-16 which has shown the lowest classification error [14]. When considering features from the VGGNet-16, we used PCA to reduce the descriptions from 4,096 to 1,000 on the second and third scales.

To compose the dictionary, we set 2,175 words for Scene15, 3,886 words for MIT67 and 6,907 words for SUN397, according to the number of regions extracted from both scales and the number of classes of each dataset. The regularization factor λ_{dl} for the dictionary learning (Eq. 1) is set to 0.1. We set the same sparsity controller value L (Eq. 2) for all configurations when executing OMP. It was set to $0.03 \times D_c$ non-zeros, where D_c is the number of columns of the dictionary. All vectors were $L2$ normalized after each step and after concatenation. We used a Linear Support Vector Machine to perform the classification.

Robustness evaluation. We verified the descriptor robustness against occlusion and noise. For this purpose, we automatically generated squares randomly positioned, to simulate occlusions (black squares) and noise (granular squares), as illustrated in Fig. 2. Experiments were performed for different sizes of windows $\frac{\mathbf{W}}{n}$, where $\mathbf{W}$ represents the dimensions of the image, with varying n values.

Table 1. Accuracy performance comparison for occlusion and noise scenarios. Our approach shows higher accuracy performance. It provides a better feature selection for scene patches instead of just max-pooling between CNN features.

Window size		Scene15				MIT67				SUN397			
		W/10	W/8	W/6	W/4	W/10	W/8	W/6	W/4	W/10	W/8	W/6	W/4
Occlusion	Herranz et al.	94.07	93.87	93.97	92.93	83.79	84.01	83.79	83.56	39.95	40.05	38.99	35.61
	Ours	94.43	94.50	94.37	93.50	84.61	84.61	84.69	84.76	64.64	64.71	64.01	62.39
	Avg. gain	**0.49**				**0.88**				**25.29**			
Noise	Herranz et al.	94.50	94.27	93.97	92.93	83.94	84.01	83.49	81.32	40.79	40.42	34.29	30.88
	Ours	95.30	94.57	94.40	93.50	84.24	85.28	84.69	82.74	65.20	65.29	57.25	50.83
	Avg. gain	**0.53**				**1.05**				**23.05**			

Fig. 2. Occlusion and noise examples. The location of windows was randomly computed, and experiments were performed for different sizes of windows.

Table 1 presents the results for occlusion and noise robustness tested against Herranz et al. [3]. Our methodology performs better in all cases, indicating that our method is less sensitive to perturbations on the image. This behaviour is even more evident for SUN397, by far the most challenging of all three datasets, where the Herranz et al.'s methodology was inferior, on average 25.29% for occlusion, and 23.05% when noise was added.

Discussion. Besides being highly robust to perturbations in the input image, one can clearly see in Table 2 that our methodology also leads the performance on Scene15 and MIT67, while performing competitively on SUN397. It is worth highlighting that our method surpasses human accuracy, which was measured for SUN397 at 68.5% [18]. To evaluate our assumption that local information can greatly benefit the task of indoor scene classification, Table 2 also presents the average accuracy achieved by a model trained solely on features from VGGNet-16 trained on Places, which composes the global portion of our methodology. On all datasets, global features by themselves showed inferior performance relative to our representation, revealing the complementary nature of local features.

We also performed a detailed performance assessment by comparing the accuracies among each indoor class on MIT67. The relative accuracies considering VGGNet-16 as a baseline are shown in Fig. 3. As we can see, for the classes that present a large number of object information (e.g., children room, emphasized in Fig. 3) we have a significant increase in accuracy. We can draw the following two observations. First, since the class of these environments strongly depends on the object configuration, the image fragments containing features of common objects can be represented by the sparse features on different scales. Second, for environments such as movie theater (Fig. 3), the proposed approach is less effective, since such environments are strongly distinguished by their overall structure, and not necessarily by the constituent objects.

Table 2. Comparing average accuracy against other approaches.

Method	SPM [5]	ScSPM [19]	MOP-CNN [20]	SFV [1]	VGGNet-16	Herranz et al. [3]	VSAD [17]	Ours
Scene15	81.40%	80.28%	-	-	94.33%	95.18%	-	**95.73%**
MIT67	-	-	68.88%	79.00%	80.88%	86.04%	86.2%	**87.22%**
SUN397	-	-	51.98%	61.72%	66.90%	70.17%	**73.0%**	71.08%

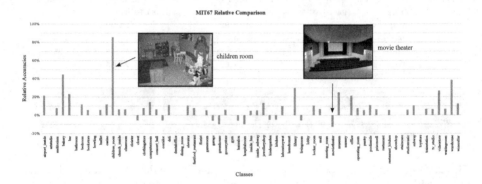

Fig. 3. Relative accuracy between VGGNet-16 and our method on MIT67. In most cases, our method leads the performance.

4   Conclusion

We proposed a robust method that combines both global and local features to compose a high discriminative representation of indoor scenes. Our method improves the accuracy of CNN features by composing local features using Sparse-Coding and a max-pooling technique, by creating an indoor scene representation based on an over-complete dictionary, which is built from several image fragments. Our sparse coding-based composition approach was capable of encoding local patterns and structures for indoor environments, and in cases of strong occlusion and noise, we lead the performance on scene recognition, which is explained by the advantages of sparse representations. Our representation outperformed VGGNet-16 in all cases, reflecting the complementary nature of local features, and outperformed the current state of the art on Scene15 and MIT67, while being competitive on SUN397.

References

1. Dixit, M., Chen, S., Gao, D., Rasiwasia, N., Vasconcelos, N.: Scene classification with semantic fisher vectors. In: IEEE Conference on Computer Vision and Pattern Recognition, pp. 2974–2983. IEEE (2015)
2. Gao, S., Tsang, I.W.H., Chia, L.T., Zhao, P.: Local features are not lonely - Laplacian sparse coding for image classification. In: IEEE Conference on Computer Vision and Pattern Recognition, pp. 3555–3561. IEEE Computer Society (2010)

3. Herranz, L., Jiang, S., Li, X.: Scene recognition with CNNs: objects, scales and dataset bias. In: IEEE Conference on Computer Vision and Pattern Recognition, June 2016
4. Jaakkola, T., Haussler, D.: Exploiting generative models in discriminative classifiers. In: Advances in Neural Information Processing Systems, vol. 11, pp. 487–493. MIT Press (1998)
5. Lazebnik, S., Schmid, C., Ponce, J.: Beyond bags of features: spatial pyramid matching for recognizing natural scene categories. In: IEEE Conference on Computer Vision and Pattern Recognition, pp. 2169–2178. IEEE Computer Society (2006)
6. Li, F.F., Perona, P.: A Bayesian hierarchical model for learning natural scene categories. In: IEEE Conference on Computer Vision and Pattern Recognition, pp. 524–531. IEEE Computer Society, Washington, DC (2005)
7. Lowe, D.G.: Distinctive image features from scale-invariant keypoints. Int. J. Comput. Vis. 60(2), 91–110 (2004)
8. Mairal, J., Bach, F., Ponce, J., Sapiro, G.: Online dictionary learning for sparse coding. In: Proceedings of the 26th Annual International Conference on Machine Learning, ICML 2009, pp. 689–696, ACM, New York (2009)
9. Russakovsky, O., Deng, J., Su, H., Krause, J., Satheesh, S., Ma, S., Huang, Z., Karpathy, A., Khosla, A., Bernstein, M., Berg, A.C., Fei-Fei, L.: ImageNet large scale visual recognition challenge. Int. J. Comput. Vis. 115(3), 211–252 (2015)
10. Oliva, A., Torralba, A.: Modeling the shape of the scene: a holistic representation of the spatial envelope. Int. J. Comput. Vis. 42(3), 145–175 (2001)
11. Oliveira, G., Nascimento, E., Vieira, A., Campos, M.: Sparse spatial coding: a novel approach to visual recognition. IEEE Trans. Image Process. 23(6), 2719–2731 (2014)
12. Quattoni, A., Torralba, A.: Recognizing indoor scenes. In: IEEE Conference on Computer Vision and Pattern Recognition, pp. 413–420 (2009)
13. Sánchez, J., Perronnin, F., Mensink, T., Verbeek, J.: Image classification with the fisher vector: theory and practice. Int. J. Comput. Vis. 105(3), 222–245 (2013)
14. Simonyan, K., Zisserman, A.: Very deep convolutional networks for large-scale image recognition. CoRR abs/1409.1556 (2014)
15. Tibshirani, R.: Regression shrinkage and selection via the lasso. J. R. Stat. Soc. Ser. B 58, 267–288 (1994)
16. Tropp, J.A., Gilbert, A.C.: Signal recovery from random measurements via orthogonal matching pursuit. IEEE Trans. Inf. Theory 53(12), 4655–4666 (2007)
17. Wang, Z., Wang, L., Wang, Y., Zhang, B., Qiao, Y.: Weakly supervised PatchNets: describing and aggregating local patches for scene recognition. IEEE Trans. Image Process. 26, 2028–2041 (2017)
18. Xiao, J., Hays, J., Ehinger, K.A., Oliva, A., Torralba, A.: SUN database: large-scale scene recognition from abbey to zoo. In: IEEE Conference on Computer Vision and Pattern Recognition, pp. 3485–3492. IEEE Computer Society (2010)
19. Yang, J., Yu, K., Gong, Y., Huang, T.: Linear spatial pyramid matching using sparse coding for image classification. In: IEEE Conference on Computer Vision and Pattern Recognition (2009)
20. Gong, Y., Wang, L., Guo, R., Lazebnik, S.: Multi-scale orderless pooling of deep convolutional activation features. CoRR abs/1403.1840 (2014)
21. Zhou, B., Lapedriza, A., Xiao, J., Torralba, A., Oliva, A.: Learning deep features for scene recognition using places database. In: Ghahramani, Z., Welling, M., Cortes, C., Lawrence, N.D., Weinberger, K.Q. (eds.) Advances in Neural Information Processing Systems 27, pp. 487–495. Curran Associates, Inc. (2014)

Deep Convolutional Neural Networks and Noisy Images

Tiago S. Nazaré, Gabriel B. Paranhos da Costa, Welinton A. Contato,
and Moacir Ponti[(⊠)]

Instituto de Ciências Matemáticas e de Computação, Universidade de São Paulo,
São Carlos, SP 13566-590, Brazil
{tiagosn,gbpcosta,welintonandrey,ponti}@usp.br

Abstract. The presence of noise represent a relevant issue in image
feature extraction and classification. In deep learning, representation is
learned directly from the data and, therefore, the classification model
is influenced by the quality of the input. However, the ability of deep
convolutional neural networks to deal with images that have a different
quality when compare to those used to train the network is still to be
fully understood. In this paper, we evaluate the generalization of models learned by different networks using noisy images. Our results show
that noise cause the classification problem to become harder. However,
when image quality is prone to variations after deployment, it might be
advantageous to employ models learned using noisy data.

1 Introduction

In real-world applications image quality may vary drastically depending on factors such as the capture sensor used and lighting conditions. These scenarios
need to be taken into consideration when performing image classification, since
quality shift directly influence its results.

Lately, deep convolutional neural networks have obtained outstanding results
in image classification problems. Nonetheless, little has been done to understand
the impacts of image quality on the classification results of such networks. In
most studies, networks were only tested on images whose quality is similar to
the training set (i.e. similar noise/blur levels). The lack of research in this topic
is not exclusive to deep learning applications. Most image classification systems
neglect preprocessing [12] and assume that image quality does not vary [5].

Given that in real-world applications image quality may vary, we evaluate
classification performance of classic deep convolutional neural networks when
dealing with different types and levels of noise. In addition, we investigate if
denoising methods can help mitigate this problem.

1.1 Related Work

We devote our efforts to investigate the effects of noisy images when using deep
convolutional neural networks in classification tasks. There are papers investigating the effect of label noise in the learning capability and performance of

© Springer International Publishing AG, part of Springer Nature 2018
M. Mendoza and S. Velastín (Eds.): CIARP 2017, LNCS 10657, pp. 416–424, 2018.
https://doi.org/10.1007/978-3-319-75193-1_50

convolutional neural networks [2]. However, this problem is not addressed in this paper. Thus, in our experiments, we assume that all labels are correct.

Some studies have already identified that image quality can hinder classification performance in systems that employ neural networks [5] and in systems that use hand-crafted features [4,9]. Recently, the development of noise-robust neural networks has been investigated. For instance, [6] presented a network architecture that can cope with some types of nosy images, while [13] designed a network that is capable of dealing with noise in speech recognition.

Dodge and Karam [5] showed that state-of-the-art deep neural networks are affected when classifying images with lower quality. In their experiments, each network was trained on images from the original dataset (with a negligible amount of noise due to the image formation process) and, then, used to classify images from the same dataset on their original state, degraded by noise and affected by blur. Their results show that classification performance is hampered when classifying images with lower quality. However, their experiments do not cover the presence of low-quality images in the training set and their impact in the learned model.

Paranhos da Costa et al. [4] extended the methodology of [5] by considering that low-quality images can also appear in the training set. In their setup, several noisy versions of a dataset are created: each version has the same images as the original dataset, wherein all images are affected by a type of noise at a fixed level. They also evaluated the effects of denoising techniques by studying restoration of noisy images. Hand-crafted features (LBP and HOG) were extracted, SVM classifiers trained with each version of the training set and, then, used to classify all versions of the test set. Even so, their study only considered two hand-crafted features (LBP and HOG).

We believe that noise makes classification more difficult due to the fact that models trained with a particular noisy/restored training set version – and tested on images with the same noise configuration – usually perform worse than a model trained and tested on the original data. Our empirical evaluation is based on [4], however there are two main differences. First, our experiments target deep neural networks, with the ability to learn from data, even from low quality data, while the previous study considers hand-crafted features and SVM classifiers. Second, we investigate if training models with a specific noise or image restoration setup can help to build models that are more resilient to changes in image quality for future data, i.e., in the test set.

2 Experiments

2.1 Experimental Setup

The **first** step in the experimental setup used in this study is to create noisy and restored versions of each one of the three publicly-available datasets selected for our experiments: MNIST, CIFAR-10 and SVHN (further information on these datasets is presented in Sect. 2.2). To do that, five copies of the original dataset are degraded by a Gaussian noise with standard deviation

$\sigma = \{10, 20, 30, 40, 50\}$, and another five copies were hindered by a salt & pepper noise using $p = \{0.1, 0.2, 0.3, 0.4, 0.5\}$, where p is the probability of a pixel being affected by the noise. Next, denoising methods are applied on each of these versions, generating 10 restored versions of each dataset, one for each noisy version.

The restored versions of the datasets affected by Gaussian noise are obtained by filtering the images with the Non-Local Means (NLM) algorithm [3]. To perform the NLM denoising, we used a 7×7 patch, a 11×11 window and a set the parameter h equal to the standard deviation of the Gaussian noise used to corrupt the dataset being restored. Regarding the salt & pepper noise, all restored versions were generated by filtering the noise images using a 3×3 median filter. Hence, we have 21 different versions of each dataset (the original dataset, 10 noisy and 10 restored versions). Since all versions contain the same images, we always use the same training-test split presented in the original dataset paper.

The **second** step is to learn a classifier for each training set version. This means that a single network architecture was trained with each dataset version, creating 21 different classifiers. Then, each classifier was tested on all versions of the test set. This process is illustrated by the diagram shown in Fig. 1. For the MNIST dataset, we selected an architecture similar to the LeNet-5 [10], while for the CIFAR-10 and the SVHN datasets, an architecture similar to the base model C of [15] was used. These architectures were implemented using the Keras library and our implementation was based on the code available on [1]. A convolutional neural network architecture was selected for each dataset.

In the **third** part of our experimental setup, we compare the learned models. We begin our analysis by comparing classification accuracies when both the training and test set have the same type and level of noise. By doing so, we want to measure how much harder classifying these datasets gets – for that network

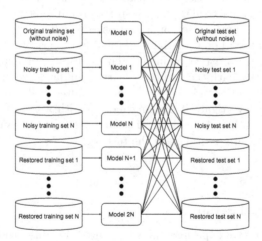

Fig. 1. Experimental setup diagram. In our experiments a different model is trained for each noise configuration, then these models are used to classify all versions of the test set. This figure was based on Fig. 2 of [4].

architecture – once the noise occurs at a particular level in the entire dataset (training and test set).

Additionally, we compare the results on the noisy versions with their restored counterparts, which allows us to measure how much the use of denoising techniques can help to improve accuracy. After, we visualize the classification results of all trained models in all versions of the test set using heatmaps. Such visualization shows how performance varies for each model. Lastly, we compute the mean (and standard deviation) of the accuracies obtained by each classifier over all test set versions. Using these values we can quantize the overall performance difference among models and compare models with regards to their resilience to different types of noisy images. This setup is illustrated by the diagram shown in Fig. 1.

2.2 Datasets

MNIST: handwritten digits [10] broadly used in deep learning experiments due to being real-world data that requires minimal pre-processing/formatting.

CIFAR-10: consisting of 60,000 color 32×32 images equally split into 10 classes [8]. This dataset is subdivided into training and test sets, which include 50,000 and 10,000 images, respectively.

SVHN: house numbers from Google Street View images [11], defines a real-world problem of recognizing digits in natural images. It is composed by 73,257 images in the training set and 26,032 images in the test set.

3 Results and Discussion

To contrast the impact of noise and denoising methods in image quality, the average Peak Signal-to-Noise Ratio (PSNR) values for each dataset version are shown in Table 1. By comparing the results when training and testing is performed in the same dataset version it is possible to analyse how noise affects classification, in particular if it makes the task more difficult by changing the parameter space learned by the network. These results are shown in Table 2, in which it is possible to notice that, in general, the presence of noise, even when restored using a denoising algorithm, increases the complexity of the classification task.

To better understand the effects of using denoising methods we plot some of the results presented in Table 2 in Fig. 3, showing the accuracies of the models trained with images affected by different levels of Gaussian noise as well salt & pepper noise, and their restored counterparts. Neither of these two figures present the results for the MNIST dataset, because, as can be seen in Table 2, the differences for this dataset were too small.

Despite the increase in PSNR when employing NLM for denoising, accuracy decreased when classifying data restored by this method. This is probably due to that fact such denoising procedure generate blurry images, removing relevant information, as can be seen in Fig. 2.

Table 1. Average PSNR for each noise level.

Noise		MNIST		CIFAR-10		SVHN	
		Noisy	Restored	Noisy	Restored	Noisy	Restored
Gaussian	$\sigma = 10$	30.93	33.18	28.25	29.88	28.15	32.65
	$\sigma = 20$	24.86	27.46	22.36	25.54	22.21	28.15
	$\sigma = 30$	21.35	24.35	18.99	22.81	18.81	25.21
	$\sigma = 40$	18.87	21.54	16.67	20.85	16.47	23.40
	$\sigma = 50$	16.96	17.84	14.95	19.49	14.74	22.26
s & p	$p = 0.1$	13.44	20.78	15.23	25.89	15.52	34.30
	$p = 0.2$	10.79	18.54	12.50	24.00	12.79	29.32
	$p = 0.3$	9.36	16.69	10.98	21.64	11.28	24.82
	$p = 0.4$	8.43	15.13	9.97	19.22	10.28	21.28
	$p = 0.5$	7.78	13.84	9.23	17.04	9.55	18.53

Table 2. Accuracy of each network when training and testing was conducted using the same dataset version.

Noise Type		MNIST		CIFAR-10		SVHN	
		Noisy	Restored	Noisy	Restored	Noisy	Restored
Original		0.9903		0.8192		0.9438	
Gaussian	$\sigma = 10$	0.9916	0.9885	0.8007	0.8016	0.9415	0.9445
	$\sigma = 20$	0.9875	0.9878	0.7727	0.7583	0.9284	0.9210
	$\sigma = 30$	0.9898	0.9867	0.7309	0.7156	0.9015	0.8793
	$\sigma = 40$	0.9890	0.9851	0.6991	0.6625	0.8849	0.8455
	$\sigma = 50$	0.9860	0.9814	0.6608	0.6277	0.8542	0.7987
s & p	$p = 0.1$	0.9799	0.9861	0.7227	0.7613	0.9308	0.9418
	$p = 0.2$	0.9793	0.9802	0.6902	0.7508	0.9193	0.9425
	$p = 0.3$	0.9753	0.9718	0.6088	0.7138	0.9095	0.9301
	$p = 0.4$	0.9641	0.9605	0.5610	0.6921	0.9057	0.9239
	$p = 0.5$	0.9437	0.9426	0.5398	0.6627	0.8889	0.9228

Fig. 2. Examples of noisy images for each dataset. The first row show the original images. The second row depict images with Gaussian noise ($\sigma = 30$) and their restored versions. Finally, the third row has images affected by salt & pepper noise ($p = 0.3$) and denoised by a median filter.

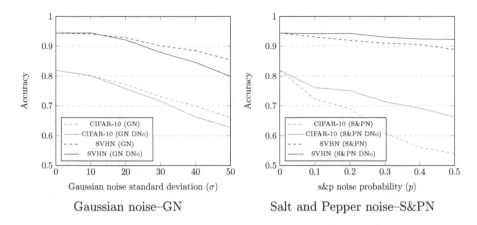

Gaussian noise–GN   Salt and Pepper noise–S&PN

Fig. 3. Comparison of the accuracy of each network for different noise parameters (a) Gaussian noise standard deviation σ (b) Salt & Pepper probability p, with and without the use of denoising algorithms (DNo) for restoration.

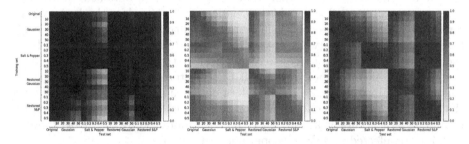

Fig. 4. Heatmaps representing the results obtained on MNIST (left), CIFAR-10 (center) and SVHN (right).

Next, results comparing models when classifying all dataset versions are presented using the heatmaps of Fig. 4. Each row in a heatmap represent a version of the training set, while each column displays the results for a version of the test set. As demonstrated in [4], models tend to achieve their best accuracy when classifying data that has the same quality as the data used to train them. Nevertheless, depending on the training set, different generalization capability is achieved. This is demonstrated by models whose results are similar to their best, even when classifying data affected by other types of noise. To compare the noise resilience of these models, Table 3 show the mean and standard deviation accuracies obtained by each classifier over all test set versions (the mean and standard deviation of each row of each heatmap).

In this comparison, the network trained with the original dataset is used as a baseline scenario, given that this network has no previous knowledge of any type of noise, while the others have already seen noisy images in some level. Therefore, networks trained with noisy images have an advantage when

Table 3. Average accuracy and standard deviation (in percentages) for each model in all test set versions.

Noise type		MNIST		CIFAR-10		SVHN	
		Noisy	Restored	Noisy	Restored	Noisy	Restored
Original		93.13 ± 9.68		43.50 ± 21.84		68.21 ± 24.64	
Gaussian	$\sigma = 10$	94.59 ± 7.35	87.68 ± 18.68	45.04 ± 21.80	48.83 ± 21.12	71.87 ± 22.47	67.41 ± 25.15
	$\sigma = 20$	91.00 ± 14.08	90.77 ± 14.18	49.52 ± 20.41	48.68 ± 21.06	73.72 ± 20.28	67.41 ± 25.15
	$\sigma = 30$	94.26 ± 8.65	88.29 ± 18.43	50.72 ± 18.57	47.75 ± 18.48	74.36 ± 18.94	61.09 ± 28.32
	$\sigma = 40$	94.49 ± 6.94	92.95 ± 9.90	51.76 ± 16.79	41.82 ± 18.07	75.62 ± 17.54	61.01 ± 28.67
	$\sigma = 50$	93.58 ± 8.16	90.72 ± 12.42	51.23 ± 14.85	38.31 ± 16.04	75.18 ± 16.76	63.51 ± 27.81
s & p	$p = 0.1$	95.85 ± 4.02	90.12 ± 12.55	56.78 ± 15.04	44.20 ± 22.47	82.97 ± 11.81	66.69 ± 24.95
	$p = 0.2$	97.13 ± 2.12	84.71 ± 19.06	56.92 ± 12.89	51.69 ± 20.08	$\mathbf{83.41 \pm 11.47}$	76.33 ± 20.13
	$p = 0.3$	$\mathbf{97.27 \pm 1.74}$	88.13 ± 15.61	49.79 ± 11.89	57.20 ± 14.97	82.26 ± 13.01	77.27 ± 19.79
	$p = 0.4$	97.11 ± 1.57	83.70 ± 18.57	42.83 ± 11.85	$\mathbf{57.83 \pm 13.71}$	82.84 ± 13.28	80.14 ± 16.04
	$p = 0.5$	96.32 ± 1.87	81.66 ± 22.40	33.21 ± 12.21	57.79 ± 11.76	80.02 ± 14.14	82.16 ± 12.89

dealing with noise in future data even when it occurs at a different level. For the MNIST dataset, models trained on the original data obtained an average accuracy of 93.13% and a standard deviation of 9.68%, while the best overall result of $97.27 \pm 1.74\%$ was obtained by the model trained using images affected by salt & pepper noise (with $p = 0.3$). On the CIFAR-10 dataset, the best average results ($57.83 \pm 13.71\%$) were obtained by the model trained with data corrupted by salt & pepper with $p = 0.4$ and restored using the median filter. The model trained using the original data obtained $43.50 \pm 21.84\%$. Lastly, in the SVHN dataset, the model trained on the original dataset obtained a overall $68.21 \pm 24.64\%$ accuracy, against $83.41 \pm 11.47\%$ obtained by the model trained with images affected by salt & pepper noise ($p = 0.2$).

Nevertheless, it is possible to notice that some models are better at generalising to other types of noise. For instance, in the MNIST dataset, most models trained with salt & pepper noise obtained were able to achieve results around 0.6 or higher, while the other model did not.

To facilitate reproducibility of our experiments, our code is publicly available at http://github.com/tiagosn/dnnnoise2017.

4   Conclusions

We analysed the behaviour of deep convolutional neural networks when dealing with different types of image quality. Our study covered images affected by s&p and Gaussian noise and their restored versions. Although noise injection in the training data is a common practice, our systematic methodology provide a better understanding of the behaviour of the models under noise conditions. The results indicate that training networks using data affected by some types of noise could be beneficial for applications that need to deal with images with varying quality, given that it seems to improve the resilience of the network to other types of noise and noise levels.

Concerning denoising methods, images restored with the median filter, when compared against images with s&p noise, were able to improve the accuracy in data with the same quality. Nevertheless, models trained with s&p noise usually obtained a better noise resilience. Restoring images with NLM resulted, for the most part, in a decrease in performance. This was probably due to the removal of relevant information caused by NLM smoothing. Hence, better results might be achieved with a different parameter choice.

As future work we intend to explore deeper models such as VGG [14] and ResNets [7]. These experiments should also include neural networks designed to be robust to noise such as in [6]. Moreover, we aim at conducting experiments in larger datasets like ImageNet.

Acknowledgment. The authors would like to thank FAPESP (grants #16/16111-4, #13/07375-0, #15/05310-3, #15/04883-0).

References

1. Arnold, T.: Stat 365/665: Data mining and machine learning: Lecture notes (transfer learning and computer vision I) (April 2016)
2. Bekker, A.J., Goldberger, J.: Training deep neural-networks based on unreliable labels. In: IEEE International Conference on Acoustics, Speech and Signal Processing (ICASSP), pp. 2682–2686. IEEE (2016)
3. Buades, A., Coll, B., Morel, J.M.: A non-local algorithm for image denoising. In: IEEE Computer Society Conference on Computer Vision and Pattern Recognition, CVPR 2005, vol. 2, pp. 60–65. IEEE (2005)
4. Paranhos da Costa, G.B., Contato, W.A., Nazare, T.S., Batista Neto, J.E.S., Ponti, M.: An empirical study on the effects of different types of noise in image classification tasks. In: XII Workshop de Visão Computacional (WVC 2016) (2016)
5. Dodge, S., Karam, L.: Understanding how image quality affects deep neural networks. In: 2016 Eighth International Conference on Quality of Multimedia Experience (QoMEX), pp. 1–6 (Jun 2016)
6. Ghifary, M., Kleijn, W.B., Zhang, M.: Deep hybrid networks with good out-of-sample object recognition. In: 2014 IEEE International Conference on Acoustics, Speech and Signal Processing (ICASSP), pp. 5437–5441. IEEE (2014)
7. He, K., Zhang, X., Ren, S., Sun, J.: Deep residual learning for image recognition. arXiv preprint arXiv:1512.03385 (2015)
8. Krizhevsky, A., Hinton, G.: Learning multiple layers of features from tiny images (2009)
9. Kylberg, G., Sintorn, I.M.: Evaluation of noise robustness for local binary pattern descriptors in texture classification. EURASIP J. Image Video Process. **2013**, 17 (2013)
10. LeCun, Y., Bottou, L., Bengio, Y., Haffner, P.: Gradient-based learning applied to document recognition. Proc. IEEE **86**(11), 2278–2324 (1998)
11. Netzer, Y., Wang, T., Coates, A., Bissacco, A., Wu, B., Ng, A.Y.: Reading digits in natural images with unsupervised feature learning (2011)
12. Ponti, M., Nazaré, T.S., Thumé, G.S.: Image quantization as a dimensionality reduction procedure in color and texture feature extraction. Neurocomputing **173**, 385–396 (2016)

13. Seltzer, M.L., Yu, D., Wang, Y.: An investigation of deep neural networks for noise robust speech recognition. In: IEEE International Conference on Acoustics, Speech and Signal Processing, pp. 7398–7402. IEEE (2013)
14. Simonyan, K., Zisserman, A.: Very deep convolutional networks for large-scale image recognition. CoRR abs/1409.1556 (2014)
15. Springenberg, J.T., Dosovitskiy, A., Brox, T., Riedmiller, M.: Striving for simplicity: the all convolutional net. arXiv preprint arXiv:1412.6806 (2014)

Spiking Hough for Shape Recognition

Pablo Negri[1,2(✉)] [iD], Teresa Serrano-Gotarredona[3],
and Bernabe Linares-Barranco[3]

[1] CONICET, Dorrego 2290, Buenos Aires, Argentina
[2] Universidad Argentina de la Empresa (UADE), Lima 717, Buenos Aires, Argentina
pnegri@uade.edu.ar
[3] CSIC, Instituto de Microelectronica Sevilla (IMSE-CNM), 41012 Sevilla, Spain

Abstract. The paper implements a spiking neural model methodology inspired on the Hough Transform. On-line event-driven spikes from Dynamic Vision Sensors are evaluated to characterize and recognize the shape of Poker signs. The multi-class system, referred as Spiking Hough, shows the good performance on the public POKER-DVS dataset.

1 Introduction

Histogram based features are a useful tool employed in Computer Vision to describe the content of an image. They are successfully used for object detection associated with different kinds of classifiers. There exists several versions, such as: HOG [4], SIFT [8], HO2L [9], etc.

Recently, a histogram based feature were proposed to characterize the shape of objects captured by an Event-Driven Dynamic Vision Sensor (DVS) [10]. This cameras are inspired on the neuromorphic behavior of the human visual system [7]. They consist of an artificial retina where each pixel captures a light change and generates a spike event. This event is defined as $\mathbf{e} = ((x,y), t, pol)$, (x,y) being the coordinates of the pixel on the grid, t the event time stamp, and pol the polarity. Polarity is a binary ON/OFF output. ON polarity informs an illumination increase, and OFF polarity is obtained when illumination decreases. An event flow composed of N consecutive events is defined as: $\mathbf{w}_N = \{\mathbf{e}_1, \ldots, \mathbf{e}_N\}$. This kind of vision sensors is considered as "frameless" providing asynchronous high temporal resolution data.

Others histograms representations using DVS events flows were proposed in the literature. Clady *et al.* [3] proposed a hand-gesture recognition framework using a histogram representation of flow motion vectors. In [5] they employed a hierarchical model architecture (HOTS) consisting of several consecutive layers of increasing detail and including a histogram representation of time-surface activations for each object class. The architecture is based on a deep neural network, similar to the Convolutional Spiking Neural Network in [11,18] for object recognition.

The proposed system is denominated Spiking Hough, because it is inspired on the Hough Transform and uses a spiking neural network approach. In [6,12]

© Springer International Publishing AG, part of Springer Nature 2018
M. Mendoza and S. Velastín (Eds.): CIARP 2017, LNCS 10657, pp. 425–432, 2018.
https://doi.org/10.1007/978-3-319-75193-1_51

the Hough transform is applied on the DVS outputs for straight edges and lane detection. On the other hand, this paper proposes a methodology which captures the spatial distribution of the events generated in the retina, and organize this information in histogram features. The histograms feed a multi-class classifier to recognize the shape of the objects.

The paper is organized as follow. Next section details the Spiking Hough approach and the classification system. Section 3 presents and discusses the results. The conclusions in Sect. 4 resume the work and propose some perspectives.

2   Spiking Hough Multi-class Classification System

The systems aims to classify the events flow into one of the four Poker signs. The pipeline is presented in Fig. 1. Incoming events spike the neurons distributed on the retina grid. The Spiking Hough receives the firing neurons, capturing the configuration of the object's shape and builds the cell histograms. The framework counts the number of incoming events, and when this number reaches N, the histograms features are evaluated with a multi-class Support Vector Machine (SVM) [15] classifier. After the classification, the histograms are reset for the next windows event.

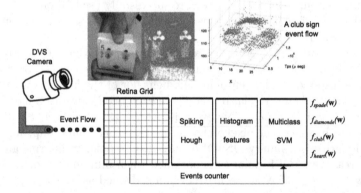

Fig. 1. Poker-DVS dataset extraction. Left image shows a RGB capture (image from [11]), and right image shows the DVS retina captured events of a *club* sign.

2.1   Spiking Neural Network Model

Figure 2 shows a neuron model [1]. In this model, dendrites are the inputs terminals to the nucleus, and the axons their outputs. The neural interfaces between dendrites and axons are denominated synapses.

The dynamic of the neuron model is controlled by input spikes, an electrical signal with short duration, arriving through the dendrites. The potential of these

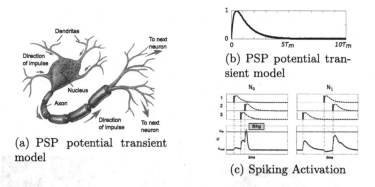

Fig. 2. (a) Neuron model [1], (b) PSP transient model and (c) Spiking neuron model (from [17]) (Color figure online)

incoming spikes are accumulated in the neuron membrane. When the potential reaches a threshold, the neuron fires a signal, an output spike, to the next neurons via the axons. Synapses control the amplitude of the spike traveling to the following neuron.

The potential of the membrane receiving an excitation from an spike has a transient behavior commonly referred as Post-Synaptic Potential (PSP). PSP can affects the membrane's potential by: Excitatory PSP (EPSP) which increments the potential, or Inhibitory PSP (IPSP) which decrements the potential. Figure 2(b) shows the PSP potential computed using the Integrate-and-Fire approach, with equation $u(t - t_i) = V_0(e^{-\frac{t-t_i}{\tau_m}} - e^{-\frac{t-t_i}{\tau_s}})$, where τ_m and τ_s denote decay time constants of the membrane integration, and V_0 normalizes the potential so that the maximum value is 1.

The spiking neural model is shown in Fig. 2(c). Two neurons, N_0 and N_1, with three afferent synapses inputs. The spikes on the input synapses are shown as blue signals. After the spike, the PSP model shows the individual afferent transient potential added the membrane potential $u_j(t)$. For N_0 all PSP signals corresponds to excitatory PSPs, with different synapses weights. In the case of N_1 synapses 1 corresponds to a IPSP which provides a negative potential to $u_1(t)$. In the example, after the three spikes of their afferent synapses, the potential of N_0 reaches threshold u_{th} and generating an output spike. Then, the potential $u_1(t)$ is reset to the u_{rest} value. Neuron N_1 did not fires, because their potential never cross the threshold in the temporal window.

Spiking neural models are well adapted to be employed with the DVS data flow. The flow of individual events feed the neurons which are trained to trigger when a special configuration is detected. The spiking model is adapted to a Hough analysis in the next section.

2.2 Hough Transform Analysis

The Hough Transform projects data from the image *xy-plane* to a new $\theta\rho-plane$. In this space, the equation of a single line passing through the point (x_0, y_0),

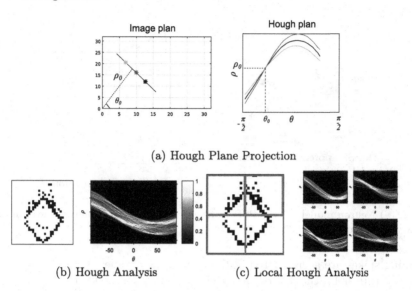

(a) Hough Plane Projection

(b) Hough Analysis   (c) Local Hough Analysis

Fig. 3. Original Hough Transform straight lines projection. Hough transform on a $\mathbf{w}_{N=200}$ windows event of a diamond sign.

Fig. 3(a), can be written as: $x \sin \theta_0 + y \cos \theta_0 = \rho_o$, which corresponds to a point in the $\theta\rho - plane$. Thus, the infinite lines which pass through (x_0, y_0) generate a curve in the Hough plane, Fig. 3(a). The Hough Transform identifies when several points lying on the same line. The curves on the Hough plane generated by the points intersects at (θ_0, ρ_0) in the $\theta\rho - plane$, which are the parameters of the line. Computationally, the Hough Transform converts the $\theta\rho - plane$ into a so-called *accumulator cell*. A cell (θ_i, ρ_i) receiving a high number of votes shows that several points in the *xy-plane* lay on the line with this parameters.

Because Hough Transform is based on a votes algorithm it is robust against noise. In Fig. 3(b), voting cells in the Hough space having maximal votes are placed at $\theta = -\pi/4$ and $\theta = \pi/4$ as expected for the diamond shape. The two maximums for each orientation corresponds to the two different ρ values of the two edges with the same orientation. In Fig. 3(c) the 32×32 pixels retina was split in four cells. This allows a local analysis of the edges, which is in fact a more discriminant way to describe the shape. The four Hough accumulators, one for each cell, are dominated for a maximum value placed at the orientation associated with the diamond shape edge.

2.3 Spiking Hough for Feature Extraction

The Spiking Hough defines a set of 12 neurons inspired on the Hough parameter space. Such space, referred as Spiking Hough Space, is discretized by four orientations and three biases, as shown Fig. 4(a). The neurons fire when the sequential events spikes at all the gray positions, and then reset their potential membrane.

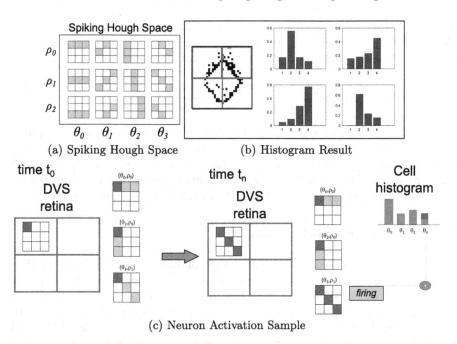

(a) Spiking Hough Space   (b) Histogram Result

(c) Neuron Activation Sample

Fig. 4. (a) Spiking Hough Space containing 12 neurons specialized on each orientation, (b) Example of the activation of a neuron and the building of the cell histogram, (c) Histogram feature result for the diamond shape using $N = 200$ events. (Color figure online)

The poker signs are contained in retina of 32×32 pixels size. Similar to Sect. 2.2, the Poker retina is subdivided into four non-overlapped cells. Each cell has 16×16 pixels size. Inside each cell, overlapped patches of 3×3 elements receive the events spikes. There are 36 overlapped patches inside each cell to cover all their pixels. To be robust against the stochastic nature of the DVS, each element in the 3×3 patch corresponds to 2×2 pixels of the retina.

Spiking Hough identifies the predominant edges configuration by firing a corresponding neurons on its space. Figure 4(c) illustrates the firing of a neuron. At time t_0 there arrives an event $\mathbf{e}_0$ at the cell lying on the 3×3 patch corner, painted in red. This event spikes the neurons with this position activated (in gray), increasing their potential. A sequence of events produce spikes on others positions of the patch. Later, at time t_n, an event $\mathbf{e}_n$ fires the neuron with (θ_3, ρ_1). Then, the firing neuron triggers and increments the 3rd. bin of the cell histogram, which correspond to the orientation of the detected edge.

2.4 Supervised Classification

The multi-class SVM classifier framework was trained using the LIBSVM library [2] and the best parameters for the linear and the RBF kernels were estimated using a 5 cross-fold validation approach. The framework was composed of the four SVM classifiers, trained using the one-against-one approach. LIBSVM uses [16] to obtain a single probability score for each class k: $f_k^{svm}(\mathbf{w}_i)$, with $k = 1, 2, 3, 4$.

$$P_k(\mathbf{w}_N^i) = \alpha f_k(\mathbf{w}_N^i) + (1 - \alpha) f_k(\mathbf{w}_N^{i-1}) \tag{1}$$

$$k^* = argmax_{k=1,2,3,4} P_k(\mathbf{w}_N^i) \tag{2}$$

For a input window $\mathbf{w}_N^i$, the probability to belong to sign k is computed with Eq. 1, where α is a memory factor. Thus $P_k(\mathbf{w}_N^i)$ uses the current and previous event window classification functions $f_k(\mathbf{w}_N^i)$ and $f_k(\mathbf{w}_N^{i-1})$ to smooth the response and become robust to noisy windows. Sample $\mathbf{w}_N^i$ is classified as in class k^* which $P_k^*(\mathbf{w}_N^i)$ produces the largest probability output on Eq. 1.

3   Experiments and Results

The Spiking Hough classification system was conducted on the 2015 Poker-DVS dataset [13]. On their website, the authors share a complete recording of the asynchronous events while they were browsing the poker cards, as well as a set of 131 individual files of cropped events. Each file has a name indicating the sign to which the flow of events corresponds. A character 'i' is added if the card is inverted. There are 30 club signs (13 inverted), 43 diamonds (8 inverted), 23 hearts, and 35 spades (10 inverted).

Given the low number of samples per class, the tests were conducted using the Leave-One-Out approach. This methodology employs all the samples of the set to train the multi-class classifier, except for one sample which is evaluated by the classifier and the result is saved in a confusion matrix. The overall performance is then obtained by computing the accuracy on the diagonal of the matrix.

The event flow of a test sample feeds the Spiking Hough system until the number of events reaches N. This event window is referred $\mathbf{w}_N^0$ and is then evaluated by the four SVM classifiers using Eqs. 1 and 2, and $\alpha = 0.5$ (which gives the best results on the tests). In this way, the output of the classification accumulates votes for each sign. This procedure is repeated to the next events of the sign, until the end of the flow, and the output of each $\mathbf{w}_N^i$ is evaluated with Eqs. 1 and 2. The test sample is finally classified by the sign that receives the highest number of votes.

Linear and Radial Basis Function (RBF) kernels are used to implement the SVM multi-classification. The results of the linear and non-linear (RBF) kernels are compared on Fig. 5. There were also tested to cell grids. One with 4 non-overlapped cells, as shows Fig. 4(b). The other configuration incorporates 4 more cells overlapping the original ones, to obtain 8 cells in total. The last configuration helps the system to be robust against little movements of the shape, and the fact that the 32×32 Poker retina is not necessarily centered all the time. The length of the event window N was also evaluated, from a minimum of 100 events to a maximum of 500. The Figure also shows the associated time delays representing the different lengths of event windows. It was calculated as the average value of all the $\mathbf{w}$ in the dataset for a specified N.

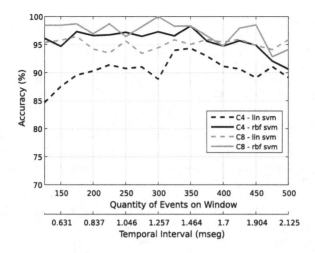

Fig. 5. Results of different classifiers and features extractions approaches.

The systems shows a good robustness, even if there exist several signs that are inverted. The non-linear RBF kernel on the configuration of 8 cells using $N = 300$ obtains an accuracy of 100%. This result outperforms the best performance published in the state of the art [14]. For this case, all the Poker signs were correctly classified. From the temporal axis, this event windows lengths corresponds on average to 1.256 *mseg*.

4   Conclusions

The Spiking Hough methodology describes *on-line* the shape of the events generated by an object moving in front of the DVS camera. The resulting histograms features show a good discriminating power for the Poker sign recognition.

Further research should be oriented to obtain a complete system which works on-line from the feature generation until the classification. Also, the system must be prepared to moving object and different scales in order to be though as a good choice to be implemented in real applications.

Acknowledgements. This work was funded by PID Nro. P16T01 (UADE, Argentine), EU H2020 grants 644096 "ECOMODE" and 687299 "NEURAM3", and by Spanish grant from the Ministry of Economy and Competitivity TEC2015-63884-C2-1-P (COGNET) (with support from the European Regional Development Fund).

References

1. Anderson, J.: An Indroduction to Neural Networks. MIT Press, Cambridge (1995)
2. Chang, C.C., Lin, C.J.: LIBSVM: a library for support vector machines. ACM Trans. IST **2**(3) (2011). http://www.csie.ntu.edu.tw/~cjlin/libsvm

3. Clady, X., et al.: A motion-based feature for event-based pattern recognition. Front. Neurosci. **10**, 594 (2017)
4. Dalal, N., Triggs, B.: Histograms of oriented gradients for human detection. In: CVPR, vol. 1, pp. 886–893 (2005)
5. Lagorce, X., et al.: HOTS: a hierarchy of event-based time-surfaces for pattern recognition. PAMI **39**(7), 1346–1359 (2017)
6. Li, X., et al.: Lane detection based on spiking neural network and hough transform. In: CISP, pp. 626–630 (2015)
7. Lichtsteiner, P., Posch, C., Delbruck, T.: A 128*128 120dB 15us latency asynchronous temporal contrast vision sensor. JSSC **43**(2), 566–576 (2008)
8. Lowe, D.: Distinctive image features from scale-invariant keypoints. Int. J. Comput. Vis. **60**(2), 91–110 (2004)
9. Negri, P.: Pedestrian detection using multi-objective optimization. In: Pardo, A., Kittler, J. (eds.) CIARP 2015. LNCS, vol. 9423, pp. 776–784. Springer, Cham (2015). https://doi.org/10.1007/978-3-319-25751-8_93
10. Negri, P.: Extended LBP operator to characterize event-address representation connectivity. In: Beltrán-Castañón, C., Nyström, I., Famili, F. (eds.) CIARP 2016. LNCS, vol. 10125, pp. 241–248. Springer, Cham (2017). https://doi.org/10.1007/978-3-319-52277-7_30
11. Pérez-Carrasco, J., et al.: Mapping from frame-driven to frame-free event-driven vision systems by low-rate rate coding and coincidence processing-application to feedforward convnets. PAMI **35**(11), 2706–2719 (2013)
12. Seifozzakerini, S., et al.: Event-based hough transform in a spiking neural network for multiple line detection and tracking using a dynamic vision sensor, pp. 94.1–94.12, September 2016
13. Serrano-Gotarredona, T., Linares-Barranco, B.: 2015 poker-DVS dataset (2015). http://www2.imse-cnm.csic.es/caviar/POKERDVS.html. Accessed 8 June 2017
14. Stromatias, E., Soto, M., Serrano-Gotarredona, T., Linares-Barranco, B.: An event-driven classifier for spiking neural networks fed with synthetic or dynamic vision sensor data. Front. Neurosci. **11**, 350 (2017)
15. Vapnik, V.: The Nature of Statistical Learning Theory. Springer, New York (1995). https://doi.org/10.1007/978-1-4757-3264-1. ISBN 9780387987804
16. Wu, T.F., Lin, C.J., Weng, R.: Probability estimates for multi-class classification by pairwise coupling. J. Mach. Learn. Res. **5**, 975–1005 (2004)
17. Zhao, B., et al.: Event-driven simulation of the tempotron spiking neuron. In: BioCAS, pp. 667–670, October 2014
18. Zhao, B., et al.: Feedforward categorization on AER motion events using cortex-like features in a spiking neural network. Neural Netw. Learn. Syst. **26**(9), 1963–1978 (2015)

Region-Based Classification of PolSAR Data Through Kernel Methods and Stochastic Distances

Rogério G. Negri[1]($\boxtimes$)(iD), Wallace C. O. Casaca[2](iD), and Erivaldo A. Silva[3](iD)

[1] Instituto de Ciência e Tecnologia, Univ. Estadual Paulista,
São José dos Campos, São Paulo, Brazil
`rogerio.negri@ict.unesp.br`
[2] Campus Experimental de Rosana, Univ. Estadual Paulista,
Rosana, São Paulo, Brazil
`wallace.coc@gmail.com`
[3] Faculdade de Ciência e Tecnologia, Univ. Estadual Paulista,
Presidente Prudente, São Paulo, Brazil
`erivaldo@fct.unesp.br`

Abstract. Stochastic distances combined with Minimum Distance method for region-based classification of Polarimetric Synthetic Aperture Radar (PolSAR) image was successfully verified in Silva *et al.* (2013). Methods like K-Nearest Neighbors may also adopt stochastic distances and then used in a similar purpose. The present study investigates the use of kernel methods for PolSAR region-based classification. For this purpose, the Jeffries-Matusita stochastic distance between Complex Multivariate Wishart distributions is integrated in a kernel function and then used in Support Vector Machine and Graph-Based kernel methods. A case study regarding PolSAR remote sensing image classification is carried to assess the above mentioned methods. The results show superiority of kernel methods in comparison to the other analyzed methods.

Keywords: PolSAR · Image classification · Region-based
Stochastic distances · Kernel function

1 Introduction

Polarimetric Synthetic Aperture Radar (PolSAR) are sensors able to record the amplitude, phase and orientation of electromagnetic waves backscattered from targets in earth surface. PolSAR image classification has been intensively investigated. Earlier studies about PolSAR image classification were developed with basis on the supervised Maximum Likelihood Classifier framework considering as probability density function the Complex Multivariate Wishart distribution [6] and on unsupervised classification process based on the eigenvalue analysis of coherency matrix [3].

The study and development of new methods for PolSAR data classification still being investigated. Region-based classification is useful for radar data, which

© Springer International Publishing AG, part of Springer Nature 2018
M. Mendoza and S. Velastín (Eds.): CIARP 2017, LNCS 10657, pp. 433–440, 2018.
https://doi.org/10.1007/978-3-319-75193-1_52

are normally analyzed using pixel-based methods. The use of stochastic distances between Complex Multivariate Wishart distributions on the Minimum Distance Classifier for region-based classification was verified in [9] and better results were achieved in comparison to pioneer method proposed in [6]. Through the integration of stochastic distances between Multivariate Gaussian distributions in kernel functions, the Support Vector Machine (SVM) and Graph-Based (GB) methods were used for region-based classification of Synthetic Aperture Radar (SAR) in [7]. The K-Nearest Neighbors (KNN) method with stochastic distances was also reported in [7]. However, investigations involving other stochastic distances and kernel machines for PolSAR data still not made.

Face to the exposed, this study analyzes the use of stochastic distances on SVM and Graph-Based kernel methods, through kernel functions, for region-based classification of PolSAR image. Comparisons with the Minimum Distance Classifier framework investigated in [9] and KNN integrate with stochastic distances between Complex Multivariate Wishart distributions are presented. Such comparisons are conducted on a PolSAR image acquired from an actual PolSAR sensor with different classification scenarios.

2   Stochastic Distances for PolSAR Data

Supposing that targets on PolSAR images has homogeneous texture, the Complex Multivariate Wishart distribution can be used to model such targets:

$$f(\mathbf{Z}; n, \mathbf{\Sigma}) = \frac{n^{3n} |\mathbf{Z}|^{n-3} e^{-nTr(\mathbf{\Sigma}^{-1}\mathbf{Z})}}{|\mathbf{\Sigma}|^n \Gamma_3(n)}, \tag{1}$$

where $\mathbf{Z}$ is a n-looks covariance matrix computed from the average of n backscatter measurements in a neighborhood. Usually, the backscatter measurements are complex scattering vector $\mathbf{z}^T = (S_{hh}\ S_{hv}\ S_{vv})$ with components as complex number representing the amplitude and phase in a transmitting-receiving linear polarization combination. Regarding the parameters, n is the equivalent number of looks and $\mathbf{\Sigma}$ represent the target mean covariance matrix computed through a set of independent samples. In addition, $\Gamma_3(n) = \pi^3 \prod_{i=0}^{2} \Gamma(n-i)$, with $\Gamma(\cdot)$ representing the gamma function.

In statistical analysis of PolSAR data stochastic distances appears as a powerful tool. These distances quantify the contrast between sets of information with basis on the dissimilarity between its probability distributions.

Let X and Y random variables following Complex Multivariate Wishart distributions with parameters $\mathbf{\Sigma}_X$ and $\mathbf{\Sigma}_Y$, respectively, both with equivalent number of looks equal to n. The Bathacharrya distance between X and Y is [5]:

$$D_B(X,Y) = n \left[\frac{\log|\mathbf{\Sigma}_X| + \log|\mathbf{\Sigma}_Y|}{2} - \log\left| \left(\frac{\mathbf{\Sigma}_X^{-1} + \mathbf{\Sigma}_Y^{-1}}{2} \right)^{-1} \right| \right]; [0, \infty]. \tag{2}$$

Re-mapping the Bathacharrya measurements from $[0, \infty]$ to $[0, 2]$, Jeffries-Matusita arises as alternative distance:

$$D_{JM}(X,Y) = 2 \left(1 - e^{-D_B(X,Y)} \right); [0, 2]. \tag{3}$$

3 Region-Based Classification and Kernel Methods

The region-based classification process consists of associating a class $\omega_j \subset \Omega, j = 1, \ldots, c$, to all pixels that compose a region $\mathcal{R}_i$. The regions $\mathcal{R}_i, i = 1, \ldots, r$, are subsets of pixels that partition the support $\mathcal{S} \subset \mathbb{N}^2$ of a given image $\mathcal{I}$. For supervised region-based classification, the decision rule is built using information from $\mathcal{D} = \{(\mathcal{R}_i, \omega_j) \in \mathcal{S} \times \Omega : i = 1, \ldots, m; j = 1, \ldots, c\}$, a set of labelled training regions. The notation $(\mathcal{R}_i, \omega_j)$ indicates that $\mathcal{R}_i$ is assigned to ω_j.

A simple way to perform region-based classification is adopt the Minimum Distance Classifier framework using stochastic distances as a measure to compare the similarity between classes and unlabeled regions [9]. We refer to this method as Minimum Stochastic Distance Classifier (MSDC). Formally, let $\mathcal{R}_i$ be an unlabeled region and let $D(f_{\mathcal{R}_i}, f_{\omega_j})$ be a stochastic distance between the distributions of the pixels in $\mathcal{R}_i$ and the class ω_j, represented by f_{ω_j} and modeled through the pixels of labeled regions assigned to ω_j in $\mathcal{D}$, an assignment $(\mathcal{R}_i, \omega_j)$ is made when the following rule is satisfied:

$$(\mathcal{R}_i, \omega_j) \Leftrightarrow j = \underset{j=1,\ldots,c}{\arg \min} \, D(f_{\mathcal{R}_i}, f_{\omega_j}). \tag{4}$$

K-Nearest Neighbors (KNN) is another method successfully adopted for region-based classification using stochastic distances [7]. For this purpose, $D(\cdot, \cdot)$ in (4) is substituted by $D_{\text{knn}}(\cdot, \cdot)$ defined as:

$$D_{\text{knn}}(f_{\mathcal{R}_i}, f_{\omega_j}) = e^{-h_j(f_{\mathcal{R}_i})}, \tag{5}$$

where $h_j(f_{\mathcal{R}_i}) = \#\left\{(\bar{\mathcal{R}}, \omega_j) \in V_k(\mathcal{R}_i)\right\}$, such that $V_k(\mathcal{R}_i)$ is the set of k training regions close to $\mathcal{R}_i$ given a distance $D(\cdot, \cdot)$. Formally,

$$V_k(\mathcal{R}_i) = \{(\bar{\mathcal{R}}_p, \omega_q) \in \mathcal{D} : 0 < D(f_{\mathcal{R}_i}, f_{\bar{\mathcal{R}}_1})$$
$$\leq D(f_{\mathcal{R}_i}, f_{\bar{\mathcal{R}}_2}) \leq \cdots \leq D(f_{\mathcal{R}_i}, f_{\bar{\mathcal{R}}_k}); \; p = 1, \ldots, k; \; q \in \{1, \ldots, c\}\} \tag{6}$$

where $\bar{\mathcal{R}}_p$ represents a new indexing of the k nearest regions of $\mathcal{D}$ based on the proximity to $\mathcal{R}_i$.

Besides MSDC and KNN, among several methods that may be adopted to perform region-based classification, kernel-based methods are an option. SVM [10] and GB [2] are examples of kernel-based methods. Such methods allows the use of kernel functions, $K : \mathcal{X}^2 \to \mathbb{R}$, which are usually adopted to improve classification performance on non-linearly separable data in the original attribute space $\mathcal{X}$ as to generalize the application in problems where the input data are non-vectorial. In special, for region-based approach where the input data are regions, the use of an adequately kernel function is a convenient alternative.

A given $K : \mathcal{X}^2 \to \mathbb{R}$ is a kernel function if it is symmetric and conforms to Mercer conditions. A straight way to define a valid kernel is considering a general model, like the radial basis model [8]:

$$K(\mathbf{x}, \mathbf{y}) = g(d(\mathbf{x}, \mathbf{y})), \tag{7}$$

where $g : \mathbb{R} \to \mathbb{R}$ is a strictly positive real function and $d : \mathcal{X}^2 \to \mathbb{R}$ is a metric.

From Eq. (7) it is possible to develop kernel functions for PolSAR region-based image classification. A reasonable choice for $g(\cdot)$ is the negative exponential function. With respect to $d(\cdot, \cdot)$, which measures the similarity between the input data, adopt a metric based on stochastic distances is convenient. Straightly considering $d(\cdot, \cdot)$ as Bathacharrya (Eq. (2)) or Jeffries-Matusita (Eq. (3)) stochastic distances will not produces valid kernel functions since such distances does not attains the triangle inequality. However, admitting $\tau \in \mathbb{R}_+$ such that $d(\mathbf{x}, \mathbf{y}) \leq \tau$ for all $\mathbf{x}, \mathbf{y} \in \mathcal{X}$, the following expression provides a metric $\overline{d}(\cdot, \cdot)$ from a distance $d(\cdot, \cdot)$:

$$\overline{d}(\mathbf{x}, \mathbf{y}) = \begin{cases} 0 \text{ if } \mathbf{x} = \mathbf{y} \\ d(\mathbf{x}, \mathbf{y}) + \tau \text{ if } \mathbf{x} \neq \mathbf{y} \end{cases}. \tag{8}$$

Once the values of D_{JM} are limited to $[0, 2]$, a metric $\overline{D_{JM}}$ is defined substituting $d(\cdot, \cdot)$ by $D_{JM}(\cdot, \cdot)$ and adopting $\tau = 2$ in Eq. (8). As result, the following kernel function is defined:

$$K_{JM}(X, Y) = e^{-\gamma \overline{D_{JM}}(X, Y)}, \tag{9}$$

where $\gamma \in \mathbb{R}_+$ is a regularization parameter.

4   Experiments and Results

In order to assess the performance of SVM and GB methods adopting the kernel function defined in Eq. (9) on region-based classification of PolSAR data, a case study regarding a multi-class remote sensing image classification under three distinct scenario was carried. The MSDC and KNN methods using the Jeffries-Matusita distance, as presented in Eqs. (4) and (5), were included in the analysis for comparison.

The PolSAR data adopted in this study corresponds to an image acquired on March 13th, 2009, by the ALOS-PALSAR sensor in a region near the Tapajos National Forest, State of Pará, Brazil. This image has approximately 20 m resolution after a 3×3 multi-look process. The following land use and land cover (LULC) types were considered: Primary Forest (PF), Regeneration (RE), Pasture (PS), Bare Soil (BS) and three types of Agriculture (A1, A2 and A3). Figure 1(a) present a color composition of the ALOS-PALSAR image. The spatial distribution of LULC samples is depicted in Fig. 1(b), where training and test samples correspond to solid and void polygons, respectively. Table 1 presents a summary about the LULC samples. The region-growing method, available in the Geographic Information System SPRING [1], was used in order to segment the image and define the regions. The segmentation parameters were chosen by visual inspection. Figure 1(c) represents the contours of the regions on the segmented study image.

As above mentioned, three distinct classification scenarios were considered. The first scenario is composed by all LULC classes identified in the study area. From the union of the agriculture classes (i.e., A1, A2 and A3) it is defined the new class called Agricultural Areas (AA), and then a second scenario it is

(a) Study image   (b) LULC samples   (c) Segmentation

Fig. 1. PolSAR image, LULC samples and segmentation used in the study. (Color figure online)

Table 1. Summary of the LULC samples.

LULC classes	Training		Testing	
	Polygons	Pixels	Polygons	Pixels
Agriculture 1 (A1)	4	3669	8	7455
Agriculture 2 (A2)	4	2902	8	6731
Agriculture 3 (A3)	3	2332	8	7049
Primary Forest (PF)	3	5430	10	29306
Pasture (PS)	5	3334	10	12866
Regeneration (RE)	5	2570	10	7307
Bare Soil (BS)	5	5384	11	13352

created with the five classes AA, PF, PS, RE and BS. The last scenario has the Agricultural Areas, High Biomass (HB) and Low Biomass (LB) classes. While HB is obtained merging PF and RE classes, LB comes from the union between PS and BS. Such scenarios represent plausible situations, once classes with similar semantic are merged from scenario 1 to define the classes of scenarios 2 and 3. Additionally, worth observe that each scenario has a specific complexity in terms of number of classes and inter/intra-class contrasts. Consequently, its use for comparison purposes allows more robust analysis of the investigated methods. It is valid mention that training and testing samples of AA, HB and LB classes arise by simple merging the sample polygons of the individual classes.

The selections of adequate number of neighbors (k – KNN), penalty (C – SVM), neighbors influence (α – GB) and kernel regularization (γ – Eq. (9)) parameters were based on a grid search process with tenfold cross-validation. The space search for each parameter was: $k \in \{3, 5, 7, 9\}, C \in \{1, 10, 100, 1000, 10000\}, \alpha \in \{0.1, 0.2, \ldots, 1.0\}$ and $\gamma \in \{0.25, 0.5, \ldots, 3.0\}$. The one-against-all multi-class strategy was adopted by SVM.

Concerning the PolSAR image and LULC classes in different scenarios, a total of 12 classification results were obtained. The accuracy of results were calculated by means of kappa agreement coefficient, regarding the LULC testing samples, and hypothesis test with 5% significance level were performed in order to compare the kappa values [4]. The analyzed methods were also compared in terms of computational time. The experiments were performed using a computer with an Intel *Core i7* processor and 16 GB of RAM running the Ubuntu Linux version 14.4 operating system. The implementations were conducted using the IDL (Interactive Data Language) programming language. The performance of analyzed methods are shown in Fig. 2.

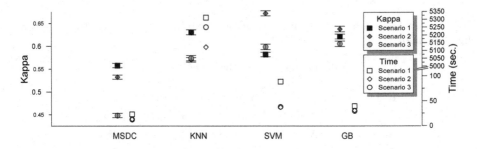

Fig. 2. Accuracy (kappa) and computational time (seconds) of analyzed methods. Error bars representing ±1 kappa standard deviation.

As initial discussion, we can observe that low kappa values were assigned to MSDC. Furthermore, the accuracy of MSDC decreases as the number of classes also decreases along the scenarios. This behavior is assigned to the scenario complexity, which increase when some LULC classes are merged giving place to new classes with higher variability.

Regarding the first scenario, the most accurate results was achieved by KNN method. Although GB presented a lower kappa coefficient in comparison to KNN, its accuracy values are statistically equivalent.

The accuracy levels presented by SVM and GB were higher in scenario 2 compared to other scenarios, where SVM was more accurate. It is worth note that KNN was superior to MSDC.

We also can observe that the variability of the classes in scenario 3 plays less effect on SVM and GB than MSDC and KNN. Furthermore, SVM and GB are statistically equivalent in this scenario.

Face to its simple algorithmic implementation, MSDC is the less expensive method in terms of computational time. SVM presents the higher computational time, which tends to increase as the amount of classes increase in reason of its multi-class classification architecture. An intermediate cost between MSDC and SVM is presented by GB method. As consequence of multiple comparisons needed to build the decision rule to classify each unlabeled region, KNN presented computational time over 5100 s.

Figure 3 depicts the classification results for each method and scenario. It can be note that, independently of method, RE and PS classes were not well discriminated in scenarios 1 and 2. In the first scenario, while MSDC and KNN were not able to discriminate BS areas, SVM and GB fails to identify A2 areas.

GB tends to classify A1 as BS and SVM frequently classifies PS as RE areas. Focusing the second scenario, MSDC was imprecise classifying AA areas as SVM the BS class and KNN pasture areas. The last scenario reveals MSDC and KNN as unable to distinguish LB and AA classes.

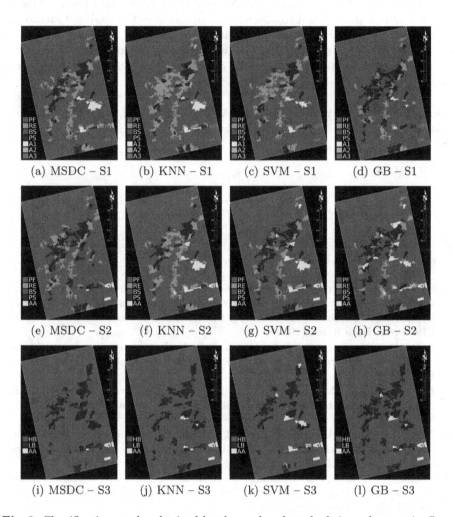

(a) MSDC – S1   (b) KNN – S1   (c) SVM – S1   (d) GB – S1

(e) MSDC – S2   (f) KNN – S2   (g) SVM – S2   (h) GB – S2

(i) MSDC – S3   (j) KNN – S3   (k) SVM – S3   (l) GB – S3

Fig. 3. Classification results obtained by the analyzed methods in each scenario. Scenarios 1, 2 and 3 are identified by S1, S2 and S3, respectively.

5   Conclusions

This study verified the performance SVM and GB on region-based classification of PolSAR image in comparison to MSDC and KNN. The Jeffries-Matusita stochastic distances between Complex Multivariate Wishart distributions was integrated in a kernel function and then adopted by SVM and GB. A case study about LULC classification using ALOS-PALSAR image was addressed.

Both SVM and GB achieved higher accuracy levels, especially on scenarios with higher intra-class variability. Regarding the first scenario, where the contrast between classes is higher, KNN provided more accurate results. Lower accuracy values are assigned to MSDC in comparison to the analyzed methods.

The better tradeoff between classification accuracy and computational cost is offered by GB. Consequently, the GB method becomes a potential alternative for region-based classification of PolSAR data. Although KNN allowed accurate results the computational time its main drawback.

Acknowledgments. The authors thank FAPESP (Proc.: 2014/14830-8) for funding this research.

References

1. Camara, G., Souza, R.C.M., Ii, F.M., Freitas, U., Garrido, J.: Spring: integrating remote sensing and GIS by object-oriented data modelling. Comput. Graph. **20**, 3 (1996)
2. Camps-Valls, G., Tatyana, V.B., Zhou, D.: Semi-supervised graph-based hyperspectral image classification. IEEE Trans. Geosci. Remote Sens. **45**, 2044–3054 (2007)
3. Cloude, S.R., Pottier, E.: An entropy based classification scheme for land applications of polarimetric SAR. IEEE Trans. Geosci. Remote Sens. **35**(1), 68–78 (1997)
4. Congalton, R.G., Green, K.: Assessing the Accuracy of Remotely Sensed Data. CRC Press, Boca Raton (2009)
5. Frery, A.C., Nascimento, A.D.C., Cintra, R.J.: Analytic expressions for stochastic distances between relaxed complex Wishart distributions. IEEE Trans. Geosci. Remote Sens. **52**(2), 1213–1226 (2014)
6. Lee, J., Grunes, M., Kwok, R.: Classification of multi-look polarimetric SAR imagery based on complex Wishart distribution. Int. J. Remote Sens. **15**(11), 2299–2311 (1994)
7. Negri, R.G., Dutra, L.V., Sant'Anna, S.J.S., Lu, D.: Examining region-based methods for land cover classification using stochastic distances. Int. J. Remote Sens. **37**(8), 1902–1921 (2016). https://doi.org/10.1080/01431161.2016.1165883
8. Schölkopf, B., Smola, A.J.: Learning with Kernels: Support Vector Machines, Regularization, Optimization, and Beyond. Adaptive Computation and Machine Learning. MIT Press, Cambridge (2002)
9. Silva, W.B., Freitas, C.C., Sant'Anna, S.J.S., Frery, A.C.: Classification of segments in PolSAR imagery by minimum stochastic distances between Wishart distributions. IEEE J. Sel. Top. Appl. Earth Obs. Remote Sens. **6**(3), 1263–1273 (2013)
10. Vapnik, V.N.: The Nature of Statistical Learning Theory. Springer-Verlag New York Inc., New York (1995)

Hand Posture Recognition Using Convolutional Neural Network

Dennis Núñez Fernández[2] and Bogdan Kwolek[1(✉)]

[1] AGH University of Science and Technology, 30 Mickiewicza, 30-059 Kraków, Poland
bkw@agh.edu.pl
[2] National University of Engineering, Av. Túpac Amaru 210, 15109 Lima, Peru

Abstract. In this work we present a convolutional neural network-based algorithm for recognition of hand postures on images acquired by a single color camera. The hand is extracted in advance on the basis of skin color distribution. A neural network-based regressor is applied to locate the wrist. Finally, a convolutional neural network trained on 6000 manually labeled images representing ten classes is executed to recognize the hand posture in a sub-window determined on the basis of the wrist. We show that our model achieves high classification accuracy, including scenarios with different camera used in testing. We show that the convolutional network achieves better results on images pre-filtered by a Gabor filter.

Keywords: Gesture recognition
Biologically inspired computer vision · Gabor filter
Convolutional neural network

1 Introduction

Hand gesture recognition is one evident way to build user-friendly interfaces between machines and their users. In the near future, hand gesture recognition technology would allow for the operation of complex machines and smart devices through only series of hand postures, finger and hand movements, eliminating the necessity for physical contact between man and machine. Gesture recognition on images from single camera is a difficult problem due to occlusions, differences in hand anatomy, variations of posture appearance, etc. In the last decade, several approaches to gesture recognition on color images were proposed [1].

In recent years, Convolutional Neural Networks (CNNs) have become the state-of-the-art for object recognition in computer vision [2]. Despite high potential of CNNs in object detection [3,4] and image segmentation [2], only few papers report promising results - a recent survey on hand gesture recognition [1] reports only one significant work [5]. Some obstacles to wider use of CNNs are high computational demands, lack of sufficiently large datasets, as well as lack of hand detectors suitable for CNN-based classifiers. In [6], a CNN has been used for classification of six hand gestures expressed by humans to control robots using

© Springer International Publishing AG, part of Springer Nature 2018
M. Mendoza and S. Velastín (Eds.): CIARP 2017, LNCS 10657, pp. 441–449, 2018.
https://doi.org/10.1007/978-3-319-75193-1_53

colored gloves. In more recent work [7], a CNN has been implemented in Theano and executed on the Nao robot. In a recent work [8], a CNN has been trained on one million of exemplars. However, only a subset of data with 3361 manually labeled frames in 45 classes of sign language is publicly available.

2   Method Overview

The proposed system for hand posture recognition operates on color RGB images. It has been designed to run in real-time with respect to low power usage, including devices without a GPU support. In order to reduce computational demands, the hand region proposals are determined at low computational cost on the basis of skin color. Given the extracted hand, a neural network based regressor is executed to estimate the wrist position, see Fig. 1. The wrist location is used to extract a sub-image with a hand, which is then fed to a CNN.

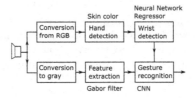

Fig. 1. Block diagram of our method for hand posture recognition.

The CNN operates on gray images of size 38×38. We trained two models for hand posture recognition: the first CNN was trained on gray images, whereas the second one was trained on gray images filtered by a Gabor filter. We show that the system achieves far better classification performance on images pre-filtered by a Gabor filter. We show that a CNN operating over pre-filtered images gives good results on both images from the camera that has been used to acquire images for training and images from a different camera, i.e. camera, whose images do not appear in the training repository.

The training of the CNN has been realized on a collection of 6000 images representing ten different hand postures. The network has been trained using Caffe [9]. The trained CNN models (Caffe Model Zoo), the training images as well as test images are available at: http://home.agh.edu.pl/~bkw/code/ciarp2017.

3   Gesture Dataset

The developed gesture set for interacting with computers is intuitive and easy to learn. It consists of ten gestures, see Fig. 2. The gestures were performed by ten persons of different nationalities. They were recorded using Kinect's RGB

Fig. 2. Hand postures.

camera. The RGB images of size 640 × 480 were recorded in the **png** image format. On the basis of manually selected wrist position the sub-images containing the hands were determined. Afterwards, the background was automatically subtracted. Finally, all images were manually refined to eliminate remaining background pixels in the vicinity of the hands. Such manually cropped and refined sub-images of size 38 × 38 were stored in the image repository.

The whole dataset consists of 6000 gray images of size 38 × 38. Having on regard that such a dataset can be too little for building powerful classification models the hands were roughly aligned in such a way that characteristic hand features (e.g. the wrist) are approximately located at pre-defined positions in the image. This means that in every class the wrists in all images are roughly located at the same position. Moreover, thanks to such an approach the recognition of gestures at satisfactory frame rates can be achieved with a simpler neural network and at a lower computational cost.

Until now, many hand datasets have been developed and made publicly available [10]. A total of 13 datasets with static hand gestures were available in the survey mentioned above at time of its publication. We found that only one set of data collections mentioned above has a sufficiently large number of examples per class, say 1000 images per class. With the discussed collection of benchmark data we selected dataset [11], albeit other dataset with large number of examples per class were also considered. As it turned out, the selected dataset does not fulfill our requirements well since it contains hands in front of background. Moreover, the hands are not properly aligned within classes. What is more, it contains the data that were shot by the same camera. Another issue is that the discussed dataset includes gestures, which are unlikely to have been designed for the control of the devices. Having the above on regard, we decided to record a hand posture dataset, which size is sufficient to perform deep learning. To the best of our knowledge, there is no other publicly available dataset of proper size, with appropriately aligned hands for learning of CNNs for real-time applications.

4   Hand Detection and Wrist Localization

At the beginning of this section, we describe how hands in RGB color images are detected. In the remainder of this section, we describe how the wrist is localized.

4.1 Hand Detection

Skin color is a powerful feature for fast hand detection. Basically, all skin color-based approaches try to learn a skin color distribution, and then use it to delineate the hands. In this work the hand has been extracted on the basis of statistical color models [12]. A model in RGB-H-CbCr color spaces has been prepared on the basis of a training set. Then, the hand probability image has been thresholded. Afterwards, after morphological closing, a connected components labeling has been executed to extract the gravity center of the blob, coordinates of the most top pixel as well as coordinates of the most left pixel of the hand region.

4.2 Wrist Localization

Modeling the statistical relation between the stochastic output y and input vector $\mathbf{x}$ is referred to as regression, which is typically expressed by an additive noise model $y = g(\mathbf{x}) + e$, where e is a random error in y that cannot be expressed by the input. Taking into account the universal function approximation ability of neural networks, the regression model can be expressed as: $y = g(\mathbf{x}, \mathbf{w}) + \epsilon$, where ϵ is a random vector. Given the binary image representing the extracted hand, we extracted a feature vector $\mathbf{x}$ of size 38 elements. It consists of x, y-coordinates of gravity center, x, y-coordinates of the most left non-zero pixel, x, y-coordinates of the most top non-zero pixel, 16-bin histogram expressing the number of pixels in nonoverlapping column pairs, 16-bin histogram expressing the number of pixels in nonoverlapping row pairs. Two NN-based regression models were trained to model the relation between the feature vector $\mathbf{x}$ and wrist positions with respect to x, y coordinates. The weights $\mathbf{w}$ of a neural network with one hidden layer consisting of ten neurons were determined using standard backpropagation algorithm.

5 Hand Posture Recognition

At the beginning of this section, we describe how the input images are preprocessed by a Gabor filter. Then, we describe the convolutional neural network.

5.1 Gabor Filter

Hubel and Wiesel demonstrated in 1962 [13] the existence of simple cells in primary visual cortex (Nobel Prize, 1981). They discovered that its receptive field comprised sub-regions, which were layered over each other to cover the entire visual field. Gabor filter (GF) is excellent approximator of the receptive fields found in the mammalian primary visual cortex (V1). In the spatial domain, the 2D Gabor filters are Gaussian kernel functions modulated by a sinusoidal waves [14]. Their frequency and orientation representations are similar to those of the human visual system. The GF has a real and an imaginary component representing orthogonal directions. The real component can be expressed as:

$$g_{\lambda,\theta,\phi,\sigma,\gamma}(x,y) = \exp\left(-\frac{x'^2 + \gamma^2 y'^2}{2\sigma^2}\right)\cos\left(2\pi\frac{x'}{\lambda} + \phi\right) \tag{1}$$

$$x' = x\cos\theta + y\sin\theta \qquad y' = -x\sin\theta + y\cos\theta \tag{2}$$

$$b = \log_2\left(\frac{\frac{\sigma}{\lambda}\pi + \sqrt{\frac{\ln 2}{2}}}{\frac{\sigma}{\lambda}\pi - \sqrt{\frac{\ln 2}{2}}}\right), \qquad \frac{\sigma}{\lambda} = \frac{1}{\pi}\sqrt{\frac{\ln 2}{2}}\cdot\frac{2^b + 1}{2^b - 1} \tag{3}$$

where λ is the wavelength of the cosine factor of the filter kernel, θ denotes the orientation of the normal to the parallel stripes of Gabor function, ϕ specifies the phase offset, γ is the spatial aspect ratio, that specifies the ellipticity of the support of the Gabor function, σ is the sigma/standard deviation of the Gaussian envelope, whereas b is related to the ratio $\frac{\sigma}{\lambda}$ and denotes the half-response spatial frequency bandwidth of Gabor filter.

5.2   Convolutional Neural Network

Unlike regular neural network, which only permits the input as vectors, convolutional neural networks allow 2 or 3-dimensional arrays at input layer. What makes convolutional neural networks distinct is that the weights are shared, that is, being different with respect to the position relative to the center pixel they are identical for different pixels in the image. Thus, it is straightforward to view a CNN as hierarchy of organized into layers a collection of local filters whose weights should be updated in a learning process. Every network layer acts as a detection filter for the presence of specific features or patterns present in the original data. The convolutions are usually followed by a non-linear operations after each layer since cascading linear convolutions would lead to a linear system. Besides, max-pooling is a mechanism that provides a form of translation invariance, which contributes towards the position independence. The CNNs are typically trained like standard neural networks using backpropagation.

In this work we utilized a convolutional neural network that is similar to the LeNet CNN [15]. The original model of LeNet relies on convolutional filters layers that are interlaced with non-linear activation functions, followed by spatial feature pooling operations, e.g. sub-sampling. It has only 60K learnable parameters out of 345 308 connections. Our convolutional neural network additionally uses ReLU activation functions, which were proposed in the Alexnet CNN [2]. The discussed work proposed also dropout regularization to selectively ignore single neurons during training, a way to avoid overfitting of the model.

Figure 3 depicts the architecture of our convolutional neural network. It consists of seven layers and takes a 38 × 38 pixel field as input.

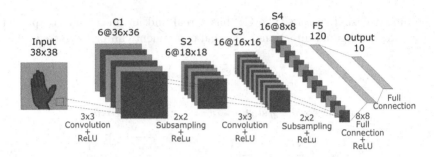

Fig. 3. Architecture of convolutional neural network.

6   Experimental Results

First, we trained CNNs on the hand posture dataset. In order to evaluate their generalization capability we additionally recorded images with the considered hand postures using a notebook camera. The camera that was utilized in image recording delivers images of size 640 × 480, i.e. the images have the same size as images acquired by Kinect RGB camera. Two volunteers of different nationalities, who did not attend in the dataset recordings performed 400 gestures.

Afterwards, we evaluated the error of wrist localization using the skin-color based algorithm for hand extraction and neural network based algorithm for wrist localization. The error was determined on 727 images (person #3) from the dataset and 400 images recorded by the notebook camera. On such an image set we manually determined the wrist positions and then used them to calculate the errors. Table 1 depicts the errors for each class together with the averages both for the Kinect and notebook camera (Dif. cam.). As we can notice, the average error for the different camera is slightly larger.

Table 1. Error of wrist localization [pix] for classes 1–10.

	1	2	3	4	5	6	7	8	9	10	Avg.
Kinect	1.2	1.3	7.8	1.4	2.1	1.8	2.6	1.3	1.2	1.4	2.2
Dif. cam.	4.6	3.4	2.9	3.5	1.7	2.7	2.7	3.8	4.6	2.6	3.2

In the next stage, for each person, we divided the images into training and testing parts. About 15% of images from each class were selected for the tests, whereas the remaining images were used in the learning of CNNs. The images with gestures performed by person #3 were not included in the training data and they were used only in person-independent tests. This way we selected 6000 gray images of size 38 × 38 for training the CNNs. All CNNs were trained using Caffe [9] with the following parameters: batch size = 64, momentum = 0.9, base learning rate = 0.001, gamma = 0.1, step size = 1000, max. iteration = 5000.

The first convolutional neural network was trained on raw gray images. Table 2 depicts the recognition performance with respect to both automatically and manually determined wrist position on test images acquired by the notebook camera and Kinect camera, i.e. test part of the hand posture dataset. As we can notice, on images acquired by Kinect the CNN achieves good classification performance. In the discussed performer-independent test the results are slightly worse for the wrist positions determined automatically. As we can observe, despite that the classification was done on images from a different camera, the CNN quite well recognizes the hand postures expressed by a different performer.

Table 2. Performance measures using CNN on raw gray images.

	Accuracy	Precision	Recall	F1 score
Kinect man.	0.950	0.945	0.960	0.949
Dif. cam. man.	0.928	0.931	0.937	0.934
Kinect aut.	0.905	0.906	0.932	0.906
Dif. cam. aut.	0.783	0.788	0.832	0.785

Table 3 depicts classification results that were obtained on images preprocessed by the Gabor filter. As we can observe, the results are far better in comparison to results presented in Table 2. The classification results achieved on images filtered by the Gabor filter are better for both the Kinect and notebook camera. In the person-independent test the CNNs achieves the classification accuracy equal to 97% if the images are taken by the same camera as in the training. If the classification is done on images acquired by different camera the classification accuracy is equal to 87%. For the discussed case the improvement with respect to classification on raw images is equal to 8.5%. The discussed results were achieved using the following parameters of the Gabor filter: $\lambda = 2$, $\theta = [0 \, \pi/4 \, \pi/2 \, 3/4\pi]$, $\phi = [0 \, \pi/2]$, $b = 1.8$, $\sigma = 0.75$, $\gamma = 0.5$.

Table 3. Performance measures using CNN with Gabor-based preprocessing.

	Accuracy	Precision	Recall	F1 score
Kinect man.	0.992	0.992	0.992	0.992
Dif. cam. man.	0.930	0.930	0.936	0.929
Kinect aut.	0.970	0.967	0.976	0.970
Dif. cam. aut.	0.868	0.868	0.882	0.866

Table 4 depicts classification accuracies that were achieved with respect to the considered hand postures. As we can notice, the CNNs achieves significantly worse results for class 6 when it operates on images taken by a different camera.

Table 4. Recognition accuracy [%] for wrist position determined automatically.

	1	2	3	4	5	6	7	8	9	10	Avg.
Kinect	100	94	100	100	90	100	83	100	100	100	97
Dif. cam.	93	95	80	95	93	63	90	88	73	100	87

Our system has been designed to operate on humanoid Robot Nao as well as ARM processor-based mobile devices. Having on regard that they are not equipped with GPUs, which can considerably reduce the processing time of CNNs, we extract low-level features using techniques from biologically inspired computer vision, which are then further processed hierarchically, and finally recognized by moderate-size CNN. The recognition of hand posture on a single sub-image with the extracted hand and the estimated wrist position is about 5 ms. The presented software has been developed in C++, Python and Matlab.

7   Conclusions

There is still strong need for RGB-based hand posture recognition, particularly for mobile devices, and even in robotics since the cameras relying on structured light typically perform poorly under natural illumination. Despite huge number of approaches [1], only few papers demonstrated complete solutions, where the images are taken by a RGB camera and both detection and classification algorithms are able to deliver promising results in real-time.

In this work we have presented CNN-based algorithm for hand posture recognition on images acquired by a single color camera. We considered ten static gestures for the operation of machines or smart devices. The CNNs have been trained on a repository of 6000 images of size 38×38. Two CNNs were trained both on raw images as well as images prefiltered by a Gabor filter. We demonstrated that CNN operating on prefiltered images has far better classification performance. We showed that our model allows for high classification accuracy even when the recognition is done on images taken by a different camera, i.e. a camera that has not been used in acquiring images for training of the CNNs.

Acknowledgment. This work was supported by Polish National Science Center (NCN) under a research grant 2014/15/B/ST6/02808.

References

1. Oyedotun, O., Khashman, A.: Deep learning in vision-based static hand gesture recognition. Neural Comput. Appl. **28**(12), 3941–3951 (2017)
2. Krizhevsky, A., Sutskever, I., Hinton, G.E.: ImageNet classification with deep convolutional neural networks. In: NIPS, pp. 1097–1105 (2012)

3. Kwolek, B.: Face detection using convolutional neural networks and Gabor filters. In: Duch, W., Kacprzyk, J., Oja, E., Zadrożny, S. (eds.) ICANN 2005. LNCS, vol. 3696, pp. 551–556. Springer, Heidelberg (2005). https://doi.org/10.1007/11550822_86

4. Arel, I., Rose, D., Karnowski, T.: Research frontier: deep machine learning-a new frontier in artificial intelligence research. Comput. Intell. Mag. **5**(4), 13–18 (2010)

5. Tompson, J., Stein, M., Lecun, Y., Perlin, K.: Real-time continuous pose recovery of human hands using convolutional networks. ACM Trans. Graph. **33**(5), 169:1–169:10 (2014)

6. Nagi, J., Ducatelle, F., et al.: Max-pooling convolutional neural networks for vision-based hand gesture recognition. In: IEEE ICSIP, pp. 342–347 (2011)

7. Barros, P., Magg, S., Weber, C., Wermter, S.: A multichannel convolutional neural network for hand posture recognition. In: Wermter, S., Weber, C., Duch, W., Honkela, T., Koprinkova-Hristova, P., Magg, S., Palm, G., Villa, A.E.P. (eds.) ICANN 2014. LNCS, vol. 8681, pp. 403–410. Springer, Cham (2014). https://doi.org/10.1007/978-3-319-11179-7_51

8. Koller, O., Ney, H., Bowden, R.: Deep Hand: how to train a CNN on 1 million hand images when your data is continuous and weakly labelled. In: IEEE Conference on Computer Vision and Pattern Recognition, pp. 3793–3802 (2016)

9. Jia, Y., Shelhamer, E., Donahue, J., et al.: Caffe: convolutional architecture for fast feature embedding. In: ACM International Conference on Multimedia, pp. 675–678 (2014)

10. Pisharady, P., Saerbeck, M.: Recent methods and databases in vision-based hand gesture recognition. Comput. Vis. Image Underst. **141**, 152–165 (2015)

11. Pisharady, P., Vadakkepat, P., Loh, A.: Attention based detection and recognition of hand postures against complex backgrounds. IJCV **101**(3), 403–419 (2013)

12. Jones, M.J., Rehg, J.M.: Statistical color models with application to skin detection. Int. J. Comput. Vis. **46**(1), 81–96 (2002)

13. Hubel, D., Wiesel, T.: Receptive fields, binocular interaction and functional architecture in the cat's visual cortex. J. Physiol. **160**(1), 106–154 (1962)

14. Petkov, N.: Biologically motivated computationally intensive approaches to image pattern recognition. Future Gener. Comput. Syst. **11**(4–5), 451–465 (1995)

15. LeCun, Y., Bottou, L., Bengio, Y., Haffner, P.: Gradient-based learning applied to document recognition. In: Proceedings of the IEEE, pp. 2278–2324 (1998)

On Semantic Solutions for Efficient Approximate Similarity Search on Large-Scale Datasets

Alexander Ocsa[1]([✉]) [iD], Jose Luis Huillca[1]([✉]), and Cristian Lopez del Alamo[2]([✉])

[1] Universidad Nacional de San Agustin, Arequipa, Peru
{aocsa,jhuillcama}@unsa.edu.pe
[2] Universidad La Salle, Arequipa, Peru
clopez@ulasalle.edu.pe

Abstract. Approximate similarity search algorithms based on hashing were proposed to query high-dimensional datasets due to its fast retrieval speed and low storage cost. Recent studies, promote the use of Convolutional Neural Network (CNN) with hashing techniques to improve the search accuracy. However, there are challenges to solve in order to find a practical and efficient solution to index CNN features, such as the need for heavy training process to achieve accurate query results and the critical dependency on data-parameters. Aiming to overcome these issues, we propose a new method for scalable similarity search, i.e., Deep frActal based Hashing (DAsH), by computing the best data-parameters values for optimal sub-space projection exploring the correlations among CNN features attributes using fractal theory. Moreover, inspired by recent advances in CNNs, we use not only activations of lower layers which are more general-purpose but also previous knowledge of the semantic data on the latest CNN layer to improve the search accuracy. Thus, our method produces a better representation of the data space with a less computational cost for a better accuracy. This significant gain in speed and accuracy allows us to evaluate the framework on a large, realistic, and challenging set of datasets.

Keywords: Multidimensional index · Approximate similarity search
Fractal theory · Deep learning

1 Introduction

The increasing availability of data in diverse domains has created a necessity to develop techniques and methods to discover knowledge from massive volumes of complex data, motivating many research works in databases, machine learning, and information retrieval communities. This has driven the development of scalable and efficient techniques to organize and retrieve this kind of data. Similarity search has been the traditional approach for information retrieval. Although several similarity search algorithms have been proposed to speed up similarity queries, most of them are either affected by the well-known "curse of

© Springer International Publishing AG, part of Springer Nature 2018
M. Mendoza and S. Velastín (Eds.): CIARP 2017, LNCS 10657, pp. 450–457, 2018.
https://doi.org/10.1007/978-3-319-75193-1_54

dimensionality". Retrieve complex data causes stability problems when the data dimensionality is too high [3].

One of the few approaches that ensure an approximate solution with sub-linear search cost for high-dimensional data is the Locality Sensitive Hashing (LSH) [1]. LSH is based on the idea that closeness between two objects is usually preserved by a random projection operation. In other words, if two objects are close together in their original space, then these two objects will remain close after a scalar projection operation [12]. However, it presents some difficulties for approximate kNN queries, in particular, related to data domain parameter dependence and quality results. Therefore, in complex domains, in particular, in high dimensional data problems, an approximate solution with a solid theoretical analysis may be the best option in many application areas because of their efficiency in time and space.

On the other hand, in Machine Learning traditionally images are often described by the hand-craft visual features. However, these hand-craft features cannot well reveal the high-level semantic meaning (labels or tags) of images, and often limit the performance of image retrieval [9]. Inspired by recent advances in Convolutional Neural Network (CNN) [8], many methods solved the problem of precision of similarity retrieval by using CNN as feature extractor and then build a compact similarity-preserving hash code for fast image retrieval. Again, hashing is widely used for large-scale image retrieval as well as video and document searches because the compact representation of hash code is essential for data storage and reasonable for query searches [14]. However, some drawbacks based on these supervised hashing methods have not been solved entirely, as follows:

- There is a trade-off between classification error and quantization error: activations of lower layers are more general-purpose [16], so training is more effective. However lower layers have larger activations maps (many nodes), which are harder to encode which leads to a compromise.
- There is a dependency on parameter values for approximate similarity search schemes based on LSH, which determine the number of hash functions and number of hash tables.

This paper proposes a novel supervised hashing technique, named Deep frActal based Hashing (DAsH), designed to perform scalable approximate similarity search. The contributions of our work are as follows. First, we introduce and define a scheme based on CNN and optimized using fractal theory. To overcome the limitation of large activations on lower layers of CNN (output of the last convolutional layer) we reduce its dimensionality using autoencoders to the optimal sub-space. Then we index this new representation with LSH scheme. Second, we present a novel method, based on fractal theory, which allow us to can find the optimal number of hash functions for an approximate similarity search scheme based on LSH.

The paper is organized as follows. Section 2 summarizes the background for this work. Section 3 describes the proposed technique and Sect. 4 reports experimental results on real and synthetic datasets. Finally, we conclude in Sect. 5.

2 Locality Sensitive Hashing

Previous work [1] has explored the idea of hashing objects and grouping them into buckets with the goal of performing approximate similarity search within buckets associated with the query element. The idea behind LSH is that if two objects are close together in their original space, then these two objects will remain close after a scalar projection operation [12]. Hence, let $h(x)$ be a hash function that maps a d-dimensional point x to a one-dimensional value. The function $h(x)$ is said to be *locality sensitive* if the probability of mapping two d-dimensional points x_1, x_2 to the same value grows as their distance $d(x_1, x_2)$ decreases.

LSH based methods report efficient results when adequate values for m (number of hash functions) and L (number of indexes) are chosen. The E^2-LSH algorithm find the best value for m and L by experimentally evaluating the cost of calculation for samples in the given dataset. Basically, the tuning parameter of LSH is chosen as a function of the dataset to minimize the running time of a query while the space requirement is within the memory bounds [13].

2.1 Fractal Theory

A fractal is characterized by the self-similarity property, i.e., it is an object that presents roughly the same characteristics when analyzed over a broad range of scales [6]. From the Fractal Theory, the Correlation Fractal Dimension $\mathfrak{D}$ is particularly useful for data analysis, since it can be applied to estimate the intrinsic dimension of real datasets that exhibit fractal behavior, i.e., exactly or statistically self-similar datasets [4]. It has been shown that, given a set of N objects in a dataset with a distance function $d(x, y)$, the average number of k neighbors within a given distance r is proportional to r raised to $\mathfrak{D}$. Thus, the pair-count $PC(r)$ of pairs of elements within distance r follows the power law:

$$PC(r) = K_p \times r^{\mathfrak{D}} \tag{1}$$

where, K_p is a proportionality constant, and $\mathfrak{D}$ is the correlation fractal dimension of the dataset. Consequently, a fractal is defined by the self-similarity property, that is the main characteristic that represents exactly or statistically the similarity between the parts to the whole fractal.

3 Deep Fractal Based Hashing - DAsH

In this section, we propose the Deep Fractal based Hashing (DAsH) designed to perform a scalable approximate search by supervised hashing by a supervised hashing scheme. As introduced in Sect. 1, our strategy is to use the fractal theory to find the optimal sub-space for the last convolutional layer output of the CNN network, and the optimal number of hash functions for LSH index as well.

Figure 1 illustrates the training process structure. The network consists of three types of layers: (1) the convolutional layers whose weights are pre-trained most of the time on Imagenet and the target dataset is fine-tuned via transfer learning [16]; (2) the fully connected layers, with the last softmax layer returning the categorical probability distribution; (3) the autoencoders layers which are used for dimensionality reduction. The Convolutional Neural Network (CNN) is trained end-to-end with the groundtruth labels. We use the output of the last convolutional layer because it has the most general-purpose representation for learning, however there is somes issues with high dimensional data. To overcome the high-dimensionality problem, we reduce to the optimal sub-space using an autoencoder. Then we index the optimal sub-space obtained by the autoencoder with LSH scheme which, as we mentioned, it is also tuned thanks to fractal theory. At the same time, we use another n-autoencoders to learning the representation of each class. After, we will use these autoencoders to improve the retrieval process.

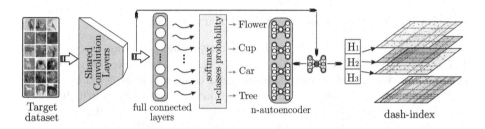

Fig. 1. DAsH: training and indexing process.

As it was showed in [11], a successful dimensionality reduction algorithm projects the data into a feature space with dimensionality close to the fractal dimensionality (FD) of the data in the original space and preserves topological properties. Thus, to find the target dimensionality (m) needed by autoencoder networks we follow the following heuristic. We start with the value at $m_1 = 2^2$, compute the FD of the new space with just that, then increment value at $m_2 = 2^3$, recompute the FD, and continue doing this until some $t(m_t = 2^t)$ where we can see a flattening in the fractal dimension, meaning that more features do not change the fractal dimensionality of the dataset.

The second step of our procedure is image retrieval via DAsH. We process the query image forwarding it through CNN aiming to obtain the strongest n classes. In contrast to existing similarity algorithms that learn similarity from the low-level feature, our similarity is the combination of semantic-level and hashing-level similarity. So, the semantic level similarity is computed firstly (the n strongest classes). After the semantic relevance checking, we will obtain the new n queries using the strongest n autoencoders. The query is transformed into new query objects $(q_1, q_2, ...q_n)$ which are hashed to locate the appropriate buckets. Once the buckets are located, the relevant candidate set is formed. Then, the elements in the candidate set are exhaustively analyzed to recover only the objects that satisfy the query condition (e.g. $d(x, q) \leq r$). This process is performed for each of the L hash tables and it is illustrated in Fig. 2.

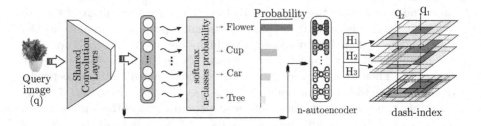

Fig. 2. DAsH. Retrieval process.

3.1   Using Fractals to Estimate LSH Parameters

To tune the LSH parameters we used a property of the correlation fractal dimension $\mathfrak{D}$, which can describes statistically a dataset. Moreover, the correlation fractal dimension $\mathfrak{D}$ can be estimated in linear time as it is depicted in [15].

We are interested to find out the resolution scale $log(r)$ at which there are approximately k objects. Considering the line with slope $\mathfrak{D}$ passing at a point defined as $<log(r), log(Pairs(k))>$ the constant K_d using the Eq. 1 is:

$$log(PC(r)) = \mathfrak{D} \times log(r) + K_p$$
$$K_p = log(Pairs(k)) - \mathfrak{D} \times log(r) \tag{2}$$

Considering another point $<log(R), log(Pairs(N))>$, the constant K_p is defined as:

$$K_p = log(Pairs(N)) - \mathfrak{D} \cdot log(R) \tag{3}$$

Now, combining Eqs. 2 and 3, we can define the radius r as:

$$r = R \cdot exp(\frac{log(Pairs(k)) - log(Pairs(N))}{\mathfrak{D}}) \tag{4}$$

Using the last Eq. 4 we find out that the optimal number of hash functions m for a Locality Sensitive Hashing (LSH) based index configured to retrieve the k nearest neighbors is proportional to the number of pairs at a distance r. This has sense, because an average number of k neighbors are within a given distance r. Then, we define:

$$m \approx log(PC(r)) \tag{5}$$

combining Eqs. 5 and 1 we obtain that $m \approx \mathfrak{D} \cdot log(r)$. Experimentally, we confirm out that the optimal m is:

$$m = (\lceil \mathfrak{D} + 1 \rceil) \cdot log(r) \tag{6}$$

4   Experiments

In this section, we are interested in answering the following question: (a) How accurate is our model in estimating the LSH parameters using the fractal dimension; (b) How does our DAsH method improve the other LSH implementations in terms of *querying performance* and *precision*. The performance of DAsH method was compared to two well-known approximate search methods, namely Multi-probe LSH [10], LSH-Forest [2], ITQ [5], and LOPQ [7]. All of the experiments were performed on a workstation with Intel core i7 3.0 GHz CPU and 64 GB RAM which is supplied with four Geforce GTX 1080 GPU.

We first conduct experiments on eight widely used datasets using hand-crafted features (audio, cities, eigenfaces, histograms, mgcounty, randomwalk, synth16d, synth6d, video)[1] to evaluate our proposed method for estimating the LSH parameters. Beside hand-crafted features, we also show the effectiveness of our methods when deep features are extracted by the deep Convolutional Neural Networks (CNN), we conduct this experiment on three datasets (MNIST[2], CIFAR-10[3], SVHN[4]) to evaluate our in terms of querying performance, meap average precision (mAP), and precision. The following describes the details of the experiments and results.

4.1   Experiment 1: Tunning LSH Parameters

LSH based methods report efficient results when adequate values for m (number of hash functions) are chosen. L (number of hash tables) is given by $L = m(m-1)/2$, see the E2LSH implementation[5]. To evaluate the effectiveness of the presented approach to tune the LSH parameters using fractal dimension, we worked on a variety of synthetic and real dataset. Table 1 summarizes the main features and parameters of the datasets, including the number of elements N, number of attributes d, their intrinsic (fractal) dimension $\mathfrak{D}$, the LSH parameters computed using two approaches: the Andoni (see footnote 5) algorithm and our proposal based on fractal dimension, and the total computation time for tune the LSH index (in seconds). The experiment results for the number of hash functions m show that the estimations given by Eq. 6 are comparable with those obtained with the E2LSH algorithm proposed by Andoni using up to $10X$ less time.

4.2   Retrieval Performance

The aim of this experiment is to measure the total time spent retrieving the k-nearest neighbor objects. The data structures being compared were tested with an specific values for queries. Thus we use $k = 1000$ when compute the Mean Average Precision (mAP) metric and $k = 25$ when compute the precision metric ($P(\%)$).

---

[1] https://github.com/joselhuillca/fractal_dataset.
[2] http://yann.lecun.com/exdb/mnist/.
[3] https://www.cs.toronto.edu/~kriz/cifar.html.
[4] http://ufldl.stanford.edu/housenumbers/.
[5] http://www.mit.edu/~andoni/LSH/.

Table 1. Optimal LSH params using exhaustive e2lsh and the fractal based method.

dataset			fractal params			e2lsh		fractalsh		
	N	d	$\mathcal{D}$	$log(R)$	$log(CR)$	m	$time(s)$	r	m	$time(s)$
audio	54387	192	6.49	-1.30	17.00	18	64.20	0.14	16	12.68
cities	5507	2	2.36	2.30	16.00	8	16.45	4.16	6	0.81
eigenfaces	11900	16	4.25	-1.40	18.00	12	42.28	0.13	13	1.57
histograms	4247	256	2.50	-0.81	16.01	6	15.62	0.22	7	6.12
mgcounty	27282	2	1.81	0.70	19.00	4	87.96	0.27	4	1.13
randomwalk	10000	1024	5.52	2.80	15.00	16	50.60	10.16	17	10.60
synth16d	10000	16	8.36	-1.40	17.00	20	27.20	0.18	18	7.51
synth6d	10000	6	4.95	-1.40	17.00	12	28.16	0.14	12	6.47
video	79094	50	7.73	-1.40	21.00	16	1205.73	0.13	19	92.54

Table 2. Mean average precision (mAP), precision, and cumulative time spent to compute mAP for different methods on the MNIST, SVHN and CIFAR-10 datasets.

	MNIST			SVHN			CIFAR-10		
	mAP	P (%)	time(s)	mAP	P (%)	time(s)	mAP	P (%)	time(s)
MpLSH [10]	0.86	0.95	19.58	0.71	0.73	8.09	0.83	0.89	122.45
ITQ [5]	0.87	0.95	1297.32	0.70	0.80	8411.26	0.84	0.88	1056.83
LOPQ [7]	0.86	0.90	119.21	0.71	0.74	613.65	0.83	0.88	528.20
LSH-F [2]	0.85	0.96	217.33	0.61	0.78	365.90	0.86	0.89	365.47
DAsH (Ours)	0.93	0.98	20.82	0.74	0.84	10.34	0.88	0.91	125.37

Table 2 show the comparison in terms of mean average precision mAP, precision, and total time (in seconds).

5   Conclusions

In this paper, we presented a new scheme to solve approximate similarity search by supervised hashing called Deep Fractal base Hashing. Our approach shows the potential of boosting querying operations when a specialized index structure is designed from end-to-end. Due to the abilities of Fractal theory to find the optimal sub-space of the dataset and the optimal number of hash functions for LSH Index, we are able to find an optimal configuration for learning and indexing process. Moreover, we defined a novel method, based on fractal theory, which allows us to can find the optimal number of hash functions for LSH indexes. We can estimate these parameters in linear time due to it depends on computing fractal dimension.

We conducted performance studies on many real and synthetic datasets. The empirical results for LSH parameters show that our method based on fractal theory is comparable with those obtained with the brute-force algorithm using up to $10X$ less time. Moreover, in retrieval performance, the DAsH method was significantly better than other approximate methods, providing up to 8% better precision maintaining excellent retrieval times.

Acknowledgements. This project has been partially funded by CIENCIA-ACTIVA (Perú) through the Doctoral Scholarship at UNSA University, and FONDECYT (Perú) Project 148-2015.

References

1. Andoni, A., Indyk, P.: Near-optimal hashing algorithms for approximate nearest neighbor in high dimensions. Commun. ACM **51**(1), 117–122 (2008)
2. Bawa, M., Condie, T., Ganesan, P.: LSH forest: self-tuning indexes for similarity search. In: Proceedings of the 14th International Conference on World Wide Web, pp. 651–660, Chiba, Japan (2005)
3. Böhm, C., Berchtold, S., Keim, D.A.: Searching in high-dimensional spaces: index structures for improving the performance of multimedia databases. ACM Comput. Surv. **33**(3), 322–373 (2001)
4. Bones, C.C., Romani, L.A.S., de Sousa, E.P.M.: Clustering multivariate data streams by correlating attributes using fractal dimension. JIDM **7**(3), 249–264 (2016)
5. Gong, Y., Lazebnik, S., Gordo, A., Perronnin, F.: Iterative quantization: a procrustean approach to learning binary codes for large-scale image retrieval. IEEE Trans. Pattern Anal. Mach. Intell. **35**(12) (2013)
6. Traina Jr., C., Traina, A.J.M., Wu, L., Faloutsos, C.: Fast feature selection using fractal dimension. JIDM **1**(1), 3–16 (2010)
7. Kalantidis, Y., Avrithis, Y.: Locally optimized product quantization for approximate nearest neighbor search. In: 2014 IEEE CVPR, pp. 2329–2336 (2014)
8. Krizhevsky, A., Sutskever, I., Hinton, G.E.: Imagenet classification with deep convolutional neural networks. In: Proceedings of the 25th International Conference on Neural Information Processing Systems NIPS 2012, pp. 1097–1105, USA (2012)
9. Li, Z., Liu, J., Tang, J., Lu, H.: Robust structured subspace learning for data representation. IEEE Trans. Pattern Anal. Mach. Intell. **37**(10) (2015)
10. Lv, Q., Josephson, W., Wang, Z., Charikar, M., Li, K.: Multi-probe LSH: efficient indexing for high-dimensional similarity search. In: Proceedings of the International Conference on Very Large Data Bases, pp. 950–961, Vienna, Austria (2007)
11. Moysey Brio, A.Z., Webb, G.M.: Chapter 7 fractal dimension. Math. Sci. Eng. **209**, 167–218 (2007). L-System Fractals
12. Ocsa, A., Sousa, E.P.M.: An adaptive multi-level hashing structure for fast approximate similarity search. JIDM **1**(3), 359–374 (2010)
13. Shakhnarovich, G., Darrell, T., Indyk, P.: Nearest-neighbor methods in learning and vision: theory and practice. In: Locality-Sensitive Hashing Using Stable Distributions, pp. 55–67. The MIT Press (2006)
14. Shen, F., Shen, C., Liu, W., Shen, H.T.: Supervised discrete hashing. In: CVPR, pp. 37–45. IEEE Computer Society (2015)
15. Traina Jr., C., Traina, A., Wu, L., Faloutsos, C.: Fast feature selection using fractal dimension. J. Inf. Data Manag. **1**(1), 3 (2010)
16. Yosinski, J., Clune, J., Bengio, Y., Lipson, H.: How transferable are features in deep neural networks? CoRR abs/1411.1792 (2014)

Self-organizing Maps for Motor Tasks Recognition from Electrical Brain Signals

Alvaro D. Orjuela-Cañón[1]([✉])([iD]), Osvaldo Renteria-Meza[2], Luis G. Hernández[2],
Andrés F. Ruíz-Olaya[1], Alexander Cerquera[1], and Javier M. Antelis[2]

[1] Universidad Antonio Nariño, Bogota D.C., Colombia
alvorjuela@uan.edu.co
[2] Tecnologico de Monterrey, Campus Guadalajara,
Av. Gral. Ramón Corona 2514, 45201 Zapopan, Jalisco, Mexico
mauricio.antelis@itesm.mx

Abstract. Recently, there has been a relevant progress and interest for brain–computer interface (BCI) technology as a potential channel of communication and control for the motor disabled, including post-stroke and spinal cord injury patients. Different mental tasks, including motor imagery, generate changes in the electro-physiological signals of the brain, which could be registered in a non-invasive way using electroencephalography (EEG). The success of the mental motor imagery classification depends on the choice of features used to characterize the raw EEG signals, and of the adequate classifier. As a novel alternative to recognize motor imagery tasks for EEG-based BCI, this work proposes the use of self-organized maps (SOM) for the classification stage. To do so, it was carried out an experiment aiming to predict three-class motor tasks (rest versus left motor imagery versus right motor imagery) utilizing spectral power-based features of recorded EEG signals. Three different pattern recognition algorithms were applied, supervised SOM, SOM+k-means and k-means, to classify the data offline. Best results were obtained with the SOM trained in a supervised way, where the mean of the performance was 77% with a maximum of 85% for all classes. Results indicate potential application for the development of BCIs systems.

Keywords: Electroencephalogram · Brain-computer interfaces
Motor imagery · Classification · Self-organized maps

1 Introduction

A Brain-Computer Interface (BCI) is an emergent technology that provides a non-muscular communication channel for people with motor disabilities [1], therefore, this technology provides a novel way to recover functionality of limbs impaired by diseases and injuries of the central nervous system. The key in a BCI is the recording and processing of the brain activity [2]. On the first hand, the non-invasive electroencephalogram (EEG) technique is the most used method to

© Springer International Publishing AG, part of Springer Nature 2018
M. Mendoza and S. Velastín (Eds.): CIARP 2017, LNCS 10657, pp. 458–465, 2018.
https://doi.org/10.1007/978-3-319-75193-1_55

record the brain activity. On the second hand, machine learning algorithms are employed to classify patterns or changes contained in the recorded EEG signals which are associated to a mental task performed by the BCI user.

The recognition of the mental task is very difficult because the EEG signals are highly noisy, moreover they are non-stationary and present strong variability across participants and time. In order to overcome these situations, BCI technology uses linear discriminant analysis (LDA), artificial neural networks (ANN) and support vector machines (SVM) which are classical supervised learning models commonly used in these systems [3]. This has resulted in several applications that have demonstrated the functionality and usability of BCI technology. However, the performance in the recognition of mental tasks from EEG signals (or classification accuracy) is prone to errors, thus novel strategies in the classification need be developed, evaluated and incorporated in BCI technologies.

As a novel alternative to recognize mental tasks for EEG-based BCI, this work proposes the use of self-organized maps (SOM) [4] to recognize motor imagery tasks from EEG signals recorded in a real BCI experiment. SOM is a (supervised or not supervised) learning method that transforms high dimensional data to a lower representation in a way that similar inputs are associated to a region that is easily separated from the regions of other similar inputs [4]. This classification model has been extensively used with biomedical and topographical data, however, its application in the recognition of mental tasks for BCI technologies is still very limited [5]. This work consists of an experimental study devised to obtain EEG signals from a participant performing motor imagery tasks. The goal is to implement and evaluate the classification performance of supervised SOM, SOM+k-means and k-means in the recognition of mental motor tasks from EEG signals recorded in a cue-based BCI experiment. Recorded signals were used to evaluate the three-class classification of *rest* versus *left motor imagery* versus *right motor imagery*. The results showed that the proposed supervised SOM provided the higher classification accuracies (mean of 77% and maximum of 85%) which were above the chance level.

The rest of this work is organized as follows: Sect. 2 describes the experiments performed to obtain EEG signals, the theoretical basis and the implementation of the SOM algorithm and the evaluation process; Sect. 3 presents and discusses the results; Sect. 4 presents the conclusions.

2 Methods and Materials

This section describes (i) the experiments carried out to obtain EEG signals from a volunteer who performed motor imagery of the upper limbs, (ii) the data analysis and the feature extraction process, (iii) the classification model based on SOM that was used to recognize motor imagery tasks from EEG signals, and (iv) the evaluation procedure and the metric used to asses performance.

2.1   EEG Data Description

Participants. The experiment was conducted with a male student from our faculty. He had no known neurological or motor disorder, moreover he had no prior experience with EEG or BCI experiments. The participant was duly informed about the aim of the study and freely signed a consent form.

Recording system. EEG signals were recorded using the Emotiv EPOC System. This is a low-cost and wireless technology useful for EEG recording and BCI experiments. This system records the brain electrical activity from 14 scalp locations at AF3, F7, F3, FC5, T7, P7, O1, AF4, F8, F4, FC6, T8, P8 and O2. These electrodes are distributed according to the international 10/20 system. The EEG signals were recorded at a sampling frequency of 128 Hz with two reference electrodes (CMS, which is located on the left side; DRL which is located on the right side). A band-pass filtering from 0.5 to 60 Hz was applied during the recording. Prior to the initiation of the experiment, a saline solution was applied to each electrode to maintain the impedance below 5 kΩ. This was cheeked with the software application provided by the Emotiv manufacturer.

Description of the experiment. The experiment consisted of a synchronous motor imagery task performed with the left hand or the right hand. The participant had to perform many repetitions or trials composed of three consecutive periods: rest, motor imagery and relax. Each of these three periods has a duration of 3 s, therefore the duration of a trial is 9 s. Figure 1 illustrates the temporal sequence of a trial. In rest period, the participant had to maintain a natural body position without performing or imaging any body movement. In the motor imagery period, the participant had to imagine opening and closing either the left or the right hand at a natural and comfortable pace. The participant was requested to avoid blinking or performing any movement during the rest and the motor imagery periods. Finally, in the relax period, the participant could move and blink.

The experiment was executed in blocks of 20 trials, therefore, a block lasted 3 min. In total, 10 blocks were recorded, that is, 200 trials were recorded which required 30 min. Note that this resulted in 200 periods of *rest*, 100 periods of *left motor imagery* and 100 periods of *right motor imagery*. For the execution of the experiment, the participant was seated if front a computer screen that displayed visual stimuli indicating to perform either, the rest, the motor imagery or the relax period. The presentation of these visual stimuli, the recording of the EEG signals from the Emotiv system and the recording of synchronization signals that marked the rest, motor imagery and relax periods were controlled by a Raspberry Pi microcomputer with an application developed in Python.

Preprocessing. After the experimental session, a four-order Butterworth-type bandpass from 2 to 40 Hz and a common average reference (CAR) filters were applied to the recorded EEG signals. Then, the filtered EEG signals were

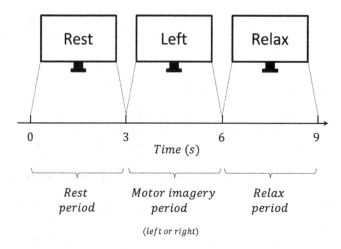

Fig. 1. Temporal sequence of a trial. The trial has a duration of 9 s and consists of three periods of 3 s each: rest (maintain a natural body position without performing any movement), motor imagery (imaging opening and closing either the left or right hand) and relax (blink and/or move any body part).

trimmed in 200 trials of 9 s containing the rest, motor imagery and relax periods, corresponding to the time intervals $[0, 3)$ s, $[3, 6)$ s and $[6, 9)$ s, respectively.

2.2   Feature Extraction and Selection

It has been established that the execution, imagination, observation or attempt to perform movements produces changes in the spectral power of the EEG in the motor-related $\alpha \sim [7, 13]$ Hz and $\beta \sim [14, 35]$ Hz frequency bands of electrodes located in the motor cortex [6]. Therefore, the spectral power of the recorded EEG signals were used as features. The Welch's averaged modified periodogram method was used to compute the spectral power. The spectral power was computed in the frequency range $[6, 40]$ Hz at a resolution of 1 Hz using Hanning-windowed windows of length 1 s with overlap of 0.5 s. Therefore, the number of spectral power values for each electrode is 35. These spectral power values were computed separately from EEG in the rest period $[0, 3)$ s (and they were labeled as 0) and from EEG in the motor imagery period $[3, 6)$ s (and they were labeled as 1 or 2 if the imagined hand was left or right, respectively). As a result, the feature vector is $\mathbf{x} \in \mathbb{R}^{n \times 1}$ with a class label $y \in \{0, 1, 2\}$ where $n = 490$ (number of electrodes × number of spectral power values).

To reduce the dimension of the feature vector, the square of the Pearson's correlation (r-squared) was employed to select the channels that present the higher discriminative power between rest and motor imagery [7]. The r-squared was computed independently for each electrode between each vector of spectral power values with the vector of class labels. Then, the four channels that presented the highest r-squared values were selected while the remaining electrodes were discharged and not used in the rest of the study. As a result, the feature vector is $\mathbf{x} \in \mathbb{R}^{n \times 1}$ with a class label $y \in \{0, 1, 2\}$ where $n = 140$ (four electrodes × thirty-five spectral power values).

2.3   Self-organizing Maps (SOM)

Supervised, semi-supervised and unsupervised learning models of artificial neural networks (ANN) can be developed by Self Organizing Maps (SOM). Inspired by the visual, aural and sensory areas of the brain, were introduced by Kohonen and Somervuo [4].

These ANN models are widely used to analyze nonlinear relations between variables in high volume of data. SOM architecture allows to represent this information in a reduced space, generally of two dimensions known as a map. This new space has descriptive characteristics in the same way as in the original space, showing properties and relations that cannot be seen before, due to the high number of dimensions [8].

For training, SOM architecture must be fixed, where the number of neurons, size of the map, type of lattice and neighborhood function information were provided. Most of these parameters obey to experimental rules with some initial information. Size of the map was computed based on the inertias relation using a Multiple Correspondence Analysis (MCA), where the ratio of the two first inertias is the same to the ratio of wide and high dimensions [9]. Then, number of neurons was adjusted according with preliminary results measured by activations of the neurons. Type of lattice was preserved as hexagonal topology due to yield the same distance between neurons, and a Gaussian neighborhood function was chosen for its good performance. SOM training is developed mainly in an unsupervised manner in a process followed by three steps: one of them competitive, then a cooperative one, and finally, an adaptive. Competition stage consists on an input vector with information extracted from each subject is presented to the map, and compared with information from synaptic weights of each neuron. Euclidian distance is used for this comparison and the neuron with weights closer to the input is defined as the best matched unit (BMU).

Cooperation step is given around the BMU, where neurons around this unit or neuron are chosen to be updated, modifying its weights through the neighborhood function. Finally, adaptive stage changes the BMU weights and its neighbors using the expression:

$$w_i(t+1) = w_i + \eta(t)h_{ij}(t)(x(t) - w_i(t)) \tag{1}$$

where $x(t)$ is the input vector, $w_i(t)$ are the weights of the map, $\eta(t)$ is a learning rate coefficient, and $h_{ij}(t)$ is the neighborhood function based on Gaussian distribution, given by:

$$h_{ij}(t+1) = exp(-d_{ij}^2/2\sigma^2(t)) \tag{2}$$

where $d_{ij}(t)$ is the Euclidian distance between the j input and the BMU, and $\sigma(t)$ is the basis of the function in the t iteration. This parameter changes during the training, beginning with a basis of four units and ending with just one unit.

Supervised Training. As the task developed by the SOM is about classification, a modification in the training was implemented. In the input vector an extra variable was included with information of the class of to which that input belongs to. This new variable makes that the training of the SOM has a guide in a supervised mode. Information of the additional variable was adjusted according to the three classes presented in the data.

Semi-supervised Training. In this case, three groups were proposed: classes one to three. This was adjusted after training, where the *k-means* algorithm was applied to create three groups based on weights between neurons. The *k-means* clustering algorithm worked developing these groups based on distances between the neurons, joining the closest neurons [10]. According to the number of activations in each group based on training data presented to the map, labels were marked for each group of neurons.

2.4   Performance Evaluation

The proposed SOM-based classification was evaluated in the three-class classification of *rest* versus *left motor imagery* versus *right motor imagery*. The cross-validation technique with ten folds was used to measure the results of the classification [11]. In each case, nine folds were combined to create the training set, and then the generalization was measured in the validation fold that was out. Performance accuracy was measured as the percentage of correct classifications which was computed separately for each class and for all classes. Performance of the two established SOM-based classification approaches, supervised SOM and SOM+*k-means*, were compared with a clustering developed just by the *k-means* algorithm, allowing to determine differences between the use of SOM before and without the use of this algorithm.

3   Results

Figure 2 shows the classification results for the three proposed approaches. Red horizontal line points out a performance of 60% (note that the chance level of the classification accuracy is 33.33% as the number of classes is 3). Results for the k-means algorithm (Fig. 2c) show that classification is not a stronghold of the method. Indeed, results just reach a performance of 51% in mean, with a maximum of 62%. When a SOM was used in combination with the k-means algorithm (Fig. 2b) is possible to see that rate of classification was improved, showing that the mapping of the network can be used in a previous task before the classification is developed. Results with this approach reach a performance of 61% in mean with a maximum of 77%. Best results were obtained with the SOM trained in a supervised way (Fig. 2a), where the mean of the performance was 77% with a maximum of 85% for all classes.

In general, results are comparable with previous studies based on the used features from the same dataset. Classification methods obtained 80% based on

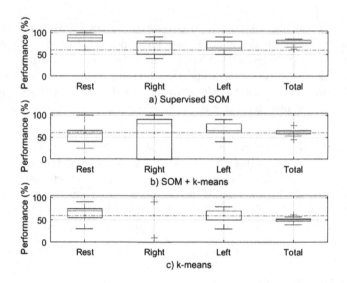

Fig. 2. Classification accuracy results obtained for each class and the total for all of them obtained with (a) Supervised SOM, (b) SOM+k-means, (c) k-means. (Color figure online)

lattice neural networks (LNN) [12] and 72% as maximum with the used of Support Vector Machines (SVM) with a radial basis function kernel [13]. Note however that is the present work we are evaluating the SOM method in a three-class classification, which is a more difficult tasks than the two-class classification of the previous studies. It is possible to see for the three approaches, that imagination of the right hand movement is one of the most difficult classes to match, where dispersion of the results and the low performance is notable (Fig. 2).

4   Conclusions

This work presented a novel classification model based on self-organized maps (SOM) to discriminate between mental motor imagery tasks from EEG signals for brain-computer interfaces (BCI). Traditionally, the SOM method is employed to carry out clustering analysis. In the present work, the SOM model was modified to perform supervised classification. The performance of this proposed supervised SOM classifier along with SOM+k-means and k-means alone were evaluated in a three-class classification scenario consisting of recognizing motor imagery tasks (rest, left hand and right hand) from spectral power-based features extracted from EEG signals. The classification results showed that the proposed supervised SOM yielded the higher accuracies which were above the chance level. On the basis of the presented results, supervised SOM is a promising classification model that can be used in real BCI settings to recognize motor imagery tasks.

Acknowledgments. The authors thank to Universidad Antonio Nariño under project 2016207 and publication PI/UAN-2017-611GIBIO for the support in this work. J.M. Antelis thanks the financial support of CONACYT through grants 268958 and PN2015-873.

References

1. Lebedev, M.A., Nicolelis, M.A.: Brain-machine interfaces: past, present and future. Trends Neurosci. **29**(9), 536–546 (2006)
2. Allison, B.Z., Dunne, S., Leeb, R., Millan, J.D.R., Nijholt, A.: Towards Practical Brain-Computer Interfaces: Bridging the Gap from Research to Real-World Applications. Springer, Heidelberg (2012). https://doi.org/10.1007/978-3-642-29746-5
3. Vega, R., Sajed, T., Mathewson, K.W., Khare, K., Pilarski, P.M., Greiner, R., Sánchez-Ante, G., Antelis, J.M.: Assessment of feature selection and classification methods for recognizing motor imagery tasks from electroencephalographic signals. Artif. Intell. Res. **6**(1), 37 (2017)
4. Kohonen, T.: Self-organization and Associative Memory, vol. 8. Springer Science & Business Media, Heidelberg (2012). https://doi.org/10.1007/978-3-642-88163-3
5. Carino-Escobar, R.I., Cantillo-Negrete, J., Gutierrez-Martinez, J.: Decodificación de imaginación motora en la señal de electroencefalografía mediante mapas auto-organizados. Revista del Centro de Investigación de la Universidad La Salle **12**, 107–125 (2016)
6. Pfurtscheller, G., Brunner, C., Schlögl, A., Lopes da Silva, F.H.: Mu rhythm (de)synchronization and EEG single-trial classification of different motor imagery tasks. NeuroImage **31**(1), 153–159 (2006)
7. Wolpaw, J., Birbaumer, N., McFarland, D., Pfurtscheller, G., Vaughan, T.: Brain-computer interfaces for communication and control. Clin. Neurophysiol. **113**(6), 767–791 (2002)
8. Haykin, S.S., Haykin, S.S., Haykin, S.S., Haykin, S.S.: Neural Networks and Learning Machines, vol. 3. Pearson, Upper Saddle River (2009)
9. Agresti, A., Kateri, M.: Categorical Data Analysis. Springer, Heidelberg (2011). https://doi.org/10.1007/978-3-642-04898-2
10. Hartigan, J.A., Hartigan, J.: Clustering Algorithms, vol. 209. Wiley, New York (1975)
11. Kohavi, R., et al.: A study of cross-validation and bootstrap for accuracy estimation and model selection. In: IJCAI, vol. 14, Stanford, CA, pp. 1137–1145 (1995)
12. Ojeda, L., Vega, R., Falcon, L.E., Sanchez-Ante, G., Sossa, H., Antelis, J.M.: Classification of hand movements from non-invasive brain signals using lattice neural networks with dendritic processing. In: Carrasco-Ochoa, J.A., Martínez-Trinidad, J.F., Sossa-Azuela, J.H., Olvera López, J.A., Famili, F. (eds.) MCPR 2015. LNCS, vol. 9116, pp. 23–32. Springer, Cham (2015). https://doi.org/10.1007/978-3-319-19264-2_3
13. Gudiño-Mendoza, B., Sanchez-Ante, G., Antelis, J.M.: Detecting the intention to move upper limbs from electroencephalographic brain signals. Comput. Math. Methods Med. **2016** (2016)

Semantic Segmentation of Color Eye Images for Improving Iris Segmentation

Dailé Osorio-Roig$^{(\boxtimes)}$, Annette Morales-González, and Eduardo Garea-Llano

Advanced Technologies Application Center (CENATAV), 7a # 21406 e/ 214 y 216,
Rpto. Siboney, Playa, 12200 Havana, Cuba
{dosorio,amorales,egarea}@cenatav.co.cu

Abstract. Iris segmentation under visible spectrum (VIS) is a topic that has been gaining attention in many researches in the last years, due to an increasing interest in iris recognition at-a-distance and in non-cooperative environments such as: blur, off-axis, occlusions, specular reflections among others. In this paper, we propose a new approach to detect the iris region on eye images acquired under VIS. We introduce the semantic information of different classes of an eye image (such as sclera, pupil, iris, eyebrows among others) in order to segment the iris region. Experimental results on UBIRIS v2 database show that the semantic segmentation improves the iris segmentation by reducing the intra-class variability, especially for the non-iris classes.

1 Introduction

Iris segmentation refers to the task of automatically locating the annular region delimited between the pupillary and limbic boundaries of the iris in a given image [1]. Results in the NICE I dataset [2] show that iris segmentation under visible spectrum (VIS) is still a very challenging problem. According to [3], non-cooperative iris recognition refers to automatically recognize at-a-distance and dealing with several factors that deteriorate the quality of an iris image [4], such as occlusions, blur, off-axis, specular reflections, poor illumination among others.

Many algorithms have been proposed for separating the iris region from the non-iris regions on images affected by the aforementioned factors. One of the main approaches consists on boundary-based methods [5–7] which generally use techniques for detecting obstructions by specular reflections as first step, and then, they use methods based on the detection of two circular boundaries or non-circular trajectories with the aim of localizing both iris boundaries. After, techniques for detecting upper and lower eyelids and eyelashes are applied. The main drawback of these algorithms is that the detection of non-iris regions such as eyelids, eyelashes and specular reflections, does not preserve entirely the spatial relationship with respect to the iris region. Moreover, these approaches only take into account as non-iris regions, the eyelids, eyelashes and specular reflections disregarding other important non-iris regions such as eyebrows, pupil, skin, hair and other factors related to problems of non-cooperative environments such

© Springer International Publishing AG, part of Springer Nature 2018
M. Mendoza and S. Velastín (Eds.): CIARP 2017, LNCS 10657, pp. 466–474, 2018.
https://doi.org/10.1007/978-3-319-75193-1_56

as glasses and lenses. Other approaches are based on a combination of region and boundary-based methods, which initially find an approximate eye area that is discriminatory and adjacent to the iris region followed by a process of detection of circular boundaries with the aim of separating the iris region of the non-iris regions. Examples of these approaches are the methods based on the localization of the sclera region, followed by a fast circular hough transform [7], which locates the iris region by identifying the sclera as the most easily distinguishable part of the eye under varying illumination [8]. The main drawback of these methods is that they only take into account the discriminatory and adjacency notion of a single region (sclera) of the set of non-iris regions present in an eye image.

In general, the methods for iris segmentation mentioned above use only low level features for locating the iris region. They do not take into account the semantic information of all classes that conform an eye image. For this reason, we believe that, using more semantic classes (namely sclera, iris, pupil, eyelids, skin, hair, glasses, eyebrows and specular reflection) and not only iris and non-iris classes, we may improve the iris segmentation by reducing the intra-class variability and increasing the inter-class variability, especially for the non-iris class. On the other hand, by semantically segmenting color eye images (employing manual annotations in the learning process) it is possible to extend the applicability of iris segmentation algorithms to other soft-biometric features, such as the sclera and periocular regions, that can be useful to improve recognition performance on color eye images [1]. To the best of our knowledge, iris segmentation have not been addressed in terms of semantic segmentation in the literature. Summarizing, our contributions are: (1) creating a set of manual annotations for eye images containing 9 semantic classes, (2) exploring the performance of semantic segmentation for eye image classes and (3) extending the use of semantic segmentation for the specific case of iris segmentation.

The remainder of this paper is organized as follows. In Sect. 2, we present a description of the proposed method. The experimental evaluation, showing our results and a comparison with other state-of-the-art methods is presented in Sect. 3. Conclusions and future work are presented in Sect. 4.

2   Proposed Method

We propose to employ a semantic segmentation approach in order to perform iris segmentation. For this task we chose a semantic segmentation algorithm named HMRF-PyrSeg [9] that was originally designed for general images. To the best of our knowledge, there are no semantic segmentation approaches specifically designed for iris segmentation.

In Fig. 1, a description of this semantic segmentation process can be seen.

2.1   Semantic Segmentation Algorithm

The first step of the algorithm is to segment the image using the Felzenszwalb's segmentation algorithm [10] (Fig. 1b). This segmented image is represented as a

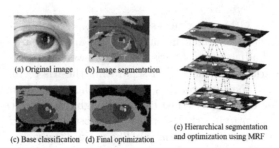

(a) Original image   (b) Image segmentation

(c) Base classification   (d) Final optimization

(e) Hierarchical segmentation and optimization using MRF

Fig. 1. General steps of the segmentation/classification process. (a) Original image, (b) initial segmentation (c) initial classification of the regions obtained by the base classifier, (d) final classification/optimization result and (e) overall example of hierarchical segmentation, classification and optimization by means of a hierarchical MRF approach (Best viewed in color).

Region Adjacency Graph (RAG), where each vertex represents a region and each edge represents the adjacency between regions. We obtain several segmentations by varying the algorithm parameters in order to avoid under-segmentation and over-segmentation problems. For all the regions in each segmentation, low-level features are extracted and a base classifier (it could be any classifier with probabilistic output) is trained taking into account the semantic classes presented in the dataset. All the image regions are classified using this base classifier (Fig. 1c). This will provide a base line classification/segmentation result for the entities present in the image.

We build a hierarchy on top of each initial segmentation, with the aim to improve it using the semantic information coming from the classification process. Given an initial segmentation, and an initial classification of its regions, we build a new segmentation level by contracting the edges of the graph, and joining the regions connected by these edges. In order to decide which edges should be contracted, we employ a criterion based on semantic information (i.e., the classes assigned to each vertex/region and their probability values), and low-level information given by the Canny edge detector to preserve important edges in the image. For more information regarding this criterion, please refer to [9]. After the new segmentation level is created, we perform the classification process on the level, and we can create a new level on top of this one using this information. This process can be repeated until no further contractions can be performed.

Once we have this segmentation hierarchy and a first classification of all its levels, a refinement is performed by means of a hierarchical Markov Random Field (MRF) approach (See Fig. 1e). Within the segmentation hierarchy, a MRF is built per level. Each MRF will receive information from the MRFs computed in adjacent levels in the hierarchy. In this way, it is possible to link spatial information within each level of segmentation and the hierarchical information between levels. Intuitively, one MRF per segmentation ensures a smoothing in terms of region labeling (classification), by taking into account the spatial relations among the regions. By using a hierarchical MRF, we are trying to add

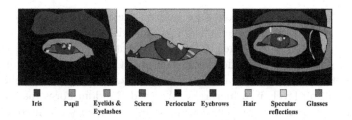

Iris   Pupil   Eyelids &   Sclera   Periocular   Eyebrows   Hair   Specular   Glasses
                 Eyelashes   reflections

Fig. 2. Examples of manual annotation of eye images (Best viewed in color).

a hierarchical smoothing, which means, for instance, that two regions wrongly classified in one level can be corrected if their father (those two regions merged as one) in the upper level was correctly classified. In this way, we can combine local an global information at different levels of segmentation. At the end of this optimization process, the regions should be annotated with more coherent classes, according to their spatial relations and context (See Fig. 1d).

After a final classification is obtained, we add to the algorithm a post-processing step for the iris regions, which consists in a morphological close followed by a connected components analysis. The objective of this step is to remove the isolated pixels or regions that are not connected to larger objects.

2.2 Manual Annotation of Eye Images

In order to obtain a ground truth for training classifiers using the semantic classes presented in eye images, we manually annotated 53 images from the UBIRIS.v2 [11] training set. These annotations were done at pixel level, i.e. each pixel is annotated with its corresponding class id. Since this annotation process is a hard and time-consuming process, we conducted a data augmentation step, where we performed a horizontal flip on the 53 original images and their corresponding manual ground truths, thus increasing the training set to 106 eye images. Examples of these manual annotations can be seen in Fig. 2. The images were annotated with 9 semantic classes: sclera, iris, pupil, eyelids and eyelashes, skin, hair, glasses, eyebrows and specular reflections. We selected these labels with the aim of obtaining the biggest semantic information describing the different regions of an eye image.

3   Experimental Results

In this section we show the experimental results of our approach. The main objective is to show that the semantic information from eye images have advantages over the iris and non-iris classification, as well as the classification of others parts of an eye image, such as: sclera, skin, eyelids/eyelashes (E&E), pupil, eyebrows, specular reflections (SpRef), glasses and hair.

The proposed method was tested on the UBIRIS.v2 dataset employed in the NICE I competition [2]. It contains eye images acquired under unconstrained conditions with subjects at-a-distance and on-the move. They present serious problems of occlusions, specular reflections, off-axis and blur.

We obtained, for each image, 10 initial segmentations (changing parameters from over-segmentation to under-segmentation). On top of each segmentation, we applied the hierarchical classification/optimization algorithm, obtaining 5 new segmentation levels in a hierarchy. We used Random Forest as base classifier and the low-level features employed to describe each segmented region are Geometric Context [12] and the texton-like features described in [13].

The first experiment was conducted to validate the semantic segmentation of all classes in terms of pixelwise annotation accuracy. For this experiment we employed 53 images (106 including the horizontal flip) with manual annotations (See Sect. 2.2). Those 53 images are a subset of the 500 UBIRIS.v2 images that were used for training in the NICE I [2] competition. We performed a hold out cross-validation, by splitting the set randomly in 70% for training and 30% for testing over 5 rounds. The results can be seen in Table 1. For the base classification, we show the results for level 3 and for optimization we show the results for level 1 in the hierarchy, since this configuration exhibited the best performance.

Table 1. Pixelwise annotation accuracy (%) for each semantic class

Algorithm	Pupil	Iris	Sclera	E&E	Eyebrows	SpRef	Skin	Glasses	Hair	Mean (Std)
Classification	56.80	**82.68**	73.92	80.09	72.28	14.55	79.38	22.92	22.19	75.00 (±2.73)
Optimization	46.97	**83.81**	71.43	**85.72**	**75.54**	12.92	77.16	**28.02**	6.65	75.36 (±2.97)
Post-processing	–	**85.81**	–	**87.33**	**77.39**	–	–	–	–	–

In Table 1 can be seen that, for some classes (iris, eyelashes & eyelids, eyebrows and glasses), the MRF optimization outperformed the base classifier, but for the other classes there was no improvement (rows 1 and 2). The reason for this might be that, for classes with small area (i.e. pupil, sclera, specular reflections, hair) the pairwise smoothing term of the MRF might induce them to change the initial label to that of the surrounding class region. Also, the specular reflections, glasses and hair classes obtained the lowest results because in the training set there are very few images annotated with theses classes. Nevertheless, it is important to notice that for the best level of the base classifier the iris class obtained the best performance with respect to the rest of the classes. The third row shows the result of post-processing the iris, eyebrows and E&E classes since these classes are the more compact ones and with better defined shapes, and they usually appear as a single connected component.

The process of optimization was able to correct some errors of the initial semantic segmentation made by the base classifier. Some of these corrected errors are shown in the Fig. 3. Notice, for instance, iris (dark blue), eyelids/eyelashes (green), specular reflections (yellow) and eyebrows (brown) regions.

The second experiment was designed for evaluating the performance of iris and non-iris classification of the proposed iris semantic segmentation method.

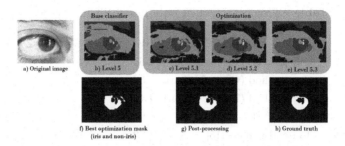

Fig. 3. Example of classification/optimization results on NICE I. (b) Level 5 of initial segmentation and base classifier result, (c–e) 3 sample levels of the optimization hierarchy built on top of Level 5, (f) mask obtained for iris/non-iris classification and (g) mask after post-processing (Best viewed in color).

We employed for training the 106 manually annotated images, and for testing we used the test set of 500 images from the NICE I competition (disjoint from the training set). We used the evaluation protocol of the NICE I competition and the average segmentation error (E_1) [2] as evaluation measure. In Table 2 we show these results for 10 initial segmentation levels with the base classifier (row 1) and their corresponding optimization (row 2).

Table 2. Comparison using E_1 (%) for different levels in classification and optimization on NICE I.

Algorithm	1	2	3	4	5	6	7	8	9	10
Base classifier	2.63	2.57	2.56	2.52	**2.51**	2.56	2.52	2.61	2.61	2.59
Optimization	2.39	2.39	2.39	2.39	**2.38**	2.49	2.52	2.55	2.68	2.69

Table 3 shows the comparison of our proposed approach with other state-of-the-art methods. It is important to notice that the number of training images used in our method is considerable smaller than other state-of-the-art methods. In Table 3 we can see that the E_1 of our method using only the base classifier was 2.51. When we used the hierarchical MRF optimization we decreased the error to 2.38, by choosing a fixed level of the optimized segmentation hierarchy. When we chose the best level of segmentation for each image, then the error was even smaller: 1.59, which outperforms most of the other methods. This last case implies selecting manually the best level of segmentation given the E_1 against the ground truth, which cannot be used in practice, but is a good indicative of the potential of our method. Automatically selecting these best levels is a good line for future improvements of our method.

Although our method was not able to outperform the one of [14], we improved some of the errors reported by that work. This can be seen in Fig. 4. In this figure we considered to show E_1 as well as the false positive and false negative

Table 3. Comparison of E_1 error in the NICE I dataset.

Method	NICE I error E^1 (%)
Our method (Base classifier)	2.51
Our method (Optimization)	2.38
Our method (Best selection)	**1.59**
Liu et al. [14]	0.90
Tan and Kumar [15]	1.72
Proenca [8]	1.87
Tan and Kumar [16]	1.90
Chai et al. [17]	2.80
Chen et al. [7]	2.97
Abdullah et al. [6]	2.95

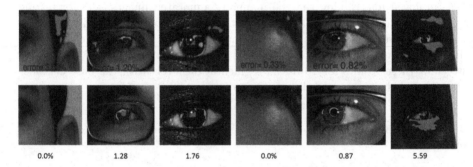

Fig. 4. Examples of classification errors. First row shows images obtained from [14] and second row shows our images. Red points denote false accept points (i.e. points labeled as non-iris by the ground truth but iris by our method). Green points denote false reject points (i.e. points labeled as iris by the ground truth but non-iris by our method) (Best viewed in color).

of those segmentations reported in [14]. We analyze these differences because this algorithm, based on multi-scale fully convolutional network (MFCNs), is the best result of iris segmentation under VIS on NICE I dataset. It can be noticed, particularly for columns 1 and 4, where there is no iris in the image, that our method was able to correctly detect that situation, while the other method detected some missclassified iris areas. Also for the case of dark skin, our method improved the recognition of these areas (see column 3).

4   Conclusions

In this paper we proposed a new iris segmentation algorithm for eye images acquired under VIS, which introduces the semantic information of the different classes of an eye image. A set of manual annotations for eye images containing 9 semantic classes was created. Experimental results showed that using the semantic information of the different classes of an eye image is a promising path for iris segmentation. It is important to note that only with 53 annotated images the proposal achieved state-of-the-art iris segmentation accuracy. Also, by using our semantic segmentation proposal, it is possible to address other segmentation tasks like sclera and periocular segmentation, which are of interest of many researches. As future work we plan to automatically select the best segmentation level for each image, as well as to work on a fusion scheme that allows merging regions of different segmentation levels.

References

1. Jillela, R.R., Ross, A.: Segmenting iris images in the visible spectrum with applications in mobile biometrics. Pattern Recogn. Lett. **57**, 4–16 (2015)
2. Proença, H., Alexandre, L.A.: The NICE.I: noisy iris challenge evaluation-part I. In: First IEEE International Conference on Biometrics: Theory, Applications, and Systems, BTAS 2007. IEEE, pp. 1–4 (2007)
3. Proenca, H., Alexandre, L.A.: Toward noncooperative iris recognition: a classification approach using multiple signatures. IEEE Trans. Pattern Anal. Mach. Intell. **29**(4) (2007)
4. Rathgeb, C., Uhl, A., Wild, P.: Iris Biometrics: From Segmentation to Template Security, vol. 59. Springer Science & Business Media, New York (2012). https://doi.org/10.1007/978-1-4614-5571-4
5. Jeong, D.S., Hwang, J.W., Kang, B.J., Park, K.R., Won, C.S., Park, D.K., Kim, J.: A new iris segmentation method for non-ideal iris images. Image Vis. Comput. **28**(2), 254–260 (2010)
6. Abdullah, M.A., Dlay, S.S., Woo, W.L., Chambers, J.A.: Robust iris segmentation method based on a new active contour force with a noncircular normalization. IEEE SMC (2016)
7. Chen, Y., Adjouadi, M., Han, C., Wang, J., Barreto, A., Rishe, N., Andrian, J.: A highly accurate and computationally efficient approach for unconstrained iris segmentation. Image Vis. Comput. **28**(2), 261–269 (2010)
8. Proenca, H.: Iris recognition: on the segmentation of degraded images acquired in the visible wavelength. IEEE Trans. PAMI **32**(8), 1502–1516 (2010)
9. Morales-González, A., García-Reyes, E., Sucar, L.E.: Improving image segmentation for boosting image annotation with irregular pyramids. In: Ruiz-Shulcloper, J., Sanniti di Baja, G. (eds.) CIARP 2013. LNCS, vol. 8258, pp. 399–406. Springer, Heidelberg (2013). https://doi.org/10.1007/978-3-642-41822-8_50
10. Felzenszwalb, P.F., Huttenlocher, D.P.: Efficient graph-based image segmentation. IJCV **59**(2), 167–181 (2004)
11. Proenca, H., Filipe, S., Santos, R., Oliveira, J., Alexandre, L.A.: The UBIRIS.v2: a database of visible wavelength iris images captured on-the-move and at-a-distance. IEEE Trans. (PAMI) **32**(8), 1529–1535 (2010)
12. Hoiem, D., Efros, A.A., Hebert, M.: Recovering surface layout from an image. IJCV **75**(1), 151–172 (2007)
13. Gould, S., Fulton, R., Koller, D.: Decomposing a scene into geometric and semantically consistent regions. In: ICCV, pp. 1–8. IEEE (2009)

14. Liu, N., Li, H., Zhang, M., Liu, J., Sun, Z., Tan, T.: Accurate iris segmentation in non-cooperative environments using fully convolutional networks. In: ICB, pp. 1–8, June 2016
15. Tan, C.W., Kumar, A.: Towards online iris and periocular recognition under relaxed imaging constraints. IEEE Trans. Image Process. **22**(10), 3751–3765 (2013)
16. Tan, C.W., Kumar, A.: Unified framework for automated iris segmentation using distantly acquired face images. IEEE Trans. Image Process. **21**(9), 4068–4079 (2012)
17. Chai, T.Y., Goi, B.M., Tay, Y.H., Chin, W.K., Lai, Y.L.: Local Chan-Vese segmentation for non-ideal visible wavelength iris images. In: TAAI, pp. 506–511. IEEE (2015)

A Single-Step 2D Thinning Scheme with Deletion of P-Simple Points

Kálmán Palágyi$^{(\boxtimes)}$ and Péter Kardos

Department of Image Processing and Computer Graphics,
University of Szeged, Szeged, Hungary
{palagyi,pkardos}@inf.u-szeged.hu

Abstract. Thinning is a frequently applied technique for producing skeletons from digital binary pictures in a topology-preserving way. Bertrand proposed a two-step thinning scheme that is based on P-simple points. In this paper, we give two sufficient conditions for topology-preserving reductions working on the three possible 2D regular grids. The new conditions combined with parallel thinning strategies and geometrical constraints yield a single-step thinning scheme that deletes solely P-simple points.

Keywords: Shape analysis · Digital topology · Regular 2D grids
Topology preservation · P-simple points

1 Introduction

Various applications of image processing and pattern recognition are based on *skeletons* [11]. *Thinning* is an iterative layer-by-layer reduction: the border points of an object that satisfy certain topological and geometric constraints are deleted until only the skeleton of that object is left [4,8].

A *binary picture* on a grid is a mapping that assigns a color of *black* or *white* to each grid element called *point* [8]. A regular partitioning of the 2D Euclidean space is formed by a tessellation of regular polygons (i.e., polygons having equal angles, and sides are all off the same length). There are exactly three polygons that can form such regular tessellations: the equilateral triangle, the square, and the regular hexagon [10]. Although 2D digital pictures sampled on the square grid are generally assumed, triangular and hexagonal grids also attract remarkable interest [8,10].

A *reduction* transforms a binary picture only by changing some black points to white ones which is referred to as *deletion* [8]. Reductions play a key role in some *topological algorithms*, e.g., thinning [4,7,8].

Topology preservation is a major concern of reductions [7,8]. In [7], Kong proposed some sufficient conditions for topology-preserving 2D reductions on the square grid, then the authors of this paper adapted his results to the hexagonal and triangular grids [5].

© Springer International Publishing AG, part of Springer Nature 2018
M. Mendoza and S. Velastín (Eds.): CIARP 2017, LNCS 10657, pp. 475–482, 2018.
https://doi.org/10.1007/978-3-319-75193-1_57

Bertrand introduced the notion of a *P-simple point* [1], and he proved that simultaneous deletion of *P*-simple points is topology-preserving. His concept can be used not only as a verification method but also as a methodology to design parallel thinning algorithms.

In this paper we prove that a reduction deletes only *P*-simple points if and only if it satisfies our previously published sufficient conditions for topology preservation [5] for the three possible regular 2D grids. Two new sufficient conditions for topology-preserving reductions are derived from our former conditions. Those new conditions provide a single-step (topology-preserving) thinning scheme with deletion of *P*-simple points.

2   Basic Notions and Results

In this paper, we use the fundamental concepts of digital topology as reviewed by Kong and Rosenfeld [8]. Despite the fact that there are other approaches based on cellular/cubical complexes [9], we consider here the 'conventional paradigm' of digital topology.

Let us denote by $\mathcal{T}$, $\mathcal{S}$, and $\mathcal{H}$ the triangular, the square, and the hexagonal grids, respectively, and throughout this paper, if we will use the notation $\mathcal{V}$, we will mean that $\mathcal{V}$ belongs to $\{\mathcal{T}, \mathcal{S}, \mathcal{H}\}$. The elements of the considered grids (i.e., regular polygons) are called *points*. Two points are 1-*adjacent* if they share an edge and they are 2-*adjacent* if they share an edge or a vertex, see Fig. 1. Let us denote by $N_j^{\mathcal{V}}(p)$ the set of points being j-adjacent to a point p in the grid $\mathcal{V}$, and let $N_j^{*\mathcal{V}}(p) = N_j^{\mathcal{V}}(p) \setminus \{p\}$ $(j = 1, 2)$.

Fig. 1. The considered adjacency relations on the three possible regular 2D grids. Points that are 1-adjacent to the central point p are marked "•", while points being 2-adjacent but not 1-adjacent to p are depicted by "○".

A sequence of distinct points $\langle p_0, p_1, \ldots, p_m \rangle$ is called a *j-path* from p_0 to p_m in a non-empty set of points $X \subseteq \mathcal{V}$ if each point of the sequence is in X and p_i is j-adjacent to p_{i-1} for each $i = 1, 2, \ldots, m$. Two points are said to be *j-connected* in a set X if there is a j-path in X between them. A set of points X is *j-connected* in the set of points $Y \supseteq X$ if any two points in X are j-connected in Y. A *j-component* of a set of points X is a maximal (with respect to inclusion) j-connected subset of X.

Let $(k, \bar{k})$ be an ordered pair of adjacency relations. Throughout this paper, it is assumed that $(k, \bar{k})$ belongs to $\{(1, 2), (2, 1)\}$. A $(k, \bar{k})$ *binary digital picture* is a quadruple $(\mathcal{V}, k, \bar{k}, B)$ [8], where set $\mathcal{V}$ contains all points of the considered grid, $B \subseteq \mathcal{V}$ denotes the set of *black points*, and each point in $\mathcal{V} \setminus B$ is said to be a *white point*. A *black component* or *object* is a k-component of B, while a *white component* is a $\bar{k}$-component of $\mathcal{V} \setminus B$.

A black point p is said to be a *border point* if p is $\bar{k}$-adjacent to at least one white point (i.e., $N_{\bar{k}}^{*\mathcal{V}}(p) \setminus B \neq \emptyset$). A border-point p is called an *isolated point* if all points in $N_k^{*\mathcal{V}}(p)$ are white (i.e., $\{p\}$ is a singleton object).

A reduction in a 2D picture is *topology-preserving* if each object in the input picture contains exactly one object in the output picture, and each white component in the output picture contains exactly one white component in the input picture [8].

A black point is said to be *simple* for a set of black points if its deletion is a topology-preserving reduction [7,8]. In [5], the authors gave both formal and easily visualized characterizations of simple points in all the considered five types of pictures on the regular 2D grids (i.e., two for $\mathcal{T}$, two for $\mathcal{S}$, and one for $\mathcal{H}$). Note that the simpleness of p is a local property (i.e., it can be decided by examining $N_2^{*\mathcal{V}}(p)$), and only non-isolated border points may be simple [5].

In [5], the authors formulated the following unified sufficient conditions for topology-preserving reductions:

Theorem 1. [5] *For any picture $(\mathcal{V}, k, \bar{k}, B)$, a reduction is topology-preserving if all of the following conditions hold.*

1. *Only simple points for B are deleted.*
2. *For any two $\bar{k}$-adjacent black points, $p, q \in B$ that are deleted, p is simple for $B \setminus \{q\}$.*
3. *If $(k, \bar{k}) = (2, 1)$, no object is deleted completely, such that is composed by two or more mutually 2-adjacent points, and is not formed by two 1-adjacent points.*

In order to construct topology-preserving reductions and provide a verification method, Bertrand proposed the notion of a *P-simple point*:

Definition 1. [1] *Let $Q \subset B$ be a set of black points in a picture. A point $q \in Q$ is called a P-simple point for Q if q is simple for $B \setminus R$ for any $R \subseteq Q \setminus \{q\}$.*

Bertrand gave a local characterization of P-simple points in $(26, 6)$-pictures on the 3D cubic grid [1], and Bertrand and Couprie reported a similar characterization in $(2, 1)$-pictures on the square grid $\mathcal{S}$ [3]. Later, the authors of this paper presented both formal and easily visualized characterizations of P-simple points in all the considered five types of 2D pictures [6].

Bertrand's approach provides the following sufficient condition for topology preservation:

Theorem 2. [1] *A reduction that deletes a subset composed solely of P-simple points is topology-preserving.*

3   Linking P-Simple Points and Our Sufficient Conditions

In this section, we show that the two kinds of conditions stated by Theorems 1 and 2 are equivalent.

Since the simpleness of a point is a local property, we can state the following proposition:

Proposition 1. *Let $Q \subset B$ be a set of points in a picture $(\mathcal{V}, k, \bar{k}, B)$. A point $q \in Q$ is P-simple for Q if for any $R \subseteq N_2^{*\mathcal{V}}(q) \cap Q$, q is simple for $B \backslash R$.*

Theorem 3. *A reduction satisfies all conditions of Theorem 1 if and only if it deletes only P-simple points.*

Proof. In order to prove the '*if*' part, let us assume that a reduction deletes the set of P-simple points $Q \subset B$ (for Q).

- Let $q \in Q$ be a point that is deleted by the reduction. Since $\emptyset \subset Q$, by Definition 1, q is simple for $B \backslash \emptyset = B$. Hence Condition 1 of Theorem 1 holds.
- Let $q_1, q_2 \in Q$ be two $\bar{k}$-adjacent points that are deleted by the reduction in question. As $\{q_2\} \subseteq Q \backslash \{q_1\}$, by Definition 1, q_1 is simple for $B \backslash \{q_2\}$, which means that Condition 2 of Theorem 1 is also fulfilled.
- Let us assume that Condition 3 of Theorem 1 is violated. Hence there is an object Q (according to Condition 3 of Theorem 1) that is completely deleted by the reduction. Consider a point $q \in Q$. Since the investigated reduction deletes only P-simple points, q is simple for $B \backslash (Q \backslash \{q\})$ by Definition 1. Since q is an isolated (non-simple black) point for $B \backslash (Q \backslash \{q\})$, we arrived at a contradiction.

Then we establish the '*only if*' part for $(1, 2)$-pictures on $\mathcal{S}$. Let us assume that a reduction satisfies all conditions of Theorem 1 and it deletes the set of points $Q \subset B$ from a picture $(\mathcal{S}, 1, 2, B)$.

In [5], the authors gave easily visualized characterizations of simple points in all the considered five types of pictures by sets of matching templates. The four base matching templates assigned to $(1, 2)$-pictures on $\mathcal{S}$ are depicted in Fig. 2.

Let $p \in Q$. Since Condition 1 of Theorem 1 holds, p is simple for B. Without loss of generality, we may investigate only the four templates shown in Fig. 2.

By Proposition 1, the following cases are to be investigated:

(a) Assume that p is matched by the template in Fig. 2a. In this case, only black point q is to be investigated. Since p is a non-simple (isolated black) point for $B \backslash \{q\}$, by Condition 2 of Theorem 1, $q \notin Q$.

(b) Assume that p is matched by the template in Fig. 2b. It can be readily seen that p is not simple for $B \backslash \{s\}$. Hence, by Condition 2 of Theorem 1, $s \notin Q$. Let us investigate the remaining two black points q and r.
   - Assume that $q \in Q$ and $r \notin Q$. Since p is matched by the template in Fig. 2a in $B \backslash \{q\}$, p remains simple after the deletion of q.
   - Assume that $r \in Q$ and $q \notin Q$. Since p is matched by the template in Fig. 2a in $B \backslash \{r\}$, p remains simple after the deletion of r.

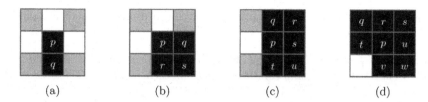

Fig. 2. The four base matching templates for characterizing a simple point p in $(1, 2)$-pictures on $\mathcal{S}$. Notations: each black template position matches a black point; each white element matches a white point; each position depicted in gray matches either a white point or a black point. All the rotated and reflected versions of these four templates also match simple points.

- Assume that $q \in Q$ and $r \in Q$. Since q is not simple for $B \backslash \{r\}$, and r is not simple for $B \backslash \{q\}$, by Condition 2 of Theorem 1, we arrived at a contradiction.

(c) Assume that p is matched by the template in Fig. 2c. It can be readily seen that p is not simple for $B \backslash \{r\}$, $B \backslash \{s\}$, and $B \backslash \{u\}$. Hence, by Condition 2 of Theorem 1, the three points r, s, and u are not in Q. Let us investigate the remaining two black points q and t.

- Assume that $q \in Q$ and $t \notin Q$. Since p is matched by the template in Fig. 2b in $B \backslash \{q\}$, p remains simple after the deletion of q.
- Assume that $t \in Q$ and $q \notin Q$. Since p is matched by the template in Fig. 2b in $B \backslash \{t\}$, p remains simple after the deletion of t.
- Assume that $q \in Q$ and $t \in Q$. Since p is matched by the template in Fig. 2a in $B \backslash \{q, t\}$, p remains simple after the deletion of $\{q, t\}$.

(d) Assume that p is matched by the template in Fig. 2d. It can be readily seen that p is not simple for $B \backslash \{q\}$, $B \backslash \{r\}$, $B \backslash \{s\}$, $B \backslash \{u\}$, and $B \backslash \{w\}$. Hence, by Condition 2 of Theorem 1, the five points q, r, s, u, and w are not in Q. Let us investigate the remaining two black points t and v.

- Assume that $t \in Q$ and $v \notin Q$. Since p is matched by the template in Fig. 2c in $B \backslash \{t\}$, p remains simple after the deletion of t.
- Assume that $v \in Q$ and $t \notin Q$. Since p is matched by the template in Fig. 2c in $B \backslash \{v\}$, p remains simple after the deletion of v.
- Assume that $t \in Q$ and $v \in Q$. Since p is matched by the template in Fig. 2b in $B \backslash \{t, v\}$, p remains simple after the deletion of $\{t, v\}$.

Since p remains simple after the deletion of each subset of Q, p is a P-simple point for Q. Hence only P-simple points are deleted.

The proofs of the '*only if*' part for the remaining four kinds of pictures can be carried out analogously with the help of matching templates given in [5]. □

4   New Sufficient Conditions

Theorem 1 (i.e., our unified sufficient conditions for topology-preserving reductions) provides a method of verifying that a thinning algorithm always preserves topology. The following new conditions allow us to directly construct deletion rules for topological algorithms.

Theorem 4. *For any picture* $(\mathcal{V}, k, \bar{k}, B)$, *a reduction* $\mathbf{R}$ *satisfies all conditions of Theorem 1 if the following conditions hold for each point* $p \in B$ *deleted by* $\mathbf{R}$:

1. p *is simple for* B.
2. *For any* $q \in N_{\bar{k}}^{*\mathcal{V}}(p) \cap B$ *that is simple for* B, p *is simple for* $B \backslash \{q\}$.
3. *If* $(k, \bar{k}) = (2, 1)$, p *does not belong to an object that is composed by two or more mutually 2-adjacent points, and is not formed by two 1-adjacent points.*

Proof. It can be readily seen that if $\mathbf{R}$ satisfies Condition i of this theorem, Condition i of Theorem 1 is also satisfied for each $i \in \{1, 2, 3\}$.   □

Let us consider the addressing schemes depicted in Fig. 3, which refer every point in the three possible regular 2D grids by a pair of coordinates. The *lexicographical order relation* '$\prec$' between two distinct points $p = (p_x, p_y)$ and $q = (q_x, q_y)$ is defined as follows: $p \prec q \Leftrightarrow (p_y < q_y) \vee ((p_y = q_y) \wedge (p_x < q_x))$. Let Q be a finite subset of $\mathcal{V}$. Then, point $p \in Q$ is said to be the *smallest element* of Q if for any $q \in Q \backslash \{p\}$, $p \prec q$.

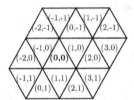 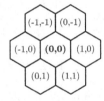

Fig. 3. Feasible addressing schemes for the three possible regular 2D grids.

Theorem 5. *For any picture* $(\mathcal{V}, k, \bar{k}, B)$, *a reduction* $\mathbf{R}$ *satisfies all conditions of Theorem 1 if the following conditions hold for each point* $p \in B$ *deleted by* $\mathbf{R}$:

1. p *is simple for* B.
2. *For any* $q \in N_{\bar{k}}^{*\mathcal{V}}(p) \cap B$ *that is simple for* B *and* $p \prec q$, p *is simple for* $B \backslash \{q\}$.
3. *If* $(k, \bar{k}) = (2, 1)$, p *is not the smallest element of an object that is composed by two or more mutually 2-adjacent points, and is not formed by two 1-adjacent points.*

Proof. It can be readily seen that if $\mathbf{R}$ satisfies Condition i of this theorem, Condition i of Theorem 1 is also satisfied for each $i \in \{1, 2, 3\}$.   □

The new sufficient conditions (i.e., Theorems 4 and 5) directly specify the following five pairs of reductions $(\mathbf{R}_{\mathrm{sym}}^{\mathcal{T},1,2}, \mathbf{R}_{\mathrm{asym}}^{\mathcal{T},1,2})$, $(\mathbf{R}_{\mathrm{sym}}^{\mathcal{T},2,1}, \mathbf{R}_{\mathrm{asym}}^{\mathcal{T},2,1})$, $(\mathbf{R}_{\mathrm{sym}}^{\mathcal{S},1,2}, \mathbf{R}_{\mathrm{asym}}^{\mathcal{S},1,2})$, $(\mathbf{R}_{\mathrm{sym}}^{\mathcal{S},2,1}, \mathbf{R}_{\mathrm{asym}}^{\mathcal{S},2,1})$, and $(\mathbf{R}_{\mathrm{sym}}^{\mathcal{H},1,2}, \mathbf{R}_{\mathrm{asym}}^{\mathcal{H},1,2}) = (\mathbf{R}_{\mathrm{sym}}^{\mathcal{H},2,1}, \mathbf{R}_{\mathrm{asym}}^{\mathcal{H},2,1})$, where $\mathbf{R}_{\mathrm{sym}}^{\mathcal{V},k,\bar{k}}$ deletes all points from a $(k,\bar{k})$ picture on $\mathcal{V}$ that satisfy all conditions of Theorem 4, and $\mathbf{R}_{\mathrm{asym}}^{\mathcal{V},k,\bar{k}}$ deletes all points from a $(k,\bar{k})$ picture on $\mathcal{V}$ that satisfy all conditions of Theorem 5.

The following proposition is an easy consequence of Theorems 3, 4, and 5:

Proposition 2. *Reductions* $\mathbf{R}_{sym}^{\mathcal{V},k,\bar{k}}$ *and* $\mathbf{R}_{asym}^{\mathcal{V},k,\bar{k}}$ *delete only* P-*simple points.*

Figure 4 is to illustrate the different behaviors of $\mathbf{R}_{\mathrm{sym}}^{\mathcal{S},2,1}$ and $\mathbf{R}_{\mathrm{asym}}^{\mathcal{S},2,1}$. It is important to emphasize the disparity between Conditions 2 and 3 of Theorems 4 and 5. Theorem 4 ensures *symmetric* conditions, since both $\bar{k}$-adjacent black points p and q are preserved if p is not simple after the deletion of q (in that case, q is not simple after the deletion of p, as well), and all 'small' objects composed by three or more mutually k-adjacent points remain unchanged. The conditions of Theorem 5 are *asymmetric*, since only the smallest elements of pairs of points and objects considered in Conditions 2 and 3 are preserved.

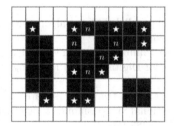 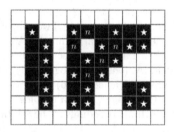

Fig. 4. Comparison of reductions $\mathbf{R}_{\mathrm{sym}}^{\mathcal{S},2,1}$ (left) and $\mathbf{R}_{\mathrm{asym}}^{\mathcal{S},2,1}$ (right) for a $(2,1)$ picture on the square grid $\mathcal{S}$. Non-simple points are marked 'n', and deleted (simple and P-simple) points are marked white stars. (Unmarked black points are simple but non-deleted points.)

5 A Single-Step 2D Thinning Scheme

In [2], Bertrand proposed a two-step (topology-preserving) thinning scheme that is based on P-simple points. One phase/reduction of the iterative thinning process is performed as follows:

1. A set of points $Q \subset B$ is (somehow) chosen and labeled.
2. All P-simple points in Q are deleted (simultaneously).

Note that Step 2 deals with *tricolor* pictures (say: the value '0' corresponds to white points, the value '1' is assigned to the elements in $B \backslash Q$, and value '2' corresponds to points in Q). Hence that two-step scheme is both space- and time-consuming.

Proposition 2 provides a single-step (topology-preserving) thinning scheme with deletion of P-simple points. The deletion rule of a reduction of the iterative thinning process can be directly constructed by combining the reduction $\mathbf{R}_{\text{sym}}^{\mathcal{V},k,\bar{k}}$ (or $\mathbf{R}_{\text{asym}}^{\mathcal{V},k,\bar{k}}$) with different thinning strategies (i.e., *fully parallel*, *subiteration-based*, and *subfield-based* [4]) and various geometric constraints (say *endpoints* [4]). The generated deletion rule is a common *Boolean function* that is to be evaluated for the neighborhood of the investigated points in binary (*two-level*) pictures. That Boolean function can be stored in a pre-calculated *look-up-table*, hence, the proposed single-step scheme can be implemented efficiently.

Acknowledgments. This work was supported by the grant OTKA K112998 of the National Scientific Research Fund.

References

1. Bertrand, G.: On P-simple points. Compte Rendu de l'Académie des Sciences de Paris, Série Math. **I**(321), 1077–1084 (1995)
2. Bertrand, G.: P-simple points: a solution for parallel thinning. In: Proceedings of the 5th International Conference on Discrete Geometry for Computer Imagery, DGCI 2005, pp. 233–242 (1995)
3. Bertrand, G., Couprie, M.: Two-dimensional parallel thinning algorithms based on critical kernels. J. Math. Imaging Vis. **31**, 35–56 (2008)
4. Hall, R.W.: Parallel connectivity-preserving thinning algorithms. In: Kong, T.Y., Rosenfeld, A. (eds.) Topological Algorithms for Digital Image Processing, pp. 145–179. Elsevier Science (1996)
5. Kardos, P., Palágyi, K.: On topology preservation in triangular, square, and hexagonal grids. In: Proceedings of the 8th International Symposium on Image and Signal Processing and Analysis, ISPA 2013, pp. 782–787 (2013)
6. Kardos, P., Palágyi, K.: Unified characterization of P-simple points in triangular, square, and hexagonal grids. In: Barneva, R.P., Brimkov, V.E., Tavares, J.M.R.S. (eds.) CompIMAGE 2016. LNCS, vol. 10149, pp. 79–88. Springer, Cham (2017). https://doi.org/10.1007/978-3-319-54609-4_6
7. Kong, T.Y.: On topology preservation in 2-D and 3-D thinning. Int. J. Pattern Recogn. Artif. Intell. **9**, 813–844 (1995)
8. Kong, T.Y., Rosenfeld, A.: Digital topology: introduction and survey. Comput. Vis. Graph. Image Process. **48**, 357–393 (1989)
9. Kovalevsky, V.A.: Geometry of Locally Finite Spaces. Publishing House, Berlin, Germany (2008)
10. Marchand-Maillet, S., Sharaiha, Y.M.: Binary Digital Image Processing - A Discrete Approach. Academic Press, Cambridge (2000)
11. Siddiqi, K., Pizer, S. (eds.): Medial Representations - Mathematics, Algorithms and Applications. Computational Imaging and Vision, vol. 37. Springer, Heidelberg (2008). https://doi.org/10.1007/978-1-4020-8658-8

Non Local Means Image Filtering Using Clustering

Alvaro Pardo[(✉)]

Department of Electrical Engineering, School of Engineering and Technologies,
Universidad Catolica del Uruguay, Montevideo, Uruguay
apardo@ucu.edu.uy

Abstract. In this work we study improvements for the Non Local Means image filter using clustering in the space of patches. Patch clustering it is proposed to guide the selection of the best patches to be used to filter each pixel. Besides clustering, we incorporate spatial coherence keeping some local patches extracted from a local search window around each pixel. Therefore, for each pixel we use local patches and non local ones extracted from the corresponding cluster. The proposed method outperforms classical Non Local Means filter and also, when compared with other methods from the literature based on random sampling, our results confirm the benefits of sampling inside clusters combined with local information.

1 Introduction

Image filtering is one of the most studied problems in image processing and probably one of the most used techniques in image processing, computer vision and related areas. In the literature we can find two broad categories of image filters: linear and non linear. More recently, non local methods attracted the attention of researchers in the area. In fact, several of the state of the art algorithms are both non local and non linear [6]. Among several image filtering applications, image denoising stands as one of the most relevant. The main goal of denoising is to remove undesired components from the image. Here we assume an additive noise model where the observed image I is the result of adding a random noise N to the ideal noiseless image I_o: $I = I_o + n$.

The main goal of image denoising is to estimate I_o while preserving its edges and details. Usually there is a tradeoff between noise reduction and detail preservation. Non linear and non local methods were a good step forward with respect to linear and local filters. In [2] Buades, Morel and Coll introduced the Non Local Means image filter (NLM) which opened a whole area of research of non linear and non local filtering methods. The underlying idea of this method is to estimate, I_{o_i} (image value at pixel i), using a weighted average of all pixels in the observed image I.

In this work we introduce a modified version of NLM that better selects the pixels to be used in weighted averaging. Instead of using all pixels from

© Springer International Publishing AG, part of Springer Nature 2018
M. Mendoza and S. Velastín (Eds.): CIARP 2017, LNCS 10657, pp. 483–490, 2018.
https://doi.org/10.1007/978-3-319-75193-1_58

the image, or restrict to points in a local search window, we propose to cluster points using local information and average only corresponding points. As we will see later, similarity is measured using patches of pixels around the pixel of interest. Also we will borrow ideas from [4] where instead of clustering a random sample of patches was proposed. In order to improve the results we will add spatial coherence to the process and show that this is a key element to produce competitive results.

2   Review of Non Local Means Image Filter

The NLM filter, as presented in [2], estimates the filtered version of the image using a weighted average of pixels in a neighboring region, $\mathcal{N}_i$, of the pixel to be filtered:

$$\hat{I}_i = \frac{1}{W_i} \sum_{j \in \mathcal{N}_i} w_{ij} I_j. \tag{1}$$

The weights w_{ij} express the similarity of pixels i and j and W_i is a normalization term. One of the key factors of NLM is that this similarity is based on the distance between patches. A patch is a square window of size $(2K+1) \times (2K+1)$ centered at the pixel of interest. If p_i and p_j are patches at pixels i and j respectively, the similarity weight between them is defined as:

$$w_{ij} = \exp(-||p_i - p_j||^2/\sigma^2). \tag{2}$$

This idea can be easily extended to average all pixels in the image and make the filter truly non local. Although this extension looks very attractive, it has some problems. The first one is obviously its computational cost which has been addressed in the literature (If the image has N pixels and the patches are of size $(2K+1) \times (2K+1)$ the computational cost is $O(N^2(2K+1)^2)$). Assuming that the computational cost is not a critical issue, the second weakness of the extension is that it does not improve the results in terms of MSE (Mean Square Error) or PSNR (Peak Signal to Noise Ratio) [1,2]. The intuition behind this problem is that weights not always discriminate non corresponding patches. If the number of non corresponding patches grows as the number of patches increases the filtered version deviates from the expected value. In the literature this problem has been addressed by several authors with several techniques that basically intend to average only patches belonging to the same *class* [5]. In Sect. 4.1 we will come back to this discussion when introducing clusters into NLM.

3   NLM and Random Sampling

In [4] Chan, Zickler and Lu proposed an interesting approach to reduce the computation cost of the NLM filter by randomly selecting a small number of patches from the whole set of patches of the image. So, in Eq. (1) instead of summing for all patches the sum considers only the subset of sampled patches (no other modifications are needed). Together with experimental results the authors present a theoretical analysis to support their proposal that was further studyed in [3]. The results of this method are analyzed in [8] where the authors show that using the same number of patches as the classical NLM filter (the number of points inside $\mathcal{N}_i$) the performance in terms of PSNR or MSE is worse than the classical approach. That is, at the same computation cost, worst denoising performance.

Random sampling is the best sampling option in order to reduce computational cost? To reduce the computational cost and at the same time select the best candidates we need to guide the sampling; a random sampling is not the best option. To guide the sampling we propose to cluster all patches and perform a sampling only within each cluster. In this way we accomplish the computational cost reduction while selecting only similar patches. For each pixel i with patch p_i belonging to cluster C_{k_i} we randomly sample patches inside this cluster. In the next section we present the details of this approach.

4 Non Local Means over Clusters of Patches

Here we present a modified version of the NLM filter that works over clusters of patches. The main idea is as follows: instead of arbitrary average patches across the whole image, to aggregate information only of close patches. For that end we cluster the patches of the image into N_c clusters and apply NLM inside each cluster. In the remainder of this section first we discuss the adaptation of the NLM filter and clustering method.

4.1 Non Local Means Using Clusters

Assume we cluster the set of patches $X = \{p_1, ..., p_N\}$ into N_c clusters. Let C_k be the cluster k, the NLM filter can be expanded as:

$$\hat{I}_i = \frac{1}{W_i} \sum_j w_{ij} I_j = \frac{1}{W_i} \sum_{k=1}^{N_c} \sum_{j \in C_k} w_{ij} I_j.$$

To include only similar patches in the weighted average the above equation is modified to use only one cluster. If the patch p_i around of pixel i belongs to cluster C_{k_i} the modified NLM filter equation is:

$$\hat{I}_i = \frac{1}{W_i} \sum_{j \in C_{k_i}} w_{ij} I_j.$$

The previous equation provides a computation cost reduction via the decrease of the number of patches to average. Of course we have to remember that we need to run a clustering algorithm before filtering. In the next sub-section we describe the clustering method used in this work.

4.2 Clustering of Patches

The set of patches of the image, $X = \{p_1, ..., p_N\}$, is clustered using a standard k-means algorithm. Before applying the clustering the patches are transformed using PCA (Principal Component Analysis). In [9] the author showed that using PCA improves the results of NLM. It is also known that PCA concentrates the noise in the components with lowest eigenvalue. In this work in all cases the patches are of size 5×5 and the patches are projected into the first 12 components.

The number of clusters was determined using spectral clustering [10]. The idea in this case is to construct a graph that encodes the structure of the patches in X. The first step is to construct the pairwise similarity matrix S. The element s_{ij} of the matrix is the similarity between patches p_i and p_j, $s_{ij} = \exp(\|p_i - p_j\|^2/\sigma^2)$. Then, the similarities are normalized using the total weight of incident arcs for each patch p_i: $d_i = \sum_j s_{ij}$. The Laplacian of the graph is defined as $L = D^{-1}S$, where $D = \mathrm{diag}(d_1, ..., d_N)$. One of the properties of the Laplacian is that the multiplicity of the eignevalue 1 gives the number of natural clusters in the data [10]. In a real case the eigenvalues are not exactly one and thresholding techniques must be applied. In our case we estimate the number of clusters with the number of eigenvalues greater than 0.8.

Once we have the clusters, instead of using all points inside the cluster, following [4], we apply a random sampling. As we said before, the main goal of this step is to select the best possible candidates. In terms of computation cost this approach, given the clusters, is the same as the random sampling proposed in [4]. We fix the number of sampled patches to keep the computational cost of the filtering process constant and evaluate the different algorithms using the MSE and the SSIM (Structural Similarity Index) [11]. The following section describes all the algorithms that will be evaluated; the proposed ones and the ones from the literature included for comparison purposes.

5   Proposed Methods

This section presents the implementation details of the proposed algorithms together with a review of the methods from the literature used for comparison.

Traditional Non Local Means (NLM): The traditional NLM implements the Eq. (1). The search window $\mathcal{N}$ was set as a square window of size 21×21 and the parameter σ used in the weight computation in (2) was set based on an estimation of the noise variance $\hat{\sigma}_n$: $\sigma = \hat{\sigma}_n$. The estimation of $\hat{\sigma}_n$, assuming Gaussian noise, can be done applying the proposal in [7].

Non Local Means with Random Sampling (NLMRS): This algorithm implements the idea presented in [4] and discussed in Sect. 3. A subset of patches is randomly sampled and all weights are computed against the selected patches. To compare all the algorithms under comparable situations we sample 21×21 patches to keep the computation cost constant. In this way, this algorithm uses the same number of patches as the NLM described in previous section.

Non Local Means with Random Sampling inside Clusters (NLMRS-C): As explained in Sect. 3 in this case we propose to apply a random sampling but restricting the samples inside each cluster. That is, given a pixel i to be processed, with patch p_i belonging to cluster C_{k_i}, this pixel is filtered using only patches from cluster C_{k_i}. As we explained before, to be able to compare filtering results we fix the number of patches used. Therefore, we sample 21×21 patches inside each cluster.

Non Local Means with Random Sampling inside Cluster and Spatial Neighbors (NLMRS-S): In Sect. 1 we discussed the desired characteristics of images filters. As we observed, spatial coherence is important. However, if we analyze the methods NLMRS-C and NLMC, described below, it is clear that they not enforce spatial coherence. If two neighboring pixels belong to different clusters they will be filtered with different patches and coherence cannot be guaranteed. This is a key difference between these methods and NLM; the later one working in a search window centered at the pixel being processed intrinsically includes spatial coherence.

To combine spatial coherence with patches selected via clustering and sampling we decided to combine local and patches sampled from the corresponding cluster. Let i be a pixel with corresponding patch p_i belonging to cluster C_{k_i} and $\mathcal{N}_i$ the corresponding search window. The set of patches to be used for the filtering operation is constructed as the union of the patches corresponding to pixels in $\mathcal{N}_i$ with patches sampled from C_{k_i}. We use a local search window $\mathcal{N}_i$ of size 15×15 and complete the remaining patches via sampling inside C_{k_i}. Hence, we use $15 \times 15 = 225$ local patches and $21 \times 21 - 15 \times 15 = 216$ sampled from the cluster trying to have a 50/50 relationship between both sets.

Non Local Means with Clusters (NLMC): This algorithm is a modification of the NLM that instead of using all pixels in the image, as explained in Sect. 1, uses only patches inside each cluster. For a given patch p_i centered at pixel i and belonging to cluster C_{k_i} it will only use corresponding patches inside C_{k_i}. This algorithm will be used for comparison purposes but will not be compared directly to NLM, NLMRS-C and NLMRS-S since it uses a different number of patches to filter each pixel.

6   Results

This section summarizes the results obtained in terms of MSE and SSIM for the algorithms detailed in previous section. We tested the algorithms over a set of six well known images in two different additive gaussian noise configurations. The following tables present the results for two noise levels, $\sigma_n = 10$ and $\sigma_n = 20$. For each image we present the MSE and SSIM for each method. The central columns contain the methods that can be directly compared while the left and rightmost columns are included for comparison purposes with the works from the literature and used as references in this work. Bold numbers highlight the best results among the algorithms under evaluation (NLMC and NLMRS are not considered).

This experiments confirm that adding spatial coherence plays an important role in the denoising quality (both objectively observing MSE and subjectively as expressed in the SSIM). Looking at the MSE NLMRS-S outperforms NLM in 4 out of the 6 images tested for both noise configurations. Barbara and House images are the two cases where NLM performs better than NLMRS-S. These images contain periodic textures that are better denoised using local patches. Since NLMRS-S combines 50% of local patches with 50% of patches sampled from the cluster it ends up using less highly similar patches for the filtering. Similar results are encountered when observing the SSIM. For $\sigma_n = 10$ we observe that NLMRS-S produces the best SSIM scores in all images. We can see that this is a promising approach to improve traditional NLM filter without increasing its computational cost. In our approach we only added a clustering stage before applying the filter. This additional step generates improvements in terms of MSE and SSIM. For future work, it would be of interest to further analyze image content to detect periodic structures to switch between NLMRS-S and NLM.

Additionally, when comparing the results NLMRS against NLMC and NLMRS-S, we confirm that better results can be obtained if some guidance is added during the sampling process. As we said before, complete random sampling is not a good strategy. Furthermore, if we compare NLMC with NLMRS-S we conclude that spatial coherence is needed to improve denoising results. Even though NLMC uses more patches to filter each pixel this does not directly translate into better MSE and SSIM scores. Only in two cases for $\sigma_n = 10$ NLMC outperforms NLMRS-S in terms of MSE (for SSIM there is no improvement) (Table 1).

In Fig. 1 we depict the results for the image Cameran for the case $\sigma_n = 10$. The clustering clearly show the different regions of the mage in terms of local patch configuration. Note how NLMRS-S preserves in a better way the details of the image (see the grass) and NLM generates smoother results (see the sky) (Table 2).

Table 1. MSE and SSIM results for $\sigma_n = 20$.

Image	NLMC		NLMRS-C		NLMRS-S		NLM		NLMRS	
	MSE	SSIM	MSE	SSIM	MSE	SSIM	MSE	SSIM	MSE	SSIM
Cameraman	92.60	0.786	163.58	0.751	**83.54**	**0.818**	88.55	0.827	269.47	0.742
Barbara	120.58	0.880	164.39	0.864	86.08	**0.922**	**73.33**	0.917	213.55	0.840
Boat	88.76	0.870	113.52	0.861	**76.85**	**0.904**	86.94	0.860	157.64	0.846
Lena	97.23	0.772	126.71	0.745	**83.85**	0.805	89.18	**0.809**	148.85	0.754
House	49.82	0.822	60.22	0.805	46.70	0.827	**42.02**	**0.840**	97.54	0.789
Goldhill	91.33	0.865	102.12	0.860	**73.24**	**0.894**	85.48	0.839	117.10	0.856

Table 2. MSE and SSIM results for $\sigma_n = 10$.

Image	NLMC		NLMRS-C		NLMRS-S		NLM		NLMRS	
	MSE	SSIM	MSE	SSIM	MSE	SSIM	MSE	SSIM	MSE	SSIM
Cameraman	40.15	0.888	136.57	0.834	**35.16**	**0.911**	50.30	0.891	230.36	0.807
Barbara	40.37	0.960	120.41	0.932	41.45	**0.967**	**30.34**	0.965	189.75	0.880
Boat	38.88	0.947	84.16	0.930	**39.23**	**0.960**	40.06	0.937	124.57	0.898
Lena	38.27	0.886	98.47	0.826	**38.74**	**0.893**	43.44	0.885	140.20	0.799
House	21.32	0.897	38.86	0.875	23.52	**0.897**	**20.57**	0.888	67.73	0.856
Goldhill	40.17	0.942	69.37	0.928	**38.91**	**0.954**	41.70	0.922	85.99	0.903

Fig. 1. Results for $\sigma_n = 10$. From left-top to bottom right: original cameraman image, visualization of patch clusters, NLM result and NLMRS-S result. Note that NLMRS-S preserves in a better way the details of the image (see the grass) and NLM generates smoother results (see the sky).

7   Conclusions

We proposed a modified version of NLM using clustering and spatial regularization that provides good results in terms of MSE and SSIM. The proposed method outperformed classical NLM in a dataset of standard images. These results allowed us to confirm that random sampling as proposed in [4] can be improved guiding the sampling using clustering. We also confirmed that spatial coherence is critical, as expected, in denoising algorithms. We managed to balance spatial coherence with non local patches obtained via clustering and sampling. We believe this opens an interesting are of analysis for NLM and similar methods.

Acknowledgements. Work partially supported by ANII and Stic-Amsud. Thanks to Andrés Almansa for valuable discussions.

References

1. Bertalmio, M., Caselles, V., Pardo, A.: Movie denoising by average of warped lines. IEEE Trans. Image Process. **16**(9), 2333–2347 (2007)
2. Buades, A., Coll, B., Morel, J.M.: A review of image denoising algorithms, with a new one. SIAM Multiscale Model. Simul. **4**(2), 490–530 (2005)
3. Chan, S.H., Zickler, T., Lu, Y.M.: Monte Carlo non-local means: random sampling for large-scale image filtering. IEEE Trans. Image Process. **23**(8), 3711–3725 (2014)
4. Chan, S.H., Zickler, T.E., Lu, Y.M.: Fast non-local filtering by random sampling: it works, especially for large images. In: ICASSP, pp. 1603–1607 (2013)
5. Mahmoudi, M., Sapiro, G.: Fast image and video denoising via nonlocal means of similar neighborhoods. IEEE Sig. Process. Lett. **12**(12), 839–842 (2005)
6. Milanfar, P.: A tour of modern image filtering: new insights and methods, both practical and theoretical. IEEE Sig. Process. Mag. **30**(1), 106–128 (2013)
7. Olsen, S.I.: Noise variance estimation in images. In: Proceedings of 8th SCIA. Citeseer (1993)
8. Pardo, A.: Comments on randomly sampled non local means image filter. In: Bayro-Corrochano, E., Hancock, E. (eds.) CIARP 2014. LNCS, vol. 8827, pp. 498–505. Springer, Cham (2014). https://doi.org/10.1007/978-3-319-12568-8_61
9. Tasdizen, T.: Principal neighborhood dictionaries for nonlocal means image denoising. IEEE Trans. Image Process. **18**(12), 2649–2660 (2009)
10. von Luxburg, U.: A tutorial on spectral clustering. Stat. Comput. **17**, 4 (2007)
11. Wang, Z., Bovik, A.C., Sheikh, H.R., Simoncelli, E.P.: Image quality assessment: from error visibility to structural similarity. IEEE Trans. Image Process. **13**(4), 600–612 (2004)

Neural Networks for the Reconstruction and Separation of High Energy Particles in a Preshower Calorimeter

Juan Pavez[1](✉), Hayk Hakobyan[2], Carlos Valle[1], William Brooks[2], Sergey Kuleshov[2], and Héctor Allende[1]

[1] Departamento de Informática, Universidad Técnica Federico Santa María, CP 110-V, Valparaíso, Chile
juan.pavezs@alumnos.usm.cl
[2] Departamento de Física, Universidad Técnica Federico Santa María, CP 110-V, Valparaíso, Chile

Abstract. Particle detectors have important applications in fields such as high energy physics and nuclear medicine. For instance, they are used in huge particles accelerators to study the elementary constituents of matter. The analysis of the data produced by these detectors requires powerful statistical and computational methods, and machine learning has become a key tool for that. We propose a reconstruction algorithm for a preshower detector. The reconstruction algorithm is in charge of identifying and classifying the particles spotted by the detector. More importantly, we propose to use a machine learning algorithm to solve the problem of particle identification in difficult cases for which the reconstruction algorithm fails. We show that our reconstruction algorithm together with the machine learning rejection method are able to identify most of the incident particles. Moreover, we found that machine learning methods greatly outperform cut based techniques that are commonly used in high energy physics.

Keywords: Machine learning · Neural networks
Reconstruction algorithm · Computational physics

1 Introduction

Event reconstruction is an important task in pattern recognition. The task is inherently statistical and consist on inferring an event given the data that it produced. It has applications in fields such as *high energy physics, nuclear medicine* and *astrophysics*.

In particular, in *high energy physics* event reconstruction is used to analyze the particles produced in particle accelerators. These accelerators collide particles at high velocity, producing a huge amount of secondary particles. The observation and study of the particles produced may help to understand the structure of matter and its fundamental properties. The particles produced at

© Springer International Publishing AG, part of Springer Nature 2018
M. Mendoza and S. Velastín (Eds.): CIARP 2017, LNCS 10657, pp. 491–498, 2018.
https://doi.org/10.1007/978-3-319-75193-1_59

collisions must be measured by detectors attached to the accelerators. These detectors quantify different properties of the particles, such as energy, momentum or interaction time.

To infer the particles that were produced, an event reconstruction algorithm is used. The reconstruction algorithm must process the information produced by the detector in order to identify the particles that originated the data. To do that, the algorithm must be able to distinguish between signal and background, cluster together data from each particle, separate overlapping particles, classify data from different particles, among others. Such tasks require powerful computational and statistical methods, and machine learning has become an important tool to solve many of these problems [1–3].

In this work we propose a complete reconstruction algorithm for a detector under construction at the Scientific and Technological Center of Valparaíso (CCTVal). More importantly, we propose the use of machine learning, particularly neural networks, to solve the problem of identifying overlapping particles. The detector and reconstruction algorithm can be used to identify close particles produced in collisions of electron ion colliders and also have applications in nuclear medicine.

2   Preshower Detector

The identification of neutral pions is an important problem in high energy physics. The task is difficult because pions decay into two photons with a very small opening angle. Consequently, the two photons arrive very close to each other to the detector, which makes very hard to distinguish between the decaying neutral pion and a high energy photon. The impossibility to identify neutral pions might affect the whole reconstruction process, since those particles can contribute important clues to the understanding of the underlying process.

The main limitation of commonly used detectors is the spatial resolution. To solve that problem a preshower calorimeter detector is proposed in [4]. This kind of detectors work by generating electromagnetic particle showers when an incident particle is detected. The showers are measured using a readout mechanism that quantifies the energy produced by the showers. For that reason, and given that the detector is used in front of the main detector, it is called preshower detector. The preshower detector is composed by a matrix of scintillating crystals, which are the ones that produce the electromagnetic showers. The signal of the showers is conducted by a set of optical fibers to the readout mechanism located on the sides of the front face of the crystal matrix. The readout system is composed by two vectors of measuring cells, producing two vectors of energy counts. The smaller transversal size of the crystals used in the preshower allows a better spatial resolution, helping with the identification of particles produced by neutral pions.

3   Reconstruction Algorithm

The outputs of the readout are two vectors of photoelectron counts, from where the energy can be computed. Each vector has 25 values, corresponding to each measuring cell. The vectors can be considered X-Y projections of the energy deposited on the crystal matrix. Using that information the reconstruction algorithm must be able to reconstruct the energy and position of the incident particles. The reconstruction process is composed by a series of steps which are resumed in Fig. 1. Each one of the steps are explained below:

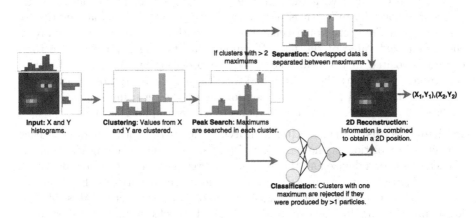

Fig. 1. Diagram of the full reconstruction process.

Clustering algorithm: The process starts with a clustering algorithm that must identify and separate each one of the incident particles. This is done by clustering together the values on the cells and assuming that each cluster corresponds to either one incident particle or two or more overlapping particles. The study of clustering algorithms is extensive and there are plenty of options to use in this step, including methods specially designed for clustering in calorimeters [5] or general pattern recognition methods [6]. We decided to use a simple, but efficient method called *topological clustering*. The method starts with a single cell, commonly the one with the maximum energy. Then, it iteratively adds the neighbors of the cells already included. Cells are only added if they are above an acceptance threshold which depends on the expected noise. Special care must be put in the process of combining clusters constructed for each one of the axes, since ambiguities may show up when two or more hits are spotted. To solve that problem we observe that clusters with similar energies in different axes were probably produced by the same particle. Using that assumption, we solved the ambiguities problem by considering all possible combinations and then choosing the ones with almost equal energy deposition.

Peak finding: In a second step, a peak finding algorithm is used to find the maximums in each one of the clusters. More than one maximums might indicate the presence of overlapping particles. We used the peak finding method proposed in [7]. The algorithm assumes that the peaks can be approximated by a normal function and that the background is piecewise linear. Then, peaks are identified using the second differences of the values, since the background is removed at the second derivative of the function.

Separation algorithm: If two or more maximums are identified in a cluster, a separation algorithm is used to set apart the overlapping particles. A separation algorithm for calorimeters is proposed in [8]. The algorithm is based on the lateral response function of the electromagnetic shower. That function relates the energy deposited in each cell with the distance from that cell to the incident particle position. Then, the function is used in an iterative algorithm that distributes the energy of the overlapping cells. The distribution is done proportionally to the normalized value of the lateral response function multiplied by the total energy deposited in the cell. The algorithm proceeds by alternating between re estimating the particle position and distributing the energy of each cell. We estimated the lateral response function for the preshower using simulations and we used the described algorithm to separate the overlapping clusters.

Position reconstruction: After identifying the energy corresponding to each particle, the position of the particle must be reconstructed. Using the location of each measuring device and the deposited energy, the reconstruction algorithm must estimate the position of each one of the detected particles. For that, we used the center of gravity of the cluster, but with logarithmic weights [9]. The logarithm accounts for the exponential decay of the electromagnetic showers.

Classification algorithm: Finally, a classification algorithm is used to reject cases in which only one maximum is observed for two incident particles. That is because the separation algorithm can only separate overlapping showers when two or more maximums are detected, however, two particles can be observed with one maximum if they are too close to each other.

4   Neural Networks for Particle Separation

In cases for which only one maximum is observed, it is hard to separate the overlapping showers. Nevertheless, it is possible to reject overlapping showers detected as single particles. In [8] the authors propose to use a cut in the second central moment of the shower, or dispersion, defined as

$$D_x = \frac{\sum E_i x_i^2}{\sum E_i} - \left(\frac{\sum E_i x_i}{\sum E_i} \right)^2. \tag{1}$$

It is observed, however, that the cut efficiency is dependent on the particle incident position respect to the center of the cell and on the energy of the particle. To solve the first issue, a parabolic cut on D_x is proposed, by defining the values

$$D_x^{corr} = D_x - D_x^{min}, \tag{2}$$
$$D_x^{min} = (\bar{x} - x_R)(\bar{x} - x_L), \tag{3}$$

where $\bar{x}$ is the first moment of the cluster and x_R, x_L are the right and left edges of the central cell of the cluster. A linear cut in D_x^{corr} is equivalent to a parabolic cut in the distribution of D_x versus $\bar{x}$. While the method is useful to solve the dependence on the incident position, it is noticed that the rejection efficiency is energy dependent even when using D_x^{corr}.

More advanced multivariate classification techniques can be used at this stage. In [10,11], the authors propose to use machine learning methods in the task of particles discrimination in calorimeters. In both papers the authors derive a set of features from the showers and then use those features to train a multivariate classification method.

For the preshower it is interesting to study the use of multivariate classification methods. Multivariate methods can: (1) Provide more complicated nonlinear cuts that might improve the rejection performance when compared to simple linear cuts. (2) Obtain a rejection efficiency that is invariant with the incident energy. The latter can be done by using the total energy as input feature. That allows the classifier to varies smoothly for different incident energies, avoiding the energy dependence of the dispersion cut. Similar ideas have been presented in [12,13]. Based on that, we propose to use a classifier trained on a set of features extracted from the cluster values. In particular, we propose to use a *multilayer perceptron* trained on 7 features extracted from the X and Y projections of the shower: the position mean and variance, the position skewness, the normalized height of the shower defined as $E_{max}/\sum E_i$ and the corrected dispersion. Each one of the features is computed for the X and Y projections obtaining a total of 14 features. Moreover, the total energy of the incident particle is included in the feature vector.

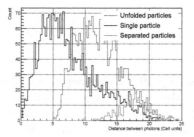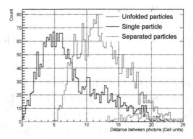

Fig. 2. Number of clusters identified as single particles, two overlapping or two separated particles given the distance between the particles. The experiments were performed for: 10 GeV + 1 GeV (left), 10 GeV + 10 GeV (right).

5   Experiments

To study the performance of the preshower detector a computational simulation of the detector was built using *Geant4* [14]. The experiments consisted on throwing photons on the front side of the preshower detector. Then, all the physics that occur in the crystals matrix and readout system were simulated. Finally, the reconstruction algorithm was applied in the simulated measures of the readout cells. Simulations were carefully calibrated in order to obtain measures that are similar to the real experimental setup. The use of simulations is needed since is the only way to know exactly the event that originated the measures.

We evaluated the performance of the reconstruction algorithm in simulated pairs of photons with uniformly distributed incident positions and with energies 10 GeV + 1 GeV and 10 GeV + 10 GeV. In Fig. 2 we show the number of particles reconstructed given the distance between both incident particles. The options are: (1) A single cluster. (2) Overlapping clusters. (3) Two separated clusters. Since all particles were simulated in pairs, the detector obtains a correct result if it identifies two overlapped or separated clusters. It is noticed that those cases are mainly observed for distances $\geq$28 mm (or equivalently 7 cell units of size 4 mm). On the other hand, for distances $\leq$20 mm most of the particles are misidentified as a single cluster.

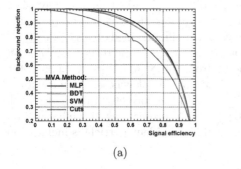
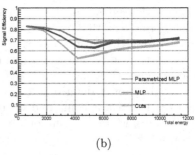

(a)   (b)

Fig. 3. (a) ROC curves for each one of the tested methods. (b) Signal efficiency for: (a) A neural network with total energy as input feature (parametrized), (b) A neural network without using the energy, (c) The cuts based method.

Next, we studied the rejection algorithm for particles with distances $\leq$20 mm, where the reconstruction algorithm fails to identify two particles. For that, we simulated single particles and pairs of particles. The algorithms were optimized to identify the cases of two particles. To evaluate the performance of the multivariate methods we will compare various machine learning algorithms (including *support vector machines, boosted decision trees* and *multilayer perceptron*) to the method based on cuts on the dispersion variable that has been previously used in [8].

For each one of the machine learning methods the features were normalized to the range $[-1, 1]$. We included the total energy for each classifier. For the *support vector machine* we used a *Gaussian* kernel with regularization $C = 1.0$. In the case of *boosted decision trees*, we used *AdaBoost* with 200 trees. We also used a maximum depth of 3 and learning rate of 0.5. For the *multilayer perceptron* we used an architecture of two hidden layers of size 15 and 5. We used *tanh* activations for the hidden layers and *sigmoid* activation for the output layer.

To train we used *stochastic gradient descent* with a learning rate of 0.02, ℓ_2 regularization and 600 epochs. The cuts were automatically selected in order to maximize the background rejection and signal efficiency, (r_B, e_S). For that, the distributiony of (r_B, e_S) distributionestimated using *monte carlo* sampling. Then, the best cut value for each one of the features was selected. It was observed that the method works poorly with many features, because of that, we used only the dispersion and corrected dispersion. All methods were implemented using TMVA [15].

The ROC curves for each one of the methods are shown in Fig. 3a. It can be seen that the best results are obtained by the *multilayer perceptron*.

Next, we studied the energy dependence of the classifiers. For that, we compared the *multilayer perceptron* with the method based on cuts and with a *multilayer perceptron* that does not includes the energy as a feature. We measured the signal efficiency for each method and for different energy ranges. Note, however, that training was made on the full range. Results are shown in Fig. 3b. While the efficiency of the cuts based method varies considerably with the energy, the performance of the classifier is more stable when the energy is used as feature.

6   Conclusions

We have presented a reconstruction algorithm for a preshower detector. We have shown that the algorithm is able to identify incident particles, separate overlapping particles and reconstruct the position and energy of the identified particles. Moreover, we have proposed a machine learning algorithm that can identify the incident particles when the reconstruction algorithm fails. We have found that the method based on a parametrized neural network outperforms the method based on cuts that is commonly used in *high energy physics*. By including the total energy of the incident particles we have reduced the dependence of the algorithm on the energy of the detected particles.

The reconstruction algorithm is general and can be extended to other calorimeter detectors. Moreover, the use of the reconstruction algorithm is not limited to *high energy physics* since these kind of detectors have important applications in other fields such as *nuclear medicine*. In future work we plan to try the algorithm on real data produced by accelerators.

Acknowledgements. This work was partially supported by the Research Project Basal FB 0821 and by Fondecyt Project 1170123.

References

1. Adam-Bourdarios, C., Cowan, G., Germain, C., Guyon, I., Kégl, B., Rousseau, D.: The Higgs boson machine learning challenge. In: NIPS 2014 Workshop on High-Energy Physics and Machine Learning, pp. 19–55 (2015)
2. Sadowski, P.J., Whiteson, D., Baldi, P.: Searching for Higgs boson decay modes with deep learning. In: Advances in Neural Information Processing Systems, pp. 2393–2401 (2014)

3. Racah, E., Ko, S., Sadowski, P., Bhimji, W., Tull, C., Oh, S.Y., Baldi, P., et al.: Revealing fundamental physics from the Daya Bay neutrino experiment using deep neural networks. In: International Conference on Machine Learning and Applications (2016)

4. Kuleshov, S., Brooks, W.: A preshower detector for forward electromagnetic calorimeters (2012). https://wiki.bnl.gov/conferences/images/b/bd/RD2012-13_Brooks_preshower_final.pdf

5. Lampl, W., Loch, P., Menke, S., Rajagopalan, S., Laplace, S., Unal, G., Ma, H., Snyder, S., Lelas, D., Rousseau, D.: Calorimeter clustering algorithms. Technical report (2008)

6. Sandhir, R.P., Muhuri, S., Nayak, T.K.: Dynamic fuzzy c-means (dFCM) clustering and its application to calorimetric data reconstruction in high-energy physics. Nucl. Instrum. Methods Phys. Res., Sect. A: Accel. Spectrom. Detect. Assoc. Equip. **681**, 34–43 (2012)

7. Morháč, M., Kliman, J., Matoušek, V., Veselský, M., Turzo, I.: Identification of peaks in multidimensional coincidence γ-ray spectra. Nucl. Instrum. Methods Phys. Res. Sect. A: Accel. Spectrom. Detect. Assoc. Equip. **443**(1), 108–125 (2000)

8. Berger, F., Bock, D., Clewing, G., Dragon, L., Glasow, R., Hölker, G., Kampert, K., Peitzmann, T., Purschke, M., Roters, B., et al.: Particle identification in modular electromagnetic calorimeters. Nucl. Instrum. Methods Phys. Res., Sect. A: Accel. Spectrom. Detect. Assoc. Equip. **321**(1), 152–164 (1992)

9. Awes, T., Obenshain, F., Plasil, F., Saini, S., Sorensen, S., Young, G.: A simple method of shower localization and identification in laterally segmented calorimeters. Nucl. Instrum. Methods Phys. Res., Sect. A: Accel. Spectrom. Detect. Assoc. Equip. **311**(1), 130–138 (1992)

10. Palma, F., Collaboration, C., et al.: Simulation studies of the expected proton rejection capabilities of CALET. In: 34th International Cosmic Ray Conference (ICRC2015), vol. 34 (2015)

11. Bellotti, R., Boezio, M., Volpe, F.: A particle classification system for the pamela calorimeter. Astropart. Phys. **22**(5), 431–438 (2005)

12. Cranmer, K., Pavez, J., Louppe, G.: Approximating likelihood ratios with calibrated discriminative classifiers. arXiv preprint arXiv:1506.02169 (2015)

13. Baldi, P., Cranmer, K., Faucett, T., Sadowski, P., Whiteson, D.: Parameterized machine learning for high-energy physics. arXiv preprint arXiv:1601.07913 (2016)

14. Agostinelli, S., Allison, J., Amako, K.A., Apostolakis, J., Araujo, H., Arce, P., Asai, M., Axen, D., Banerjee, S., Barrand, G., et al.: GEANT4 - a simulation toolkit. Nucl. Instrum. Methods Phys. Res., Sect. A: Accel. Spectrom. Detect. Assoc. Equip. **506**(3), 250–303 (2003)

15. Hoecker, A., Speckmayer, P., Stelzer, J., Therhaag, J., von Toerne, E., Voss, H., Backes, M., Carli, T., Cohen, O., Christov, A., et al.: TMVA-toolkit for multivariate data analysis. arXiv preprint physics/0703039 (2007)

Noisy Character Recognition Using Deep Convolutional Neural Networks

Sirlene Peixoto[1], Gabriel Gonçalves[2], Andrea Bianchi[1], Alceu De S. Brito[3], William Robson Schwartz[2] (iD), and David Menotti[4(✉)] (iD)

[1] Federal University of Ouro Preto, Ouro Preto, MG, Brazil
[2] Federal University of Minas Gerais, Belo Horizonte, MG, Brazil
[3] Pontifical Catholic University of Paraná, Curitiba, MG, Brazil
[4] Federal University of Paraná, Curitiba, MG, Brazil
menotti@inf.ufpr.br

Abstract. Due to degradation and low quality in noisy images, such as natural scene images and CAPTCHAs (Completely Automated Public Turing test to tell Computers and Humans Apart) based on text, the character recognition problem continues to be extremely challenging. In this work, we study two convolutional neural network approaches (filter learning and architecture optimization) to improve the feature representations of these images through deep learning. We perform experiments in the widely used Street View House Numbers (SVHN) dataset and a new dataset of CAPTCHAs created by us. The approach to learn filter weights through back-propagation algorithm using data augmentation technique and the strategy of adding few locally-connected layers to the Convolutional Neural Network (CNN) has obtained promising results on the CAPTCHA dataset (97.36% of accuracy for characters and 85.4% for CAPTCHAs) and results very close to the state-of-the-art regarding the SVHN dataset (97.45% of accuracy for digits).

Keywords: Deep learning · Convolutional network
Noisy character recognition · CAPTCHAs · Street house view numbers

1 Introduction

Automatically character recognition in noisy images such as natural scene images and CAPTCHAs based on text is still a difficult task and a recurrently researched subject. Such images usually have complex background, uneven illumination, distorted characters, multiple sources and varying sizes, which lead to misrepresentation of the data. In general, the performance of machine learning methods is highly related to the recognition system ability to extract features to maximize and minimize intra- and inter-class similarities, respectively.

Over the past few years numerous studies show the effectiveness of deep learning techniques in various fields of challenging applications. Deep learning techniques aim to gradually map the representation of raw pixel into more abstract representations. Besides success in recognizing handwritten text [11],

© Springer International Publishing AG, part of Springer Nature 2018
M. Mendoza and S. Velastín (Eds.): CIARP 2017, LNCS 10657, pp. 499–507, 2018.
https://doi.org/10.1007/978-3-319-75193-1_60

deep learning techniques have also been applied in face recognition [9], speech recognition [13], scene recognition [12] problems, among others.

Aiming at improving the feature representation and therefore the classification task, many studies have applied deep learning for character recognition in noisy images [2,7,10]. In [7], the authors compare and show the supremacy of two deep learning techniques over two hand-designed descriptors using optimized classifiers in SVHN database. In [10], the authors included a multi-stage characteristic approach and different pooling methods (Lp pooling) in the architecture of a traditional convolutional neural network, achieving an accuracy rate of 95.1% in the SVHN database. More recently, Goodfellow et al. [2] proposed an approach that integrates the three traditional steps of a recognition system (location, segmentation and recognition) by using a deep convolutional neural network, achieving an accuracy of 97.84% in the SVHN dataset.

Despite the importance of the aforementioned contributions, the recognition of noisy characters is still an open problem. In this paper, following the same direction than the literature, we apply deep learning to address such a challenging problem. However, the contribution of this paper is fourfold: (i) evaluation of two different approaches to learn/generate the filter weights during the convolution operation of the CNN, named Learning Filters and Architecture Optimization; (ii) evaluation of locally-connected layer and data augmentation for training the proposed CNN; (iii) comparison of the proposed deep learning recognition method with other approaches based on hand-designed descriptors; and (iv) a new dataset composed of noisy character-based CAPTCHA images, which is currently available for research purposes upon request.

According to the experiments based on the well-known SVHN dataset and the proposed CAPTCHA image dataset, we observe that the CNN presents higher accuracy when compared to the hand-crafted approaches evaluated. These results also indicate that the application of the two investigated techniques reduces the error rate in about 2% in the SVHN dataset. We achieved an error rate in this dataset near to the one obtained in [2] using fewer convolutional layers (three against nine). In the CAPTCHA dataset, very promising results were achieved, i.e. 97.36% of accuracy for characters and 85.4% for CAPTCHAs. Moreover, the approach that learns filter weights through back-propagation algorithm using data augmentation technique and the strategy of adding few locally-connected layers to the CNN has obtained promising results for the CAPTCHA dataset.

2 Theoretical Background

A CNN is composed of layers that usually perform a series of four sequential linear and nonlinear operations [8], such as:

1. Convolution: calculates the intensity of a given pixel according to the weighted sum of its neighbors pixels.
2. Activation: maps the output of convolution operation aiming to limit the "neuron" output.

3. Spatial Pooling: reduces the resolution of the map features. It reduces the sensitivity of the output map displacements and other forms of distortion, selecting invariant features that improve the generalization performance.
4. Normalization: usually creates competition between maps. Aims further strengthening the robustness of the CNN output feature vector.

Through these operations, CNN maps the input image to a higher-level representation that can be viewed as a multi-band image. In other words, the CNN gradually transforms the raw pixel in representations highlighting abstract concepts. This multi-band representation is later applied to a classification task.

In this work, we discuss two kinds of CNNs available in the literature that apply different approaches to learn/generate the filter weights of the convolution operation. While one uses a supervised learning algorithm [3], the other optimizes the architecture employing random filters with zero mean and unit norm [4,8]. Both approaches are briefly detailed in the next sections.

2.1 Learning Filters (LF)

In the LF approach, the convolution layer indeed performs two required operations: convolution and activation, and two additional operations: normalization and pooling. The order of these last two operations is interchangeable. The learning of the filters of the convolution layers as well as the weights of the other layers are estimated in a supervised fashion through back-propagation with the stochastic gradient descent rule to update the filters and their weights.

While the feature representation of this approach is performed by the convolutional layers, the classifier consists of one (or more) fully connected layer with n output values (in which n is the number of classes), and a multinomial logistic regression layer. Multinomial logistic regression is a classification method that generalizes logistic regression to multi-class problems by softmax function. In particular, we use the CUDA-convnet library implemented in C++/CUDA by Krizhevsky [3][1] to evaluate the LF approach.

2.2 Architecture Optimization (AO)

Each convolutional layer of a CNN is defined by n hyperparameters such the number and size of the filters related to the convolution operation, among others [8]. Considering c convolutional layers and the possible values of each of the $n \times c$ hyperparameters (the *search space*), there are a large number of possible architectures to be evaluated. An effective way to explore such *search space* is by employing the AO approach as described in [5], in which some few thousands of randomly generated architectures are evaluated and the best found convolutional topology is returned, this process is illustrated in Fig. 1.

To make the AO practical, the filter weights used to evaluate the candidate architectures are randomly generated from a uniform distribution $U(0,1)$ and normalized to zero mean and unit variance [5]. The prediction is performed by linear support vector machines (SVMs).

---

[1] https://code.google.com/p/cuda-convnet/.

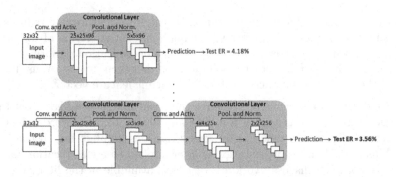

Fig. 1. AO: randomly generates thousands of architectures, evaluates the candidate architectures and returns the best convolutional architecture found.

3   Experimental Results

In this section, we present the experiments undertaken to evaluated the proposed approaches for noisy character recognition. After describing the used datasets, we present and analyze the results obtained.

3.1   Datasets

The Street View House Numbers Dataset. This image dataset consists of 73,257 digits for training, 26,032 digits for testing, and 531,131 additional digits for training. Here, we combine the training and additional sets for training. The digit images are organized into 10 classes and available in two formats, as follows: (1) original images with character level bounding boxes; and (2) images centered around a single character (cropped digits). We use the latter format as illustrated on the left of Fig. 2[2].

CAPTCHA Dataset. This dataset consists of 12,000 images of CAPTCHAs composed of 6 characters. The characters are segmented by fixed position in the images and they belong to 36 classes (26 upper case letters and 10 digits). It is divided in training and testing sets of 8k/48k and 2k/12k CAPTCHAs/characters, respectively. Two samples of this database are shown on the right in Fig. 2. This dataset is currently available for research purposes upon request.

Fig. 2. Samples of the datasets: one from SVHN and two from CAPTCHAs

---

[2]   http://ufldl.stanford.edu/housenumbers/.

3.2   Experiments

We now evaluate the effectiveness of the LF and AO approaches described in Sect. 2 for character recognition. We execute experiments applying these approaches in the SVHN and CAPTCHA datasets. More specifically, we present results for the use of locally-connected layers in CNN and the use of data augmentation in the learning process. We also compare the effectiveness of our deep learning-based approach with others based on hand-designed descriptors.

Architecture Optimization. Due to the memory space required for both feature representation and the extra space required by the linear SVM for learning in the AO approach, it was necessary to restrict the size of training set for both datasets, SVHN and CAPTCHA, to 100 samples per class, such that the best convolutional architectures, i.e., AO-SVHN and AO-CAPTCHA, respectively, could be obtained in dozens of processing hours. The hyperparameter values of the AO-SVHN and AO-CAPTCHA CNNs are detailed in Table 1, in which the value x^i in tuple $\{x^1, x^2, \ldots, x^n\}$ corresponds to the hyperparameter value of convolutional layer i and the absence of any operation in the i-th convolution layer is representing by '-'. Note that the input image size in both datasets is always 32×32 pixels.

Our first experiment evaluates the impact in the accuracy when varying the training set size to learn the linear SVM, once the architecture optimization has been performed using few samples per class due to memory constraints[3]. These results are presented in Table 2. It is observed that the error rate decreases in the SVHN and CAPTCHA datasets as the number of samples per class increases.

Learning Filters. Based on standard publicly convolutional architectures available in the CUDA-convnet library and on our knowledge domain, we define and evaluate various architectures based on LF approach. To train our CNNs, we followed the methodology of CUDA-convnet library[4]. Those performing better in the SVHN (LF-SVHN) and CAPTCHA (LF-CAPTCHA) datasets are detailed

Table 1. Hyperparameter values of the best architectures.

Operation	Convolution		Pooling		Normalization
CNN	Filters	Size	Size	Stride	Size
AO-SVHN	$\{64, 128\}$	$\{5, 3\}$	$\{3, 3\}$	$\{2, 1\}$	$\{3, -\}$
LF-SVHN	$\{64, 128, 192\}$	$\{3, 3, 3\}$	$\{3, 3, 3\}$	$\{2, 2, 2\}$	$\{9, 9, 9\}$
AO-CAPTCHA	$\{192, 192\}$	$\{3, 3\}$	$\{3, 5\}$	$\{2, 1\}$	$\{-, -\}$
LF-CAPTCHA	$\{32, 64\}$	$\{5, 5\}$	$\{3, 3\}$	$\{2, 2\}$	$\{9, 9\}$

---

[3] Our system is a Intel i7 Core, 32 GB RAM with NVIDIA GTX Geforce Titan Black GPU and the largest training set possible to be evaluated was around 20k samples.

[4] https://code.google.com/p/cuda-convnet/wiki/Methodology.

Table 2. AO results: evaluating the training set size to learn the final linear SVM.

CN	# training		# testing	test error
	Per class	Total	Images	Rate (%)
AO-SVHN	100	1,000	26,000	24.28
AO-SVHN	560	5,600	26,000	17.13
AO-SVHN	2,000	20,000	26,000	14.03
AO-CAPTCHA	100	3,600	12,000	17.85
AO-CAPTCHA	560	20,160	12,000	5.45

in Table 1. The obtained test error rates are 4.29% and 3.12% for the SVHN and CAPTCHA datasets, respectively.

Locally-Connected Layer. Locally-connected layer is similar to a convolutional layer but the former does not share weights as the latter does. That is, a different set of filter weights is applied to each portion of the input map/image, which can help to better discriminate that particular region of the character image, since different regions of a character have different local statistics. As the CUDA-convnet library supports this type of operation/layer, we investigated the influence on the effectiveness by inserting one, two, and three locally-connected layers in a CN, atop the last convolution layer. Table 3 reports the test error rate (ER) obtained, in which L corresponds to the number of locally-connected layers added. The smallest error rates were found for $L = 2$ and $L = 1$ in the SVHN and CAPTCHA datasets, respectively. Hereinafter, we call these architectures LF-LL-SVHN and LF-L-CAPTCHA.

Table 3. Test Error Rates (ER) related to the number of locally-connected layers added (L).

CNN	L	Test ER (%)
LF-SVHN	1	3.95
LF-SVHN	2	3.84
LF-SVHN	3	3.95
LF-CAPTCHA	1	2.73
LF-CAPTCHA	2	2.91
LF-CAPTCHA	3	2.98

Table 4. Data augmentation. Test Error Rate (ER) for different values of crop borders.

CNN	crop_border	Test ER (%)
LF-LL-SVHN	1	2.75
LF-LL-SVHN	2	**2.55**
LF-LL-SVHN	4	2.75
LF-L-CAPTCHA	1	**2.64**
LF-L-CAPTCHA	2	2.91
LF-L-CAPTCHA	4	4.01

Data Augmentation. The data augmentation technique strikes overfitting by artificially expanding the dataset. Our data augmentation consists of generating horizontally translated images from the original images. Let I be an image of $n \times n$ pixels. From I, squared sub-regions (and their horizontal reflections) of

Table 5. Comparison of the proposed CNNs with classifiers based on hand-designed descriptors

Dataset	Test error rates (%)			
	Oblique Random Forest	Linear SVM	RBF SVM	Proposed CNN
SVHN	21.9	43.5	39.7	2.55
CAPTCHA	13.8	37.6	24.2	2.64

$n - 2 \times crop\_border$ order are generated, in which $2 \times crop\_border$ specifies the width of the cropped region in pixels. We investigated the influence of data augmentation on the effectiveness of the LF-LLL-SVHN and LF-L-CAPTCHA by varying the $crop\_border$ parameter. According to Table 4, different values of $crop\_border$ obtain better results for the two datasets.

Comparison with Other Approaches Based on Hand-Designed Descriptors. To compare the deep learning approaches with other approaches based on hand-designed descriptors, we perform experiments in both datasets using the hand-designed descriptor *Histograms-of-Oriented-Gradients* (HOG) [1] with 8 bins and 4 different cell sizes: 16×16, 8×8, 4×8 and 8×4. All cell have stride set up to 50%. To the prediction, we use three classifiers. The first is the *Oblique Random Forest* (oRF) method with SVM on each node [6]. The oRF is an approach that utilizes another classifier to split the data on each nodes of its trees instead of perform using just one feature as custom Random Forest does. The other two classifiers are SVM with a Linear kernel and SVM with a RBF kernels. The results are reported in Table 5. The smallest error rates are obtained using the oRF classifier, but its effectiveness are far away from the ones obtained by the CNNs using learning filters.

Another important topic to be described is the processing time to perform these classifications. In the case of the Oblique Random Forest with a SVM on each node, which proved to be the best classifier of all three baselines, it takes more than twenty hours to execute the learning phase in both datasets, reaching up to fourteen hours in the case of the SVHN dataset. All SVMs utilized on each node are Linear and the parameter C is 0.01, which was determined empirically. The linear SVM has a learning-time of 1 h to SVHN dataset and about 30 min to the other dataset. The SVM with RBF kernel executes the learning in three hours in SVHN dataset, ninety minutes in the Captcha dataset. On the other hand, the CNN LF-LL-SVHN and CNN LF-L-CAPTCHA with data augmentation takes in average nine hours and one hour, respectively, to perform the learning, which is a much smaller time compared to the ORF baseline.

3.3   Discussion

The results show that deep learning approaches are far superior than the ones using hand-designed descriptors in the two datasets of characters evaluated.

The LF approach has obtained the smallest error rates in both datasets. We believe that the use of complete and large sets of training samples contributed greatly to this result. For the SVHN dataset, the more complex, the LF-SVHN did not converged during the training (underfitting) when we used training sets of size similar to the ones used in AO (100, 560 and 2000 samples per class). When training the LF-CAPTCHA with the first two sets of training (100 and 560 samples per class), we obtained the respective error rates of 29.50% and 4.09% and the AO-CAPTCHA obtained 17.85% and 5.45%, respectively (see Table 2). Note that AO approach was able to use 3.5% and 42% of the entire training sets of the SVHN and CAPTCHA databases, respectively.

4   Conclusions

In this paper we employed deep learning approaches and hand-designed descriptors for character recognition problem in noisy images. In our experiments, we use the CAPTCHA dataset, proposed by us, and the widely used SVHN dataset. Our experiments demonstrated that the most promising results have been obtained by LF approach. Furthermore, the use of data augmentation and the addition of locally-connected layers to CN reduce the error rates.

For future work, we intend to independently evaluate the influence of the classifiers on the results obtained by deep learning representations, since AO employs linear SVM and the classifier used by the LF approach is based on the fully connected layers and softmax regression.

Acknowledgments. The authors thank UFPR, PUCPR, UFOP, UFMG, FAPEMIG, CAPES and CNPq (Grants #428333/2016-8 & # 307010/2014-7) for supporting this work.

References

1. Dalal, N., Triggs, B.: Histograms of oriented gradients for human detection. In: IEEE Conference on Computer Vision and Pattern Recognition (CVPR), vol. 1, pp. 886–893 (2005)
2. Goodfellow, I.J., et al.: Multi-digit number recognition from street view imagery using deep CNNs. In: International Conference on Learning Representation (2014)
3. Krizhevsky, A., Sutskever, I., Hinton, G.E.: ImageNet classification with deep convolutional neural networks. In: NIPS, pp. 1097–1105 (2012)
4. Menotti, D., Chiachia, G., Falcao, A., Oliveira Neto, V.: Vehicle license plate recognition with random convolutional networks. In: 2014 27th SIBGRAPI Conference on Graphics, Patterns and Images (SIBGRAPI), pp. 298–303 (2014)
5. Menotti, D., et al.: Deep representations for iris, face, and fingerprint spoofing detection. IEEE Trans. Inf. Forensics Secur. **10**(4), 864–879 (2015)
6. Menze, B.H., Kelm, B.M., Splitthoff, D.N., Koethe, U., Hamprecht, F.A.: On oblique random forests. In: Gunopulos, D., Hofmann, T., Malerba, D., Vazirgiannis, M. (eds.) ECML PKDD 2011. LNCS (LNAI), vol. 6912, pp. 453–469. Springer, Heidelberg (2011). https://doi.org/10.1007/978-3-642-23783-6_29

7. Netzer, Y., et al.: Reading digits in natural images with unsupervised feature learning. In: NIPS Workshop on Deep Learning and Unsupervised Feature Learning, pp. 1–9 (2011)
8. Pinto, N., et al.: A high-throughput screening approach to discovering good forms of biologically inspired visual representation. PLoS **5**(11), e1000579 (2009)
9. Schroff, F., Kalenichenko, D., Philbin, J.: FaceNet: a unified embedding for face recognition and clustering. arXiv preprint arXiv:1503.03832 (2015)
10. Sermanet, P., Chintala, S., LeCun, Y.: Convolutional neural networks applied to house numbers digit classification. In: ICPR, pp. 3288–3291 (2012)
11. Vajda, S., Rangoni, Y., Cecotti, H.: Semi-automatic ground truth generation using unsupervised clustering and limited manual labeling: application to handwritten character recognition. Pattern Recogn. Lett. **58**, 23–28 (2015)
12. Yuan, Y., Mou, L., Lu, X.: Scene recognition by manifold regularized deep learning architecture. IEEE Trans. Neural Netw. Learn. Syst. **26**(10), 1–12 (2015)
13. Zhang, X., Trmal, J., Povey, D., Khudanpur, S.: Improving deep neural network acoustic models using generalized maxout networks. In: IEEE International Conference on Acoustics, Speech and Signal Processing (ICASSP), pp. 215–219 (2014)

A Flexible Framework for the Evaluation of Unsupervised Image Annotation

Luis Pellegrin[1] ⓘ, Hugo Jair Escalante[1](✉) ⓘ, Manuel Montes-y-Gómez[1],
Mauricio Villegas[2], and Fabio A. González[3]

[1] Computer Science Department, Instituto Nacional de Astrofísica,
Óptica y Electrónica (INAOE), Tonantzintla, Mexico
{pellegrin,hugojair,mmontesg}@inaoep.mx
[2] SearchInk, Koppenplatz 10, 10115 Berlin, Germany
[3] Computing Systems and Industrial Engineering Department, MindLab,
Universidad Nacional de Colombia, Bogotá DC, Colombia

Abstract. Automatic Image Annotation (AIA) consists in assigning keywords to images describing their visual content. A prevalent way to address the AIA task is based on supervised learning. However, the unsupervised approach is a new alternative that makes a lot of sense when there are not manually labeled images to train supervised techniques. AIA methods are typically evaluated using supervised learning performance measures, however applying these kind of measures to unsupervised methods is difficult and unfair. The main restriction has to do with the fact that unsupervised methods use an unrestricted annotation vocabulary while supervised methods use a restricted one. With the aim to alleviate the unfair evaluation, in this paper we propose a flexible evaluation framework that allows us to compare coverage and relevance of the assigned words by unsupervised automatic image annotation (UAIA) methods. We show the robustness of our framework through a set of experiments where we evaluated the output of both, unsupervised and supervised methods.

1 Introduction

AIA consists in describing an image by keywords related to its visual content. Traditional AIA methods are based on supervised learning, where labeled images are used to train models in which labels are taken as classes, then the model can assign labels from the learnt models to new images. Recently, advances in deep learning have increased considerably the performance in tasks such as image classification [1] and object recognition [2]. However, supervised approaches are still limited to assign only the labels available in the training dataset.

As an alternative to the supervised approach to AIA, the so called unsupervised approach has recently emerged [3]. Here, the dataset with labeled images is replaced by a reference collection of image-text pairs where the image is immersed in textual information. An unsupervised method uses this unstructured information in order to annotate new images. This is a more challenging

© Springer International Publishing AG, part of Springer Nature 2018
M. Mendoza and S. Velastín (Eds.): CIARP 2017, LNCS 10657, pp. 508–516, 2018.
https://doi.org/10.1007/978-3-319-75193-1_61

task because the labels to assign need to be mined from the textual part. However, the scenario has as advantage the availability and easy access to huge collections of images with associated text from different sources such as wikipedia, social networks and, in general, the web. Furthermore, any word mined from the text in the reference collection can be used to describe the visual content of images.

These two approaches of AIA work in different scenarios. On the one hand, supervised approaches are competitive working with a fixed number of labels, which makes them appropriate for specific domains. However, supervised methods rely on huge labeled datasets which are only possible for a limited number of labels, e.g. ImageNet (http://www.image-net.org/) formed by labeled images with presence or absence of 1000 object categories. On the other hand, unsupervised approaches are not restricted to a fixed number of words to assign because these can be mined from the reference collection. Notwithstanding, both approaches are evaluated under a supervised scenario using a fixed vocabulary of labels, e.g. see ImageCLEF (http://imageclef.org/). In Fig. 1 the current evaluation is exemplified, the output of three annotation systems are compared, two of them unsupervised. Unsupervised methods labeling images with words extracted from a free vocabulary. We can see that even though all assigned words from unsupervised methods seem to be relevant to annotate the evaluated image, only those words matching exactly with the ground truth are taken into account by supervised evaluation metrics.

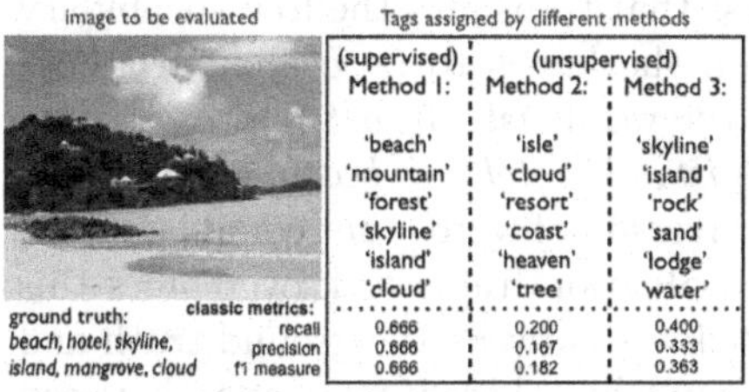

Fig. 1. Current evaluation framework.

	ground truth					
output	island	beach	mangrove	skyline	hotel	cloud
isle	0.923	0.615	0.400	0.571	0.500	0.400
cloud	0.428	0.428	0.285	0.400	0.352	1.000
resort	0.588	0.571	0.434	0.666	0.947	0.500
coast	0.615	0.769	0.400	0.571	0.500	0.400
heaven	0.625	0.615	0.400	0.714	0.500	0.400
tree	0.470	0.470	0.916	0.444	0.500	0.500
match	isle	coast	tree	heaven	resort	cloud

Fig. 2. Semantic relatedness similarity among labels

Due to this unfair evaluation, the effectiveness of unsupervised methods is not really known, since classic metrics do not capture the relevance of output words from free-vocabularies. With the aim to provide a more suitable evaluation that serves to encourage new developments of unsupervised automatic image annotation (UAIA) methods, in this paper we propose a flexible evaluation framework that allows us to compare coverage and relevance of the assigned labels from a free-vocabulary. We introduce new metrics that adapt classic ones such as recall, precision and F1, to deal with approximate matchings between predicted labels and the ground truth. The proposed framework defines a scenario where it is possible to evaluate the assignments from free-label vocabularies, it makes possible to use labeled or unlabeled datasets for training phase, and taking advantage of available resources for AIA, where we only need a labeled test set. We used the proposed framework for performing a comparison between supervised and unsupervised AIA methods. The experimental results show that the proposed framework is a good alternative for evaluating supervised and unsupervised AIA methods under fair conditions by taking into account the semantic relatedness of assigned and ground truth labels.

As far as we know there is not much work on evaluation of UAIA methods. The most related contributions are directed to automatic description generation [4]. However, this task goes beyond of describing isolated parts of the image. It involves generating a textual description that verbalizes the content of the image. For evaluating descriptions, there exists measures that compute a score indicating the semantic similarity between the description generated by a system and a set of human-written reference descriptions. One of the most representative measures is Meteor (http://www.cs.cmu.edu/~alavie/METEOR/) that allows matchings using exact and synonyms, exhibiting higher correlation with human judgments.

The remainder of the paper is organized as follows: Sect. 2 describes our evaluation framework; Sect. 3 reports experimental results. Finally, Sect. 4 presents some conclusions of this work.

2   Evaluation Framework for UAIA

To overcome the limitation in evaluation, many systems that follow an unsupervised approach have to perform a process that translates the free vocabulary extracted from the reference collection into the labels to be evaluated (those present in the ground truth). For instance, the words *isle, resort*, and *coast* from Method 2 in Fig. 1 need to be translated to *island, hotel* and *beach*, respectively. However, this kind of translation limits the diversity of words that could be mined on the reference collection. Our proposed evaluation framework overcomes this limitation by measuring the *semantic* relatedness between the ground truth and the words assigned by the UAIA method, even when there is no overlap between both sets of words. In this sense, the assigned labels by the system can refer to everything related with the visual content in the image. The proposed evaluation framework follows two steps:

1. *Measuring the semantic relatedness.* This is calculated between all words in the ground truth and all output words from the UAIA method. The idea is to assign a score that captures the semantic relatedness of each word pair. In this way, it is possible to identify a best approximated matching for each word in the ground truth. For instance, in Fig. 2, the semantic relatedness between the output words of an unsupervised method and all words in the ground truth is calculated, where a score of 1 means exact matching.
2. *Measuring coverage and relevance.* Having the best approximated matching from the previous step, we conceived a way to measure coverage and relevance by adapting three classic measures: recall, precision and F1 measure. More details about the adapted metrics will be provided in Subsect. 2.1.

2.1 Measuring Coverage and Relevance in UAIA Methods

In order to decide whether a word from the ground truth is covered and which words assigned by system are relevant, we have adapted classic metrics by establishing a threshold α and taking into account the similarity between words. This α regulates the hardness matching among words, an α value of 1 is an exact match, whereas a value of 0.2 refers to a matching with low semantic relatedness. Note that with the definition of α is possible to relax the matching among words as much as necessary. Accordingly, the new metrics for evaluation of UAIA methods are as follows:

- Recall$_\alpha$ (coverage). Unlike classic recall, for this adaptation to reach a maximum score implies that for each word from the ground truth, there exists a matching with a word from the output system, such that this matching is greater than α.

$$R_\alpha = \frac{\sum_{w_i \in gt} 1_{(\exists w_j : w_j \in outs \land sim(w_i, w_j) \geq \alpha)}}{|gt|} \tag{1}$$

where gt refers to words from ground truth, and $outs$ are the output words given by the annotation system, $sim(w_i, w_j)$ expresses the semantic relatedness between two words and can be estimated by either one of the two approaches defined in Subsect. 2.2.

- Precision$_\alpha$ (relevance). In this measure, the effectiveness of the output words is given by the number of words that surpass the α threshold:

$$P_\alpha = \frac{\sum_{w_j \in outs} 1_{(\exists w_i : w_i \in gt \land sim(w_i, w_j) \geq \alpha)}}{|outs|} \tag{2}$$

The precision$_\alpha$ complements the coverage with a score that expresses the quality of the words given by the system in relation to the quantity.

- F1$_\alpha$. This metric combines recall$_\alpha$ and precision$_\alpha$ by using a harmonic mean:

$$F1_\alpha = 2 \cdot \frac{R_\alpha \cdot P_\alpha}{R_\alpha + P_\alpha} \tag{3}$$

Furthermore, the relaxation defined by the α threshold can be used to adapt other metrics such as MAP (Mean Average Precision) that is usually used to evaluate the ranked list of words that are output of the AIA methods. Accordingly the AP (Average Precision) is defined as:

- AP$_\alpha$. This adapted metric makes use of the approximated matching defined by the α threshold in P$_\alpha$, so it is not necessary to define a function as relevance indicator unlike the original metric:

$$AP_\alpha = \frac{\sum_{k=1}^{|outs|} P_\alpha(k)}{|gt|} \tag{4}$$

where k is the word to evaluate and

$$P_\alpha(k) = \frac{\sum_{j=1}^{k} 1_{(\exists w_i : w_i \in gt \land sim(w_i, w_j) \geq \alpha)}}{|k|} \tag{5}$$

Retaking the example of Fig. 1, using the proposed evaluation metrics. When the scores are reported at $\alpha = 0.7$, the method1 obtains $R_\alpha = 0.66$, $P_\alpha = 1$ and $F1_\alpha = 0.79$, whilst method2 obtains $R_\alpha = 1$, $P_\alpha = 1$ and $F1_\alpha = 1$ and method3 obtains $R_\alpha = 0.5$, $P_\alpha = 0.66$ and $F1_\alpha = 0.57$. Under this new evaluation, both unsupervised methods obtained competitive results and they can be compared with the scores obtained by the supervised method, even when the words do not exactly correspond with the words of the ground truth.

2.2   Measuring the Semantic Relatedness

Measuring semantic relatedness between two words is a well known task in natural language processing. This relatedness is generally established by context around the words, and it can be produced by knowledge-based approaches, e.g. using a lexical database like WordNet (https://wordnet.princeton.edu/), or by statistical-based approaches, e.g. distributed representations [5]. We describe these two approaches in the following paragraphs:

Knowledge-Based Approach. For instance, WuP metric [6] calculates relatedness using the depths of two words in WordNet by counting their number of nodes, along with the depth of the lower common subsumer (lcs).

Statistical-Based Approach. Distributed representations [5] allow representing words by vectors in a $\mathbb{R}^r$ space. The final purpose is to build accurate representations of words that semantically associated words with similar meaning by using context around the words. A cosine similarity between word vectors is used.

3   Experiments and Results

This section is divided into two parts. The first part describes the used dataset, the evaluated methods, and the experimental settings. While in the second part, the obtained results are reported and discussed.

3.1   Dataset, Methods and Experimental Settings

Dataset. We considered a benchmark used in the large concept annotation task at ImageCLEF for year 2013 [7]. This dataset was created with the aim of being exclusively used for UAIA methods, it provides a reference collection with 250,000 documents. Each document is composed by an image, and a web page that provides textual information. We used the recently released test set for evaluation. This contains 2000 images annotated with 116 different concepts.

Evaluated AIA Methods. We used two UAIA methods [3] that were developed to annotate any word extracted from a reference collection. Both methods exploit the information differently, also they are flexible to use visual descriptors and fixed the number of words that can be assigned as annotations. In addition, we have evaluated a supervised method [8] that uses a model trained by context dependent support vector machines (SVM) for each concept in the ground truth. This method has obtained the best results in the considered dataset. Following a traditional classification problem this supervised method uses the development set for training models for each concept.

Experimental Settings

- We used the statistical-based approach for calculation of semantic relatedness. The learned vectors were obtained with word2vec [5] using Wikipedia for training. We performed the same experiments using the WuP metric from WordNet obtaining similar results.
- In order to reduce the low variability over the semantic relatedness similarity values, the reported results are normalized values using z score. This helps to follow a more proportional and intuitive scale close to human perception. This scores are then mapped to $[0, 1]$.
- In order to report a fair comparison between supervised and unsupervised methods, the quantity of words used by unsupervised methods is equal to the maximum of words labeled by the supervised system (20 labels).

3.2 Evaluating Proposed Metrics

Table 1 shows the results reported for ImageCLEF 2013 dataset on the test set using classic metrics for different AIA methods under two scenarios. On top are reported results obtained by considering a free vocabulary, while on the bottom part are reported results obtained by considering only labels in the test set.

Table 1. Classic metrics used in AIA - TWO SCENARIOS

Scenario	Metrics	Methods		
		Supervised	Local	Global
Free Vocabulary	Recall	—	0.141	0.110
	Precision	—	0.101	0.018
	F1	—	0.103	0.030
	MAP	—	0.135	0.018
In Set Labels	Recall	0.470	0.439	0.456
	Precision	0.462	0.252	0.134
	F1	0.425	0.287	0.194
	MAP	0.443	0.362	0.269

For this last case, unsupervised methods (called *Local* and *Global*) reveal competitive recall. However, using free vocabulary, note that the scores reported by unsupervised methods have been drastically reduced due to almost null overlap with the ground truth. In the following, we evaluate the same systems with our proposed metrics, under the free vocabulary scenario. Our aim is to show that output words could be relevant although these do not have an exact overlap with words from the ground truth. Figure 3 shows the $F1_\alpha$, we can see at $\alpha = 0.8$ that the unsupervised methods have similar results. However, at the lowest levels there exist a large separation, thus making possible a fairer comparison. For knowing how good are the words at different levels of α threshold see examples in Subsect. 3.3. In Fig. 4, the MAP_α allows us to measure the ranking given by annotation systems. From this figure we can see that the difference between the unsupervised methods at $\alpha = 0.6$ is reduced, this indicates that both systems are competitive including at top the most relevant words for annotating images.

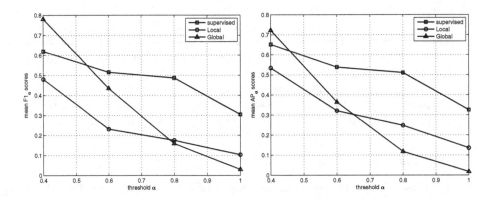

Fig. 3. Evaluation of $F1_\alpha$ measure.   **Fig. 4.** Evaluation of MAP_α measure.

Recapitulating on the last two figures, these have shown that the supervised method maintains its performance and it can be better compared to unsupervised methods without misplace its competitiveness.

3.3   How Relevant Are the Words at Different α Thresholds?

This subsection presents a qualitative analysis of the words obtained at different levels of the α threshold. Figure 5 show images with their ground truth and some of the words generated by unsupervised methods. Each column shows the words between the ranges of the α thresholds, the idea is to allow the reader to judge the relevance when using a free vocabulary for labeling images.

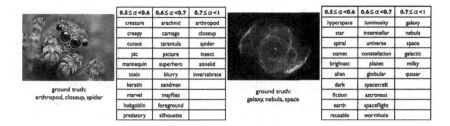

Fig. 5. Words obtained at different levels of α threshold

4 Conclusions

In this paper, we have proposed a flexible framework for evaluation of UAIA methods. It aims at mitigating the unfair evaluation of UAIA methods with respect to supervised methods, and to boost new developments. In this sense, our proposed framework allows us to compare the output labels of UAIA systems without restricting the number of assigned words or fixing the vocabulary of annotation. The scenario of evaluation is that it can used on labeled or unlabeled datasets, taking advantage of available resources. The preliminary results show that it is possible to compare supervised and unsupervised methods. In this regard, we have showed that the performance of supervised methods is not affected or misplaced. Besides, we showed that unsupervised approaches can be competitive, also when evaluating supervised and unsupervised the difference is reduced. Furthermore, it is possible relaxing the relevance of the output words from unsupervised methods in order to increase diversity of image annotation.

Acknowledgment. This work was supported by CONACYT under project grant CB-2014-241306 (Clasificación y recuperación de imágenes mediante técnicas de minería de textos). And, also by the CONACyT with scholarship No. 214764.

References

1. Krizhevsky, A., Sutskever, I., Hinton, G.E.: Imagenet classification with deep convolutional neural networks. In: Pereira, F., Burges, C., Bottou, L., Weinberger, K. (eds.) Advances in Neural Information Processing Systems 25, pp. 1097–1105 (2012)
2. Girshick, R., Donahue, J., Darrell, T., Malik, J.: Rich feature hierarchies for accurate object detection and semantic segmentation. In: CVPR, pp. 580–587 (2014)
3. Pellegrin, L., Escalante, H.J., Montes-y Gómez, M., González, F.A.: Local and global approaches for unsupervised image annotation. Multimedia Tools Appl. **76**(15), 16389–16414 (2017). https://doi.org/10.1007/s11042-016-3918-9. ISSN 1573-7721
4. Bernardi, R., Çakici, R., Elliott, D., Erdem, A., Erdem, E., Ikizler-Cinbis, N., Keller, F., Muscat, A., Plank, B.: Automatic description generation from images: a survey of models, datasets, and evaluation measures. CoRR abs/1601.03896 (2016)
5. Mikolov, T., Sutskever, I., Chen, K., Corrado, G., Dean, J.: Distributed representations of words and phrases and their compositionality. In: NIPS, pp. 3111–3119 (2013)

6. Wu, Z., Palmer, M.: Verbs semantics and lexical selection. In: ACL, pp. 133–138 (1994)
7. Villegas, M., Paredes, R.: 2013 imageCLEF WEBUPV collection. Zenodo. http:// doi.org/10.5281/zenodo.257722. Accessed 06 Nov 2017
8. Sahbi, H.: CNRS - telecom paristech at imageCLEF 2013 scalable concept image annotation task: winning annotations with context dependent SVMS. In: CLEF 2013 Evaluation Labs and Workshop, Online Working Notes, pp. 1–12 (2013)

Unsupervised Local Regressive Attributes for Pedestrian Re-identification

Billy Peralta[1]([✉]), Luis Caro[1], and Alvaro Soto[2]

[1] Catholic University of Temuco, Temuco, Chile
{bperalta,lcaro}@uct.cl
[2] Pontifical Catholic University of Chile, Santiago, Chile
asoto@ing.puc.cl

Abstract. Discovering of attributes is a challenging task in computer vision due to uncertainty about the attributes, which is caused mainly by the lack of semantic meaning in image parts. A usual scheme for facing attribute discovering is to divide the feature space using binary variables. Moreover, we can assume to know the attributes and by using expert information we can give a degree of attribute beyond only two values. Nonetheless, a binary variable could not be very informative, and we could not have access to expert information. In this work, we propose to discover linear regressive codes using image regions guided by a supervised criteria where the obtained codes obtain better generalization properties. We found that the discovered regressive codes can be successfully re-used in other visual datasets. As a future work, we plan to explore richer codification structures than lineal mapping considering efficient computation.

Keywords: Attribute discovery · Pedestrian re-identification
Unsupervised learning

1 Introduction

An important task in computer vision is the recognition of specific people considering semantic attributes. As an application, we could search for people with similar characteristics to a particular person, or we could find visual similarities between two people. An usual scheme is considering attributes as binary variables, where the attribute codes gives the answer to question: *Has the image such attribute?* On the other hand, we can consider multiple-valued attributes where they are more informative than binary attributes which should lead to better visual recognizers.

For such reason, we propose to use a variation of relative attributes (Parikh and Grauman, [1]), where the attributes consider an order relationship instead

B. Peralta—This author was partially funded by FONDECYT Initiation grant 11140892.

© Springer International Publishing AG, part of Springer Nature 2018
M. Mendoza and S. Velastín (Eds.): CIARP 2017, LNCS 10657, pp. 517–524, 2018.
https://doi.org/10.1007/978-3-319-75193-1_62

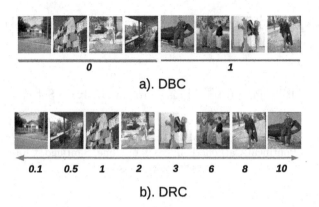

Fig. 1. (a) DBC roughly divides data considering the presence of the attribute, which can be seen as "green-cloth". (b) On the other hand, DRC estimates the degree of a particular objectness, in this case "green-cloth-ness" of image. Note that DRC codification is only guided by a joint optimization of codes, therefore it obtains a richer codification than the given by DBC.

of a binary criteria. In our work, we associate an arbitrary ordinal codification to the set of codes located on particular image regions; this codification implies that it exists a hidden and meaningful regressive localized relationship between classes. These ordinal codifications are not evident; but for example we can imagine a 't-shirt degree' for the mid part of a clothed person. Our goal is learning pedestrian-specific mid-level features as regression codes. In contrast to *Discriminative Binary Codes* (DBC) [2], we associate the classifiers to some specific region of image and they are more expressive as we use real values to codes. On the other hand, [1] uses a attribute-labelled datasets and they use a discrete code based on ranks; in our case, we have not access to attribute information and we also obtain a more flexible codification than discrete codification. As we use regression in our proposed codification, we call our technique as *Discriminative Regression Codes* (DRC).

The attribute-based techniques show similarities to mid-level features based techniques, however, they are different in that while the first obtains a compressed coding of data, the second simultaneously obtains a classifier with a mid-level codification. For example, Lobel et al. [3] propose to jointly learn mid-level and top-level features and the classifier for object recognition tasks. Their optimization model is oriented to discriminate between different classes using mid-level features with a max-margin approach. In contrast, a standard attribute learning model is oriented to the discovery of attributes, where the codification of samples with the same class must be similar and the codification of samples with different classes must be different; that is, the class information is used laterally. Moreover, an image is codified using a fixed set of attributes, which is expected to be smaller than the original features sets.

The difference between DBC and DRC techniques can be visualized in Fig. 1. While DBC learns a binary code in a unsupervised fashion, we observe that it loses valuable information. On the other hand, DRC is designed to assign a code oriented to measure a particular attribute of an image. In the figure, this attribute is the "green-cloth-ness" degree. In this paper, we make the following contributions: (1) We present a method to learn unsupervised discriminative regressive codes (DRC); and (2) we apply our method in person re-identification problems considering a transfer learning approach.

2   Our Approach

2.1   Proposed Method

We assume a training set (x_i, y_i) with N records where the index $i \in 1, \ldots, N$. Each training record is represented by a low-level d visual features, $x_i \in \Re^d$, and the labels $y_i \in 1, \ldots, C$ where C is the number of classes. Our objective is to represent each low-level feature x_i by an intermediate attribute representation $R^i = [R_1^i, R_2^i, \ldots, R_K^i]$ as we consider K regression codes with $K \ll d$ which generates a compressed data representation. To train each k-th regression code, we need to learn the output for each training record. The output is given by the prediction of regression given by support vector machine (SVM) models over each training record. Our proposed method mainly comprehends two stages: (i) Attribute initial selection and (ii) Attribute training.

(i) Attribute initial selection
Initially, we consider a simple strategy that consists of randomly assigning regression codes for all training records, but we found that this method has poor results i.e. typically with less of 5% to 10% of accuracy respect to the best found method of initialization. This reveals that it is necessary a careful initialization in order to obtain an reliable model.

Our procedure assumes that the number of codes is L. First, we apply PCA to training dataset and store the L principal components. Then, we apply these components to data in order to obtain a codification with L components. Finally, we normalize the data considering the range $[0, 1]$. We believe that there are better ways to initialize the codes and we left them as future research avenues. It is relevant because the initialization appears to be a decisive factor of accuracy.

(ii) Attribute training
Considering as initial code values given by PCA procedure, we jointly train all them using the class information. This modelling is challenging because the attributes are expected to be interdependant. Particularly, we propose the following optimization based on max-margin models for regression SVMs:

$$\min_{w,\xi,\gamma,L,B} \frac{1}{2} \sum_{c \in 1:C} \sum_{m,n \in c} d(R_m, R_n) + \lambda_1 \sum_{s \in 1:K} \|w^s\|^2$$

$$+ \lambda_2 \sum_{\substack{s \in 1:K \\ i \in 1:N}} (\xi_i^{s+} + \xi_i^{s-}) - \lambda_3 \sum_{\substack{c' \in 1:C \\ p \in c'}} \sum_{\substack{c'' \in 1:C \\ q \in c'' \\ c' \neq c''}} d(R_p, R_q)$$

s.t.

$$R_i^s - w^s diag(Z_s)x_i \leq \epsilon - \xi_i^{s+}, \ \forall i \in X, \forall s \in 1:K$$
$$w^s diag(Z_s)x_i - R_i^s \leq \epsilon - \xi_i^{s-}, \ \forall i \in X, \forall s \in 1:K$$
$$R_i^s = w^s diag(Z_s)x_i, \ \forall i \in 1:N$$
$$\xi_i^{s+}, \xi_i^{s-} > 0, \ \forall i \in 1:N, \forall s \in 1:K \tag{1}$$

In the previous equation appears some terms which will be explained as following. $\sum_{c \in 1:C} \sum_{m,n \in c} d(R_m, R_n)$ represents the intraclass mid-level regressive distance. The idea is that in the same group, the regression distance between mid-level codifications should be small. In the case of $\sum_{s \in 1:K, i \in 1:N} (\xi_i^{s+} + \xi_i^{s-})$, it represents the slack from a typical max-margin regression model. The term $\sum_{c' \in 1:C, p \in c'} \sum_{c'' \in 1:C, q \in c'', c' \neq c''} d(R_p, R_q)$ measures the interclass mid-level distance i.e. this terms represents the discriminativeness of regressive attributes classifiers. The intuition is that records of different classes should have different codifications.

Finally, the variable $Z_s \in \Re^d$ represents the region of data where the $s - th$ attribute is localized. Z selects some of them by considering a binary vector with γ active features ($\gamma < d$). We use the diagonalization operation $diag(Z_s)$ obtaining a matrix of dimensionality $d * d$ in order to appropriately interact with w^s and x^i as they have dimensionalities $1 * d$ and $d * 1$, respectively. Given that the possible number of combinations to explore is exponential (d^γ), we consider a fixed set of regions.

For the optimization of variable R, we consider as metric the classic Euclidean distance. In this case, we also apply a stochastic gradient optimization over R. The algorithm firstly initializes the regressive codes R following the procedure given in previous Subsection. Then, we proceed by iterating sequentially the following three steps. **First**, we optimize R for fixed W_i and ξ_i^s using stochastic gradient descent; this step improves the code discriminativity considering separation between different classes and it is started with the current value of R. **Second**, we update W_i and ξ_i^s by fixing R and Z and training linear SVMs over them. **Third**, the resulting set of SVMs is used to predict an improved version of regressive codes R, and we continue this process until to reach convergence of resulting codifications or the energy function is not minimized.

While this optimization procedure does not guarantee cost function descent in each iteration due to non-convex nature, our experiments show that we get convergence of solutions in practice. Furthermore, we find that the accuracy in test set is typically increased with the number of iterations. We fix the parameters for all datasets; λ_1 is setted to $1e-3$, λ_2 with λ_3 are setted to 0.1 where λ_3 is normalized for the size of categories.

3   Experiments

Our method is tested considering four public re-identification datasets and one new re-identification datasets, which corresponds to typical pedestrian datasets. The public pedestrian datasets are ETHZ, CAVIAR, ILIDS and VIPER datasets, while the private dataset is DCC. ETHZ dataset consists on 146 persons with an average of 30 images for person. CAVIAR dataset consists on 72 persons with an average of 10 images by person. ILIDS dataset consists on 119 persons with 3 images by person. Both, CAVIAR and ILIDS have image under multiple views. VIPER datasets consists on 632 persons with 2 images for person. Our own pedestrian dataset, DCC, is obtained from an indoor environment in an university campus and consists on 16 persons and an average of 30 images by person. Before to extract the feature, we rescale the images to dimensions 128×64 in order to uniformize the analysis of images. We use CAFFE features [4] which are based on convolutional neural networks. In detail, we apply the public CAFFE feature extractor over the images and apply it over a sliding window with dimensions 100×50 and we use a grid of 3×3 overlapping regions obtaining $L = 9$ regions (Table 1).

Table 1. Datasets details.

Dataset name	# images	# subjects	Type
ETHZ	4500	150	Pedestrian
CAVIAR	720	72	Pedestrian
ILIDS	357	119	Pedestrian
DCC	480	16	Pedestrian
VIPER	1264	632	Pedestrian

3.1   Transfer Learning

Our main experiment is oriented to a transfer learning task, which consists on learning attributes from a known set of classes (training dataset), and then testing such learned attributes over a new set of a-priori unknown classes (testing dataset). We consider all datasets, except VIPER because it has only two images by class, therefore, is not feasible to follow the proposed evaluation scheme.

First, we assume the access to low-level feature dataset of N records, $D = [d_1, \ldots, d_N]$ with $d_i \in R^L$. In particular, we first divide the data using $\tau_1\%$ of classes for training dataset D_{tr} and $(1 - \tau_1)\%$ of classes for testing dataset D_{te}. In this part, we sort each class according to number of samples where the most populated classes belongs to training set and the less populated to testing set.

This is justified because the new classes (pedestrians) in operation can have few samples and we have no control over them. Then, we initialize DCR using a validation dataset. In this respect, we divide the original training dataset in two by using $\tau_2\%$ for a new training dataset D'_{tr} and the remaining $(1 - \tau_2)\%$ for a new validation dataset D'_{val} considering a stratified scheme, which means that we consider the class proportions. Note that in this case, this division does not consider class information, in contrast to the previous separation given by τ_1. We optionally also test the use of feature selection over initial codes given by PCA considering a method based on Random Forest [5]. Next, we optimize the regressive codes of DRC using the complete original training dataset D_{tr}. We consider to use all available information in order to find generalizable regressive codes. Subsequently, we use the learned codes in the testing dataset D_{te}.

We test both methods based on attributes, DBC and DRC; and a baseline based on SVM over original CAFFE features. Both methods are based on the automatic discovering of attributes, which is essentially hard because the attributes are not known in advance. The number of attributes for both algorithms is 128 as this value has obtained good results in previous experiments [2]. We test two sets of initial codes, one given by the first codes of PCA and the other by feature selection based on random forest [5]. In the case of SVM, we use all original features. Due to the good discriminativity of sparse CAFFE features, the results using full original features are better than DRC and DBC, however, the training and testing of this model requires the 36824 features and the attribute discovery models only 128 after lineal dimensionality reduction. The average test time for SVM over original features is five times slower than DRC in our experiments, which can be crucial in re-identification applications.

Table 2. Accuracy performance for DBC and DCR algorithms (mean and standard deviation) considering 128 codes. The use of feature selection in initialization by (s). We also show the results with full original CAFFE features.

Dataset	ETHZ	CAVIAR	ILIDS	DCC
All	*99.4 (0.3)*	*94.1 (2.1)*	*51.0 (4.6)*	*99.9 (0.1)*
DBC	93.2 (1.3)	78.4 (4.1)	35.2 (4.2)	95.1 (0.6)
DRC	97.6 (0.6)	84.6 (3.4)	42.0 (4.8)	99.9 (0.2)
DRC(s)	98.1 (0.5)	88.7 (2.2)	45.3 (4.5)	99.9 (0.1)

Table 2 shows the results for each dataset by comparing accuracies. Our technique is able to outperform the binary method DBC by an average of 6% considering 128 codes. We note that high values of variance correspond to datasets with few exemplars by individuals, mainly in ILIDS dataset which has variance over 4% and has three images by person. We also note the use of feature selection over initial codes also improves the results, for example, in ILIDS dataset the increasing is 3%. On the other hand, the original CAFFE features are very discriminative as we can see in the results; however, the use of full features requires much more computing processing (five times slower).

3.2   Pedestrian Re-identification

We apply our method in re-identification tasks assuming a transfer learning setting. For space reason, we show results in two public datasets, which are pedestrian datasets. These datasets are VIPER and ILIDS, where both datasets have been previously detailed. These datasets have the characteristic that have only few examples by person, therefore, are natural to be applied transfer learning considering other pedestrian datasets. Again, we use the CAFFE features over 9 overlapping regions given by a 3×3 grid and obtain 36864 sparse features.

We compare our technique with some state-of-the-art methods considering transfer learning setting for pedestrian re-identification. We also add PCA-CAFFE features in order to slightly boost our method. Tables 3 and 4 show the results for datasets VIPER and ILIDS, respectively, by comparing rate recognition according CMC curves using 256 codes for DRC. Our technique is able to compete these methods although these methods usually use some information of test datasets. On the other hand, our method is simple and competitive because it is based on simple linear transformations that are ready to use in test settings. For example, [6] calibrates these models using small partitions of test datasets; furthermore, they use a convolutional neural network for re-identification problem that require specialized GPU hardware in order to learn model.

Table 3. CMC score according to ranking for VIPER dataset. The used rankings are 1%, 10%, 20% and 30%. Although our method does not uses any test dataset information, it is able to be competitive with state-of-the-art methods.

Method	1%	10%	20%	30%
DTRSVM [7]	10.9	28.2	37.7	44.9
DML [8]	16.2	45.8	57.6	64.2
CNN [6]	18.6	48.5	60.6	68.5
DRC	16.4	40.4	57.7	67.7

Table 4. CMC score according to ranking for ILIDS dataset. The used rankings are 1%, 5%, 10% and 20%. Our method is competitive with some state-of-the-art methods.

Method	1%	5%	10%	20%
MCC [9]	12.0	33.7	48.0	67.0
LMNN [10]	23.7	45.4	57.3	70.9
Xing [11]	23.2	45.2	56.9	70.5
CNN [6]	32.4	55.2	67.8	81.5
DRC	27.1	51.6	61.7	72.7

4   Conclusions

The discovering of attributes is an attractive area because it can help to automatic visual recognition by proposing new attributes. Moreover, the attributes are associated to semantic information therefore they can have good generalization properties. We proposed a method considering a different idea in relation to typical binary attribute: what happens if we want to learn a regression of t-shirts? We apply our method to re-identification problem with results which are competitive respect to state-of-the-art methods under the same setting. Five public datasets were tested to evaluate the performance of these codes. Extensive experiments show that the learned representations have good generalization. Although the results are encouraging, we think that the linear mapping restricts our model from finding more interesting patterns. As future work, we will explore richer codification structures than lineal mapping.

References

1. Parikh, D., Grauman, K.: Relative attributes. In: Proceedings of IEEE International Conference on Computer Vision and Pattern Recognition, pp. 503–510 (2011)
2. Rastegari, M., Farhadi, A., Forsyth, D.: Attribute discovery via predictable discriminative binary codes. In: Fitzgibbon, A., Lazebnik, S., Perona, P., Sato, Y., Schmid, C. (eds.) ECCV 2012. LNCS, vol. 7577, pp. 876–889. Springer, Heidelberg (2012). https://doi.org/10.1007/978-3-642-33783-3_63
3. Lobel, H., Vidal, R., Soto, A.: Hierarchical joint max-margin learning of mid and top level representations for visual recognition. In: Proceedings of International Conference on Computer Vision (2013)
4. Jia, Y., Shelhamer, E., Donahue, J., Karayev, S., Long, J., Girshick, R., Guadarrama, S., Darrell, T.: Caffe: convolutional architecture for fast feature embedding. In: Proceedings of the ACM International Conference on Multimedia, MM 2014, pp. 675–678. ACM, New York (2014)
5. Genuer, R., Poggi, J.M., Tuleau-Malot, C.: Variable selection using random forests. Pattern Recogn. Lett. **31**(14), 2225–2236 (2010)
6. Hu, Y., Yi, D., Liao, S., Lei, Z., Li, S.Z.: Cross dataset person re-identification. In: Jawahar, C.V., Shan, S. (eds.) ACCV 2014. LNCS, vol. 9010, pp. 650–664. Springer, Cham (2015). https://doi.org/10.1007/978-3-319-16634-6_47
7. Ma, A., Yuen, P., Li, J.: Domain transfer support vector ranking for person re-identification without target camera label information. In: Proceedings of International Conference on Computer Vision (2013)
8. Yi, D., Lei, Z., Li, S.: Deep metric learning for practical person re-identification. In: Proceedings of International Conference on Pattern Recognition (2014)
9. Globerson, A., Roweis, S.T.: Metric learning by collapsing classes. In: Advances in Neural Information Processing Systems (2005)
10. Weinberger, K.Q., Blitzer, J., Saul, L.K.: Distance metric learning for large margin nearest neighbor classification. In: Advances in Neural Information Processing Systems, pp. 1473–1480 (2005)
11. Xing, E., Ng, A., Jordan, M., Russell, S.: Distance metric learning, with application to clustering with side-information. In: Advances in Neural Information Processing Systems, vol. 15 (2003)

A Deep Boltzmann Machine-Based Approach for Robust Image Denoising

Rafael G. Pires[1], Daniel S. Santos[2], Gustavo B. Souza[1],
Aparecido N. Marana[2], Alexandre L. M. Levada[1],
and João Paulo Papa[2(✉)]

[1] Department of Computing, UFSCar - Federal University of São Carlos,
São Carlos, Brazil
rafapires@gmail.com, gustavo.botelho@gmail.com,
alexandre.levada@gmail.com
[2] Department of Computing, UNESP - Univ Estadual Paulista,
Av. Eng. Luiz Edmundo Carrijo Coube, 14-01, Bauru, SP 17033-360, Brazil
danielfssantos1@gmail.com, {nilceu,papa}@fc.unesp.br

Abstract. A Deep Boltzmann Machine (DBM) is composed of a stack of learners called Restricted Boltzmann Machines (RBMs), which correspond to a specific kind of stochastic energy-based networks. In this work, a DBM is applied to a robust image denoising by minimizing the contribution of some of its top nodes, called "noise nodes", which often get excited when noise pixels are present in the given images. After training the DBM with noise and clean images, the detection and deactivation of the noise nodes allow reconstructing images with great quality, eliminating most of their noise. The results obtained from important public image datasets showed the validity of the proposed approach.

1 Introduction

During the image acquisition process, some level of noise is usually added to the real data mainly due to physical limitations of the acquisition sensor, and also regarding imprecisions during the data transmission and manipulation. Therefore, the resultant image needs to be processed in order to attenuate its noise without losing details present at high frequencies areas, being the field of image processing that addresses such issue called "image restoration". In this context, some well-known image restoration methods, such as inverse and Wiener filter, regularization, projection-based [1,2] and *Maximum a Posteriori* probability techniques have been developed over the last decades [3]. Despite of machine learning being a well consolidated research field dating back to the 60's, only in the last years it has been employed to address the problem of image restoration [4,5].

Recently, deep learning techniques have been considered a page-turner due to the outstanding results in a number of computer vision-related problems,

J. P. Papa—The authors are grateful to Capes, CNPq grant #306166/2014-3, and FAPESP grants #2014/16250-9, #2014/12236-1, and #2016/19403-6.

© Springer International Publishing AG, part of Springer Nature 2018
M. Mendoza and S. Velastín (Eds.): CIARP 2017, LNCS 10657, pp. 525–533, 2018.
https://doi.org/10.1007/978-3-319-75193-1_63

such as face and object recognition, just to name a few. In the last years, some works have addressed the problem of image restoration using such approaches, such as Keyvanrad et al. [6], which employed Deep Belief Networks (DBN) to smooth noise in images. Tang et al. [7] proposed the Robust Boltzmann Machine (RoBM), which allows Boltzmann Machines to be more robust to image corruptions. The model is trained in an unsupervised fashion with unlabeled noisy data, and can learn the spatial structure of the occluders. Compared to some standard algorithms, the model has been significantly better for denoising face images.

Xie et al. [8] used deep networks pre-trained with auto-encoders for image inpainting and denoising, and Tang et al. [9] employed Restricted Boltzmann Machines (RBMs) for the very same purpose of image denoising. Later on, Yan and Shao [10] used Deep Belief Networks [11] (DBNs) to identify blur type and parameters in natural images. Recently, a new RBM-based architecture was proposed, called Deep Boltzmann Machines (DBMs) [12]. This new approach have presented some great results in many areas outperforming the DBNs, since in the training phase of that network not only bottom-up information is considered, but also top-down influences. As a matter of fact, our approach is based on the work by Keyvanrad et al. [6], a new deep learning-based approach for robust image denoising using DBMs is proposed. We showed the proposed approach outperforms DBNs and standard DBMs in the context of image denoising.

2   Restricted Boltzmann Machines

Restricted Boltzmann Machines [13] are energy-based stochastic neural networks composed of two layers of neurons (visible and hidden), in which the learning phase is conducted by means of an unsupervised fashion. A naïve architecture of a Restricted Boltzmann Machine comprises a visible layer $\mathbf{v}$ with m units and a hidden layer $\mathbf{h}$ with n units. Additionally, a real-valued matrix $\mathbf{W}_{m \times n}$ models the weights between the visible and hidden neurons, where w_{ij} stands for the weight between the visible unit v_i and the hidden unit h_j.

At first, let us assume both $\mathbf{v}$ and $\mathbf{h}$ as being binary-valued units. In other words, $\mathbf{v} \in \{0,1\}^m$ e $\mathbf{h} \in \{0,1\}^n$, thus leading to the so-called Bernoulli-Bernoulli Restricted Boltzmann Machine, since both units follow a Bernoulli distribution. The energy function of an RBM is given by:

$$E(\mathbf{v}, \mathbf{h}) = -\sum_{i=1}^{m} a_i v_i - \sum_{j=1}^{n} b_j h_j - \sum_{i=1}^{m} \sum_{j=1}^{n} v_i h_j w_{ij}, \tag{1}$$

where $\mathbf{a}$ e $\mathbf{b}$ stand for the biases of visible and hidden units, respectively.

The probability of a joint configuration $(\mathbf{v}, \mathbf{h})$ is computed as follows:

$$P(\mathbf{v}, \mathbf{h}) = \frac{1}{Z} e^{-E(\mathbf{v}, \mathbf{h})}, \tag{2}$$

where Z stands for the so-called partition function, which is basically a normalization factor computed over all possible configurations involving the visible and hidden units. Similarly, the marginal probability of a visible (input) vector is given by:

$$P(\mathbf{v}) = \frac{1}{Z} \sum_{\mathbf{h}} e^{-E(\mathbf{v},\mathbf{h})}. \tag{3}$$

Since the RBM is a bipartite graph, the activations of both visible and hidden units are mutually independent, thus leading to the following conditional probabilities:

$$P(v_i = 1|\mathbf{h}) = \phi \left(\sum_{j=1}^{n} w_{ij} h_j + a_i \right), \tag{4}$$

and

$$P(h_j = 1|\mathbf{v}) = \phi \left(\sum_{i=1}^{m} w_{ij} v_i + b_j \right). \tag{5}$$

Note that $\phi(\cdot)$ stands for the sigmoid function.

Let $\Theta = (W, a, b)$ be the set of parameters of a RBM, which can be learned though a training algorithm that aims at maximizing the product of probabilities given all the available training data $\mathcal{V}$, as follows:

$$\arg\max_{\Theta} \prod_{v \in \mathcal{V}} P(\mathbf{v}). \tag{6}$$

One of the most used approaches to solve the above problem is the Contrastive Divergence (CD) [13], which basically ends up performing Gibbs sampling using the training data as the visible units, instead of random inputs.

2.1 Deep Boltzmann Machines

Salakhutdinov and Hinton [12] presented the DBM, which aims at improving the inference during the learning process, since it now considers both directions of interaction among adjacent layers. Salakhutdinov and Hinton [14] proposed the use of a variable inference method called "Mean-Field" to enhance the DBM learning procedure. This technique approximates the posterior distributions inferred from the observed data of the estimates based on isolated network segments. The training process of a DBM consists in minimizing the total energy of the system according to the parameters found through partial inferences made through the mean-fields (MF).

Roughly speaking, the idea is to find an approximation $Q^{MF}(\mathbf{h}|\mathbf{v};\boldsymbol{\mu})$ that best represents the true distribution of the hidden layers, i.e. $P(\mathbf{h}|\mathbf{v})$. This approximation is computed through the following factored distribution:

$$Q^{MF}(\mathbf{h}|\mathbf{v};\boldsymbol{\mu}) = \prod_{l=1}^{L} \left[\prod_{k=1}^{F_l} q(h_k^l) \right], \tag{7}$$

where L stands for the number of hidden layers, F_l represents the number of nodes in the hidden layer l, and $q(h_k^l = 1) = \mu_k^l$. The goal is to find the parameters of the mean-field $\boldsymbol{\mu} = \{\boldsymbol{\mu_1}, \boldsymbol{\mu_2}, \ldots, \boldsymbol{\mu^L}\}$ according to following equations:

$$\mu_k^1 = \phi \left(\sum_{i=1}^{m} w_{ik}^1 v_i + \sum_{j=1}^{F_2} w_{kj}^2 \mu_j^2 \right), \tag{8}$$

which represents the interaction between the first hidden layer and your previous visible layer where ϕ is a sigmoid function. Similarly, the interactions between hidden layers $l-1$ and l are given as follows:

$$\mu_k^l = \phi \left(\sum_{i=1}^{F_{l-1}} w_{ik}^l \mu_i^{l-1} + \sum_{j=1}^{F_{l+1}} w_{kj}^{l+1} \mu_j^{l+1} \right), \tag{9}$$

where w_{ij}^l stands for the weight between node i from hidden layer $l-1$ and node j from hidden layer l. Finally, the mean-field parameters for the hidden layer at the top of the DBM are calculated by:

$$\mu_k^L = \phi \left(\sum_{i=1}^{F_{L-1}} w_{ik}^L \mu_i^{L-1} \right). \tag{10}$$

3   Proposed Approach

The goal of this work is to propose a new DBM-based denoising approach, which learns how to deactivate some of its nodes in order to smooth the noise levels from images. After training the DBM with the clean (noise-free) and noisy training images together, we used a criterion called relative activity [6] (ψ^*), which is defined as the difference among the mean activation values of the upper most hidden nodes. For the sake of explanation, after training the DBM using the mean-field procedure, we take first the clean images and propagate them upwards. For each clean image, we store the activation field of the top layer in order to compute the mean activation field regarding all clean images, hereinafter called ψ_{clean}. Further, we conduct the very same procedure for the noisy images to estimate their mean activation field at the top layer[1], hereinafter called ψ_{noisy}. Therefore, one has two mean activation fields at the very top layer: one for the clean and another for the noisy images. Thus, the aforementioned relative activity is computed as the difference between the mean activation fields of the clean and noisy images, i.e., $\psi^* = |\psi_{clean} - \psi_{noisy}|$.

Further, one needs to find out the so-called "noise nodes", which stand for the nodes of the upper most hidden layer that are activated upon the presence

---

[1] Notice the procedure over the noisy images do not consider the activation field already computed for the clean images.

of noisy structures, i.e., such nodes become more "excited" when they are presented to noisy elements. Basically, we thresholded the relative activity values as follows:

$$\psi_i^* = \begin{cases} \psi_i^{clean} & \text{if } \psi_i^* > T \\ \psi_i^* & \text{otherwise,} \end{cases} \tag{11}$$

where T stands for the threshold value[2]. Figure 1a illustrates the aforementioned process.

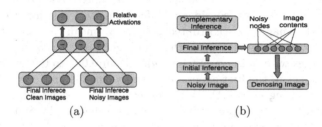

(a)   (b)

Fig. 1. Illustration of the proposed DBM for denoising purposes: (a) difference between the mean activations fields of the clean and noisy images, and (b) proposed DBM-based image denoising approach.

Finally, concerning the denoising step, given a noisy image, we perform a bottom-up pass until we reach the top layer (final inference). Then, the noisy nodes will be in charge of replacing their corresponding nodes of the noisy image. Since the noisy nodes were deactivated in the previous step, the reconstructed image (top-down step) is now much cleaner than before. This procedure is depicted in Fig. 1b.

4   Experiments and Discussion

We used the following image databases to evaluate the performance of the proposed approach: MNIST [15], Semeion [16], and Caltech 101 Silhouettes [17]. In this work, we employed a neural architecture composed of 3 layers containing 784-1000-500-250 nodes. Recall that such architecture has been empirically chosen[3,4].

The main idea is to train the network in a way that is possible to learn the mapping between clean and noisy images. The MNIST database contains 60,000

---

[2] In this paper, we evaluated $T \in [0.1, 0.9]$ with steps of 0.1.

[3] Since we have 28×28 images concerning all datasets, the visible layer has $28 \times 28 = 784$ nodes. Also, since we are using Bernoulli-based DBM/DBN models, we employed a min-max normalization of the grayscale values of the images' pixels.

[4] Since Semeion database images are 16×16-sized, we centered them into a 28×28-black-squared window in order to have all images used in the experiments with the very same size.

training images, as well as $10,000$ testing images. In the experiments, we used a subset of $20,000$ images for training composed of $10,000$ noiseless images and their respective noisy versions $(10,000)$. The noisy images were generated by means of a Gaussian noise with zero mean and two different variance values: $\sigma \in \{0.1, 0.2\}$, since we conducted two different experiments. In regard to the test phase (denoising), we used all the $10,000$ testing images corrupted with a zero-mean Gaussian noise and variance levels of $\sigma \in \{0.1, 0.2\}$.

Since Semeion database contains $1,400$ training images only, we increased the training set size to $22,400$ as follows: we kept the original $1,400$ images, and further we generated $1,400$ more images that are noisy versions of the original ones. Once again, since we considered different noise levels for two separated experiments, the images were corrupted with a zero-mean Gaussian distribution and variance $\sigma \in \{0.1, 0.2\}$. After that, we generated $9,800$ more Gaussian-corrupted images with noise variance ranging from 0.001 to 0.007 with steps of 0.001 with respect to the first experiment (i.e., when the first corrupted images were generated using $\sigma = 0.1$). With respect to the second experiment (i.e., when the images were corrupted using $\sigma = 0.2$), we generated $9,800$ more images corrupted with a Gaussian noise with variance ranging from 0.201 to 0.207 with steps of 0.001.

Finally, Caltech 101 Silhouettes database contains $4,100$ training images and $2,307$ testing data. Regarding this dataset, we also increased the number of training images to $24,600$ as follows: $4,100$ clean images and their corresponding noisy versions (Gaussian noise with zero-mean and variance of $\sigma \in \{0.1, 0.2\}$ for both experiments). Also, we generated $4,100$ more images corrupted with a zero-mean Gaussian noise (variances of 0.001 and 0.002) considering $\sigma = 0.1$, and variances of 0.201 and 0.207 considering $\sigma = 0.2$.

The proposed DBM-based approach was compared against a similar DBN (i.e. a DBN with "noisy nodes" as proposed by Keyvanrad et al. [6]) standard DBNs and DBMs, as well as against the well-known Wiener Filter. In order to evaluate the performance of the proposed method, the Peak signal-to-noise ratio (PSNR) between the noise-free image and its respective restored version is computed. Table 1 presents the parameters used for each technique. These values have been empirically chosen.

Table 1. Parameters used considering both DBMs and DBNs.

Parameters	DBM pretraining	Mean-Field	DBN
Learning rate	0.05	0.00005	0.05
Momentum (min)	0.5	0.1	0.5
Momentum (max)	0.9	0.5	0.9
Weight Decay	0.001	0.0005	0.001
# Epochs	100	20	100
Batch size	100	100	100
# Mean-Field updates		30	

Table 2. PSNR results concerning the image denoising procedure.

	No-denoising	DBM	Proposed DBM	DBN	DBN [6]	Wiener
$\sigma = 0.1$						
MNIST	12.888	17.738	**19.280 (0.4)**	17.184	18.002 (0.5)	15.765
Caltech	13.036	16.005	**16.302 (0.3)**	14.017	14.857 (0.5)	15.139
Semeion	13.051	17.081	**19.104 (0.3)**	16.290	18.231 (0.5)	15.423
$\sigma = 0.2$						
MNIST	10.163	13.275	16.230 (0.4)	15.651	**16.747 (0.6)**	13.282
Caltech	10.216	13.60	**13.920 (0.4)**	12.782	13.637 (0.4)	12.640
Semeion	10.173	14.675	**17.367 (0.4)**	14.800	16.253 (0.4)	12.744

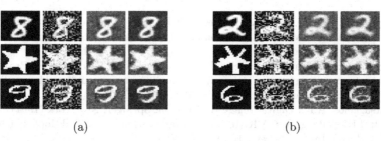

(a)   (b)

Fig. 2. Example images considering MNIST (first row), Caltech (second row) and Semeion (third row) databases. (a) First experiment, from left to right: original, noisy, standard DBM, and proposed DBM-based denoised images; (b) Second experiment from left to right: original, noisy, DBN considering MNIST, DBM considering both Caltech and Semeion, DBN [6] considering MNIST, and proposed DBM for Caltech and Semeion.

Table 2 presents the results, being the best ones in bold. The values in parenthesis stand for the best thresholds used to find the "noisy nodes". As one can observe, the proposed approach obtained the best results in all cases whereas Gaussian noise with zero-mean and variance of 0.1. In regard to Gaussian noise with zero-mean and variance of 0.2, the proposed approach obtained the best results in two out three datasets, namely Caltech and Semeion. In regard to the MNIST, the best result was obtained by the DBN, but being closely followed by the proposed approach. More interestingly, the DBM with "noisy nodes" outperformed both the standard DBM and DBN in all situations. Also, the proposed approach obtained better results than Wiener Filter, which is considered one of the best approaches for image denoising.

Figure 2 displays some example images from the databases. Clearly, the images denoised by the proposed DBM seem to have less noise levels than the images filtered by standard DBM. The content of the number itself seems to be similar among the images, but its surroundings have been better restored by the proposed DBM.

5 Conclusion

In this work, a new DBM-based approach for robust image denoising has been proposed. The idea is to learn how to turn off nodes that are often activated when noisy images are presented to the network. The experiments in three public datasets showed the proposed approach obtained better results in two out three situations, but producing images with lower reconstruction errors than standard DBNs, DBMs and Wiener Filter in all datasets. In regard to future works, we aim at working with gray-scale images, as well as how to learn "noisy nodes" at different layers, not only in the top one.

References

1. Papa, J.P., Fonseca, L.M.G., Carvalho, L.A.S.: Projections onto convex sets through particle swarm optimization and its application for remote sensing image restoration. Pattern Recognit. Lett. **31**(13), 1876–1886 (2010)
2. Pires, R.G., Pereira, D.R., Pereira, L.A.M., Mansano, A.F., Papa, J.P.: Projections onto convex sets parameter estimation through harmony search and its application for image restoration. Nat. Comput. **15**(3), 493–502 (2016)
3. Katsaggelos, A.K.: Digital Image Restoration. Springer, New York (1991)
4. Paik, J.K., Katsaggelos, A.K.: Image restoration using a modified Hopfield network. IEEE Trans. Image Process. **1**, 49–63 (1992)
5. Sun, Y., Yu, S.Y.: A modified Hopfield neural network used in bilevel image restoration and reconstruction. In: International Symposium on Information Theory Application, vol. 3, pp. 1412–1414 (1992)
6. Keyvanrad, M.A., Pezeshki, M., Homayounpour, M.A.: Deep belief networks for image denoising. arXiv preprint arXiv:1312.6158 (2013)
7. Tang, Y., Salakhutdinov, R., Hinton, G.: Robust Boltzmann machines for recognition and denoising. In: IEEE Conference on Computer Vision and Pattern Recognition, Providence, Rhode Island, USA (2012)
8. Xie, J., Xu, L., Chen, E.: Image denoising and inpainting with deep neural networks. In: Pereira, F., Burges, C.J.C., Bottou, L., Weinberger, K.Q. (eds.) Advances in Neural Information Processing Systems, vol. 25, pp. 341–349. Curran Associates Inc. (2012)
9. Tang, Y., Salakhutdinov, R., Hinton, G.E.: Robust Boltzmann machines for recognition and denoising. In: IEEE Conference on Computer Vision and Pattern Recognition, CVPR 2012, pp. 2264–2271. IEEE (2012)
10. Yan, R., Shao, L.: Image blur classification and parameter identification using two-stage deep belief networks. In: 24th British Machine Vision Conference, pp. 1–11 (2013)
11. Osindero, S., Hinton, G.E., Teh, Y.W.: A fast learning algorithm for deep belief nets. Neural Comput. **18**(7), 1527–1554 (2006)
12. Salakhutdinov, R., Hinton, G.E.: Deep Boltzmann machines. In: AISTATS, vol. 1, p. 3 (2009)
13. Hinton, G.E.: Training products of experts by minimizing contrastive divergence. Neural Comput. **14**(8), 1771–1800 (2002)

14. Salakhutdinov, R., Hinton, G.E.: An efficient learning procedure for deep Boltzmann machines. Neural Comput. **24**(8), 1967–2006 (2012)
15. Corinna, C., LeCun, Y., Burges, C.J.C.: The MNIST database of handwritten digits (1998)
16. Lichman, M.: UCI machine learning repository (2013)
17. Holub, A., Griffin, G., Perona, P.: Caltech-256 object category dataset (2007)

Discovering Bitcoin Mixing Using Anomaly Detection

Mario Alfonso Prado-Romero[1](✉)(iD), Christian Doerr[2],
and Andrés Gago-Alonso[1]

[1] Advanced Technologies Application Center (CENATAV),
7a # 21406, Rpto. Siboney, Playa, 12200 Havana, Cuba
{mprado,agago}@cenatav.co.cu
[2] Delft University of Technology (TU Delft),
Mekelweg 4, 2628CD Delft, The Netherlands
c.doerr@tudelft.nl

Abstract. Bitcoin is a peer-to-peer electronic currency system which
has increased in popularity in recent years, having a market capitalization of billions of dollars. Due to the alleged anonymity of the Bitcoin
ecosystem, it has attracted the attention of criminals. Mixing services are
intended to provide further anonymity to the Bitcoin network, making
it impossible to link the sender of some money with the receiver. These
services can be used for money laundering or to finance terrorist groups
without being detected. We propose to model the Bitcoin network as
a social network and to use community anomaly detection to discover
mixing accounts. Furthermore, we present the first technique for detecting Bitcoin accounts associated to money mixing, and demonstrate our
proposal effectiveness on real data, using known mixing accounts.

Keywords: Bitcoin · Bitcoin mixing · Anomaly detection

1 Introduction

Bitcoin is an electronic currency designed to be decentralized and to provide
anonymity to users [1]. In the Bitcoin network, accounts are identified by public
keys and each user can own multiple accounts. All transactions in the Bitcoin
system are stored in a public ledger called the blockchain which is the mechanism
used by the system to prevent double spending. Low transaction fees, easiness to
create accounts and anonymity have all influenced in the fast increase in Bitcoin
popularity with a current market capitalization of around 14 billion dollars in
2017[1]. Despite the alleged privacy, there is an understanding amongst Bitcoin
more technical users that anonymity is not a primary design goal of the system
[2]. It is possible to use the blockchain to trace money from one user to another,
and it has been demonstrated that the identity of the users can be uncovered
using information external to the Bitcoin network.

---

[1] https://blockchain.info/charts/market-cap.

© Springer International Publishing AG, part of Springer Nature 2018
M. Mendoza and S. Velastín (Eds.): CIARP 2017, LNCS 10657, pp. 534–541, 2018.
https://doi.org/10.1007/978-3-319-75193-1_64

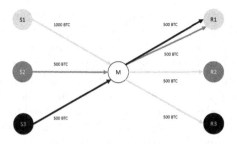

Fig. 1. Bitcoin mixing example

In recent years, many services intended to provide further transaction anonymization have emerged such as BitLaundry, BitFog and the Send Shared functionality of Blockchain.info. These are known as Mixing Services and some of them routinely handle the equivalent of 6-digit dollar amounts [3]. The idea behind Mixing Services is to be an intermediary in user transactions. They take the money of many senders and then for each one, send the desired amount of money to the receiver using money coming from other senders. The goal of Bitcoin mixing is to make it impossible to link the sender with the actual receiver of the money. The use of these services imply some inconveniences for the users, like a delay of many days in the transactions, the payment of an extra fee for the operation, and the risk of having their money stolen by a fraudulent mixing service. Bitcoin has always attracted the attention of the criminal world due to its decentralized nature [4] and Mixing Services can be used for money laundering or to finance terrorist groups without being detected.

Figure 1, displays an example of mixing where senders $S1$, $S2$ and $S3$ want to transfer money to the receivers $R1$, $R2$ and $R3$ respectively. To avoid being related to the receivers, the senders use a mixing service M which transfers the desired amount to $R2$ and $R3$ using the money from $S1$, and the bitcoins from $S2$ and $S3$ are sent to $R1$. This is a basic example, actual Mixing Services use many evasion techniques to avoid money tracing. Two of the most used tactics for this end are delaying transactions, to avoid be linked by time, and splitting the money into small transactions, to make impossible to relate the transactions by amount. Also, a common practice is to use many accounts for moving the money before performing the actual mixing.

Tracing money through mixing services has been demonstrated to be an extremely difficult task in most cases [3]. In the other hand, the discovery of Bitcoin mixing accounts is still possible and worthwhile in its own. Mixing Services create new accounts regularly, and the whole set of accounts belonging to them is not known by users. While tracing is capable of discovering some mixing accounts used by a particular person of interest, other accounts from the service remain unknown. Mixing detection can identify which accounts from the network are related to these services. Once the mixing accounts are discovered, users related to them could be identified and further analyzed to determine if they are involved in criminal activities. Existing works for detecting malicious activities in Bitcoin [3,5–7] are not focused in identifying unknown mixing accounts.

Anomaly detection refers to the problem of finding patterns in data that do not conform to expected behavior [8]. Due to the inconveniences of mixing sites, they are not used by the majority of people in the Bitcoin network, for this reason mixing accounts and its users are an anomaly from the perspective of the network topology. We take advantage of this property and use anomaly detection to identify mixing accounts. Most existing anomaly detection techniques are for vector data [9] and cannot be used for this task. From the techniques designed to work in graphs [10], only a small number considers communities of elements [11–14] and they are not designed to identify bitcoin mixing accounts. The main contributions of our work are:

– **We tackle the problem of Bitcoin mixing detection:** Many works have focused in studies of anonymity and criminal activities in Bitcoin. To the best of our knowledge this is the first work focused on detecting mixing accounts, which can be helpful to uncover potential money laundering.
– **We discover mixing accounts using community outlier detection:** We propose to build the Bitcoin user network where members are structured in communities. Then, we use community anomaly detection to identify mixing accounts. Furthermore, we present the first algorithm for this purpose.
– **We validate our approach using real data:** We test our algorithm in real data and the results demonstrate the effectiveness of our proposed approach for identifying mixing accounts.

The remainder of this paper is structured as follows: In Sect. 2, our proposal is presented. In Sect. 3, the effectiveness of our proposal is demonstrated on real data and the results are analyzed. Finally, in Sect. 4, the work is concluded and some open challenges are discussed.

2 Discovering Bitcoin Mixing

People normally have a tendency to use the services they known and like, and to exchange money with the same group of known people. We believe this behavior is also present in the Bitcoin network.

Bitcoin transactions are linked to each other, and can be naturally modeled as a network where the transactions will be the vertices, and the money stream among them will be represented by the edges. This graph, called the transaction network, can be used to trace money or to identify double-spending, but it is not very useful to recognize user behavior. Also the user behavior is spread across many accounts. Due to this, our first step is to merge accounts into user identities to build a user network following the process described in [2].

The user network behaves as a social network where users are organized in communities. The idea of mixing is to merge money from different users, and for each sender, give the desired amount of money to the target receiver using the money coming from another sender. For this reason, mixing violates the community structure of the network relating users that have nothing in common. As we mention in the introduction, the people using mixing sites are a minority of Bitcoin users. We affirm that users having many more inter-community connections compared to the rest of users belonging to its same community are probably mixing sites. We propose a new algorithm named InterScore designed to identify this kind of outliers. Due to unavailability of labeled data and the difference between user groups, our algorithm finds the communities in the network and analyzes each element in its community in an unsupervised fashion. As result, it returns an outlier ranking of Bitcoin users.

Definition 1 (Outlier ranking). *An outlier ranking from a graph G is a set $R = \{(v,r)|v \in V, r \in [0,1]\}$ of tuples, each one containing a vertex from G and its outlierness score.*

The input of our algorithm is a user graph G_U. In a first stage, the Louvain community detection method [15] is used on G_U to identify groups of related users, returning a clustering C of vertices from G_U. Any state-of-the-art graph clustering algorithm could be used in this stage. The Louvain method was selected based mainly in its performance and applicability in large graphs.

In the second stage, our algorithm iterates over each community $C_i \in C$ and for each vertex calculates the number of inter-community links it has, using a function $l : V \to \mathbb{R}$. Then, for each community C_i is calculated the mean difference among the number of inter-community links from its elements as defined below:

$$IMD(C_i) = \frac{\sum_{v_j \in C_i} \sum_{v_k \in C_i, v_j \neq v_k} |l(v_j) - l(v_k)|}{|C_i|} \tag{1}$$

Once the inter-community links mean difference is calculated for each community, our algorithm iterates over the elements of each C_i and determines its anomaly score using the next function:

$$r(v, C_i) = \frac{\sum_{u \in C_i, u \neq v} d(v, u, C_i)}{|C_i|} \tag{2}$$

where $d : V \times V \times 2^V \to \{0,1\}$ is a function that determines if the inter-community links difference between two vertices is greater than its community mean. The function is defined as below:

$$d(v, u, C_i) = \begin{cases} 0 & |l(v) - l(u)| \leq IMD(C_i), \\ 1 & |l(v) - l(u)| > IMD(C_i) \end{cases} \tag{3}$$

Intuitively, the score function measures with how percent of the community the user has a difference in the amount of inter-community links greater than the mean difference for that community. This function adaptively ranks users outlierness according to their context, and detects anomalies that cannot be identified from a global point of view. In the Algorithm 1, the steps of the InterScore method can be observed in more detail.

Algorithm 1. InterScore

Input: A users network G_U
Output: An outlier ranking R of the vertices from G_U
1 $R \leftarrow \emptyset$
2 $C \leftarrow \text{Clustering}(G_U)$
3 **foreach** *community* $C_i \in C$ **do**
4 $\quad$ $m \leftarrow IMD(C_i)$
5 $\quad$ **foreach** *user* $v \in C_i$ **do**
6 $\quad\quad$ $r_v \leftarrow 0$
7 $\quad\quad$ **foreach** *user* $u \in C_i$ *with* $u \neq v$ **do**
8 $\quad\quad\quad$ **if** $|l(u) - l(v)| > m$ **then**
9 $\quad\quad\quad\quad$ $r_v \leftarrow r_v + 1$
10 $\quad\quad\quad$ **end**
11 $\quad\quad$ **end**
12 $\quad\quad$ $r_v \leftarrow \dfrac{r_v}{|C_i|}$
13 $\quad\quad$ $R \leftarrow R \cup \{r_v\}$
14 $\quad$ **end**
15 **end**
16 **return** R

The InterScore algorithm has two fundamental stages, the community detection stage and the anomaly detection stage. In the former, we use the Louvain algorithm whose exact computational complexity is unknown because it is an heuristic, but the authors said it appears to be $O(V \log(V))$ based in the experiments. The outlierness score used in the last stage has a $O(V^2)$ complexity. As a result, the InterScore algorithm has a computational complexity of $O(V^2)$. It is important to mention that the Bitcoin network is sparse, so calculate the inter-community links of all vertices is less expensive than $O(V^2)$ in practice. Also, the anomaly score is calculated only in relation to the users in the same community, for this reason is less expensive when the number of communities is higher.

3   Experiments

We use two subsets from the blockchain in our analysis, the first one contains all transactions ranging from 2012-09-01 to 2012-10-01, and the second one, all transactions from 2013-04-19 to 2013-05-31. The former subset will be referred as the 2012 data set and the later as 2013 data set. It is important to note that the algorithms for Bitcoin mixing detection should be capable of finding interesting results using only a subset of the data because the size of Bitcoin network growths exponentially.

As ground truth we use six accounts, identified in [3], as involved in money mixing. The authors found these accounts while trying to trace money through three known mixing services BitLaundry, BitFog and the Send Shared functionality of Blockchain.org. The problem of tracing money through this services is complex and the authors cannot guarantee all six transactions are related to the mentioned mixing services.

We propose two different methods for determining the inter-community links of a user. The first one will be called Relative Inter Links and consist in for each neighbor u of the analyzed node v, adds 1 if the community of u is different from that of v. The second will be called Total Inter Links and it is a very similar process, but instead of increase 1, increases in the number of transactions between the users v and u. The former method is less sensitive to misclassified users during the community detection stage, but the later is better in identifying a common practice among Mixing Services of dividing transactions in smaller ones.

The user networks for the 2012 and 2013 data set where built. The result was a graph with $412,330$ vertices and $885,808$ edges in the former, and with $942,204$ vertices and $2,835,807$ edges in the later. These values evidenced the sparse nature of the user network.

The results of our algorithm can be observed in Table 1. The advantages of our proposal in identifying candidate mixing accounts, for being further analyzed, can be appreciated. The number of elements identified as mixing services in both data sets is less than a 0.6% of total users. Furthermore all known mixing accounts are identified by our algorithm demonstrating the effectiveness of our approach.

Table 1. InterScore performance

Database	Interlink counting function	Total users	Discovered users	Known mixing users	Known mixing users identified
2012	Relative Inter Links	412330	100	2	2
2012	Total Inter Links	412330	133	2	2
2013	Relative Inter Links	942204	2114	4	4
2013	Total Inter Links	942204	5422	4	4

The ground truth used in this section are four transactions and two accounts reported in [3] as involved in money mixing. As a threshold in the outlier ranking, we use the outlierness score of the element from the ground truth with the lowest score. It is important to mention that a transaction could have many sender and receiver accounts. We say our algorithm detects a known mixing user if it identifies a user that owns at least one account involved in a transaction from the ground truth. Not all accounts involved in mixing transactions are identified as anomalous. Some of them are accounts used only to move money and made it harder to trace. Also, some new mixing accounts that at the moment of the analysis have been used only by users in the same community are hard to identify. Despite the previous cases, at least one account involved in each transaction for the ground truth was identified by our algorithm.

It is interesting to analyze the difference in the number of detected elements depending on the function used to count the inter-community links. The function that counts the total number of links detects more elements, especially in the 2013 data set. This increase can be the result of some elements misclassified by the community detection algorithm due to the increase in the complexity of the network. Also, could indicate an increase in the activity of mixing services. The last explanation is possible due to an increase in the number of Bitcoin users and the fact that the number of detected elements greatly increased using both functions. Additionally, the function that counts the total number of inter-community links better captures the common mixing sites behavior of dividing big transactions into many smaller ones, being capable of detecting more mixing accounts. It should be mentioned that we cannot ensure that all detected elements are mixing accounts, but cannot ensure neither that those of them which are not known mixing accounts are normal users.

In Table 2, we show the scores assigned by our algorithm to each account from the ground truth. In general, these accounts get very high scores. It is curious that, in the 2012 data set, the accounts get the same score no matter the link counting function used. In the 2013 data set the results vary accordingly to the function used. This behavior indicates that in 2012 the Mixing Services did not split the senders money into smaller transactions, a practice that seems common in 2013. An increase in the complexity of Mixing Services behavior could be an explanation to the difference in the number of detected elements we got using different link counting functions in 2013 data set.

Table 2. Ground truth accounts outlierness score

Account key	Outlierness score (total)	Outlierness score (relative)	Database
1E8QctAG6oeM6sw9tZccxvUPRg9dhf9VDm	0.999972676849	0.999972676849	2012
1Bw7ohts4BHVPAMBiSAqWeELjTUV1ZY87g	0.999977892736	0.999977892736	2012
13udyfBcdA2PUDCFM69VYDEHRRFnqkjEkx	0.9996	0.999266666667	2013
1MsmThtteKPu6fWxwn2SMDEnmJex3vKSBk	0.996448195994	0.999341002258	2013
152Yd71xMDzVntyD86xWgag5jfeNmyhMyS	0.999980617713	0.999995154428	2013
1KdPv6GWpg6eoj6cxcV65uc1NwufvhtGGQ	0.999995154428	0.999995154428	2013

An interesting result is that we identify as anomalous accounts belonging to all known mixing transactions. The approach used by [3] was focused in trace money through mixing sites, and the fact that we find these same elements using a different approach is indicative of the association of these transactions with Bitcoin mixing.

4   Conclusions

We discussed the problem concerning the detection of mixing accounts in the Bitcoin network. We modeled the Bitcoin user network as a social network where members are associated in communities and mixing sites behave as community anomalies. Furthermore, we proposed the first algorithm to detect Bitcoin mixing accounts and demonstrated its effectiveness on real data.

We will focus on some challenges in future work. First, our algorithm can be naturally parallelized to increase the performance. Also, there is information about the direction of transactions and about the accounts that could be interesting to include in our method analysis. Finally, it is important to mention that the same idea used for detecting Bitcoin mixing can be used to identify spammers or even bots in botnets, being interesting domains for future work.

References

1. Nakamoto, S.: Bitcoin: a peer-to-peer electronic cash system (2008). http://www.cryptovest.co.uk/resources/Bitcoin%20paper%20Original.pdf
2. Reid, F., Harrigan, M.: An analysis of anonymity in the Bitcoin system. In: Altshuler, Y., Elovici, Y., Cremers, A., Aharony, N., Pentland, A. (eds.) Security and privacy in social networks, pp. 197–223. Springer, Heidelberg (2013). https://doi.org/10.1007/978-1-4614-4139-7_10
3. Möser, M., Böhme, R., Breuker, D.: An inquiry into money laundering tools in the Bitcoin ecosystem. In: eCrime Researchers Summit (eCRS) (2013)
4. Christin, N.: Traveling the silk road: a measurement analysis of a large anonymous online marketplace. In: Proceedings of the 22nd International Conference on World Wide Web (2013)
5. Bonneau, J., Miller, A., Clark, J., Narayanan, A., Kroll, J.A., Felten, E.W.: Sok: research perspectives and challenges for Bitcoin and cryptocurrencies. In: 2015 IEEE Symposium on Security and Privacy (SP) (2015)
6. Möser, M., Böhme, R., Breuker, D.: Towards risk scoring of Bitcoin transactions. In: International Conference on Financial Cryptography and Data Security (2014)
7. Spagnuolo, M., Maggi, F., Zanero, S.: BitIodine: extracting intelligence from the bitcoin network. In: Christin, N., Safavi-Naini, R. (eds.) FC 2014. LNCS, vol. 8437, pp. 457–468. Springer, Heidelberg (2014). https://doi.org/10.1007/978-3-662-45472-5_29
8. Chandola, V., Banerjee, A., Kumar, V.: Anomaly detection: a survey. ACM Comput. Surv. (CSUR) **41**, 15 (2009)
9. Hodge, V., Austin, J.: A survey of outlier detection methodologies. Artif. Intell. Rev. **22**, 85–126 (2004)
10. Akoglu, L., Tong, H., Koutra, D.: Graph based anomaly detection and description: a survey. Data Min. Knowl. Disc. **29**, 626–688 (2015)
11. Gao, J., Liang, F., Fan, W., Wang, C., Sun, Y., Han, J.: On community outliers and their efficient detection in information networks. In: Proceedings of the 16th ACM SIGKDD International Conference on Knowledge Discovery and Data Mining (2010)
12. Müller, E., Iglesias Sánchez, P., Mülle, Y., Böhm, K.: Ranking outlier nodes in subspaces of attributed graphs. In: 2013 IEEE 29th International Conference on Data Engineering Data Engineering Workshops (ICDEW) (2013)
13. Perozzi, B., Akoglu, L., Iglesias Sánchez, P., Müller, E.: Focused clustering and outlier detection in large attributed graphs. In: Proceedings of the 20th ACM SIGKDD International Conference on Knowledge Discovery and Data Mining (2014)
14. Prado-Romero, M.A., Gago-Alonso, A.: Detecting contextual collective anomalies at a glance. In: Proceedings of the 23rd International Conference on Pattern Recognition (ICPR) (2016)
15. Blondel, V.D., Guillaume, J.L., Lambiotte, R., Lefebvre, E.: Fast unfolding of communities in large networks. J. Stat. Mech: Theory Exp. **2008**, P10008 (2008)

A Scale-Space Approach for Multiscale Shape Analysis

Lucas Alexandre Ramos[1]([✉])[ID], Aparecido Nilceu Marana[1][ID],
and Luis Antônio de Souza Junior[2][ID]

[1] UNESP - São Paulo State University, Bauru, SP, Brazil
`l.a.ramos@amc.uva.nl`
[2] UFSCar - Federal University of São Carlos, São Carlos, SP, Brazil

Abstract. Currently, given the widespread of computers through society, the task of recognizing visual patterns is being more and more automated, in particular to treat the large and growing amount of digital images available. Two well-referenced shape descriptors are BAS (Beam Angle Statistics) and MFD (Multiscale Fractal Dimension). Results obtained by these shape descriptors on public image databases have shown high accuracy levels, better than many other traditional shape descriptors proposed in the literature. As scale is a key parameter in Computer Vision and approaches based on this concept can be quite successful, in this paper we explore the possibilities of a scale-space representation of BAS and MFD and propose two new shape descriptors SBAS (Scale-Space BAS) and SMFD (Scale-Space MFD). Both new scale-space based descriptors were evaluated on two public shape databases and their performances were compared with main shape descriptors found in the literature, showing better accuracy results in most of the comparisons.

Keywords: Image analysis · Scale-space · Multiscale
Shape analysis · BAS · MFD

1 Introduction

Shape analysis is considered one of the most important areas in image analysis, since it is able to describe an object preserving its most relevant information. Shape classification is a fundamental problem in Computer Vision, which has many applications such as, object detection, classification and image retrieval.

When analyzing an image, it is of utmost importance to consider that certain information only makes sense under certain viewing conditions, such as the scale. Nevertheless, choosing the appropriate scales of observation is not a trivial task, which motivated the development of scale-space filters used to create a multiscale image representation [5,15]. Some multiscale shape descriptor that

---

L. A. Ramos—The authors thank São Paulo Research Foundation (FAPESP) for the financial support (grant 2014/10611-0).

M. Mendoza and S. Velastín (Eds.): CIARP 2017, LNCS 10657, pp. 542–549, 2018.
https://doi.org/10.1007/978-3-319-75193-1_65

can be found in the literature are Multiscale Fractal Dimension [2], Multiscale Hough Transform Statistics [10], Multiscale Fourier Descriptor [4], and Curvature Scale-Space [14].

It is well known that multiscale/multiresolution methods are regarded to be consistent with plausible models of the human visual system, therefore, approaches based on this concept can be promising [9]. In this paper, we proposed two new shape descriptors: SBAS, a scale-space version of BAS (Beam Angle Statistics) proposed by [1], and SMFD, a scale-space version of MFD (Multiscale Fractal Dimension) proposed by [2]. Experimental results obtained on two public shape database images showed that both SBAS and SMFD presented better recognition rates than many traditional shape description methods, such as: Zernike Moments [13], BAS [1], HTS (Hough Transform Statistics) [12], MFD [2], TS (Tensor Scale) [7], FD (Fourier Descriptors) [16] and CS (Contour Salience) [14].

2   Scale-Space BAS and MFD

In scale-space theory, the characteristics of interest describe a continuous path in the representation, allowing a consistent manipulation of structures present in different scales. One of the main reasons to represent information at multiple levels is that the successive simplification removes unwanted details, such as noise and insignificant structures. Also, the scale reduction is directly related with information reduction, which implies in a reduction in processing time and increase in computational efficiency [15]. An important property of the scale-space theory is that the transformation to a coarser level should not introduce new structures, thus, structures present in a coarser scale should be present in all others refined scales [15].

In this paper, the image representation in coarse scales was performed by convolving the image with a low-pass filter, the 2D Gaussian kernel, defined by Eq. 1.

$$G(x, y) = \frac{1}{2\pi\sigma^2} e^{-\frac{x^2+y^2}{2\sigma^2}}. \tag{1}$$

where σ is the standard deviation of the Gaussian distribution. The σ value represents the 2D Gaussian kernel's width, and is referred here as the scale parameter. The convolution with the Gaussian kernel causes blurring of the image's border, reducing the information and creating a coarser level of it as the scale (σ) increases. For each scale i, the scale parameter is defined by $\sigma = 2^i$. Figure 1 illustrates the process of Gaussian kernel convolution. One can notice that the higher the value of σ, the coarser the image and the blurrier the borders.

After the convolution, the resulting images are binarized as shown in Fig. 2, so their border can be extracted. Once the borders are blurred and the images are binarized the BAS and MFD methods can be applied for each image from each scale and the feature vectors of each scale are concatenated in a single feature vector making the new SBAS and SMFD shape descriptors, respectively.

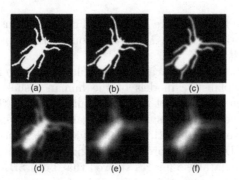

Fig. 1. Examples of images obtained by the convolution with the 2D Gaussian kernel: (a) $\sigma = 2$, (b) $\sigma = 4$, (c) $\sigma = 8$, (d) $\sigma = 16$, (e) $\sigma = 32$, (f) $\sigma = 64$.

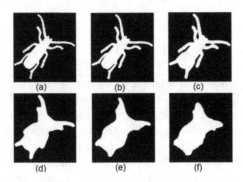

Fig. 2. Binarized images after the convolution with the 2D Gaussian kernel: (a) $\sigma = 2$, (b) $\sigma = 4$, (c) $\sigma = 8$, (d) $\sigma = 16$, (e) $\sigma = 32$, (f) $\sigma = 64$.

3   Experimental Results

In order to evaluate the performance of the proposed methods, SBAS and SMFD, they were applied on two public and well known shape datasets: Kimia-216 [11] and MPEG-7 Part B [3]. Then, their results were compared with results obtained with some well-referenced shape description methods: Zernike Moments [13], BAS [1], HTS (Hough Transform Statistics) [12], MFD [2], TS (Tensor Scale) [7], FD (Fourier Descriptors) [16], and CS (Contour Salience) [14].

The performance comparisons were based on the following metrics:

Precision x Recall: The precision is the fraction of retrieved instances that are relevant, while the recall is the fraction of relevant instances that are retrieved [8];

Multiscale Separability: The Multiscale Separability indicates how clusters of different classes are distributed in the feature space. The more separated the clusters, the better is the descriptor [14];

Bulls-Eye Score: The Bulls-Eye score is calculated as follows: given a dataset $(S_{c,n})$, where c is the number of classes in S and n the number of images per class, each image in $(S_{c,n})$ is used as a query and the number of correct images in the top $2n$ matches is computed. A perfect score is achieved when $c.n^2$ positive cases are found across all the dataset [6].

3.1   Experiments on MPEG-7 Part B Shape Dataset

The MPEG-7 Part B dataset is composed by 1400 images divided into 70 classes of 20 images each, with white silhouette and black background.

The Precision x Recall curves for the MPEG-7 Part B dataset obtained with the proposed shape descriptors (SBAS and SMFD), and with the other shape descriptors (Zernike Moments, BAS, HTS, MFD, TS, FD, and CS) are presented in Fig. 3. One can observe that SBAS presented the best Precision x Recall results (highest curve), followed by its monoscale version BAS. Although the SMFD did not present top results, its performance was better than the results of its monoscale version (MFD).

Precision x Recall

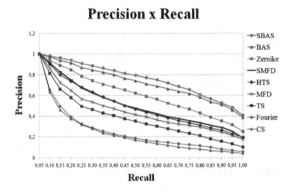

Fig. 3. Precision x Recall curves for the MPEG-7 Part B dataset.

For the shape descriptors that presented the best Precision x Recall curves (SBAS, BAS, Zernike Moments, SMFD, HTS, and MFD), we also calculated their Multiscale Separability curves. Figure 4 presents such curves. One can observe that SBAS also presented the best result according to this measure. The SMFD and Zernike Moments presented very similar results, both outperforming BAS, MFD and HTS.

Finally the Bulls-Eye score was calculated for the methods that presented the best performances according to the Precision x Recall and Multiscale separability results. Table 1 presents the Bulls-Eye score for each method. One can observe that also for this measure SBAS presented the best results, followed by its monoscale version BAS [1] and Zernike Moments [13]. The SMFD method showed better results than its monoscale version MFD [2] and than HTS [12].

From all these results we conclude that the proposed scale-space approach for shape recognition significantly improved the accuracy of the monoscale shape recognition approach.

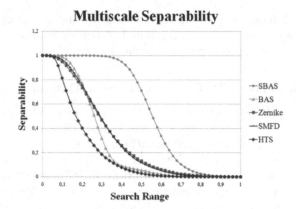

Fig. 4. Multiscale separability curves for the MPEG-7 Part B dataset [3].

Table 1. Bulls-Eye scores for each method using the MPEG-7 Part B dataset [3].

Shape descriptor	Bulls-Eye score
SBAS	0.80336
BAS	0.76646
Zernike M	0.70220
SMFD	0.57750
HTS	0.56014
MFD	0.52568

3.2 Experiments on Kimia-216 Shape Dataset

The Kimia-216 shape dataset [11] is composed by 216 images divided into 18 classes of 12 images each. This dataset is simpler than MPEG-7 Part B, because it is composed by fewer classes, and they do not present many transformations as the MPEG-7 Part B classes do.

The Precision x Recall curves for the Kimia-216 dataset obtained with the proposed shape descriptors (SBAS and SMFD), and with the other shape descriptors (Zernike Moments, BAS, HTS, MFD, TS, FD, and CS) are presented in Fig. 5. One can observe that SBAS presented best Precision x Recall results for most parts of the curve, followed closely by its monoscale version BAS. Although the SMFD did not present top results, its performance in this dataset was also improved when compared to its monoscale version MFD, likewise in MPEG-7 Part B dataset.

For the shape descriptors that presented the best Precision x Recall curves (SBAS, BAS, Zernike Moments, HTS, SMFD and MFD), we also calculated their Multiscale Separability curves. Figure 6 presents such curves. One can observe that SBAS also presented the best results according to this measure, followed by Zernike Moments. The SMFD did not present the top results, but it was significantly better than BAS and MFD.

Precision X Recall

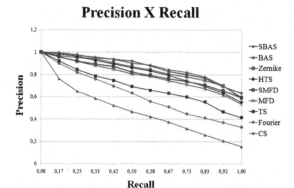

Fig. 5. Precision x Recall curves for the Kimia-216 dataset [11].

Multiscale Separability

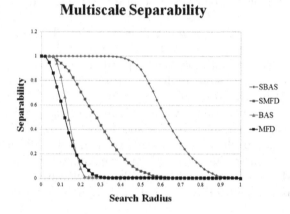

Fig. 6. Multiscale separability curves for the Kimia-216 dataset [11].

Finally the Bulls-Eye score was calculated for the methods that presented the best performances according to the Precision x Recall and Multiscale separability results. Table 2 presents the Bulls-Eye score from each method. One can observe that also for this measure SBAS presented the best results, followed by its monoscale version BAS. The SMFD method showed better results than its monoscale version MFD.

Likewise in the MPEG-7 Part B dataset, from all the results obtained on Kimia-216 dataset, we conclude that the proposed scale-space approach for shape recognition significantly improved the accuracy of the monoscale shape recognition approach.

Table 2. Bulls-Eye scores for each method using the Kimia-216 dataset [11].

Shape descriptor	Bulls-Eye score
SBAS	0.89426
BAS	0.88850
HTS	0.88542
SMFD	0.85725
MFD	0.85494

4   Discussion and Conclusion

In this paper we presented two new shape description methods, SBAS and SMFD. Experiments carried out on MPEG-7 Part B dataset showed that the SBAS presented the best results among several well-referenced shape description methods in the literature, such as: Zernike Moments [13], BAS [1], HTS (Hough Transform Statistics) [12], MFD [2], TS (Tensor Scale) [7], FD (Fourier Descriptors) [16] and CS (Contour Salience) [14], for all three evaluation metrics used in this work (Precision x Recall, Multiscale Separability and Bulls-Eye). While the SMFD did not present so good results, it performed better than its monoscale version, the MFD shape descriptor.

Regarding the results obtained with the Kimia-216 dataset, SBAS presented the best Multiscale Separability results and Bulls-Eye score. The SMFD also presented better results than its monoscale version, but it did not outperform other methods. It is important to notice that the Kimia-216 dataset is a simpler dataset than MPEG-7 Part B and the results obtained by the methods in the Kimia-216 dataset are already very good, making it harder to obtain relevant improved results.

From the obtained results, one can observe that the new descriptor SBAS showed better results than all methods compared in this paper, improving the monoscale version of BAS [1] accuracy in approximately 5.3% according to the Bulls-Eye score for the MPEG-7 Part B dataset, and in 0.65% for the Kimia-216 dataset.

Therefore, the results obtained in this paper suggests that the proposed scale-space approach for shape recognition can significantly improve the accuracy of any shape description method already proposed in literature that does not explores the scale-space. In this work we assessed the BAS and MFD shape descriptors. However, since the proposed multiscale approach is applied in the pre-processing stage, it can be applied in any other shape description method.

References

1. Arica, N., Vural, F.T.Y.: BAS: a perceptual shape descriptor based on the beam angle statistics. Patt. Recogn. Lett. **24**(9–10), 1627–1639 (2003)
2. Florindo, J.B., Backes, A.R., Castro, M., Bruno, O.M.: A comparative study on multiscale fractal dimension descriptors. Pattern Recog. Lett. **33**(6), 768–806 (2012)

3. Jeannin, S., Bober, M.: Description of Core Experiments for MPEG-7 Motion/Shape (1999)
4. Kunttu, L., Lepist, L., Rauhamaa, J., Visa, A.: Multiscale fourier descriptor for shape-based image retrieval. In: 17th ICPR, pp. 765–768 (2004)
5. Lindeberg, L.: Scale-Space Theory in Computer Vision. Kluwer Academic Publishers, Norwell (1994)
6. Liu, M., Vemuri, B.C., Amari, S., Nielsen, F.: Shape retrieval using hierarchical total Bregman soft clustering: overview and proposals. IEEE Trans. Pattern Anal. Mach. Learn. **34**(12), 2407–2419 (2012)
7. Miranda, P.A.V., Torres, R.S., Falcão, A.X.: TSD: a shape descriptor based on a distribution of tensor scale local orientation. In: 27th SIBGRAPI, pp. 139–146 (2005)
8. Müller, H., Müller, W., Squire, D.M., Maillet, S.M., Pun, T.: Performance evaluation in content-based image retrieval: overview and proposals. Pattern Recog. Lett. **22**(5), 593–601 (2001)
9. Phongsuphap, S., Takamatsu, R., Sato, M.: Multiscale image analysis through the surface shape operator. J. Electron. Imaging **9**(3), 305–316 (2000)
10. Ramos, L.A., de Souza, G.B., Marana, A.N.: Shape analysis using multiscale hough transform statistics. Progress in Pattern Recognition, Image Analysis, Computer Vision, and Applications. LNCS, vol. 9423, pp. 452–459. Springer, Cham (2015). https://doi.org/10.1007/978-3-319-25751-8_54
11. Sebastian, T., Klein, P., Kimia, B.: Recognition of shapes by editing their shock graphs. IEEE Trans. PAMI **26**(5), 550–571 (2004)
12. Souza, G.B., Marana, A.N.: HTS and HTSn: new shape descriptors based on Hough transform statistics. CVIU **127**, 43–56 (2014)
13. Teague, M.R.: Image analysis via the general theory of moments. J. Opt. Soc. Am. **70**(8), 920–930 (1980)
14. Torres, R.S., Falco, A.X.: Contour salience descriptors for effective image retrieval and analysis. Image Vision Comput. **25**, 3–13 (2007)
15. Witkin, A.: Scale-space filtering: a new approach to multi-scale description. In: 9th IEEE ICASSP (1984)
16. Zhang, D., Lu, G.: A comparative study of fourier descriptors for shape representation and retrieval. In: 5th Asian Conference on Computer Vision, Patterns and Images, pp. 646–651 (2002)

An Unsupervised Approach for Eye Sclera Segmentation

Daniel Riccio[1,2]([email]) [id], Nadia Brancati[2], Maria Frucci[2], and Diego Gragnaniello[2]

[1] University of Naples "Federico II", Naples, Italy
daniel.riccio@unina.it
[2] Institute for High Performance Computing and Networking,
National Research Council of Italy (ICAR-CNR), Naples, Italy
{nadia.brancati,maria.frucci,diego.gragnaniello}@cnr.it

Abstract. We present an unsupervised sclera segmentation method for eye color images. The proposed approach operates on a visible spectrum RGB eye image and does not require any prior knowledge such as eyelid or iris center coordinate detection. The eye color input image is enhanced by an adaptive histogram normalization to produce a gray level image in which the sclera is highlighted. A feature extraction process is involved both in the image binarization and in the computation of scores to assign to each connected components of the foreground. The binarization process is based on clustering and adaptive thresholding. Finally, the selection of foreground components identifying the sclera is performed on the analysis of the computed scores and of the positions between the foreground components. The proposed method was ranked 2nd in the Sclera Segmentation and Eye Recognition Benchmarking Competition (SSRBC 2017), providing satisfactory performance in terms of precision.

Keywords: Sclera segmentation · Gray level clustering
Feature extraction

1 Introduction

The sclera is a white region in the eye, around the eyeball. It contains randomly distributed blood vessels, which have recently been investigated as a biometric trait in an eye recognition system for personal identification [5,7]. The first step for an eye recognition system based on the extraction of the sclera vessels is sclera segmentation. However, uncontrolled human posing, multiple iris gaze directions, and illumination variations result in many challenges in sclera segmentation. For these reasons, some sclera vessel identification methods adopt manual sclera segmentation [9,10].

Several methods for sclera segmentation have been proposed in the literature. A semi-automatic technique is proposed in [4], where the output of an automatic clustering method is refined by manual intervention. Some fully automatic methods are robust to illumination variations and to occluded sclera regions and these

© Springer International Publishing AG, part of Springer Nature 2018
M. Mendoza and S. Velastín (Eds.): CIARP 2017, LNCS 10657, pp. 550–557, 2018.
https://doi.org/10.1007/978-3-319-75193-1_66

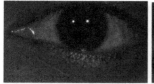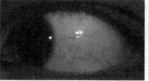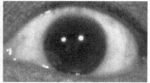

Fig. 1. Examples of images with different issues

are based on color image segmentation in the HSV color space [13,14]. An adaptive approach to sclera segmentation based on active contours has been presented in [3]. The same authors propose a sclera segmentation method based on a fusion technique which processes each pixel of the eye image in different color spaces [2]. A method based on multiple classifier architecture is presented in [12].

In this paper, a fully automatic unsupervised sclera segmentation method is proposed. This approach has been designed to take into account the different iris gaze directions and illumination variations in eye images. It allows you to obtain a correct segmentation independently of whether or not the sclera appears divided into many parts, whether only one part of the sclera is visible or whether the sclera completely surrounds the iris (see Fig. 1).

The color image is converted into a gray level image, where the sclera is highlighted. Next, the image is binarized by means of a clustering and an adaptive thresholding. For each foreground component of the binary image, a score is computed taking into account the shape and geometric features of the component. Finally, the sclera is detected by analyzing the score information and relative positions of the candidate components.

Our algorithm was submitted to the Sclera Segmentation and Eye Recognition Benchmarking Competition (SSRBC 2017) [1,6]. The dataset of SSRBC 2017 included eye images acquired in the visible spectrum and having a high degree of variation in illumination and eye pose.

The rest of the paper is organized as follows: in Sect. 2, the proposed method is presented; Sect. 3 describes the obtained results; and finally in Sect. 4 some conclusions are drawn.

2   The Method

The sclera is a white area so that its pixels have a high intensity in each channel of the RGB color space. Skin pixels can also have such a feature, but the intensity in the red channel (R) is generally higher than one associated in the blue (B) and green (G) channels. To discriminate the color information associated with the sclera from a characterizing skin region, the color input image is converted into a gray level image by merging the three channels with a low contribution of the red channel.

The input image I is decomposed into R, G and B channels and the intensities of the each channel C are mapped with a range from 0 to 1 by means of a non-linear transformation based on a quasi-sigmoid function defined in [8]:

$$C' = 128 \left(1 + \frac{1 - b^{\frac{C}{1.5\mu}}}{ab^{\frac{C}{1.5\mu}} + 1} \right) \tag{1}$$

with a and b being pre-defined constant terms with values of $a = 2 + \sqrt{3}$ and $b = 7 - 4\sqrt{3}$, respectively. The choice of the μ value assumes a fundamental rule in order to avoid the possibility that intensity values greater than μ are predominantly mapped onto high values in the range $[0, 1]$. We assume:

$$\mu = min(nR, nG, nB) \tag{2}$$

where

$$nR = mean(R) + \sigma(R)/2$$
$$nG = max(G) + \sigma(G)/2$$
$$nB = max(B) + \sigma(B)/2$$

where $\sigma(R)$ $(\sigma(G), \sigma(B))$ represents the standard deviation of intensity values in the R (G, B) channel(s). Since the mean value of R is considered, differently from G and B for which the maximum value is computed, the red channel contributes with a low weight to the calculation of μ. Next, a gray level image I_Q is obtained by combining the normalized channels R', G' and B' according to the following formula:

$$I_Q = B' + G' - R' \tag{3}$$

Finally, a full scale histogram stretch of I_Q in the range $[0, 255]$ is performed. Note that the red channel R' is subtracted in the combination of normalized channels of I in such a way as to highlight the sclera. Figure 2 shows the difference between the gray level image obtained using a classic weighted average method to convert a color image in grayscale and the one computed according to (3) for a running example.

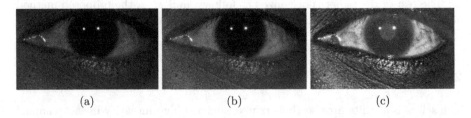

      (a)   (b)   (c)

Fig. 2. (a) Running example, (b) gray level image according to a classic weighted average method, and (c) gray level image according to (3). (Color figure online)

The second step of the method is based on an incremental clustering algorithm [11] to divide I_Q into a set of regions characterized by gray levels with similar information content. For each couple q_i and q_j of gray levels in I_Q, a distance measure $d(q_i, q_j)$ is computed by summing the following parameters:

- the actual difference between q_i and q_j defined as $log(|q_i - q_j| + 1)$;
- the weighted difference between q_i and q_j, given by

$$1/|I_Q| \cdot (|w(q_i)q_i - w(q_j)q_j|) \tag{4}$$

where $w(q_i)$ $(w(q_j))$ represents the occurrence of $q_i(q_j)$ in I_Q; and
- the *joint sparsity index* defined as:

$$\sigma(q_i, q_j) = \frac{min\{\sigma(q_i), \sigma(q_j)\}}{max\{\sigma(q_i), \sigma(q_j)\}} \tag{5}$$

where $\sigma(q_i)$ $(\sigma(q_j))$ is the standard deviation of the coordinates of the pixels with a gray level q_i (q_j).

The assignment criterion of the gray level is based on the thresholding of the distance measure, which has to be smaller than a fixed value t. In our experiments t has been set to 1.5. This incremental clustering algorithm produces h clusters (with $0 < h < 255$), such that indexes assigned to the cluster increase proportionally to gray level of the pixel they include. A new image I'_Q is obtained, by assigning to each pixel p, with gray level g_i in I_Q, the label associated with the cluster including g_i. Figure 3a shows I'_Q in false color.

Thresholding of I'_Q allows an assignment of non sclera pixels to the background. In particular, each pixel p of I'_Q is assigned to the foreground if its value is larger than t', where $t' = avg(I'_Q) + 0.75 \cdot \sigma(I'_Q)$. Let I''_Q be the obtained binarized image (see Fig. 3b).

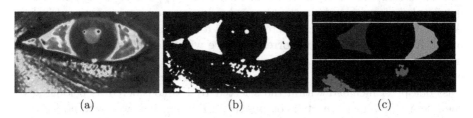

(a)   (b)   (c)

Fig. 3. (a) Clustered image, (b) binary image, and (c) candidate connected components. The green component has the highest score value, the blue component falls inside the region of interest (ROI). (Color figure online)

In the following section, only connected components of the foreground pixels will be taken into account.

The third step of the method involves the computation of a score for each connected component F_k. Such a score is computed on the basis of the following parameters:

- the Proximity of F_k to the center of the image computed as:

$$\alpha(F_k) = 1/2 \cdot \delta/\sqrt{W^2 + H^2}$$

where δ is the Euclidean distance between the center of F_k and the center of the image. The values W and H are the width and height of the image.

- the Density $\omega(F_k)$ computed as the ratio between the area of F_k and the area of the convex hull of F_k.
- the Compactness of F_k defined as:

$$\gamma(F_k) = 1/|F_k| \cdot \left(\sum_{i,j} g(i,j)/bg(i,j)\right)$$

where $g(i,j)$ is the gray level in I_Q of the pixel p of F_k with coordinates (i,j), while

$$bg(i,j) = \begin{cases} g(i,j) + \alpha(F_k) \ if \ p \ is \ a \ border \ pixel \\ \alpha(F_k) \ otherwise \end{cases}$$

Note that before the computation of $\gamma(F_k)$, a process is performed to fill holes included in F_k.

Finally, the score of F_k is given by:

$$s(F_k) = \gamma(F_k) \cdot \omega(F_k)$$

The foreground connected component of I_Q'' with the highest score is considered as belonging to the sclera (the green connected component in Fig. 3c). Let F_m be the selected connected component. Since in an eye image the sclera can be divided into many regions, as in the case of the running example, a further selection of the foreground connected components is performed. Precisely, only connected components F_k with $k \neq m$ such that $s(F_k) > 0.70 \cdot s(F_m)$ are candidates to belong to the sclera (the blue and red connected components in Fig. 3c). A region of interest (ROI) is selected in such a way that the bottom and the top of the ROI are positioned in correspondence with the lowest border pixel and highest border pixel of F_m, respectively (see Fig. 3c). The width of the ROI is equivalent to the width of the image. Any candidate component F_k having at least one pixel included within the selected ROI is considered as belonging to the sclera (the blue connected component in Fig. 3c). It is worth to notice that all parameters have been set experimentally so to obtain the best segmentation performance.

Finally, a process to fill the holes of the detected sclera regions is performed.

3   Results

The experiments have been performed on an Intel Pentium i7-6900K@3.2 GHz with 8 GB of RAM. Our Matlab implementation of the method requires 0.34 s to segment an eye image with resolution 1295×576 pixels.

Some results of our method can be observed in Fig. 4, where in each line the input image, the ground truth and the result of our method are shown from left to right.

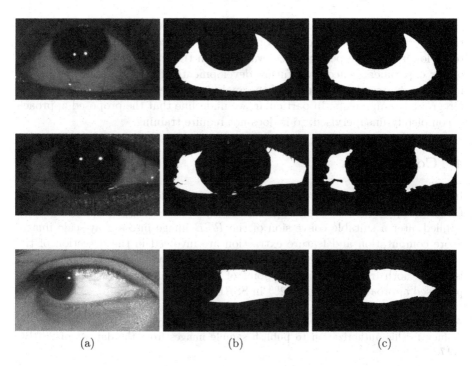

Fig. 4. Results of our method: (a) original image, (b) ground truth, (c) results of the proposed method.

Table 1. Results of SSRBC 2017

Rank	Precision in %	Recall in %
1	95.34	96.65
2	85.59	64.60
3	76.45	62.89
4	55.72	88.16
5	53.91	49.92

An evaluation of the proposed method has been performed by a team of the SSRBC 2017 organization [1,6] in terms of precision and recall. Precision was considered the most important measure for the performance evaluation. The results of the competition are shown in Table 1.

Our algorithm obtained a satisfactory result in terms of precision and it was ranked in second position in the classification.

The tuning dataset provided by SSRBC 2017 to the participants in competition contained 120 eye images from 30 individuals. The images were acquired under different illumination conditions and contained noise such as reflections or occluded sclera regions. We obtained 95.47% in terms of accuracy, 98.80% in specificity, 82.32% in recall and 90.36% in precision. Although the evaluation of the performance of our algorithm on the whole dataset of SSRBC 2017 proved to be lower than the performance evaluated on the tuning data set, the results of our experiments encourage future developments of the proposed method by improving, particularly in terms of the analysis of the scores and positions of the foreground components. In particular, we underline that the proposed approach is completely unsupervised, so it does not require training.

4   Conclusions

In this work, we have presented an unsupervised sclera segmentation method for visible spectrum eye images. The method is based on a gray level clustering, applied after a suitable conversion of the RGB image into a gray-scale image. Score computation and feature extraction are involved in the detection of the components representing the sclera. The method provides results that are satisfactory both from a qualitative point of view and in terms of precision. The proposed approach was ranked 2nd in SSRBC 2017.

Acknowledgements. We would like to thank Dr. Abhijit Das of the University of Sydney for his authorization to publish sample images from the dataset of SSRBC 2017.

References

1. IJCB 2017. http://www.ijcb2017.org/ijcb2017/competitions.php
2. Alkassar, S., Woo, W., Dlay, S., Chambers, J.: Enhanced segmentation and complex-sclera features for human recognition with unconstrained visible-wavelength imaging. In: 2016 International Conference on Biometrics (ICB), pp. 1–8. IEEE (2016)
3. Alkassar, S., Woo, W.L., Dlay, S.S., Chambers, J.A.: Robust sclera recognition system with novel sclera segmentation and validation techniques. IEEE Trans. Syst. Man Cybern.: Syst. **47**(3), 474–486 (2017)
4. Crihalmeanu, S., Ross, A., Derakhshani, R.: Enhancement and registration schemes for matching conjunctival vasculature. In: Tistarelli, M., Nixon, M.S. (eds.) ICB 2009. LNCS, vol. 5558, pp. 1240–1249. Springer, Heidelberg (2009). https://doi.org/10.1007/978-3-642-01793-3_125
5. Das, A., Pal, U., Ballester, M.A.F., Blumenstein, M.: Fuzzy logic based selera recognition. In: 2014 IEEE International Conference on Fuzzy Systems (FUZZ-IEEE), pp. 561–568. IEEE (2014)
6. Das, A., Pal, U., Ballester, M.A.F., Blumenstein, M.: Multi-angle based lively sclera biometrics at a distance. In: 2014 IEEE Symposium on Computational Intelligence in Biometrics and Identity Management (CIBIM), pp. 22–29. IEEE (2014)
7. Das, A., Pal, U., Blumenstein, M., Ballester, M.A.F.: Sclera recognition-a survey. In: 2013 2nd IAPR Asian Conference on Pattern Recognition (ACPR), pp. 917–921. IEEE (2013)

8. De Marsico, M., Riccio, D.: A new data normalization function for multibiometric contexts: a case study. In: Campilho, A., Kamel, M. (eds.) ICIAR 2008. LNCS, vol. 5112, pp. 1033–1040. Springer, Heidelberg (2008). https://doi.org/10.1007/978-3-540-69812-8_103

9. Derakhshani, R., Ross, A.: A texture-based neural network classifier for biometric identification using ocular surface vasculature. In: International Joint Conference on Neural Networks, IJCNN 2007, pp. 2982–2987. IEEE (2007)

10. Derakhshani, R., Ross, A., Crihalmeanu, S.: A new biometric modality based on conjunctival vasculature. In: Proceedings of Artificial Neural Networks in Engineering (ANNIE), pp. 1–8 (2006)

11. Jain, A.K., Murty, M.N., Flynn, P.J.: Data clustering: a review. ACM Comput. Surv. (CSUR) **31**(3), 264–323 (1999)

12. Radu, P., Ferryman, J., Wild, P.: A robust sclera segmentation algorithm. In: 2015 IEEE 7th International Conference on Biometrics Theory, Applications and Systems (BTAS), pp. 1–6. IEEE (2015)

13. Thomas, N.L., Du, Y., Zhou, Z.: A new approach for sclera vein recognition. In: SPIE Defense, Security, and Sensing, p. 770805. International Society for Optics and Photonics (2010)

14. Zhou, Z., Du, E.Y., Thomas, N.L., Delp, E.J.: A new human identification method: sclera recognition. IEEE Trans. Syst. Man. Cybern.-Part A: Syst. Hum. **42**(3), 571–583 (2012)

An Approach to Clustering Using the Expectation-Maximization and Selection of Attributes ReliefF Applied to Water Treatment Plants process

Fábio Cosme Rodrigues dos Santos[1,2], André Felipe Henriques Librantz[1(✉)], and Renato José Sassi[1]

[1] Nove de Julho University - UNINOVE, São Paulo, SP 01504-000, Brazil
librantz@uninove.br
[2] Basic Sanitation Company of the State of São Paulo - SABESP, São Paulo, SP 05429-010, Brazil

Abstract. The water treatment process contains several physico-chemical parameters relevant to decision making and the water quality scenarios' identification. Some scenarios are evident and can be observed without the application of mathematical or statistical techniques, however some of these scenarios are difficult to distinguish, and it is necessary to use computational intelligence techniques for solution. In this context, the paper aims to show the application of the expectation-maximization (EM) algorithm for data clusters of the coagulation process and the ReliefF algorithm to determine the importance of the physico-chemical parameters, using the WEKA tool to analyze historical dataset of a water treatment plant. The results were favorable to the scenarios' identification and to determine the relevance of the parameters related to the process.

Keywords: Clustering · Expectation-maximization · ReliefF
Water treatment · Coagulant dosage

1 Introduction

Fresh water available in rivers or dams contains micro-organisms and impurities that may be harmful to human health. When captured, it is adduced to the treatment plant with the objective of removing all undesirable substances, in order to meet the potability requirements defined by legislation [5,12].

According to [11], the water treatment process has several physical-chemical parameters that must be monitored and controlled in order to guarantee the quality of the final product. One of the more complex subsystems is coagulation, which consists in destabilizing the dirt particles, so they are retained in the later processes of the water treatment plant [4].

© Springer International Publishing AG, part of Springer Nature 2018
M. Mendoza and S. Velastín (Eds.): CIARP 2017, LNCS 10657, pp. 558–565, 2018.
https://doi.org/10.1007/978-3-319-75193-1_67

The coagulation subsystem has as main parameters the raw water quality and intermediate subsystems' parameters, in this order, making possible the measurement of the efficacy of the coagulation process. The reference dosage of coagulant in water is determined by bench-testing the treatment process or by measuring the electrical charge every time the raw water quality scenario has changes [2, 17].

Although the scenarios represent droughts and rainy seasons, it is not easy to classify them, especially when there is a possibility of adjusting the chemical dosage reference values as a result of the transition between the periods that most affect quality of raw water [2].

One way to find possible scenarios in water treatment plant is the use of computational intelligence techniques, specifically data clustering techniques [6]. According to [13], the use of artificial neural network architecture for clustering Self Organizing Maps (SOM) showed good results in the detection of contamination problems in wellsprings and river, allowing the development of corrective measures to avoid health problems.

In [5] the possibility of internal problems detection in the water treatment plant caused by equipment faults or wellspring conditions was shown, also using the SOM network, allowing to anticipate measures so that the quality of the treated water is kept within the limits settled down.

Several segments of the sanitation area have applied SOM network as a technique for solving problems, presenting favorable results, such as [7] in validation and reconstruction of information in databases with the presence of incoherent values and [3] in the identification of water quality and process behavior.

Other techniques are used in water treament plants to study process behaviors by clustering information. In [14] k-means was used as part of their study of a coagulant dosage prediction model, while [1] used the expectation-maximization technique in order to determine the appropriate amount of clusters and their characteristics of information from lakes in Alberta city, in Canada.

In [16] an evolutionary algorithm and expectation-maximization were applied to improving detection of pipe bursts and other events in water distribution systems. The results obtained have shown that the use of these strategies could improve the performance in terms of event detection performance.

In [19] the authors proposed a polynomial function fitted to the historic flow measurements based on a weighted least-squares method in conjunction with an expectation-maximization algorithm for automatic burst detection in the U.K. water distribution networks. This approach can automatically select useful data from the historic flow measurements, which may contain normal and abnormal operating conditions in the distribution network, e.g., water burst.

In [8] the EM technique was used in conjunction with a Bayesian model with the purpose of predicting leakage in water distribution networks, showing the possibility of minimizing the risks of water loss.

Thus, the objective of this work was to propose a clustering approach and the relevance of each physical-chemical parameter determination, using the expectation-maximization technique and the ReliefF algorithm respectivily, in the coagulation process data from a water treatment plant located in the metropolitan region of São Paulo, Brazil.

2   Theoretical Background

2.1   Expectation-Maximization (EM) Algorithm

The expectation-maximization algorithm can be considered a data mining technique. It aims to find the maximum likelihood parameters in a base of information considered incomplete, allowing the application such as pattern recognition with neural networks [10].

The basic algorithm idea consists in the representation of a problem by two information called x and y. The term Y can be considered as an observed data random vector y with probability density function, denoted by $g\left(y|\varphi\right)$, where $\varphi = [\varphi_1, ..., \varphi_d]^T$ represents the unknown parameter vector. The term X corresponds to the random vector of the total data vector x, denoted as a probability density function $g_c\left(x|\varphi\right)$. Thus, φ can be the maximization parameter in Eq. 1 [9].

$$logL_c(\varphi) = log\, g_c\left(x|\varphi\right) \tag{1}$$

According to [9], there are two steps for each algorithm iteration, as follows:

- Expectation ou E-Step: aimed to finding the clusters' probabilities:

$$Q\left(\varphi|\varphi^{(i)}\right) = E_{\varphi^{(i)}}\left[logL_c\left(\varphi\right)|y\right] \tag{2}$$

- Maximization ou M-Step: corresponds to the maximization $M(\varphi^{(i)})$ through Eqs. 3 and 4:

$$Q\left(\varphi^{(i+1)}|\varphi^{(i)}\right) \geq Q\left(\varphi|\varphi^{(i)}\right) \tag{3}$$

$$M\left(\varphi^{(i)}\right) = \arg\max_{\varphi} Q\left(\varphi|\varphi^{(i)}\right) \tag{4}$$

where: i is the number of iterations.

2.2   ReliefF Algorithm

The ReliefF algorithm has the purpose of qualifying attributes according to proximity and distinction. The basic operation idea is to obtain the updated vector with the attributes' quality [15].

The algorithm starts resetting the quality vector $W[x]$ and randomly selects an instance R and searches for the nearest neighbor of the same class, called H and other classes, known as M. Then, it is verified which class instance has the most relevance and, finally, updates the quality vector, according to Eq. 5:

$$[x] = W\left[x\right] - \textstyle\sum_{=1}^k \frac{dist(x,R,H)}{m.k} + \sum_{C \neq class(R)} \frac{\frac{P()}{1-P(class(R))}\sum_{j=1}^k dist\left(x,R,M\right)}{m.k} \tag{5}$$

where: C is the class identification, $P(C)$ class probability, $1 - P(Class(R))$ the sum of probabilities of the different classes and m the number of iterations.

The dist function calculates the difference between the atribute's values and the two instances (M and H), with a range from 0 to 1. After the finalization, we have the vector of weights $W[x]$ from all the attributes and the highest weight is considered as the most relevant atribute.

3 Materials e Methods

3.1 Characteristics of the Studied Process

The Alto Cotia water treatment plant (WTP) has a nominal production capacity of 1.25 m^3/s and is located in the metropolitan region of São Paulo. This WTP has two coagulant dosage application points due to an expansion to increase treatment capacity, known as dosing system by gravity and dosing system by pumping. Data were collected from SABESP's laboratory management system on the period from 2010 to 2012, totalizing 6686 records. Table 1 shows the maximum and minimum values of each parameter.

Table 1. Maximum and minimum values of each parameter

Parameter	Maximum	Minimum
Raw water turbidity (NTU)	23.90	3.19
Raw water color (uC)	135	25
Clarified water turbidity (NTU)	1.80	0.90
Filtered water turbidity (NTU)	0.60	0.20
Residual aluminum of clarified water (mg/L)	0.54	0.06
Residual aluminum of treated water (mg/L)	0.20	0.05
Coagulant dosage by gravity system (mg/L)	46.57	8.36
Coagulant dosage by pumping system (mg/L)	48.90	7.29

The physical-chemical parameters that refer to the raw water quality are gross water turbidity and gross water color. The effectiveness of the coagulation process is measured by physicochemical parameters: clarified water turbidity, residual aluminum of clarified water, filtered water turbidity and residual aluminum of treated water. The dosage references of coagulant are: coagulant dosage by gravity system and coagulant dosage by pumping system. These variables are considered as outputs, as they are values used in the water treatment process.

3.2 Computational Experiments

The computational experiments were carried out with WEKA software, version 3.8.10 x64 [18]. All the algorithms used in the work are available in this tool.

The experiments were divided into the following steps: (i) Pre-processing of data: consists on the elimination of inconsistent and null values of the database collected at Sabesp's laboratory management system; (ii) Use of the ReliefF algorithm in the database with preprocessing: the attribute selection algorithm was applied to verify the importance of collected data without the clustering techniques application, having as reference, the coagulant dosage variables the coagulant dosage by gravity system and the coagulant dosage by pumping system; (iii) Data clustering: use the EM algorithm for clustering, with an automatic cluster quantification function enabled, in the WEKA tool. In this stage the data mining technique was applied; (iv) Clusters' selection: the selection of clusters was based on information that represented optimized values on coagulation process of water treatment plant; (v) Use of the ReliefF algorithm in the selected database: applied the attribute selection algorithm again in the information, which is contained in the selected clusters, also, having as systemic coagulant dosage variables by gravity and coagulant dosage by pumping system, applied feature selection algorithm again to the information; (vi) Attribute selections' comparison: the comparison of the importance of the attributes between the clustered database and the database with preprocessing was performed.

4   Results and Discussions

The pre-processed database, with 6686 registers, were processed by WEKA clustering algorithm, in 6965.12 seconds. The WEKA generated 20 clusters for the submitted database. The quantities of records in each cluster and their percentage are presented in Table 2.

Table 2. Number of records per cluster

Cluster	Record amount	Percentage	Cluster	Record amount	Percentage
0	222	3%	11	633	9%
1	358	5%	12	147	2%
2	142	2%	13	56	1%
3	258	4%	14	223	3%
4	169	3%	15	188	3%
5	545	8%	16	357	5%
6	421	6%	17	608	9%
7	299	4%	18	249	4%
8	453	7%	19	518	8%
9	647	10%	Mean	334	5%
10	193	3%			

It can be seen in Table 2 that some clusters contain different numbers in comparison with the mean, demonstrating the existence of several scenarios into varied clusters, which may be related to the raw water quality or the efficacy of the coagulation process by measuring the turbidity parameters of clarified and filtered water.

Table 3 shows the means and standard deviations of each attribute grouped in clusters obtained by the EM algorithm. The data presented in Table 3 shows that the clustering algorithm considered the raw water parameters and dosages as relevant to define the clusters. Thus, it can be observed that clusters 0, 2, 6, 12, 13, 14 and 16, highlighted in red, present high values of dosages in both systems (by gravity and pumping) for very close values on turbidity and color of raw water, which represents inconsistent information.

Therefore, the data selected as optimized values of the coagulant dosage process are presented in the clusters 1, 3, 4, 5, 7, 8, 9, 10, 11, 15, 17, 18 and 19. From the total of 6686 data, 5118 are used, representing a reduction of 23.45 % in the pre-processed database.

The relevance of attributes or parameters was evaluated by the ReliefF algorithm in the pre-processed and mined data, to verify if there was a change in the significance of the attributes after the selection of clusters with optimized values of the process. Table 4 shows the most relevant parameters in relation to the dosing variables by gravity and pumping systems.

Table 3. Mean values and standard deviations of each attribute and each cluster

Cluster	Raw water color		Raw water turbidity		Clarified water turbidity		Filtered water turbidity		Aluminum of clarified water		Aluminum of filtered water		Dosage System Gravity		Dosing System Pumping	
	Mean	Standard deviation	Mean	Standard deviation	Mean	Standard deviation	Mean	Standard deviation	Mean	Standard deviation	Mean	Standard deviation	Mean	Standard deviation	Mean	Standard deviation
00	35.58	3.43	3.89	0.33	1.04	0.13	0.40	0.08	0.28	0.05	0.19	0.03	23.17	2.30	22.99	2.18
01	77.87	8.70	11.09	1.89	1.36	0.22	0.31	0.08	0.19	0.04	0.11	0.03	26.10	1.98	26.32	2.21
02	45.48	5.30	5.54	0.46	0.97	0.05	0.38	0.06	0.21	0.03	0.14	0.02	22.64	2.65	22.66	2.49
03	61.31	5.94	7.76	0.91	1.21	0.21	0.32	0.06	0.16	0.02	0.09	0.02	21.34	1.41	21.01	1.70
04	81.65	10.60	11.37	2.04	1.21	0.19	0.35	0.07	0.20	0.04	0.12	0.02	34.04	1.91	33.80	1.52
05	54.32	4.09	6.99	0.75	1.19	0.17	0.39	0.08	0.23	0.04	0.14	0.03	19.98	1.08	19.80	1.38
06	37.72	4.45	4.35	0.80	1.17	0.16	0.43	0.07	0.27	0.05	0.17	0.03	18.37	0.98	17.85	1.23
07	35.35	3.13	3.73	0.23	1.27	0.17	0.47	0.05	0.20	0.04	0.10	0.02	11.14	0.77	9.48	0.60
08	56.49	5.59	7.02	0.73	1.16	0.16	0.36	0.06	0.18	0.03	0.11	0.02	17.64	1.22	17.48	1.32
09	38.37	2.89	4.52	0.46	1.20	0.14	0.43	0.05	0.21	0.02	0.14	0.02	14.96	1.52	15.18	1.36
10	110.99	12.01	17.77	2.84	1.46	0.20	0.33	0.08	0.22	0.06	0.12	0.03	39.94	4.65	39.94	5.04
11	48.08	4.23	5.72	0.62	1.25	0.19	0.40	0.07	0.20	0.03	0.12	0.02	15.45	1.51	16.07	1.16
12	59.89	5.19	7.60	0.75	1.04	0.09	0.33	0.07	0.20	0.03	0.13	0.02	27.54	2.07	27.62	2.02
13	80.33	10.72	11.22	1.80	1.38	0.21	0.45	0.04	0.32	0.06	0.20	0.02	36.42	2.36	35.72	1.93
14	58.21	7.47	7.11	1.47	1.21	0.19	0.46	0.05	0.26	0.05	0.18	0.03	24.49	1.48	24.98	1.45
15	36.47	3.39	4.33	0.44	1.51	0.16	0.45	0.05	0.24	0.05	0.12	0.03	13.95	1.26	12.90	1.20
16	48.18	6.53	5.49	0.72	1.12	0.16	0.37	0.07	0.19	0.02	0.13	0.01	11.44	1.34	12.05	1.36
17	36.94	2.15	4.00	0.20	1.04	0.10	0.34	0.06	0.16	0.01	0.10	0.01	11.32	1.18	12.00	1.35
18	64.59	6.41	8.91	1.14	1.42	0.18	0.42	0.06	0.21	0.02	0.12	0.02	20.23	2.36	19.89	2.04
19	39.27	2.89	4.57	0.50	1.07	0.11	0.34	0.06	0.17	0.03	0.11	0.02	13.87	1.78	14.44	1.72

The data importance in the dosing process of coagulant in the lack of data mining shows, according to Table 4, that the quality of the raw water (highlighted in green) is fundamental for the control of the dosage. In addtion, the dependence between the dosages of the systems in both scenarios, pre-processing and clustering stages, is noticed.

In applying the data mining methods, the parameters that monitor the efficacy of the coagulation process are presented in the sequence of the treatment process, i.e. first the clarified water and subsequently the filtered and treated water (highlighted in bold). This shows that the process must be monitored and controlled, according to the treatment steps, following the parameters of the clarified, filtered and treated water. Thus, contrary to what is reported in the pre-processed data, the control of the clarified water will prevent problems being transferred to the filtration process and consequently damages the quality of the treated water.

Table 4. Relevance of physico-chemical parameters

Relevance	Pre-processed data		Mined data	
1	Dosing system by gravity	Dosing system by pumping	Dosing system by gravity	Dosing system by pumping
2	Dosing system by pumping	Dosing system by gravity	Dosing system by pumping	Dosing system by gravity
3	Raw water turbidity	Raw water turbidity	Raw water color	Raw water color
4	Raw water color	Raw water color	Raw water turbidity	Raw water turbidity
5	Residual aluminum clarified water	Residual aluminum clarified water	**Clarified water turbidity**	**Clarified water turbidity**
6	Residual aluminum treated water	Residual aluminum treated water	**Residual aluminum clarified water**	**Residual aluminum clarified water**
7	Clarified water turbidity	Clarified water turbidity	**Filtered water turbidity**	**Filtered water turbidity**
8	Filtered water turbidity	Filtered water turbidity	**Residual aluminum treated water**	**Residual aluminum treated water**

5   Conclusion

The proposed approach showed that it was possible to select the optimized values of the process by means of clustering and to identify that the optimization done by the water treatment technicians is not uniform, evidencing empirical evaluations.

Moreover, the obtained results point that is possible to developing models using computational intelligence resources to obtain more adequate training database, avoiding that undesirable data applied to learning step of intelligent algorithms, e.g. artificial neural networks, can lead to wrong output of the model.

For further investigatons, information from clusters with stored values can be submitted to metaheuristics classifiers to improve the data selection, which may be more representative in comparison to the choices made by expert evaluations of the water treatment process.

Acknowledgments. The authors would like to thank São Paulo Research Foundation (FAPESP, grant#2016/02641-1), Basic Sanitation Company of the State of São Paulo - (SABESP) and Nove de Julho University (UNINOVE) for their supports.

References

1. Akbar, T.A., Hassan, Q.K., Achari, G.: A methodology for clustering lakes in alberta on the basis of water quality parameters. Clean - Soil Air Water **39**(10), 916–924 (2011)
2. Baxter, C.W., Stanley, S.J., Zhang, Q.: Development of a full-scale artificial neural network model for the removal of natural organic matter by enhanced coagulation. Aqua **48**(4), 129–136 (1999)

3. Bieroza, M., Baker, A., Bridgeman, J.: New data mining and calibration approaches to the assessment of water treatment efficiency. Adv. Eng. Softw. **44**(1), 126–135 (2012)
4. Heddam, S., Bermad, A., Dechemi, N.: ANFIS-based modelling for coagulant dosage in drinking water treatment plant: a case study. Environ. Monit. Assess. **184**(4), 1953–1971 (2012)
5. Juntunen, P., Liukkonen, M., Lehtola, M., Hiltunen, Y.: Cluster analysis by self-organizing maps: an application to the modelling of water quality in a treatment process. Appl. Soft Comput. **13**(7), 3191–3196 (2013)
6. Kalteh, A., Hjorth, P., Berndtsson, R.: Review of the self-organizing map (SOM) approach in water resources: analysis, modelling and application. Environ. Model. Softw. **23**(7), 835–845 (2008)
7. Lamrini, B., Lakhal, E.K., Le Lann, M.V., Wehenkel, L.: Data validation and missing data reconstruction using self-organizing map for water treatment. Neural Comput. Appl. **20**(4), 575–588 (2011)
8. Leu, S.S., Bui, Q.N.: Leak prediction model for water distribution networks created using a bayesian network learning approach. Water Resour. Manag. **30**(8), 2719–2733 (2016)
9. McLachlan, G.J., Krishnan, T.: The EM Algorithm and Extensions, 2nd edn. Wiley, Hoboken (2008)
10. North, B., Blake, A.: Using expectation-maximisation to learn dynamical models from visual data. Image Vis. Comput. **17**(8), 611–616 (1999)
11. Ogwueleka, T., Ogwueleka, F.: Optimization of drinking water treatment processes using artificial neural network. Niger. J. Technol. **28**(1), 16–25 (2009)
12. Olanweraju, R.F., Muyibi, S.A., Salawudeen, T.O., Aibinu, A.M.: An intelligent modeling of coagulant dosing system for water treatment plants based on artificial neural network. Aust. J. Basic Appl. Sci. **6**(1), 93–99 (2012)
13. Olawoyin, R., Nieto, A., Grayson, R.L., Hardisty, F., Oyewole, S.: Application of artificial neural network (ANN) self-organizing map (SOM) for the categorization of water, soil and sediment quality in petrochemical regions. Expert Syst. Appl. **40**(9), 3634–3648 (2013)
14. Park, S., Bae, H., Kim, C.: Decision model for coagulant dosage using genetic programming and multivariate statistical analysis for coagulation/flocculation at water treatment process. Korean J. Chem. Eng. **25**(6), 1372–1376 (2008)
15. Robnik-Šikonja, M., Kononenko, I.: Theoretical and empirical analysis of ReliefF and RReliefF. Mach. Learn. **53**(1), 23–69 (2003)
16. Romano, M., Kapelan, Z., Savić, D.A.: Evolutionary algorithm and expectation maximization strategies for improved detection of pipe bursts and other events in water distribution systems. J. Water Resour. Plann. Manag. **140**(5), 572–584 (2007)
17. Siti Rozaimah, S.A., Pasilatun Adawiyah, I., Mohd. Marzuki, M., Rakmi, A.R.: Pattern recognition of fractal profiles in coagulation-flocculation process of wastewater via neural network. J. Inst. Eng. **68**(4), 17–19 (2007)
18. WEKA: Homepage. http://www.cs.waikato.ac.nz/ml/weka. Accessed 09 Oct 2016
19. Ye, G., Fenner, R.A.: Weighted least squares with expectation-maximization algorithm for burst detection in U.K. water distribution systems. J. Water Resour. Plann. Manag. **140**(4), 417–424 (2014)

A Two-Step Neural Network Approach to Passage Retrieval for Open Domain Question Answering

Andrés Rosso-Mateus[1(✉)], Fabio A. González[1], and Manuel Montes-y-Gómez[2]

[1] MindLab Research Group, Universidad Nacional de Colombia, Bogotá, Colombia
{aerossom,fagonzalezo}@unal.edu.co
[2] Computer Science Department,
Instituto Nacional de Astrofísica, Óptica y Electrónica, Puebla, Mexico
mmontesg@ccc.inoep.mx

Abstract. Passage retrieval is an important subtask of question answering. Given a question and a set of candidate passages, the goal is to rank them according to their relevance to the question. This work presents a two-stage approach for solving this problem. Both stages are based on convolutional neural network architecture with a reduced set of parameters. In the first stage the network is used to identify the degree of similarity between question and candidate answers, then, in the second stage, the result of the first stage is used to re-rank the answers according to their similarity with the initial best-ranked answer in such a way that the most similar candidate answers are moved up. This approach is analogous to a pseudo-relevance feedback strategy. The experimental results suggest that the proposed method is competitive with the state-of-the-art methods, achieving a remarkable performance in three evaluation datasets.

Keywords: Question answering · Passage retrieval · Answer ranking
Pseudo-relevance feedback

1 Introduction

Question answering over open domain data is the focus of next generation web search engines [4]. This task considers three main subtasks: (*i*) information retrieval, which collects the most relevant documents to the given question; (*ii*) passage retrieval, which selects the passages from the returned documents more likely to contain the answer to the question; and (*iii*) answer extraction, which analyzes the selected passages and extracts the exact answer to the question.

This work focuses on the passage retrieval subtask: given a question q and a set of candidate passages $\{s_1, s_2, \ldots s_n\}$, the goal is to rank the passages according to their similarity with the question.

Most of the state-of-the-art approaches exhibit a good performance in ranking the first candidate passage. That is, the first-ranked candidate passage frequently

© Springer International Publishing AG, part of Springer Nature 2018
M. Mendoza and S. Velastín (Eds.): CIARP 2017, LNCS 10657, pp. 566–574, 2018.
https://doi.org/10.1007/978-3-319-75193-1_68

contains a valid answer to the posed question. Based on this observation, in this paper, we propose a passage retrieval method based on a two-stage ranking approach. In the first stage, the passages (referred to as answers from now on) are ranked according to their similarity to the question. This initial ranking is generated by a convolutional neural network, which is applied to a matrix encoding query−answer term similarities, and returns a score that indicates the degree of similarity between the query and a candidate answer. In the second stage, passages are re-ranked based on their similarity to the first passage in the initial ranking. To generate this new ranking, a convolutional neural network is also applied but this time to a matrix encoding first_answer−passage term similarities. This strategy is analogous to the pseudo-relevance feedback method used in information retrieval where the highest ranked results are used to expand the query.

The paper is organized as follows. Section 2 presents some state-of-the-art methods for passage retrieval. Section 3 describes the method and the proposed architecture. Section 4 depicts the experimental setup in detail. Section 5 discusses the results achieved by the method in the two evaluation datasets, and finally, Sect. 6 presents our conclusions and future work directions.

2   Related Work

The Association for Computational Linguistics (ACL) has maintained a rank with the most successful methods for passage retrieval [1]. This list includes methods based on pure linguistic techniques, on statistical approaches, and more recently on deep learning.

Based on the hypothesis that questions can be generated from correct answers, Yu and the team of DeepMind [16] proposed a transformation model where question and answers are represented in the same space and their distance is used to get their similarity score. Le and Mikolov [6] proposed a paragraph vector (PV), which learns how to represent a variable size sentence in a vector. The learned vectors are then used to measure the similarity between question and answers. Severyn and Moschitti [11] presented a convolutional neural network method (CNN), for ranking pairs of short texts. The goal is to learn an optimal representation of text pairs and a similarity function. A pairwise word interaction model (Pairwise CNN), is presented by He and Lin [5]. They used a novel deep neural network architecture (BiLSTM) aimed to capture the relation between sequences of terms of two sentences read in both directions (left-right and right-left). They also proposed an attention mechanism to give importance to some component terms. Finally, Yang et al. [14] presented an attention-based model where the importance of terms is learned based on their correlations seen at the training phase. All these methods based on deep-learning techniques do not use any query expansion strategy to enhance the evaluation of question and answers similarities. The only method similar in spirit to ours, is the one proposed by Riezler et al. [10], which uses a Statistical Machine Translation (SMT) technique for expanding questions with synonyms. Although it achieved good results, it could not outperform state of the art results.

3   Model Description

The method proposed in this paper is detailed in Fig. 1, each of those steps is going to be detailed in the next section. The whole process consists of two stages: the training phase where the similarity model is obtained, and the testing phase where the calculated model is used to rank the question-answer pairs. During training: (1) question-answer pairs (qa-pairs) are pre-processed, (2) the similarity matrix between qa-pairs terms is calculated, (3) a convolutional neural network model is trained to predict the relevance of the answer to the question. Once the model is built it can be used to predict the rank order of candidate answers. At testing time, for a particular question, the model is applied to predict the relevance score of the set of candidate answers: (4) answers are ranked according to their scores, (5) answers are re-ranked according to their similarity with the highest ranked answer at step (4), producing a new ranking of the answers.

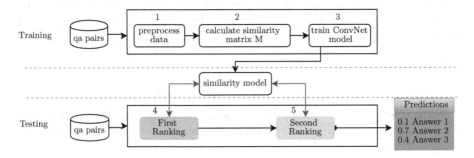

Fig. 1. Process of ranking and re-ranking qa pairs.

3.1   Step 1. Preprocess Data

Questions and candidate answers are processed using: tokenization to delimit terms; lowercasing to standardize the terms; pos-tagging, using the nltk pos-tagger [2], to extract syntactical information that will be used in salience weighting; and transforming terms to a word2vec vector representation [9], to make possible their semantic similarity comparison.

3.2   Step 2. Calculate Similarity Matrix

The similarity matrix M represents the semantic relatedness of the i-th question term and the j-th answer term according to a similarity measure. Each element $M_{i,j}$ of this matrix is a composition of a similarity score and a salience score as described by the Eq. 1.

$$M_{i,j} = scos(q_i, a_j) * sal(q_i, a_j) \tag{1}$$

Similarity Score. The similarity score for a question-answer pair terms (q_i, a_j) is calculated by means of the cosine distance between their word2vec vectors as indicated by Formula 2.

$$scos(q_i, a_j) = 0.5 + \frac{q_i \cdot a_j}{2 \, \|q_i\|_2 \, \|a_j\|_2} \tag{2}$$

In the case that there does not exist the word2vec representation for one of the terms, their similarity is measured based on their distance in Wordnet [13]. In particular, we use as similarity measure the edge distance between the first common concept related with q_i and a_j. If there is not a common concept between the terms, then we calculate the Levenshtein distance between the words [7], defined as the number of operations (insertions and eliminations of characters) needed to transform q_i to a_j.

Salience Weighting. As not all terms are equally informative for measuring text similarities [3,8], we consider weighting the terms from the question and the answer based on part of speech functions: verbs, nouns, and adjectives are considered to be the most relevant. We model this information through a salience score.

The salience score is calculated as follows. If both terms are relevant then their score is 1. If only one of the terms is important then the score is 0.6, in the case none of them is relevant the score is 0.3. The salience function is defined in the Formula 3.

$$sal(q_i, a_j) = \begin{cases} 1 & if \; imp(q_i) + imp(a_j) = 2 \\ 0.6 & if \; imp(q_i) + imp(a_j) = 1 \\ 0.3 & if \; imp(q_i) + imp(a_j) = 0 \end{cases} \tag{3}$$

Where $imp(q_i)$ and $imp(a_j)$ are the evaluation of importance weighting function for every question and answer term. The related function returns 1 if the term is a verb, noun or adjective, otherwise, returns 0.

Finally, we sort the calculated matrix M leaving the most related terms in the top left cell, and if the number of rows or columns exceeds 40, the remaining data is truncated. This step provides an invariable representation of the similarity patterns that can be exploited by the convolutional network.

3.3   Step 3. Convolutional Model

Convolutional neural networks (CNN) are a popular method for image analysis thanks to their ability to capture spatial invariant patterns. In the proposed method, they play a similar role, but instead of receiving an input image the CNN receives the similarity matrix M. The hypothesis is that it will be able to identify term-similarity patterns that help to determine the relevance of a question-answer pair. Patterns identified by the CNN are sub-sampled by a pooling layer. The output of the pooling layer feeds a fully-connected layer. Finally, the output of the model is generated by a sigmoid unit. This output corresponds to a score, **simScore**(q, a), that can be interpreted as a degree of relatedness between the question q and the answer a.

The architecture of the convolutional model is depicted in Fig. 2.

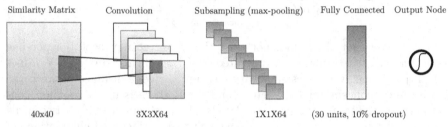

Similarity Matrix   Convolution   Subsampling (max-pooling)   Fully Connected   Output Node

40x40   3X3X64   1X1X64   (30 units, 10% dropout)

Fig. 2. Convolutional neural network model architecture.

3.4   Step 4 and 5. Two Ranking Stages

During the testing phase, a new query along with candidate answers are presented to the method. The candidate answers $(a_1, a_2, \ldots, a_k)$ are ranked using the CNN model producing the first rank of them. Based on the premise that the first candidate answer, a^*, is expected to be highly correlated with the question q, a second score, $simScore(a^*, a_k)$, is calculated by comparing each candidate answer with the highest ranked answer. A new ranking is calculated by using a new score corresponding to a linear combination of the first and second score as is shown in Eq. 4.

$$finalScore(q, a_k) = (1 - \alpha) * simScore(q, a_k) + \alpha * simScore(a^*, a_k) \quad (4)$$

As we are introducing a weighting term α to scale the second score, we calculated this term based on the exploration carried out in validation partition, which gives 0.32 as the optimal value.

This strategy promotes candidate answers which share similar terms with the highest ranked answer. This is a strategy analogous to pseudo-relevance feedback in information retrieval [10], where the original query is extended with terms from the highest ranked documents.

4   Experimental Setup

4.1   Test Datasets

The proposed method was compared to baseline and state-of-the-art methods using two information retrieval performance measures Mean Reciprocal Rank (MRR) and Mean Average Precision (MAP). MAP is defined as the mean of the average precision scores for each query over a set of Q queries and MRR evaluates the relative ranks of correct answers in the candidate sentences of a question [15]. To evaluate the method, two standard data sets for the passage retrieval task were used. The baseline and state-of-art methods were evaluated over the same dataset using the same experimental setup.

- **TrecQA:** was provided by Mengqiu Wang and collects data from Text REtrieval Conference (TREC) QA track (8–13). It was first used in [12]. The dataset has two partitions. In TRAIN partition the correctness of answer was carried out manually while in TRAIN-ALL the correctness of candidate answer sentences, was identified by pattern regular expressions matching answer, which induce noise in data, the statics of the related dataset are presented in Table 1.
- **Wiki QA:** is a dataset released in 2015 by Microsoft Research Group [15], that contains Question-Answer pairs for open domain. The Microsoft research group collected Bing Search Engine query logs and extract the questions the user submit from May of 2010 to July of 2011, and the answers are sentences of Wikipedia summary page, Table 2.

Table 1. TrecQA dataset

Split	#Questions	#Pairs
TRAIN ALL	1,229	53,417
TRAIN	94	4,718
DEV	82	1,148
TEST	95	1,517

Table 2. WikiQA dataset

Split	#Questions	#Pairs
TRAIN	2,118	20,358
DEV	296	2,716
TEST	633	6,156

4.2 Baseline Models

Three baseline models were implemented to evaluate the performance of the proposed method. (1) Word Count, which is a word matching method that counts the number of non-stopwords that occur both in the question and in the answer sentences. (2) Weighted Word Count, a modified approach which weights the word counts using semantical information [15]. (3) DeepMind model [16], a semantic parsing method based on similarity metric learning and latent representations.

The list of comparative methods are the following: Word Count, Weighted Word Count, DeepMind model [16], Paragraph Vector (PV) [6], Attention-Based Model (aNMM) [14], Convolutional Neural Network Method (CNN) [11], Pairwise Word Interaction Model (Pairwise CNN) [5], and the proposed model without rerank (This Work) and with rerank (This Work Rerank).

5 Results

Table 3 summarizes the results of all the evaluated methods applied to both TrecQA and WikiQA datasets. In the case of TrecQA two configurations were evaluated: the TRAIN partition and the TRAIN ALL partition, which were described in Subsect. 4.1.

In TrecQA dataset, the proposed method presents the best performance of all the evaluated methods. This is consistent in both configurations. Also, we can observe that the use of re-ranking improves the method performance in terms of

MAP. The main reason is that in most cases the first ranked answer is relevant we can be evidenced by the high value of the MRR measure.

In WikiQA dataset, the best result is obtained by the Pairwise CNN method [5], however the proposed method has a competitive performance that clearly outperforms the other evaluated methods. As evidenced by the overall performance of all the methods, this dataset seems to be more challenging. One problem of this dataset is that it contains several questions without a valid answer in the dataset. The re-ranking strategy produces an important improvement for TrecQA dataset, while it did not improve the performance for the WikiQA dataset. Our conclusion is that the lower MRR in this dataset means that the top ranked answer is less likely to be relevant and thus it has less probability of improving the ranking of relevant answers.

In general, we can say that the proposed method exhibits a very competitive performance when compared to state-of-the-art methods. However, its main strength is the fact that it is simpler than the other methods. This can be objectively measured by counting the number of parameters that the learning algorithm has to adjust during training. Table 4 shows the number of parameters

Table 3. Overview of results QA answer selection task datasets. We also include the results of the base line models. ('-' is Not Reported)

| Method | TREC | | TREC | | WikiQA | |
| | TRAIN ALL | | TRAIN | | | |
	MAP	MRR	MAP	MRR	MAP	MRR
BaseLines						
Word Count	0.6402	0.7021	0.6402	0.7021	0.4891	0.4924
Weighted Word Count	0.6512	0.7223	0.6512	0.7223	0.5099	0.5132
DeepMind model	0.6531	0.6885	0.6689	0.7091	0.5908	0.5951
aNMM [14]	0.7385	0.7995	0.7334	0.8020	-	-
CNN [11]	0.7459	0.8078	0.7329	0.7962	-	-
Pairwise CNN [5]	0.7588	0.8219	-	-	**0.7090**	**0.7234**
PV [6]	-	-	-	-	0.5110	0.5160
This work	0.7644	**0.8414**	0.7605	0.8344	0.6368	0.6614
This work (Rerank)	**0.7737**	0.8403	**0.7750**	**0.8350**	0.6351	0.6583

Table 4. Number of parameters

Split	# of parameters
aNMM [14]	14,000
CNN [11]	100,000
Pairwise CNN (2016)[5]	1.7 million
This work	**3,198**

for some of the evaluated methods. Clearly the proposed method has orders of magnitude less parameters than the other methods. This has a positive impact on the amount of computational resources which are required during training and testing.

6   Conclusions

This work presents a novel method for question-answer ranking based on convolutional neural networks and a pseudo-relevance-feedback-inspired re-ranking strategy. The experimental results shows that the proposed method is competitive when compared to state-of-the-art methods, despite being a simple model with a reduced set of parameters. Taking into account the lack of complexity of the proposed model, the obtained results are very promising.

The experimental evaluation shows an improvement of 2% in the MAP score when the proposed method was applied. Such results suggest that the first ranked answer contains information that can help to rank the subsequent answers.

The future work will focus on exploiting the contextual relationships of terms as well as the inclusion of attention mechanisms that have shown very good results in other text analysis tasks.

References

1. ACL, Association for Computational Linguistics: Question Answering (State of the art) (2007). http://www.aclweb.org/aclwiki/index.php. Accessed 1 May 2017
2. Bird, S.: NLTK: the natural language toolkit. In: Proceedings of the COLING/ACL on Interactive Presentation Sessions, vol. 1, pp. 69–72. ACL (2006)
3. Dong, L., Wei, F., Zhou, M., Xu, K.: Question answering over freebase with multi-column convolutional neural networks. In: ACL, vol. 1, pp. 260–269 (2015)
4. Etzioni, O.: Search needs a shake-up. Nature 476(7358), 25–26 (2011)
5. He, H., Lin, J.J.: Pairwise word interaction modeling with deep neural networks for semantic similarity measurement. In: HLT-NAACL, vol. 1, pp. 937–948 (2016)
6. Le, Q.V., Mikolov, T.: Distributed representations of sentences and documents. In: ICML, vol. 14, pp. 1188–1196 (2014)
7. Levenshtein, V.I.: Binary codes capable of correcting deletions, insertions, and reversals. In: Soviet Physics Doklady, vol. 10, pp. 707–710 (1966)
8. Liu, F., Pennell, D., Liu, F., Liu, Y.: Unsupervised approaches for automatic keyword extraction using meeting transcripts. In: Proceedings of Human Language Technologies, vol. 1, pp. 620–628. ACL (2009)
9. Mikolov, T., Chen, K., Corrado, G.S., Dean, J.: Efficient estimation of word representations in vector space (2013)
10. Riezler, S., Vasserman, A., Tsochantaridis, I., Mittal, V., Liu, Y.: Statistical machine translation for query expansion in answer retrieval. In: ACL (2007)
11. Severyn, A., Moschitti, A.: Learning to rank short text pairs with convolutional deep neural networks. In: 38th ACM SIGIR (2015)
12. Wang, M., Smith, N.A., Mitamura, T.: What is the jeopardy model? A quasi-synchronous grammar for QA, vol. 7, pp. 22–32 (2007)

13. Wu, Z., Palmer, M.: Verbs semantics and lexical selection. In: 32nd Proceedings ACL, vol. 1, pp. 133–138. Association for Computational Linguistics (1994)
14. Yang, L., Ai, Q., et al.: aNMM: ranking short answer texts with attention-based neural matching model. In: Proceedings of the 25th ACM ICIKM. ACM (2016)
15. Yang, Y., Yih, W., Meek, C.: WikiQA: a challenge dataset for open-domain question answering. In: EMNLP, vol. 1, pp. 2013–2018 (2015)
16. Yu, L., Hermann, K.M., Blunsom, P., Pulman, S.: Deep learning for answer sentence selection. In: NIPS Deep Learning Workshop (2014)

Ensembles of Multiobjective-Based Classifiers for Detection of Epileptic Seizures

Fernando S. Beserra, Marcos M. Raimundo, and Fernando J. Von Zuben[(⊠)]

LBiC/DCA/FEEC - University of Campinas, Campinas, SP, Brazil
{fbeserra,marcos,vonzuben}@dca.fee.unicamp.br

Abstract. This paper proposes multiobjective-based classifiers to detect epileptic seizures using ensemble approaches, transfer-learning methods, and three alternative feature extraction techniques. Two aspects of the problem were investigated: (1) the relative merit of distinct proposals to synthesize an ensemble of classifiers, considering all the three feature extraction techniques; (2) the potential of an ensemble composed of transfer-learned classifiers. The blend approaches with the best performance detected all test seizures, with a high proportion of correctly detected samples inside the seizure interval and high proportion of time intervals correctly classified as non-seizures.

Keywords: Ensemble of classifiers · Transfer learning
Multiobjective optimization · Epilepsy

1 Introduction

Biological signals tend to impose big challenges to traditional machine learning approaches. It is common knowledge that data acquisition of such signals is often subject to (i) high levels of noise, (ii) the inherent trade-off between signal quality and the level of invasiveness, and (iii) the fact that reproducibility is not always guaranteed. In this work, we focus on the problem of seizure detection in epileptic brain recordings. Seizures, being scarce events, pose a defiance to machine learning algorithms both regarding biological variability of their types and also regarding the formation of datasets with balanced sample classes. In the literature, great detection performances have been reported in recent works, using classifiers such as Support Vector Machines (SVMs) [12,14], Extreme Learning Machines [13,17] and Deep (Convolutional) Neural Networks [10,15].

Concerning the nature of the features that are usually passed to such classifiers, there is no consensus: energies of subbands of the Fourier spectrum have been shown to be descriptive of seizure events [12] as well as wavelet expansions [8], entropic measures, and other linear and nonlinear features [5]. More recently, there is an increasing interest in extracting features capable of unveiling the interdependence between multiple channels' recordings, forming synchronization graphs [3]. All these extraction methods can be used to generate explanatory features that a learning machine would make use to detect seizures. Depending on

© Springer International Publishing AG, part of Springer Nature 2018
M. Mendoza and S. Velastín (Eds.): CIARP 2017, LNCS 10657, pp. 575–583, 2018.
https://doi.org/10.1007/978-3-319-75193-1_69

particular characteristics of the state of the patient under evaluation, some feature extraction methods may outperform others. These considerations motivate us to construct ensemble approaches, which are capable of weighting multiple points of view, automatically, giving preference to distinct classifiers depending on how each sample is positioned in the feature space.

Aiming at constructing a robust classifier, our main strategy consists of an ensemble that aggregates learning machines trained with three different feature extraction strategies. The training of these classifiers, done by a Multiobjective Optimization (MOO) solver, the Noninferior Set Estimation method (NISE [2]), further provides to the ensembles' units proper regularization strengths. Addressing strategies for information sharing among patients, we adopted a transfer learning method, which uses MOO to share data from a source patient to a target one. Furthermore, we resort to ensembles to aggregate these transfer-learned classifiers.

2   Feature Extraction

The data analyzed in this work were extracted from electroencephalographic (EEG) recordings in the Physionet Database [6,11]. Among its available patients and channels, here we work with the following subset of channels $\mathbb{C}$ present in all patients $\mathbb{P} = \{1, 2, 3, 4, 5, 6, 7, 8, 9, 10, 11, 12, 14, 20, 21, 22, 23\}$, for uniformity across patient's settings.

Whenever convenient, the EEG recording at a given channel $c \in \mathbb{C}$ and time t will be denoted by $data(t)[c]$. The set of extraction methods are defined by $E \equiv \{gph, fou, wlt\}$ corresponding respectively to synchronization graph-based, Fourier transform-based, wavelet expansion-based extractions.

Wavelet-based features are extracted by a wavelet transform of each data channel. Let $h_i(t) = data(t)[i]$, $t = 0, \ldots, (N-1)\delta t$ be the recording of the i-th channel at time $t = T$, where $\delta_t = 1/256$ is the sampling time precision. Then, a wavelet transform of $h_i(t)$ is defined as

$$W_i(n, a) = \sum_{j=0}^{N-1} h_i(j\delta t) f\left(\frac{(j-n)\delta_t}{a}\right) \tag{1}$$

where $f(\cdot)$ is the Sombrero Wavelet [8]: $f(t) = \sqrt{\frac{\delta_t}{a}} \frac{2}{\sqrt{3}} \pi^{-1/4}(1 - t^2)e^{-t^2/2}$. The adopted a-scales are $A = \{0.031, 0.033, 0.037, 0.049, 0.080, 0.165, 0.392, 1\}$.

The complete generation of a feature vector concatenates two shifts of one second of the W coefficients obtained at t and $t-1$. At each of these time instants, one second of data is used for the transform and the $N = 256$ coefficients are reduced to a pair of means and standard deviations, per channel and per a-scale.

Synchronization graph-based features were inspired by the works of Dhulekar et al. [3] and Kramer et al. [7]. This procedure divided windows of 10 s into twenty intervals of the form $I(n) = [t - (1 + n/20), t - L(1 + n/20) + 1](s)$. For each of these subintervals and for every pair of electrodes, the Pearson correlation coefficients between these electrodes' signals were computed, the maximum absolute values of these correlations were retained and used to create a weighted graph, which is then converted to three unweighted undirected graphs, using $\tau = [0.4, 0.5, 0.6]$ as discretization thresholds.

The metrics extracted from each of these graphs were composed of all those outlined at Sect. 3.1 of reference [3], adding : number of $\lambda = 1$ eigenvalues of the Laplacian Matrix; average connected component size; adjacency matrix spectral radius; adjacency matrix trace; adjacency matrix energy; clustering coefficient; eccentricity; radius; number of edges; normalized Laplacian Energy; the ratio between the first non-zero and the largest eigenvalue of the normalized Laplacian; the second largest eigenvalue of the Laplacian matrix.

Finally, the features of the kind **Fourier transform-based** are extracted following the procedure defined in Shoeb et al. [11]. They have information of the channels at times $[t - 4, t + 2]$, for a feature characterizing time t.

3   Learning Machines

Multiobjective optimization is an expanding field in machine learning; this methodology helps to: select meta-parameters for classifiers, obtain simultaneously accurate and simple models [9]; develop classifiers with high accuracy and sensitivity [4]; and search for accurate and diverse ensembles [1]. In this work, we use an adaptive approach that employs a weighted method, called NISE [2], to tune a scalarization parameter, finding accurate and simple models for single-task classifiers, being both accurate on the source and target patients.

3.1   Multinomial Logistic Regression

We approach the detection of epileptic seizures in a binary classification fashion. Despite this fact, we aim to construct a general framework for multiclass problems supported by multiobjective optimization. Seeking to use a simple but robust statistical model, we resort to a regularized multinomial regression:

$$\min_{\theta} \quad -\sum_{i,k} \mathbf{y}_i^k \ln \left(\frac{e^{\theta_k^\top \mathbf{x}_i}}{\sum_\kappa e^{\theta_\kappa^\top \mathbf{x}_i}} \right) + \lambda \sum_k ||\theta_k||^2 \equiv \mathbf{w}_1 l(\theta, \mathbf{X}, \mathbf{Y}) + \mathbf{w}_2 r(\theta). \quad (2)$$

where $\mathbf{w}_1 = \frac{1}{1+\lambda}$, $\mathbf{w}_2 = \frac{\lambda}{1+\lambda}$, $\mathbf{X}, \mathbf{Y}$ are the data matrixes ($\mathbf{x}, \mathbf{y}$ its vectorial representation), θ is the parameter vector, i, k are the sample and label indexes. Equation 2 shows the following equivalence: since NISE finds efficient solutions by varying $\mathbf{w}$, it also determines models with different regularization parameters λ.

3.2   Multinomial Logistic Regression for Transfer Learning

Considering a source patient $s \in \mathbb{P}$ and a target patient $t \neq s$, $t \in \mathbb{P}$, transfer learning consists of a strategy that conveys knowledge extracted from s to t.

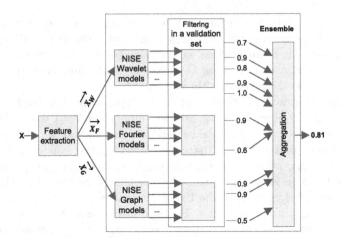

Fig. 1. Representation of the ensemble operation.

Denoting $(\mathbf{X}^s, \mathbf{Y}^s)$ and θ^s as the data and the parameters of the source learner and $(\mathbf{X}^t, \mathbf{Y}^t)$ and θ^t as the data and the parameters of the target one, it is possible to define a multiobjective problem, in the weighted form, to make trade-off transfers between a single source task learning function and a specific target task learning function, based on the following formulation:

$$\min_{\theta} \quad \mathbf{w}_1 \left(l(\theta, \mathbf{X}^s, \mathbf{Y}^s) + \lambda r(\theta) \right) + \mathbf{w}_2 \left(l(\theta, \mathbf{X}^t, \mathbf{Y}^t) + \lambda r(\theta) \right) \tag{3}$$

The regularization parameter λ is obtained by a cross-validation in the sour-ce-task (a particular patient) using multinomial classifiers.

3.3   Model Selection and Ensembles

The construction of an ensemble takes three steps: generation, filtering, and aggregation [18]. The **generation** of a set of classifiers is explored in different ways, further explained in Sect. 3.4. Considering the harmonic mean between sensitivity and specificity calculated in the validation set as the evaluation metric, we tested two types of **filtering**: (a) selection of the best classifier (wta) and (b) selection of the ten best classifiers (elt). Finally, the **aggregation** is done by averaging the distributions produced by the components selected in the filtering step.

3.4   Proposed Methodologies

Single-task predictor generated by multiobjective optimization. Considering a unique target-task $t \in P$ and a unique feature extraction $e \in E \equiv \{gph, fou, wlt\}$, with P and E defined in Sect. 2, the models are trained using NISE on a multinomial model (Sect. 3.1). This procedure generates a set of 25 models, where the predictor is selected using the harmonic mean between sensitivity and specificity in a validation set. Here, three predictors are created: STNISE_gph, STNISE_fou and STNISE_wlt, corresponding to single-task learning strategies using, respectively, graph, Fourier and wavelet feature extractions.

Ensemble of single-task learned models. For every feature extraction $e \in E$ on target-task $t \in P$, the models are trained using NISE on a multinomial model (Sect. 3.1). Then, the 25 models for every extraction $e \in E$ are gathered to compose an ensemble, selected and aggregated according to the content of Sect. 3.3. This pipeline is depicted in Fig. 1 where each NISE box represents the training of an extraction $e \in E$ to a target-task $t \in P$. Here, two predictors are created: STNISE_wta and STNISE_elt.

Ensemble of transfer-learned models. For every feature extraction $e \in E$ and for every transfer for a unique source-task $s \in P$ to a unique target-task $t \in P$, the models are trained using NISE on transferring multinomial model (Sect. 3.2). This pipeline might also be represented in Fig. 1 where each NISE box represents the training of an extraction $e \in E$, all using a transfer from a source-task $s \in P$ to a unique target-task $t \in P$. Then, the 25 models for every source-task $s \in P$ and every extraction $e \in E$ are gathered to compose an ensemble, selected and aggregated as described in Sect. 3.3. Here, two predictors are created: TLNISE_wta and TLNISE_elt.

4   Experiments

4.1   Definition of Training, Validation and Test Sets for Source and Target Patients

The feature extraction stage generated a set of pairs $(\boldsymbol{x}_p(t), y_p(t))$, with time discretized in windows of size $\Delta t = 1\,\mathrm{s}$, and labels generated according to [16], where only the first 20 s of a seizure are marked with $y = 1$ and time instants outside a seizure receive label $y = 0$.

Target patients $p \in P$ had their data divided into training, validation and out-of-time test sets, by assigning to the test sets all the samples whose time instants were $t \geq t^p_{seiz} = t^*_p - 5$ min, t^*_p being the lowest time index of a sample occurring after the n^*_p-th last seizure, $n^*_p = max(1, \lfloor n^{seiz}_p/4 \rfloor)$, n^{seiz}_p denoting the number of annotated seizures for patient p. All patients were tested in a set containing a single seizure interval mark, except for patients $6, 12, 14, 20$, whose test sets contained $6, 10, 2, 2$ annotated test seizures, respectively. Training and validation data consisted of a random split $(70\%/30\%)$ of the samples whose extraction times were $t < t^p_{seiz}$ and, due to the large amount of data, we retained only a random quarter of the $y = 0$-labeled samples. Source patients $p \in P$ had their non-seizure training samples with times $t < t^p_{seiz}$ again subsampled by a quarter, while their validation data consisted of all samples with associated $t \geq t^p_{seiz}$.

Table 1. Friedman rank and average values for SEN, SPE, LAT, AUC and UND metrics.

	SEN		SPE		LAT		AUC		UND
	Rnk	Mean ± Std	Rnk	Mean ± Std	Rnk	Mean ± Std	Rnk	Mean ± Std	Sum
SVM_fou	5.7	0.77 ± 0.24	4.0	0.87 ± 0.31	6.3	3.96 ± 4.65	3.9	0.89 ± 0.23	0
SVM_gph	5.7	0.81 ± 0.21	7.4	0.83 ± 0.26	5.2	2.23 ± 3.96	6.5	0.88 ± 0.15	0
SVM_wlt	6.5	0.74 ± 0.25	8.1	0.80 ± 0.23	6.7	3.23 ± 3.64	8.1	0.81 ± 0.20	2
stNISE_fou	3.9	0.86 ± 0.17	4.5	0.86 ± 0.21	3.9	1.89 ± 3.23	3.4	0.95 ± 0.08	0
stNISE_gph	7.6	0.61 ± 0.24	7.2	0.89 ± 0.14	7.9	5.73 ± 4.63	7.3	0.88 ± 0.09	1
stNISE_wlt	6.5	0.72 ± 0.27	5.6	0.91 ± 0.17	5.7	2.47 ± 3.31	5.7	0.91 ± 0.09	1
stNISE_wta	4.5	0.80 ± 0.20	4.5	0.90 ± 0.18	4.5	2.80 ± 3.52	6.8	0.85 ± 0.11	0
stNISE_elt	4.9	0.82 ± 0.17	4.7	0.96 ± 0.06	5.5	2.84 ± 3.72	3.2	0.96 ± 0.07	0
tlNISE_wta	4.9	0.80 ± 0.20	4.5	0.92 ± 0.15	4.6	2.45 ± 3.39	7.5	0.86 ± 0.10	0
tlNISE_elt	4.9	0.82 ± 0.18	4.5	0.92 ± 0.16	4.8	2.55 ± 3.58	2.7	0.96 ± 0.07	0

4.2   Performance Metrics and Developed Comparison Benchmarks

Jointly with STNISE_{gph,fou,wlt} single-task baseline methodologies, three other comparison benchmarks were developed: SVM_gph, SVM_fou and SVM-_wlt, characterized by the results of the best chosen configurations of Support Vector Machines, in a validation set, with a regularization parameter varying in $C \in \{10^{-3}, 10^{-2}, \ldots, 10^{+2}\}$ and kernels belonging to sigmoid or radial basis function, for each of the feature extraction methods treated individually.

The metrics used to evaluate the performance were sensitivity, denoted by SEN, the ratio of true positives to the total of positives; specificity, denoted by SPE, the ratio of true negatives to the total of negatives; latency, denoted by LAT, the required time to detect a given seizure since its onset mark; the area under the receiver operating characteristic curve, denoted by AUC; and the number of fully undetected seizures, i.e., cases where all the 20 positive instances of a seizure were missed, across all patients, denoted by UND.

5   Results

The first column of each metric corresponds to the average Friedman rank, used in the Friedman test, and the second corresponds to the mean ± standard deviation, except for the undetected seizures metric (UND) which indicates the sum of the undetected seizures across all patients. For all metrics, lower ranks indicate better methods, higher means of SEN, SPE and AUC values indicate better methods, and for latency and UND, lower values indicate better methods. The obtained results for all the detailed metrics are exhibited in Table 1.

To make more assertive comparisons, we use a Friedman test, first to detect whether all the methods are similar, or reject the null hypothesis. If the null hypothesis is rejected, as a *post-hoc* stage, we used the Finner test. The latter test compares pairs of methods and, when the similarity hypothesis is rejected,

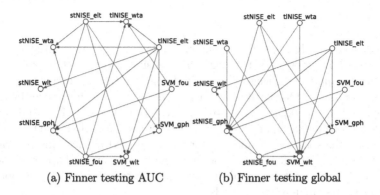

(a) Finner testing AUC   (b) Finner testing global

Fig. 2. Graphs denoting the results of a Finner *post-hoc* test, indicating the methods hierarchy obtained for AUC and the global comparison of metrics.

the method with lower rank is considered better than the method with higher rank. Both tests consider a significance level of $t = 0.01$.

The null hypotheses were rejected by the Friedman test for SPE, LAT, AUC and a global test, that compared all metrics, for all the patients. The Finner *post-hoc* test for the metrics AUC and global are in Fig. 2(a) and (b), in a directed graph format, where the arrow goes from the better method to the (pairwise) worse method. This *post-hoc* test also indicates that SVM_fou performed better than SVM_wlt for SPE, and for LAT, stNISE_fou performed better than stNISE_gph.

6   Analysis

From the statistical point of view, the evaluation of AUC and global metrics, presented at Fig. 2a and b, it is possible to infer that the top three algorithms, those with the highest out-degrees, are tlNISE_elt, stNISE_elt and stNISE_fou. These algorithms keep a good performance on SPE and LAT metrics, not being statistically worse than any method. Additionally, analyzing the performances *per se* at Table 1, these classifiers have different facets. stNISE_fou is specialized in sensitivity and latency but keeping a fair specificity, stNISE_elt and tlNISE_elt having a higher specificity keeping a reasonable sensitivity and latency. Moreover, none of these algorithms had completely missed seizures, since their UND value is 0. Interestingly, the model selection approaches (stNISE_wta and tlNISE_wta) did not achieve the performance level obtained by ensemble approaches (*_elt), even though selecting from the same set of classifiers.

Analyzing classifiers with specific feature extractions, when it is evaluated the in-degree and out-degree on graphs of Fig. 2 for both SVM and stNISE, it is possible to see that the Fourier feature extraction produces the best classifiers, and the graph feature extraction produces the worst classifiers.

7   Conclusion

This work successfully implemented an ensemble strategy that aggregates classifiers produced by different feature extractions using multiobjective optimization to conduct the model systhesis. The methodologies with ensemble aggregation of multiple classifiers (stNISE_elt) have achieved the best performance, being followed only by the classifiers using Fourier extraction. We have also implemented an ensemble of classifiers produced by transfer learning (tlNISE_elt), and this learning machines showed similar performance to stNISE_elt. This equivalence draws attention to the fact that at least no negative transfer happened, in this scenario characterized by data abundance, where transfer learning is not supposed to surpass single-task methods.

Acknowledgments. This research was supported by FAPESP, processes 16/19080-2 and 14/13533-0, and by CNPq, process 309115/2014-0.

References

1. Chandra, A., Yao, X.: DIVACE: diverse and accurate ensemble learning algorithm. In: Yang, Z.R., Yin, H., Everson, R.M. (eds.) IDEAL 2004. LNCS, vol. 3177, pp. 619–625. Springer, Heidelberg (2004). https://doi.org/10.1007/978-3-540-28651-6_91
2. Cohon, J.L.: Multiobjective Programming and Planning. Academic Press, New York (1978)
3. Dhulekar, N., Nambirajan, S., Oztan, B., Yener, B.: Seizure prediction by graph mining, transfer learning, and transformation learning. In: Perner, P. (ed.) MLDM 2015. LNCS (LNAI), vol. 9166, pp. 32–52. Springer, Cham (2015). https://doi.org/10.1007/978-3-319-21024-7_3
4. Fernández Caballero, J.C., Martinez, F.J., Hervas, C., Gutierrez, P.A.: Sensitivity versus accuracy in multiclass problems using memetic Pareto evolutionary neural networks. IEEE Trans. Neural Netw. **21**(5), 750–770 (2010)
5. Giannakakis, G., Sakkalis, V., Pediaditis, M., Tsiknakis, M.: Methods for seizure detection and prediction: an overview. In: Sakkalis, V. (ed.) Modern Electroencephalographic Assessment Techniques. N, vol. 91, pp. 131–157. Springer, New York (2014). https://doi.org/10.1007/7657_2014_68
6. Goldberger, A.L., Amaral, L.A.N., Glass, L., Hausdorff, J.M., Ivanov, P.C., Mark, R.G., Mietus, J.E., Moody, G.B., Peng, C., Stanley, H.E.: Physiobank, physiotoolkit, and physionet. Circulation **101**(23), e215–e220 (2000)
7. Kramer, M.A., Kolaczyk, E.D., Kirsch, H.E.: Emergent network topology at seizure onset in humans. Epilepsy Res. **79**, 173–186 (2008)
8. Latka, M., Was, Z., Kozik, A., West, B.J.: Wavelet analysis of epileptic spikes. Phys. Rev. E **67**, 052902 (2003)
9. de Miranda, P.B.C., Prudencio, R.B.C., de Carvalho, A.C.P.L.F., Soares, C.: Combining a multi-objective optimization approach with meta-learning for SVM parameter selection. In: 2012 IEEE International Conference on Systems, Man, and Cybernetics (SMC), pp. 2909–2914 (2012)
10. Mirowski, P., Madhavan, D., LeCun, Y., Kuzniecky, R.: Classification of patterns of EEG synchronization for seizure prediction. Clin. Neurophysiol. **120**(11), 1927–1940 (2009)

11. Shoeb, A.H.: Application of machine learning to epileptic seizure onset detection and treatment. Ph.D. thesis, Harvard-MIT Division of Health Sciences and Technology (2009)
12. Shoeb, A.H., Guttag, J.V.: Application of machine learning to epileptic seizure detection. In: Proceedings of the 27th International Conference on Machine Learning (ICML-10), pp. 975–982 (2010)
13. Song, Y., Crowcroft, J., Zhang, J.: Automatic epileptic seizure detection in EEGs based on optimized sample entropy and extreme learning machine. J. Neurosci. Methods **210**, 132–146 (2012)
14. Subasi, A., Kevric, J., Abdullah Canbaz, M.: Epileptic seizure detection using hybrid machine learning methods. Neural Comput. Appl., 1–9 (2017). https://doi.org/10.1007/s00521-017-3003-y
15. Swami, P., Gandhi, T.K., Panigrahi, B.K., Bhatia, M., Santhosh, J., Anand, S.: A comparative account of modelling seizure detection system using wavelet techniques. Int. J. Syst. Sci. Oper. Logist. **4**(1), 41–52 (2017)
16. Van Esbroeck, A., Smith, L., Syed, Z., Singh, S., Karam, Z.: Multi-task seizure detection: addressing intra-patient variation in seizure morphologies. Mach. Learn. **102**(3), 309–321 (2016)
17. Yuan, Q., Zhou, W., Liu, Y., Wang, J.: Epileptic seizure detection with linear and nonlinear features. Epilepsy Behav. **24**(4), 415–421 (2012)
18. Zhou, Z.H.: Ensemble Methods: Foundations and Algorithms, 1st edn. Chapman and Hall/CRC, London (2012)

Class Confusability Reduction in Audio-Visual Speech Recognition Using Random Forests

Gonzalo D. Sad, Lucas D. Terissi$^{(\boxtimes)}$, and Juan C. Gómez

Laboratory for System Dynamics and Signal Processing,
Universidad Nacional de Rosario CIFASIS-CONICET, Rosario, Argentina
{sad,terissi,gomez}@cifasis-conicet.gov.ar

Abstract. This paper presents an audio-visual speech classification system based on Random Forests classifiers, aiming to reduce the intra-class misclassification problems, which is a very usual situation, specially in speech recognition tasks. A novel training procedure is proposed, introducing the concept of Complementary Random Forests (CRF) classifiers. Experimental results over three audio-visual databases, show that a good performance is achieved with the proposed system for the different types of input information considered, *viz.*, audio-only information, video-only information and fused audio-video information. In addition, these results also indicate that the proposed method performs satisfactorily over the three databases using the same configuration parameters.

Keywords: Speech recognition · Audio-visual speech · Random forests

1 Introduction

It is known that human speech perception is a multimodal process which makes use of information not only from what we hear (acoustic) but from what we see (visual). For instance, a significant role in spoken language communication is played by lip reading. This is essential for the hearing-impaired people, and is also important for normal listeners in noisy environments to improve the intelligibility of the speech signal. Because of this, significant research effort have been devoted to the development Audio Visual Speech Recognition Systems (AVSRS) where the acoustic and visual information (mouth movements, facial gestures, etc.) during speech are taken into account [4].

In the past decades, different kind of models and classifiers have been proposed in the literature for speech recognition tasks. In general, all these methods can be classified into two large groups: generative models and discriminative models. In generative models, like Hidden Markov Models (HMM) [2], usually each word or phoneme is represented by a model, which is trained with examples of the class to be represented. In contrast, if discriminative models are used, like Random Forests (RF) [9], Support Vector Machine (SVM) [10], Deep Neural Networks (DNN) [7], etc., usually a single classifier is trained which internally models all classes of the dictionary, using all examples of all classes. In this paper,

© Springer International Publishing AG, part of Springer Nature 2018
M. Mendoza and S. Velastín (Eds.): CIARP 2017, LNCS 10657, pp. 584–592, 2018.
https://doi.org/10.1007/978-3-319-75193-1_70

another way of using the data in the training stage of Random Forests to produce a new classifier, namely Complementary Random Forests (CRF), is introduced, aiming of improving the recognition performance when the problem at hand suffer of intra-class misclassification problems, which is a very common situation, specially in speech recognition tasks. The performance of the proposed system is evaluated over three audio-visual databases, considering different scenarios, *viz.*, considering only audio information, only video information (lip-reading), and fused audio-visual information, respectively.

The rest of this paper is organized as follows. First, a preliminary analysis of the class confusability problem is given in Sect. 2 and the classification based on absent classes is explained in Sect. 3. The description of the proposed system is given in Sect. 4, and the databases used for the experiments are described in Sect. 5. In Sect. 6 experimental results are presented and the performance of the proposed strategy is analyzed. Finally, some concluding remarks are included in Sect. 7.

2   Class Confusability

Class confusability or intra-class confusion is a very common source of error in classification and recognition tasks. This phenomenon can be observed by analyzing the resulting confusion matrix. If the classifier has p classes, the confusion matrix is square matrix of dimension $p \times p$. The rows of the confusion matrix represent the system decisions, and the columns represent the true observation labels. Ideally, the values of off-diagonal entries are zeros, which means that none of the samples have been misclassified. Usually, the classifier makes some misclassification and non zero values appear in the off-diagonal entries. In many cases, these values are not randomly distributed, *i.e.*, the total error is distributed between few off-diagonal elements, while the rest have very low values. The position of these predominant elements (rows and columns), determine which are the classes that are most likely to be confused. Thus, the most conflicted classes can be easily found by ordering in descending order the off-diagonal elements. The row and column of the resulting first element determine the pair of most conflicting classes. In a similar way, the row and column of the resulting last element determine the pair of less conflicting classes. In this paper, the M most conflicting classes will be determined by taking the position (row and column) of the first M elements.

3   Classification Based on Absent Classes

In this paper, another way of using the data in the training stage of Random Forests to produce a new classifier, namely Complementary Random Forests (CRF), is introduced. The main idea is to detect absent classes by means of redefining the classes inside the Random Forests classifier. For each particular word in the vocabulary, a new class is defined using all the instances of the words in the vocabulary excepts the corresponding to the one being represented.

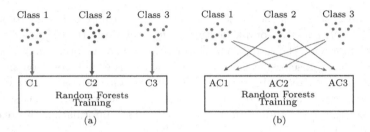

Fig. 1. Training procedure of the classifiers for the case of 3 classes. (a) Normal models (Random Forests). (b) Complementary models (Complementary Random Forests).

Then, a Random Forests classifier is trained using the traditional training procedure, but using this new set of classes. Then, given an observation sequence associated to the ith-word in the vocabulary, if the output probabilities of this Random Forests classifier are computed, it would be reasonable to expect that the minimum value would correspond to the ith-class. This is because the data used in the training of the ith-class does not include instances of the ith-word, whereas the rest of the classes do so. Based on this, in the recognition stage the probabilities corresponding to the new set of classes are computed and the recognized word corresponds to the class giving the minimum value. This can be interpreted as detecting absent classes, since the class giving the minimum probability is the only one that does not include data from the i-class in the training stage. This decision rule is expressed in the following equation

$$i = \underset{j}{\mathbf{argmin}}\ P_j\left(O|\lambda_{CRF}\right),\tag{1}$$

where i is the recognized class, λ_{CRF} is the classifier which has been trained with the new set of classes, which will hereafter be referred to as *complementary classifier*, and P_j indicates the probability of the new class j. Figure 1 schematically depicts the training procedure for the classifiers based on normal classifiers (RF) and complementary classifiers (CRF), for the case of a vocabulary with three different classes (N = 3).

4   Proposed Approach

Speech is inherently a time varying signal, where the utterances of the same word usually have different temporal durations. On the other hand, Random Forests classifiers require fixed-length input data. For that reason, audio-visual speech information must be represented with a fixed-length structure. For this task, the method proposed in [9] based on Discrete Wavelet Transform (DWT) is employed in this paper. This method is independent of the type and length of the input data. Each time varying parameter is resampled and then a multilevel decomposition is performed using DWT. Finally, the resulting DWT approximation coefficients are used to represent in a compact and fixed-length form the input parameters.

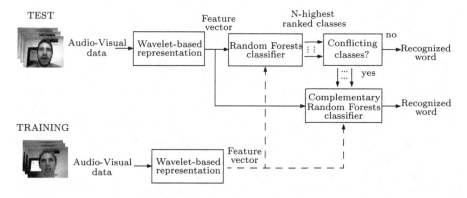

Fig. 2. Schematic representation of the proposed AV speech classification system.

A schematic representation of the proposed speech classification scheme is depicted in Fig. 2. In a first stage, wavelet-based audio-visual feature vectors are computed from the input audio-visual speech data. In the second stage, a classification based on Random Forests (RF) is performed, and the N most likely classes are pre-selected using the output probabilities. At this point, the possible solutions are reduced to these N classes. Then, if none of these pre-selected classes are considered as conflicted ones, the recognized word corresponds to the highest ranked class of this N classes, *i.e.*, the one with the highest probability. Otherwise, if any pair of these N classes is considered as conflicted ones, the system advance to a third stage.

In the third and last stage, the Complementary Random Forests classifier (λ_{CRF}) is formed, using only these N classes. This classifier will hereafter be referred to as *N-class complementary classifier*. Finally, the same input feature vector used in stage one is used as input to this new *N-class complementary classifier* and the recognized word corresponds to the class which gives the minimum probability.

5   Audio-Visual Databases

The performance of the proposed classification scheme is evaluated over three isolated word audio-visual databases, *viz.*, a database compiled by the authors, hereafter referred to as AV-UNR database, the Carnegie Mellon University (AV-CMU) database (now at Cornell University) [1], and the AVLetters database.

(I) *AV-UNR* database: The AV-UNR database consists of videos of 16 speakers, pronouncing a set of ten words (*up, down, right, left, forward, back, stop, save, open* and *close*) 20 times. The audio features are represented by the first eleven non-DC Mel-Cepstral coefficients, and its associated first and second derivative coefficients. Visual features are represented by three parameters, *viz.*, mouth height, mouth width and area between lips.

(II) *AV-CMU* database: The AV-CMU database [1] consists of ten speakers, with each of them saying the digits from 0 to 9 ten times. The audio features are represented by the same parameters as in AV-UNR database. To represent the visual information, the weighted least-squares parabolic fitting method proposed in [2] is employed in this paper. Visual features are represented by five parameters, *viz*, the focal parameters of the upper and lower parabolas, mouth's width and height, and the main angle of the bounding rectangle of the mouth.

(III) *AVLetters* database: The AVLetters database [6] consists of three repetitions by each of 10 talkers, five males and five females, of the isolated letters A-Z, resulting in a total of 780 utterances. In this case, only visual information is available. To represent the visual information associated with the utterance of each word, the method based on local spatiotemporal descriptors proposed in [10] is employed in this paper. As it is reported in [10], the best representation of the words in AVLetters database is obtained by computing local spatiotemporal descriptors using $2 \times 5 \times 3$ spatiotemporal blocks.

6   Experimental Results

The performance of the proposed system is evaluated separately at different scenarios, *viz.*, considering only audio information, only video information (lip-reading), and fused audio-visual information, respectively. These evaluations are carried out over the audio-visual databases described in Sect. 5. The audio signal is partitioned in frames with the same rate as the video frame rate. For the case of considering audio-visual information, the audio-visual feature vector at frame t is composed by the concatenation of the acoustic parameters with the visual ones. As described in Sect. 2, the confusion matrix is analyzed to detect which classes are conflicting or confused. For each database, this procedure is carried out forming the confusion matrix by evaluating a Random Forests classifier, randomly selecting (70%) of the database for training and using the remaining (30%) for testing, and taking the 4 most conflicting pair of classes. These classes are used to determine if there are conflicting classes in the second stage of the system proposed in Fig. 2.

In all the experiments presented in this work, the tuning parameters of the proposed classification system are remained fixed, except for the case of the number of randomly selected splitting variables to be considered at each node of the RF classifiers, hereafter denoted as α. For the wavelet-based representation, the normalized length of the resampled time functions is set to 256, the wavelet resolution level to 3, and the db4 is used as mother wavelet. Regarding the complementary models, 3-*class complementary classifiers* were considered ($N = 3$). For the RF classifiers, the number of trees to grow is set to 2000, while different values of parameter α are considered in the range $[2, 10]$.

To evaluate the recognition rates under noisy acoustic conditions, experiments with additive Babble noise, with SNRs ranging from -10 dB to 40 dB, were performed. Multispeaker or Babble noise environment is one of the most challenging noise conditions, since the interference is speech from other speakers. This noise is uniquely challenging because of its highly time evolving structure and its similarity to the desired target speech [5]. In this paper, Babble noise

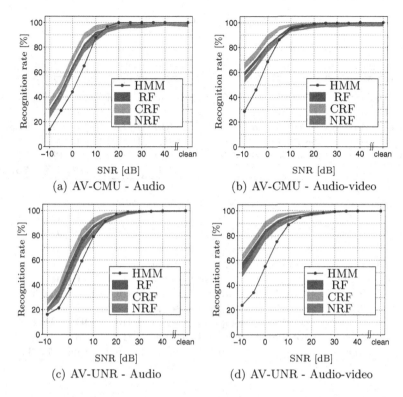

Fig. 3. Recognition based on acoustic information (first column) and audio-visual information (second column). Recognition rates obtained over the AV-CMU (first row) and AV-UNR (second row) databases for different SNRs for the cases of considering Babble noise. The color areas correspond to the recognition rates obtained by selecting α in the range from 2 to 10.

samples were extracted from *NOISEX-92* database, compiled by the Digital Signal Processing (DSP) group at Rice University [3].

In order to obtain statistically significant results, at each experiment a D-fold cross-validation (CV) is performed over the whole data to compute the recognition rates. For the cases of AV-CMU and AV-UNR databases, at each fold, one speaker is used for testing and the remaining ones for training, resulting in a speaker independent evaluation. Thus, a 10-fold CV is used for the AV-CMU database (10 speakers) and a 16-fold CV for the AV-UNR database (16 speakers). For the case of AVLetters database, in order to fairly compare our results with the ones obtained by other works in the literature [6,7,10], the training set was the first two utterances of each of the letters from all talkers (520 utterances) and the test set was the third utterance from all talkers (260 utterances). For comparison purposes, these experiments are also performed with other 5 recognition approaches, *viz.*, one based on traditional Hidden Markov Models (HMMs), and other 4 using variants of the proposed system: using (i) only the first Random

Table 1. Lip-reading results.

(a) AV-UNR

Classifier	Features	Accuracy
HMM	MSP [8]	70.16 %
RF	MSP [8]	85.21 % - 88.67 %
NRFf	MSP [8]	82.21 % - 85.83 %
NFR	MSP [8]	85.52 % - 88.15 %
CRFf	MSP [8]	88.12 % - 90.16 %
CRF	MSP [8]	**90.31 % - 91.97 %**

(b) AV-CMU

Classifier	Features	Accuracy
HMM [2]	PLCM [2]	61.17 %
HMM	PLCM [2]	57.79 %
RF	PLCM [2]	69.59% - 71.65%
NRFf	PLCM [2]	65.73% - 67.84%
NRF	PLCM [2]	68.03% - 71.34%
CRFf	PLCM [2]	70.12% - 72.34%
CRF	PLCM [2]	**72.57% - 75.02%**

(c) AVLetter

Classifier	Features	Accuracy
HMM [6]	Multiscale Spatial Analysis [6]	44.60 %
HMM [10]	Local Spatiotemporal Descriptors [10]	57.30 %
SVM [10]	Local Spatiotemporal Descriptors [10]	58.85 %
Deep learning [7]	Video-Only Deep Autoencoder [7]	64.40 %
RF	Local Spatiotemporal Descriptors [10]	61.12% - 65.38%
NRFf	Local Spatiotemporal Descriptors [10]	59.57% - 60.85%
NRF	Local Spatiotemporal Descriptors [10]	60.94% - 63.27%
CRFf	Local Spatiotemporal Descriptors [10]	64.23% - 67.87%
CRF	Local Spatiotemporal Descriptors [10]	**69.50% - 72.34%**

Forests classifier (RF), (ii) eliminating the conflicting classes detection block and always evaluating the complementary models (CRFf), (iii) replacing the *N-class complementary classifier* by a *N-class classifier*, *i.e.*, a traditional Random Forests classifier but trained using only the N pre-selected classes (NRF) and (iv) the same system than NRF, but eliminating the conflicting classes detection block, and always evaluating the second classifier (NRFf).

The recognition rates obtained at different SNRs over the AV-CMU and AV-UNR databases, using audio-only information and audio-visual information are depicted in Fig. 3. As expected, for all the cases, the performance in the recognition task deteriorates as the SNR decreases. It is clear that, for both databases and both types of inputs, the proposed recognition system (CRF) performs better than the ones based on HMMs and the 4 variants of the proposed system. This is more notorious at low and middle SNRs. The results for the lip-reading scenario for AV-UNR, AV-CMU and AVLetters databases, are shown in Table 1(a), (b) and (c), respectively. Also, the results obtained with others methods proposed in the literature are shown. As can be observed, the proposed system (CRF) performs satisfactorily and better than the other approaches, over the three databases, independently of the value used for parameter α. The results show that CRF gives better results than NRF, which highlights the importance of using the models based on absent classes. In addition, the results obtained with CRFf and NRFf are worst than the ones obtained with CRF and NRF, respectively, which means that selecting the conflicting classes is important. It is also important to note that the efficiency of the proposed scheme is slightly affected by using different values for parameter α. In contrast to other methods described in the literature, there is no need to perform an optimization of the system for each particular database.

7   Conclusions

An audio-visual speech classification system based on wavelets, two stages of Random Forests classifiers using a novel training procedure, and class confusability detection, have been proposed in this paper. In particular, the proposed system aims to improve the recognition performance when the problem at hand suffers of intra-class misclassification problems, which is a very common situation, specially in speech recognition tasks. The performance of the proposed recognition scheme was evaluated over three different isolated word audio-visual databases, considering different types of inputs, *viz.*, only audio information, only video information (lip-reading), and fused audio-visual information, respectively. Also, acoustic noisy conditions were considered. Experimental results show that good performance is achieved with the proposed system over the three databases, performing better than other methods reported in the literature. In all the experiments (different databases and types of input information), the proposed system was evaluated using the same tuning parameter. This is an important advantage of the proposed approach in comparison to other methods that necessarily require a usually time consuming optimization stage of the classifiers' metaparameters.

References

1. Advanced Multimedia Processing Laboratory. Carnegie Mellon University, Pittsburgh, PA. http://chenlab.ece.cornell.edu/projects/AudioVisualSpeechProcessing/
2. Borgström, B., Alwan, A.: A low-complexity parabolic lip contour model with speaker normalization for high-level feature extraction in noise-robust audiovisual speech recognition. IEEE Trans. Syst. Man Cybern. **38**(6), 1273–1280 (2008)
3. Digital Signal Processing Group, Rice University: NOISEX-92 Database, Houston, Rice
4. Katsaggelos, A.K., Bahaadini, S., Molina, R.: Audiovisual fusion: challenges and new approaches. Proc. IEEE **103**(9), 1635–1653 (2015)
5. Krishnamurthy, N., Hansen, J.: Babble noise: modeling, analysis, and applications. IEEE Trans. Audio Speech Lang. Process. **17**(7), 1394–1407 (2009)
6. Matthews, I., Cootes, T., Bangham, J.A., Cox, S., Harvey, R.: Extraction of visual features for lipreading. IEEE Trans. Pattern Anal. Mach. Intell. **24**, 2002 (2002)
7. Ngiam, J., Khosla, A., Kim, M., Nam, J., Lee, H., Ng, A.: Multimodal deep learning. In: Proceedings of the 28th International Conference on Machine Learning (ICML-11), pp. 689–696 (2011)
8. Terissi, L., Gómez, J.: 3D head pose and facial expression tracking using a single camera. J. Univ. Comput. Sci. **16**(6), 903–920 (2010)

9. Terissi, L.D., Sad, G.D., Gómez, J.C., Parodi, M.: Audio-visual speech recognition scheme based on wavelets and random forests classification. In: Pardo, A., Kittler, J. (eds.) CIARP 2015. LNCS, vol. 9423, pp. 567–574. Springer, Cham (2015). https://doi.org/10.1007/978-3-319-25751-8_68
10. Zhao, G., Barnard, M., Pietikäinen, M.: Lipreading with local spatiotemporal descriptors. IEEE Trans. Multimedia **11**(7), 1254–1265 (2009)

Exploring Image Bit Planes for Video Shot Boundary Detection

Anderson Carlos Sousa e Santos$^{(\boxtimes)}$ and Helio Pedrini (iD)

Institute of Computing, University of Campinas, Campinas, SP 13083-852, Brazil
anderson.santos@ic.unicamp.br

Abstract. The wide availability of digital content and the advances in multimedia technology have leveraged the development of efficient mechanisms for storing, indexing, transmitting, retrieving and visualizing video data. A challenging task is to automatically construct a compact representation of video sequences to help users comprehend the most relevant information present in their content. In this work, we develop and evaluate a novel method for detecting abrupt transitions based on bit planes extracted from the video frames. Experiments are conducted on two public datasets to demonstrate the effectiveness of the proposed method. Results are compared against other approaches of the literature.

Keywords: Bit planes · Cut detection · Video shot boundary
Video analysis

1 Introduction

The availability of several mobile devices, such as digital cameras, cell phones, and tablets, has enabled people to generate a large amount of multimedia content. The growth of digital data, particularly video streaming, has contributed to the advance of many knowledge domains, for instance, entertainment, education, telemedicine, robotics, surveillance and security.

In contrast to computationally expensive and time consuming task of manual annotation, the development of automatic and scalable strategies for storing, indexing, transmitting and retrieving multimedia data [15,18] is crucial to manage such massive growth of digital content.

Shot boundary detection plays an important role in temporal video segmentation [6,9,10,12,19], whose purpose is to partition the video content into meaningful units that constitute the most representative keyframes, known as shots. The summary of a video can be constructed from a set of keyframes that represent the shots.

Two categories of video transitions between shots are commonly defined: gradual and abrupt transitions. A gradual transition represents a smooth change over several frames, whereas an abrupt transition corresponds to a cut between one frame of a shot and its adjacent frame in the next shot.

© Springer International Publishing AG, part of Springer Nature 2018
M. Mendoza and S. Velastín (Eds.): CIARP 2017, LNCS 10657, pp. 593–600, 2018.
https://doi.org/10.1007/978-3-319-75193-1_71

There are various challenges associated with the video shot boundary detection task, such as illumination variability, camera motion, diversity of video genres, as well as the inherent subjectivity of the segmentation process. Although different video shot boundary detection methods have been proposed in the literature [2,3,8,16], two common steps are generally performed: (i) a similarity or dissimilarity measure is computed for each pair of consecutive frames and (ii) a cut is detected if the measure is higher than a specified threshold.

This work investigates and evaluates a novel shot boundary detection approach based on bit planes extracted from the video frames. An adaptive thresholding scheme is employed to determine if a transition is an abrupt shot boundary. Experiments are conducted on two public video benchmarks to show the effectiveness of the proposed method. Results are compared to other approaches available in the literature.

This paper is organized as follows. The proposed shot boundary detection method is detailed in Sect. 2. Experimental results are presented and discussed in Sect. 3. Finally, some final remarks and directions for future work are included in Sect. 4.

2 Methodology

The proposed video shot boundary detection method is based on the bit planes that compose a grayscale image. Since a pixel in a grayscale frame typically requires 1 byte of storage, 8 binary images can be produced by taking each bit plane independently. Figure 1 illustrates the bit planes extracted as binary images from a grayscale image (video frame). Each of these images describes different details about the pixel intensities, so a set of these images are used to extract features of the frame.

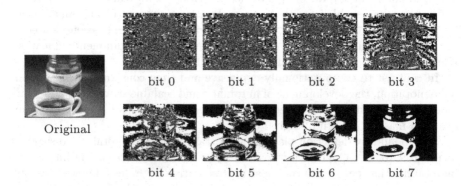

Fig. 1. Grayscale image and its decompositions in binary bit plane images.

Algorithm 1 describes how the features are extracted. For each bit plane image, a horizontal and vertical projection are calculated. These projections differ from the standard approach to summation of foreground pixels [7], whose purpose is to describe an object present in the image, since the information here about background is also relevant.

Algorithm 1. Projection

---

    input : Image f of size $N \times M$ pixels
               Set of bit planes B
    output: Feature vector W

1 **for** $b \in B$ **do**
2     $f_b \leftarrow$ Extract bit plane b
3     $V \leftarrow \emptyset$
4   **for** $l \in N$ **do**
5         $V_l = \sum_{j=0}^{M} -(-1)^{f(l,j)}$
6     $H \leftarrow \emptyset$
7   **for** $c \in M$ **do**
8         $H_c = \sum_{i=0}^{N} -(-1)^{f_b(i,c)}$
9     $W_b \leftarrow \{H, V\}$
10 **return** W

The background computation is important in the projection process since there are often cases with no foreground pixels, whose sum would result in zero. To avoid this, a transformation is applied to convert the binary image values from $[0, 1]$ to $[-1, 1]$. Thus, the projections correspond to the difference between the number of foreground and background pixels along each line and column, assigning positive values where the foreground exceeds the background, zero where they are equal, and negative values otherwise.

The concatenation of both horizontal and vertical projections constitutes the feature vector for a bit plane image. In order to compute the frame dissimilarities, a correlation-based distance (Eq. 1) between feature vectors is applied between equivalent bit planes for frames t and $t - 1$. The total dissimilarity is calculated as the average distance in the set of bit planes. Algorithm 2 summarizes such procedure.

$$correlation(u, v) = 1 - \frac{(u - \overline{u}) \cdot (v - \overline{v})}{\|(u - \overline{u})\|_2 \|(v - \overline{v})\|_2} \tag{1}$$

where u and v are two feature vectors.

Finally, the vector of dissimilarities for the entire video is subject to a thresholding method [14], which is locally adaptive. It normalizes the vector into the range $[0, 1]$, takes a moving window over the vector, computes the median value

---

Algorithm 2. Dissimilarity

---

   input : Video K
              Set of bit planes B
   output: Dissimilarity vector D

1   $D \leftarrow \emptyset$
2   **for** $f_i \in K$ **do**
3       $w^{i-1} \leftarrow projection(f_{i-1}, B)$
4       $w^i \leftarrow projection(f_i, B)$
5   **for** $b \in B$ **do**
6          $d_b \leftarrow correlation(w_b^{i-1}, w_b^i)$

7       $D_i \leftarrow \dfrac{\sum d_b}{|B|}$

8 **return** D

---

within the window, and determines a shot boundary if the center of the window is the maximum value within the window and is greater than the median plus a fixed α. Figure 2 shows an example of dissimilarity vector after applying the thresholding process.

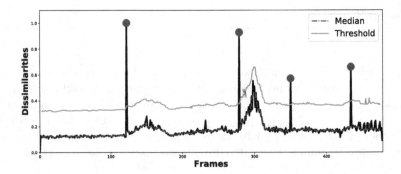

Fig. 2. Illustration of the thresholding technique.

3   Experimental Results

In this section, we present the experimental setup and analyze the results obtained on two data sets with our shot boundary detection method based on bit planes.

3.1   Data Sets

The TRECVID'2002 [17] consists of 18 video sequences with variability in quality, length, production style and noise level. The minimum number of cuts in a video is 18 and the maximum is 163, there is an average of 83 cuts per video. The duration of the longest video is 28 min 48 s, whereas the shortest is 6 min 32 s. This data set was used as benchmark for the TREC Video Retrieval Evaluation competition in the shot boundary detection task. Unlike some recent competition data sets, this benchmark is freely available.

The VIDEOSEG'2004 [20] is composed of 10 video sequences from different genres, sizes, digitization quality and production effects. The video with maximum number of cuts is 87, minimum number of cuts is 0, where the average number of cuts is 28.

3.2   Evaluation Metrics

The evaluation guidelines available for the TRECVID competition are followed in this work, such that the results are reported in terms of precision, recall and harmonic mean (F_{score}), as expressed in Eqs. 2, 3 and 4, respectively. The precision indicates the capacity of the method in detecting only the real cuts, the recall indicates the capacity in finding all existing cuts, whereas the F_{score} is the harmonic mean between the other two metrics.

$$\text{Precision} = \frac{\sum_{f_i \in V} S(i) \in Cut \wedge i \in True\ Cut}{\sum_{f_i \in V} S(i) \in Cut} \qquad (2)$$

$$\text{Recall} = \frac{\sum_{f_i \in V} S(i) \in Cut \wedge i \in True\ Cut}{\sum_{f_i \in V} i \in True\ Cut} \qquad (3)$$

$$F_{score} = 2 \frac{\text{Precision} \times \text{Recall}}{\text{Precision} + \text{Recall}} \qquad (4)$$

where V is a Video and S a detection set.

As established in the TRECVID competition, we set a tolerance of $+5$ and -5 for the frame number of the detected transition. This is done due to possible changes in the numeration of a frame caused by different video encoders.

3.3   Parameter Settings

Similarly to other approaches [13,14], the parameters used in the thresholding stage are kept. The window size is set to 7 and $\alpha = 0.2$. The feature extraction method and the distance calculation require as parameters only the set of bit planes.

An empirical experiment is carried out to determine the most appropriate set with respect to F_{score}. A search is executed in the TRECVID'2002 data set by evaluating all combinations out of the 8 planes. Table 1 shows the best results for each number of planes used. It is important to highlight that the results reported for VIDEOSEG'2004 are based on the same set as chosen for TRECVID'2002.

As expected, the most significant bits occur in the combinations with higher accuracy. Nonetheless, the bit plane 6 presented the best results among the

Table 1. Results for different combinations of bit planes.

# Bit planes	F_{score} (%)	
	TRECVID'2002	VIDEOSEG'2004
1 {6}	83.04	82.24
2 {6, 7}	**85.75**	82.94
3 {5, 6, 7}	85.51	83.22
4 {0, 5, 6, 7}	84.97	94.62
5 {0, 2, 5, 6, 7}	83.48	94.35
6 {0, 1, 4, 5, 6, 7}	80.06	94.04
7 {0, 1, 2, 4, 5, 6, 7}	77.95	94.33
8 {0, 1, 2, 3, 4, 5, 6, 7}	74.13	**94.92**

most significant ones, such as bit plane 7. This can be explained by the fact that the more significant the bit is, the more susceptible it is with respect to small contrast variations, which also justify the favoritism of bit 0 in place of other most significant bits. When the three most significant bits are present, bit 0 works to regulate the average and avoid false transitions caused by illumination changes.

In order to have a trade-off in terms of accuracy for both data sets and have a single best set of bit planes, the results presented in the next section consider the set of 4 bit planes ({0, 5, 6, 7}).

3.4   Results

Table 2 shows the results on TRECVID'2002 data set when the dissimilarities between frames were calculated through color histograms with 32 bins and cross-correlation between pixels. Both baseline methods were implemented in our framework with an adaptive threshold, such that only the features and distances were varied.

Table 3 shows comparative results for the VIDEOSEG'2004 data set. The results for the baseline methods were extracted from the literature.

For the TRECVID'2002 data set, our method surpassed the other methods in terms of precision, while maintained a proper recall rate. The precision rate is particularly difficult in this data set since the video sequences have several different types of transitions and glitch effects, which may be confused as cuts. For the VIDEOSEG'2004 data set, our method did not obtain the best precision or recall rate individually, however, it achieved a proper trade-off between both, resulting in the highest F_{score} measure.

Table 2. Video cut detection results for TRECVID'2002.

Method	Precision (%)	Recall (%)	F_{score} (%)
Color Histogram (CH)	78.34	91.50	83.72
Normalized Cross-Correlation (NCC)	75.05	94.99	80.45
Our method	81.29	90.81	**84.97**

Table 3. Comparative results for VIDEOSEG'2004.

Method	Precision (%)	Recall (%)	F_{score} (%)
Feature-tracking [20]	87.35	96.06	90.79
Visual rhythm [5]	85.72	96.07	89.75
Discrete Cosine Transform (DCT) [1]	93.94	89.81	91.54
Minimum ratio [11]	97.10	86.30	90.95
Color Co-occurrence Matrices (CCM) [4]	83.22	87.26	83.58
Our method	94.30	95.30	**94.62**

4   Conclusions

The decomposition of a grayscale image into bit planes was explored in this work for the purpose of video shot boundary detection. The bit planes are employed in the process as binary images, where feature vectors are extracted from them by means of vertical and horizontal projections and used in the computation of dissimilarity between adjacent video frames.

Experiments were conducted on two data sets to assess the proposed methodology. It shows that features based on the bit planes image are relevant to compute similarity between images. It was capable to outperform other approaches present in the literature of cut detection, and even when using the same settings and conditions it surpass the histogram of colors.

Acknowledgments. The authors are thankful to São Paulo Research Foundation (grant FAPESP #2014/12236-1) and Brazilian Council for Scientific and Technological Development (grant CNPq #305169/2015-7 and scholarship #141647/2017-5) for their financial support.

References

1. Almeida, J., Leite, N.J., da S. Torres, R.: Rapid cut detection on compressed video. In: San Martin, C., Kim, S.-W. (eds.) CIARP 2011. LNCS, vol. 7042, pp. 71–78. Springer, Heidelberg (2011). https://doi.org/10.1007/978-3-642-25085-9_8
2. Apostolidis, E., Mezaris, V.: Fast shot segmentation combining global and local visual descriptors. In: IEEE International Conference on Acoustics, Speech and Signal Processing, pp. 6583–6587 (2014)

3. Birinci, M., Kiranyaz, S.: A perceptual scheme for fully automatic video shot boundary detection. Signal Process.: Image Commun. **29**(3), 410–423 (2014)
4. Cirne, M.V.M., Pedrini, H.: VISCOM: a robust video summarization approach using color co-occurrence matrices. Multimed. Tools Appl. **77**(1), 857–875 (2018). https://link.springer.com/article/10.1007%2Fs11042-016-4300-7
5. Guimarães, S., Patrocínio, Z., Paula, H., Silva, H.: A new dissimilarity measure for cut detection using bipartite graph matching. Int. J. Semant. Comput. **03**(02), 155–181 (2009)
6. Huang, T.S.: Image Sequence Analysis, vol. 5. Springer Science & Business Media, Heidelberg (1981). https://doi.org/10.1007/978-3-642-87037-8
7. Jain, R., Kasturi, R., Schunck, B.G.: Machine Vision. McGraw-Hill Inc., New York (1995)
8. Jiang, X., Sun, T., Liu, J., Chao, J., Zhang, W.: An adaptive video shot segmentation scheme based on dual-detection model. Neurocomputing **116**, 102–111 (2013)
9. Koprinska, I., Carrato, S.: Temporal video segmentation: a survey. Signal Process. Image Commun. **16**(5), 477–500 (2001)
10. Ngan, K.N., Li, H.: Video Segmentation and Its Applications. Springer Science & Business Media, New York (2011). https://doi.org/10.1007/978-1-4419-9482-0
11. Pal, G., Acharjee, S., Rudrapaul, D., Ashour, A.S., Dey, N.: Video segmentation using minimum ratio similarity measurement. Int. J. Image Min. **1**(1), 87–110 (2015)
12. Piramanayagam, S., Saber, E., Cahill, N.D., Messinger, D.: Shot boundary detection and label propagation for spatio-temporal video segmentation. In: Proceedings of SPIE, vol. 9405, pp. 94050D–94050D-7 (2015)
13. Sousa e Santos, A.C., Pedrini, H.: Adaptive video transition detection based on multiscale structural dissimilarity. In: Bebis, G., et al. (eds.) ISVC 2016. LNCS, vol. 10073, pp. 181–190. Springer, Cham (2016). https://doi.org/10.1007/978-3-319-50832-0_18
14. Sousa e Santos, A.C., Pedrini, H.: Video temporal segmentation based on color histograms and cross-correlation. In: Beltrán-Castañón, C., Nyström, I., Famili, F. (eds.) CIARP 2016. LNCS, vol. 10125, pp. 225–232. Springer, Cham (2017). https://doi.org/10.1007/978-3-319-52277-7_28
15. Tekalp, A.M.: Digital Video Processing, 2nd edn. Prentice Hall Press, Upper Saddle River (2015)
16. Tippaya, S., Sitjongsataporn, S., Tan, T., Chamnongthai, K., Khan, M.: Video shot boundary detection based on candidate segment selection and transition pattern analysis. In: IEEE International Conference on Digital Signal Processing, pp. 1025–1029. IEEE (2015)
17. TRECVID: TRECVID Data Availability (2017). http://trecvid.nist.gov/trecvid.data.html
18. Veltkamp, R., Burkhardt, H., Kriegel, H.P.: State-of-the-Art in Content-Based Image and Video Retrieval, vol. 22. Springer Science & Business Media, Heidelberg (2013). https://doi.org/10.1007/978-94-015-9664-0
19. Verma, M., Raman, B.: A hierarchical shot boundary detection algorithm using global and local features. In: Raman, B., Kumar, S., Roy, P.P., Sen, D. (eds.) Proceedings of International Conference on Computer Vision and Image Processing. AISC, vol. 460, pp. 389–397. Springer, Singapore (2017). https://doi.org/10.1007/978-981-10-2107-7_35
20. Whitehead, A., Bose, P., Laganiere, R.: Feature based cut detection with automatic threshold selection. In: Enser, P., Kompatsiaris, Y., O'Connor, N.E., Smeaton, A.F., Smeulders, A.W.M. (eds.) CIVR 2004. LNCS, vol. 3115, pp. 410–418. Springer, Heidelberg (2004). https://doi.org/10.1007/978-3-540-27814-6_49

Convolutional Network for EEG-Based Biometric

Thiago Schons[1(✉)], Gladston J. P. Moreira[1]🆔, Pedro H. L. Silva[1],
Vitor N. Coelho[2], and Eduardo J. S. Luz[1]

[1] Computing Department, Universidade Federal de Ouro Preto,
Ouro Preto, Minas Gerais, Brazil
thiagoschons2@gmail.com
[2] Department of Computer Science, Universidade Federal Fluminense,
Niterói, Rio de Janeiro, Brazil

Abstract. The global expansion of biometric systems promotes the emergence of new and more robust biometric modalities. In that context, electroencephalogram (EEG) based biometric interest has been growing in recent years. In this study, a novel approach for EEG representation, based on deep learning, is proposed. The method was evaluated on a database containing 109 subjects, and all 64 EEG channels were used as input to a Deep Convolution Neural Network. Data augmentation techniques are explored to train the deep network and results showed that the method is a promising path to represent brain signals, overcoming baseline methods published in the literature.

1 Introduction

Humankind has urged for safety in all spheres of our society, thus, as technology evolves in this direction, also evolves efforts to overcome security systems. In this context, current biometric systems are in constant development, and new forms of capture discriminant and robust traits among people are desirable. The present work deals with the use of electroencephalogram (EEG) signals for *biometry task*, since the EEG is difficult to fake or steal.

The seminal work presented in [8] showed the feasibility of using the EEG to biometric task, and since that, many approaches using EEG have been proposed such as in [2] where authors performed biometric verification on Physionet EEG database. Authors concluded that the best frequency band for EEG biometric is the gamma band (30–50 Hz), where they reported 4.4% of equal error rate (EER). Their approach is based on phase synchronization, in which the Eigenvector Centrality obtained from every node (subject) is the feature vector. Signals on resting condition are considered for the analyses in two scenarios: eyes open and eyes closed.

In [12], four different task conditions, related to signal motor movement and imagery tasks are investigated. A novel wavelet-based feature was used to extract EEG feature. Experiments were conducted in Physionet EEG data and a mixture

© Springer International Publishing AG, part of Springer Nature 2018
M. Mendoza and S. Velastín (Eds.): CIARP 2017, LNCS 10657, pp. 601–608, 2018.
https://doi.org/10.1007/978-3-319-75193-1_72

of data, from different sessions, is used for training. Only nine electrodes are considered and the lowest EER achieved is 4.5%.

Several machine learning and pattern recognition techniques were investigated aiming to identify a person by means of EEG signals, however, to the best of our knowledge, deep learning based methods as Convolutional Neural Networks (CNN) [7] have not been evaluated yet. Deep learning has been used to represent patterns in several computer vision and patterns recognition problems, and outstanding results have been reported [1,5].

In this work, a novel approach for EEG representation based on deep learning is proposed. The approach is also evaluated on the Physionet database, and data augmentation techniques are explored to train a deep convolutional neural network. Results show that the use of CNN in EEG biometrics is a promising path, outperforming baseline methods by lowering the EER from 4.4% to 0.19% in the best scenario.

The remainder of this paper is organized as follows. Section 2 contains the approach with the methodology and a description of the database used. In Sect. 3, we show the experimental results and a discussion about it. Finally, in Sect. 4, the conclusions are presented.

2   Approach

In this section, the Physionet EEG database is described as the proposed method, based on the convolutional network, along with the required pre-preprocessing steps.

2.1   Physionet EEG Database

The Physionet EEG Database [3] is a popular benchmark in the literature for biometric with EEG and it is public available[1]. The records were acquired from 109 different subjects, using 64 electrodes in the region of the scalp to store the EEG signals (see Fig. 1), each sampled at 160 Hz. The database was created by the developers of the BCI2000 instrumentation system[2] and maintained by Physionet. There are 14 different acquisition sessions for each subject, each one with different motor/imagery tasks considered during recording. Among those 14 sessions, there are two one-minute (60 or 61 s per record) baseline runs (one with eyes open (EO), one with eyes closed (EC)). The others sessions are related to four kinds of tasks of three two-minute runs.

2.2   Methodology

Data pre-preprocessing: To further investigate the feasibility of the method, only resting state EEG data is considered. Thus, the baseline sessions - data captured where the subject is with EO and EC - are used during experiments.

---

[1] http://physionet.org/pn4/eegmmidb/.
[2] http://www.bci2000.org.

All EEG recording signals are band-pass filtered in 3 frequency bands. The first band covering from delta to gamma frequencies (1–50 Hz), the second band are related to low to high beta (10–30 Hz), and the third one preserves the range of gamma (30–50 Hz) frequency. A total of 61 s of raw wave are used for training and test.

Data Augmentation: In order to follow a baseline method evaluation protocol, proposed in [2], EEG data is divided into segments of 12 s window (1920 samples), i.e., 5 segments per subject for each record. Although 5 segments per subject are not enough data to train a deep convolutional neural network, data augmentation technique is proposed here to overcome this issue. The rationale for the data augmentation is to consider a large overlap between segments and therefore multiplying the number of segments. The new augmented data is created sliding a 12 s (1920 samples) window overall record signal (9600 or 9760 samples), shifting from 0.125 to 0.125 s, a sliding window strategy with 20 samples per step [9]. This technique yields 42696 new instances for training.

Convolutional Neural Network: The architecture of a typical convolutional neural network is structured as a series of stacked operations, beginning with convolutional layers, followed by activation with Rectified Linear Units (ReLu), pooling, normalization and finally fully connected layers (FC) [7].

For this work, three CNN architectures have been investigated. One with small receptive fields in the first convolutional layer inspired by [10] and two others with large receptive fields on the first convolutional layers inspired by [6, 13]. Note that filters are one-dimensional and proportionally adapted to EEG raw signal (see Fig. 2). The width and depth of the networks have been empirically evaluated based on validation error.

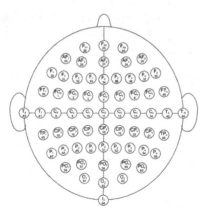

Fig. 1. Positions of electrodes on scalp. Source: http://physionet.org/pn4/eegmmidb/.

After the learning process, last three layers are removed (Softmax, Dropout, and FC4 as seen in Table 1) and the new network output is used as a feature vector for a 12 s EEG segment, which will be used for verification task.

In verification task, the performance of methods is expressed in terms of Detection Error Trade-off (DET) curves, which show the trade-off between type I error (false acceptance error - FAR) and type II error (false rejection error - FRR). To construct the DET curve, all instances from testing dataset are compared to each other, in an all-against-all scheme. Verification task can be modeled as the Eq. 1, where S is the function that measures the similarity between two feature vectors (X_1 and X_2) and t is a predefined threshold [4]. The value $S(X_1, X_2)$ is the similarity or matching score between the biometric measurements with Euclidean distance. A person's identity is claimed and classified into genuine when pairs are similar and impostor, otherwise. After that genuine (intra-class) and impostor (inter-class) distribution curves are generated from similarities scores.

$$(X_1, X_2) \in \begin{cases} genuine, & \text{if } S(X_1, X_2) \geq t \\ impostor, & otherwise \end{cases} \tag{1}$$

3   Experimental Results and Discussion

Experiments were conducted on an Intel (R) Core i7-5820K CPU @ 3.30 GHz 12-core machine, 64 GB of DDR4 RAM and one GeForce GTX TITAN X GPU. The MatConvNet library is used for the convolutional networks [11] linked to NVIDIA CuDNN.

Data segmentation for experiments was performed following the evaluation proposed in [2], where the window size consists of 12 s (as detailed in Sect. 2.2).

Data augmentation is used on data from EO session (training data), yielding 384 or 392 (from 60 and 61 s) segments of size 1920 samples (12 s) per subject. For evaluation, five segments of 12 s are extracted for each subject from EC session, i.e., no overlapping during the test (See Fig. 3).

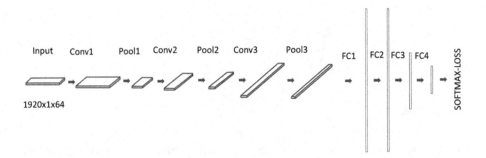

Fig. 2. Deep learning model.

Table 1. Architecture for EEG biometry.

Name	Type	Input size	Number of filters	Filter size/stride/pad	Relu	Norm
Network arch						
Conv1	Conv	1×1920	96	$1 \times 11/1/0$	Yes	Yes
Pool1	Max pooling	1×1910	N/A	$1 \times 2/4/0$	No	No
Conv2	Conv	1×478	128	$1 \times 9/1/0$	Yes	Yes
Pool2	Max pooling	1×470	N/A	$1 \times 2/2/0$	No	No
Conv3	Conv	1×235	256	$1 \times 9/1/0$	Yes	Yes
Pool3	Max pooling	1×227	N/A	$1 \times 2/2/0$	No	No
FC1	Full. conn.	1×113	4096	$1 \times 113/1/0$	Yes	No
FC2	Full. conn.	1×1	4096	$1 \times 1/1/0$	Yes	No
FC3	Full. conn.	1×1	256	$1 \times 1/1/0$	No	Yes
FC4	Full. conn.	1×1	109	$1 \times 1/1/0$	No	No
Drop	Dropout	1×1	2	N/A	No	No
Cost	Softmax	N/A	N/A	N/A	N/A	N/A

During training, the input signal, represented by a 12 s length EEG time series, is feed-forwarded through network layers. Each layer represents one or more CNN operations: convolutional filter; pooling; stride; rectification (RELU); normalization (L2 Norm). Convolutional stride and padding are set to one. Pooling layer performs a max-pooling operation, and when there is down-sampling (*stride* > 1), it happens in conjunction with pooling. The stack of layers are followed by *Fully-Connected* (FC) layers and the last FC layer is for classification. These FC layers can be seen as multi-layer-perceptron (MLP) network.

The final layer is a soft-max loss one. The FC layer with 1×1 filter size is used for dimension reduction and rectified linear activation. The network architecture is presented in Table 1.

For training the network, three learning rates of value $L = [0.01, 0.001, 0.0001]$ are distributed over the epochs, mini batches are set to size 100, and a momentum coefficient of 0.9 is considered during all training. Filter weights are randomly initialized and stochastic gradient descent is used for optimization. The dropout operation is placed before the last layer with 10% to minimize overfitting.

During the training phase, 90% of the data is reserved for training and 10% for validation as shown in Fig. 3. The CNN are trained for over 60 epochs.

Evaluation is carried in verification mode and the metric used to report results is *Equal Error Rate* (EER) which, in turn, is defined as the point where the *False Acceptance Rate* (FAR) is equal to the *False Rejection Rate* (FRR). FAR and FRR are generated from intra-class and inter-class pairs comparison. The present protocol produces 1086 genuine (intra-class) pairs and 146610 impostors (inter-class) pairs. In Fig. 4, DET curve shows the relationship between FAR, FRR, EER by means of a threshold variation.

Fig. 3. The distribution of ECG segments used for training and testing.

Table 2. EER obtained for the specified frequency bands.

Frequency band	EER
01–50 Hz	11.2%
10–30 Hz	6.25%
30–50 Hz	0.19%

The DET curves in Fig. 4 depicts the performance for the detailed experiments. The curve related to 30–50 Hz resulted in an overall performance of 0.19% EER as shown in Table 2, overcoming results published in the literature. As shown in Fraschini et al. [2], the best frequency band for EEG biometrics is the gamma band. The results presented here confirm the findings in [2] regarding

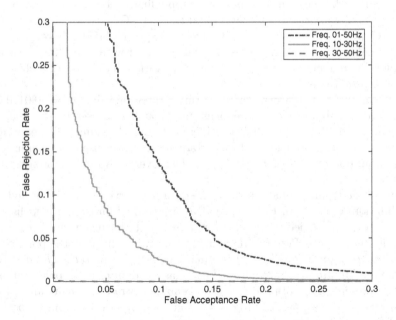

Fig. 4. DET curve for proposed experiments.

frequency band, however, the discrepancy of results with other frequency bands was greater here. More experiments are needed to investigate whether this phenomenon extends to other tasks (T1–T4) or even other databases.

Results presented in Table 3 compares the proposed method with state-of-the-art approaches. As can be noticed, the proposed method significantly reduced the EER. The usage of all 64 EEG channels shows the robustness of the method since it was able to handle all electrodes, even if not all of them effectively contribute to the identification of individuals [12].

Table 3. Comparison with related works.

Reports	Features	Train-Test	Electrodes	Subjects	EER(%)
Fraschini et al. [2]	Eigenvector centrality	EO-EC	64	109	4.40
Yang et al. [12]	Wavelet coefficients	T1–T4	9	108	4.50
Proposed work	CNN	EO-EC	64	109	0.19

4 Conclusions

In this work, the use of CNN in an EEG-based biometric system is investigated for the first time. When compared to the baseline methods presented in the literature (under the Physionet EEG database), EEG data represented by the CNN model showed a lower EER for person recognition (verification mode).

The contribution of this paper is the proposed deep CNN architecture and the data augmentation technique, which is of paramount importance in the training process. The sliding window strategy for generating new training samples allowed the deep network architecture to learn efficiently even with reduced data.

Results showed that the proposed EEG-based biometric system can be a promising method for future real-world applications since researchers are developing hardware to facilitate embedding a CNN model, such FPGA-based deep learning acceleration and NVIDIA TX1[3].

Acknowledgements. The authors thank UFOP and funding Brazilian agencies CNPq, Fapemig and CAPES. We gratefully acknowledge the support of NVIDIA Corporation with the donation of the Titan X Pascal GPU used for this research.

References

1. Collobert, R., Weston, J., Bottou, L., Karlen, M., Kavukcuoglu, K., Kuksa, P.: Natural language processing (almost) from scratch. J. Mach. Learn. Res. **12**, 2493–2537 (2011). http://dl.acm.org/citation.cfm?id=1953048.2078186

---

[3] http://www.nvidia.com/.

2. Fraschini, M., Hillebrand, A., Demuru, M., Didaci, L., Marcialis, G.L.: An EEG-based biometric system using eigenvector centrality in resting state brain networks. IEEE Signal Process. Lett. **22**(6), 666–670 (2015)
3. Goldberger, A.L., Amaral, L.A.N., Glass, L., Hausdorff, J.M., Ivanov, P.C., Mark, R.G., Mietus, J.E., Moody, G.B., Peng, C.K., Stanley, H.E.: PhysioBank, PhysioToolkit, and PhysioNet: components of a new research resource for complex physiologic signals. Circulation **101**(23), e215–e220 (2000)
4. Jain, A.K., Ross, A., Prabhakar, S.: An introduction to biometric recognition. IEEE Trans. Cir. Sys. Video Technol. **14**(1), 4–20 (2004)
5. Krizhevsky, A., Sutskever, I., Hinton, G.E.: Imagenet classification with deep convolutional neural networks. In: Proceedings of the 25th International Conference on Neural Information Processing Systems, NIPS 2012, pp. 1097–1105. Curran Associates, Inc., Red Hook (2012). http://dl.acm.org/citation.cfm?id=2999134.2999257
6. Krizhevsky, A., Sutskever, I., Hinton, G.E.: Imagenet classification with deep convolutional neural networks. In: Pereira, F., Burges, C.J.C., Bottou, L., Weinberger, K.Q. (eds.) Advances in Neural Information Processing Systems 25, pp. 1097–1105. Curran Associates, Inc. (2012). http://papers.nips.cc/paper/4824-imagenet-classification-with-deep-convolutional-neural-networks.pdf
7. LeCun, Y., Bengio, Y., Hinton, G.: Deep learning. Nature **521**(7553), 436–444 (2015)
8. Poulos, M., Rangoussi, M., Chrissikopoulos, V., Evangelou, A.: Person identification based on parametric processing of the EEG. In: The 6th IEEE International Conference on Electronics, Circuits and Systems, Proceedings of ICECS 1999, vol. 1, pp. 283–286 (1999)
9. Reitermanov, Z.: Data splitting. In: Proceedings of Contributed Papers, WDS 2010, pp. 31–36 (2010)
10. Simonyan, K., Zisserman, A.: Very deep convolutional networks for large-scale image recognition. CoRR abs/1409.1556 (2014). http://arxiv.org/abs/1409.1556
11. Vedaldi, A., Lenc, K.: Matconvnet - convolutional neural networks for MATLAB. In: Proceeding of the ACM International Conference on Multimedia (2015)
12. Yang, S., Deravi, F., Hoque, S.: Task sensitivity in EEG biometric recognition. Pattern Anal. Appl., 1–13 (2016)
13. Zeiler, M.D., Fergus, R.: Visualizing and understanding convolutional networks. In: Fleet, D., Pajdla, T., Schiele, B., Tuytelaars, T. (eds.) ECCV 2014. LNCS, vol. 8689, pp. 818–833. Springer, Cham (2014). https://doi.org/10.1007/978-3-319-10590-1_53

Skeleton Pruning Based on Elongation and Size of Object's Limbs and Boundary's Convexities

Luca Serino$^{(\boxtimes)}$ ⓘ and Gabriella Sanniti di Baja

Institute for High Performance Computing and Networking, CNR, Naples, Italy
{luca.serino,gabriella.sannitidibaja}@cnr.it

Abstract. We present a new pruning method able to remove peripheral branches of the skeleton of a 2D object without altering more significant branches. Pruning criteria take into account elongation and size of the object's parts associated with skeleton branches. Only peripheral branches associated with scarcely significant object's limbs and boundary's convexities are removed, so that the object can be recovered satisfactorily starting from the pruned skeleton. Since by removing peripheral branches, new peripheral branches can be created, pruning is iterated until the skeleton structure becomes stable. The algorithm does not require fine tuning of the parameters and the obtained results are satisfactory.

Keywords: Shape representation · Shape analysis · Skeleton · Pruning

1 Introduction

The skeleton is a well known representation system useful in the framework of shape analysis. A wide literature is available as concerns different skeletonization methods, devised in the continuous and in the digital space, and the use of the skeleton in several application fields [1]. Ideally, the skeleton is characterized by unit thickness, centrality, homotopy, recoverability and significance. This means that the skeleton should consist exclusively of curves placed in the middle of the object, and should be characterized by the same topological features as the object; moreover, skeleton pixels should be labeled with their distance from the complement of the object so that the object can be recovered by the union of the discs centered on the skeleton pixels and having radii equal to the corresponding distance values; finally skeleton branches should be found only in correspondence of significant limbs and strong boundary convexities of the object. Actually, whichever skeletonization algorithm is considered, the obtained skeleton S unavoidably includes a number of peripheral branches not all obtained in correspondence of individually meaningful object protrusions or boundary's convexities, thus making skeleton pruning an indispensable step for any skeletonization algorithm.

Pruning can be accomplished by preventing the creation of non significant branches via a preliminary filtering process that identifies suitable anchor points on the tips of significant object's convexities from which skeleton branches will originate [2, 3]. More typically, pruning is accomplished during a post-processing phase aimed at removing scarcely significant branches.

© Springer International Publishing AG, part of Springer Nature 2018
M. Mendoza and S. Velastín (Eds.): CIARP 2017, LNCS 10657, pp. 609–617, 2018.
https://doi.org/10.1007/978-3-319-75193-1_73

In the literature, different criteria have been suggested to distinguish branches corresponding to significant parts of the object from those whose removal does not affect the representative power of the skeleton. See for instance [4], where pruning criteria based on propagation velocity, maximal thickness, radius function, axis arc length, and ratio between boundary and axis length are presented. A pruning criterion based on the number of slices of object pixels that would not be recovered starting from the pruned skeleton is discussed in [5]. Other pruning methods involve contour partitioning via discrete curve evolution and bending potential ratio [6, 7], where pruning is either performed during a post-processing step, or is embedded into the skeleton computation process. Sequential and parallel modalities to perform pruning are discussed in [8], while criteria to remove concatenations of branches without altering the topology of S or reducing its representative power are described in [9].

In this paper, we present a pruning method based on elongation and size of object's protrusions and boundary convexities. The method can be applied to any skeleton, provided that its pixels are labeled with their distance from the complement of the object. In this paper we refer to the skeleton computed by the algorithm [5], where the <3, 4> distance [10] is used to label skeleton pixels. Elongation and size of the object part associated with a skeleton branch are evaluated by resorting to polygonal approximation of the skeleton branch in a 3D space, where the three coordinates of any skeleton pixel are its spatial coordinates and distance label. The criterion adopted to evaluate the goodness of the suggested method is based on the reconstruction ability of the skeleton. The higher is the percentage of pixels of the input object recovered by the pruned skeleton, the better the object is represented by the pruned skeleton.

2 Basic Notions

We work with binary images, where the object consists of the pixels with value 1, while the background consists of the pixels with value 0. The 8-connectedness and the 4-connectedness are used for the object and the background, respectively. Since the skeleton S is a subset of the object, the 8-connectedness is used also for S.

Given two pixels p and q, their <3, 4> distance $d(p, q)$ is the length of a minimal 8-connected path linking p to q, where unit moves towards horizontal/vertical neighbors and diagonal neighbors along the path are respectively weighted 3 and 4.

For the sake of simplicity, let us suppose that only one 8-connected object exists in the image at hand, so that its skeleton S consists of exactly one 8-connected component. Pixels of S are classified as end points, normal points, or branch points, depending on whether they have one, two, or more than two neighbors in S.

Each branch of S, which is a curve entirely consisting of normal points with the exception of the two pixels delimiting the curve, which can be end points or branch points, maps a region of the input object. A branch is termed peripheral branch if one delimiting pixel is an end point and the other is a branch point. To avoid altering the topology of S, only peripheral branches can be pruned. Thus, skeletons rid of peripheral branches do not undergo pruning.

When a skeleton branch is pruned, its delimiting branch point is not removed; it can be transformed into a new end point or into a normal point, or can maintain the status of branch point, depending on whether all or only part of the peripheral branches sharing it are pruned. Once the initial peripheral branches of S have been checked against the pruning criteria and possibly removed, new peripheral branches might have been created in S due to the transformation of some branch points into end points or normal points. Then, pruning is applied again. The process is iterated until the skeleton becomes stable.

3 The Method

At each iteration, the following main tasks are performed: (1) polygonal approximation of peripheral skeleton branches, (2) computation of elongation and size of the corresponding object's parts, and (3) removal of scarcely significant branches. Pruning is iterated until the skeleton reaches a stable structure.

Polygonal approximation
We resort to the split type approach [11] to compute the polygonal approximation of peripheral skeleton branches. The extremes of the current skeleton branch are taken as the two starting vertices from which the process recursively splits the branch. The Euclidean distance of all normal points of the skeleton branch from the straight line joining the two vertices is computed. The point at the largest distance is taken as a new vertex, provided that such a distance overcomes an a priori fixed threshold θ. Otherwise the process terminates. When a new vertex is detected, the branch is divided into two curves, to each of which the split type algorithm is applied. Splitting continues as far as new vertices are detected. At the end of the process, the skeleton branch is approximated by a number of curves that, in the limits of the adopted tolerance, are straight segments and is represented by the ordered sequence of the detected vertices. To explain the reason for which polygonal approximation is performed in a 3D space, where the three coordinates of any skeleton pixel are its spatial coordinates and distance label, let us consider the object in Fig. 1 left and its skeleton S in Fig. 1 middle, where different colors denote different distance values. We may note that the geometry of S reflects the geometry of the object: curvature changes along the skeleton correspond to curvature changes along the object boundary of the object. Thus, polygonal approximation of S in the 2D space would divide it into straight segments corresponding to parts of the object whose boundary is characterized by the absence of curvature changes. We may also note that the different distance values of the pixels of S take into account the changes in width of the object. Thus, by approximating S in the 3D space, we divide it into straight segments along which distance labels are either constant or monotonically increase/decrease. Each so found segment is interpreted as the spine of a simple region whose boundary is rid of curvature changes and whose width is either constant or linearly increases/decreases. From an operative point of view, the computation of the Euclidean distance d of a point C from the straight line joining two points A and B in the 3D space is done by using the following expression:

$$d^2 = ||AC||^2 - P_{ABC} * P_{ABC}/||AB||^2 \qquad (1)$$

where $||AB||$ is the norm of the vector AB, and P_{ABC} is the scalar product between vectors AB and AC. We have experimentally found that to obtain a quite faithful approximation of S with a reasonably small number of vertices, the best value for the threshold is in the average $\theta = 3.5$. Such a value has been used for all the examples shown in this paper. As an example, see Fig. 1 right, where the found vertices are shown in black.

Fig. 1. From left to right, an object, its skeleton (different colors denote different distance values), and the vertices (black) resulting in the skeleton after polygonal approximation in 3D. (Color figure online)

Size and elongation

The advantage provided by polygonal approximation is that elongation and size of the object's parts associated with skeleton branches can be computed immediately once spatial coordinates and distance labels of the vertices are available. In fact, a simple region is shaped either as a rectangle (Fig. 2 middle left), when the distance labels are constant along the spine (Fig. 2 left), or as a trapezium (Fig. 2 right), when distance labels linearly increase/decrease along the spine (Fig. 2 middle right), and is delimited by two half discs centered on the two vertices of the spine. A concatenation of simple regions is associated to a skeleton branch approximated by a number of segments.

Fig. 2. Straight line skeleton segments and their corresponding simple regions.

We remark that any vertex along a skeleton branch (except the end point and the branch point) is common to two successive spines. We also remark that pruning does not remove the branch point delimiting a skeleton branch. Thus, the evaluation of the size of the object's part that would be lost by removing the corresponding skeleton branch is done as follows. For each simple region in the concatenation forming the object's part associated with the *i-th* peripheral skeleton branch, the half disc centered on the more external vertex is added to the area of the rectangle/trapezium, while the half disc centered on the more internal vertex is subtracted. The total area A_i associated to the i-th skeleton branch is the sum of the so computed contributions in correspondence with the spines polygonally approximating the i-th skeleton branch.

The elongation E_i of the object's part that would be lost by removing the i-th skeleton branch is computed by adding to the radius of the disc centered on the end point of the i-th skeleton branch the height of each rectangle/trapezium in the concatenation of simple regions and by subtracting the radius of the disc centered on the branch point. The height of any rectangle/trapezium is computed as the distance between the two vertices delimiting the corresponding spine.

Pruning

To decide whether the i-th peripheral skeleton branch can be pruned, we should compare the elongation E_i and the size A_i of the corresponding object's part with two thresholds. We automatically compute the value for the threshold on elongation in terms of the average elongation AvE of all the skeleton branches that result to be peripheral at the current iteration of the process, and regard the i-th peripheral branch prunable with respect to elongation if the following condition is satisfied:

$$E_i < e \times AvE_i \qquad (2)$$

where e is a constant.

We set the value of the threshold on size in terms of the ratio between the size A_i of the object part that would be lost by removing the i-th skeleton branch and the size D_i of the disc centered on the branch point of the i-th skeleton branch, and regard the i-th skeleton branch prunable with respect to size if the following condition is satisfied:

$$A_i/D_i < a \qquad (3)$$

where a is a constant.

We take into account that often S is affected by a large number of "short" noisy peripheral branches, whose contribution to AvE significantly reduces the real significance of this parameter. See for example Fig. 3 left. Thus, at each iteration of pruning we perform a pre-processing step to remove "short" noisy branches. We regard a skeleton branch as "short" if it consists of at most four pixels. A close up of a portion of S is given in Fig. 3 middle, where "short" branches are shown in red. This choice is done since any such a branch would be definitely approximated by a single spine, so that during pre-processing we don't perform polygonal approximation of "short" branches and directly take their delimiting end points and branch points as the extremes of the corresponding spines. Obviously, a "short" branch is not necessarily noisy and might still be important for a quite faithful object recovery. Thus, we consider the i-th "short" branch as noisy and remove it only if condition (3) is satisfied for it. We point out that a number of pixels different from four can characterize the length of "short" noisy branches, depending e.g., on acquisition noise and image resolution. However, even if some "short" branches are longer than four pixels and hence are not removed during pre-processing, we still considerably trim S so that the successive computation of AvE is done in a more reliable manner.

After "short" noisy branches have been pruned, polygonal approximation of the current peripheral branches of S is done and E_i, A_i and D_i are computed for each branch. The average elongation AvE of all peripheral branches is also computed.

In general, S consists both of noisy peripheral branches and of significant peripheral branches. If the majority of peripheral branches are noisy, a larger value for the threshold on elongation can be used, while a smaller value is necessary if S has a limited number of noisy branches, so as to avoid pruning together with noisy branches also some significant branches. We evaluate whether a smaller or a larger threshold value should be used by exploiting the criterion adopted to measure the goodness of pruning, which is based on the reconstruction ability of the skeleton. If at the beginning of a given iteration of pruning the number of pixels recovered by the un-pruned skeleton, #Unpr, and the number of pixels recovered by S if all its peripheral branches are removed, #Prun, have similar size, we could argue that all peripheral branches of S are noisy. Clearly, in general not all peripheral branches of S are noisy, so that we need to set a threshold τ_{iter} on the ratio #Prun/#Unpr to decide whether we can use larger threshold value. To take into account that pruning is iterated until prunable branches are found, we also compare the ratio between #Prun and the number of pixels in the input object, #Obj, with another, smaller, threshold τ_{obj}. Then, if #Prun/#Obj > τ_{obj} and #Prun/#Unpr > τ_{iter}, a larger value for the threshold on elongation can be used by setting $e = e_1 > 1$ in condition (2). Otherwise, $e = 1$.

We have experimentally found that in the average satisfactory results are obtained by setting $\tau_{obj} = 0.80$, $\tau_{iter} = 0.85$, $e_1 = 6$ and $a = 0.6$.

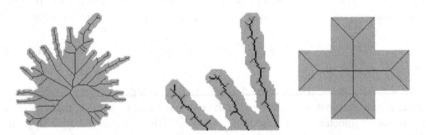

Fig. 3. Left, a noisy skeleton with many "short" peripheral branches; middle, a close up with "short" branches in red; right, a skeleton with peripheral branches with similar elongation. (Color figure online)

Finally, if the peripheral branches of S are all characterized by very similar elongation (see for example the skeleton in Fig. 3 right), we regard all the skeleton branches as significant and no pruning at all is accomplished. We have experimentally verified that if all peripheral branches are characterized by elongation that differs from AvE for at most 10%, pruning should not be accomplished.

Summarizing, at each iteration the following tasks are performed:

(1) Peripheral branches including at most four pixels are identified and each of them is removed if $A_i/D_i < a$;
(2) Peripheral branches are identified and undergo polygonal approximation, during which for each of them elongation E_i, size A_i and area of the disc D_i for the associated object's part are computed. The average elongation AvE is also computed;
(3) If the elongation of all peripheral branches differs from AvE for at most 10%, all peripheral branches are significant and pruning terminates;
(4) If $\#Prun/\#Obj > \tau_{obj}$ and $\#Prun/\#Unpr > \tau_{iter}$, the i-th peripheral branch is pruned provided that $A_i/D_i < a$ and $E_i < e_1 \times AvE_i$, else the i-th peripheral branch is pruned provided that $A_i/D_i < a$ and $E_i < AvE_i$.

The tasks (1)–(4) are repeated until peripheral branches are pruned. In general, at most four iterations are enough to obtain the final result. An example where skeleton stability is reached in two iterations is shown in Fig. 4.

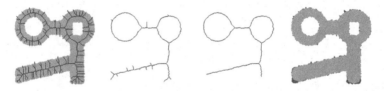

Fig. 4. From left to right, the un-pruned skeleton superimposed on the input object, the skeleton resulting after the 1st and after the 2nd iteration, and the object recovered by the pruned skeleton (object pixels that are not recovered by the pruned skeleton are shown in red). (Color figure online)

We have tested our pruning method on a large set of binary images, taken from shape repositories such as tosca.cs.technion.ac.il/book/resources_data.html, cs.rug.nl/svcg/Shapes/SkelBenchmark, vision.lems.brown.edu/faculty/kimia, cs.toronto.edu/∼dmac/ShapeMatcher/. We point out that our method is robust in case of noisy objects and when the input object appears at a different scale or in a different pose. In fact, the average value of the ratio between input pixels recovered by the pruned skeleton and input pixels recovered by the un-pruned skeleton is about 0.96, showing that the representative power of the pruned skeleton is not significantly modified with respect to that of the un-pruned skeleton.

A few examples showing the performance of our pruning method can be appreciated with reference to Fig. 5. We remark that the same values of the parameters involved in the pruning algorithm have been computed for all the images independently of image resolution, object size and position, and noise.

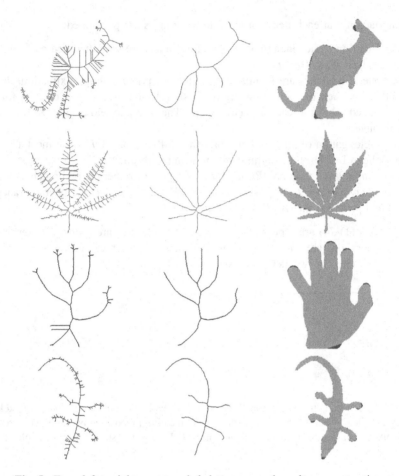

Fig. 5. From left to right, un-pruned skeletons, pruned results, reconstructions.

4 Conclusion

We have introduced a pruning method that removes scarcely significant branches of the skeleton of 2D objects without altering significant branches. The two main pruning criteria take into account elongation and size of the object's parts associated with skeleton branches. To guarantee topology preservation, only peripheral branches are possibly removed. Since removal is done only for branches associated with scarcely significant object's parts, the object can be recovered satisfactorily starting from the pruned skeleton. Pruning is iterated since removal of peripheral branches may create new peripheral branches. The algorithm does not require fine tuning of the parameters and the obtained results are satisfactory.

References

1. Siddiqi, K., Pizer, S.M. (eds.): Medial Representations: Mathematics, Algorithms and Applications. Springer, Berlin (2008). https://doi.org/10.1007/978-1-4020-8658-8
2. Serino, L., Sanniti di Baja, G.: Selecting anchor points for 2D skeletonization. In: Kamel, M., Campilho, A. (eds.) ICIAR 2011. LNCS, vol. 6753, pp. 344–353. Springer, Heidelberg (2011). https://doi.org/10.1007/978-3-642-21593-3_35
3. Postolski, M., Couprie, M., Janaszewski, M.: Scale filtered Euclidean medial axis and its hierarchy. Comput. Vis. Image Underst. **129**, 89–102 (2014)
4. Shaked, D., Bruckstein, A.M.: Pruning medial axes. Comput. Vis. Image Underst. **69**(2), 156–169 (1998)
5. Sanniti di Baja, G.: Well-shaped, stable and reversible skeletons from the (3, 4)-distance transform. VCIR **5**, 107–115 (1994)
6. Bai, X., Latecki, L.J., Liu, W.-Y.: Skeleton pruning by contour partitioning with discrete curve evolution. IEEE Trans. PAMI **29**(3), 449–462 (2007)
7. Shen, W., Bai, X., Hu, R., Wang, H., Latecki, L.J.: Skeleton growing and pruning with bending potential ratio. Pattern Recogn. **44**, 196–209 (2011)
8. Frucci, M., Sanniti di Baja, G., Arcelli, C., Cordella, L.P.: On the strategy to follow for skeleton pruning. In: De Marsico, M., Fred, A. (eds.) ICPRAM 2013, pp. 263–266. SCITEPRESS, Lisboa (2013)
9. Serino, L., Sanniti di Baja, G.: A new strategy for skeleton pruning. Pattern Recogn. Lett. **76**, 41–48 (2016)
10. Borgefors, G.: Distance transformations in digital images. Comput. Vis. Graph. Image Process. **34**(3), 344–371 (1986)
11. Ramer, U.: An iterative procedure for the polygonal approximation of plane curves. CGIP **1**, 244–256 (1972)

Volume Rendering by Stochastic Neighbor Embedding-Based 2D Transfer Function Building

Walter Serna-Serna(✉)(iD), Andres M. Álvarez-Meza,
and Álvaro-Ángel Orozco-Gutierrez

Automatics Research Group, Universidad Tecnológica de Pereira, Pereira, Colombia
{wserna,andres.alvarez1,aaog}@utp.edu.co

Abstract. Multidimensional transfer functions (MDTF) allow studying a volumetric data in a space built from features of interest. Thus, a transfer function (TF) can be defined as a region in a feature space that assigns optical properties to each voxel supporting volume rendering. Since voxels belonging to different objects can share feature similarities, segmentation of individual volume structures is not a straightforward task. We present a TF building approach from a 2D low-dimensional space using dimensionality reduction (DR). Namely, we carried out a Stochastic Neighbor Embedding (SNE)-based DR from MDTF domains. The outcomes show how our proposal, termed SNETF, outperform state-of-the-art approaches that use DR techniques in TF domains. The experiments were performed in a synthetical volume and in a standard volumetric tomography. Our method achieved a higher separability among objects on the new 2D space preserving the original distances between voxel samples. Thus, it was possible to get 3D representation of an object of interest into a given volume, which is an important fact for the next step in automating the generation of TF for volume rendering.

Keywords: Volume rendering · Transfer functions
Sthocastic neighbor embedding

1 Introduction

Information in a volumetric data is described by 3D features. It is necessary to analyze the elements of a volume through variables like the shape, scale, orientation, boundaries, curvature, the density of materials, in order to estimate how is a volumetric object represented by the data, and further, to compare the object with other ones of the same nature. However, visual inspection of a volume is not a simple task due to the data must be displayed by projection or slices, in consequence, a spatial dimension will be lost [2].

To visualize fragmented information systematically is the paradigm in volumetric exploration. Related to this, clinical application in diagnosis and surgical planning from medical images is an actively developing field [9]. Here, the use

© Springer International Publishing AG, part of Springer Nature 2018
M. Mendoza and S. Velastín (Eds.): CIARP 2017, LNCS 10657, pp. 618–626, 2018.
https://doi.org/10.1007/978-3-319-75193-1_74

of segmentation and classification techniques is imperative. However, the original data is not enough to separate structures and additional features must be extracted from it. A Transfer Function (TF) in direct volume rendering (DVR) pretends to find a visual solution through three stages: the multi-dimensional (MD) feature space construction, the TF delimitation that is the selection of clustered voxels on the feature space, and the voxel mapping to optical properties for DVR [6]. Furthermore, the most implemented way to delimit the TF is manual, however, an MD feature space is not usable for visual exploration.

The advantage of an MD feature space is the separability of particular structures with information that is merged in other low-dimensional (LD) spaces. Nonetheless, due to the TF-based volume render is not a completely automated methodology, the TF building is not feasible in high-dimensional (HD) domains. Often, Dimensionality Reduction (DR) is performed to produce an LD feature space where the user can sketch the TFs. These DR techniques seek to preserve relevant information of HD features through discriminative measures to be represented as intelligible data visualization [8]. An unsupervised approach was presented in [7] using Self-Organizing Maps (SOM), where a lattice compounded of neurons are adjusted to the data structure in the MD-Space considering sample concentration. Later, all the samples are projected to the lattice which is considered the LD representation where the TF is built. PCA has been used to preserve variance in orthogonal projections of the data, Local Linear Embedding (LLE) exploits the relation between a point and its neighborhood, and Isomap conserves the geodesic distances between points [3]. Nevertheless, from a general perspective in DR tasks, these methods have not shown better outcomes that older techniques [5]. This is the case in volume rendering with MD-TF, where a complete isolation of volumetric objects has not been achieved.

In this paper, is exploited the capability of SNE preserving similarities in the construction of LD feature spaces as a method to extract relevant information of a multiple TF domains combination for DVR. In particular, we implement derivative-based features as the gradient and second derivative, likewise statistical measures as the mean and standard deviation, and boundary-related features as the low-value and high-value from the LH-histogram. All these features are used to compound an HD domain where is expected that the final topology allows a better isolation of volumetric objects. First, a representation of each voxel (with its multiple features) is randomly generated in a 2D space, and a Gaussian-based similarity is computed among voxels in the HD and LD spaces to measure both probability distributions. Later, the LD representations of voxels are iteratively optimized to find the lower difference between distributions. Here, a Kullback-Leibler divergence is employed as the cost function in the manner proposed by [5], where the minimization is solved for the LD space by a gradient descent-based algorithm. In order to compare the performance of the SNE-based method, PCA and SOM techniques are implemented. Finally, several TF are manually built over the LD space of each DR method looking for specific volume objects of two datasets used in the experiments.

The rest of this paper is organized as follows. Section 2 describes how the divergence is computed among HD and LD similarities in SNE to optimize the LD representation of data. Section 3 presents the proposed experiments and discusses the results. Finally, Sect. 4 resumes the conclusions.

2   Theoretical Background

Let $\boldsymbol{V} \in \mathbb{R}^{w \times h \times l}$ be an input volume holding $N = w \cdot h \cdot l$ voxels $v_r \in \mathbb{R}$, where $\boldsymbol{r} = (i, j, k)$ is a 3D coordinate with range $i \in \{1, 2, \ldots, w\}, j \in \{1, 2, \ldots, h\}$, and $k \in \{1, 2, \ldots, l\}$. Our SNETF approach includes the following main stages: (i) HD feature estimation from MDTF to code volume structures, (ii) DR based on SNE algorithm to find a 2D LD space coding discriminative patterns, and (iii) a TF building by user region sketching. Following, each stage is described in detail.

Feature estimation from MDTF. We represent $\boldsymbol{V}$ through M mappings to gather the feature set $\Psi = \{\boldsymbol{S}^{(m)} \in \mathbb{R}^{w \times h \times l \times Q_m}\}_{m=1}^{M}$, where $\boldsymbol{S}^{(m)}$ is the volume data transformed by the m-th function $f^{(m)} : \mathbb{R} \to \mathbb{R}^{Q_m}$, which intends to enhance some specific properties of the voxels. In this respect, each voxel in $\boldsymbol{V}$ is mapped to a Q_m-dimensional representation given by $f^{(m)}(v_r|\theta_m) \mapsto \boldsymbol{s}_r^{(m)}$, where θ_m corresponds to the function parameters, and $\boldsymbol{s}_r^{(m)} \in \boldsymbol{S}^{(m)}$. After voxel concatenation, we devise the HD feature space matrix $\boldsymbol{X} \in \mathbb{R}^{N \times P}$, where $P = \sum_{m=1}^{M} Q_m$, and each element $\boldsymbol{x}_n \in \mathbb{R}^P$ preserves the coordinates (i, j, k) in $\boldsymbol{V}$ with $n = i + j \cdot w + k \cdot w \cdot h$.

Dimensionality reduction based on SNE. In turn, to extract a discriminative LD space matrix $\boldsymbol{Y} \in \mathbb{R}^{N \times U}$ ($U = 2$ for TF building purposes) we carried out an SNE projection as a probabilistic approach to weigh the geometrical relation among voxels in $\boldsymbol{X}$ [5]. In such a way, let $\boldsymbol{D} \in \mathbb{R}^{N \times N}$ and $\boldsymbol{B} \in \mathbb{R}^{N \times N}$ be a HD and a LD space distance matrices with elements $d_{ij} = ||\boldsymbol{x}_i - \boldsymbol{x}_j||_2$ and $b_{ij} = ||\boldsymbol{y}_i - \boldsymbol{y}_j||_2$, respectively, where $|| \cdot ||$ stands for the 2-norm and $\boldsymbol{y}_i \in \mathbb{R}^U$. Later, to estimate the probability distribution of the samples in both spaces, a Gaussian-based similarity is employed as follows: $\gamma_{nn'}^{X} = \exp(-d_{nn'}^2/\sigma_n^X)/\sum_{q=1}^{N} \exp(-d_{nq}^2/\sigma_n^X)$, $\gamma_{nn'}^{Y} = \exp(-b_{nn'}^2/\sigma_n^Y)/\sum_{q=1}^{N} \exp(-b_{nq}^2/\sigma_n^Y)$; $\forall q \neq n$, $\sigma_n^X, \sigma_n^Y \in \mathbb{R}^+$, and $\gamma_{nn'}^{X}, \gamma_{nn'}^{Y} \in [0,1]$. Hence, to compute $\boldsymbol{Y}$, the SNE algorithm minimizes a Kullback-Leibler divergences between the probability distributions, yielding:

$$\boldsymbol{Y}^* = \arg\min_{\boldsymbol{y}_n} \sum_{n=1}^{N} \sum_{n'=1}^{N} \gamma_{nn'}^{X} \log(\gamma_{nn'}^{X}/\gamma_{nn'}^{Y}). \tag{1}$$

Equation 1 is solved by a gradient descent-based algorithm [5]. Remarkably, the LD space $\boldsymbol{Y}$ in SNE optimally preserves the neighborhood identity, allowing the identification of relevant data clusters to support further TF building.

TF building. Finally, a TF is built on the LD space by sketching a region $\Lambda \in \boldsymbol{Y}$ to render (label) an object of the input volume $\boldsymbol{V}$ as $\hat{\boldsymbol{V}} = \{v_r : \boldsymbol{y}_n \in \Lambda; r = (i, j, k), n = i + j \cdot w + k \cdot w \cdot h\}$. Figure 1 summarizes the proposed SNETF.

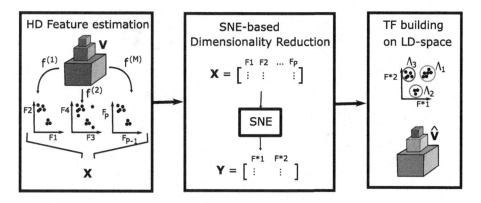

Fig. 1. SNETF pipeline for volume rendering.

3   Experiments and Results

We test our SNETF approach as a tool for estimating a relevant LD space from MDTF to guide volume rendering. Indeed, we built a 2D TFs by manually clustering the voxels in Δ from Y that successfully emphasize the region of interest in V. So, both synthetic and real-world databases are studied as follows:

Synthetic data results. We evaluate a synthetic data $V \in \mathbb{R}^{50 \times 50 \times 50}$ holding three 3D spheres (SV) of different intensity and radius sizes (see Figs. 2(a), (b), (c)). Then, the following $M = 3$ MDTF are computed for each voxel: derivative that assesses voxel intensity variations to reveal volume boundaries [4], statistical-based function that quantifies homogeneity/heterogeneity of the object densities [1], and Low-High (LH)-based histogram that code volume boundaries as concentrated clusters [10]. Each mapping holds the following features: intensity, gradient magnitude, and second derivative ($Q_1 = 3$); mean and standard deviation ($Q_2 = 2$); low value and high value ($Q_3 = 2$). Also, the kernel size of the derivative function is fixed as 3 [4]. Moreover, the neighborhood size to compute the statistical mapping is customized for each voxel following the approach in [1], which uses a spherical neighborhood that keeps growing as long as the elements within it show a normal distribution while new neighbors preserve a similarity-based threshold. Besides, the step size in LH-based histogram is set as 1 [10]. Since the input volume comprise a huge number of voxels, we randomly selected $N = 2500$ samples before building the HD feature matrix X with $P = 7$ [11]. Furthermore, the background was removed previously by a thresholding stage to increase the probability to get samples from small structures. Note that each column in X is normalized within the range $[0, 1]$. In turn, the SNE approach is carried out over X to compute the LD feature space matrix Y. For concrete testing, $U = 2$ to perform visual inspection analysis. The σ_n^X and σ_n^Y values are found when the distribution entropy over a number κ of neighbors is equal to $log(\kappa)$, where κ is empirically fixed as 90. Moreover, the complementary voxels after random selection are mapped to Y through a k-nearest neighbor

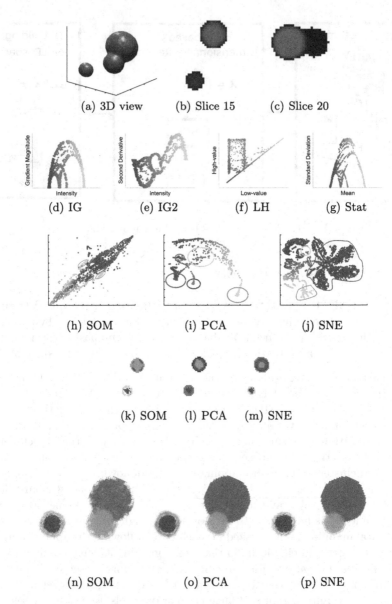

Fig. 2. SV results. First row: input volume. Second row: multi-domain features. Third row: LD spaces. Fourth row: 2D-Segmented slices. Fifth row: volume rendering. (Color figure online)

interpolation. Here, the number of neighbors is empirically fixed as $\vartheta = 15$. Furthermore, the PCA and SOM-based volume rendering techniques are utilized as benchmark [3,7]. For the SOM approach, a rectangular lattice of 32×32 neurons is trained in X, fixing the Euclidean metric as the topological distance over the lattice, and new samples are projected to the LD space.

Figures 2(d), (e), (g) shows the traditional 2D visualizations from the considered MDTF: Intensity vs Gradient (IG), intensity vs Second derivative (IG2), low vs high intensities (LH), and Mean vs standard deviation histogram (Stat). Likewise, Figs. 2(i), (h), (j) show the LD spaces generated by PCA, SOM, and SNE techniques. The projected samples are colored according to the original voxel intensities trying to provide a visual relation with the input volume. The same colormap is used in all the spaces so that the clustering can be compared among techniques. Related to PCA algorithm, the reduced space shows a similar distribution to the IG histogram (see Fig. 2(d)). Although some bifurcations appear due to the interpolation process, both the arches and their intersections remain in Y. Hence, the PCA-based projection favors the segmentation of the internal voxels of the spheres due to such voxels are represented as the arch lowest part. However, it is not possible to segment each sphere boundaries individually without overlapping them, which is the same problem for IG-based histograms. Concerning the SOM algorithm, the generated LD space does not show a representative separation of the volumes. In consequence, finding a suitable TF is not a straightforward task because of the voxels overlap. Therefore, a lot of voxels are left out, and that is why the spheres are incomplete in the original space (see Figs. 2(k) and (n)). In contrast, the introduced SNETF reaches a separation among objects, even boundaries. As shown in Fig. 2(j), a path of points can be seen between objects representing boundaries without intersections between paths. Although SNE spreads the interior regions of the spheres in comparison to the PCA-based projection, a suitable discrimination among objects is obtained.

Real data results. As a second test, we evaluate the well-known tomography of a tooth $V \in \mathbb{R}^{256 \times 256 \times 161}$ (TV) [6], which exhibits the pulp, the dentin, and the enamel of a tooth (see Figs. 3(a) to (c)). In the same way, as in the synthetic data, seven features are computed, and a set of $N = 10000$ voxels are selected, excluding the background. Besides, each feature is normalized to the range $[0, 1]$. In turn, we compute the LD space by the SNE approach fixing $\kappa = 120$, and the interpolation method is performed with $\vartheta = 25$. Again, the PCA and SOM-based methods are computed in the same conditions as in the synthetic data. It is important to notice that the tomographic images are cylindrical shaped, but these are padded in a rectangular volume. So the cylindrical background is taken as an additional object. In this case, we build four TFs: a red TF involves the pulp, a yellow TF represents the enamel, a green TF is related to the tooth dentin, and a magenta TF is used to study boundaries (see Figs. 3(h) to (j)). For SOM and PCA renderings the pulp is merged with the cylinder edge that can be seen as a red circle boundary in the slice views in Figs. 3(k) and (l). In contrast, our proposal allows to separate the pulp from this cylindrical edge. Furthermore, the boundaries of the dentin-pulp and the background-dentin are wrapped into the same TF. In contrast, the SNETF isolate such boundaries but spreads the interior regions of the objects. Lastly, note that an object can present more than

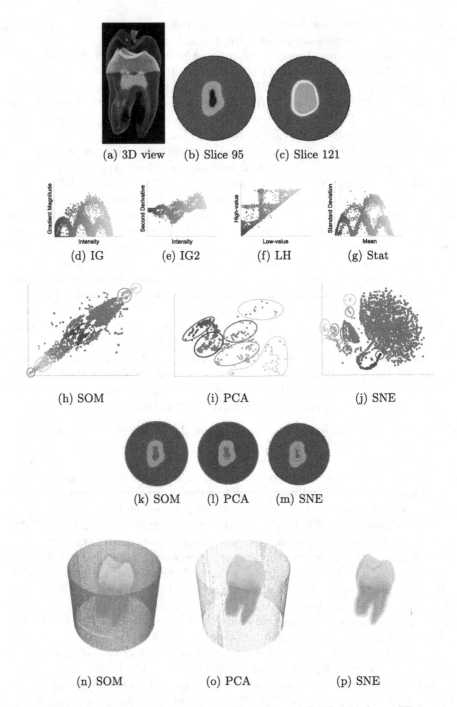

Fig. 3. TV results. First row: input volume. Second row: MDTF. Third row: LD spaces. Fourth row: 2D-Segmented slices. Fifth row: volume rendering. (Color figure online)

one cluster in the LD space for all considered techniques [10]; nonetheless, the SNETF does not present a significant mixing between objects.

4   Conclusions

A new volume rendering approach based on DR is introduced by TF building from 2D spaces. In particular, we represent an input volume as a HD feature space combining different TF domains. Then, a similarity preservation method, termed SNE, is employed to extract a suitable 2D space to favor TF construction. Moreover, our method, called SNETF, uses an interpolation scheme based on k-nearest neighbors to project complementary voxels after training the DR mapping from a randomly selected subset of input voxels. Experimental results show a high separability among volume objects in the 2D LD space and the capability to conserve the boundaries as isolated transitions without intersections. Moreover, SNETF preserves the cluster distributions, even for large datasets holding millions of voxels. As future work, authors plan to incorporate semi-supervised information to avoid the spread of internal objects. Also, spatial information from input volume can be used to join/separate the clusters representing individual structure in the LD space.

Acknowledgements. Under grants provided by the project (111074455778) "Desarrollo de un sistema de apoyo al diagnstico no invasivo de pacientes con epilepsia fármaco-resistente asociada a displasias corticales cerebrales: Método costoefectivo basado en procesamiento de imágenes de resonancia magnética", funded by Colciencias.

References

1. Haidacher, M., Patel, D., Bruckner, S., Kanitsar, A., Gröller, M.E.: Volume visualization based on statistical transfer-function spaces. In: IEEE PacificVis (2010)
2. Kehrer, J., Hauser, H.: Visualization and visual analysis of multifaceted scientific data: a survey. IEEE TVCG **19**(3), 495–513 (2013)
3. Kim, H.S., Schulze, J.P., Cone, A.C., Sosinsky, G.E., Martone, M.E.: Dimensionality reduction on multi-dimensional transfer functions for multi-channel volume data sets. Inf. Vis. **9**(3), 167–180 (2010)
4. Kniss, J., Kindlmann, G., Hansen, C.: Multidimensional transfer functions for interactive volume rendering. IEEE TVCG **8**(3), 270–285 (2002)
5. Lee, J.A., Renard, E., Bernard, G., Dupont, P., Verleysen, M.: Type 1 and 2 mixtures of Kullback-Leibler divergences as cost functions in dimensionality reduction based on similarity preservation. Neurocomputing **112**, 92–108 (2013)
6. Ljung, P., Krüger, J., Groller, E., Hadwiger, M., Hansen, C.D., Ynnerman, A.: State of the art in transfer functions for direct volume rendering. Comput. Graph. Forum **35**, 669–691 (2016)
7. de Moura Pinto, F., Freitas, C.M.: Design of multi-dimensional transfer functions using dimensional reduction. In: Proceedings of the 9th Joint Eurographics/IEEE VGTC conference on Visualization, pp. 131–138. Eurographics Association (2007)

8. Peluffo-Ordóñez, D.H., Lee, J.A., Verleysen, M.: Recent methods for dimensionality reduction: a brief comparative analysis. In: ESANN (2014)
9. Preim, B., Botha, C.P.: Visual Computing for Medicine: Theory, Algorithms, and Applications. Newnes, Oxford (2013)
10. Sereda, P., Bartroli, A.V., Serlie, I.W., Gerritsen, F.A.: Visualization of boundaries in volumetric data sets using LH histograms. IEEE Trans. Vis. Comput. Graph. **12**(2), 208–218 (2006)
11. Tzeng, F.Y., Ma, K.L.: A cluster-space visual interface for arbitrary dimensional classification of volume data. In: IEEE TCVG Conference on Visualization (2004)

Text Localization in Born-Digital Images of Advertisements

Dirk Siegmund$^{(\boxtimes)}$ ⓘ, Aidmar Wainakh, Tina Ebert, Andreas Braun,
and Arjan Kuijper

Fraunhofer Institute for Computer Graphics Research (IGD),
Fraunhoferstraße 5, 64283 Darmstadt, Germany
{dirk.siegmund,aidmar.weinakh,tina.ebert,andreas.braun,
arjan.kuijper}@igd.fraunhofer.de

Abstract. Localizing text in images is an important step in a number of applications and fundamental for optical character recognition. While born-digital text localization might look similar to other complex tasks in this field, it has certain distinct characteristics. Our novel approach combines individual strengths of the commonly used methods: stroke width transform and extremal regions and combines them with a method based on edge-based morphologically growing. We present a parameter-free method with high flexibility to varying text sizes and colorful image elements. We evaluate our method on a novel image database of different retail prospects, containing textual product information. Our results show a higher f-score than competitive methods on that particular task.

1 Introduction

Optical character recognition (OCR) is a widely used application describing the task of converting images of typed, handwritten or printed text into machine readable text. One challenge in OCR is the localization of text within an image. While born-digital (BD) text localization might look similar to other complex tasks of this field, they present certain distinct characteristics. They originate from materials in digital form and carry salient semantic information such as advertisements or security-related information. Therefore, extracting text information from BD images enhances the semantic relevance of its content for indexing and retrieval. One application example is retail advertisement, where brochures and prospects are the most common marketing media elements. To make their offline advertisements accessible on new media devices like tablets and smart-phones, internet platforms digitize and index them [1,8]. An important component in this challenging task is the localization of text on these images (see Fig. 1). They are available free of charge to readers as a printed edition or as e-paper on the internet. Mainly they contain product names, with its textual description, price and images. Usually, OCR systems extract these kind of text by following the steps: text localization, text segmentation and text recognition. Text localization is critical to the overall system performance and is influenced by varying image background, text color, the used font and layout (see Sect. 3).

© Springer International Publishing AG, part of Springer Nature 2018
M. Mendoza and S. Velastín (Eds.): CIARP 2017, LNCS 10657, pp. 627–634, 2018.
https://doi.org/10.1007/978-3-319-75193-1_75

When BD images are getting used, the performance of known OCR methods drop drastically. Main difficulty in theses methods are the definition of color clusters and parameters for local thresholding. We found additional challenges and features that reduced the appliance of state of the art methods: extremal regions (ER) and stroke width transform (SWT) to the recognition of BD advertisement images. To overcome these problems, we (1) provide a new technical viewpoint for the localization of text in born-digital advertisement images. Our method combines the strength of SWT and ER with a novel method based on morphology growing and edge-detection. (2) We present an entropy based color clustering method that is flexible to colorful text elements. (3) A novel database is presented, containing advertisement prospects from different retailers.

In Sect. 2 we will give a brief overview about related methods. An overview of our database is given in Sect. 3, where we also focus on the special properties of our database and the differences to other data-sets of BD images. Our method is introduced in Sect. 4, where we explain, how we addressed the identified challenges. In Sect. 5 we show our results in comparison with other methods.

2   Related Work

Research on BD images is focused mostly on applications like image-spam filtering, trying to classify images in emails as legit or spam [5]. In these tasks defining the extent of text is more important, then recognizing its content [6]. We found following methods developed for these domains not necessary transferable to our task. Nevertheless, known general techniques on text localizing are the detection of ER across several image projections (channels), as proposed by Neumann and Matas [10]. In their approach, a two phase classifications is presented using the features: area, bounding box, perimeter and Euler number for classification with AdaBoost. In the second phase a SVM classifier is used for classification of the hole area ratio, convex hull ratio, and the number of outer boundary inflexion points as features. Moreover, a clustering algorithm groups ER's into text lines, where finally an OCR classifier recognizes the text. Epshtein [4] and proposed using SWT to detect text in natural scenes images. A method proposed by Cho et al. [3] exploited the similarity between characters in order to detect text. Their approach is based on the fact that the cohesive characters compose a word or sentence are sharing similar properties such as spatial location, size, color, and stroke width. It consists of finding ER's, applying double threshold classifier and hysteresis tracking. This work achieved a recall harmonic mean of 82.17%, exceeding the winner in ICDAR 2013 RRC (Challenge 2). Yu et al. presented an approach [13] that localizes text, based on edge analysis. The authors used over-segmentation of edges and recombined them using a neighbor map, where they applied an edge filter, using stroke width, angle between corresponding points, shape feature, and the variance of morphological operation. Moreover, a text line filter was applied, which considers that characters in the same line having similar color, stroke width, height, and width.

3   Database

BD images are digitally created images that may show any content as textual information. As many of them are used online, low resolution and compression artifacts (see Fig. 1c) may occur. Especially in advertisements, there is a variety of text sizes and fonts (see Fig. 1e and f), which sometimes have complex and unusual designs. In comparison to real-scene images, the amount of text can be quite large and often contains text that is of no particular interest (like brand names on the pictured products as in Fig. 1d). Sometimes, the background is a complex image with high contrast, which makes the text detection more difficult (see Fig. 1a). Also, the text is spread over the whole page in small groups and is not placed in one well defined block, as it is in pure text documents (see Fig. 1b). Another challenge is the variety of colors, although it is often confined to a maximum number of three because of design reasons. This is helpful in determining the number of required color clusters. Other beneficial characteristics are the usually high text to background contrast and the good readability, which by contrast is challenging in real-scene images. Also, there are no illumination problems and no perspective transformations.

Fig. 1. (a) High contrast background (b) different alignment (c) compression artifacts (d) undesired text elements (e) variety of text sizes (f) special fonts. (Color figure online)

3.1   Advertisement Database

We created a novel advertisement database, containing a number of 78 different advertisement prospects. This included advertisements from local retailers, like beverage shops (see Fig. 1 left) and supermarkets with challenging content. In all cases, the total image size is 1200×1860 pixel (px) and a combination of product pictures, product information and prizes is shown. The layout differs for every page, and there are both, single color backgrounds and complex figure backgrounds. A large variety in text sizes, fonts and colors is also included on each page. The ground truth was created manually by creating bounding boxes, covering text elements line-wise and drawing them on a binary mask.

4   Our Approach

We present a novel approach for the localization of text in images of advertisement prospects. We combined useful aspects of the three known text localization methods: Stroke Width Transform [4] and Extremal Regions [9] and Edge Based Morphological Growing [7] into a novel method in order to localize the different text elements on the images of our data-set.

4.1   Edge-Based Morphological Growing

This approach uses morphological operators in order to merge nearby detected edges to form connected components [7].

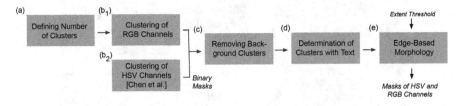

Fig. 2. Process of edge-based morphological growing.

Advertisement images are most often very colorful but still provide good readability and contrast of the relevant information. But in our tests, analyzing advertisements of many companies, we noticed that relevant product information like: product name, price and its description show not more than two or three colors. Thus, we cluster the image colors in two different ways by using k-mean to reduce the complexity of the image (Fig. $2b_1$ and b_2). As the use of different color models represents different color-perception, we use RGB and HSV color model and combine their advantages in later processing.

1. Entropy based Color Clustering (RGB): Images contain different amount of colors, which is reflected by the image entropy. In Formula 1 we propose using the 'Shannon entropy' to define the amount of k color-clusters. First, we convert the image to HSV color space and calculate the entropy E of the H channel. Second, we assign E divided by 2 to the formula on the right, to find

k number of clusters where $\alpha = 0.2$ and $\beta = 2.4$. The variables α and β are defined by testing iteratively on our data-set.

$$E(I) = -\sum_{C=1}^{C=3} \sum_{i=1}^{i=2^8} p(i) \log_2 (p(i)), k = e^{\left(\frac{E(I)-\alpha}{\beta}\right)} \tag{1}$$

2. Centroid based Color Clustering (HSV): We use the approach proposed by Chen et al. [2] operating across different dimension of a HSV image. The initialization of centroids and thereby the number of clusters is calculated using histogram quantization.

After having clusters from these two methods we exclude their main background cluster (see Fig. 2c) by choosing the one with most non zero pixel. Thereafter, we calculate the amount of text in each cluster using the software 'Google Tesseract' [12]. We choose the three clusters with the highest amount of text (Fig. 2d) for further processing. In the last processing step (see Fig. 2e) we detect edges using the 'Sobel Edge Detector' and apply morphological closing horizontally. Since characters are rich of edges, the closing operation will lead to merging them. The size of the closing operation element was chosen by testing over the full data-set. Finally, we detect connected components as candidates for text lines using structural component analysis. Since the output of this approach contains a lot of false positive, we filter components based on their geometric properties extent and size. Second, we filter smaller isolated boxes, since most likely text is gathered in lines and blocks. In the last step binary masks are generated from both, the RGB and HSV clustered masks containing the remaining components.

4.2   Stroke Width Transform

Stroke width transform has shown its advantages in different applications of this field [11]. We use the approach of Epshtein [4] because it represents a suitable image operator representing the stroke line of text. To define stroke width for each image pixel and group pixel that have similar stroke width. Each group of pixels is considered as a candidate letter because usually letters have same stroke width. Non-letter groups get filtered out to aggregate text elements. SWT shows limitation in detecting especially the smaller characters. We extended this method by a measurement of the intensity of characters in a specific position in the image. Based on that, we decide whether there is a missed character at that position. Most likely the missed character is surrounded by several detected characters within the same line or the upper/lower lines.

4.3   Extremal Regions

We employed the ER text localization method proposed by Neumann and Matas (NM) [9], as they showed good results on similar tasks. A corresponding program flow chart is given in Fig. 3, where in (a), a combination of red (R), green (G),

Fig. 3. Sequence of ER detection.

blue (B), and an intensity gradient (∇) is used to calculate channels, that are to be processed individually.

In (b) 1st and 2nd stage classifiers are generated. In the first stage, extremal regions are detected by using incrementally computable descriptors, which are then classified as text or non-text in the second stage. These two ER filters are both applied to the previous calculated NM channels, and return ER objects and their bounding boxes. Afterward, in the overlapping boxes are merged to avoid redundancy (Fig. 3c). Especially when trying to localize text boxes of similar size, such as prices, the previously found boxes need to be filtered (d). For this, the most occurring box height is determined by using a histogram of heights. Then, the average area of boxes with a height around this value is calculated, and used to filter out over- and undersized boxes. In the last step a binary mask is created from the remaining set of bounding boxes.

4.4 Combination of Masks

We integrate these methods, using their individual strength on different text sizes. We distinguish between: small (<60px height) and big (>60px height) texts elements. We use ER to detect the big text, because it shows the highest amount of false-positive while we used morphology to detect small text. In addition, SWT is used to detect any text size since it is not text size sensitive.

First combination: In our experiments, we noticed that the masks resulting from 'Edge-Based Morphological Growing' RGB and HSV clustering rarely contain the same non-text boxes. Therefore, we intersect elements on both masks and verify them by using the SWT mask. Boxes that were excluded intersecting RGB and HSV mask, get included again, in case they are found in SWT mask. Then we add the ER mask since it is responsible of detecting big texts.

Second combination: In this combination we used SWT as a base mask and added to it the ER mask, containing the big text elements. As SWT has limitation in detecting small texts, we add a intersection mask of the RGB and HSV resulting from the 'Edge-Based Morphological Growing' method.

5 Results

The evaluation was performed in terms of text element-level localization on our data-set. To quantify our results we used three parameters: recall rate, precision, and f-score. ER led to very high false positive rate as shown in Table 1. The number of the detected elements was 5795, which is more than double of the

Table 1. Results of proposed combination of methods.

Approach	#Elements	Recall%	Precision%	f-score%
RGB	3736	69.95	50.98	57.91
HSV	3992	52.72	52.40	50.29
SWT	1788	73.84	70.95	71.44
ER	5795	71.04	50.35	57.95
Combination 1	2732	75.21	65.53	68.88
Combination 2	2005	88.54	67.63	75.91

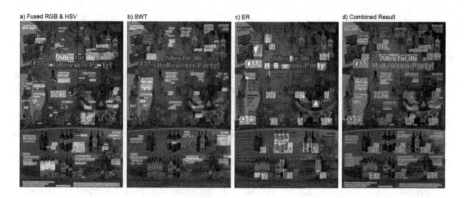

Fig. 4. Example, showing individual results (a–c) and final results (d).

ground truth elements of 2077. Our evaluation showed that most of these falsely detected elements are of small size, while the approach achieved better detection and precision rate for big elements. Edge-Based Morphological Growing showed its strengths especially in detecting areas of small font size. We noticed low recall rate and low precision, due to the limited filtering criteria that is available in this approach when using both RGB and HSV clusters. The main criteria to filter out the non-text blobs was the extent, which is the ratio between the size, white pixel and their bounding box. SWT reached a high recall rate in case of big text while its performance decreases when the text becomes smaller. We presented two combinations of the presented methods, as shown in Table 1. In combination 1, the morphology approach is mainly used to detect small texts but has a high false positive rate. Therefore more elements were detected in combination 1 compared to combination 2. On the other hand, we noticed better precision in combination 2 as it depends on SWT, which strictly verifies the intersection of the masks gathered by the edge-based morphology growing method (Fig. 4).

6   Conclusion

We presented a novel localization method for born-digital images. We showed the characteristics of this specific use-case and explored methods that are used in similar tasks. Our presented method adapts several strengths of these methods and extends them to be more robust against challenges of advertisement images. We introduced a novel image database, containing advertisement prospects on which we evaluated our method. As a result we showed that our method overcomes limitation of competitive methods. Our pipeline is able to detect text, without any parameter, regardless of its scale, color and font.

Acknowledgment. This work was supported by the German Federal Ministry of Education and Research (BMBF) as well as by the Hessen State Ministry for Higher Education, Research and the Arts (HMWK) within CRISP.

References

1. Bonial International GmbH: Kaufda (2017). http://www.kaufda.de/
2. Chen, T.W., Chen, Y.L., Chien, S.Y.: Fast image segmentation based on K-Means clustering with histograms in HSV color space. In: 2008 IEEE 10th Workshop on Multimedia Signal Processing, pp. 322–325 (2008)
3. Cho, H., Sung, M., Jun, B.: Canny text detector: fast and robust scene text localization algorithm. In: Proceedings of the IEEE Conference on Computer Vision and Pattern Recognition, pp. 3566–3573 (2016)
4. Epshtein, B.: Detecting text in natural scenes with stroke width transform, pp. 2963–2970 (2010)
5. Gonzalez, A., Bergasa, L.M., Yebes, J.J., Bronte, S.: Text location in complex images. In: 2012 21st International Conference on Pattern Recognition (ICPR), pp. 617–620. IEEE (2012)
6. Hanif, S.M., Prevost, L.: Text detection and localization in complex scene images using constrained adaboost algorithm. In: 2009 10th International Conference on Document Analysis and Recognition, ICDAR 2009, pp. 1–5. IEEE (2009)
7. Khan, N., Puri, S.: A study on text detection techniques of printed documents. In: Proceedings of the 2016 IEEE International Conference on Wireless Communications, Signal Processing and Networking, WiSPNET 2016, pp. 2478–2482 (2016)
8. marktguru Deutschland GmbH: Marktguru (2017). http://info.marktguru.de/
9. Neumann, L., Matas, J.: Real-time scene text localization and recognition. In: 2012 IEEE Conference on Computer Vision and Pattern Recognition (CVPR), pp. 3538–3545. IEEE (2012)
10. Neumann, L., Matas, J.: Real-time lexicon-free scene text localization and recognition. IEEE Trans. Pattern Anal. Mach. Intell. **38**(9), 1872–1885 (2016)
11. Siegmund, D., Ebert, T., Damer, N.: Combining low-level features of offline questionnaires for handwriting identification. In: Campilho, A., Karray, F. (eds.) ICIAR 2016. LNCS, vol. 9730, pp. 46–54. Springer, Cham (2016). https://doi.org/10.1007/978-3-319-41501-7_6
12. Smith, R.: An overview of the Tesseract OCR engine. In: Proceedings of the Ninth International Conference on Document Analysis and Recognition, ICDAR 2007, Washington, DC, USA, vol. 02, pp. 629–633. IEEE Computer Society (2007)
13. Yu, C., Song, Y., Zhang, Y.: Scene text localization using edge analysis and feature pool. Neurocomputing **175**, 652–661 (2016)

Deep Learning Techniques Applied to the Cattle Brand Recognition

Carlos Silva[1(✉)], Daniel Welfer[2], and Juliano Weber[3]

[1] Federal University of Pampa, Alegrete, Brazil
carlos.al.silva@live.com
[2] Federal University of Santa Maria, Santa Maria, Brazil
daniel.welfer@ufsm.br
[3] Federal Institute Farroupilha, Santo Ângelo, Brazil
juliano.weber@iffarroupilha.edu.br

Abstract. The automatic recognition of cattle brandings is a need of the government organizations responsible for controlling this activity. To help this process, this work presents a method that consists in using Deep Learning techniques for extracting features from images of cattle branding and Support Vector Machines for their classification. This method consists of six stages: (a) selection of a database of images; (b) selection of a pre-trained CNN; (c) pre-processing of the images, and application of the CNN; (d) extraction of features from the images; (e) training and classification of images (SVM); (f) evaluation of the results obtained in the classification phase. The accuracy of the method was tested on the database of a City Hall, where it achieved satisfactory results, comparable to other methods reported in the literature, with 91.94% of Overall Accuracy, and a processing time of 26.766 s, respectively.

Keywords: Deep learning · Convolutional neural network
Recognition images · Support Vector Machines · Cattle brand

1 Introduction

The use of computer tools to aid in the analysis and recognition of images is a matter of interest for the most renowned research centers in the world. The use of computing for image analysis and recognition is in constant development, generating many benefits to society in the most diverse areas of knowledge. Concerning the recognition of cattle branding in particular, a very traditional activity and of great social-economic relevance for Latin American countries, including Brazil, there is no specific and appropriately consolidated tool for this end.

The use of brands or symbols on cattle presupposes the public recognition of its property by an individual or group. Livestock has a relevant role in social formation, which is still today an activity of great importance in cultural expressions associated to it, as it is linked to the culture and the countryside way of life, and it also has a role in the affirmation and building of individual and group identities [1]. Generally, cattle brand records involve books with the drawings of the brands and the identification of

M. Mendoza and S. Velastín (Eds.): CIARP 2017, LNCS 10657, pp. 635–642, 2018.
https://doi.org/10.1007/978-3-319-75193-1_76

their owner. Currently, brand recording in Brazil is performed by town offices, generally without a more effective systematization and without instituted renewals.

In face of the presented scenario, this work intends to present and assess a tool that performs automatic cattle brand recognition, with the goal of replacing the manual control of cattle branding performed today, in order to potentially decrease the possibility of duplicate records, reducing waiting times for the recording of new brands, improving governmental administration regarding the brand archive under its care and aid security officials in preventing cattle raiding crimes.

Our work focused on the development of algorithms for the recognition of cattle branding for the São Francisco de Assis City Hall, in Rio Grande do Sul, Brazil. Therefore, the employees in the Cattle Branding Record Section and in the Data Processing Center of this township validated the suggested tool.

The remaining work is organized as follows: in Sect. 2, we describe related works. In Sect. 3, we present the materials and methods used in the article. Section 4 describes the results and discussions obtained by the application of the recommended methods. Lastly, in Sect. 5, final considerations are described.

2   Related Work

In general, we could not find any works in the literature review that report the use of convolutional neural networks for the recognition of cattle branding images.

Sanchez *et al.* [2] present a tool for recognition of cattle branding that use Hu and Legendre moments for extracting features of images in a grey scale, and also a classifier of k-nearest neighbors (k-NN). The authors used Hu and Legendre moments to source features that were not prone to rotation, translation, and scale transformations. The peak percentage of correct classification presented by the authors was 99.3%, with a significant decrease in accuracy as the number of classified images increased, however. No classification model is created. Instead, for each new object to be classified, training data are scanned, and the suggested classifier becomes computationally expensive.

Convolutional neural networks have been used for many years in image recognition, and they have obtained great success in recognizing characters in the research made by LeCun *et al.* [3] Recent studies using convolutional neural networks known as Deep Convolutional Neural Networks (CNN) obtained the new state of the art in object recognition based in CIFAR-10 and NORB [4]. Generally, CNNs are trained under supervision, but some research revealed that pre-training CNNs with filters obtained in a non-supervised way yields better results [5]. In the research conducted by Sermanet *et al.* [6] they present a framework using CNNs to perform recognition, localization, and detection of images. An important feature of CNNs is their ability to be reused and refined for different image bases. In the research conducted by Razavian *et al.* [7] a pre-trained CNN named *Overfeat* [6] was used to perform the extraction of a descriptor from different image bases whose CNN was not originally trained. In this case, the descriptors are then sorted by an SVM linear classifier. These results indicate a performance compatible with the state of the art, even when compared with algorithms that use manually-segmented images, a procedure that is unnecessary when CNNs are used and trained specifically in the analyzed basis. Usage of deep learning is also described in the research made by Constante *et al.* [8]. In this paper, the method was used to sort strawberries and obtained Overall Accuracy results of 92.5% in the "Extra" category; 90% in the "Consumption" category; 90% in the "Raw material" category; and 100% in the "Alien objects" category.

Differently from the research presented by Sanchez *et al.* [2], the work we propose here intends to show results that can be generalized or reproduced, deploying state-of-the-art techniques for feature extraction and statistical classification of digital images, such as Convolutional Neural Networks (CNN) and Support Vector Machines (SVM) to achieve superior performances in image recognition in databases with a greater number of records, but with a lower computational cost and processing time.

3   Materials and Methods

The images from cattle branding presented in this research were provided by the São Francisco de Assis City Hall, in Rio Grande do Sul. We used 30 cattle branding images, each one of them composed by 45 sub-images (samples), totaling 1,350 samples from original images, but with size and orientation variations. We intended to identify patterns with the greatest independence possible from these variable factors. The images were provided in high resolution in the Portable Network Graphics format at a size of 600 × 600 pixels.

For the implementation of the proposed tool, as well as image database storage, algorithm processing and viewing of the results we used a personal computer with a video card that supports CUDA parallel computing platform with a 5.0 compute capability version. Furthermore, we used the MATLAB software with the Neural Network, Parallel Computing and Statistics and Machine Learning libraries and the pre-trained convolutional network model obtained from the open source library VLFeat.org [9].

The proposed method consists of six steps, which are: image database selection; selection of pre-trained CNN model; pre-image processing and application of CNN; extraction of features of the images; training and classification of images by Support Vector Machines; and, finally, evaluation of the classification results. Figure 1 shows a summarized flowchart of the proposed method.

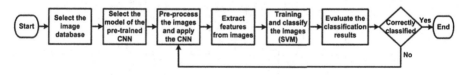

Fig. 1. Summarized flowchart of the proposed method

3.1   Image Database Selection

The brands, and brand codes used are presents in Fig. 2.

Fig. 2. Images of cattle branding used in this article

3.2   Model Selection of a Pre-trained CNN

Convolutional neural networks exhibit a relatively unexploited property known as knowledge transfer. By applying this concept, we selected the CNN model of the open source VLFeat library. The option for a pre-trained model was made due to the fact that the dataset is not large enough to effectively train the CNN. Moreover, the time needed to train a model from scratch would preclude the use of the proposed tool, since it needs speed regarding the task of recognizing cattle brandings. The use of the above mentioned pre-trained CNN model did not directly influence the Overall Accuracy of the cattle branding images.

3.3   Image Pre-processing and CNN Application

The adopted method is a neural network with five convolutional layers. Pre-processing is performed for grey-shaded images, which conducted by replicating images 3 times in order to create a RGB image. The 1st convolutional layer has as its input the 3 color channels (RGB). Each convolution performs an application of the non-linear activation function ReLU and a reduction by the Maxpooling. The last layers are composed of totally-connected neurons. Figure 3 shows the proposed convolutional neural network architecture.

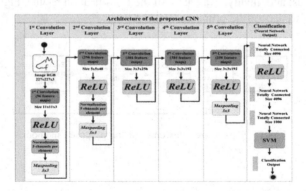

Fig. 3. Proposed convolutional neural network architecture

3.4   Image Feature Extraction

The set of filters learned by the CNN during training is responsible for the sensing of features in the new image at the moment of a query. In the first filter level, we can observe some lines and orientations used for that sensing. Figure 4 shows the learned filters in the first convolutional layer using the RGB color space. There are 96 individual sets that represent the 96 filters used in that layer. It is possible to observe how the areas with horizontal, vertical and diagonal bulges are highlighted after the first convolution.

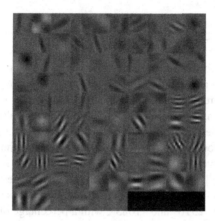

Fig. 4. Convolutional 1st layer filters from the executed experiment

After the first convolutions, it was possible to perform the extraction of image features for training the classifier.

3.5   Training and Classification of Images with Support Vector Machines

The model for automatic learning adopted in the presented work was the Support Vector Machine (SVM) supervised classifier. Support Vector Machine is a classification algorithm known for its success in a wide range of applications. SVMs are one of the most popular approaches for data modeling and classification. Its advantages include their outstanding capacity for generalization, concerning the ability to correctly sort the samples that are not in the feature space used for training. SVMs are used to classify and recognize patterns in several types of data; they are also employed in a wide range of applications, such as face recognition, clinical diagnosis, industrial process monitoring, and image processing and analysis [10].

In the proposed research, we randomly separated the set of images in two parts, where one of these parts was used for the training stage and the other for the test stage, thus eliminating polarization in the results. The final result is the average of the result obtained in the test stage. The percentage division used here was 30% for training and 70% for test.

3.6   Assessment of Classification Results – Confusion Matrix

The confusion matrix of a classifier indicates the number of correct classifications versus the predictions made in each case over a group of examples. In this matrix, the lines depict the actual cases and the columns depict the predictions made by the model. Through the confusion matrix, it is possible to find information related to the number of correctly and incorrectly classified images for each group of samples. This is an A × A matrix, where A is the amount of categories to which we apply the classifier. In our case, the experiment conducted included 30 brands, so we have a 30 × 30 confusion matrix in this situation.

4   Results and Discussions

Through the results found we were able to assess the proposed method. The assessment of experiment results was performed based on the recognition rate obtained in the confusion matrix rendered from the classification attained in the test stage. Furthermore, the total processing time of the proposed method was also checked.

Figure 5 presents the confusion matrix for the best result obtained in the experiments, whose recognition rate was up to 91.94%. Through the analysis of the main diagonal we can observe that the correct classification rate is emphasized in three brandings "864", "1085", and "2542", in which the percentage of correctness reached 100%, or 31 correct hits. We can also observe that the brandings with the lowest correct classification rates were "2516" and "1083", with a percentage of 77.42% and 80.64%, in this order.

The hypothesis of wrong classification of cattle branding as shown in the confusion matrix may also be associated to the complexity of the samples, since some of the images present similar features among themselves. In general, the image samples of brandings with a better descriptive power and better quality correctly classified more images, since they encompass more characteristics when compared to brandings with worse sample quality, and, consequently, less extracted features. The capacity of recognizing patterns in an image over a set of images depends on the amount of a priori information available about the object in question.

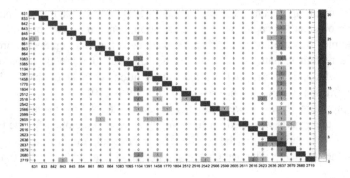

Fig. 5. Confusion matrix obtained in the test stage

Taking into account the confusion matrix shown in Fig. 5, we can observe that, through the sum of main diagonal, we can find the total number of correctly classified cattle branding, more precisely, 855, while the sum of the other values points to incorrectly classified brandings, that is, 75.

The processing time of the algorithm based on the number of cattle branding samples is shown in Fig. 6. The processing times of the proposed method were measured for the classification of five sample groups, respectively. Each group contained 270; 540; 810; 1,080, and 1,350 images. The processing times for the classification of the images in each group was 9.616 s; 14.162 s; 18.114 s; 22.320 s, and 26.766 s, in this order. When analyzing the graphic in Fig. 6, we can understand that the processing time observed in the Y-axis, varies in direct proportion to the increase of samples of cattle branding classified in the X-axis, i.e., a pattern of linear growth of the function is observed, even if the growth rate is not exactly a constant number.

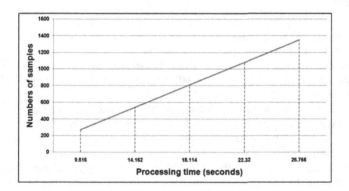

Fig. 6. Processing time of the algorithm according to number of samples

The results obtained with the experiments performed with 30 cattle brandings and 1,350 image samples, used both for training and test, obtained an Overall Accuracy of 91.94%, an error rate of 8.06%, and a total processing time of 26.766 s. The Overall Accuracy rate was obtained through the calculation of the arithmetic average of brands properly classified from the confusion matrix, and the total processing time was obtained with the use of the MATLAB software which, at the end of code processing, performs the breakdown of the algorithm processing speed.

5   Conclusion

In this research, we presented an automated method for the recognition of cattle branding. The experiments performed in this project used a Deep Learning technique of Convolutional Neural Network (CNN) to extract features and an SVM supervised classifier.

The usage of the suggested method delivered an Overall Accuracy of 91.94% and an algorithm processing time of 26.766 s for the 30 assessed brands, in a total of 1,350 samples used for training and test. The deployed method has effectively and efficiently performed the recognition of different cattle brands, although it used a pre-trained CNN. Its main limitation, however, was the need for a high number of sample images to train the classifier, as CNN needs these images in order to extract the features from convolutional layers.

As future researches, we strive for the expansion of the image base used for training and test, with a main goal of obtaining new accuracy and performance measurements of the proposed method. In time, we intend to conduct new experiments through the use of an untrained model for the implemented CNN to extract features, in order to compare with results obtained by the pre-trained model. Finally, we will use a Sofmax layer at the output of the CNN to sort cattle brandings to compare the results of this new model with the ones that were obtained by using SVM.

References

1. Arnoni, R.: Os registros e catálogos de marcas de gado da região platina. Memória em Rede, Pelotas, **3**, 1–10 (2013). http://guaiaca.ufpel.edu.br
2. Sanchez, G., Rodriguez, M.: Cattle marks recognition by hu and legendre invariant moments. ARPN J. Eng. Appl. Sci. **11**, 607–614 (2016). http://www.arpnjournals.org/jeas/research_papers/rp_2016/jeas_0116_3379.pdf
3. LeCun, Y., Boser, B., Denker, J.S., Henderson, D., Howard, R.E., Hubbard, W., Jackel, L. D.: Handwritten digit recognition with a back-propagation network. In: Touretzky, D. (ed.) Advances in Neural Information Processing Systems, pp. 396–404. Morgan Kaufmann, Denver (1990)
4. Ciregan, D., Meier, U., Schmidhuber, J.: Multi-column deep neural networks for image classification. In: Proceedings of the 25th IEEE Conference on Computer Vision and Pattern Recognition, pp. 3642–3649. IEEE Press, Providence (2012). https://doi.org/10.1109/cvpr.2012.6248110
5. Kavukcuoglu, K., Sermanet, P., Boreau, Y.L., Gregor, K., Mathieu, M., LeCun, Y.: Learning convolutional feature hierarchies for visual recognition. In: Advances in Neural Information Processing Systems, pp. 1090–1098. IEEE Press, Vancouver (2012)
6. Sermanet, P., Eigen, D., Zhang, X., Mathieu, M., Fergus, R., LeCun, Y.: Overfeat: integrated recognition, localization and detection using convolutional networks. arXiv preprint arXiv: 13126229 (2013)
7. Razavian, A.S., Azizpour, H., Sullivan, J., Carlsson, S.: CNN features off-the-shelf: an astounding baseline for recognition. In: Proceedings of the IEEE Conference on Computer Vision and Pattern Recognition Workshops, pp. 512–519. IEEE Press, Columbus (2014). https://doi.org/10.1109/cvprw.2014.131
8. Constante, P., Gordon, A., Chang, O., Pruna, E., Acuna, F., Escobar, I.: Artificial vision techniques for strawberry's industrial classification. IEEE Lat. Am. Trans. **14**, 2576–2581 (2016). https://doi.org/10.1109/TLA.2016.7555221
9. Vlfeat. Biblioteca open source VLFeat (2016). http://www.vlfeat.org/matconvnet/models/beta16/imagenet-caffe-alex.mat
10. Tchangani, P.: Support vector machines: a tool for pattern recognition and classification. Stud. Inform. Control J. **14**, 99–110 (2005)

Efficient Transfer Learning for Robust Face Spoofing Detection

Gustavo B. Souza[1](✉)⬵, Daniel F. S. Santos[2]⬵, Rafael G. Pires[1]⬵,
Aparecido N. Marana[2]⬵, and João P. Papa[2]⬵

[1] UFSCar - Federal University of São Carlos, São Carlos, SP, Brazil
gustavo.botelho@gmail.com, rafapires@gmail.com
[2] UNESP - São Paulo State University, Bauru, SP, Brazil
{nilceu,papa}@fc.unesp.br, danielfssantos1@gmail.com

Abstract. Biometric systems are synonym of security. However, nowadays, criminals are violating them by presenting forged traits, such as facial photographs, to fool their capture sensors (spoofing attacks). In order to detect such frauds, handcrafted methods have been proposed. However, by working with raw data, most of them present low accuracy in challenging scenarios. To overcome problems like this, deep neural networks have been proposed and presented great results in many tasks. Despite being able to work with more robust and high-level features, an issue with such deep approaches is the lack of data for training, given their huge amount of parameters. Transfer Learning emerged as an alternative to deal with such problem. In this work, we propose an accurate and efficient approach for face spoofing detection based on Transfer Learning, i.e., using the very deep VGG-Face network, previously trained on large face recognition datasets, to extract robust features of facial images from the Replay-Attack spoofing database. An SVM is trained based on the feature vectors extracted by VGG-Face from the training images of Replay database in order to detect spoofing. This allowed us to work with such 16-layered network, obtaining great results, without overfitting and saving time and processing.

1 Introduction

Despite the higher security of biometric systems, criminals are already violating them by presenting forged traits, e.g., facial photographs, to their capture sensors, process known as spoofing attack [1]. In this sense, countermeasures techniques must be integrated to the traditional biometric systems in order to prevent such frauds.

Antispoofing methods proposed so far use, in general, handcrafted features extracted at the moment of identification, e.g., presence of facial movements, to detect whether there is a real or synthetic biometric trait being presented to the

Authors are grateful to grants #2014/12236-1 and #2017/05522-6 from FAPESP, grant #306166/2014-3 from CNPq, and grant #88881.132647/2016-01 from CAPES.

sensor. Such handcrafted methods, however, are shown to be not good enough, especially in challenging scenarios, due to their raw data-based approaches.

Recently, deep neural networks have gained attention due to their greater results in many complex tasks [2]. However, despite of being more robust, the training of such architectures is computationally expensive and usually there are no public databases with enough images to avoid overfitting, especially when working with deeper architectures (which usually allow better results). To address such problem, Transfer Learning has been recently proposed, in which the deep architectures are trained on large datasets, even from other domains, and then a classifier, or the own network, is just fine-tuned based on the images of the smaller database of the problem being addressed [2].

In this work, we present a novel efficient method for spoofing detection in face-based biometric systems by means of high-level and robust features, extracted from images by the very deep VGG-Face [3] network, a well-referenced 16-layered Convolutional Neural Network (CNN) model, previously trained on large face recognition datasets. By using such trained network in face spoofing detection (smaller databases available), overfitting is avoided since only a Support Vector Machine is trained on the spoofing images, time is saved and accuracy tends to be preserved due to the similar domain of face recognition and face spoofing detection applications. Results on the traditional Replay-Attack [4] spoofing dataset show that the proposed method, besides of efficiency, presents great results, close to other state-of-the-art techniques, in terms of attack detection.

2   Face Spoofing Detection

In attacks to biometric systems, criminals usually generate synthetic samples of biometric traits of legal users, such as printed facial photographs or latex and gelatin fingers, to fool their capture sensors [5]. Figure 1 shows images of real and fake faces (protographs) presented to a real capture camera. After capture, it is difficult, even for humans, to differentiate between the real and fake ones.

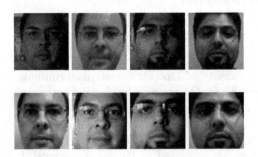

Fig. 1. Real (top) and fake (bottom) faces from the Replay-Attack [4] spoofing dataset.

Among the main biometric traits, face is a promising one especially due to its convenience and low cost of acquisition, being very suitable to a wide variety of environments. However, despite all these advantages, face recognition systems are the ones that most suffer with spoofing attacks since they can be easily fooled even with common printed photographs obtained in the worldwide network. In this context, attack detection methods for face recognition systems are essential.

Antispoofing techniques for face recognition systems have been proposed based on different principles. Nevertheless, spoofing detection is still an open question [6]. As mentioned, most of techniques are based on simple rules (hand-crafted features) in order to detect attacks, e.g., facial movements. However, criminals can easily identify these rules and improve attacks: moving the facial photographs in a specific way. In this sense, algorithms able to work with high-level and non trivially generated features are necessary. Among them, the deep learning-based methods simulate the deep structures of neurons in human brain and have outperformed state-of-the-art techniques in many areas [2].

3   Convolutional Neural Networks (CNN)

Convolutional Neural Networks (CNN) [7] are deep learning architectures con-stituted of layers in which different kind of filters (convolution and sampling) are applied to the input data, initially two-dimensional images. The result of a given layer serves as input to the above one until the top of the network is reached. As shown in Fig. 2, besides convolution and sampling, layers with neurons fully connected can be included at the top of the network for classification, usually performing signal rectification or applying the softmax normalization function, in this case, transforming their input values into output probabilities. At the top layers of the network, high-level representation of the original image are obtained, more robust than the raw pixels information for many applications [1].

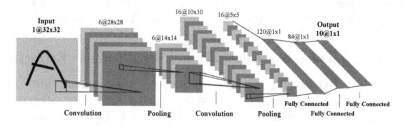

Fig. 2. Illustration of the Lenet-5 [7] model of CNN, with convolution, pooling and fully connected layers. The last one presents softmax neurons for classification of the input images. Sizes of the feature maps generated in each layer are shown at top.

Despite of obtaining high level of abstraction over the input data and, usually, more accurate results, a common problem in working with deep neural networks consists in the fact that, due to their huge amount of free parameters to be optimized (weights of connections, biases, etc.), a huge amount of samples are required to train the architecture in an effective way and avoid overfitting [8].

Besides of demanding time and processing, for many problems there are no public databases with such dimensions.

In order to deal with such issues, still preserving the robustness of the very deep neural networks in problems with lack of training data, Transfer Learning, which consists in applying trained models on large datasets to extract high-level features from the images of other databases (usually smaller), was proposed [2]. The weights of the trained deep network are just fine-tuned based on the training set of the given problem. Other more efficient approach consists in extracting the high-level features from the samples of the target dataset using the trained deep network and training only a classifier based on such features, which is less computationally expensive and also allows great results [9].

An interesting and public available trained model of a very deep CNN, as denoted by the own authors, which can be used in Transfer Learning, is VGG-Face [3] network. It presents 16 layers with convolution, pooling, linear rectification and softmax operations. Each operation has its own set of parameters which are already optimized over the Labeled Faces in the Wild (LFW) [10] and YouTube Face [11] databases (huge datasets) for the face recognition task, problem similar to ours. Details of VGG-Face architecture can be found in [3].

4   Proposed Approach

In most of cases, the very deep neural networks, already trained and used to extract features on other smaller datasets, are trained on data of a different domain of the problem being addressed. In this work, as said, we use a pre-trained model, the very deep VGG-Face [3] network, on a similar domain problem (face recognition), tending to avoid at maximum a decrease in performance. The proposed approach for face spoofing detection presents the following training steps: (i) Face Detection and Normalization; (ii) Deep Features Extraction; (iii) Support Vector Machine (SVM) Training; and (iv) Threshold Calibration.

4.1   Face Detection, Normalization and Deep Features Extraction

After loading VGG-Face [3] model, each frame of the training videos of the assessed Replay-Attack [4] dataset passed through a preprocessing step: the face in each frame was cropped from the original image following the coordinates provided by the own authors of the dataset in order to avoid segmentation errors. We incremented or decremented, by few pixels, the size of the provided rectangles in order to crop only 112×112-sized squares, to facilitate their resizing to 224×224 pixels, dimensions required to feed the VGG-Face [3] network.

Actually, we regularly sampled and worked with only 1 of each group of 8 frames of the videos in order to speed up the feature extraction and SVM Training, obtaining almost $11,650$ training facial images. Such resultant 224×224 facial images were also normalized (subtraction of the mean image of the databases on which the VGG-Face was trained) and presented to VGG-Face, performing a foward pass until the layer "Pool5", last layer before the fully connected ones in the VGG-Face [3] model (almost at the top of the network).

Given an input image presented to the network, the resultant 512 feature maps with size of 7×7 generated as output of the layer "Pool5" were also regularly sampled (a quarter of them) and their values, after concatenation, were taken as the feature vector of such training image.

4.2 SVM Training and Threshold Calibration

Given all high-level feature vectors extracted by VGG-Face [3] from the training images (from the real and fake training videos of the Replay [4] dataset), and their respective labels, their values were normalized to the range $[0; 1]$ and a radial basis function SVM classifier was trained. Given all training vectors, we performed the grid-search for finding the parameters c and g of the SVM kernel using a 5-fold cross-validation scheme. After then, the SVM was trained with such parameters and the whole set of training vectors.

Actually, a probabilistic version of the SVM classifier [12] was used. After training, given a facial image from a video being analyzed, the classifier outputed a probability indicating the similarity of such face with the real and fake classes. In order to calibrate the threshold τ, used to determine whether a test face is real or fake given the probability generated by the SVM to it, we used the videos of the development set of the Replay-Attack [4] dataset, varying τ from 0 to 1.

Given each development video, we also sampled it in regularly spaced positions, using the ratio of 1 frame in each 8, also obtaining almost 11, 650 facial images, which feeded the VGG-Face [3] model, after face cropping and normalization, in order to extract their feature vectors and generate their probabilities of belonging to each class. For each facial image, if its output probability was higher than threshold τ being considered, we classified it as real, fake otherwise.

Finally a votation scheme was applied to determine the final class of each development video, i.e., given its classified frames, if at least half of them were considered fake, the video was classified as attack, otherwise the video was taken as real. We varied τ in the interval $[0; 1]$, in steps of 0.01, and calculated the False Acceptance Rate (FAR) and False Rejection Rate (FRR) of the method based on the incorrect classified development videos. We fixed the τ parameter when obtained the same values for FAR and FRR (EER - Equal Error Rate).

4.3 Testing

The same process done for the training and development videos was applied to the test videos after fixing τ. For each test video we also sampled its frames (considering all training videos, a total of 15, 500 frames were sampled), cropped and normalized the faces in them, and passed such images through the VGG-Face [3] network in order to extract their feature vectors and also classify them by comparing their SVM output probabilities with the τ value. If greater than τ, the face was considered as real, and fake otherwise. A votation scheme also was applied to determine the final class of the test videos: real or fakes. Again, if at least half of the classified frames from a given video were considered as fake, the video was classified as attack, and real otherwise.

5   Experiments, Results and Discussion

The proposed approach for face spoofing detection was assessed on the traditional Replay-Attack [4] dataset using a computer with an Intel i7-4790K CPU (8 cores), 32 GB of RAM memory and an NVIDIA GTX980 GPU with 4 GB of RAM (our best computer). Such database contains color videos with frames with 320×240-sized frames of real and fake faces, divided in training (360 videos), development (other 360 videos, only for threshold estimation) and testing sets (with 480 videos). The real videos from all sets present 375 frames while the attack videos present 240 or 230 frames, depending on the type of attack.

The fake videos were made by presenting many kinds of synthetic facial images to the capture cameras: high-definition printed photographs, videos or even photographs exhibited on displays of mobile devices, etc. Figure 1, in Sect. 2, shows examples of real and fake faces present in the frames of the training set of videos from such database. As said, it is difficult to differentiate between them, especially due to high intraclass variability and high interclass similarity.

In our experiments we used the Caffe [13] framework with Matlab R2014b interface (MatCaffe) in order to load the trained VGG-Face model and to perform the foward pass of the detected and normalized facial image from the videos. Given the feature vectors of the faces in the training frames (from all training videos), we used the LibSVM [12] library in order to normalize their values in the range $[0; 1]$ and train a radial basis function kernel SVM. As said, the grid-search was performed in a 5-folds cross-validation scheme and the SVM was trained with the best found kernel parameters ($c = 32.0$ and $g = 0.0078$). After training the SVM (probabilistic version), we used the videos from the delvelopment set to optimize the τ threshold.

After finding the optimal threshold $\tau = 0.18$, we classified the sampled frames from the testing videos and used a votation scheme, as said, to find a final classification for each of them (based on the classes of their sampled frames). As in literature, the results in terms of Half-Total Error Rate (HTER) of the proposed approach and of other state-of-the-art methods are shown in Table 1. The lower the HTER, the better the method. We also provided our Accuracy (ACC) result, measure of performance not mentioned in the other works.

It is important to observe in Table 1 that our efficient approach, based on an already trained very deep CNN on similar domain databases, presents an HTER result close to the state-of-the-art methods: with an HTER of 16.62%, it outperforms the LBP+LDA technique and is close to the other methods. Such methods are computationally expensive: LBP-TOP, e.g., extracts multiple LBP histogramas for lots of small sets of frames of the videos (a slide window is deslocated over all the frames) and concatenates such histograms generating large feature vectors to be processed by an SVM. In such method, in order to classify each video, a votation scheme also is performed.

Table 1. Results (%) on the Replay-Attack database regarding Half-Total Error Rate (HTER) of the proposed method and other state-of-the-art techniques. We also provided our Accuracy (ACC) result. The best values are highlighted.

Method	ACC	HTER
Proposed approach	**93.95**	16.62
Diffusion speed model [14]	-	12.50
LBP+LDA [4]	-	17.17
LBP+SVM [4]	-	16.16
LBP-TOP+SVM [15]	-	**7.60**
Non-linear diffusion [16]	-	10.00

Our architecture only requires to train an SVM based on robust feature vectors directly extracted by the already trained very deep VGG-Face [3] model, being efficient and saving processing time and hardware consumption: the feature extraction, based on the layer "Pool5" of VGG-Face, took about only 0.4 s per frame. It is important to note that the obtained HTER and ACC results of our method can be considerably improved by considering more frames of the videos instead of 1 in each 8 of them, as well as using all the 7×7 output feature maps from layer "Pool5" to represent the images. Also, as done in other works, we could use the top fully connected layers of the VGG-Face [3] network or even a combination of feature maps from many of its layers, including the bottom ones, in order to better represent faces (which present larger dimensions).

The other mentioned approach for Transfer Learning, i.e., fine-tune of the own deep network in the database of the problem, changing the size of the softmax layer according to the number o classes existent, in our case fake and real, also can be performed to improve even more the results. Despite this, it is important to note that all these mentioned approaches require more processing time and computer resources (especially memory space) than the proposed architecture, and, due to this reason, it was not possible to execute them given our hardware restrictions. However, having more powerfull equipaments, they can be performed easily.

6   Conclusion

In this work, we presented an efficient and also robust approach for face spoofing detection based on Transfer Learning by applying an already trained model of a very deep Convolutional Neutal Network (CNN), the VGG-Face network, on the Replay-Attack spoofing database. The proposed approach, due to its work with a deep architecture, is very robust, presenting results close to the state-of-the art techniques on the traditional Replay dataset. Besides, since we just need to train an SVM based on the high-level features extracted by VGG-Face from the training facial images, just performing a foward pass of such images into the trained model, it saves much processing time and also hardware resources, especially in training. Also, since such VGG-Face was trained on images of a similar domain (facial images for face recognition) than our given problem (facial images for face spoofing detection), robustness of the high-level features extracted are strongly preserved. Overfitting is also avoided given the lack of huge amounts of training data for face spoofing detection, since we do not need to train the deep VGG-Face model for robust feature extraction. Based on all this, the proposed approach is very suitable for real applications or in cases where there are not available high-tech computers servers, as ours.

References

1. Menotti, D., Chiachia, G., Pinto, A., Schwartz, W., Pedrini, H., Falcao, A., Rocha, A.: Deep representations for iris, face, and fingerprint spoofing attack detection. IEEE Trans. Inf. Forensics Secur. **10**(4), 864–879 (2015)
2. Bengio, Y.: Deep learning of representations: looking forward. Stat. Lang. Speech Process. **7978**, 1–37 (2013)
3. Parkhi, O.M., Vedaldi, A., Zisserman, A.: Deep face recognition. In: Proceedings of the British Machine Vision Conference (2015)
4. Chingovska, I., Anjos, A., Marcel, S.: On the effectiveness of local binary patterns in face anti-spoofing. In: Proceedings of International Conference of BIOSIG, USA, pp. 1–7 (2012)
5. Ratha, N.K., Connell, J.H., Bolle, R.M.: An analysis of minutiae matching strength. In: Bigun, J., Smeraldi, F. (eds.) AVBPA 2001. LNCS, vol. 2091, pp. 223–228. Springer, Heidelberg (2001). https://doi.org/10.1007/3-540-45344-X_32
6. Patel, K., Han, H., Jain, A.K.: Cross-database face antispoofing with robust feature representation. In: You, Z., Zhou, J., Wang, Y., Sun, Z., Shan, S., Zheng, W., Feng, J., Zhao, Q. (eds.) CCBR 2016. LNCS, vol. 9967, pp. 611–619. Springer, Cham (2016). https://doi.org/10.1007/978-3-319-46654-5_67
7. LeCun, Y., Bottou, L., Bengio, Y., Haffner, P.: Gradient-based learning applied to document recognition. Proc. IEEE **86**(11), 2278–2324 (1998)
8. Krizhevsky, A., Sutskever, I., Hinton, G.E.: Imagenet classification with deep convolutional neural networks. In: Advances in Neural Information Processing Systems, pp. 1106–1114 (2012)
9. Montavon, G., Orr, G.B., Müller, K.-R. (eds.): Neural Networks: Tricks of the Trade. LNCS, vol. 7700, 2nd edn. Springer, Heidelberg (2012). https://doi.org/10.1007/978-3-642-35289-8
10. Huang, G., Ramesh, M., Berg, T., Miller, E.: Labeled faces in the wild: a database for studying face recognition in unconstrained environments. Technical report 07–49, University of Massachusetts, Amherst (2007)
11. Wolf, L., Hassner, T., Maoz, I.: Face recognition in unconstrained videos with matched background similarity. In: Proceedings of the IEEE Conference on Computer Vision and Pattern Recognition (2011)
12. Chang, C., Lin, C.: LIBSVM: a library for support vector machines. ACM Trans. Intell. Syst. Technol. **2**, 1–27 (2011)
13. Jia, Y., Shelhamer, E., Donahue, J., Karayev, S., Long, J., Girshick, R., Guadarrama, S., Darrell, T.: Caffe: convolutional architecture for fast feature embedding. arXiv preprint arXiv:1408.5093 (2014)

14. Kim, W., Suh, S., Han, J.: Face liveness detection from a single image via diffusion speed model. IEEE Trans. Image Process. **24**(8), 2456–2465 (2015)
15. de Freitas Pereira, T., Anjos, A., De Martino, J.M., Marcel, S.: *LBP–TOP* based countermeasure against face spoofing attacks. In: Park, J.-I., Kim, J. (eds.) ACCV 2012. LNCS, vol. 7728, pp. 121–132. Springer, Heidelberg (2013). https://doi.org/10.1007/978-3-642-37410-4_11
16. Alotaibi, A., Mahmood, A.: Deep face liveness detection based on nonlinear diffusion using convolution neural network. Sig. Image Video Process. **11**(4), 713–720 (2017)

A Kernel-Based Optimum-Path Forest Classifier

Luis C. S. Afonso[1] (ID), Danillo R. Pereira[2] (ID), and João P. Papa[3(✉)] (ID)

[1] Department of Computing, UFSCar - Federal University of São Carlos,
São Carlos, Brazil
sugi.luis@dc.ufscar.br
[2] University of Western São Paulo, Presidente Prudente, Brazil
danilopereira@unoeste.br
[3] School of Sciences, UNESP - São Paulo State University, Bauru, Brazil
papa@fc.unesp.br

Abstract. The modeling of real-world problems as graphs along with the problem of non-linear distributions comes up with the idea of applying kernel functions in feature spaces. Roughly speaking, the idea is to seek for well-behaved samples in higher dimensional spaces, where the assumption of linearly separable samples is stronger. In this matter, this paper proposes a kernel-based Optimum-Path Forest (OPF) classifier by incorporating kernel functions in both training and classification steps. The proposed technique was evaluated over a benchmark comprised of 11 datasets, whose results outperformed the well-known Support Vector Machines and the standard OPF classifier for some situations.

Keywords: Optimum-Path Forest · Kernel · Support Vector Machines

1 Introduction

The nature of real-world problems has driven their modeling to structured objects where samples are usually represented by nodes and their relationship are represented by connections (edges), similarly to a graph [1,2]. Among these problems, one can mention studies in bioinformatics, social networks, and data mining in documents, just to name a few, in which it is necessary to identify structures or elements that share similar properties.

Another very common characteristic present in real-world problems is the non-linear distribution, which requires complex models to identify the different existing patterns in the dataset [3]. In this matter, kernels have been proposed as a tool to overcome this issue by providing a way to map the data into a space of higher dimension, where the input data may be linearly separable.

Support Vector Machines (SVM) is perhaps the most well-known technique that makes use of kernels for data embedding into higher dimensional feature spaces [4]. There are also the kernel-PCA (Principal Component Analysis) [5],

The authors are grateful to Capes, CNPq grant #306166/2014-3, and FAPESP grants #2014/16250-9, #2014/12236-1, and #2016/19403-6.

© Springer International Publishing AG, part of Springer Nature 2018
M. Mendoza and S. Velastín (Eds.): CIARP 2017, LNCS 10657, pp. 652–660, 2018.
https://doi.org/10.1007/978-3-319-75193-1_78

and the kernel-Fisher discriminant analysis [6], just to mention a few. Although the mapping can take a feature space to a very high dimension that increases the computing cost, both training and classification steps can be performed by means of a dot product, the so-called *kernel trick*. Such procedure consists in computing the dot product without the need of mapping the samples to a higher dimensional space. The combination of the these two worlds, i.e., representations based on graphs and non-linear distributions, has motivated studies in the application of kernels in structured data [1,7]. In this research area, it is typical to find two different approaches, i.e., applications that make use of kernels to identify similarity among graphs, while others focus on the similarity among the graph nodes.

One of the first works on kernels among graphs was proposed by Gärtner et al. [8] and Borgwardt et al. [9] using the random walk, and marginalized kernels by Tsuda et al. [10], Kashima et al. [11,12] and Mahé et al. [13]. The basic idea of random-walk graph kernels is to perform a random walk in a pair of graphs and sum up the number of matching walks [2] on both graphs. The marginalized kernel is defined as the inner product of the count vectors averaged over all possible label paths [11]. Regarding kernels in graphs, interesting works were proposed by Kondor and Lafferty [1] and Smola and Kondor [14]. In [1], the authors proposed the diffusion kernels, which is a special class of exponential kernels based on the heat equation.

Graph-based machine learning techniques can be noticed as well. The Optimum-Path Forest (OPF) framework [15–18] is a graph-based classification algorithm where samples are represented as graph nodes and each node can be also weighted by a probability density function that takes into account its k-nearest neighbors, and the edges are weighted by a distance function. The partitioning using the OPF algorithm is performed by a competition process, in which prototypes try to conquer the remaining samples by offering them paths with optimum cost. The result is a set of trees (forest) rooted at a set of prototypes, where each tree may represent a single class or cluster. The OPF algorithm has been applied in a large variety of applications, such as network security [19], Parkinson's disease identification [20], clustering [21,22], and biometrics [23,24], just to name a few, and it has showed competitive results against some state-of-art and well-known machine learning algorithms.

The current OPF algorithm implementation works naturally with non-linear situations, but it does not map samples from one space to another. In this paper, we take one step further by modifying the OPF algorithm to work with kernels on graphs in order to improve its training and classification results, since such approach has not been applied so far. The performance of the proposed approach is assessed under three different kernels, and it is compared against the original OPF algorithm and the well-known SVM in 11 different datasets. The remainder of this paper is organized as follows. Sections 2 and 3 present the theoretical background concerning OPF and its kernel-based variant, respectively. Section 4 discusses the methodology and experiments, and Sect. 5 states the conclusions.

2   Optimum-Path Forest

In this section, we explain the OPF working mechanism considering the first proposed version [16,17]. Roughly speaking, the OPF classifier models the problem of pattern recognition as a graph partition in a given feature space. The nodes are represented by the feature vectors and the edges connect all pairs of them, defining a full connectedness graph. The partition of the graph is performed through a competition process among some key samples (prototypes), which offer optima paths to the remaining nodes of the graph. Each prototype sample defines its own optimum-path tree (OPT), and the collection of all OPTs defines an optimum-path forest, which gives the name to the classifier.

Let $\mathcal{Z} = \mathcal{Z}_1 \cup \mathcal{Z}_2$ be a dataset labeled with a function λ, in which $\mathcal{Z}_1$ and $\mathcal{Z}_2$ stand for the training and test sets, respectively, such that $\mathcal{Z}_1$ is used to train a given classifier and $\mathcal{Z}_2$ is used to assess its accuracy. Let $\mathcal{S} \subseteq \mathcal{Z}_1$ be a set of prototype samples. Essentially, the OPF classifier creates a discrete optimal partition of the feature space such that any sample $\mathbf{s} \in \mathcal{Z}_2$ can be classified according to this partition.

The OPF algorithm may be used with any *smooth* path-cost function which can group samples with similar properties [25]. Papa et al. [16,17] employed the path-cost function f_{max}, which is computed as follows:

$$f_{max}(\langle \mathbf{s} \rangle) = \begin{cases} 0 & \text{if } \mathbf{s} \in \mathcal{S}, \\ +\infty & \text{otherwise} \end{cases}$$
$$f_{max}(\pi \cdot \langle \mathbf{s}, \mathbf{t} \rangle) = \max\{f_{max}(\pi), d(\mathbf{s}, \mathbf{t})\}, \tag{1}$$

in which $d(\mathbf{s}, \mathbf{t})$ denotes the distance between samples $\mathbf{s}$ and $\mathbf{t}$, and a path π is defined as a sequence of adjacent samples. As such, we have that $f_{max}(\pi)$ computes the maximum distance among adjacent samples in π, when π is not a trivial path.

The OPF algorithm assigns one optimum path $P^*(\mathbf{s})$ from $\mathcal{S}$ to every sample $\mathbf{s} \in \mathcal{Z}_1$, forming an optimum path forest P (a function with no cycles that assigns to each $\mathbf{s} \in \mathcal{Z}_1 \backslash \mathcal{S}$ its predecessor $P(\mathbf{s})$ in $P^*(\mathbf{s})$ or a marker *nil* when $\mathbf{s} \in \mathcal{S}$. Let $R(\mathbf{s}) \in \mathcal{S}$ be the root of $P^*(\mathbf{s})$ that can be reached from $P(\mathbf{s})$. The OPF algorithm computes for each $\mathbf{s} \in \mathcal{Z}_1$, the cost $C(\mathbf{s})$ of $P^*(\mathbf{s})$, the label $L(\mathbf{s}) = \lambda(R(\mathbf{s}))$, and the predecessor $P(\mathbf{s})$.

2.1   Training

In the training phase, the OPF algorithm aims to find the set $\mathcal{S}^*$, that is the optimum set of prototypes, by minimizing the classification errors for every $\mathbf{s} \in \mathcal{Z}_1$ through the exploitation of the theoretical relation between minimum-spanning tree (MST) and optimum-path tree (OPT) for f_{max} [26]. The training essentially consists in finding $\mathcal{S}^*$ from $\mathcal{Z}_1$ and an OPF classifier rooted at $\mathcal{S}^*$.

By computing a MST, we obtain a connected acyclic graph whose nodes are all samples of $\mathcal{Z}_1$ and the arcs are undirected and weighted by the distances d between adjacent samples. The spanning tree is optimum in the sense that the sum of its arc weights is minimum as compared to any other spanning tree in the complete graph. In the MST, every pair of samples is connected by a single path which is optimum according to f_{max}. That is, the minimum-spanning tree contains one optimum-path tree for any selected root node. The optimum prototypes are the closest elements of the MST with different labels in $\mathcal{Z}_1$ (i.e., elements that fall in the frontier of the classes). After finding the prototypes, the competition process takes place to build the optimum-path forest.

2.2 Classification

For any sample $\mathbf{t} \in \mathcal{Z}_2$, we consider all arcs connecting $\mathbf{t}$ with samples $\mathbf{s} \in \mathcal{Z}_1$, as though $\mathbf{t}$ were part of the training graph. Considering all possible paths from $\mathcal{S}^*$ to $\mathbf{t}$, we find the optimum path $P^*(\mathbf{t})$ from $\mathcal{S}^*$ and label $\mathbf{t}$ with the class $\lambda(R(\mathbf{t}))$ of its most strongly connected prototype $R(\mathbf{t}) \in \mathcal{S}^*$. This path can be identified incrementally by evaluating the optimum cost $C(\mathbf{t})$ as

$$C(\mathbf{t}) = \min\{\max\{C(\mathbf{s}), d(\mathbf{s}, \mathbf{t})\}\}, \ \forall \mathbf{s} \in \mathcal{Z}_1. \tag{2}$$

Let the node $\mathbf{s}^* \in \mathcal{Z}_1$ be the one that satisfies Eq. 2 (i.e., the predecessor $P(\mathbf{t})$ in the optimum path $P^*(\mathbf{t})$). Given that $L(\mathbf{s}^*) = \lambda(R(\mathbf{t}))$, the classification simply assigns $L(\mathbf{s}^*)$ as the class of $\mathbf{t}$. An error occurs when $L(\mathbf{s}^*) \neq \lambda(\mathbf{t})$.

3 Proposed Approach

The proposed kernel-based OPF, hereinafter called kOPF, works similarly to SVM algorithm, in which samples are mapped into a feature space of higher dimension. In the SVM context, such mapping is performed as an attempt to make the data linearly separable. The OPF, on the other hand, naturally works with non-linear data. Therefore, the main idea of this work is to evaluate OPF's behavior under such assumption of samples' separability in higher dimensional spaces.

Let $\Phi(\cdot, \cdot)$ be a kernel function that generates a new dataset $\mathcal{X} = \Phi(\mathcal{Z})$. Given a sample $\mathbf{p} \in \mathcal{Z}$, such that $\mathbf{p} \in \Re^n$, its new representation $\hat{\mathbf{p}} \in \mathcal{X}$ is defined as follows.

$$\hat{\mathbf{p}} = (\Phi_1, \Phi_2, \ldots, \Phi_{|\mathcal{Z}_1|}), \tag{3}$$

where $\Phi_i = (\mathbf{p}, \mathbf{s}_i)$, $\mathbf{s}_i \in \mathcal{Z}_1$. Notice that $\hat{\mathbf{p}} \in \Re^{|\mathcal{Z}_1|}$, which means the new sample $\mathbf{p}$ contains as many dimensions as the number of training samples.

In short, $\Phi(\mathbf{p}, \mathbf{s}_i)$ makes use of a distance function (i.e., Euclidean, Mahalanobis, among others) to compute a term that replaces the norm in kernel functions, such as Radial Basis Function (RBF) and Sigmoid, for instance. The aforementioned term corresponds to the distance between a sample to be mapped $\mathbf{p}$ and a training sample $\mathbf{s}$. Basically, the mapping performed by kOPF is carried out by computing a feature vector, where each component has the distance value from the sample to be mapped (either training or testing sample) to a different training sample applied to a kernel function. It is important to highlight that large training sets may cause a significant increase on the size of the new feature vector, since the size is dependent on the number of training samples $|\mathcal{Z}_1|$.

4   Methodology and Experiments

The kOPF classifier has both its performance and accuracy assessed by means of 11 public benchmarking datasets[1] that provide different classification scenarios. The implementation of our proposed approach is developed over the LibOPF [27], being standard OPF and SVM used as baselines for the experiments. With respect to SVM, we used the well-known LibSVM[2]. Table 1 presents detailed information from the datasets.

Table 1. Information about the datasets used in the experiments.

Dataset	No. samples	No. features	No. classes
boat (bt)	100	2	3
cone-torus (ct)	400	2	3
data1 (d1)	1,423	2	2
data2 (d2)	283	2	2
data3 (d3)	340	2	5
data4 (d4)	698	2	3
data5 (d5)	1,850	2	2
mpeg7-BAS (m-B)	1,400	180	70
mpeg7-Fourier (m-F)	1,400	126	70
petals (ps)	100	2	4
saturn (sn)	200	2	2

Since the kernel function can influence the final accuracy, we evaluated its impact by applying three different kernel functions for kOPF as follows:

- Identity: $\Phi(\mathbf{p}, \mathbf{s}) = \|\mathbf{p}, \mathbf{s}\|$
- RBF: $\Phi(\mathbf{p}, \mathbf{s}) = e^{-(\gamma \|\mathbf{p}, \mathbf{s}\|^2)}$
- Sigmoid: $\Phi(\mathbf{p}, \mathbf{s}) = \tanh(\gamma \|\mathbf{p}, \mathbf{s}\| + C)$

where $\|\mathbf{p}, \mathbf{s}\|$ denotes the Euclidean distance. Notice the Identity kernel is parameterless. The SVM was also evaluated using three different kernel functions: linear, RBF and Sigmoid. The situations in which parameters C and γ are required, it is performed their optimization using the intervals $C = [-32, 32]$ and $\gamma = [0, 32]$ with steps equals to 2 for both of them.

---

[1] https://github.com/jppbsi/LibOPF.
[2] https://www.csie.ntu.edu.tw/~cjlin/libsvm.

The classification experiments were conducted by means of a hold-out process using 15 runs, in which both training and testing sets were randomly generated in each run and always having a number of samples equals to 50% of the dataset size. The experiments also evaluated the impact in the accuracy rate when features are normalized. Tables 2 and 3 present the mean accuracy results considering non-normalized and normalized datasets, respectively. The accuracy rates were computed using the accuracy measure proposed by Papa et al. [16], which considers unbalanced data. The best results according to the Wilcoxon signed-rank test with significance 0.05 are shown in bold.

Table 2. Mean classification rates for non-normalized features.

Dataset	OPF	kOPF			SVM		
		Identity	RBF	Sigmoid	Linear	RBF	Sigmoid
bt	98.5 ± 0.0	96.8 ± 3.0	$\mathbf{100.0 \pm 0.0}$	99.1 ± 1.2	76.9 ± 1.8	$\mathbf{100.0 \pm 0.0}$	76.9 ± 1.8
ct	86.2 ± 0.0	84.3 ± 2.6	84.5 ± 2.7	84.2 ± 2.4	74.9 ± 0.2	$\mathbf{86.5 \pm 1.1}$	74.4 ± 0.4
d1	99.5 ± 0.0	$\mathbf{99.4 \pm 0.2}$	67.2 ± 8.6	68.5 ± 14.5	94.8 ± 0.8	$\mathbf{99.4 \pm 0.3}$	94.0 ± 0.8
d2	$\mathbf{99.3 \pm 0.0}$	98.1 ± 1.1	57.0 ± 2.6	62.0 ± 6.1	97.1 ± 0.5	98.2 ± 0.2	81.3 ± 3.6
d3	99.6 ± 0.0	$\mathbf{99.7 \pm 0.4}$	64.0 ± 3.6	71.1 ± 5.4	98.4 ± 0.8	$\mathbf{99.7 \pm 0.4}$	97.9 ± 0.7
d4	$\mathbf{100.0 \pm 0.0}$	$\mathbf{100.0 \pm 0.0}$	60.8 ± 2.6	67.8 ± 6.4	$\mathbf{100.0 \pm 0.0}$	$\mathbf{100.0 \pm 0.0}$	50.9 ± 1.3
d5	$\mathbf{100.0 \pm 0.0}$	$\mathbf{100.0 \pm 0.0}$	62.5 ± 3.2	67.5 ± 9.3	50.0 ± 0.0	$\mathbf{100.0 \pm 0.0}$	50.0 ± 0.0
m-B	89.4 ± 0.0	89.0 ± 0.30	50.1 ± 0.1	50.1 ± 0.1	87.9 ± 0.2	$\mathbf{90.5 \pm 0.2}$	50.0 ± 0.0
m-F	73.0 ± 0.0	$\mathbf{73.0 \pm 0.5}$	64.4 ± 0.6	66.2 ± 0.5	69.0 ± 0.8	$\mathbf{73.0 \pm 0.5}$	68.1 ± 0.5
ps	98.7 ± 0.0	$\mathbf{100.0 \pm 0.0}$	99.2 ± 0.6	98.9 ± 1.0	99.6 ± 0.6	$\mathbf{100.0 \pm 0.0}$	$\mathbf{100.0 \pm 0.0}$
sn	$\mathbf{93.0 \pm 0.0}$	90.2 ± 2.0	80.8 ± 2.6	81.8 ± 4.0	42.7 ± 3.7	91.3 ± 1.9	49.0 ± 3.3

Table 3. Mean classification rates for normalized features.

Dataset	OPF	kOPF			SVM		
		Identity	RBF	Sigmoid	Linear	RBF	Sigmoid
bt	97.1 ± 0.0	99.7 ± 0.6	99.4 ± 1.2	$\mathbf{100.0 \pm 0.0}$	76.9 ± 1.8	99.5 ± 0.7	76.5 ± 1.2
ct	89.3 ± 0.0	86.7 ± 1.7	88.0 ± 1.2	$\mathbf{89.9 \pm 0.5}$	75.9 ± 1.6	$\mathbf{89.4 \pm 0.5}$	76.1 ± 1.6
d1	$\mathbf{99.3 \pm 0.0}$	99.0 ± 0.3	99.0 ± 0.3	99.0 ± 0.3	94.7 ± 0.4	$\mathbf{99.3 \pm 0.2}$	94.7 ± 0.4
d2	97.3 ± 0.0	96.9 ± 1.5	97.1 ± 1.2	97.0 ± 1.3	$\mathbf{98.8 \pm 0.9}$	98.6 ± 0.6	$\mathbf{98.6 \pm 0.6}$
d3	100.0 ± 0.0	98.4 ± 1.4	$\mathbf{100.0 \pm 0.0}$	98.7 ± 0.8	99.5 ± 0.4	99.3 ± 0.7	99.5 ± 0.4
d4	100.0 ± 0.0	100.0 ± 0.0	100.0 ± 0.0	100.0 ± 0.0	100.0 ± 0.0	100.0 ± 0.0	100.0 ± 0.0
d5	100.0 ± 0.0	100.0 ± 0.0	100.0 ± 0.0	100.0 ± 0.0	50.0 ± 0.0	100.0 ± 0.0	50.7 ± 0.9
m-B	88.8 ± 0.0	$\mathbf{90.8 \pm 0.3}$	53.2 ± 0.5	56.3 ± 0.4	87.7 ± 0.9	89.9 ± 0.6	87.6 ± 0.9
m-F	62.9 ± 0.0	62.5 ± 0.4	59.2 ± 0.7	61.1 ± 0.8	64.7 ± 0.7	$\mathbf{65.4 \pm 0.5}$	64.3 ± 0.9
ps	98.7 ± 0.0	$\mathbf{100.0 \pm 0.0}$	99.5 ± 0.6	99.5 ± 1.0	$\mathbf{100.0 \pm 0.0}$	99.6 ± 0.6	$\mathbf{100.0 \pm 0.0}$
sn	91.0 ± 0.0	$\mathbf{91.8 \pm 0.2}$	79.2 ± 4.8	84.6 ± 0.8	52.3 ± 1.7	90.7 ± 4.5	50.7 ± 3.4

In the non-normalized feature scenario, SVM-RBF achieved the best results (or similar) in 9 out of 11 datasets, followed by the kOPF (kOPF-Identity and kOPF-RBF) with 7 out of 11 (being kOPF-Identity the best in 6 out of 11).

The standard OPF obtained the best results in 5 out of 11 datasets. Considering normalized features, the kOPF (kOPF-Identity, kOPF-RBF and kOPF-Sigmoid) obtained the best results (or similar) in 8 out of 11 datasets (being kOPF-Identity the best in 5 out of 11), followed by SVM (SVM-Linear, SVM-RBF and SVM-Sigmoid) with 7 out of 11 (being SVM-RBF the best in 6 out of 11). The OPF obtained the best results in only 4 out of 11 datasets.

In both (non-normalized and normalized) scenarios, the proposed kOPF outperformed the traditional OPF in most datasets, and for normalized features kOPF outperformed the SVM in some datasets as well. The results are quite interesting, since kOPF was able to improve OPF and outperforming SVM in some datasets. Considering some other datasets, although kOPF did not outperform both OPF and SVM, their results were considerably close.

The experiments also comprised the analysis of computational load required by each technique in each dataset. The results showed OPF and kOPF require a considerably small computational load in the training phases when compared against SVM. The high training time consumption turn the SVM prohibitive in real-time learning systems, specially if the training set is very dynamic over time. In this situation, both OPF and kOPF seems to be the most suitable approach. Due to lack of space, it was not possible to insert the computational load results of each technique, but OPF-based approaches have been around 200 times faster than SVM for training in the larger datasets.

5   Conclusions

This paper introduced a kernel-based OPF, which is a modification made over the standard OPF classifier that allows the usage of different kernel functions for both learning and classification. In our proposed approach, the mapping makes use of distance metric whose results are applied to kernel functions, such as RBF and Sigmoid. The main goal of such modification is to improve the accuracy rate.

The evaluation using 11 benchmark datasets and three different kernels showed the proposed approach achieved very interesting results, in which the application of kernel functions improved the accuracy rate of the traditional OPF, and even outperformed the well-known SVM when features were normalized. In summary, kOPF achieved satisfactory results and is an interesting option for classification, specially when training sets are very dynamic due to its low computational load for training purposes.

References

1. Kondor, R.I., Lafferty, J.D.: Diffusion kernels on graphs and other discrete input spaces. In: Proceedings of the Nineteenth International Conference on Machine Learning, ser. ICML 2002, pp. 315–322. Morgan Kaufmann Publishers Inc., San Francisco (2002)
2. Vishwanathan, S.V.N., Schraudolph, N.N., Kondor, R., Borgwardt, K.M.: Graph kernels. J. Mach. Learn. Res. **11**, 1201–1242 (2010)

3. Hofmann, T., Schölkopf, B., Smola, A.J.: Kernel methods in machine learning. Ann. Stat. **36**(3), 1171–1220 (2008)
4. Burges, C.J.: A tutorial on support vector machines for pattern recognition. Data Min. Knowl. Discov. **2**(2), 121–167 (1998)
5. Schölkopf, B., Smola, A., Müller, K.-R.: Nonlinear component analysis as a kernel eigenvalue problem. Neural Comput. **10**(5), 1299–1319 (1998)
6. Mika, S., Rtsch, G., Weston, J., Schlkopf, B., Mller, K.-R.: Fisher discriminant analysis with kernels (1999)
7. Gärtner, T.: A survey of kernels for structured data. SIGKDD Explor. Newsl. **5**(1), 49–58 (2003)
8. Gärtner, T., Flach, P., Wrobel, S.: On graph kernels: hardness results and efficient alternatives. In: Schölkopf, B., Warmuth, M.K. (eds.) COLT-Kernel 2003. LNCS (LNAI), vol. 2777, pp. 129–143. Springer, Heidelberg (2003). https://doi.org/10.1007/978-3-540-45167-9_11
9. Borgwardt, K.M., Ong, C.S., Schnauer, S., Vishwanathan, S.V.N., Smola, A.J., Kriegel, H.-P.: Protein function prediction via graph kernels. Bioinformatics **21**, i47 (2005)
10. Tsuda, K., Kin, T., Asai, K.: Marginalized kernels for biological sequences. Bioinformatics **18**, S268 (2002)
11. Kashima, H., Tsuda, K., Inokuchi, A.: Marginalized kernels between labeled graphs. In: Proceedings of the Twentieth International Conference on Machine Learning, pp. 321–328. AAAI Press (2003)
12. Kashima, H., Tsuda, K., Inokuchi, A.: Kernels for Graphs. MIT Press, Cambridge (2004)
13. Mahé, P., Ueda, N., Akutsu, T., Perret, J.-L., Vert, J.-P.: Extensions of marginalized graph kernels. In: Proceedings of the Twenty-First International Conference on Machine Learning, ser. ICML 2004, p. 70. ACM, New York (2004)
14. Smola, A.J., Kondor, I.R.: Kernels and regularization on graphs. In: Proceedings of the Annual Conference on Computational Learning Theory (2003)
15. Rocha, L.M., Cappabianco, F.A.M., Falcão, A.X.: Data clustering as an optimum-path forest problem with applications in image analysis. Int. J. Imaging Syst. Technol. **19**(2), 50–68 (2009)
16. Papa, J.P., Falcão, A.X., Suzuki, C.T.N.: Supervised pattern classification based on optimum-path forest. Int. J. Imaging Syst. Technol. **19**(2), 120–131 (2009)
17. Papa, J.P., Falcão, A.X., Albuquerque, V.H.C., Tavares, J.M.R.S.: Efficient supervised optimum-path forest classification for large datasets. Pattern Recogn. **45**(1), 512–520 (2012)
18. Papa, J.P., Fernandes, S.E.N., Falcão, A.X.: Optimum-path forest based on k-connectivity: theory and applications. Pattern Recogn. Lett. **87**, 117–126 (2017)
19. Pereira, C.R., Nakamura, R.Y., Costa, K.A., Papa, J.P.: An optimum-path forest framework for intrusion detection in computer networks. Eng. Appl. Artif. Intell. **25**(6), 1226–1234 (2012)
20. Spadoto, A.A., Guido, R.C., Papa, J.P., Falco, A.X.: Parkinson's disease identification through optimum-path forest. In: 2010 Annual International Conference of the IEEE Engineering in Medicine and Biology, pp. 6087–6090, August 2010
21. Pisani, R., Riedel, P., Ferreira, M., Marques, M., Mizobe, R., Papa, J.: Land use image classification through optimum-path forest clustering. In: 2011 IEEE International Geoscience and Remote Sensing Symposium, pp. 826–829, July 2011
22. Afonso, L., Papa, J., Papa, L., Marana, A., Rocha, A.: Automatic visual dictionary generation through optimum-path forest clustering. In: 2012 19th IEEE International Conference on Image Processing, pp. 1897–1900, September 2012

23. Chiachia, G., Marana, A.N., Papa, J.P., Falcao, A.X.: Infrared face recognition by optimum-path forest. In: 2009 16th International Conference on Systems, Signals and Image Processing, pp. 1–4, June 2009

24. Papa, J.P., Falcao, A.X., Levada, A.L.M., Correa, D.C., Salvadeo, D.H.P., Mascarenhas, N.D.A.: Fast and accurate holistic face recognition using optimum-path forest. In: 2009 16th International Conference on Digital Signal Processing, pp. 1–6, July 2009

25. Falcão, A., Stolfi, J., Lotufo, R.: The image foresting transform theory, algorithms, and applications. IEEE Trans. Pattern Anal. Mach. Intell. **26**(1), 19–29 (2004)

26. Allène, C., Audibert, J.-Y., Couprie, M., Keriven, R.: Some links between extremum spanning forests, watersheds and min-cuts. Image Vis. Comput. **28**(10), 1460–1471 (2010). Image Analysis and Mathematical Morphology

27. Papa, J.P., Suzuki, C.T.N., Falcão, A.X.: LibOPF: a library for the design of optimum-path forest classifiers (2014). http://www.ic.unicamp.br/~afalcao/libopf/

A Possibilistic c-means Clustering Model with Cluster Size Estimation

László Szilágyi[1]($\boxtimes$) and Sándor M. Szilágyi[2]

[1] Computational Intelligence Research Group,
Sapientia - Hungarian Science University of Transylvania, Tîrgu Mureş, Romania
lalo@ms.sapientia.ro
[2] Department of Informatics, Petru Maior University, Tîrgu Mureş, Romania

Abstract. Most c-means clustering models have serious difficulties when facing clusters of different sizes and severely outlier data. The possibilistic c-means (PCM) algorithm can handle both problems to some extent. However, its recommended initialization using a terminal partition produced by the probabilistic fuzzy c-means does not work when severe outliers are present. This paper proposes a possibilistic c-means clustering model that uses only two parameters independently of the number of clusters, which is able to correctly handle the above mentioned obstacles. Numerical evaluation involving synthetic and standard test data sets prove the advantages of the proposed clustering model.

Keywords: Fuzzy c-means clustering
Possibilistic c-means clustering · Cluster size sensitivity
Outlier sensitivity

1 Introduction

The introduction of the fuzzy logic into c-means clustering opened new horizons toward fine data partitioning, but also rose several obstacles that proved difficult to handle simultaneously. The early probabilistic clustering models introduced by Dunn [6] and generalized by Bezdek [4] are strongly influenced by outlier data and uneven sized clusters. A solution to the problem of outliers was given by Dave [5], who relaxed the probabilistic constraint by introducing an extra cluster that attracted noisy data, but this method still creates clusters of equal diameter. The possibilistic c-means clustering algorithm [8] addresses the uneven sized clusters as well, but frequently creates coincident clusters [3]. Several solutions have been proposed to enable the probabilistic clustering models to correctly handle clusters of different weight and/or diameter (e.g. [10,11]), but these are still sensitive to outlier data due to their probabilistic constraints. Leski [9] recently proposed the so-called fuzzy c-ordered means algorithm, which achieved high

The work of L. Szilágyi was supported by the Institute for Research Programs of the Sapientia University. The work of S. M. Szilágyi was supported by UEFISCDI Romania through grant no. PN-III-P2-2.1-BG-2016-0343, contract no. 114BG.

M. Mendoza and S. Velastín (Eds.): CIARP 2017, LNCS 10657, pp. 661–668, 2018.
https://doi.org/10.1007/978-3-319-75193-1_79

robustness, but dropped the classical alternative optimization scheme of the fuzzy c-means algorithm.

In this paper we propose a possibilistic fuzzy c-means clustering approach, which employs an extra noise cluster whose prototype is situated at constant distance from every input vector, and an adaptation mechanism that enables the algorithm to handle different cluster diameters.

2   Background

The fuzzy c-means algorithm. The conventional fuzzy c-means (FCM) algorithm partitions a set of object data $\mathbf{X} = \{\mathbf{x}_1, \mathbf{x}_2, \ldots, \mathbf{x}_n\}$ into a number of c clusters based on the minimization of a quadratic objective function, defined as:

$$J_{\text{FCM}} = \sum_{i=1}^{c} \sum_{k=1}^{n} u_{ik}^{m} ||\mathbf{x}_k - \mathbf{v}_i||_{\mathbf{A}}^{2} = \sum_{i=1}^{c} \sum_{k=1}^{n} u_{ik}^{m} d_{ik}^{2}, \tag{1}$$

where $\mathbf{v}_i$ represents the prototype or centroid of cluster i $(i = 1 \ldots c)$, $u_{ik} \in [0, 1]$ is the fuzzy membership function showing the degree to which vector $\mathbf{x}_k$ belongs to cluster i, $m > 1$ is the fuzzyfication parameter, and d_{ik} represents the distance (any inner product norm defined by a symmetrical positive definite matrix $\mathbf{A}$) between $\mathbf{x}_k$ and $\mathbf{v}_i$. FCM uses a probabilistic partition, meaning that the fuzzy memberships assigned to any input vector $\mathbf{x}_k$ with respect to clusters satisfy the probability constraint $\sum_{i=1}^{c} u_{ik} = 1$. The minimization of the objective function J_{FCM} is achieved by alternately applying the optimization of J_{FCM} over $\{u_{ik}\}$ with $\mathbf{v}_i$ fixed, $i = 1 \ldots c$, and the optimization of J_{FCM} over $\{\mathbf{v}_i\}$ with u_{ik} fixed, $i = 1 \ldots c, k = 1 \ldots n$ [4]. Obtaining the optimization formulas involves zero gradient conditions of J_{FCM} and Langrange multipliers. Iterative optimization is applied until cluster prototypes $\mathbf{v}_i$ $(i = 1 \ldots c)$ converge.

Relaxing the probabilistic constraint. The relaxation of the probabilistic constraint was a necessity provoked by the outlier sensitivity of the FCM algorithm. Here we need to mention two different ways the constraint was eliminated. Krishnapuram and Keller [8] introduced the possibilistic c-means (PCM) algorithm, which optimizes

$$J_{\text{PCM}} = \sum_{i=1}^{c} \sum_{k=1}^{n} \left[t_{ik}^{p} d_{ik}^{2} + (1 - t_{ik})^{p} \eta_i \right], \tag{2}$$

constrained by $0 \leq t_{ik} \leq 1 \ \forall i = 1 \ldots c, \forall k = 1 \ldots n$, and $0 < \sum_{i=1}^{c} t_{ik} < c \ \forall k = 1 \ldots n$, where $p > 1$ represents the possibilistic exponent, and parameters η_i are the penalty terms that control the diameter of the clusters. The iterative optimization algorithm of PCM objective function is derived from the zero gradient conditions of J_{PCM}. In the probabilistic fuzzy partition, the degrees of membership assigned to an input vector $\mathbf{x}_k$ with respect to cluster i depends on the distances of the given vector to all cluster prototypes: $d_{1k}, d_{2k}, \ldots, d_{ck}$. On the other hand, in the possibilistic partition, the typicality value u_{ik} assigned to input vector $\mathbf{x}_k$ with respect to any cluster i depends on only one distance: d_{ik}. PCM efficiently suppresses the effects of outlier data, at the price of frequently producing coincident cluster prototypes. The latter is the result of the highly independent cluster prototypes [3].

On the other hand, Dave [5] introduced a noise cluster into the FCM algorithm, which by definition is situated at a constant distance d_0 from any input vector $\mathbf{x}_k$ $(k = 1 \dots n)$. Thus, the objective function becomes

$$J_{\text{Dave}} = \sum_{i=0}^{c} \sum_{k=1}^{n} u_{ik}^m d_{ik}^2 = J_{\text{FCM}} + \sum_{k=1}^{n} u_{0k}^m d_0^2, \qquad (3)$$

where the noise cluster is the one with index 0. The probabilistic constraint becomes $\sum_{i=0}^{c} u_{ik} = 1 \ \forall k = 1 \dots n$, thus the degrees of membership of any input vector with respect to the real clusters does not sum up to 1 anymore. Outliers will be attributed with high degrees of membership towards the noise class, making the algorithm insensitive to outlier data, without producing coincident clusters. However, this approach cannot handle clusters of different size or diameter.

Several fuzzy-possibilistic mixture partition models have been proposed to deal with the coincident clusters of PCM (e.g. [12,13]). The most recent additive mixture model proposed by Pal et al. [13] called possibilistic-fuzzy c-means (PFCM) clustering minimizes

$$J_{\text{PFCM}} = \sum_{i=1}^{c} \sum_{k=1}^{n} [a u_{ik}^m + b t_{ik}^p] d_{ik}^2 + \sum_{i=1}^{c} \eta_i \sum_{k=1}^{n} (1 - t_{ik})^p, \qquad (4)$$

constrained by the conventional probabilistic and possibilistic conditions of FCM and PCM, respectively. Here a and b are two tradeoff parameters that control the strength of the possibilistic and probabilistic term in the mixed partition. All other parameters are the same as in FCM and PCM. This clustering model was found accurate and robust, but still sensitive to outlier data [14].

Fuzzy c-means with various cluster diameters. Komazaki and Miyamoto [7] presents a collection of solutions how the FCM algorithm can adapt to different cluster sizes and diameters. From the point of view of this paper, it is relevant to mention the FCMA algorithm by Miyamoto and Kurosawa [11], which minimizes

$$J_{\text{FCMA}} = \sum_{i=1}^{c} \sum_{k=1}^{n} \alpha_i^{1-m} u_{ik}^m d_{ik}^2, \qquad (5)$$

subject to the probabilistic constraint of the fuzzy memberships u_{ik} $(i = 1 \dots c, k = 1 \dots n)$, and of the extra terms α_i $(i = 1 \dots c)$: $\sum_{i=1}^{c} \alpha_i = 1$. The optimization algorithm of J_{FCMA} can be derived from zero gradient conditions using Lagrange multipliers. Each iteration updates the probabilistic memberships u_{ik} $(i = 1 \dots c, k = 1 \dots n)$, the cluster prototypes $\mathbf{v}_i$ $(i = 1 \dots c)$, and the cluster diameter terms α_i $(i = 1 \dots c)$ as well. The algorithm stops when cluster prototypes stabilize.

3   Methods

In the following, we propose a fuzzy c-means clustering model with relaxed probabilistic constraint, which employs an extra noise cluster whose prototype is situated at constant distance from every input vector, and an adaptation mechanism to different cluster diameters. The proposed objective function is:

$$J = \sum_{i=0}^{c} \sum_{k=1}^{n} \alpha_i^{1-m} u_{ik}^m d_{ik}^2, \tag{6}$$

where $d_{ik} = \|\mathbf{x}_k - \mathbf{v}_i\|$ $(\forall i = 1 \ldots c, \forall k = 1 \ldots n)$, $m > 1$ is the fuzzy exponent, and $d_{0k} = d_0$ $\forall k = 1 \ldots n$ with d_0 predefined constant. Variables α_i $(i = 1 \ldots c)$ satisfy the probabilistic constraint $\sum_{i=0}^{c} \alpha_i = 1$, while fuzzy memberships are also constrained to the probabilistic rule, similarly to the FCM algorithm: $\sum_{i=0}^{c} u_{ik} = 1$ for any $k = 1 \ldots n$. The minimization formulas of the objective function given in Eq. (6) are obtained using zero gradient conditions and Lagrange multipliers. Let us consider the functional

$$\mathcal{L} = J + \sum_{k=1}^{n} \lambda_k \left(1 - \sum_{i=0}^{c} u_{ik}\right) + \lambda_\alpha \left(1 - \sum_{i=0}^{c} \alpha_i\right), \tag{7}$$

where $\lambda_1 \ldots \lambda_n$ and λ_α are the Lagrange multipliers. The zero gradient conditions with respect to the fuzzy memberships u_{ik} $(\forall i = 0 \ldots c, \forall k = 1 \ldots n)$ imply

$$\frac{\partial \mathcal{L}}{\partial u_{ik}} = 0 \Rightarrow \alpha_i^{1-m} m u_{ik}^{m-1} d_{ik}^2 = \lambda_k, \tag{8}$$

and so

$$u_{ik} = \left(\frac{\lambda_k}{m}\right)^{1/(m-1)} \alpha_i d_{ik}^{-2/(m-1)}. \tag{9}$$

According to the probabilistic constraint of fuzzy memberships, we have:

$$\sum_{j=0}^{c} u_{jk} = 1 \Rightarrow 1 = \left(\frac{\lambda_k}{m}\right)^{1/(m-1)} \sum_{j=0}^{c} \alpha_j d_{jk}^{-2/(m-1)}. \tag{10}$$

Equations (9) and (10) allows us to eliminate the Lagrange multiplier λ_k from the formula of u_{ik}:

$$u_{ik} = \frac{u_{ik}}{1} = \frac{\left(\frac{\lambda_k}{m}\right)^{1/(m-1)} \alpha_i d_{ik}^{-2/(m-1)}}{\left(\frac{\lambda_k}{m}\right)^{1/(m-1)} \sum_{j=0}^{c} \alpha_j d_{jk}^{-2/(m-1)}} = \frac{\alpha_i d_{ik}^{-2/(m-1)}}{\sum_{j=0}^{c} \alpha_j d_{jk}^{-2/(m-1)}}. \tag{11}$$

The optimization formula for α_i $(i = 0 \ldots c)$ is obtained similarly. We start from the zero crossing of the partial derivative

$$\frac{\partial \mathcal{L}}{\partial \alpha_i} = 0 \Rightarrow \sum_{k=1}^{n} \left[\alpha_i^{-m}(1-m) u_{ik}^m d_{ik}^2\right] = \lambda_\alpha, \tag{12}$$

Algorithm 1. The proposed algorithm

Data: Input data $\mathbf{X} = \{\mathbf{x}_1, \mathbf{x}_2, \ldots, \mathbf{x}_n\}$
Data: Number of clusters c, fuzzy exponent m, distance d_0, threshold ε
Result: Possibilistic partition u_{ik} $(\forall i = 0 \ldots c, \forall k = 1 \ldots n)$
Initialize $\mathbf{v}_i^{(\text{new})}$ $(i = 1 \ldots c)$ as random input vectors that have several input vectors in their close neighborhood, to avoid outliers
$\alpha_i \leftarrow 1/(c+1), \forall i = 0 \ldots c$
repeat
$\quad$ $\mathbf{v}_i^{(\text{old})} \leftarrow \mathbf{v}_i^{(\text{new})}, \forall i = 1 \ldots c$
$\quad$ Update partition u_{ik}, $(i = 0 \ldots c, k = 1 \ldots n)$, according to Eq. (11)
$\quad$ Obtain new cluster prototypes $\mathbf{v}_i^{(\text{new})}$, $(i = 1 \ldots c)$, according to Eq. (17)
$\quad$ Update α_i values for any $i = 0 \ldots c$, according to Eq. (16)
until $\sum_{i=1}^{c} ||\mathbf{v}_i^{(\text{new})} - \mathbf{v}_i^{(\text{old})}|| < \varepsilon$;

---

which implies

$$(1-m)\alpha_i^{-m} \sum_{k=1}^{n} u_{ik}^m d_{ik}^2 = \lambda_\alpha \Rightarrow \alpha_i^m = \left(\frac{1-m}{\lambda_\alpha}\right)\left[\sum_{k=1}^{n} u_{ik}^m d_{ik}^2\right], \quad (13)$$

and so we get

$$\alpha_i = \left(\frac{1-m}{\lambda_\alpha}\right)^{1/m}\left[\sum_{k=1}^{n} u_{ik}^m d_{ik}^2\right]^{1/m}. \quad (14)$$

On the other hand, the probabilistic constraint $\sum_{j=0}^{c} \alpha_j = 1$ implies:

$$\sum_{j=0}^{c} \alpha_j = 1 \Rightarrow 1 = \left(\frac{1-m}{\lambda_\alpha}\right)^{1/m} \sum_{j=0}^{c}\left[\sum_{k=1}^{n} u_{jk}^m d_{jk}^2\right]^{1/m}. \quad (15)$$

Equations (14) and (15) allows us to eliminate the Lagrange multiplier λ_α from the formula of α_i:

$$\alpha_i = \frac{\alpha_i}{1} = \frac{\left(\frac{1-m}{\lambda_\alpha}\right)^{1/m}\left[\sum_{k=1}^{n} u_{ik}^m d_{ik}^2\right]^{1/m}}{\left(\frac{1-m}{\lambda_\alpha}\right)^{1/m}\sum_{j=0}^{c}\left[\sum_{k=1}^{n} u_{jk}^m d_{jk}^2\right]^{1/m}} = \frac{\left[\sum_{k=1}^{n} u_{ik}^m d_{ik}^2\right]^{1/m}}{\sum_{j=0}^{c}\left[\sum_{k=1}^{n} u_{jk}^m d_{jk}^2\right]^{1/m}}. \quad (16)$$

The update formula of cluster prototypes $\mathbf{v}_i$ is obtained as:

$$\frac{\partial \mathcal{L}}{\partial \mathbf{v}_i} = 0 \Rightarrow \sum_{k=1}^{n}(-2)\alpha_i^{1-m} u_{ik}^m(\mathbf{x}_k - \mathbf{v}_i) = 0$$

$$\Rightarrow \sum_{k=1}^{n} u_{ik}^m \mathbf{x}_k = \mathbf{v}_i \sum_{k=1}^{n} u_{ik}^m \quad (17)$$

$$\Rightarrow \mathbf{v}_i = \frac{\sum_{k=1}^{n} u_{ik}^m \mathbf{x}_k}{\sum_{k=1}^{n} u_{ik}^m}.$$

If a defuzzyfied partition is desired, any input vector $\mathbf{x}_k$ can be assigned to cluster number $\arg\max_i\{u_{ik}, i = 0 \ldots c\}$. Vectors belonging to cluster number 0 are detected outliers. The proposed algorithm is summarized in Algorithm 1.

4   Results and Discussion

The proposed method was evaluated on three different data sets, and its behavior compared to FCM [4] and PFCM [13]. The first data set consisted of two groups of randomly generated two-dimensional input vectors, situated inside the circle with center at $(0, 1)$ and radius 1.2, and the circle with center at $(0, -1)$ and radius 0.6, respectively. Each group contained 100 vectors. The input vectors are exhibited in Fig. 1, together with the partitions created by the proposed algorithms and its counter candidates. Adding an extra input vector situated at $(\delta, 0)$ and treating δ as a parameter allowed us to evaluate the sensitivity of the clustering to outliers. FCM and PFCM were able to provide two valid clusters up $\delta = 132$ and $\delta = 158$, respectively. The proposed method was not influenced by high values of δ, it assigned the extra vector to the third (outlier) class.

The second data set employed by the numerical evaluation of the algorithms was the IRIS data set [1], which consist of 150 labeled feature vectors of four dimensions, organized in three groups that contain fifty vectors each. It is a reported fact, that conventional clustering models like FCM produce 133–134 correct decisions when classifying IRIS data. PFCM produced the best reported accuracy with 140 correct decisions using $a = b = 1, m = p = 3$, and initializing v_i with terminal FCM prototypes [13]. Under less advantageous circumstances, PFCM reportedly produced 136–137 correct decisions. Initially we normalized the IRIS data set, and included an outlier situated at $(\delta, \delta, \delta, \delta)$, where δ was a variable parameter. We tested the clustering models for various values of the algorithm parameters and outlier positions δ. Figure 2 shows the outcome of tests. FCM produces three valid clusters in case of $\delta < 9$, PFCM can correctly handle the outlier for values of δ up to 12, while the proposed method can deal with the outlier situated at any distance.

The third numerical test employed the WINE data set [2], which consist of 178 labeled feature vectors of 13 dimensions, organized in three groups of uneven cardinality. The WINE data set was initially normalized, and an outlier was included at the position $(\delta, \delta, \ldots, \delta)$, where δ represents a variable parameter. We tested the clustering models for various values of the algorithm parameters and outlier positions δ. Figure 3 exhibits the test results. FCM and PFCM can produce three valid clusters in case of $\delta < 6$, and $\delta < 7$, respectively, while the proposed method has no difficulty with handling correctly an outlier situated at any distance.

Based on all numerical tests, we can assert that the proposed clustering model is able to perform better than its counter candidates, as: (1) it creates valid clusters according to cluster validity indexes proposed for the characterisation of such partitions; (2) it produces more correct decisions than previous algorithms; (3) it can correctly handle severe outlier vectors; (4) it is able to adapt the diameter of clusters to the input data to some considerable extent; (5) it does not produce coincident clusters; (6) it uses a reduced number of parameters (2 instead of the $c + 1$ parameters of PCM). Results reported in Figs. 1, 2 and 3 were obtained using the following parameter settings: $m = 2$ for FCM and the

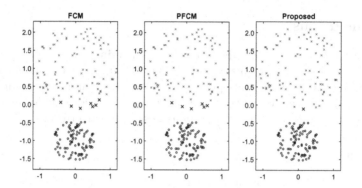

Fig. 1. The case of two groups of same cardinality but different diameter

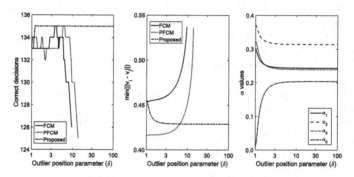

Fig. 2. Evaluation using the IRIS data set: (left) correct decisions out of 150 vs. δ, (middle) CVI vs. δ, (right) final α_i values of the proposed algorithm vs. δ

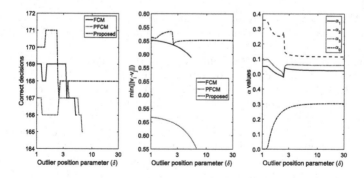

Fig. 3. Evaluation using the WINE data set: (left) correct decisions out of 150 vs. δ, (middle) CVI vs. δ, (right) final α_i values of the proposed algorithm vs. δ

proposed method, $m = p = 2$ and $a = b = 1$. Possibilistic penalty terms η_i were not established based on a terminal FCM partition, as recommended by their authors, because in many cases the terminal FCM partition was severely damaged by the presence of the outlier. Parameters η_i and d_0 were used as constants whose values were set empirically to favor correct and fine partitions.

5   Conclusions

This paper proposed a possibilistic c-means clustering model with a reduced number of parameters (two instead of $c + 1$) that can robustly handle distant outlier data. The advantageous properties of the algorithms were numerically validated using synthetic and standard test data sets. A formula for an optimal d_0 distance may further enhance the robustness of the algorithm.

References

1. Anderson, E.: The irises of the Gaspe Peninsula. Bull. Am. Iris Soc. **59**, 2–5 (1935)
2. Asuncion, A., Newman, D.J.: UCI Machine Learning Repository. http://archive. ics.uci.edu/ml/datasets.html
3. Barni, M., Capellini, V., Mecocci, A.: Comments on a possibilistic approach to clustering. IEEE Trans. Fuzzy Syst. **4**, 393–396 (1996)
4. Bezdek, J.C.: Pattern Recognition with Fuzzy Objective Function Algorithms. Plenum, New York (1981)
5. Dave, R.N.: Characterization and detection of noise in clustering. Pattern Recogn. Lett. **12**, 657–664 (1991)
6. Dunn, J.C.: A fuzzy relative of the isodata process and its use in detecting compact well-separated clusters. Cybern. Syst. **3**(3), 32–57 (1973)
7. Komazaki, Y., Miyamoto, S.: Variables for controlling cluster sizes on fuzzy c-means. In: Torra, V., Narukawa, Y., Navarro-Arribas, G., Megías, D. (eds.) MDAI 2013. LNCS (LNAI), vol. 8234, pp. 192–203. Springer, Heidelberg (2013). https:// doi.org/10.1007/978-3-642-41550-0_17
8. Krishnapuram, R., Keller, J.M.: A possibilistic approach to clustering. IEEE Trans. Fuzzy Syst. **1**, 98–110 (1993)
9. Leski, J.M.: Fuzzy c-ordered-means clustering. Fuzzy Sets Syst. **286**, 114–133 (2016)
10. Lin, P.L., Huang, P.W., Kuo, C.H., Lai, Y.H.: A size-insensitive integrity-based fuzzy c-means method for data clustering. Pattern Recogn. **47**(5), 2024–2056 (2014)
11. Miyamoto, S., Kurosawa, N.: Controlling cluster volume sizes in fuzzy c-means clustering. In: SCIS and ISIS, Yokohama, Japan, pp. 1–4 (2004)
12. Pal, N.R., Pal, K., Bezdek, J.C.: A mixed c-means clustering model. In: Proceedings of IEEE International Conference on Fuzzy Systems (FUZZ-IEEE), pp. 11–21 (1997)
13. Pal, N.R., Pal, K., Keller, J.M., Bezdek, J.C.: A possibilistic fuzzy c-means clustering algorithm. IEEE Trans. Fuzzy Syst. **13**, 517–530 (2005)
14. Szilágyi, L.: Fuzzy-possibilistic product partition: a novel robust approach to c-means clustering. In: Torra, V., Narakawa, Y., Yin, J., Long, J. (eds.) MDAI 2011. LNCS (LNAI), vol. 6820, pp. 150–161. Springer, Heidelberg (2011). https://doi. org/10.1007/978-3-642-22589-5_15

Randomized Neural Network Based Signature for Classification of Titanium Alloy Microstructures

Jarbas Joaci de Mesquita Sá Junior[1,2]($\boxtimes$),
André R. Backes[3], and Odemir Martinez Bruno[1]

[1] São Carlos Institute of Physics, University of São Paulo,
PO Box 369, São Carlos, SP 13560-970, Brazil
bruno@ifsc.usp.br, jarbas_joaci@yahoo.com.br
[2] Department of Computer Engineering,
Campus de Sobral - Universidade Federal do Ceará,
Rua Estanislau Frota, S/N, Centro, Sobral, Ceará 62010-560, Brazil
[3] School of Computer Science, Universidade Federal de Uberlândia,
Av. João Naves de Ávila, 2121, Uberlândia, MG 38408-100, Brazil
arbackes@yahoo.com.br

Abstract. This paper presents the application of the randomized neural network based signature, an innovative and powerful texture analysis algorithm, to a relevant problem of metallography, which consists of classifying zones of titanium alloys Ti-6Al-4V into two categories: "alpha and beta" and "alpha + beta". The obtained results are very promising, with accuracy of 98.84% by using LDA, and accuracy of 98.64%, precision of 99.11% for "alpha and beta", and precision of 98.09% for "alpha + beta" by using SVM. This performance suggests that this texture analysis method is a valuable tool that can be applied to many other problems of metallography.

Keywords: Randomized neural network based signature
Titanium alloy · Texture · Metallography

1 Introduction

Texture analysis has playing a preponderant role in computer vision since the birth of this research field. Although a texture image does not have a unique definition, it is an attribute easily perceived by humans. In general words, textures are images composed of sub-patterns, which can be pixels, regions or visual attributes (shape, color etc.) [1]. Obviously, such concept does not address a vast range of natural textures, such as images of smoke, mammography, bark, water etc., which present a persistent stochastic pattern [2].

Throughout the years, increasingly accurate methods have been proposed for extracting signatures from texture images. As a consequence, these methods have also been applied successfully in several domains, providing expert systems for

M. Mendoza and S. Velastín (Eds.): CIARP 2017, LNCS 10657, pp. 669–676, 2018.
https://doi.org/10.1007/978-3-319-75193-1_80

many human activities. In the medicine field, for instance, we can cite: the paper [3] performed a combination of texture measures for meningioma classification, the papers [4,5] applied several LBP variants to classify a database of images from the cervix (*pap-smear* dataset [6]); in the biology field: the paper [7] applied Gabor wavelets to analyze texture images of leaf surfaces; and in industry area: the paper [8] proposed to apply a texture extraction method based on LBP and GLCM to classify tea leaves in order to improve real-time controls of tea production lines.

This paper aims to apply a recent and powerful texture analysis method to the analysis of images of titanium alloys, which is a very relevant problem of metallography. For this purpose, this work is organized as follows: Sect. 2 briefly describes what is metallography and the study of titanium alloys; Sect. 3 describes the randomized neural network based signature used for classifying the titanium alloy images; Sect. 4 shows all the details of the performed experiments; Sect. 5 presents the obtained results and a discussion on them; and, finally, Sect. 6 establishes some remarks about this work.

2   Metalography and Titanium Alloy

Metallography can be defined as the study of the physical properties of metals based on the inspection of their constituents. This inspection can be performed by naked eye, or, more accurately, by using optical or electron microscopes, which allow us to examine the metal microstructures [9]. In the context of this work, we focus on the microstructures of an important category of metal called "titanium alloy Ti-6Al-4V", which has been used, for instance, in aerospace industry and medical prostheses. This type of alloy presents two phases, called "alpha" and "beta", which are hexagonal and cubic structures, respectively, and determine the physical properties of the metal.

When titanium alloys Ti-6Al-4V are subjected to Friction Stir Welding (FSW) [10], which is a solid-state joining technique, the resulting product presents microstructure zones that can be grouped into two classes. The first

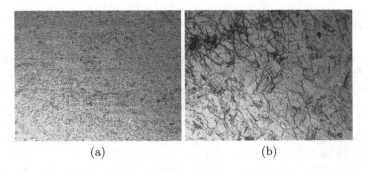

(a)   (b)

Fig. 1. Samples of titanium alloy Ti-6Al-4V: (a) - class Alpha and Beta; (b) - class Alpha + Beta (images provided by the authors of the paper [9]).

class, called here "alpha and beta", is composed of noise-like microstructures, which presents grains of alpha and beta. The second class, called here "alpha + beta", is composed of crystal-like microstructures presented in the deformed weld zones [9]. Because these two categories of microstructures have different physical properties, classifying them correctly is an important problem in metallography. Figure 1 shows one example for each class.

3   Texture Analysis Using Randomized Neural Network

3.1   Random Neural Network

A randomized neural network [11–14] is a type of neural network with only two layers (hidden and output) and a very fast learning algorithm when compared to the traditional backpropagation approach. Basically, the weights of the hidden neurons are randomly determined according to a uniform or a Gaussian distribution, and the weights of the output neurons are calculated using a least-squares solution. Next, a variant of this neural network, which was used in [15], is briefly explained.

Let $X = [x_1, x_2, \ldots, x_N]$ (each x_i has a constant $x_{i0} = -1$ to connect it to the biases of the neurons) be a matrix containing N feature vectors used for training and $D = [d_1, d_2, \ldots, d_N]$ be a matrix with their respective label vectors. First, it is necessary to assign random values (uniform or Gaussian) to the weights of the hidden layer. These weights can be arranged in a matrix W, where each line represents the weights of a determined hidden neuron $q \in \{1, 2, \ldots, Q\}$ (the weights w_{q0} of the first column are the biases of each hidden neuron and Q is the number of hidden neurons).

$$W = \begin{pmatrix} w_{10} & w_{11} & \cdots & w_{1p} \\ w_{20} & w_{21} & \cdots & w_{2p} \\ \vdots & \vdots & \vdots & \vdots \\ w_{Q0} & w_{Q1} & \cdots & w_{Qp} \end{pmatrix} \tag{1}$$

The next step is to calculate $Z = \phi(WX)$ (ϕ is the activation function, which can be, for instance, the tangential or hyperbolic function) in order to obtain the output of the hidden neurons for each feature vector x_i. This new matrix $Z = [z_1, z_2, \ldots, z_N]$, where each z_i corresponds to each x_i, can be used for obtaining the weights of the output neurons. For this purpose, a constant -1 is again added to each vector z_i to connect it to the biases of the output neurons. Next, the weights of the output neurons are arranged in a matrix M, where each line is the weights of a specific output neuron. Finally, the weights of the output layer (arranged in a matrix M) can be computed by the following equation

$$M = DZ^T(ZZ^T)^{-1}. \tag{2}$$

3.2 Randomized Neural Network Based Signature

To use the weights of the output neurons as an image signature, the first step is to divide the image into windows $K \times K$ (K is odd) and consider the central pixel as a label vector d_i and the neighboring pixels as the feature vector x_i. For this purpose, the paper [15] adopted $K = \{3, 5, 7\}$, which determined a circular neighborhood composed of pixels with distance less than or equal to $\sqrt{2}, \sqrt{8}$ and $\sqrt{13}$, respectively.

In order to make the method robust to rotation, a specific fixed mask is multiplied by every rotation of the neighboring pixels, thus resulting in several linear combinations (for instance, eight linear combinations for neighborhood $\sqrt{2}$). Thus, the linear combination with the minimum value determines the order of the pixels in the feature vector x_i. The paper [15] proposes the masks $[2^{1.0}, 2^{2.0}, 2^{3.0}, \ldots, 2^{8.0}]$, $[2^{1.0}, 2^{1.6}, 2^{2.2} \ldots, 2^{7.6}]$, and $[2^{1.0}, 2^{1.4}, 2^{1.8} \ldots, 2^{7.0}]$ for the windows $3 \times 3, 5 \times 5$ and 7×7, respectively.

The next step of the algorithm is to assign values to the matrix W. In every training phase of a randomized neural network, different random values are assigned to the weights of the hidden layer, so that the outputs of the neural network are not necessary the same for the same input vectors. In an image such effect is undesirable because a texture analysis method must provide the same signature for the same input image. Thus, the paper [15] used the classical linear congruent generator (LCG) [16,17] with fixed values of "seed" and "adjustment parameters" to generate pseudorandom uniform numbers for the matrix W. Lastly, all the weights of the matrix W and each line of the matrix X are normalized to zero mean and unit variance.

Taking into account all the presented procedures, the feature vector is obtained by the equation $f = DZ^T(ZZ^T)^{-1}$. Based on this feature vector, it is possible to construct two image signatures. The first is

$$\Theta(Q)_{K_1, K_2, \ldots, K_n} = [f_{K_1}, f_{K_2}, \ldots, f_{K_n}], \tag{3}$$

where K is the window size used to divide the image.

The second signature can be expressed as the concatenation of the previous signature for different numbers of hidden neurons, that is,

$$\Psi_{Q_1, Q_2, \ldots, Q_m} = [\Theta(Q_1)_{K_1, \ldots, K_n}, \Theta(Q_2)_{K_1, \ldots, K_n}, \ldots, \Theta(Q_m)_{K_1, \ldots, K_n}]. \tag{4}$$

More details on this algorithm can be found in the paper [15].

4 Experiments

Our database was kindly provided by the authors of the paper [9] and consists of 30 images 1079×816 pixel size equally divided into two classes ("alpha and beta" and "alpha + beta"). To construct a new database for our experiments, we followed the same procedure of the paper [9], that is, we divided each image into a grid of 13×10 windows 80×80 pixel size and discarded the residual

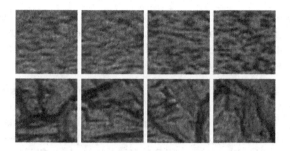

Fig. 2. Samples 80 × 80 pixel size of titanium alloy Ti-6Al-4V: first line - class Alpha and Beta; second line - class Alpha + Beta (images provided by the authors of the paper [9]).

pixels of the borders. Also, we converted the images into grayscale. Thus, each class contains 1950 images 80 × 80 pixel size. Figure 2 shows some samples of both classes.

In the experiments, we used the algorithm proposed by Sá Junior and Backes [15] with the same parameter values ($K = \{3, 5, 7\}$, W determined by LCG and $Q = \{19, 39\}$ for the second type of signature, that is, $\Psi_{19,39}$). We also compared the method with other very important grayscale texture analysis methods, which are:

- **Gabor wavelets:** we used the same parameter values adopted in the paper [18], that is, 6 rotations, 4 scales, and minimum and maximum frequencies of 0.05 and 0.4, respectively. The paper [18] proposes a mathematical rule to obtain σ_x and σ_y of the Gabor filters according to the parameters above. The feature vector is constructed using the mean and standard deviation of the magnitude of the transform coefficients, thus resulting in a signature with 48 attributes.
- **Co-occurrence matrices:** this is a classical algorithm that computes the "co-occurrence" of pairs of pixels at determined orientation and distance [19]. For the experiments, we used the most common distances ($d = 1$ and $d = 2$) and orientations ($0°, -45°, 90°$ and $135°$) and non-symmetric matrices. The descriptors extracted were energy and entropy, totaling 16 attributes.
- **Tourist walk** [20,21]: this method considers each image pixel as a city. A tourist starting at determined city must visit the neighboring cities according to a deterministic rule of going to the nearest or farthest city not visited in the last μ time steps. Because of this, the tourist's path has two elements: a transient time t (new cities are visited) and a cycle of period p (the tourist is trapped in a cycle and, therefore, visits the same cities). The next step is to construct a joint distribution map of transient times p and periods p for a determined memory μ. To obtain the feature vector, we adopted the strategy of concatenation of histograms presented in the paper [20] using $m = 4$ and memories $\mu = \{0, 1, 2, \ldots, 5\}$ for the rule of going to the nearest and farthest city. The resulting signature has 48 attributes.

- **Wavelet descriptors** [22,23]: we used the multilevel 2D wavelet and performed three dyadic decompositions using Daubechies 4. Next, we computed energy and entropy from each vertical, horizontal and diagonal details, totaling a feature vector of 18 attributes.

For the classification, we used a linear discriminant analysis (LDA) with a leave-one-out cross validation strategy [24]. We also performed a comparison "similar" to that of the paper [9] by using a radial basis SVM as classifier and the **precision** as performance measure. For this purpose, we used the LIBSVM [25], which is a public library for SVM. We adopted as SVM parameters the default values of C and γ of this library (the paper [9] adopted $C = 1$ and $\gamma = 1$, which give the best results). The validation strategy adopted was the *holdout* (50% of the samples for training and the remainder for testing) with 100 repetitions, according to the paper [9].

Unfortunately, it was not possible to recover the same database used in the experiment of the paper [9] (the authors used 20 images - 10 images per class - from an original database with more than 150 images 1079×816 pixel size to construct a database of 2600 images 80×80 pixel size). However, we believe that our imperfect comparison can still confirm the high performance of the randomized neural network method.

5   Results and Discussion

Table 1 shows the comparison of the randomized neural network based signature against all the compared texture analysis methods proposed in this paper. The results clearly confirm the high performance of the randomized neural network method, since it surpassed the second best method (Gabor wavelets) in 2.28% of samples correctly classified, that is, 89 more images. When we consider the remainder methods, the difference of accuracy becomes absolutely evident. For instance, the tourist walk method, which presents a state-of-the-art approach, obtained only 86.00% of accuracy, that is, 12.84% (501 images) less than the randomized neural network based signature.

The only apparent drawback of the randomized neural network based signature is its high number of features (180) when compared to the other methods.

Table 1. Comparison results of different methods applied to the titanium alloy Ti-6Al-4V database - LDA classifier with leave-one-out cross validation.

Methods	No of descriptors	Accuracy (%)
Randomized neural network signature	180	98.84
Gabor wavelets	48	96.56
Wavelet descriptors	18	88.90
Tourist walk	48	86.00
Co-occurrence matrices	16	86.26

It happens because the same parameter values of the paper [15] were adopted in order to avoid tuning the parameters of the randomized neural network, which would result in an unfair comparison with the other approaches. However, if we use the signature Ψ_{19}, which contains only 60 features, the obtained accuracy is 98.00%, that is, a result almost equal to that obtained by the signature $\Psi_{19,39}$ and equally overcoming all the compared methods.

When we performed the experiment "similar" to that of the paper [9], the results were: accuracy of 98.64%, precision of 99.11% for "alpha and beta", and precision of 98.09% for "alpha + beta". Our results of precision were superior to the best values presented in the paper [9], which obtained approximately 84.00% of precision for "alpha and beta" and almost 95.00% for "alpha + beta". This performance demonstrates that the randomized neural network based signature is very suitable for classifying samples of titanium alloy Ti-6Al-4V, and, therefore, can be a useful tool for metallography.

6   Conclusion

This paper presented an application of a very discriminative state-of-the-art texture analysis method to an important problem of metallography. The results obtained are very promising, surpassing several compared methods as well as the results of a recent work that addresses the same problem. Thus, the obtained performance suggests that the randomized neural network based signature can be a valuable tool for other relevant problems of metallography.

Acknowledgments. Jarbas Joaci de Mesquita Sá Junior thanks CNPq (National Council for Scientific and Technological Development, Brazil) (Grant: 152054/2016-2 and 453835/2017-1) for the financial support of this work. André R. Backes gratefully acknowledges the financial support of CNPq (Grant #302416/2015-3) and FAPEMIG (Foundation to the Support of Research in Minas Gerais) (Grant #APQ-03437-15). Odemir M. Bruno gratefully acknowledges the financial support of CNPq (307797/2014-7 and 484312/2013-8) and FAPESP (14/08026-1). We also thank the authors of the paper [9] for kindly providing the titanium alloy images used in this paper.

References

1. Backes, A.R., Martinez, A.S., Bruno, O.M.: Texture analysis based on maximum contrast walker. Pattern Recogn. Lett. **31**(12), 1701–1707 (2010)
2. Kaplan, L.M.: Extended fractal analysis for texture classification and segmentation. IEEE Trans. Image Process. **8**(11), 1572–1585 (1999)
3. Al-Kadi, O.S.: Texture measures combination for improved meningioma classification of histopathological images. Pattern Recogn. **43**(6), 2043–2053 (2010)
4. Nanni, L., Lumini, A., Brahnam, S.: Local binary patterns variants as texture descriptors for medical image analysis. Artif. Intell. Med. **49**(2), 117–125 (2010)
5. Nanni, L., Lumini, A., Brahnam, S.: Survey on LBP based texture descriptors for image classification. Expert Syst. Appl. **39**(3), 3634–3641 (2012)

6. Jantzen, J., Norup, J., Dounias, G., Bjerregaard, B.: Pap-smear benchmark data for pattern classification. In: Proceedings of the NiSIS 2005, Albufeira, Portugal. NiSIS, pp. 1–9 (2005)
7. Casanova, D., de Mesquita Sá Junior, J.J., Bruno, O.M.: Plant leaf identification using Gabor wavelets. Int. J. Imaging Syst. Technol. **19**(1), 236–243 (2009)
8. Tang, Z., Su, Y., Er, M.J., Qi, F., Zhang, L., Zhou, J.: A local binary pattern based texture descriptors for classification of tea leaves. Neurocomputing **168**, 1011–1023 (2015)
9. Ducato, A., Fratini, L., La Cascia, M., Mazzola, G.: An automated visual inspection system for the classification of the phases of Ti-6Al-4V titanium alloy. In: Wilson, R., Hancock, E., Bors, A., Smith, W. (eds.) CAIP 2013. LNCS, vol. 8048, pp. 362–369. Springer, Heidelberg (2013). https://doi.org/10.1007/978-3-642-40246-3_45
10. Mishra, R., Ma, Z.: Friction stir welding and processing. Mat. Sci. Eng.: R: Rep. **50**(12), 1–78 (2005)
11. Schmidt, W.F., Kraaijveld, M.A., Duin, R.P.W.: Feedforward neural networks with random weights. In: Proceedings of the 11th IAPR International Conference on Pattern Recognition, vol. II, Conference B: Pattern Recognition Methodology and Systems, pp. 1–4 (1992)
12. Pao, Y.H., Takefuji, Y.: Functional-link net computing: theory, system architecture, and functionalities. IEEE Comput. J. **25**(5), 76–79 (1992)
13. Pao, Y.H., Park, G.H., Sobajic, D.J.: Learning and generalization characteristics of the random vector functional-link net. Neurocomputing **6**(2), 163–180 (1994)
14. Huang, G.B., Zhu, Q.Y., Siew, C.K.: Extreme learning machine: theory and applications. Neurocomputing **70**(1–3), 489–501 (2006)
15. de Mesquita Sá Junior, J.J., Backes, A.R.: ELM based signature for texture classification. Pattern Recogn. **51**, 395–401 (2016)
16. Lehmer, D.H.: Mathematical methods in large scale computing units. Ann. Compt. Lab. Harv. Univ. **26**, 141–146 (1951)
17. Park, S.K., Miller, K.W.: Random number generators: good ones are hard to find. Commun. ACM **31**(10), 1192–1201 (1988)
18. Manjunath, B.S., Ma, W.Y.: Texture features for browsing and retrieval of image data. IEEE Trans. Pattern Anal. Mach. Intell. **18**(8), 837–842 (1996)
19. Haralick, R.M.: Statistical and structural approaches to texture. Proc. IEEE **67**(5), 786–804 (1979)
20. Backes, A.R., Gonçalves, W.N., Martinez, A.S., Bruno, O.M.: Texture analysis and classification using deterministic tourist walk. Pattern Recogn. **43**(3), 685–694 (2010)
21. Campiteli, M.G., Martinez, A.S., Bruno, O.M.: An image analysis methodology based on deterministic tourist walks. In: Sichman, J.S., Coelho, H., Rezende, S.O. (eds.) IBERAMIA/SBIA -2006. LNCS (LNAI), vol. 4140, pp. 159–167. Springer, Heidelberg (2006). https://doi.org/10.1007/11874850_20
22. Daubechies, I.: Ten Lectures on Wavelets. Society for Industrial and Applied Mathematics, Philadelphia (1992)
23. Chang, T., Kuo, C.J.: Texture analysis and classification with tree-structured wavelet transform. IEEE Trans. Image Process. **2**(4), 429–441 (1993)
24. Fukunaga, K.: Introduction to Statistical Pattern Recognition, 2nd edn. Academic Press, San Diego (1990)
25. Chang, C.C., Lin, C.J.: LIBSVM: a library for support vector machines. ACM Trans. Intell. Syst. Technol. **2**, 27:1–27:27 (2011)

Pap-smear Image Classification Using Randomized Neural Network Based Signature

Jarbas Joaci de Mesquita Sá Junior[1,2(✉)], André R. Backes[3], and Odemir Martinez Bruno[1]

[1] São Carlos Institute of Physics, University of São Paulo, PO Box 369, São Carlos, SP 13560-970, Brazil
jarbas_joaci@yahoo.com.br, bruno@ifsc.usp.br
[2] Department of Computer Engineering, Campus de Sobral - Universidade Federal do Ceará, Rua Estanislau Frota, S/N, Centro, Sobral, Ceará 62010-560, Brazil
[3] School of Computer Science, Universidade Federal de Uberlândia, Av. João Naves de Ávila, 2121, Uberlândia, MG 38408-100, Brazil
arbackes@yahoo.com.br

Abstract. This paper presents a state-of-the-art texture analysis method called "randomized neural network based signature" applied to the classification of *pap-smear* cell images for the Papanicolaou test. For this purpose, we used a well-known benchmark dataset composed of 917 images and compared the aforementioned image signature to other texture analysis methods. The obtained results were promising, presenting accuracy of 87.57% and AUC of 0.8983 using LDA and SVM, respectively. These performance values confirm that the randomized neural network based signature can be applied successfully to this important medical problem.

Keywords: Randomized neural network · *Pap-smear* database
Texture analysis

1 Introduction

Texture is among the most important attributes in computer vision and has been the focus of intensive research throughout the years. In a concise term, we can define texture as an arrangement of sub-patterns, which can be pixels, regions or other visual attributes [1]. Obviously, such definition is quite restrict and does not encompass a great variety of images (for instance, smoke, mammograms, fire, water etc.), which present a persistent stochastic pattern with a cloud-like appearance [2].

Even though texture lacks a formal definition, it is a feature easily understood by the human visual system. Such importance has motivated the development of many techniques for the analysis and recognition of texture patterns, making

© Springer International Publishing AG, part of Springer Nature 2018
M. Mendoza and S. Velastín (Eds.): CIARP 2017, LNCS 10657, pp. 677–684, 2018.
https://doi.org/10.1007/978-3-319-75193-1_81

this a field of intense research [3]. Among the many techniques available, there are those that describe the image texture using second-order statistics [4,5], spectral analysis (e.g., Fourier and Gabor filters) [6–8], local binary patterns [9], gravitational systems [10] and agents walking over the texture pattern [1].

Medical image analysis is a field of intense research, with many approaches being developed over the years. For instance, [11] proposed LBP variants as texture descriptors for medical image analysis, which were evaluated in different medical datasets, such as cell phenotype image classification, neonatal facial images classification of pain states and detection of abnormal smear cells. In [12], a fuzzy clustering algorithm was proposed for brain tumor segmentation. The authors stated that the fuzzy clustering enables many cases of uncertainty to be considered during the segmentation process. Breast cancer risk assessment has been the focus of many studies on texture analysis. In [13], two types of texture features are proposed to assess breast cancer risk: textons based on local pixel intensities and features based on oriented tissue structures. In [14], background intensity independent texture features were proposed for mammogram classification. Another topic of intense research is the identification and segmentation of melanocytic skin lesions. Machine learning techniques were used in [15] to select the parameters of a classification framework of melanocytic lesions. The paper [16] presented an approach using a feature learning scheme and normalized graph cuts for skin lesion image segmentation.

This paper proposes to apply a recent and very discriminative texture analysis method to a relevant medical problem, which consists of classifying *pap-smear* cells to discover pre-cancerous or cancerous stages in the cervix. Section 2 briefly describes the randomized neural network and how to use it to obtain an image signature from its neuron weights. Section 3 presents the *pap-smear* database, the other texture analysis methods used for comparison and the classification procedure. Section 4 discusses the obtained results, and, finally, Sect. 5 presents some remarks about this work.

2   Randomized Neural Network and its Texture Signature

A randomized neural network [17–20] is a recent proposal of neural network that has only two neuron layers and a very fast training procedure. In the hidden layer, the weights of the neurons are randomly determined according to a uniform or Gaussian distribution. These weights can be arranged in a matrix

$$W = \begin{pmatrix} w_{10} & w_{11} & \cdots & w_{1p} \\ w_{20} & w_{21} & \cdots & w_{2p} \\ \vdots & \vdots & \vdots & \vdots \\ w_{Q0} & w_{Q1} & \cdots & w_{Qp} \end{pmatrix}, \tag{1}$$

where each line represent the weights of a determined hidden neuron q, p is the number of attributes of an input vector x, and Q is the total of hidden neurons.

Let $X = [x_1, x_2, \ldots, x_N]$ and $D = [d_1, d_2, \ldots, d_N]$ be matrices representing the input vectors x_i and their respective labels d_i (N is the number of feature vectors). Then, after inserting a new first line composed of -1 into X (for bias), we can provide the output of the hidden neurons according to the equation $Z = \phi(WX)$, where $\phi(.)$ is a transfer function (in general, logistic or hyperbolic function).

Next, we create a matrix $Z = [z_1, z_2, \ldots, z_N]$ representing the output of the hidden neurons for each input feature vector x_i. Again, we insert a new first line composed of -1 into Z (for bias) and the objective is to solve $D = MZ$, where M represents the weights of the output neurons. The matrix M can be easily obtained after some simple matrix operations, according to the following equation

$$M = DZ^T(ZZ^T)^{-1}. \tag{2}$$

2.1 Randomized Neural Network Texture Signature

The random neural network texture signature is proposed in the paper [21] and consists of using image pixels as input and label data in order to train a randomized neural network. Next, the weights of the output neuron layer of this trained network are used as the image signature. For this purpose, the image is divided into overlapping windows $K \times K$ ($K = \{3, 5, 7\}$). For each window, its border pixels are used as input feature vector x_i and its central pixel is used as the respective scalar label d_i. Thus, we have 8-, 16- and 24-dimensional feature vectors x_i for the aforementioned window sizes, respectively.

The next step is to determine the values of the matrix W. For this, the paper [21] adopted the Linear Congruent Generator (LCG) [22, 23] to produce pseudorandom values in a uniform distribution. The parameter values for the "seed" and other adjustment parameters are based on the value Q (number of hidden neurons) and p (dimensionality of the input feature vector). All the values of W and each line of the matrix X are normalized to have zero mean and unit variance. Finally, the logistic transfer function is used in all the neurons.

Once these fundamental procedures are determined, it is possible to construct two signatures based on Eq. 2, which becomes a vector $f = DZ^T(ZZ^T)^{-1}$ because D is also a vector. The first signature considers only one value Q for multiples values K, as follows

$$\Theta(Q)_{K_1, K_2, \ldots, K_n} = \left[f_{K_1}, f_{K_2}, \ldots, f_{K_n}\right]. \tag{3}$$

The second signature, which consists of the concatenation of the previous signature for different values Q, is determined according to the following equation

$$\Psi_{Q_1, \ldots, Q_m} = \left[\Theta(Q_1)_{K_1, \ldots, K_n}, \ldots, \Theta(Q_m)_{K_1, \ldots, K_n}\right]. \tag{4}$$

A detailed description of the randomized neural network based signature can be found in the paper [21].

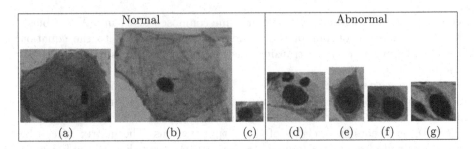

Fig. 1. Cells of the *pap-smear* database: (a) superficial squamous epithelial, (b) intermediate squamous epithelial, (c) columnar epithelial, (d) mild squamous non-keratinizing
dysplasia, (e) moderate squamous non-keratinizing dysplasia, (f) severe squamous non-
keratinizing dysplasia, (g) squamous cell carcinoma in situ intermediate [24].

3   Experiments

3.1   *Pap-smear* database

The *pap-smear* database [24] is a collection of 917 cell images extracted from cer-
vices. The images were obtained at the Herlev University Hospital and were classified into 7 groups, which are: normal superficial squamous epithelial (74 cells);
normal intermediate squamous epithelial (70 cells); normal columnar epithelial
(98 cells); mild squamous non-keratinizing dysplasia (182 cells); abnormal moderate squamous non-keratinizing dysplasia (146 cells); abnormal severe squamous
non-keratinizing dysplasia (197 cells); and abnormal squamous cell carcinoma in
situ intermediate (150 cells). These cell images can also be classified into two
groups: normal cells (242 images) and abnormal cells (675 images). In our experiments, all the images were converted into grayscale. Moreover, we addressed
only the 2-class problem, since the 7-class problem is still a challenge for texture
analysis methods. Figure 1 shows one sample of each class.

3.2   Classification Procedure

In the randomized neural network texture signature, we used the same parameter
values adopted in the paper [21], that is, $Q = \{19, 39\}$, $K = \{3, 5, 7\}$ for the
second signature (Eq. 4) in order to establish a fair comparison with the other
texture methods, in which we used parameter values according to either their
respective papers or the common use. At this point, it is important to mention
that, even though the paper [21] proposes a strategy to make the method more
robust to rotation, we did not use it for two reasons: first, there is no orientation
in the *pap-smear* cells; second, the method is faster without this strategy.

In order to assess the performance of the method, we compare it to other classical and recent texture analysis. They are: Co-occurrence matrices [5], Wavelets
descriptors [25,26], Tourist Walk [27], Discrete Cosine Transform (DCT) [28],
Lacunarity 3D [29], Local binary patterns (LBP) [9], Gray Level Difference
Matrix (GLDM) [30,31] and Complex Network Texture Descriptor (CNTD) [32].

For classification, we used the Linear Discriminant Analysis [33], which is a classical statistical classifier that creates hyperplanes among the groups based on the their centroid vectors and the covariance matrix of all the samples. As strategy validation, we adopted the leave-one-out cross-validation, which uses one sample for testing the remainder for training. This process is repeated N times (N is the number of samples), each time with a different sample for testing. The performance measure is the average of the N accuracies.

We also obtained the AUC (Area Under the ROC Curve) [34] to compare it to the highest AUC values obtained in two recent papers [11,35], which compared several LBP variants applied to the *pap-smear* database. To assess the randomized neural network signature, we used the same procedures present in these two works: a Linear Support Vector Machine (SVM) as classifier and the 5-fold cross-validation. The paper [11] does not mention the parameter values used, but the paper [35] uses the default parameter values of the LIBSVM [36], which is a public library for SVM. Thus, for a fair comparison, we also used the default parameter values of this library. Moreover, because 5-fold-cross-validation is not a deterministic strategy, we performed 101 validation runs and adopted the median AUC value as the performance measure of the randomized neural network signature.

4   Results and Discussion

Table 1 shows the comparison of the randomized neural network signature with other grayscale texture analysis methods. As one can see, the neural network approach surpasses all the compared methods in terms of accuracy. One disadvantage of the method is its excessively large number of descriptors. However, it is important to notice that its accuracy is 1.20% superior to the second best method (wavelet descriptors). This percentage represents 11 more images correctly classified by the method, thus corroborating its efficiency and ability to discriminate *pap-smear* samples, a challenging database in which any improvement is desirable.

Table 2 shows the median AUC obtained by the neural network signature and all the compared approaches, as well as the highest AUC values present in two recent papers. As it is possible to notice, the randomized neural network signature obtained the second best result among all the methods. Although this performance is already impressive, it is important to emphasize that ENS and MAG1 provide the highest AUC values of the papers [11] and [35], respectively. Thus, considering that the paper [11] applied nine LBP variants to the *pap-smear* database, and the paper [35] performed more than 50 tests on this same dataset, our obtained result acquires an even higher perspective and demonstrates that the randomized neural network descriptors are very discriminative in *pap-smear* cell images.

Table 1. Comparison of different texture analysis methods applied to the *pap-smear* database.

Methods	No of descriptors	Accuracy (%)
Randomized neural network signature	180	87.57
Wavelet descriptors	18	86.37
Tourist walk	48	85.82
Co-occurrence matrices	16	79.83
DCT	8	78.63
Lacunarity 3D	12	78.08
LBP	54	83.53
GLDM	60	83.10
CNTD	108	81.90

Table 2. Comparison of AUC (Area Under the ROC Curve) of different texture analysis methods applied to the *pap-smear* database

Methods	AUC
Randomized neural network signature	0.8983
ENS (Highest AUC in [11])	0.8840
MAG1 (Highest AUC in [35])	0.9080
Wavelet descriptors	0.8628
Tourist walk	0.8452
Co-occurrence matrices	0.7594
DCT	0.8280
Lacunarity 3D	0.7996
LBP	0.8642
GLDM	0.8759
CNTD	0.8637

5   Conclusion

This paper presented the application of a very discriminative texture analysis method to the highly relevant medical problem of classifying *pap-smear* cells. The randomized neural network texture signature obtained a high performance in this problem, surpassing all the compared methods (LDA experiment) and presenting the second best AUC value, which is comparable to the highest results of two recent papers that address the same problem. Thus, it is possible to affirm that the randomized neural network signature is suitable for the *pap-smear* problem, and, therefore, adds a new tool to the computer vision research focused on the Papanicolaou test.

Acknowledgments. Jarbas Joaci de Mesquita Sá Junior thanks CNPq (National Council for Scientific and Technological Development, Brazil) (Grant: 152054/2016-2 and 453835/2017-1) for the financial support of this work. André R. Backes gratefully acknowledges the financial support of CNPq (Grant #302416/2015-3) and FAPEMIG (Foundation to the Support of Research in Minas Gerais) (Grant #APQ-03437-15). Odemir M. Bruno gratefully acknowledges the financial support of CNPq (307797/2014-7 and 484312/2013-8) and FAPESP (14/08026-1).

References

1. Backes, A.R., Martinez, A.S., Bruno, O.M.: Texture analysis based on maximum contrast walker. Pattern Recogn. Lett. **31**(12), 1701–1707 (2010)
2. Kaplan, L.M.: Extended fractal analysis for texture classification and segmentation. IEEE Trans. Image Process. **8**(11), 1572–1585 (1999)
3. Bhosle, V.V., Pawar, V.P.: Texture segmentation: different methods. Int. J. Soft Comput. Eng. **3**, 69–74 (2013)
4. Zwiggelaar, R.: Texture based segmentation: automatic selection of co-occurrence matrices. In: ICPR, vol. I, pp. 588–591 (2004)
5. Haralick, R.M.: Statistical and structural approaches to texture. Proc. IEEE **67**(5), 786–804 (1979)
6. Manjunath, B.S., Ma, W.Y.: Texture features for browsing and retrieval of image data. IEEE Trans. Pattern Anal. Mach. Intell. **18**(8), 837–842 (1996)
7. Dawood, H., Dawood, H., Guo, P.: Efficient texture classification using short-time Fourier transform with spatial pyramid matching. In: SMC, pp. 2275–2279. IEEE (2013)
8. Li, C., Huang, Y., Zhu, L.: Color texture image retrieval based on Gaussian copula models of Gabor wavelets. Pattern Recogn. **64**, 118–129 (2017)
9. Ojala, T., Pietikäinen, M., Mäenpää, T.: Multiresolution gray-scale and rotation invariant texture classification with local binary patterns. IEEE Trans. Pattern Anal. Mach. Intell. **24**(7), 971–987 (2002)
10. Sá Junior, J.J.M., Backes, A.R.: A simplified gravitational model to analyze texture roughness. Pattern Recogn. **45**(2), 732–741 (2012)
11. Nanni, L., Lumini, A., Brahnam, S.: Local binary patterns variants as texture descriptors for medical image analysis. Artif. Intell. Med. **49**(2), 117–125 (2010)
12. Ananthi, V.P., Balasubramaniam, P., Kalaiselvi, T.: A new fuzzy clustering algorithm for the segmentation of brain tumor. Soft. Comput. **20**(12), 4859–4879 (2016)
13. Li, X.Z., Williams, S., Bottema, M.J.: Texture and region dependent breast cancer risk assessment from screening mammograms. Pattern Recogn. Lett. **36**, 117–124 (2014)
14. Li, X.Z., Williams, S., Bottema, M.J.: Background intensity independent texture features for assessing breast cancer risk in screening mammograms. Pattern Recogn. Lett. **34**(9), 1053–1062 (2013)
15. Capdehourat, G., Corez, A., Bazzano, A., Alonso, R., Musé, P.: Toward a combined tool to assist dermatologists in melanoma detection from dermoscopic images of pigmented skin lesions. Pattern Recogn. Lett. **32**(16), 2187–2196 (2011)
16. Flores, E.S., Scharcanski, J.: Segmentation of melanocytic skin lesions using feature learning and dictionaries. Expert Syst. Appl. **56**, 300–309 (2016)

17. Schmidt, W.F., Kraaijveld, M.A., Duin, R.P.W.: Feedforward neural networks with random weights. In: Proceedings of the 11th IAPR International Conference on Pattern Recognition, Conference B: Pattern Recognition Methodology and Systems, vol. II, pp. 1–4 (1992)
18. Pao, Y.H., Takefuji, Y.: Functional-link net computing: theory, system architecture, and functionalities. IEEE Comput. J. **25**(5), 76–79 (1992)
19. Pao, Y.H., Park, G.H., Sobajic, D.J.: Learning and generalization characteristics of the random vector functional-link net. Neurocomputing **6**(2), 163–180 (1994)
20. Huang, G.B., Zhu, Q.Y., Siew, C.K.: Extreme learning machine: theory and applications. Neurocomputing **70**(1–3), 489–501 (2006)
21. Sá Junior, J.J.M., Backes, A.R.: ELM based signature for texture classification. Pattern Recogn. **51**, 395–401 (2016)
22. Lehmer, D.H.: Mathematical methods in large scale computing units. Ann. Comput. Lab. Harvard Univ. **26**, 141–146 (1951)
23. Park, S.K., Miller, K.W.: Random number generators: good ones are hard to find. Commun. ACM **31**(10), 1192–1201 (1988)
24. Jantzen, J., Norup, J., Dounias, G., Bjerregaard, B.: Pap-smear benchmark data for pattern classification. In: Proceedings of the NiSIS 2005, NiSIS, pp. 1–9 (2005)
25. Daubechies, I.: Ten Lectures on Wavelets. Society for Industrial and Applied Mathematics, Philadelphia (1992)
26. Chang, T., Kuo, C.J.: Texture analysis and classification with tree-structured wavelet transform. IEEE Trans. Image Process. **2**(4), 429–441 (1993)
27. Backes, A.R., Gonçalves, W.N., Martinez, A.S., Bruno, O.M.: Texture analysis and classification using deterministic tourist walk. Pattern Recogn. **43**(3), 685–694 (2010)
28. Ng, I., Tan, T., Kittler, J.: On local linear transform and Gabor filter representation of texture. In: International Conference on Pattern Recognition, pp. 627–631 (1992)
29. Backes, A.R.: A new approach to estimate lacunarity of texture images. Pattern Recogn. Lett. **34**(13), 1455–1461 (2013)
30. Weszka, J.S., Dyer, C.R., Rosenfeld, A.: A comparative study of texture measures for terrain classification. IEEE Trans. Syst. Man. Cybern. **6**(4), 269–285 (1976)
31. Kim, J.K., Park, H.W.: Statistical textural features for detection of microcalcifications in digitized mammograms. IEEE Trans. Med. Imaging **18**(3), 231–238 (1999)
32. Backes, A.R., Casanova, D., Bruno, O.M.: Texture analysis and classification: a complex network-based approach. Inf. Sci. **219**, 168–180 (2013)
33. Webb, A.R.: Statistical Pattern Recognition, 2nd edn. Wiley, Chichester (2002)
34. Fawcett, T.: An introduction to ROC analysis. Pattern Recogn. Lett. **27**(8), 861–874 (2006)
35. Nanni, L., Lumini, A., Brahnam, S.: Survey on LBP based texture descriptors for image classification. Expert Syst. Appl. **39**(3), 3634–3641 (2012)
36. Chang, C.C., Lin, C.J.: LIBSVM: A library for support vector machines. ACM Trans. Intell. Syst. Technol. **2**, 27:1–27:27 (2011)

Estimation of Pedestrian Height Using Uncalibrated Cameras

Alejandro Valdés-Camejo, Guillermo Aguirre-Carrazana,
Raúl Alonso-Baryolo, Annette Morales-González$^{(\boxtimes)}$ [iD],
Francisco J. Silva-Mata, and Edel García-Reyes

Advanced Technologies Application Center, 7a # 21812 b/ 218 and 222,
Rpto. Siboney, Playa, 12200 Havana, Cuba
{avaldes,gaguirre,rbaryolo,amorales,fsilva,egarcia}@cenatav.co.cu

Abstract. The height of a person can be used as a Soft Biometrics feature in surveillance scenarios. The automatic estimation of pedestrian height have been addressed mostly on calibrated cameras. We are proposing a new method for real height estimation in videos from uncalibrated cameras. Our proposal computes the horizon line within the scene and then, the relative height of each person is obtained. We employ the real height distribution of a population to provide the final height value. In this process, it has been included an evaluation of the silhouette's quality, in order to improve the results. Experiments were conducted in uncontrolled scenarios, showing a good performance of our method.

Keywords: Height estimation · Video surveillance
Silhouette quality · Shape matching · Uncalibrated camera

1 Introduction

Detection, localization and identification of people are some of the hardest problems in video-surveillance research field with an increasing interest in last years. Currently, the use of physical and behavioral traits has been introduced as a recognition technique which is known as Soft Biometrics [1]. This can be used to filter large amounts of data or to identify and re-identify individuals.

This paper focuses on determining human height, which is a geometric information that can be automatically estimated in scenarios from video-surveillance. Unlike biometric features, this information can be obtained from videos taken at long distances where people walk in any directions. Height estimation of people can be used both as Soft Biometrics and as a person tracking feature. The first case eliminates some possible subjects having considerably different heights than the target, and focus on determining distinctive identification features. It can be used for temporal and spatial correspondence analysis for person tracking.

A significant number of works tackle the problem of object height estimation. Some of them are based on calibrated cameras and they use their intrinsic and extrinsic parameters for estimation [2,3]. On the other hand, there are methods based on uncalibrated cameras [4–8]. Generally, these methods perform a

© Springer International Publishing AG, part of Springer Nature 2018
M. Mendoza and S. Velastín (Eds.): CIARP 2017, LNCS 10657, pp. 685–692, 2018.
https://doi.org/10.1007/978-3-319-75193-1_82

camera's auto-calibration based on scene geometry or object tracking [6,9]. The main problem of these methods is that they require information about extrinsic parameters of the camera or measures in 3D world as reference for the estimation.

We propose a new method for estimating real pedestrian height using uncalibrated cameras. This work is based on the algorithm of Richardson et al. [10] who proposed a method to estimate the image horizon from multiple tracked objects and an estimation of relative height. Our differences w.r.t that work, and main contributions are: (1) we introduce an algorithm to evaluate the human silhouette quality within the processes of horizon detection and height estimation, and (2) a new method to estimate real human height, that obtains the real height value of pedestrians in uncalibrated cameras without using any prior knowledge about the scene geometry or camera parameters.

2   Estimation of the Horizon and Relative Height

An important step for object height estimation is the image horizon estimation. For this task, Richardson et al. [10] established the condition that the camera X-axis must be parallel to the ground plane of the 3D-world, which implies that the horizon can be uniquely defined by a single value in the image Y-axis. Nevertheless, if this condition is not met, they proposed a method to find the angle between the camera's X-axis and the ground plane in order to rotate the video. Also a flat ground plane is assumed. Under this condition, when an object moves in a way that its position in the image Y-axis ascends, it becomes smaller. The image horizon is defined as the position where the moving object becomes infinitesimally small.

Richardson et al. [10] demonstrated that a linear relationship exists between the vertical image position (y_0) and its height (h_0) on the image (See Fig. 1b). In other words, for a moving object in the scene, the equation $y = mh + n$ (height equation) is met. Parameters m, n from this equation can be estimated by performing a linear regression from a set of values $\{(h_i, y_i)\}$ extracted from the track of a moving object (See Fig. 1a). In the equation, n is the vertical position in the image where the object's height becomes zero. Therefore, n represents the horizon according to the definition (See Fig. 1b).

The information obtained from multiple tracked objects is necessary for the horizon estimation since the tracking points from an object can be partially affected by occlusion, segmentation, identity changes and others. A robust method based on the Hough transform is presented in [10], as a voting procedure. In the height equation the value of n obtained for each object is considered as a vote. A histogram is created with these values on the y axis of the image. The resulting probability function for the horizon will have a sharp peak at y_h (the most voted value).

If the image horizon (y_h) is found, for a pair (y_i, h_i) from a person, it is possible to estimate this person's height in pixels for all possible positions using the line that joins $(0, y_h)$ to (y_i, h_i) (height line). Being A and B two objects at the scene, a relationship between objects' height and the slope of their height

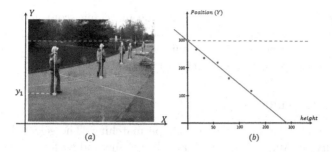

(a)   (b)

Fig. 1. (a) Multiple instances of a tracked person in a video scene. For each instance, a line is plotted representing the size of the object and their position (the lower point of the line). (b) Graphic of size vs. position shows the measurements extracted for the persons and their linear relationship.

lines can be established: A is higher than B if and only if the line's slope of A is bigger than the line's slope of B. Note that, if A is larger than B (See Fig. 2b) then for two points $p_a(x_A, y_1)$ and $p_b(x_B, y_1)$ at y_1 position, the relationship $(y_1 - y_h)/x_B < (y_1 - y_h)/x_A$ is met. The backward implication is similar. The slopes of these lines are a good estimation value of its relative height within the scene.

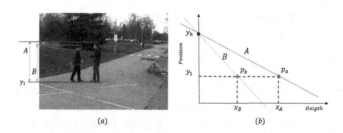

(a)   (b)

Fig. 2. (a) Instances from two pedestrians of different sizes at the same position. (b) Graphic of height vs position showing the relationship between the height and position for these two pedestrians.

3   Silhouette Quality Process

In the work of Richardson et al. [10], all the points extracted from each moving object in the video, are used to estimate the horizon and height. They consider several types of objects: cars, people, animals, among others. People are the only interest in our work. Also, as was mentioned before, the points might be affected by segmentation and tracking problems which cause significant variations on the line parameters. For those reasons, in the present work we propose to introduce a process that allows to select silhouettes belonging to people. The process consists on a shape matching algorithm to compare the frame silhouettes to a set of prototypes. As a result, depending on the matching score, it is possible to select several person's silhouettes with good quality. In our case, a good quality silhouette must be well segmented (with no missing body parts). For horizon and height estimation we only consider these good silhouettes.

3.1   Shape Matching Algorithm

Currently, there are many works for shape matching. The selected method is known as Turning Angle [11]. This is an efficient method with good results, as it is shown in the survey conducted in [12]. Also, in our case, since the persons are always standing, this method can be even faster by omitting the starting point selection in the matching process. In this method, a description of the shape is created based on its contour. The angle between the shape slope at a given point and the X-axis is calculated for each contour point in an ordered way. The shape representation is defined as the sequence of contour angles (Fig. 3). To compare two representations the Dynamic Time Warping method is used.

Fig. 3. Representation of the *Turning Angle* method.

3.2   Prototype Selection

The silhouettes from CASIA-B database [13] were used for prototype selection. In this database there are people walking in 11 different angles under controlled environments. We selected 20 people ensuring the presence of men and women. Then, an algorithm for clustering was used to split the silhouette sets into groups according to their similarity and we selected its representative element as prototype. The algorithm used was k-medoids, since it is simple, fast and provides a representative object for each cluster. The distance among silhouettes used for the k-medoids algorithm was the edit distance. It was calculated over Freeman chains extracted from silhouettes contours. We selected 50 prototypes (Fig. 4). This number was chosen as a trade-off between having enough samples and faster processing times.

Fig. 4. Prototypes examples

4 Estimation of the Real Height

The real height estimation proposed in this work is based on the relative height sampling and its comparison with a known height distribution. For example, it is well known that the Cuban population heights has a normal distribution with $\mu = 1.68$ m for men and $\mu = 1.56$ m for women [14]. The overall mean is 1.61 m.

Once the image horizon is found, we propose to store a certain amount of relative heights calculated from good silhouettes. Then, the normal distribution parameters are calculated. Under the assumption that relative heights follow the same distribution that real height, we propose to make a distribution match by their means. The real height of a person will be the corresponding value in the real height distribution given that person's relative height.

5 Experimental Results

There is not a unified criterion in literature for evaluating the results in this topic. Moreover, there are no public datasets or real height data to compare with other works. Authors usually present the results in their own databases under controlled environments [5,15].

Fig. 5. Results of the proposed method on a video from PETS 2009 dataset [16]. (a) Results when not using the silhouette quality step and (b) results using the silhouette quality step. The green line represents the horizon estimation and the pink marks to the left of each image are the horizon votes. Bounding boxes for each tracked object are shown. The bounding box is blue if the blob silhouette is selected for estimation (good human silhouette) and pink if not. (Best viewed in color)

The accuracy in the horizon estimation is very subjective, the exact value depends on a technical opinion. Nevertheless, in the majority of the analyzed videos, the estimated value falls within a neighborhood close to the real apparent horizon. The influence of introducing the silhouette's quality process can be observed in Fig. 5. The horizon estimation in Fig. 5b shows better results than Fig. 5a and the vote distribution is more compact. Note that, in the case of

Fig. 5b, the bounding boxes corresponding to two superposed persons, or to an incomplete person, were classified as bad, and not selected for estimation.

To test the height estimation we captured several videos under uncontrolled scenarios. The video scene is a parking lot where people walk towards the camera (See Fig. 6). The camera resolution is 625 × 500. To estimate the horizon and relative height distribution we considered the first 40 people in the video. Then, the height estimations of other 30 people were compared with their measured heights. People were measured wearing shoes. The results are shown in Table 1 split into men and women.

Fig. 6. Video sequence from our dataset and our height estimation

Table 1. Results of the estimation algorithm (1) for men and (2) for women: In table: E.H is estimated height; M.H. is measured height and Diff is the difference between them

Person	E.H. (m)	M.H. (m)	Diff(cm)
1	1.73	1.73	0
2	1.70	1.67	3
3	1.73	1.73	0
4	1.74	1.72	2
5	1.80	1.79	1
6	1.80	1.81	1
7	1.75	1.74	1
8	1.76	1.74	2
9	1.80	1.81	1
10	1.81	1.78	3
11	1.78	1.76	2
12	1.81	1.80	1
13	1.73	1.71	2
14	1.75	1.75	0
15	1.75	1.76	1
16	1.71	1.70	1
17	1.72	1.70	2
18	1.71	1.71	0
19	1.75	1.72	3
20	1.82	1.81	1

(1)

Person	E.H. (m)	M.H. (m)	Diff(cm)
1	1.60	1.61	1
2	1.74	1.70	4
3	1.65	1.64	1
4	1.69	1.66	3
5	1.66	1.64	2
6	1.62	1.60	2
7	1.66	1.66	0
8	1.57	1.59	2
9	1.67	1.63	4
10	1.69	1.69	0

(2)

The automatic estimations for the 30 people are in a difference range of 4 cm w.r.t. the real value. This corresponds to a 100% of accuracy if we consider an error of ±5 cm. Also in the range of difference of 1 cm there are 16 out of 30 correct measurements (53%). The range of height estimation was [1.60–1.82]. The overall mean error was 1.53 cm. The mean error for men was 1.35 cm and for women 1.90 cm. In our model, 90% of the estimated values variability is explained by the linear relation with observed values. The F statistic value is 166.81 and is statistically significant with p = 0.0001. There is evidence to refuse the null hypothesis and thus it is possible to set up a model from linear regression. This method was tested in a PC Intel Core I7 with 8 GB RAM. The processing time of 1 frame was 25 ms, therefore, we are able to process 40 frames per second, achieving real time conditions.

In order to understand these results we provide further considerations. First, human walking is a cyclic set of different poses. When legs are open the height is lower than when they are closed. For estimation we considered the values corresponding to the pose where the person appears more straight. Second, the height of blobs is affected by footwear and hairstyle, that explains why the error in women is a slightly larger than in men. Some of the women used for testing were wearing high shoes or up hairstyles. The measurement of the real height and the capture of the test videos were done in different dates. Despite the quality method, some errors can occur. If parts of a silhouette are selected as good, it decreases the distribution mean and tends to overestimate the heights.

The main limitation of the present work is that it needs a large number of people for a good approximation of the population height distribution. If the number is too small then there is no guarantee that the sample mean will be close to the population mean. Nevertheless, the distribution parameters computation can be updated with the relative height of each new tracked person that appears in the scene. Also, the view angle must be oblique enough to notice differences on the silhouette height when its position changes.

6   Conclusions

In this work we presented a method to estimate the height of pedestrians in uncalibrated cameras. The method does not require information about the scene or camera parameters. The height estimation was based on existing techniques, but this work incorporates a step for evaluating human silhouette quality. This helps to overcome some segmentation problems that affect the results. The accuracy of the proposed method has been experimentally shown in outdoor environments, under uncontrolled lighting conditions. An error mean of 1.53 cm was achieved for 30 test subjects from our database. We believe that real height estimation can be a very useful feature for people tracking, and in future work it can be used as a step for camera calibration.

References

1. Dantcheva, A., Velardo, C., Dangelo, A., Dugelay, J.L.: Bag of soft biometrics for person identification. Multimedia Tools Appl. **51**(2), 739–777 (2011)
2. Hansen, D.M., Mortensen, B.K., Duizer, P.T., Andersen, J.R., Moeslund, T.B.: Automatic annotation of humans in surveillance video. In: Fourth Canadian Conference on Computer and Robot Vision, CRV 2007, pp. 473–480. IEEE (2007)
3. Kispál, I., Jeges, E.: Human height estimation using a calibrated camera. In: Proceedings of the CVPR (2008)
4. Arigbabu, O.A., Ahmad, S.M.S., Adnan, W.A.W., Yussof, S., Iranmanesh, V., Malallah, F.L.: Estimating body related soft biometric traits in video frames. Sci. World J. **2014** (2014)
5. Jung, J., Kim, H., Yoon, I., Paik, J.: Human height analysis using multiple uncalibrated cameras. In: 2016 IEEE International Conference on Consumer Electronics (ICCE), pp. 213–214. IEEE (2016)
6. Jung, J., Yoon, I., Lee, S., Paik, J.: Object detection and tracking-based camera calibration for normalized human height estimation. J. Sens. **2016**, 8347841:1–8347841:9 (2016)
7. Criminisi, A., Zisserman, A., Van Gool, L., Bramble, S., Compton, D.: A new approach to obtain height measurements from video. In: Proceedings of SPIE, vol. 3576 (1998)
8. De Angelis, D., Sala, R., Cantatore, A., Poppa, P., Dufour, M., Grandi, M., Cattaneo, C.: New method for height estimation of subjects represented in photograms taken from video surveillance systems. Int. J. Legal Med. **121**(6), 489–492 (2007)
9. O'Gorman, L., Yang, G.: Orthographic perspective mappings for consistent wide-area motion feature maps from multiple cameras. IEEE Trans. Image Process. **25**(6), 2817–2832 (2016)
10. Richardson, E., Peleg, S., Werman, M.: Scene geometry from moving objects. In: 2014 11th IEEE International Conference on Advanced Video and Signal Based Surveillance (AVSS), pp. 13–18. IEEE (2014)
11. McConnell, R., Kwok, R., Curlander, J.C., Kober, W., Pang, S.S.: ψ-s correlation and dynamic time warping: two methods for tracking ice floes in SAR images. IEEE Trans. Geosci. Remote Sens. **29**(6), 1004–1012 (1991)
12. Mingqiang, Y., Kidiyo, K., Joseph, R.: A Survey of Shape Feature Extraction Techniques. INTECH Open Access Publisher (2008)
13. Yu, S., Tan, D., Tan, T.: A framework for evaluating the effect of view angle, clothing and carrying condition on gait recognition. In: 18th International Conference on Pattern Recognition, ICPR 2006, vol. 4, pp. 441–444. IEEE (2006)
14. Centro de estudios de población y desarrollo, Cálculos de peso y talla promedio de la población por provincias y Cuba, Oficina Nacional de Estadística (2008)
15. Shao, J., Zhou, S.K., Chellappa, R.: Robust height estimation of moving objects from uncalibrated videos. IEEE Trans. Image Process. **19**(8), 2221–2232 (2010)
16. Ferryman, J., Shahrokni, A.: Pets2009: dataset and challenge. In: 2009 Twelfth IEEE International Workshop on Performance Evaluation of Tracking and Surveillance (PETS-Winter), pp. 1–6. IEEE (2009)

Semi-supervised Online Kernel Semantic Embedding for Multi-label Annotation

Jorge A. Vanegas[1]($\boxtimes$), Hugo Jair Escalante[2]⑩, and Fabio A. González[1]⑩

[1] MindLab Research Group, Universidad Nacional de Colombia, Bogotá, Colombia
{javanegasr,fagonzalezo}@unal.edu.co
[2] Instituto Nacional de Astrofísica, Optica y Electrónica (INAOE),
Puebla, Mexico
hugojair@ccc.inaoep.mx

Abstract. This paper presents a multi-label annotation method that uses a semantic embedding strategy based on kernel matrix factorization. The proposed method called Semi-supervised Online Kernel Semantic Embedding (SS-OKSE) learns to predict the labels of a document by building a semantic representation of the document features that takes into account the labels, when available. A remarkable characteristic of the algorithm is that it is based on a kernel formulation that allows to model non-linear relationships. The SS-OKSE method was evaluated under a semi-supervised learning setup for a multi-label annotation task, over two text document datasets and was compared against several supervised and semi-supervised methods. Experimental results show that SS-OKSE exhibits a significant improvement, showing that a better modeling can be achieved with an adequate selection/construction of a kernel input representation.

Keywords: Semantic representation · Semi-supervised learning
Learning on a budget · Multi-label annotation

1 Introduction

The automatic multi-label annotation problem presents several applications in areas as diverse as text and music categorization and classification, semantic labeling of images and videos, medical diagnosis, and functional genomics, among others [11]. Several methods transform the problem of multi-label learning to a conventional classification problem (i.e., a set of binary classification problems solved independently). Unfortunately this kind of approaches present two principal drawbacks, first, usually they do not scale well when the number of labels and/or instances increase, and second, these approaches do not have into account the possible strong correlations between the labels. Another important issue is that these approaches require a significant amount of labeled data to achieve a reasonable generalization performance. In multi-label learning, this issue is more evident than in single-class classification since manually assigning multiple labels

© Springer International Publishing AG, part of Springer Nature 2018
M. Mendoza and S. Velastín (Eds.): CIARP 2017, LNCS 10657, pp. 693–701, 2018.
https://doi.org/10.1007/978-3-319-75193-1_83

is more time-demanding than assigning unique global labels. A possible strategy to deal with a large number of labels and the lack of annotated instances is to find a compact representation of them by using, for instance, a dimensionality reduction method. This approach is followed by multi-label latent space embedding methods, which have recently shown competitive results. There are several strategies to construct the latent semantic space, and most of them proposed supervised and semi-supervised extensions. The important thing about this kind of methods, is that this compact semantic representation also can be modeled by the discriminative structure of not only labeled but also unlabeled data. In this paper, we present a method for multi-label annotation based on semantic embedding that finds a common semantic space based on the kernel feature representation of an instance and its corresponding labels that model a mapping between the feature representation and the annotation labels. The proposed method has three important characteristics: (1) the method is formulated as a semi-supervised learning algorithm that learns to construct a common semantic representation not only from labeled instances but also from unlabeled ones, (2) despite being based on kernels, the method scales well to deal with large datasets thanks to a budget restriction which allows tackling one of the main problems of kernel-based methods, that is the scalability in terms of number of training instances, and (3) the method is formulated as an on-line learning algorithm, based on stochastic gradient descent, which allows it to deal with large collections of data, achieving a significant reduction in memory requirements and computational load. Additionally, the latter characteristic allows the efficient implementation of the method in dataflow GPU frameworks such as Theano and TensorFlow, which are used for efficient training and simulation of deep neural networks.

The rest of this paper is organized as follows: Sect. 2 discusses the related work; Sect. 3 formally introduce the details of the proposed multi-label annotation method; Sect. 4 presents the experimental evaluation; and, finally, Sect. 5 presents some concluding remarks.

2   Multi-label Annotation Based on Semantic Embedding Methods

The existing methods for multi-label classification problems can be grouped into two main categories [11]: (1) problem transformation methods, which transform the multi-label classification problem into several single-label classification or regression problems and (2) algorithm adaptation methods, which extend specific learning algorithm in order to handle multi-label data directly. In the group of algorithm adaptation methods, we can find several adaptations to classical discriminative methods. For instance, Andrews et al. [1] proposed two extensions for Support Vector Machine (SVM) for multi-instance learning methods the miSVM and MISVM. miSVM treats instance labels as unobserved variables and maximizes the margin on instances. MISVM, in contrast, define a new concept of bag margin maximization that maximizes the difference between individual patterns. Unfortunately, these methods present two main problems: first,

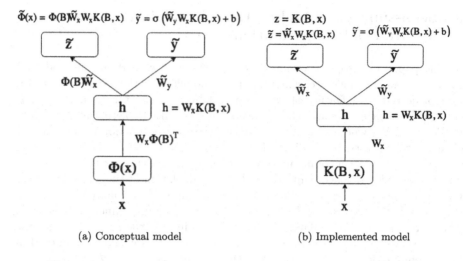

$$\tilde{\Phi}(x) = \Phi(B)\tilde{W}_x W_x K(B,x) \quad \tilde{y} = \sigma\left(\tilde{W}_y W_x K(B,x) + b\right)$$

$$\tilde{z} \qquad \tilde{y}$$

$$\Phi(B)\tilde{W}_x \qquad \tilde{W}_y$$

$$h \qquad h = W_x K(B,x)$$

$$W_x \Phi(B)^T$$

$$\Phi(x)$$

$$x$$

(a) Conceptual model

$$z = K(B,x)$$
$$\tilde{z} = \tilde{W}_x W_x K(B,x) \quad \tilde{y} = \sigma\left(\tilde{W}_y W_x K(B,x) + b\right)$$

$$\tilde{z} \qquad \tilde{y}$$

$$\tilde{W}_x \qquad \tilde{W}_y$$

$$h \qquad h = W_x K(B,x)$$

$$W_x$$

$$K(B,x)$$

$$x$$

(b) Implemented model

Fig. 1. Conceptual model and the actual implementation of SS-OKSE.

Kernel-based methods usually do not scale well due to the high computational complexity caused by the kernel matrix that grows quadratically with the number of training instances, and second, these methods cannot be extended for a semi-supervised learning and require ground-truth labels for all training documents. Topic models can overcome the second problem by modeling a compact semantic representation that can be modeled only by the discriminative structure of the data. Classical semantic embedding models are usually unsupervised, but several extensions to add supervision have been proposed, for instance, several matrix factorization based methods have been extended to improve the semantic representation by taking advantage of label information [2,7]. Under the same approach, several probabilistic topic models have been extended. For instance, many works have extended the classical Latent Dirichlet Allocation (LDA) [4] to add supervised and semi-supervised information, such as Semi-supervised LDA [8], Maximum Entropy Discrimination Latent Dirichlet Allocation (MedLDA) [12], Partially Labeled LDA (PLLDA) and more recently the Semi-supervised Multi-label Topic Model (MLTM) [9]. Probabilistic topic models like LDA have the advantage of incorporating prior knowledge to guide the topic modeling process to improve both the quality of the resulting topics and of the topic labeling, but unfortunately are very computational demanding, making them prohibited for large scale problems.

In this paper we propose the Semi-supervised Online Kernel Semantic Embedding (SS-OKSE) method that is formulated for large scale problems by tackling two main issues: first, the method presents a budget restriction that reduces the computational complexity caused by the kernel matrix, and second, the proposed method is formulated as an on-line learning algorithm which allows it to deal with large collections of data.

3   Semi-supervised Online Kernel Matrix Factorization for Multi-label Annotation

In this paper we propose a multi-label latent space embedding method that constructs an intermediate semantic space modeled by the input representation projected in a feature space generated by a kernel function, this strategy can be seen as a matrix factorization problem in the kernel feature space $(\Phi(X) \simeq F_\Phi H)$, where $X \in \mathbb{R}^{n \times l}$ describes the entire collection composed by l elements represented by $n-$dimensional vectors, $F_\Phi \in \mathbb{R}^{n \times r}$ is the basis matrix and $H \in \mathbb{R}^{r \times l}$ is the encoding matrix that represents all the input instances in a low $r-$dimensional space. Unfortunately, the calculation of this factorization is infeasible due to F_Φ depends explicitly on the mapping function $\phi(\cdot)$ (a mapping to a very highly dimensional space or even to an infinite-dimensional space). Therefore, instead of calculating directly F_Φ, we impose the restriction that the column vectors of F_Φ lie within the space of $\Phi(X)$, this is, F_Φ is composed of linear combinations of the X points in the feature space $(F_\Phi = \Phi(X)\tilde{W}_x)$.

$$\Phi(X) \simeq \Phi(X)\tilde{W}_x H, \quad \Phi(X) \approx \Phi(B)\tilde{W}_x H \tag{1}$$

This restriction avoid the necessity of evaluating the data in the feature space, additionally, only a reduced number $b \ll l$ of representative instances are used to construct the basis matrix (we construct a budget kernel matrix $B \in \mathbb{R}^{b \times l}$ instead of the full kernel matrix $X \in \mathbb{R}^{l \times l}$) This mitigates the high computational cost of constructing the huge kernel matrix. In this paper, we propose not to construct an explicit representation in the semantic space but learn a mapping from the feature representation to this semantic space, and again, using only the b representative points to model the restricted kernel feature space $(H = W_x \phi(B)^T \phi(X) = W_x K(B, X))$. In this manner, the model learns two transformations what allows to project the original data representation to the lower-dimensional semantic space and at the same time to reconstruct from this semantic representation the original data in the feature space.

$$\Phi(X) \approx \Phi(B)\tilde{W}_x W_x K(B, X) \tag{2}$$

Additionally to the original feature representation, we want the semantic representation to also lie the label representation $Y \in \mathbb{R}^{m \times k}$, where m is the total number of possible labels and k a reduced number of annotated instances (i.e., $k \ll l$), as follows:

$$Y \approx \sigma\left(\tilde{W}_y H\right) = \sigma\left(\tilde{W}_y W_x K(B, X)\right) \tag{3}$$

where $W_y \in \mathbb{R}^{m \times r}$ is another transformation matrix that projects from the semantic representation to the label space, and finally, an additional non-linear function σ is used to add more flexibility to the model. Putting all these restriction together, the final model can be represented as is shown in Fig. 1a.

Loss function. The final loss function to be minimized forces the feature reconstruction by defining a squared minimum error and learns the binary label reconstruction by imposing a binary cross entropy function:

$$\min_{W_x, \tilde{W}_x, W_y} J_i\left(W_x, \tilde{W}_x, W_y\right) = \frac{\alpha}{2} \left\| \Phi\left(x_i\right) - \Phi\left(B\right) \tilde{W}_x W_x K\left(B, x_i\right) \right\|_F^2$$

$$+ \beta \sum_{i=0}^{k} \log\left(1 + \exp\left(-y \cdot \tilde{W}_y W_x K\left(B, x_i\right)\right)\right)$$

$$+ \frac{\lambda_1}{2} \left\|W_x\right\|_F^2 + \frac{\lambda_2}{2} \left\|\tilde{W}_x\right\|_F^2 + \frac{\lambda_3}{2} \left\|\tilde{W}_y\right\|_F^2 \quad (4)$$

where α and β control the relative importance of reconstructing the feature and label representation, respectively, and $\lambda_{1,2,3}$ control the relative importance of the regularization terms, which penalize large values for the Frobenius norm of the transformation matrices. Unfortunately, solve this problem directly is infeasible due to the function $\phi(\cdot)$ performs a mapping to a highly-dimensional space or even to an infinite-dimensional space, but, the first term of the loss function can be rewritten in terms of kernel matrices and employ the kernel trick [6].

$$\frac{\alpha}{2} \left\| \Phi\left(x_i\right) - \Phi\left(B\right) \tilde{W}_x W_x K\left(B, x_i\right) \right\|_F^2 = \frac{\alpha}{2} \left(K\left(x_i, x_i\right) - 2K\left(x_i, B\right) \tilde{W}_x W_x K\left(B, x_i\right) + \right.$$

$$\left. K\left(B, x_i\right)^T W_x^T \tilde{W}_x^T K\left(B, B\right) \tilde{W}_x W_x K\left(B, x_i\right)\right)$$

Finally, we can make a change of variables a redefine the first term of the loss function ($\frac{\alpha}{2}\left(1 - 2z_i^T \tilde{z}_i + \tilde{z}_i^T K(B, B)\tilde{z}_i\right)$) and the structure (Fig. 1b), so that can be easily implemented in some deep learning framework.

Prediction. Once the parameters have been learned (coding and decoding matrices), we can use the model to predict the label representation $\tilde{y}$ from the feature representation x of a new unlabeled document by forward propagating it through the network.

Implementation details. The proposed method was implemented in the Keras [5] Framework, a high-level neural networks API written in Python and capable of running on top of either TensorFlow or Theano [10] libraries. The optimization is performed by stochastic gradient descent (SGD) with the RMSProp optimizer.

4   Experiments and Results

In this section, the proposed SS-OKSE algorithm will be evaluated in a multi-label annotation task under a semi-supervised setup. In order to compare our algorithm, we use the same experimental setup proposed by Soleimani and Miller [9], where the performance of our method is compared against several supervised (PLLDA, miSVM and MISVM) and semi-supervised (ssLDA, and MLTM) methods in two different datasets of text documents:

Delicious. This dataset is composed of tagged web pages from the social bookmarking site delicious.com [13]. We adopt the subset proposed in [9] where the top 20 common tags are used as a class labels, constructing a subset portioned in 8350 documents for training and 4000 documents for testing. The documents are represented in a bag-of-word representation composed by a codebook of 8500 unique words obtained after applying Porter stemming and stopword removal.

Ohsumed. This collection contains medical abstracts from the MeSH categories of the year 1991 [13]. The specific task in this dataset is to categorizing 23 cardiovascular diseases categories. It is composed by 11122 training and 5388 test documents. Almost half of the documents have more than one label.

Experimental setup. For a fair comparison for all methods, where in some of them a suitable selection of the threshold is not trivial, the ROC AUC (Area Under the Curve) is used as the evaluation metric. Micro-ROC and Macro-ROC AUC are reported separately. In Micro-ROC, TPR and FPR are computed globally. In Macro-ROC, the ROC AUC is computed for each class across all documents and then the average is taken over all classes. While Micro-ROC may be dominated the bigger classes, Macro-ROC gives equal weight to all classes and better reveals performance on rare classes. For each dataset, a $(1 - p)$ fraction is randomly selected from the training documents and their labels are removed. Then, for semi-supervised models, both labeled and unlabeled documents are used for training. But, for the purely supervised methods, only the remaining labeled documents are used for training. The annotation experiment is performed for different label proportions $p \in 0.01, 0.05, 0.1, 0.3, 0.6, 0.8, 0.9$ (five experiments are executed and the average is reported).

Determining the hyperparameters. Our model has 7 hyperparameters (α, β, λ_1, λ_2, λ_3, b, r). To properly determine the values of these hyperparameters, we randomly extract 20% of instances from the training set to validate the performance under a random (uniform) exploration in 30 different hyperparameter configurations trained with the remained 80% of training instances. The best configuration was chosen to be evaluated with the test partition. (this strategy have shown similar results than grid search while requires much fewer computation resources [3]).

Multi-label annotation performance. Figure 2a presents the performance in ROC AUC in the Delicious dataset for different proportions of labeled documents. As we can see, the proposed SS-OKSE using a linear kernel presents competitive results against the other semi-supervised algorithms, and using a histogram intersection (hi) kernel presents the best performance showing that a suitable kernel can help to model a good semantic space where non-labeled instances can be exploited. This becomes evident by achieving a performance very close to the maximum possible just with only 10% of labeled instances. Another important result is that this performance is obtained only with a budget composed by 500 instances randomly selected, this is only about the 6% of the training instances. Figure 2b presents the same previous experiment for the Delicious dataset. In Micro-ROC our method presents competitive results but the results in Macro-ROC suggest that the classes with the highest number of instances are dominating the learning. Still, it is important to emphasize that this result is obtained by using a budget of 2500 instances randomly selected (22% of the training instances) (Table 1).

Time-consuming and complexity. The main advantage of the SS-OKSE method is its highly scalable formulation. Table 2 shows the comparison between the proposed method and the state-of-the-art topic model method (we use the implementation supplied by the authors [9]). In both datasets SS-OKSE obtains remarkable speedups in comparison with MLTM under the same conditions (same computer with CPU device on a single core), but also thanks to its implementation, SS-OKSE is capable of running transparently in GPU devices achieving even more dramatic speedups, showing the ability of the proposed method to work on large scale datasets.

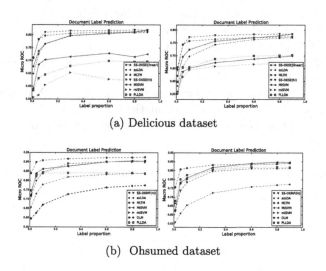

(a) Delicious dataset

(b) Ohsumed dataset

Fig. 2. Performance comparison in multi-label annotation.

Table 1. N: number of unique words, m: number of classes, l: number of instances, cardinality: average number of labels per instance.

			Training set		Test set	
Name	n	m	l	Cardinality	l	Cardinality
Delicious	8520	20	8251	2.89	3983	2.91
Ohsumed	13117	23	11122	1.65	5388	1.64

Table 2. Training time comparison (in minutes). For CPU device, only one core is using. Experiment running in a Intel(R) Xeon(R) CPU E5-2640 v3 @ 2.60 GHz with 16 cores, 64 GB RAM and an Nvidia Titan X.

Delicious					Ohsumed				
MLTM	SS-OKSE				MLTM	SS-OKSE			
CPU	CPU	Speedup	GPU	Speedup	CPU	CPU	Speedup	GPU	Speedup
961.3	43.5	22.1	5.1	186.9	1026	160.3	8.3	24.7	41.5

5 Conclusions

We have presented the SS-OKSE, a novel semi-supervised kernel semantic embedding method that uses a budget restriction to tackle the memory issue and computation time associated to kernel methods. The main advantage of using a budget in our method is that it allows us to save memory, since it is not necessary to store the complete kernel matrix, but a significantly smaller matrix defined by a budget keeping low computational requirements in large-scale problems. SS-OKSE is able to take advantage of annotated data to model a semantic low-dimensional space that preserves the separability of the original classes, and additionally, has the ability to exploit unlabeled instances for modeling the manifold structure of the data and use it to improve its performance in multi-label annotation tasks. The results confirm that the ability of the proposed method for modeling non-linearities can over-improve the performance in the multi-label annotation task.

Acknowledgment. Jorge A. Vanegas thanks for doctoral grant supports Colciencias 617/2013.

References

1. Andrews, S., Tsochantaridis, I., Hofmann, T.: Support vector machines for multiple-instance learning. In: Advances in Neural Information Processing Systems, pp. 577–584 (2003)
2. Beltrán, V., Vanegas, J.A., González, F.A.: Semi-supervised dimensionality reduction via multimodal matrix factorization. In: Pardo, A., Kittler, J. (eds.) Progress in Pattern Recognition, Image Analysis, Computer Vision, and Applications. LNCS, vol. 9423, pp. 676–682. Springer, Cham (2015). https://doi.org/10.1007/978-3-319-25751-8_81

3. Bergstra, J., Bengio, Y.: Random search for hyper-parameter optimization. J. Mach. Learn. Res. **13**, 281–305 (2012)

4. Blei, D.M., Ng, A.Y., Jordan, M.I.: Latent dirichlet allocation. J. Mach. Learn. Res. **3**(Jan), 993–1022 (2003)

5. Chollet, F.: Keras (2015). https://github.com/fchollet/keras

6. Chollet, F., et al.: Keras. GitHub (2015). https://github.com/keras-team/keras

7. Lee, H., Yoo, J., Choi, S.: Semi-supervised nonnegative matrix factorization. IEEE Sig. Process. Lett. **17**(1), 4–7 (2010)

8. Mcauliffe, J.D., Blei, D.M.: Supervised topic models. In: Advances in Neural Information Processing Systems, pp. 121–128 (2008)

9. Soleimani, H., Miller, D.J.: Semi-supervised multi-label topic models for document classification and sentence labeling. In: CIKM 2016, pp. 105–114. ACM (2016)

10. Theano Development Team. Theano: a Python framework for fast computation of mathematical expressions. arXiv e-prints, arXiv:1605.02688, May 2016

11. Tsoumakas, G., Katakis, I., Vlahavas, I.: Mining multi-label data. In: Maimon, O., Rokach, L. (eds.) Data Mining and Knowledge Discovery Handbook, pp. 667–685. Springer, Boston (2009). https://doi.org/10.1007/978-0-387-09823-4_34

12. Zhu, J., Ahmed, A., Xing, E.P.: Medlda maximum margin supervised topic models. J. Mach. Learn. Res. **13**(Aug), 2237–2278 (2012)

13. Zubiaga, A., García-Plaza, A.P., Fresno, V., Martínez, R.: Content-based clustering for tag cloud visualization. In: Social Network Analysis and Mining, ASONAM 2009, pp. 316–319. IEEE (2009)

Ultrasonic Backscatter Signal Processing Technique for the Characterization of Animal Lymph Node

Julian A. Villamarín[1], Daniela A. Montilla[2], Olga M. Potosi[1],
Luis F. Londoño[3], Fabian G. Muñoz[1], and Edgar W. Gutierrez[1(✉)]

[1] Antonio Nariño University, Popayán, Cauca, Colombia
{jvilla22, opotosi, fabian.munoz, edgargu}@uan.edu.co
[2] Valparaiso University, Valparaiso, Chile
daniela.montilla@postgrado.uv.cl
[3] Polytechnique Jaime Isaza Cadavid, Medellín, Antioquía, Colombia
lflondono@elpoli.edu.co

Abstract. Quantitative ultrasonic characterization of biological soft tissues has become in recent years an essential tool in the non-invasive-non-destructive assessment of physical properties of the microstructure of tissues, due to the potential for estimating acoustic parameters associated to density characteristics, distribution and heterogeneity of histological samples, as well as making the construction of improved quantitative images that support processes of clinical diagnosis. This paper presents the implementation of computational methods based on spectral analysis techniques for the construction of parametric ultrasonic images of animal suprascapular lymph node, which is an important tissue for the analysis of animal health risk or animal health. The computational algorithms were implemented based on the estimation of the acoustic attenuation coefficient dependent of the frequency and integrated backscatter coefficient (IBC). These computational procedures automatically processed 400 ultrasonic echoes acquired in a region of interest of 4 cm^2 for each sample of lymph node, which it was exposed to an incident ultrasonic field of 2.25 MHz with bandwidth of 1 MHz @ −3 dB. The results allowed parametric identification of nodule structures as germinal nodules, which are hardly identified in conventional qualitative ultrasound images. Finally ultrasonic parametric characterization of biological study samples provides potential quantitative indicators, which are so much accurate in the estimation of histonormality.

Keywords: Backscattering · Lymph node · Ultrasound

1 Introduction

The lymph nodes are soft structures surrounded of connective tissue, which comprises an important part of the immune system, acting as a mechanical filter of the lymph (extracellular fluid responsible of collecting and returning interstitial fluid to the blood), reducing the transit of pathogenic microorganisms and infectious agents, which can alter the immune system among others [1, 2]. Due to its direct contact with invading

© Springer International Publishing AG, part of Springer Nature 2018
M. Mendoza and S. Velastín (Eds.): CIARP 2017, LNCS 10657, pp. 702–709, 2018.
https://doi.org/10.1007/978-3-319-75193-1_84

pathogenic bacteria and microorganisms, the lymph node can be used as a biological reference structure to support the surveillance of animal health risks and consequently animal health [3]. In this context, the exploration of new tools aimed to the evaluation of the normality condition in biological tissue, especially in animal benefit plants, are a subject of great industrial, scientific and technological interest, due to the importance that it deserves the development of instruments and methods involved in the processes of risk control of zoonoses (infectious diseases or vectors that can be shared between animals and humans), in favor of preventing public health problems, like the overall purposes of the Food and Agriculture Organization of the United Nations (FAO) and the World Health Organization (WHO), in the control of diseases related to the development of biological products, food and products of animal origin, in addition to the eradication of zoonoses and the post mortem inspection of meat with commercial purposes [4]. Conventionally, the most common methods used to establish the normality of biological tissue animal and to infer in possible alterations are based on visual inspection processes, performed subjectively by a specialist in veterinary science, with procedures that include the analysis and verification of size, color and physical characteristics of the microstructure among others [5, 6]. Other procedures manual and expensive are the histological techniques used in the processing of biological tissue, in which, the disadvantage is the use of chemical reagents (hematoxylin-eosin) and invasive procedures such as tissue dehydration, staining, paraffin aggregation and microtome cutting [7–9]. In particular, the processing techniques of biological tissues have a negative effect on the environment, in addition to being susceptible to systematic error, for the variability intra-observer and inter-observer, which in some cases causes diagnostic errors. In this way, the current techniques and methods used in the histological characterization by animal benefit plants, are based in costly protocols framed in the regulations of the Ministry of Health and Social Protection, in the national context, showing the need to promote the development of support instrumentation for the diagnosis, prevention, surveillance and control of welfare of the animal health as sustainable support to public health.

In this sense, the implementation and exploration of alternative methods for the characterization of biological tissues on the basis of non-invasive non-destructive methods, has stimulated the development of techniques based in the exhibition of acoustic fields [10, 11], with potential advantages in performing analyzes, without altering the physical and chemical characteristics of tissues with an optimal relationship cost-benefit [12]. Beside this, several studies show a wide range of applications by the techniques of quantitative ultrasonic characterization, especially in soft tissues such as heart, liver, kidney among others, assuming advantages compared to tissue processing technologies like systems x-rays or technologies of characterization of laser (which presents limitations in the characterization of opaque medium with high concentrations of particles). In order to propose alternative instruments and methods to support tissue analysis, the present study evidences the implementation of a system of quantitative ultrasonic characterization of animal lymph node samples, that allows to estimate acoustic indicators associated to the normality of the tissue, with the digital processing of ultrasound signals of backscattering, based in techniques of spectral analysis and the estimation of parameters associated with the loss of acoustic energy and power of backscatter.

2 Theoretical Fundament

The ultrasound use as tools in the noninvasive and nondestructive characterization (low power), it has driven the analysis of tissue properties, heterogeneous structures and the assessment of liquids and materials. This is because the ultrasound pressure variations during propagation process in interrogated media provide records of ultrasound backscattering signals, which contribute to estimates of quantitative parameters associated to properties of the characteristic media such as density spatial distribution and concentration heterogeneities and so on.

The acoustic attenuation is the combination of physical effects such as reflection, dispersion and absorption during the mechanical propagation processes in elastic media waves. In procedures of acoustic tissue characterization describes the energy loss when it is exposed to an ultrasound field, proving that the attenuation estimation can be used like a quantitative indicator to differentiate states of histonormality in biological media. For purposes of this study, the absorption is not considered because the characterizations experiments were performed using low power and indeed there is no temperature increase of the sample study. Thus, the attenuation effects are strictly caused by reflection and scattering process related to acoustic impedance variations from spatial distribution in the media. The mathematical model that describes the acoustic attenuation is obtained from the general dÀlembert solution unidimensional [12, 13]. One important specific solution in the frequency domain from which the estimation of the attenuation can be derived as described in (1):

$$\log\left[\frac{S_f(f)}{S_i(f)}\right] = -\beta f^n x, \tag{1}$$

where the power law $\alpha = \beta f^n$ is the attenuation coefficient (expressed in dB/cm), which relates the frequency - dependent attenuation coefficient β (expressed in dB/cm MHz), the frequency f and the index n that describes the frequency dependence, usually considered linear for soft biological tissues. $S_f(f)$ is the power spectral density for ultrasonic backscatter signals from sample study and $S_i(f)$ is the power spectrum of the ultrasonic incident wave and x is the propagation medium distance, equivalent to twice x because the characterization methods is performed in pulse-echo mode.

On the other hand, the acoustic scattering is a phenomenon of multidirectional propagation caused by interaction mechanisms between incident waves and heterogeneity characteristics from the propagation medium [17]. For experimental effects of ultrasonic soft tissues characterization, the scattering processes can be described in terms of the backscattering coefficient $\sigma_b(f, x)$. This parameter is used for measuring the power of backscatter that generates a volume of biological tissue. Its mathematical model is expressed in (2) according to [13]:

$$\sigma_b(f, x) = \frac{R^2}{A_0 \Delta x} w(f, x) \tag{2}$$

where R is the focal length of the transducer, A_0 is the effective radiation area of the transducer, Δx is the propagation distance (tissue thickness) and $w(f, x)$ is the rate between the spectral average power density from backscattering ultrasonic signal on the sample of study and the average power spectrum from a reference medium (material with known acoustic properties). Considering that $\sigma_b(f,x)$ can be integrated into the bandwidth of the ultrasonic transducer, the integrated backscattering coefficient (IBC) is the most used parameter to describes structural features of tissues from the average of spectral energy backscattered per unit of volume and incident energy unit for a frequency range. The mathematical model is seen in (3):

$$IBC = \sum_{f=\min}^{f=\max} \frac{\sigma_b(f,x)}{f_{\max} - f_{\min}} \tag{3}$$

where $f_{\min} - f_{max}$ represent the bandwidth of the ultrasonic transducer.

3 Ultrasonic Tissue Characterization System

The ultrasound characterization system for the exploration of acoustic indicators from animal lymph node can be seen in the Fig. 1. The system comprises a cartesian microcontroller positioner with millimetric resolution located at the top of an acoustic tank, a pulser - receiver system (*Olympus* pulse generator *PR 5072*) to emission of an ultrasonic field in pulse–echo mode with pulse repetition frequency of 100 kHz, an ultrasonic transducer *Panametrics-NDT* with resonance frequency of 2.25 MHz and 1 MHz @ −3 dB bandwidth, an oscilloscope for acquisition and signals digitization.

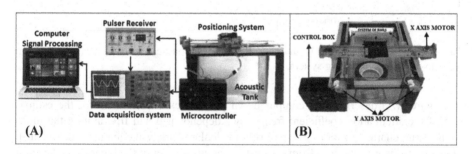

Fig. 1. The implemented system is composed by: **(A)** System of emission, acquisition and signal processing. **(B)** The acoustic transducer positioning system (cartesian robot).

The acquisition of the ultrasonic signals was carried out using an oscilloscope *Gw Instek G,* with rate sampling of 25 m/s, which allows the storage backscatter signals to enable its posterior processing by computational algorithms in the workstation. The implemented computational procedures for processing of ultrasonic signals were performed in MATLAB version R2014A.

3.1   Biological Samples and Ultrasonic Signal Processing

The bovine supra-mammary lymph nodes selected for the ultrasonic characterization were collected with the consent of veterinary services of the animal slaughterhouse. In this study lymphatic cells in a medullary region from four nodules were acoustically inspected. The delimited RoI was of 2 × 2 cm.

On the other hand, an immersion method was used aiming the ultrasonic characterization by the system described in Fig. 1. Lymph node samples were exposed to a 2.25 MHz ultrasonic field inside acoustic tank, which uses water as coupling medium for the optimal ultrasonic energy transfer to the tissue. The focal length transducer was estimated at 12 cm according to far field technical specifications. In this way, four hundred twenty ultrasonic backscattered signals were stored during the inspection space of 4 cm^2 for each lymph node. The signals were recorded by the oscilloscope used as signal acquisition system. An example of ultrasonic backscattered signal acquired from RoI is seen in Fig. 2.

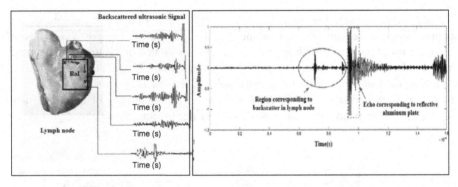

Fig. 2. Signals obtained from lymph node RoI during acoustic inspection area of 2 × 2 cm using to cartesian positioning system.

Figure 3(A) describes the computational procedures implemented for the estimation of the attenuation coefficient frequency - dependent β and IBC aiming the ultrasonic tissue characterization. The algorithm calculates firstly the angular coefficient β on the transducer bandwidth from the best linear adjustment of the logarithmic spectral subtraction between the tissue and reference normalized spectra respectively, showing spectral energy loss because biological tissues produce effects of low pass filters. The β estimation for each node allowed generating a data matrix, which is used to construct a parametric ultrasonic image taking into account first the logarithmic data compression and a spline interpolation. On the other hand, Fig. 3(B) represents the construction of IBC parametric images from the following steps: (1) Enlistment of ultrasonic backscatter signals acquired from tissue cross section during acoustic inspection; (2) Spectral calculation by the periodogram technique from tissue backscatter signals fragmented with an overlapping rate of 50% on the RoI. The ultrasonic signal partition in RoI is performed considering a partition size equivalent to the temporal pulse width from the ultrasonic incident signal. Subsequently it is selected in order to calculate the fast fourier transform (FFT); (3) IBC calculations use theoretical models previously described; (4) For generating IBC parametric images, each data matrix IBC is interpolated by spline method and is fitted with a meshgrid, according to the dimensions of the RoI of the tissue, which was inspected with a step size of 1 mm per backscatter signal collected.

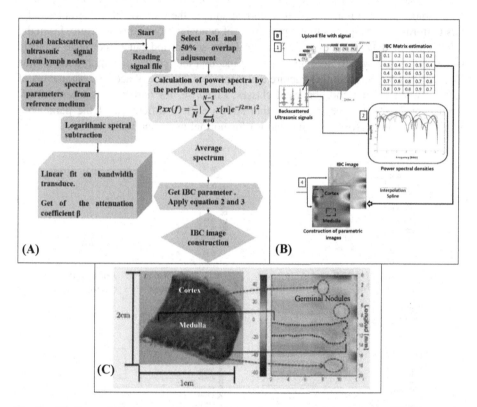

Fig. 3. (A) Computational procedures used to estimate spectral parameters. (B) Parametric image construction from IBC. (C) Histological plate of lymph node supra mammary Inspected (left). Parametric IBC image of cross-section of tissue (right).

Figure 3 shows histological plaque generated for a nodule by a processing and paraffin inclusion together with a staining of hematoxylin eosin. Figure 3 also shows a parametric image generated by the estimation algorithm of IBC parameter. In Fig. 3 (C), two structurally different sections are observed. In the upper left of the parametric image is evidence subtly small nodular areas corresponding probably effects of germinal nodules that are located preferably in the lymph node cortex. In the second instance in Fig. 3(C) also can be displayed in the middle of the spatial distribution, medullary sinusoids of a lighter color and including medullary trabeculae. The density variations of structural tissue are better detailed by generation of ultrasound parametric images ultrasound, which are more sensitive to the characteristics of the tissue microstructure, in contrast with the generation of ultrasound images in mode B, because in the latter it is lost information by the transformation of high-frequency backscatter signals to low-frequency signals, representing the qualitative characteristics of an irradiated sample. Figure 3(C) shows the medullary sinuses that presents a lymph node and are coherent with the structures seen by visual inspection in a histologic cross-section of the node. In the middle region of the image is possible to appreciate a region more uniform that for distance corresponds to the medulla of the sample study, this uniformity corresponds to the properties of echogenicity of this structure. From the calculation of data matrices, statistical analyzes are carried out to correlate acoustic parameters associated to the regions of the medulla and cortex of lymph nodes. Thus Fig. 4 shows for the medulla structure a percentage difference of acoustic parameters of 33% between normal and abnormal classes. Likewise, for cortex exist a percent difference of 24.9% for these classes, that shows a clear difference of the parametric acoustic indicators (IBC and attenuation) for normal and abnormal structures.

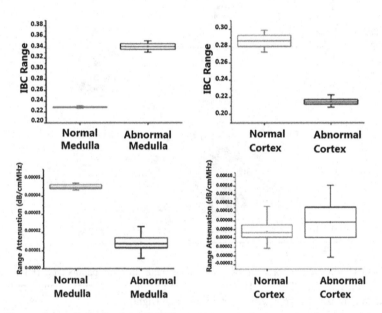

Fig. 4. Comparison of distribution of IBC parameters and attenuation between normal and abnormal lymph node samples.

The results obtained from ANOVA variance test show a statistically significant difference through the acceptance of the alternative hypothesis with an acceptance rate of 95%, which demonstrates that pixel values of parametric images in regions of medulla and cortex are useful to discern between nodules considered histologically normal and abnormal. Based on the parametric ultrasonic images different structures of the lymph nodes can be identified due to differences in echogenicity of the same.

4   Conclusions

The present study demonstrates how parametric characterization techniques by ultrasound, favor non-invasive and non-destructive analysis in the evaluation of animal soft tissue structures. The parametric ultrasound images of IBC allow a better visualization of structures such as cortex and medulla of lymph nodes in comparison with conventional ultrasound images in B mode, which difficult to identify in detail medullar structures or germinal centers. Finally, it is convenient to mention that the techniques of parametric ultrasonic characterization can contribute to technify process related with of histological analysis and the minimization of contamination risks evidencing the acoustic exposimetry techniques like an unconventional potential tool in analysis of biological tissues.

References

1. Miura, K., Nasu, H., Yamamoto, S.: Scanning acoustic microscopy for characterization of neoplastic and inflammatory lesions of lymph nodes. Sci. Reports **3**, 1–10 (2013)
2. Abdel, E., Abu, M., Mohammed, A.: The utility of multi-detector CT in detection and characterization of mesenteric lymphadenopathy with histopathological confirmation (2016)
3. Chemat, F., Huma, Z., Khan, M.: Applications of ultrasound in food technology: processing, preservation and extraction. Ultrason. Sonoche. **18**, 813–835 (2011)
4. Engels, D.: The control of neglected zoonotic diseases: zoonoses – prevention and control. World Health Organization, pp. 1–48 (2015)
5. Sedano, E.; Neira, C.: Calidad y Control de Calidad en el Laboratorio de Procedimientos Histológicos del Departamento de Patología **59** (1998). ISSN 1025-5583
6. Kiernan, J.A.: Histological and Histochemical Methods: Theory and Practice, 3rd edn. Arnold Publisher, London (2002)
7. Guzmán, X., Ramírez, P., López, S.: Manual de procedimientos estándares para el análisis histológico e histopatológico en organismos acuáticos. Instituto Nacional de Ecología INECC, pp. 1–22 (2009)
8. Badía, M., Ruiz, E., González, C.A.: Técnicas en Histología y Biología Celular. Elsevier Masson, Madrid (2009)
9. Ghoshai, G., Luchies, A.C., Blue, J.P., Oelze, M.L.: Temperature dependent ultrasonic characterization of biological media. J. Acoust. Soc. **43**, 2203–2211 (2011)
10. Cullinane, M., Reid, G., Dittrich, R., Kaposzta, Z., Babikian, V., Droste, D., Grossett, D.: Quantitative ultrasound spectroscopic imaging for characterization of disease extent in prostate cancer patients. Trans. Oncol. **8**, 25–34 (2014)
11. Oliveira, L., Maia, J., Gamba, H., Gehewht, P., Pereira, W.: Image quality evaluation of B mode ultrasound equipament. Revista Brasileira de Engenharia Bioméd. **26**(1), 11–24 (2010)
12. Herd, T., Hall, T., Jingfeng, J., Zagzebskib, J.: Improving the statistics of quantitative ultrasound techniques with deformation compounding: an experimental study. Ultrasound Med. Biol. **37**(12), 2066–2074 (2011)
13. Mamou, J., Oelze, M.: Quantitative Ultrasound in Soft Tissues. Springer, Heidelberg (2013). https://doi.org/10.1007/978-94-007-6952-6

A Distributed Shared Nearest Neighbors Clustering Algorithm

Juan Zamora[1], Héctor Allende-Cid[1(✉)], and Marcelo Mendoza[2,3]

[1] Pontificia Universidad Católica de Valparaíso, Valparaíso, Chile
hector.allende@pvcv.cl
[2] Universidad Técnica Federico Santa María, Santiago, Chile
[3] Centro Científico y Tecnológico de Valparaíso, Valparaíso, Chile

Abstract. Current data processing tasks require efficient approaches capable of dealing with large databases. A promising strategy consists in distributing the data along several computers that partially solves the undertaken problem. Then, these partial answers are integrated in order to obtain a final solution. We introduce the *Distributed Shared Nearest Neighbor* based clustering algorithm (D-SNN) which is able to work with disjoint partitions of data producing a global clustering solution that achieves a competitive performance regarding centralized approaches. Our algorithm is suited for large scale problems (e.g, text clustering) where data cannot be handled by a single machine due to memory size constraints. Experimental results over five data sets show that our proposal is competitive in terms of standard clustering quality performance measures.

Keywords: Clustering · Distributed algorithm
Shared Nearest Neighbors

1 Introduction

As a consequence of the explosive growth of the web, the integration of search engines into personal computers and mobile devices, and the wide use of social networks, the task of clustering text data for document organization has become a key aspect for web systems design. Nowadays the generation of large amounts of documents surpasses the computational capacity of personal computers and even the one of high performance computers. As an example, it is estimated that the amount of web pages indexed in the web is higher than 45 billion[1]. Therefore, it is of great interest to develop algorithmic techniques able to automatically organize, classify and summarize document collections distributed in multiple machines, and that also perform efficiently in modern hardware, particularly within parallel processing frameworks in multi-core architectures.

In real scenarios such as *Collection Selection* [3] for distributed search engines, in which for a given query the computer node containing the most

---

[1] http://www.worldwidewebsize.com/, last access at 9th June 2017.

© Springer International Publishing AG, part of Springer Nature 2018
M. Mendoza and S. Velastín (Eds.): CIARP 2017, LNCS 10657, pp. 710–718, 2018.
https://doi.org/10.1007/978-3-319-75193-1_85

suitable sub-collection must be selected to answer it, the challenges related to scalability and efficiency of clustering methods have become very important [9]. Traditional algorithms often assume that the whole dataset is loaded into main memory (RAM) and thus every document can be accessed at any time with no access latency. In scenarios where the size of the collection is much bigger than the amount of available RAM either because of the number of documents or the length of each document, this loading process is unfeasible.

There are three approaches successfully applied to the construction of clustering algorithms able to process large volumes of data. The first one introduces constraints on the number of passes allowed on a document [11]. The second one exploits multi-core architectures to perform parallel processing of the data [12]. The last one combines the computational power of a single machine together with the scalable storage capability of a distributed system by partitioning the data into several independent machines connected through a network [10]. Our proposal is focused on the last approach.

This work is structured as follows: In Sect. 2 we present the most relevant works related to our proposal. In Sect. 3 we present D-SNN. Then in Sect. 4 we present the experimental setup and the results obtained with several datasets. Finally in Sect. 5 we present concluding remarks and future directions of our work.

2   State of the Art

Over the years there have been several contributions in the Clustering task. We will focus mainly in parallel and distributed approaches for large datasets.

2.1   Main Contributions on Parallel Clustering Algorithms

Xu *et al.* [13] present a parallel version of DBSCAN (PDBSCAN). The authors present the 'shared-nothing' architecture with multiple computers interconnected through a network. A fundamental component of a shared-nothing system is its distributed data structure. They introduce the dR*-tree, a distributed spatial index structure in which the data is spread among multiple computers and the indexes of the data are replicated on every computer. A performance evaluation shows that PDBSCAN offers nearly linear speed-up and has excellent scale-up and size-up behavior. Dhillon and Modha [5] present an algorithm that exploits the inherent data-parallelism in the k-means algorithm. They analytically show that the speed-up and the scale-up of the algorithm approach the optimal as the number of data points increases. The implementation of this proposal is done on an IBM POWER parallel SP2 with a maximum of 16 nodes. On typical test data sets, nearly linear relative speed-ups are observed, for example, 15.62 on 16 nodes, and essentially linear scale-up in the size of the data set and in the number of clusters desired. For a 2 gigabyte test data set, the implementation drives the 16 node SP2 at more than 1.8 Gigaflops.

Another scalable approach based on secondary memory consists in designing algorithms able to work within the MapReduce framework. In this context we highlight the contributions made by Ene et al. [6] in which they tackle the K-Median problem by using MapReduce. Das et al. [4] present an approach to filter recommendations for users of Google News in order to generate personalized recommendations in a collaborative way. They generate recommendations using three approaches: collaborative filtering using MinHash clustering, probabilistic Latent Semantic Indexing (pLSI), and co-visitation counts. The recommendations are combined from different algorithms using a linear model. The authors claim that the approach is content agnostic and consequently domain independent, making it easily adaptable for other applications and languages with minimal effort. Another parallel approach for K-Means is presented by [1] and it is called K-Means++. The initialization of the k means algorithm is crucial, and the authors claim that the proposed algorithm obtains an initial set of centers that is probably close to the optimum solution. A major downside of k-means++ is its inherent sequential nature, which limits its applicability to massive data: one must make k passes over the data to find a good initial set of centers. In this work the authors show how to drastically reduce the number of passes needed to obtain, in parallel, a good initialization. This is unlike prevailing efforts on parallelizing k-means that have mostly focused on the post-initialization phases of k-means. The proposed initialization obtains a nearly optimal solution after a logarithmic number of passes, and then shows that in practice a constant number of passes suffices.

2.2   Distributed Approaches for Clustering Based on Prototypes

Forman and Zhang [8] describe a technique to parallelize a family of center-based data clustering algorithms. The central idea of the proposed algorithm is to communicate only sufficient statistics, yielding linear speed-up with good efficiency. The proposed technique does not involve approximation and may be used in conjunction with sampling or aggregation-based methods, such as BIRCH, to lessen the quality degradation of their approximation or to handle larger data sets. The authors demonstrate that even for relatively small problem sizes, it can be more cost effective to cluster the data in-place using an exact distributed algorithm than to collect the data in one central location for clustering.

Qi et al. [14] propose an approximate K-Median clustering technique that works over streaming data, i.e. data is continuously collected. They propose a suite of algorithms for computing $(1 + \epsilon)$-approximate k-median clustering over distributed data streams under three different topology settings: topology-oblivious, height-aware, and path-aware. The proposed algorithms reduce the maximum per node transmission to $polylogN$ (opposed to $\Omega(N)$ for transmitting the raw data). The authors perform simulations on a distributed stream system with both real and synthetic datasets composed of millions of data points. In practice, the algorithms are able to reduce the data transmission to a small fraction of the original data. The results indicate that the algorithms are scalable with respect to the data volume, approximation factor, and the number of sites.

Balcan *et al.* [2] provides novel algorithms for distributed clustering for two popular prototype-based strategies, k-medians and k-means. These algorithms have provable guarantees and improve communication complexity over existing approaches. The proposed algorithm reduces the problem of finding a clustering with low cost to the problem of finding a coreset of small size. The authors provide a distributed method for constructing a global coreset which improves over the previous methods by reducing the communication complexity, and which works over general communication topologies. Experimental results on large scale data sets show that this approach is feasible.

3   Proposal

In this work we present D-SNN, a distributed clustering algorithm based on *Shared-Nearest-Neighbor* (SNN) clustering [7]. This method automatically identifies the number of underlying groups in each data partition along with a set of representative points for each one. We pose that this method is able to deal with collections arbitrarily distributed across a Master/Worker architecture as depicted in Fig. 1. The algorithm operates in two stages: The first one starts by processing the data in each partition, producing a set of density-based samples, i.e. *core-points*, and finally, transmitting back a sample set of these *core-points* to the master node. In the second stage, the master node joins the sample points received and then starts a centralized SNN clustering over them. As the master node works over worker samples avoiding the allocation of the whole dataset in memory, the algorithm is able to handle large-scale data collections. Finally, the set of representative points is labeled in the master node consolidating a set of *core-points* that summarizes the overall collection.

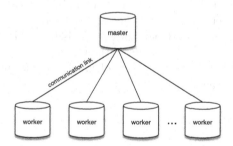

Fig. 1. Network architecture employed by the proposed algorithm

D-SNN Stages
The first stage of D-SNN starts by working over each data partition. Data partitions can be obtained by randomly partitioning and distributing the dataset into several worker nodes. In this first stage, each worker n_i starts identifying

in its partition $\mathcal{D}_i$ the core-points. The procedure starts by computing the k-nearest neighbors of each data point in $\mathcal{D}_i$. To deal with high dimensional data, we use in this step the cosine similarity. Note that this function may be replaced with a better one depending on the nature of the data. Then, we calculate the size of the shared neighborhood for every pair of distinct points in $\mathcal{D}_i$, a value that is indexed by the function named $\mathrm{SNN}_k(p,q)$. Then, for each data point in $\mathcal{D}_i$ we calculate its density which corresponds to the number of data points whose shared neighborhood is greater than Eps. Finally, the core point condition is checked. Each point is marked as a core point if its density is greater than MinPts. The collection of core points is returned by the algorithm to the cluster labeling process. Note that the similarity function $\mathrm{SNN}_k(p,q)$ is also returned.

Then the algorithm continues with a labeling process, which works as follows. Initially, all points are unlabeled. Then, we assign the same cluster label to each core point that belongs to the same connected component, i.e. the collection of points that mutually has at least a Eps similarity in the SNN space. Accordingly, a cluster is defined as a high density region of data points in the *SNN* space. Once all core-points are labeled, a weighted sample from each cluster is drawn. The weight of a data point follows the expression: $\dfrac{1 - (N_l/N_c)}{2 \cdot N_l}$ where N_c denotes the number of labeled core-points and N_l the number of points labeled with label l. The expression shown above denotes that the sampling weight of a core-point is inverse to its group size. The aim of this function is to build a sample of core-points able to represent both small and large groups alike without a bias to larger clusters which is a serious problem for clustering algorithms operating under unbalanced data scenarios.

Finally when all worker nodes transmit their sample of core-points to the master node and then this node joins all these points into a single data set $\mathcal{S}$. Then the procedure SnnCorePoints is applied over $\mathcal{S}$, obtaining a new core-point set $\mathcal{C}$, and the same labeling step applied in the initial stage in each worker is performed over this set. After this step, for each point $q \in \mathcal{S}$ not chosen for the core-point set $\mathcal{C}$, its nearest neighbor $p \in \mathcal{C}$ is found. If the SNN similarity between these points is lower than Eps, then q is marked as noise, otherwise label l_p is assigned to q. Finally, the set of labeled points in $\mathcal{S}$ (i.e. discarding the noisy subset) is returned.

4   Experiments

We assess the performance of D-SNN comparing its effectiveness against a centralized versions of Kmeans and DBSCAN. The highest performance in terms of V-measure was reported for each algorithm and it was calculated by choosing the algorithm parameters from a grid (i.e. Number of clusters for Kmeans and Epsilon/MinPts for DBSCAN). Additionally, as Kmeans and D-SNN have a randomized initialization, 10 runs were executed for each one and the mean and standard deviation of each quality measure was reported. Additionally, the machines used for running the centralized algorithms and the D-SNN workers have 2 cores and 6 GB RAM.

Three document collections previously used in the document clustering task were employed, namely 20-Newsgroups (20NG), Topic-Detection and Tracking (TDT) and Reuters (Reuters). The 20-Newsgroups collection is nowadays a standard dataset for text categorization. Documents contained in this collection are categorized into 20 groups. For the experimentation we chose only the largest 10 categories thus leaving us with 9917 documents. The TDT corpus consists of data taken from two newswires, two radio programs and two television programs. It consists of 11201 documents classified into 96 categories. For the experimentation we considered only the largest 30 categories thus leaving us with 9394 documents in total. The Reuters corpus contains 21578 documents grouped into 135 clusters. Reuters corpus is quite imbalanced, with some large clusters having more than 300 times larger than some small ones. For the experimentation we used only the documents labeled in a single category and we also chose the largest 30 categories. This left us with 8067 documents in total. Note that all these collections are quite sparse and also their document vectors are spanned onto high dimensional term spaces with tens of thousands of features (Table 1).

Table 1. Data sets employed in our experiments.

Dataset	Description	Instances	$|\mathcal{V}|$	Classes	%NNZ
20NG	20-newsgroup data	9917	26214	10	37.8%
TDT	Topic detection and tracking	9394	36771	30	29.6%
Reuters	Reuters	8067	18933	30	42.6%

Document preprocessing

In order to build a vector representation for each document, a classic feature extraction procedure for text data was performed. That is stopword removal and *Term-Frequency/Inverse Document-Frequency* (Tf-Idf) weighting scheme to generate the document vectors. As a document similarity function we used the Cosine similarity due to its good performance on high dimensional data collections.

Clustering quality measures

In this work, only external validation measures were employed. We used **Adjusted-Mutual-Information** and **V-Measure**. These scores have positive values within $[0, 1]$. For all of these measures, larger values denote a better clustering quality.

Evaluation mechanism

Lets first recall the last part of the proposed method: Within the final stage, several representative points are collected from all the nodes. Then these points are clustered by the master node and this labeled subset of the original data is reported as the result of the proposed method. Therefore, the quality scores reported further on involve only the output labels generated by the algorithm.

Results

The results of our experiments are shown in Table 2. Bold fonts indicate the best result for each configuration.

Table 2. Performance attained by the two methods. The values surrounded by parenthesis denote the standard deviation of each measure along the execution runs.

	Dataset	V	AMI
Kmeans	20NG	0.5856 (0.0206)	0.5764 (0.0215)
	TDT	0.7576 (0.0178)	0.7061 (0.0178)
	Reuters	0.5028 (0.0113)	0.4076 (0.0097)
DBSCAN	20NG	0.2897	0.0997
	TDT	0.3379	0.1510
	Reuters	0.3043	0.1321
D-SNN	20NG	0.5695 (0.0134)	0.5524 (0.0185)
	TDT	0.7038 (0.0291)	0.6812 (0.0251)
	Reuters	0.3841 (0.0169)	0.3461 (0.0139)

Analysis of the results

Table 2 shows that D-SNN obtains a comparable performance in almost all the experiments although D-SNN used randomly generated data partitions distributed along several machines. Our experimental results suggest that the core-point extraction procedure together with the sampling strategy followed to transmit the representatives to the master node enable a clear discrimination of each group in comparison to two classic centralized techniques (A partitional and a density based methods). None of the algorithms showed high performance scores due to the challenge imposed by the high dimensional nature and sparsity of text data and also because of the asymmetric sizes of the classes specially in the *Reuters* dataset.

5   Concluding Remarks

In this work, a distributed clustering method able to deal with massive text collections was proposed. In order to assess its utility especially for recovering the group structure underlying each collection, its performance was contrasted against DBSCAN and Kmeans which is widely used as a baseline clustering algorithm for text categorization. We showed that D-SNN attained a comparable performance to Kmeans on several high dimensional text data sets.

Three real text collections were employed with the aim to show the weaknesses and strengths of the shared-nearest-neighbor approach in an isolated way from the strengths of the proposed method. The results indicate that D-SNN, besides its power to process big collections in a distributed fashion, maintains the good properties of Centralized-SNN and at the same time improves the capability of dealing with the low discrimination power of the underlying similarity measure.

Currently, we are extending and testing D-SNN. The evaluation of D-SNN on a large distributed cluster will alow us to evaluate efficiency measures. An implementation of D-SNN on Spark will allow us to achieve this goal.

Acknowledgments. Juan Zamora is supported by a postdoctoral project from Pontificia Universidad Católica de Valparaíso. Héctor Allende-Cid is supported by project FONDECYT initiation into research 11150248. Marcelo Mendoza was supported by project Basal FB0821.

References

1. Bahmani, B., Moseley, B., Vattani, A., Kumar, R., Vassilvitskii, S.: Scalable k-means ++. Proc. VLDB Endow. (PVLDB) **5**, 622–633 (2012)
2. Balcan, M.F., Ehrlich, S., Liang, Y.: Distributed k-means and k-median clustering on general topologies. In: Advances in Neural Information Processing Systems 26 (NIPS 2013), pp. 1–9 (2013)
3. Crestani, F., Markov, I.: Distributed information retrieval and applications. In: Serdyukov, P., Braslavski, P., Kuznetsov, S.O., Kamps, J., Rüger, S., Agichtein, E., Segalovich, I., Yilmaz, E. (eds.) ECIR 2013. LNCS, vol. 7814, pp. 865–868. Springer, Heidelberg (2013). https://doi.org/10.1007/978-3-642-36973-5_104
4. Das, A., Datar, M., Garg, A., Rajaram, S.: Google news personalization: scalable online collaborative filtering. In: Proceedings of the 16th International Conference on World Wide Web, pp. 271–280. ACM (2007)
5. Dhillon, I.S., Modha, D.S.: A data-clustering algorithm on distributed memory multiprocessors. In: Zaki, M.J., Ho, C.-T. (eds.) LSPDM 1999. LNCS (LNAI), vol. 1759, pp. 245–260. Springer, Heidelberg (2002). https://doi.org/10.1007/3-540-46502-2_13
6. Ene, A., Im, S., Moseley, B.: Fast clustering using MapReduce. In: KDD, pp. 681–689 (2011)
7. Ertöz, L., Steinbach, M., Kumar, V.: Finding clusters of different sizes, shapes, and densities in noisy, high dimensional data. In: Proceedings of the SIAM International Conference on Data Mining, pp. 47–58 (2003)
8. Forman, G., Zhang, B.: Distributed data clustering can be efficient and exact. ACM SIGKDD Explor. Newslett. **2**(2), 34–38 (2000)
9. Mendoza, M., Marín, M., Gil-Costa, V., Ferrarotti, F.: Reducing hardware hit by queries in web search engines. Inf. Process. Manag. **52**(6), 1031–1052 (2016)
10. Sarnovsky, M., Carnoka, N.: Distributed algorithm for text documents clustering based on k-means approach. Adv. Intell. Syst. Comput. **430**, 165–174 (2016)
11. Yi, J., ZShang, L., Wang, J., Jin, R., Jain, A.K.: A single-pass algorithm for efficiently recovering sparse cluster centers of high-dimensional data. In: Proceedings of the 31st International Conference on Machine Learning, Beijing, China, vol. 3, pp. 2112–2127 (2014)

12. Zhang, J., Wu, G., Hu, X., Li, S., Hao, S.: A parallel clustering algorithm with MPI – MKmeans. J. Comput. **8**(1), 10–18 (2013)
13. Xu, X., Jäger, J., Kriegel, H.: A fast parallel clustering algorithm for large spatial databases. In: Guo, Y., Grossman, R. (eds.) High Performance Data Mining, pp. 263–290. Springer, Boston (1999). https://doi.org/10.1007/0-306-47011-X_3
14. Qi, Z., Jinze, L., Wei, W.: Approximate clustering on distributed data streams. In: Proceedings - International Conference on Data Engineering, pp. 1131–1139 (2008)

Automatic Classification of Optical Defects of Mirrors from Ronchigram Images Using Bag of Visual Words and Support Vector Machines

Daniel Zapata[1], Angel Cruz-Roa[2(⊠)]📷, and Andrés Jiménez[3,4]📷

[1] Dynamic Systems and GITECX Research Groups, Universidad de los Llanos,
Villavicencio, Colombia
daniel.zapata@unillanos.edu.co

[2] GITECX Research Group, Universidad de los Llanos, Villavicencio, Colombia
aacruz@unillanos.edu.co

[3] Dynamic Systems Research Group, Universidad de los Llanos,
Villavicencio, Colombia
ajimenez@unillanos.edu.co

[4] UN-Robot Research Group, Universidad Nacional de Colombia, Bogotá, Colombia
afjimenezlo@unal.edu.co

Abstract. The Ronchi test is known as a procedure that is able to generate visual patterns called ronchigrams. These patterns could be used to determine optical characteristics on the surface of mirrors, particularly to quantify and qualify optical aberrations and deformations. This paper presents an automatic method to detect these optical errors of mirrors using the Ronchi test by classifying ronchigram images using bag of visual words (BoVWs) for image representation and support vector machines (SVM) for ronchigrams classification. The ronchigram image data set was obtained from the optical manufacture laboratory of lenses and mirrors at Universidad de los Llanos. The BoVWs approach used was based on Scale-Invariant Feature Transform (SIFT) as visual words and a Linear SVM was trained for automatic classification of ronchigrams into optical defects. The classification performance achieved was 0.69% in terms of accuracy measure. These results shows that our proposed approach can be used to detect optical defects of mirrors with high precision in a real scenario of ronchigrams obtained from mirrors during the manufacture process of a optical laboratory.

Keywords: Bag of visual words · Image processing
Machine learning · Optical defects · Mirrors · Ronchi test
Support Vector Machine

1 Introduction

During the mirrors manufacture, the phase of optical test allows to identify optical defects which could be corrected to obtain the images of optical systems

© Springer International Publishing AG, part of Springer Nature 2018
M. Mendoza and S. Velastín (Eds.): CIARP 2017, LNCS 10657, pp. 719–726, 2018.
https://doi.org/10.1007/978-3-319-75193-1_86

sharper. These defects are considered as deformations that affect the response of the optical system resulting into image degradation. One of the optical tests performed in mirrors is known as *Ronchi test* [8]. This method was proposed by the Italian physicist Vasco Ronchi in 1922 when he published about the formation of fringe or interference patterns, a.k.a. *ronchigrams* [8]. These interference patterns are generated by placing a grid near the center of curvature of a mirror, which are related to optical aberrations (i.e. errors in the shape of the surface). The study of phase shift is one of the procedures used for ronchigrams analysis, Zernike polynomials are generally used to quantify these aberrations reflected by the mirror under test. This technique has been used in concave mirrors [1], spherical surfaces using a LCD screen to show vertical and horizontal grids [2,3,6], with interferometry, linear polarization modulation [10] and for optical tests in flat mirrors [4]. However, for the correct interpretation of these ronchigrams, it is required precision of the measurement tools, to process information of images or electromagnetic signals, and application of knowledge from mathematics, physics and engineering [7]. For ronchigram image analysis, several methods have been developed to detect aberrations by applying image processing techniques into space and frequency domains [10]. The procedure to obtain the ronchigram images, according to the Ronchi test, is performed by sampling the wavefront of the response of the optical system, in such a way that benchmarks can be compared with those resulting from the test. However, the ronchigram image acquisition requires a specific device and the subsequent image processing techniques for analysis, that is also related with the type of camera, lighting and kind of algorithms or procedures.

This paper is organized as follows: Sect. 2 explains the details of the proposed method to detect optical defects based on BoVWs and SVMs, Sect. 3 describes the experimental evaluation, including the data set description, the experimental setup and results. Finally, in Sect. 4 we present the main conclusions and future work.

2   Methodology

The proposed approach used a methodology which starts with a SIFT-based Bag of Visual Words (BoVW) image representation and Support Vector Machines (SVM) for automatic classification of Ronchigram images. During classification process, a SVM model was trained, applying linear kernel functions, different variations of regularization factor and local feature extraction methods using Scale-Invariant Feature Transform (SIFT) algorithm, together with global image representation of "Bag of Words" (BoVW). In this way, to select the results with better performance of the classifier and choose the implementation with the solution of parameters more efficient, this algorithm is integrated as MARC (Module of Analysis of Ronchigrams) module, into the ANGMAR - Image Processing V1.0 software of SAPRULL [9].

2.1   Ronchigram Image Acquisition

The Optical Center at Universidad de los Llanos produces optical components such as a simple magnifying glass or even an objective mirror of an astronomical reflecting telescope. In order to evaluate the quality of the lenses and mirrors an assisted system is used. This system consists of a Computer-Assisted Ronchi Interferometer (CARI) module and a Ronchigram Computer Analysis (RCA) module. The practical arrangement identifies three primordial elements: (i) white light source, (ii) periodic grid (4 lines/mm) and the digital camera (Exmor R®CMOS sensor, type 1/2.3 (7.82 mm), 20.4 MP). Figure 1 shows the system used to acquire ronchigrams images [9].

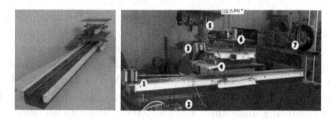

Fig. 1. Computer-aided system for Ronchi test - SAPRULL. 1. rail (1 m), 2. power source (12 V), 3. NEMA motors, 4. circuit set - long moving platform, 5. white light source, portable laser, 6. camera (CMOS), grid and slit - short motion platform, 7. mirror and bracket. Source: [9].

2.2   Local Image Feature Extraction (SIFT)

In this stage, the detection of interest points and their description by a local feature representation of each ronchigram image was developed. The implementation used SIFT points and features (Scale Invariant Feature Transform), that is an algorithm to detect and represent local features of a given image. An important aspect of this approach is that it generates a large number of features that densely cover the image across the entire range of scales and locations. An image of size 256×256 pixels will give rise to a certain number of stable characteristics (this number depends on the image content and the variation of existing parameters in the image), see Fig. 2.

Extraction and description of interest points by SIFT for the image gives rise to different results as seen in the right ronchigram of Fig. 2, where 60 key points (points of interest) were extracted. The center Ronchigram has 292 points of interest and one on the left has 82 key points. These key points represent features that allow the creation of the visual dictionary of visual words in representation of the images of each class, then, to be compared and based on the euclidean distance of their characteristic vectors can find coincidence images that allow their clustering using the K-mean algorithm for dictionary construction.

Fig. 2. Local feature extraction of a Ronchigram image using SIFT. Source: Authors. (Color figure online)

2.3   Image Representation via Bag of Visual Words (BoVW)

The Ronchigram Images were represented using the Histogram of Bag of Visual Words approach, which starts with: (i) the local features extraction using the Scale Invariant Feature Transform (SIFT), to build (ii) the visual dictionary (vocabulary) using the clustering algorithm known as K-means, in order to represent the ronchigram images as a histogram of the frequency of bag of visual words of the dictionary. Figure 3 depicts the BoVW approach for Ronchigram image representation. The vocabulary construction of the *BoVW* was done using the *K-means* clustering algorithm. Each cluster represents a visual word of the *visual* vocabulary that represents, according to the occurrence, the visual content distribution in the image. The *BoVW* method stats with the detection and description of keypoints for each image, using *SIFT* descriptors. The vocabulary was built with 50.000 keypoints with a length of 1.024 visual words.

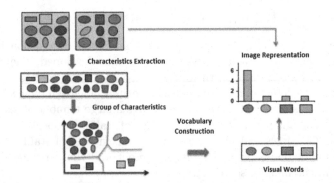

Fig. 3. Ronchigram image representation using BoVW [12].

2.4   Image Classification via Support Vector Machine (SVM)

For the automatic classification of ronchigram images from the BoVW Image Representation, a Support Vector Machine (SVM) was trained. SVM is one of the common and successful methods used in supervised machine learning in several tasks, which is used either for regression and classification tasks. This method uses the "kernel trick", which allows the computation of the original data representation into a high dimensional space. This high dimensional space is known as "feature space" ',' where data is expected to be linearly separable by hyperplanes learned during training process for the classification task, independently of the kernel function applied in training. For instance, SVM has been successfully applied in regression applications like analysis of data, to determinate fraud or not-fraud in banking transactions [5] or in Multi-class classification to recognize hand-written digits from 0 to 9 [11]. Here, the original set of ronchigrams is obtained from the Ronchi Test applied to 14 parabolic mirrors, which were selected with alterations on their surfaces (optical aberrations). The original data set of 2.554 Ronchigrams and 5 classes, was divided into 70% for training set and 30% for testing set, using stratified sampling per class (Fig. 4).

Fig. 4. Ronchigrams of the library. Source: Authors.

3   Experimental Evaluation

3.1   Ronchigram Image Dataset Description

The data set was obtained from mirrors samples without defects and mirrors that present borders and/or center defects. For each sample a digitized image was obtained by the Ronchi test according to the process described in Sect. 2.1. In Fig. 5, we can appreciate the variations of the number of fringes when the ronchigram was acquired for different places with respect to the optical focus (0.5 cm, 1.0 cm, 1.5 cm, 2.0 cm). This behavior is validate by the expert in optics aberrations from the optical laboratory of the Universidad de los Llanos. These patterns highlight the importance of the relation between optical aberrations at different focal lengths for optical aberration differentiation. Table 1 describes the data set distribution per class for training and testing of the SVM classifier.

3.2   Experimental Setup

Once the data set of ronchigram images was acquired according to description of Subsect. 2.1, the SVM classifier was trained using 5-fold cross-validation in training to obtain the best parameters. The kernel function used was linear because it is simplest and obtain best classification performance.

The performance measures of precision $Pr = TP/(TP + FP)$, recall (Rc $Rc = TP/(TP + FN)$ and accuracy $Acc = (TP + TN)/(TP + TN + FP + FN)$, were used to evaluate the ronchigrams classification. Where true-positive

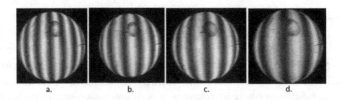

Fig. 5. Ronchigrama with movement inside of the focus, (a) 0.5 cm, (b) 1.0 cm, (c) 1.5 cm, (d) 2.0 cm. Source: Authors.

Table 1. Ronchigram image data set. Source: Authors

Classes	Training	Testing
1. Spherical (no defects)	561	240
2. SobreCorr3 (center defects)	464	199
3. SobreCorr9 (center defects)	342	146
4. Sobre9SubCorr3 (borders and center defects)	337	145
5. SubCorr9 (borders defects)	84	36

samples (TP) are those samples in test data set that are correctly classified in the corresponding class, false-positive samples (FP) are those samples that were incorrectly classified in the class but actually are from other classes, true-negative samples (TN) are those samples that are not the class of interest and were correctly classified as other class, and false-negative samples (FN) are those samples that were incorrectly classified as other classes different of the class of interest. In addition, the Receiver operating characteristic (ROC) curve was calculated to analyze the classification performance per class.

3.3   Results

The best combination of SVM parameters was obtained by grid-search in training data set using: 5-fold cross-validation, exploring between "*Linear*" and "*Radial*" *Kernels* and using a range of 10 different values for the regularization parameter C between 0.0001 and 0.02 values. The best parameters were for "*Linear*" *Kernel* and $C = 0.001$ achieving a classification performance of: precision (0.71), recall (0.68), accuracy (0.69) and execution time (1.013 s). The execution time varies according to the number of parameters that are initialized for validation and also of the CPU characteristics where the program was executed. Figure 2 shows three example images of ronchigrams with the points of interest (red circles) detected using SIFT algorithm detection and description for local features. The main advantage of these kind of local feature detector and description is that are invariant to scale, rotation, and size. Particularly, similar ronchigram image patterns reveals similar points of interest with the same local feature representation, which is useful to detect similar optical defects for the posterior classification process.

Figure 6 presents two confusion matrices of the classification performance from the trained SVM model over the test set of ronchigram images. The ice colorbar represents that dark blue colors are higher values whereas white colors are lower values. In this case, the values of the right confusion matrix correspond to the accuracy performance measure and left matrix the number of samples. This figure shows that each confusion matrix are presenting good performance of the SVM models with a diagonal values closed to the optimal. Using the accuracy measure, the best results of SVM classification in test data set found was for $C = 0.001$ achieving an average accuracy of 0.69. Finally, ROC curve is shown in Fig. 7 in order to analyze the classification performance per each class which is higher for most of classes.

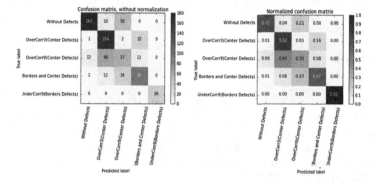

Fig. 6. Confusion matrices of the multi-class classification performance. (Color figure online)

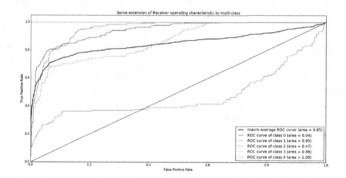

Fig. 7. ROC curve of the multi-class classification performance.

4 Conclusions and Future Work

The proposed approach presented in this paper for automatic classification of optical defects from Ronchigram images using SIFT-based Bag of Visual Words and Support Vector Machines was successfully applied achieving high classification performances. Our approach includes the image acquisition process where samples of each class corresponds to optical defects of mirrors used during the Ronchi test validated by an expert. The ronchigrams vary in terms of number of fringes but the defect is kept. However, images acquired from different distances did not alter the classification performance results. The used of SVM models requires the systematic exploration of hyperparameters through a grid search using cross-validation for each kind of kernel evaluated. This could be a time consuming task, which could suggest the exploration of alternative methods with less number of parameters with similar classification performance. The most robust results were obtained when both measures, precision and recall, were took into account. Future work includes to increase the number of image per each optical defect to train better models. In fact, data-driven and state-of-the-art methods based on convolutional neural networks will be evaluated.

References

1. Arenas, G., Doria, O., Narvaéz, A., Nunez, M.: Prueba de ronchi para un espejo cóncavo. Rev. Colomb. de Física **38**(2), 601–604 (2006)
2. Cordero, A., González, J.: Surface evaluation with Ronchi test by using Malacara formula, genetic algorithms, and cubic splines. In: IODC. SPIE (2010)
3. Cordero, A., González, J., Robledo, C.I., Leal, I.: Local and global surface errors evaluation using ronchi test, without both approximation and integration. Appl. Opt. **50**(24), 4817–4823 (2011)
4. González, M.M., Ochoa, N.A.: The Ronchi test with an LCD grating. Opt. Commun. **191**(3), 203–207 (2001)
5. Hidalgo, S.: Random forests para detección de fraude en medios de pago. Master's thesis, Univ. Autónoma de Madrid, Escuela Politécnica Superior, Dpto. Ingeniería Informática (2014)
6. Jiménez, J., Dávila, A.: Ronchi test visibility as a function of illumination source size and duty cycle of the Ronchi ruling. Optica pura y aplicada **46**(1), 29–32 (2013)
7. López, A.J., Pelayo, M.P., Forero, Á.R.: Enseñanza del procesamiento de imágenes en ingeniería usando Python (2015)
8. Malacara, D.: Optical Shop Testing, vol. 59. Wiley, Hoboken (2007)
9. Prieto, M., Forero, A.: Diseño e implementación de un sistema asistido por computador de la prueba de ronchi en el taller de óptica de la universidad de los llanos (2015)
10. Sánchez, J., Silva, G., Castro, J., Sasian, J.: 1λ Ronchi tester to obtain the wave front aberration in converging optical systems using phase shifting interferometry. Opt. Pura Apl. **45**(4), 461–473 (2012)
11. Scikit-learn developers: an introduction to machine learning with scikit-learn. An introduction to machine learning with scikit-learn (2010–2016)
12. Valveny, E., González, J.B.R.: Clasificación de imágenes: cómo reconocer el contenido de una imagen? (2010)

Author Index

Printed in the United States
By Bookmasters